THE ROUTLEDGE HANDBOOK OF PHILOSOPHY OF SCIENTIFIC MODELING

Models and modeling have played an increasingly important role in philosophy, going back to the nineteenth century. While philosophical interest in models has been remarkably lively over the last two decades, there are still many underexplored questions. *The Routledge Handbook of Philosophy of Scientific Modeling* is an outstanding reference source and guide to this fast-growing area and is the first volume of its kind. Comprised of 40 specially commissioned chapters by an international team of contributors, the *Handbook* is organized into five clear parts:

- Historical and General Perspectives
- Philosophical Accounts of Modeling
- Methodological Aspects: Model Construction, Evaluation, and Calibration
- Related Topics
- Modeling in the Wild

Within these parts, the *Handbook* covers a diverse range of topics, including historical perspectives on modeling, the relationship between models, theories, representation, idealization, and understanding, and related topics like big data, simulation, and statistical and computational modeling. Different kinds of models are discussed, for example, network models, financial models, and climate and synthetic models.

The Routledge Handbook of Philosophy of Scientific Modeling is essential reading for students and scholars of philosophy of science, formal epistemology, and philosophy of social sciences. It is also a valuable resource for those in related fields such as computer science and information technology.

Tarja Knuuttila is a Professor of Philosophy of Science at the University of Vienna, Austria. She has developed an artifactual account of models. Knuuttila focuses, in her research, on scientific modeling, interdisciplinarity, and the modal dimension of science with a special focus on synthetic biology, engineering sciences, and economics.

Natalia Carrillo is an Associate Researcher at the Institute of Philosophical Research at the National Autonomous University of Mexico (UNAM). She is interested in philosophical problems at the intersection of philosophy of science and technology, especially artifactuality and abstraction in modeling practices, and the role of analogies and metaphors in science.

Rami Koskinen is a Researcher at the University of Vienna, Austria, with an interest in the general philosophy of science, philosophy of biology, and epistemology. He has been investigating modal reasoning in the sciences, modeling in synthetic biology, and the question of multiple realizability.

ROUTLEDGE HANDBOOKS IN PHILOSOPHY

Routledge Handbooks in Philosophy are state-of-the-art surveys of emerging, newly refreshed, and important fields in philosophy, providing accessible yet thorough assessments of key problems, themes, thinkers, and recent developments in research.

All chapters of each volume are specially commissioned and written by leading scholars in the field. Carefully edited and organized, *Routledge Handbooks in Philosophy* provide indispensable reference tools for students and researchers seeking a comprehensive overview of new and exciting topics in philosophy. They are also valuable teaching resources as accompaniments to textbooks, anthologies, and research-orientated publications.

ALSO AVAILABLE:

THE ROUTLEDGE HANDBOOK OF CONTEMPORARY EXISTENTIALISM
Edited by Kevin Aho, Megan Altman, and Hans Pedersen

THE ROUTLEDGE HANDBOOK OF ESSENCE IN PHILOSOPHY
Edited by Kathrin Koslicki and Michael J. Raven

THE ROUTLEDGE HANDBOOK OF POLITICAL PHENOMENOLOGY
Edited by Nils Baratella, Steffen Herrmann, Sophie Loidolt, Tobias Matzner, and Gerhard Thonhauser

THE ROUTLEDGE HANDBOOK OF EMBODIED COGNITION, SECOND EDITION
Edited by Lawrence Shapiro and Shannon Spaulding

THE ROUTLEDGE HANDBOOK OF THE PHILOSOPHY OF SCIENTIFIC MODELING
Edited by Tarja Knuuttila, Natalia Carrillo, and Rami Koskinen

For more information about this series, please visit: https://www.routledge.com/Routledge-Handbooks-in-Philosophy/book-series/RHP

THE ROUTLEDGE HANDBOOK OF PHILOSOPHY OF SCIENTIFIC MODELING

Edited by Tarja Knuuttila, Natalia Carrillo, and Rami Koskinen

Routledge
Taylor & Francis Group
LONDON AND NEW YORK

Cover image credit: © Andrey_A / Getty Images

First published 2025
by Routledge
4 Park Square, Milton Park, Abingdon, Oxon OX14 4RN

and by Routledge
605 Third Avenue, New York, NY 10158

Routledge is an imprint of the Taylor & Francis Group, an informa business

British Library Cataloguing-in-Publication Data
A catalogue record for this book is available from the British Library

Library of Congress Cataloging-in-Publication Data
Names: Knuuttila, Tarja, editor. | Carrillo, Natalia, 1984– editor. |
Koskinen, Rami, editor.
Title: The Routledge handbook of philosophy of scientific modeling / edited
by Tarja Knuuttila, Natalia Carrillo, and Rami Koskinen.
Description: Abingdon, Oxon; New York, NY: Routledge, 2024. |
Series: Routledge handbooks in philosophy |
Includes bibliographical references and index.
Identifiers: LCCN 2024015958 (print) | LCCN 2024015959 (ebook) |
ISBN 9781032071510 (hardback) | ISBN 9781032071541 (paperback) |
ISBN 9781003205647 (ebook)
Subjects: LCSH: Science—Mathematical models—Philosophy. |
Mathematical models—Philosophy.
Classification: LCC Q175.32.M38 R68 2024 (print) | LCC Q175.32.M38
(ebook) | DDC 501/.1—dc23/eng/20240412
LC record available at https://lccn.loc.gov/2024015958
LC ebook record available at https://lccn.loc.gov/2024015959

ISBN: 978-1-032-07151-0 (hbk)
ISBN: 978-1-032-07154-1 (pbk)
ISBN: 978-1-003-20564-7 (ebk)

DOI: 10.4324/9781003205647

Typeset in Sabon
by codeMantra

CONTENTS

PART 5
Modeling in the wild 427

ACKNOWLEDGEMENTS

This work has been made possible by a grant from the European Research Council (ERC) under the European Union's Horizon 2020 research and innovation programme (Grant agreement No. 818772). We are grateful for this support.

We would like to express our sincerest gratitude to all the authors of the *Handbook*. It is thanks to your contributions that the *Handbook* has become a comprehensive and exciting collection on the philosophy of scientific modeling—a collection we both as editors and philosophers are proud to put our names on. We appreciate the amount of care and expertise that you have put into the individual entries investigating a myriad of topics related to scientific modeling from the perspectives of the philosophy of science, epistemology, metaphysics, and beyond. We also thank you for your patience and understanding during the editorial process of the *Handbook*—it has been a pleasure working with all of you!

Huge thanks also to our superb assistants at the University of Vienna, Meghan Bohardt, Konstantin Eckl and Alexander Gschwendtner: without your effort, managing a project of this scale would simply not have been possible. We also want to thank Tony Bruce and Adam Johnson from Routledge for the opportunity to produce this volume and for shepherding us through this process.

Finally, our heartfelt thanks, as always, belong to Sami, Pau, and Anni.

Tarja Knuuttila, Natalia Carrillo, and Rami Koskinen
Vienna, Austria, December 2023

CONTRIBUTORS

N. Emrah Aydinonat is a Docent and a University Researcher at TINT, Faculty of Social Science, University of Helsinki. His research focuses on the philosophy of economics, modeling, and scientific explanation.

Alessandra Basso is a Newton International Fellow at the Department of History and Philosophy of Science, University of Cambridge. Her research focuses on the philosophy of economics and social sciences, as well as the philosophy of measurement, particularly as it relates to morally laden concepts.

Alisa Bokulich is a Professor of Philosophy at Boston University and served from 2010 to 2023 as Director of the BU Center for Philosophy & History of Science. She is also an Associate Faculty member of the History of Science Department at Harvard University. Her research is primarily focused on scientific modeling, data, and explanation in the physical sciences, especially philosophy of the geosciences.

Mieke Boon is a Professor of Philosophy of Science in Practice at the University of Twente with a background in chemical engineering. Her research interests include philosophy of science for the engineering sciences including conceptual modeling in AI and machine learning technology, as well as science and engineering education.

Jarosław Boruszewski is a Researcher at the Faculty of Philosophy, Adam Mickiewicz University, Poznań. His research focuses on philosophy of information, logical semiotics, modeling, and humanities methodology.

Marcel Boumans is a Professor at Utrecht University, focusing on the history and philosophy of science. His main research interest is understanding empirical research practices in economics from a philosophy of science-in-practice perspective, the practices of measurement and modeling, and the role of mathematics in the social sciences.

Stefan Buijsman is an Assistant Professor of Philosophy of AI at TU Delft, focusing on explainability and the information that different stakeholders need to responsibly develop and use AI systems.

Natalia Carrillo is a Researcher in the Philosophy of Science and Technology at the National Autonomous University of Mexico (UNAM) with a background in mathematics. Her research interests include scientific modeling, metaphor and abstraction in science, and the epistemic role of artifactuality.

Alejandro Cassini teaches philosophy of science at the Department of Philosophy of the University of Buenos Aires. He does research on topics of general philosophy of science, philosophy of experimentation, and history and philosophy of modern physics.

Leticia Castillo Brache is a PhD candidate at Boston University. She is interested in the philosophy of climate science, paleoclimate proxies, and issues related to justice and values in climate adaptation and mitigation, especially for minority populations.

Sanjay Chandrasekharan is an Associate Professor at the Homi Bhabha Centre for Science Education, Tata Institute of Fundamental Research. His research interests include learning sciences, new computational media, science cognition, model-based imagination and reasoning in science and engineering, and philosophy of scientific modeling.

Adrian Currie is a Senior Lecturer at the Department of Sociology, Philosophy and Anthropology, University of Exeter. His research interests include philosophy of science, philosophy of biology, philosophy of historical science, social epistemology, history and philosophy of science, and creativity.

Juan M. Durán is an Assistant Professor at the Faculty of Technology, Policy and Management, TU Delft. His research focuses on the philosophy of science and ethics of computer-based science and engineering (computer simulations, AI, and big data).

Catherine Z. Elgin is a Professor of the Philosophy of Education at the Harvard Graduate School of Education. She is an epistemologist with an interest in aesthetics and the philosophy of science.

Roman Frigg is a Professor of Philosophy in the Department of Philosophy, Logic and Scientific Method at the London School of Economics. His research interests lie in the general philosophy of science and philosophy of physics.

Axel Gelfert is a Professor of Philosophy at the Technical University of Berlin. His research interests include social and applied epistemology, general philosophy of science, scientific models and scientific practice, history of philosophy, and philosophy of technoscience.

Till Grüne-Yanoff is a Professor of Philosophy at the KTH Royal Institute of Technology (KTH) in Stockholm. His research focuses on the philosophy of science, the philosophy of economics, and decision theory.

Catherine Herfeld is a Professor of Philosophy and History of Economics at the Leibniz University Hannover. Her research interests cover topics in history, methodology, and philosophy of economics. Currently, she is particularly interested in the questions of why and how models are transferred across different domains and in which way such model transfers can lead to scientific progress.

Wybo Houkes is a Professor of Philosophy of Science and Technology at Eindhoven University of Technology. His research interests include the philosophy of technical artifacts, theories of cultural evolution, the philosophy of scientific modeling, and analyses of technological knowledge.

Jennifer S. Jhun is an Assistant Professor of Philosophy at Duke University, as well as a Faculty Fellow at the Center for the History of Political Economy. She works mainly in the history and philosophy of science, especially of economics.

Tarja Knuuttila is a Professor of Philosophy of Science at the University of Vienna. She has developed a novel artifactual account of models. Knuuttila focuses, in her research, on scientific modeling, interdisciplinarity, and the modal dimension of science with a special focus on synthetic biology, engineering sciences, and economics.

Rami Koskinen is a Philosopher of Science at the University of Vienna with an interest in the general philosophy of science, philosophy of biology, and epistemology. He has been studying modal reasoning in the sciences, modeling in synthetic biology, and the question of multiple realizability.

Insa Lawler is an Assistant Professor of Philosophy at UNC Greensboro (UNCG). Her research focuses on the epistemology of scientific inquiry, with a focus on understanding and its relation to truth.

Johannes Lenhard holds the Heisenberg Professorship "Philosophy in Science and Engineering" at RPTU, Kaiserslautern. His research mainly addresses the following question: How does using computers change the methodology and epistemology of the sciences?

Arnon Levy is an Associate professor of philosophy at the Hebrew University of Jerusalem. He works in the philosophy of science and philosophy of biology. His research has focused on scientific explanation and modeling in biology. More recently, he has begun working on the role(s) of values in science.

Chia-Hua Lin is an Assistant Professor of Philosophy at Fairfield University. Her research focuses on topics in general philosophy of science with an emphasis on the interdisciplinary applications of modeling frameworks.

Chiara Lisciandra is an Assistant Professor at Utrecht University. Her primary research interests are general philosophy of science (models, explanations, interdisciplinary science), social philosophy (norms, game theory), and philosophy of economics (methodology of behavioral/experimental/development economics).

Andrea Loettgers is a Senior Researcher at the University of Vienna. She is a historian and philosopher of science with a background in physics. She has a longstanding interest in scientific modeling in physics, neurobiology, astrobiology, and systems and synthetic biology.

Sergio F. Martínez is a Philosopher at the National Autonomous University of Mexico (UNAM). His interests include general philosophy of science, naturalized philosophy of science, and modeling the co-evolution of material culture and social cognition as embodied in artifacts.

K. K. Mashood is a Reader at Homi Bhabha Centre for Science Education. His research interests include alternative conceptions in physics, concept inventories, modeling in science education, and conceptual change.

Mary S. Morgan is the Albert O. Hirschman Professor of History and Philosophy of Economics at the Department of Economic History, London School of Economics. Her research interests include history, philosophy, and sociology of science focused on economics and statistics, models, measurements, experiments, observations, and "traveling facts."

Nancy J. Nersessian is the Regents' Professor Emerita at the Georgia Institute of Technology and a Research Associate at the Department of Psychology, Harvard University. Her research focuses on creativity, innovation, and conceptual change in science, especially the creation of novel modeling practices in science and bioengineering.

Krzysztof Nowak-Posadzy is a Researcher at the Faculty of Philosophy, Adam Mickiewicz University, Poznań. His research focuses on philosophy of economics, cultural semiotics, and humanities methodology.

Patricia Palacios is an Associate Professor in philosophy of science at the University of Salzburg. Her areas of specialization are general philosophy of science, philosophy of complexity sciences, and philosophy of physics.

Wendy S. Parker is a Professor of Philosophy at Virginia Tech. Her research focuses on topics in general philosophy of science and philosophy of climate science/meteorology.

Francesca Pero is a Philosopher of Science at the University of Florence. Her research interests include scientific representation and modeling.

Ilkka Pättiniemi is a Philosopher of Science with a background in theoretical physics. He has worked mainly on the philosophy of physics and the foundations of quantum mechanics, which has led him to questions regarding the epistemology of modality in science and philosophy.

Christopher Pincock is a Professor of Philosophy at the Ohio State University. His research focuses on the philosophy of science, the philosophy of mathematics, and the history of analytic philosophy.

Charles Rathkopf is a Permanent Research Associate in Philosophy and Neuroscience at the Jülich Research Center and a lecturer at the University of Bonn. He researches the philosophy of mind and the philosophy of science, especially on topics relating to neuroscience and artificial intelligence.

Collin Rice is an Assistant Professor in the Philosophy Department at Colorado State University. His primary areas of research are on the use of idealizations and modeling in scientific practice and the nature of scientific explanation and understanding (especially in biology).

Kristina Rolin is a University Lecturer in Research Ethics at Tampere University. Her main areas of research are philosophy of science, social epistemology, and research ethics. She is interested in the role of values in science—including social science, medicine, engineering, scholarship in the humanities, epistemic injustice, trust, epistemic responsibility, and collective knowing.

Julia Sánchez-Dorado is a Postdoctoral Researcher in the area of Logic and Philosophy of Science at the University of Sevilla (Spain) and a Fellow at the Institute for Cultural Inquiry, Berlin. Her research interests include modeling practices in geosciences and engineering, the problems of creativity and abstraction, and the relationship between art and science.

Dunja Šešelja is a Professor of Philosophy at the Institute for Philosophy II, Ruhr University Bochum, and serves as a co-editor in chief of the European Journal for Philosophy of Science. Her areas of expertise are social epistemology and philosophy of science, in particular, formal modeling of inquiry, social epistemology of scientific disagreements, and integrated history and philosophy of science.

Ylwa Sjölin Wirling is a Researcher at the University of Gothenburg and a Pro Futura Scientia Fellow at the Swedish Collegium of Advanced Study. She works on epistemology, philosophy of science, and metaphilosophy.

Rawad el Skaf is an Assistant Professor (RTD-A) in Logic and Philosophy of Science in the Department of Mathematics at the Politecnico di Milano. His research interests concern topics in the history and philosophy of science, especially surrogative reasoning and scientific tools, beyond (direct) laboratory experiments.

Michael T. Stuart is a Lecturer in Philosophy at the Department of Philosophy, University of York. He works on the philosophy of science, scientific imagination, and artificial intelligence.

Mauricio Suárez is a Professor in Logic and Philosophy of Science at the Complutense University of Madrid, a life member of Clare Hall Cambridge, and a research associate at the Centre for Philosophy of Natural and Social Science, LSE. His main research interests lie in the history and philosophy of natural science (physics, chemistry, and the life sciences).

Eran Tal is an Associate Professor of Philosophy at McGill University and Canada Research Chair in Data Ethics. He specializes in the philosophy of science, the philosophy of measurement, and the ethics of big data and artificial intelligence.

Krist Vaesen is an Associate Professor in the Philosophy of Innovation at Eindhoven University of Technology. His current research interests include theories of cultural and technological evolution, foundational issues in human origins research, the philosophy of scientific models (e.g., models of innovation), scientific pluralism, science and research policy, and the history of 20th-century Anglo-American philosophy.

Rasmus Grønfeldt Winther is a Professor of Humanities at the University of California, Santa Cruz, and an Affiliate Professor of Transformative Science in the GeoGenetics Section of Globe Institute at the University of Copenhagen. He works primarily in the history and philosophy of science, biology, and race, and has interests in history, literature, anthropology, feminism, and multiculturalism.

INTRODUCTION

Scientific models in the philosophy of science

Tarja Knuuttila, Natalia Carrillo and Rami Koskinen

Modeling cuts across sundry scientific practices, contributing to theorizing, experimentation, prediction, measurement, scientific instrumentation, and science education. Beyond the sciences, modeling plays a crucial role in citizen engagement with science and public policy decision-making. It plays a major role in the efforts to address the huge challenges of the 21st century, including but not limited to climate change, shortage of natural resources, loss of biodiversity, and economic forecasting in increasingly unforeseeable situations. The diversity of scientific models is astounding; side-by-side mathematical and scale models, technological advances such as rapidly expanding big data, computational and synthetic approaches, and generative AI are pushing modeling toward new frontiers, redefining the epistemic agency between humans and scientific instruments. A discussion of what we can achieve through modeling, and how we should manage model-based practices, is critical for ensuring a good and responsible use of this epistemic resource. The chapters of *The Routledge Handbook of Philosophy of Scientific Modeling*, written by experts in various areas of the philosophy of science, seek to provide enduring philosophical insights and useful analyses for understanding modeling in its multiplicity.

The philosophical discussion on modeling

The philosophical interest in modeling within the philosophy of science has heterogeneous beginnings, testifying to a variety of theoretical, formal, and practical aspirations that appear to have different goals. While scientists such as Maxwell, Thomson, Helmholtz, Hertz, and Boltzmann addressed mechanical models and analogies in the 19th and beginning of the 20th centuries, the philosophical discussion started to bloom first in the 1950s, only to explode at the turn of the 21st century. It seems fair to say that within the philosophical discussion for most of the 20th century, models remained subordinate to theories. In the last decades, however, the situation has definitely changed. In the present philosophical discussion, modeling now occupies the center stage, even to the extent that Morrison (2007) has asked: where have all the theories gone?

Already Wartofsky (1966) paid attention to what he called the "model muddle," referring to the proliferation of a wildly heterogeneous assortment of things called models—both

DOI: 10.4324/9781003205647-1

within the sciences as well as within the philosophy of science. In scientific research, especially technological developments have added new kinds of inhabitants to the ever-increasing diversity of models, including, among others, mathematical models, scale models, general circulation models, agent-based simulations, network models, model organisms, large language models, and synthetic models. Philosophers, for their part, have taken notice of the growing importance and diversification of scientific models, coining and analyzing an amazing assortment of model types, such as idealized models, toy models, minimal models, exploratory models, analog models, fictional models, and caricature models, to name just a few (for a more comprehensive list and discussion, see Frigg 2023 and Frigg and Hartmann 2020). There have been several attempts to domesticate the wilderness of model kinds and types through categorization, such efforts extending also to the question of the ontology of models (e.g., Black 1962, Achinstein 1968, Weisberg 2013, Gelfert 2016, Frigg 2023). However, as the different model types identified by philosophers typically address some particular uses of models, the question concerning the ontology of models tends to intersect with their functional qualities (Gelfert 2016, Frigg 2023).

This *Handbook* introduces the amazing variety of scientific models and the rich philosophical discussion on them. It also addresses other philosophical topics that relate to modeling, both long-established and more recent, contributing not only to our knowledge of modeling but also to those topics themselves. Before going into the contents of the *Handbook*, we will shortly discuss the two main ways in which models have been comprehended within the philosophy of science.

Syntactic and semantic views on theories and models

The very different ways in which models have been approached within the philosophy of science may puzzle a newcomer to the field. On the one hand, there have been attempts to establish what scientific models *are*, within a formal framework (Bailer-Jones 1999, 32). The syntactic and semantic views on theories, inspired by mathematical logic, are both attempts of this kind, although the place of models in them is quite different. On the other hand, the present discussion of modeling tends to focus on the pragmatic and cognitive roles of models in scientific enterprise, without any explicit interest in defining models.

According to the syntactic view of theories, promoted by logical empiricists, a scientific theory is an uninterpreted or partially interpreted formalism, a syntactic structure consisting of a set of axioms and theorems. The axioms would be formulations of scientific laws, specifying the relationships between scientific terms. The theory, as a syntactic structure, is explicated in terms of its logical form. To interpret such a theory would be to specify a model for it, which makes all the axioms of the theory true. The interpretation provided by a model could supply, as Ernest Nagel put it, some flesh to the skeletal structure in terms of more or less familiar conceptual or visualizable materials (Nagel 1961, 90).

The semantic conception of models contested this "linguistic" view of theories, replacing the focus on the syntactic formulation of the theory and starting rather with the theory's models, which are non-linguistic entities. According to the semantic view, theories are not assemblages of propositions or statements but families of models that can be described or characterized by a number of different linguistic formulations (Suppe 1977, 221). These models would be akin to models in mathematical model theory: Suppes (1961, 65) suggested that the "meaning of the concept of model is the same in mathematics and in the empirical sciences." The semantic approaches consider models as structures that can be defined either by the use of set-theoretical predicates (e.g., Suppes 1961) or by the use of suitable mathematical language (van Fraassen 1980). Van Fraassen's version of the semantic approach comes closer to physical theories, considering non-relativistic theories in terms of systems of physical entities developing in time. A cluster of models is united by a common state space with a domain of objects and their trajectories in that space.

Within the semantic approach, especially Ronald Giere focused on scientific modeling, his account becoming increasingly aligned with practice-oriented approaches (e.g., Giere 1999; for a discussion on practice-oriented approaches, see below). While he considered scientific models to be abstract entities, he did not think that the concept of a model from formal logic and mathematics was suitable for scientific practice. Giere (1988) developed his account of models on the basis of classical mechanics as presented in advanced textbooks. He proposed that, for example, the "linear oscillator" is a cluster of models of varying degrees of specificity. Such models are not true or false with respect to the world; the role of the theory is rather to claim a "good fit" between the models and some important types of real systems. Consequently, for Giere, in contrast to most adherents of the semantic conception, the relationship between models and their target systems is not primarily that of isomorphism (or some other kind of structure-preserving mapping) but similarity. Also, the links between the models of a theory are relations of similarity, since according to Giere, nothing in the structure of a model itself determines whether it belongs to a given family of models or not. It would be up to the scientific community to judge whether the resemblance is sufficient.

That a conception of theories should provide an approach to scientific models already seems somewhat paradoxical at the outset and has been challenged by philosophers studying modeling from the perspective of scientific practice. In response, some adherents of the semantic approach have responded by invoking the so-called "partial structures" view. It addresses the concern that several kinds of models used in science are not set-theoretical models, but instead material or iconic (French and Ladyman 1999). The partial structures approach seems to accommodate such possibilities as it requires only partial isomorphisms between the model and the modeled (see e.g., Bueno 1997, da Costa and French 2003). However, French and Ladyman (1999) also claim that it is important to keep in mind that what is at stake is whether the set-theoretical account can adequately describe scientific models used in scientific practice. Consequently, the philosopher "'represents' the theory at the meta-level of the philosophy of science in terms of set theory and also 'represents' the way the theory latches onto the world via the formal notion of (partial) isomorphism" (French 2017, 3324). This is a different aim than what motivates those philosophers who are interested in the models that are constructed and used in actual scientific practices. It also implies that the notion of a model in mathematical logic is not the notion that is employed by working scientists.

A practice-oriented approach to models

The philosophical discourse surrounding models has traditionally been driven by practical concerns. Even those who advocated for a semantic view of theories, such as van Fraassen (1980), perceived their approach as offering a more realistic account of theories. Instead of reconstruction, however, the practice-oriented approach focused on different aspects of actual scientific practice. During the 1950s and 1960s, the examination of models gained traction within the philosophy of science as several scholars addressed topics such as theory reconstruction, theory change, and scientific discovery (Bailer-Jones 1999, 31). Achinstein (1968), Black (1962), Hesse (1966), and Hutten (1954) drew comparisons between models and analogies or metaphors in their efforts to comprehend the functioning of models in scientific reasoning. Black and Achinstein developed taxonomies of models with the aim of capturing the range of models employed in scientific practice. Numerous subjects and issues addressed throughout this early philosophical discourse on models remain pertinent still today. One central contribution was also methodological. Already Hutten (1954) anticipated the importance of case studies and the exploration of specific models. He recommended that philosophers "follow the scientists" as closely as possible in order to "avoid forcing science into a pre-conceived scheme" (81) and "illustrating [the] scientific method by means of old-fashioned and very simplified examples" (284). The current discourse on modeling has indeed adhered to this recommendation, with the influential collection *Models as Mediators* (Morrison and Morgan 1999) playing a significant role in this development (see below).

The physicists of the 19th century discussed mechanical models, either concretely constructed or imagined, functioning as illustrations or "working models" that would provide mechanical analogies to the physical phenomena of interest (Boltzmann 1902, Hon and Goldstein 2021). Likewise, Black (1962) and Achinstein (1968) started by considering three-dimensional physical objects, which Black thought were the "standard cases" of models in the literal sense of the word. Achinstein paid attention to the manipulability or "workability" of physical models (which he called representational models). According to him "representational models, although used in all the sciences, are particularly central in engineering. Instead of investigating an object directly, the engineer may construct a representation of it, which can be studied more readily" (1968, 209). Morrison and Morgan (1999) emphasized the importance of manipulability, extending it to models more generally. Already in 1953, Hesse (1953) claimed that mathematical formalisms may be thought of as models and that they functioned in much the same manner as mechanical models.

Morrison and Morgan's view of models as mediators that serve as investigative tools draws on this prior practice-oriented tradition, as well as Nancy Cartwright's work. Models occupy the middle ground between theory and the world (or data), in both Cartwright's account of models as bridges and Morrison and Morgan's models as mediators. Cartwright (1983) used models to argue that the fundamental laws of physics do not describe natural regularities. She believes that there is a disconnect between general theoretical physics principles and the messiness and complexity of facts that phenomenological laws seek to convey. Models are tasked with filling that gap:

> The route from the theory to reality is from theory to model, and then from model to the phenomenological law. The phenomenological laws are indeed true of the objects of reality—or might be; but the fundamental laws are true only of objects in the model.
>
> *(Cartwright 1983, 4)*

For a model to function as a bridge between theory and data, it has to include some genuine properties of the objects modeled. Yet models also contain traits of convenience and fiction. Morrison and Morgan (1999) also emphasize the incorporation of "additional elements" into models. This is precisely what allows models to connect disparate realms, but it is also what allows them to be "at least" partially autonomous. Models can operate as investigative instruments precisely because of their partial autonomy.

The crucial contribution of Morrison and Morgan's (1999) approach is their attention to scientists' active engagement with models. Morrison and Morgan scrutinize it from the standpoints of construction, functioning, representing, and learning. Models have traditionally been considered representations, a perspective that has in fact been shared by both the structuralist and many practice-oriented approaches to models. However, Morrison and Morgan's notion of representation makes it less a truthful depiction that corresponds to some particular natural or social system than an ongoing investigation of both theories and empirical findings (see also Wimsatt 2007). Morrison and Morgan's emphasis on learning is critical; scientists learn by building and manipulating models. Weisberg (2013) also highlights the independent nature of models. He considers modeling a particular kind of theoretical practice that is characterized by indirect representation with which he refers to how modelers construct and analyze hypothetical systems, i.e., models, without necessarily attempting to establish any links between them and some determinable real-world systems.

The emphasis on modeling as a particular theoretical strategy raises the question of its historical antecedents (Godfrey-Smith 2006). According to Hon and Goldstein (2021), Maxwell's concept of a "working model" indicates a shift toward a modeling approach. The working model, according to Maxwell, was "a medium capable of producing the mechanical phenomena observed" (Maxwell [1859/1890] 1965, 162). According to Hon and Goldstein, Maxwell's working model was not just a tool for understanding but also a research instrument for examining the imagined process at the microlevel. Suárez (2024) also traces the origin of what he calls the "modeling attitude"—a set of methodological commitments and style of inquiry—to the 19th century. For Suárez, the modeling attitude consists of the construction of analogical, idealized, fictional, or artifactual scenarios within models, for multiple purposes. Giorgio Israel (1993), a historian of science, places the origin of the modeling approach in the early 20th century when a new notion about the relationship of mathematics to reality was born. The classical notion of the uniqueness of mathematical representation gave way to the modeling approach, which employs the same abstract mathematical representations across a wide range of domains (see also Knuuttila and Loettgers 2023). Modeling endeavors of this type concentrate on formal structures capable of describing a wide range of isomorphic occurrences or similar patterns. Hon and Goldstein, and Suárez, as well as Israel, may have various types of models in mind, but one thing they appear to agree on is the exploratory aspect of modeling (Gelfert 2016, Massimi 2022).

The contents of the *Handbook*

The *Handbook* contains a total of 40 chapters, all specially written for the present collection by leading scholars from around the world. The chapters are divided into five thematic parts that go from the general to the particular:

1 Historical and General Perspectives
2 Philosophical Accounts of Modeling
3 Methodological Aspects: Model Construction, Evaluation, and Calibration
4 Related Topics
5 Modeling in the Wild

Part 1, Historical and General Perspectives, places the philosophical discussion of modeling in perspective. First, it offers a historical overview of how the modeling attitude emerged in 19th-century physics, addressing how these scientists interpreted the utility of models for scientific study. Second, the link between theories and models is investigated: how this relationship was interpreted across semantic and syntactic accounts and how models are perceived to relate to theories in contemporary debates. Third, the practice-oriented approaches that pushed modeling to the forefront of the philosophy of science are examined. These three historical viewpoints set the stage for the *Handbook*'s subsequent themes.

Part 2, Philosophical Accounts of Modeling, addresses a variety of subjects central to the contemporary philosophical debates on modeling. The crucial question is the epistemic role of models. Models are commonly regarded as representations, but what this implies and whether the representational viewpoint is sufficient to account for models' varied epistemic roles is debatable. Other philosophers choose to begin on a different note, arguing that models may represent, but that does not always explain their epistemic value for scientific endeavors. The proposal that models should be viewed as epistemic artifacts is an example of such an account. A related epistemological topic is idealization, traditionally dealing with the problem of why scientists misrepresent features of the target systems in their models. Idealization appears to call into question the representational perspective on models, and de-idealization offers one answer. While de-idealization is sometimes associated with idealization, it also emerges as a topic of its own. In the next entry, target systems are attended to. They are assumed to be what models are about, and how they are retrieved or constructed is an important aspect of modeling that has received less attention. Minimal models and computer simulations are discussed towards the end of Part 2. This part concludes by discussing two classical philosophy of science topics that are also connected to modeling: scientific laws and explanation.

Part 3, Methodological Aspects: Model Construction, Evaluation, and Calibration, investigates a variety of methodological issues encountered in modeling practices. The first chapter discusses the concept of robustness and how it pertains to normative modeling considerations. This leads to a debate on model evaluation—how to determine whether a model is adequate for a given purpose. Another critical concern is how scientists determine which mathematical forms to utilize in their models. The role of models in statistical inference is also addressed. The final topic covered is model transfer, which describes how and by what means formalized models can be moved and applied from one scientific domain to another.

Part 4, Related Topics, examines philosophical issues that are pertinent but not exclusive to modeling. First, the notions of representation-as and exemplification as they apply to modeling are addressed. Two key philosophical debates on models are discussed in the chapters that follow: (1) whether understanding is factive or not, and how various answers to this question affect the epistemological role of models, and (2) to what extent and how models might provide us with access to possibilities. Moreover, models have been compared to a variety of objects, and our understanding of them has grown as a result. This part of the *Handbook* discusses how models relate to and differ from thought experiments, maps, metaphors, and narratives, and it recovers the origins and evolution of these comparisons. Subsequently, the crucial issue of how values are incorporated into modeling is tackled. Interdisciplinarity and modeling, as well as learning through modeling, are discussed in the final two chapters.

While the chapters in Parts 1–4 cover a wide range of issues that have been discussed in the context of modeling in general, the purpose of the final Part 5, Modeling in the Wild, is to explore and examine different types of models. The chapters in this part address several epistemological, ontological, and methodological challenges related to modeling *in situ*. The cases studied include modeling in statistics, climate science, machine learning, biomedical and engineering sciences, synthetic biology, paleosciences, economics, formal language theory, and neuroscience. The question of why network models can be applied to such a wide range of natural and social phenomena is also discussed, relating to the pervasiveness of certain formal or other templates in modeling practices more generally.

The increasing prominence of modeling in science has generated considerable momentum, making the future of scientific modeling highly exciting. This is particularly evident in the transformations of modeling due to advancements in technology, encompassing improved experimental equipment, enhanced processing capacity, and novel computational approaches, including generative AI. We believe that the massive crowdsourcing effort represented by *The Routledge Handbook of Philosophy of Scientific Modeling* will continue to contribute pointers, philosophical insights, and valuable analyses for understanding modeling in the time to come.

References

Achinstein, Peter. 1968. *Concepts of Science*. Baltimore: Johns Hopkins Press.

Bailer-Jones, Daniela M. 1999. "Tracing the Development of Models in the Philosophy of Science." In *Model-Based Reasoning in Scientific Discovery*, edited by Lorenzo Magnani, Nancy J. Nersessian, and Paul Thagard, 23–40. Boston, MA: Springer US.

Black, Max. 1962 *Models and Metaphors: Studies in Language and Philosophy*. Ithaca, NY: Cornell University Press. https://doi.org/10.7591/9781501741326.

Boltzmann, Ludwig. 1902. "Model." *Encyclopaedia Britannica*. 10th Edition, 788–791. Cambridge: Cambridge University Press.

Bueno, Otavio. 1997. "Empirical Adequacy: A Partial Structure Approach." *Studies in the History and Philosophy of Science* 28: 585–610.

Cartwright, Nancy. 1983. *How the Laws of Physics Lie*. Oxford: Oxford University Press.

Costa, Newton C. A. da, and Steven French. 2003. *Science and Partial Truth: A Unitary Approach to Models and Scientific Reasoning*. 1st edition. Oxford; New York: Oxford University Press.

van Fraassen, Bas C. 1980. *The Scientific Image*. Clarendon Library of Logic and Philosophy. Oxford: Oxford University Press.

French, Steven. 2017. "(Structural) Realism and Its Representational Vehicles." *Synthese* 194(9): 3311–3326.

French, Steven, and James Ladyman. 1999. "Reinflating the Semantic Approach." *International Studies in the Philosophy of Science* 13(2): 103–121.

Frigg, Roman. 2023. *Models and Theories. A Philosophical Inquiry*. London: Routledge.

Frigg, Roman, and Stephan Hartmann. 2020. "Models in Science." *The Stanford Encyclopedia of Philosophy*. http://plato.stanford.edu/entries/models-science/.

Gelfert, Axel. 2016. *How to Do Science with Models: A Philosophical Primer*. SpringerBriefs in Philosophy. Springer International Publishing. https://doi.org/10.1007/978-3-319-27954-1.

Giere, Ronald N. 1988. *Explaining Science: A Cognitive Approach*. Chicago: University of Chicago Press.

———. 1999. *Science without Laws*. Chicago and London: The University of Chicago Press.

Godfrey-Smith, Peter. 2006. "The Strategy of Model-Based Science." *Biology and Philosophy* 21(5): 725–740. https://doi.org/10.1007/s10539-006-9054-6.

Hesse, Mary B. 1953. "Models in Physics." *The British Journal for the Philosophy of Science* 4(15): 198–214.

———. 1966. *Models and Analogies in Science*. Notre Dame, IN: University of Notre Dame Press.

Hon, Giora, and Bernard R. Goldstein. 2021. "Maxwell's Role in Turning the Concept of Model into the Methodology of Modeling." *Studies in History and Philosophy of Science* 88: 321–333. https://doi.org/10.1016/j.shpsa.2021.03.010.

Hutten, E. H. 1954 "The Role of Models in Physics." *British Journal for the Philosophy of Science* 4: 284–301.

Israel, Giorgio. 1993. "The Emergence of Biomathematics and the Case of Population Dynamics A Revival of Mechanical Reductionism and Darwinism." *Science in Context* 6(2): 469–509.

Knuuttila, Tarja, and Andrea Loettgers. 2023. "Model Templates: Transdisciplinary Application and Entanglement." *Synthese* 201: 200. https://doi.org/10.1007/s11229-023-04178-3

Massimi, Michela. 2022. *Perspectival Realism*. New York: Oxford University Press.

Maxwell, J. C. [1859/1890] 1965. "On the stability of the motion of Saturn's rings." Macmillan: London. Reprinted in The scientific papers of James Clerk Maxwell.W.D. Niven (ed.), 2 vols. Cambridge: University Press. Reprinted, two volumes bound as one. New York: Dover, 1: 288–376.

Morgan, Mary S., and Margaret Morrison, eds. 1999 *Models as Mediators: Perspectives on Natural and Social Science*. Cambridge; New York: Cambridge University Press.

Morrison, Margaret. 2007. "Where Have All the Theories Gone?" *Philosophy of Science* 74(2): 195–228.

Nagel, Ernst. 1961. *The Structure of Science - Problems in the Logic of Scientific Explanation*. London: Routledge & Kegan Paul.

Suárez, Mauricio. 2024. *Inference and Representation: A Study in Modeling Science*. Chicago University Press.

Suppe, Frederic. 1977. *The Structure of Scientific Theories*. 2nd edition. Urbana: University of Illinois Press.

Suppes, Patrick. 1961. "A Comparison of the Meaning and Uses of Models in Mathematics and the Emprical Sciences." In *The Concept and the Role of the Model in Mathmatics and the Natural and Social Sciences*, edited by Hans Freudenthal, 163–177. Dordrecht: Reidel.

Wartofsky, Marx W. 1966. "The Model Muddle: Proposals for an Immodest Realism." In *Models: Representation and the Scientific Understanding*, edited by Marx W. Wartofsky, 1–11. Boston Studies in the Philosophy of Science. Dordrecht: Springer Netherlands. https://doi.org/10.1007/978-94-009-9357-0_1.

Weisberg, Michael. 2013. *Simulation and Similarity: Using Models to Understand the World*. Reprint edition. Oxford: Oxford University Press.

Wimsatt, William C. 2007. *Re-Engineering Philosophy for Limited Beings*. Cambridge, MA: Harvard University Press.

PART 1

Historical and general perspectives

1
THE EMERGENCE OF THE MODELLING ATTITUDE

Mauricio Suárez

1. A History of the Modelling Attitude

An 'attitude', or a 'stance', is a set of loose methodological and heuristic commitments, a style of doing science. It is not a thesis or a set of propositions explicitly defining the nature of science or its aim (Chakravartty 2004; Rowbottom 2011). The modelling attitude is the mode of scientific work that relies on the construction, development, and application of models; it does so to achieve the plurality of aims pursued by science. It need not be defined as a thesis about scientific knowledge: it is merely a methodological stance, a commitment to a mode of work.

Philosophical discussions about stances or attitudes are by now, of course, rather entrenched, and postulating a stance, or attitude, in the study of the nature and aims of science is a respected view. Arthur Fine (1984/1987) proposed a natural ontological attitude, and Bas Van Fraassen (2002) advanced an empirical stance. Both intended their views as viable hermeneutics in a project of understanding science. The aim of this chapter is more modest: it aims to defend that a large part of the present-day scientific work in the physical sciences answers to a 'modelling attitude'. It does not claim that this is the (only) hermeneutics suitable for natural science, or science in general; in fact, it makes no claims regarding the appropriate interpretational stance on science, taken as a whole. Rather, it approaches stances and attitudes as primarily part of the scientists' own methodological practices and only derivatively sees them as informing philosophical debates and narratives. Just as philosophical realism is born out of internal scientific disputes regarding the atomic hypothesis, so is the modelling attitude born out of scientific modelling methodology. Moreover, both are interconnected fin-de-siècle developments.

Indeed, the modelling attitude has a history, (Suárez, 2014, 2024) which sees it emerge in full force in the nineteenth century, in the wake of both British Victorian physics and the German theory of models or *Bildtheorie*. The main contention of this chapter is that there are interesting insights in this history that are relevant to the contemporary debate regarding modelling and the nature of representation. The story commences at a perhaps unsuspected place and time, the Scottish Enlightenment at the beginning of the nineteenth century.

 DOI: 10.4324/9781003205647-3

2. The 'Relativity of Knowledge' in the Scottish Enlightenment

The roots of 'analogy' and its use in British Victorian nineteenth-century science lie in the Scottish Enlightenment (see Davie 1961; Olson 1975; Harman 1998; Siegel 1991; Smith and Wise 1989). They can be located more precisely in some common-sense philosophical views regarding the nature of knowledge that derive from the practice of mathematical abstraction. Outstanding amongst this is the so-called relativity of knowledge, a thesis regarding the comparative nature of knowledge (hence in no way related to our contemporary forms of epistemic relativism).

The Scottish abstract school of mathematics was in many ways shaped over the generations by Robert Simson's (1756) commentary on Euclid's Elements – a book that went through many editions and was in print in the US until the end of the nineteenth century. In a much-discussed passage in the book, Simson develops the concept of a surface by abstraction, a process carried out entirely in the mind. First, consider a solid geometric object in physical space shaped as a rectangular block. Then, imagine the solid block divided into two halves, right down the middle. Had the surface in between any thickness, it would belong to either half. Yet, it cannot be part of either half because, if we imagine that half being removed, the surface still exists in the remaining half. By reduction, it follows that the surface has no thickness and belongs to neither half – it is rather an abstraction. We apprehend the nature of a plane, or surface, only when we split the real block in our mind, into two imaginary situations, and compare them. We can continue this process of abstraction to generate cognate results regarding one-dimensional lines as the intersection of planes and non-dimensional points as the intersection of lines.

While the nature of mathematical abstraction is involved and has roots in medieval concepts and doctrines that cannot be discussed here (see Davie, 1961, 127–149), one feature stands out for our purposes. The method of abstraction is a way to infer a result about a real physical object and its properties based on a piece of reasoning that is carried out in some imaginary situations involving this object. The analogy, or comparison, between such imaginary situations yields knowledge of the nature of the object or its properties. While this is a method envisaged for abstract mathematical (geometrical) properties, it is not hard to see how it could be implemented to obtain empirical knowledge regarding physical properties too.

The method of abstraction was one central ingredient in the intellectual milieu that saw William Thomson (Lord Kelvin, 1824–1907) and James Clerk Maxwell (1831–1879) develop analogy as a method for scientific discovery. The other key ingredient was the cognate thesis in Scottish common-sense philosophy that all knowledge is the result of apt comparison, the so-called "relativity thesis" (Davie 1961; Olson 1975, chap. 12; see also related discussions in Harman 1991, chap. 2). This opposed atomistic theories of knowledge, according to which knowledge can be exclusively of a given object. The Scottish common-sense tradition emphasised the way all knowledge of an object is the product of a comparison of that object with something else. Thus, the only means to achieve genuine knowledge of the world necessarily involves likeness, comparison, or analogy. The word 'analogy' was common in the nineteenth century, and its use is widespread even now (as a simple Google search shows) but, as we shall see, its meaning shifts into what we nowadays refer to as 'model'. By the time Boltzmann writes his 1902 entry for the Encyclopaedia Britannica, he has no need for 'analogy' and employs 'model' instead. A genealogical study of 'model' thus turns out what was considered a method, and an activity, involving analogical reasoning. It is a contention of this chapter that this genealogy is not a mere accident, but the history of the modelling attitude informs our current modelling methodologies (as well as, arguably, other features of our contemporary scientific culture) and merits philosophical attention.

3. Kelvin, Maxwell, and the Uses of Analogy

James Clerk Maxwell (1831–1879), Edinburgh-born and educated at its Academy and University, completed three full courses until he left for Cambridge in 1850. William Thomson (1824–1907) grew up in Glasgow and was linked to the city throughout his life. In 1892, he was elevated Baron Kelvin after the river that runs through the city and university. Maxwell was mentored by the physicist James David Forbes and the philosopher Sir William Hamilton, within the broad-based liberal Scottish educational system. Thomson was taught by his father, the reformist mathematician James Thomson, and the radical professor of astronomy John Pringle Nichol, who in turn had been trained at Aberdeen's King's College. All of these are habitual localities in the history of Scottish common-sense philosophy and abstract mathematics, and all mentors and tutees were willing partakers in both traditions.

Maxwell, in particular, was strongly imbued with the relativity thesis, including Thomas Reid's tenet that analogical reasoning was an unavoidable – however regrettable, in Reid's view – component of scientific reasoning (Olson 1975, chaps. 2 and 3). He went on to develop his own philosophical views in a paper delivered in 1856 at the Apostles in Cambridge (Maxwell 1856b/1890). The paper is a disquisition on the nature of analogy, and it shows that the term had a somewhat more general meaning than we ascribe it today, rather closer to our current generic notion of 'model' (see Cat 2001, for an insightful account of analogy and metaphor in Maxwell's thought). His central question concerns whether analogies are in mind or nature. This would nowadays be rendered as a question regarding whether models are realistic renditions of their targets or not. His response is revealing. Maxwell acknowledges that there exist objects endowed with properties and holding an array of properties and relations to each other. In our conventional contemporary terms, he is thus a kind of metaphysical realist. Yet, he also claims there to be a distinct kind of necessity that applies to thoughts – there are laws amongst thoughts that can only be said to apply to objects by means of some comparison or likeness. This induces a method for surrogative reasoning, which, according to Maxwell, is a typical inclination of any student of analogy ('modeller'): "Whenever they [men] see a relation between two things they know well, and think they see there must be a similar relation between things well known, they reason from the one to the other" (Maxwell 1856b/1890, 382).

We can take this to be a statement for the modelling attitude in the Victorian era. The mechanical models of the aether so dear to 'the Maxwellians' (Hunt 1991) are fine examples of Maxwell's view of analogy as reasoning via the perceived shared relations amongst distinct systems of objects. Mechanical models were taken to bear informative likenesses to the electromagnetic aether, and they were thus employed by Victorian physicists such as George Francis Fitzgerald, Oliver Heaviside, or Oliver Lodge to infer a diverse range of properties of electromagnetic radiation. Maxwell even took care to fill in the concept of reasoning employed as follows: "A reason or argument is a conductor by which the mind is led from a proposition to a necessary consequence of that proposition" (1856/1890, 379). As we shall see, the notion of a 'conductor' (itself a useful analogy) turns out to be critical to the development of a modelling attitude in the nineteenth century.

Maxwell himself famously put all these ideas to use in his development of Faraday's experimental findings on electromagnetic induction in a full electromagnetic theory, culminating in his celebrated *Treatise on Electricity and Magnetism* (Maxwell 1873). This development essentially took place at Cambridge, where Maxwell moved in 1850 for further studies, graduating in 1854, as a Trinity College fellow. He would return to Cambridge in 1871 as the new Cavendish professor, after short hiatuses at Aberdeen and King's College London. Thomson had also been a graduate student at Cambridge a decade earlier, and Cambridge provided both men with formidable formal skills through its mathematics tripos.[1] It was exposure to Cambridge that turned them into what Siegel (1991) calls 'deep theory modellers'. In Scottish common-sense philosophy, analogy is essentially a heuristic for research and discovery: however, one that could mislead if taken at face value. Analogy is to be employed, but not to be trusted too much, and Reid in particular disparaged against any realist interpretation. Under the influence of John Herschel's theory of errors and William Whewell's consilience of induction, both Maxwell and particularly Thomson became wedded to a more realistic form of analogy relying on classical mechanics.

This is perhaps best exemplified in Maxwell's two most important contributions on the road to a comprehensive electromagnetic theory. In the earlier 'On Faraday's Lines of Force' (Maxwell 1856a/1890), Maxwell exhibits a characteristically 'Scottish' attitude: he compares electrical and magnetic phenomena with the flow of an incompressible fluid through a porous medium, and he uses the comparison merely as a provisional template for investigating such phenomena. Anticipating a role for fictional assumptions in science, Maxwell even claims that the incompressible fluid is 'imaginary':

> The substance here treated of must not be assumed to possess any of the properties of ordinary fluids except those of freedom of motion and resistance to compression. It is not even a hypothetical fluid [...] It is merely a collection of imaginary properties [...]. The use of the word 'Fluid' will not lead us into error, if we remember that it denotes a purely imaginary substance.
>
> *(Maxwell, 1856a/1890, 160)*

Partly inspired by Thomson's (1847) and Rankine's (1855) molecular vortices theory of elasticity, Maxwell's attitude changed in the years leading up to 1861. Analogy became more than merely a useful heuristic. It developed into a magnifying glass for probing into the world, a window on the underlying laws of apparently detached and distinct phenomena. By the time he published 'On Physical Lines of Force' (Maxwell 1860/1890), the analogical source itself had changed: rather than modelling the induction in currents as a flow, the aether was then represented as molecular vortices in rotational motion, in terms of the famous vortices and idle-wheels model. The tiny counter-rotating 'idle-wheels' were introduced to account coherently within mechanics for such rotational motion (see the famous figure 2 in plate VIII in Maxwell's 1860/1890). This model is a mixture of heuristically useful assumptions, such as the idle-wheels, and what Maxwell called 'real' analogies, namely the molecular vortices themselves. The 'relativity of knowledge' drives all these attempts to illustrate electric and magnetic phenomena by means of mechanical models made up of elastic solids or fluids (Harman 1998, 71–80; Siegel 1991, chaps 2 and 3).

The turn away from useful mechanical models towards deep theory was to be completed in Maxwell's *Treatise*, in 1873, where Maxwell finally developed the theory of electromagnetism that bears his name. The so-called Maxwell equations were a somewhat later development arising out of mainly the work of Oliver Heaviside, but essentially all the main empirical results and theoretical concepts employed by 'the Maxwellians' (Hunt 1991) were already formulated in *Treatise*. This includes the radical insight that light is a transverse wave in the electromagnetic aether, as well as the famous equivalence of the speed of light with the inverse of the square of the ratio of electrostatic and electrodynamic units. The full history of the development is rich, and there is no space to broach it in detail here (Hunt 1991; Siegel 1991; Harman 1998; Cat 2001; Nersessian 2008). The main lesson for our purposes concerns the use of mechanical models in arriving at these theoretical developments. While there is some debate regarding how necessary the analogies are methodologically to arrive at the full electromagnetic theory (Hon and Goldstein 2020), it is undeniable that in Maxwell's own reasoning, the vortices and idle-wheels model plays a key role, particularly in the derivation of the displacement current (see Harman 1998; and particularly Siegel 1991, chap. 4).

4. Helmholtz and the Origins of *Bildtheorie*

Roughly at the same time as Thomson and Maxwell developed an English-speaking modelling attitude, Hermann von Helmholtz (1821–1894) established his 'Berlin school' of physics and in so doing set up a distinct German-speaking variant of the nineteenth-century modelling attitude. Helmholtz's account of *Bilder* was essentially driven by his sign theory. The 'Bildtheorie' – literally the 'theory of images' – is not merely an account of scientific representation: it is also the name of a movement in scientific modelling practice that emerged in fin-de-siècle Austria and Germany. While it is expressly inspired by the English-speaking modellers – most prominently by Thomson and Maxwell's analogies between fluid mechanics, heat, and electricity – it also has its own roots in Neo-Kantian empiricism. Thus, although the Bildtheorie emerges most explicitly in the writings of Heinrich Hertz (1857–1894) and Ludwig Boltzmann (1844–1906) towards the end of the century, it is really to their mentor Hermann von Helmholtz (1821–1894) that we must look to searching for its intellectual and historical sources.[2]

According to Buchwald (1993), 'Helmholtzianism' is an open-ended set of methodological maxims for the practice of experimental science. At the core of this practice is the requirement to actively intervene experimental setups to obtain anomalous results or effects. These would be described in terms of the ascription of dynamic states to systems, together with functions operating on these states representing interaction potentials. The evolution of the states is therefore the key to the result of the interaction, and Helmholtz assumed everything else was essentially redundant or derivative, including charges, currents, or forces. Thus, contrary to what is sometimes supposed, Helmholtz was never entirely at ease with action-at-a-distance theories such as those of Wilhelm Weber and Gustav Fechner (or their equivalent over in Britain, such as those taught by Rouch and the other Cambridge coaches until well into the 1890s, as described in Warwick (2003)). Rather, he followed Franz Neumann in not presupposing any account of charges or currents, or the forces supposedly acting on them at a distance. Thus, Helmholtz – and Neumann – postulate a potential function between any two charges whose shape depends on their distance. The energy of the system is thereby determined without making any further assumptions regarding the nature of the system of charges itself, or the forces operating, other than the system that can be ascribed a 'state', which figures in a potential 'function' that fully describes its interaction properties. As Buchwald (1994, 15) puts it: "It could [...] be said of Helmholtz that after the early 1870s nothing was clear to him until it could be formulated in terms of interaction energies".

This is relevant to our present purposes for three reasons. First, it belies the thought that Helmholtz was initially resistant to field theories such as Maxwell's. On the contrary, Helmholtzianism is essentially neutral on whether fields mediating inductive currents on conductors exist, or instead forces acting at a distance displacing charges and thereby setting up such currents. There was no transition in Helmholtz from an action-at-a-distance to a field-theoretic account of electromagnetism because Helmholtz was never wedded to an action-at-a-distance theory to begin with.[3] The models Helmholtzians and Maxwellians countenanced were in fact similar from the start. This is perhaps not surprising, since Helmholtz and Thomson corresponded regularly and read each other's work avidly (Smith and Wise 1987, 1989). Furthermore, Helmholtz's (1870) proof that Fechner–Weber theories entail the predictions of the Maxwell displacement current model was a noted milestone on both sides of the channel (Buchwald 1993).

Second, Helmholtz's initial training was in medicine, and he started as a sort of Neo-Kantian, committed to the principle of causality and a style of causal realist explanation (Heidelberger 1993; Turner 1993). Yet, starting with his work on the physiology of perception in the 1860s, he progressively veered off towards a generic form of empiricism (Eckert 2006, 19; Patton 2010). Thus, Helmholtz moved away from the idea that perceptions are 'copies' of the objects perceived towards the view that they are signs instead, standing in the same conventional relation a name stands to its bearer. Helmholtz's 'sign theory' is a direct predecessor of the *Bildtheorie*: it identifies perceptions with representational signs, which can be operated upon in accordance with certain rules of inference. And indeed, at roughly this time, Helmholtz begins to employ the term 'Bild' to refer to the discovered laws of science (Schiemann 1998, 25). Hertz and Boltzmann inherited the insight that models are sign systems endowed with internal rules of inference.[4]

Third, and finally, Helmholtz's characteristic neutrality on issues of ontology is inherited by both Hertz and Boltzmann and turns out to be at the heart of the German-speaking modelling school. The principal lesson that Hertz and Boltzmann derived from their work in Helmholtz's laboratory is that the most appropriate representations must abstract away from the concrete material details of systems and instead focus on dynamic states and their potential and interaction functions. Once the appropriate dynamic models are adopted, ontological disputes will prove beside the point. Are there really forces in nature, or just masses? Do atoms exist, or are they just packets of energy? These are ontological disputes that are beyond the purview of scientific models per se but rather belong to the domain of interpretation. Hertz's attempt to derive a representation of mechanics devoid of forces and Boltzmann's attachment to atoms do not have the dogmatic character of a believer (in potentials and atoms, respectively) so much as that of a sceptic regarding forces and energentism, respectively. In both cases, they are attempts at justifying introducing alternative scientific models.[5]

5. Hertz and Boltzmann: Conformity and Information

There is one critical difference between the mentor and mentees, though: where Helmholtz upheld the principle of 'sign constancy' (Schiemann 1998), Hertz and Boltzmann allowed for multiple alternative representations. Hertz (1894) puts it with characteristic clarity:

> The images which we may form of things are not determined without ambiguity by the requirement that the consequents of the images must be the images of the consequents. Various images of the same objects are possible, and these images may differ in various respects. (1894, 3)

Ultimately, it is this multiplicity of models of phenomena – and their underdetermination by both experimental evidence and dynamic presuppositions – that gives rise to both Hertz's and Boltzmann's unusual scientific views at the time (D'Agostino 1990; De Regt 1999, 2005).

Heinrich Hertz's full formulation of the *Bildtheorie* came in early enough in his astonishingly deep *Principles of Mechanics*, where he famously wrote:

> We form for ourselves images or symbols of external objects; and the form which we give them is such that the necessary consequents of the images in thought are always the images of the necessary consequents in nature of the things pictured. (1894, 3)

This is through and through a Helmholtzian insight. There is first the idea that models are symbolic representations endowed with certain rules of inference (symbolic or logical necessity). There is then the thought that such models are related to the systems represented not by standing as copies of them, but only in the way in which conventional signs stand for their bearers – merely, at best, by exhibiting correlations between their consequents. The laws of nature and the rules of *Bilder* answer to different sorts of necessity (natural or physical; and logical or symbolic, respectively), but the consequences of rules and laws must correspond to each other. Thus, Hertz goes on to write: "The images that we here speak of are our conceptions of things. With the things themselves they are in conformity in one important respect, namely, in satisfying the above-mentioned requirement" (1894, 3).

It can then be argued, following Hertz, that 'conformity' is the only necessary condition on *Bilder*, the only defining condition on a scientific model or representation. It is not, however, the only virtue that a model can have. Hertz lists another four desirable properties in a *Bild*, namely *permissibility*, *correctness*, *distinctness*, and *appropriateness*. These conditions, Hertz argues, are not always fulfilled in every model. In fact, they often militate against each other, so that they must be traded wisely within their context of use. Thus, in practice, no model possesses them all, and most models struggle to possess one of them at all. Hertz's introduction of these conditions is interesting for what it lets in as desirable virtues of a model, but even more so for what it leaves out: 'conformity' is not taken to be an optional virtue, but the only necessary condition on any model.[6]

Thus, 'permissibility' is coherence "with the laws of our thought" (Hertz 1894, 2), which on the face of it appears to be a requirement of consistency or non-contradiction. Yet, Hertz is clear that a model may be contradictory, yet conform. And if a model conforms, it remains indeed a model. This makes room for models of fictional or impossible worlds, which may be 'impermissible' in this terminology, but are nonetheless allowed if they conform. 'Correctness' is the requirement of consistency with the properties of the target system, since an incorrect model, according to Hertz, is one whose "essential relations contradict the relations of external things" (1894, 2). Again, a model may be grossly 'incorrect', or inaccurate, or even an artefact, in the sense of being built to purpose but not necessarily truth-apt, yet conform and hence remain a model. According to Hertz, 'distinctness' is the requirement that a *Bild* provides an accurate rendition of every aspect of the target; we would nowadays refer to this roughly as 'completeness', and obviously, it is not a plausible requirement on any model. Thus, many models are highly streamlined, idealised, abstract, or 'indistinct' in Hertz's terminology, yet of course, they remain models if they conform to their targets. Finally, 'appropriateness', according to Hertz, is a measure of simplicity. A minimal model is 'appropriate' if it does not make or contain superfluous claims regarding its target system. Another way to put Hertz's thought is that an appropriate model lacks any properties that have no role at all in the sorts of inferences that the model promotes with respect to its target. Hertz most clearly does not think that every scientific representation is appropriate: his *Principle of Mechanics* is a forceful argument to the effect that the standard representation of mechanics in terms of forces acting at a distance is inappropriate, at least when compared to his own much more streamlined and scarce representation in terms merely of mass and potentials. More generally, it seems indeed clear that the conformity of a scientific model in no way requires its appropriateness: most models are far from minimal, and they contain elements that are extraneous to their representational tasks.

Hertz's discussion, I argue, is a *tour de force* and sets the stage for the ensuing modelling attitude. Nevertheless, Hertz's *Principles of Mechanics* remained controversial, and Hertz's untimely death in 1894 curtailed this work. So it was down to a devoted admirer, Ludwig Boltzmann, working in Vienna, to promote the *Bildtheorie* most firmly. The high peak of the German-speaking school of modelling may well be signalled by the publication of Boltzmann's *Popularen Schriften* in 1905.[7] Boltzmann's goals for modelling are also arguably less lofty than Hertz's, imbued instead with characteristic Viennese pragmatism and empiricism. The modelling attitude is, in Boltzmann's hands, what results from the application of principles of economy of thought to scientific theorising: "As the facts of science increase in number, the greatest effort had to be observed in comprehending them [models] and in conveying them to others" (Boltzmann 1902, 2).

Boltzmann also added a requirement of informational gain to Hertz's minimal condition of conformity. In discussing the models in thermodynamics that he was so instrumental in establishing, he wrote: "If for one of the elements [in the model] a quantity which occurs in calorimetry be chosen – for example, entropy – information is also gained about the behaviour of the body when heat is taken in or abstracted" (1902, 2). A model must show conformity to its target, but not any conformity will do: the model must provide us with relevant new information about that target. It is this combination of minimal conformity and informational gain that makes a model scientific – and a valuable instrument for surrogate reasoning regarding its target. Together, these two requirements bring into relief Maxwell's notion of a 'conductor' as an instrument for reasoning, which was reviewed in the first part of this chapter. It is not a coincidence: Boltzmann was arguably led to the informational gain requirement through Maxwell's analogies, which he had studied very closely (Klein 1973). Furthermore, it is through these two conditions that we can ultimately understand the work that analogy and metaphor can do for us in scientific inquiry. Reasoning by analogy requires both a degree of conformity (to make it possible to inquire into the nature of an object or system by means of a comparison to other systems or objects) and a measure of informativeness, the capacity of the source of the analogy to enlighten us regarding aspects of the target that had not been considered before.

In setting such a minimal bar on acceptable *Bilder*, Hertz paved the way for the underdetermination of theoretical models – and hence for pluralism, as the thesis that more than one model is often available for any phenomena, effect, or process of interest. And in the insistence that the logical necessity in a *Bild* is distinct from the natural necessity in the phenomena pictured or in their represented causes, he opened up models to the normative practices that sanction the rules of reasoning within *Bilder*, beyond those of logical consequence or necessity. Hertz's 'conformity' appears similar in this regard to the cognate notion of 'conformation' in Helen Longino's celebrated *The Fate of Knowledge* (Longino 2001). Both notions are attempts to set a lower bar for scientific representation, thus widening its scope and generating room for underdetermination and genuine pluralism. Moreover, they both seek to do this by grounding the activity of modelling in our socially sanctioned surrogate inferential practices, thus placing greater emphasis on the communal sets of norms required to functionally set and maintain representations. Yet, 'conformation' is not 'conformity'. According to Longino (2001, 117) 'conformation' is "a general term for a family of epistemological success concepts, including truth, but also isomorphism, homomorphism, similarity, fit, alignment and other such notions". In terms of the recent debates over representation, Longino advances a general noun for the variety of conditions of accuracy or adequacy of scientific representation, not the conditions for representation *per se*. By contrast, I shall argue, Hertz's 'conformity', like Maxwell's analogy, is a minimal requirement on the conceptually prior obtaining of representation, however erroneous, false, or inaccurate.

6. The Philosophical Reception of the Modelling Attitude

The modelling attitude in science reached a high peak at the turn of the century, as signalled by Boltzmann's entry in the Encyclopaedia Britannica (Boltzmann 1902). It is a *fin-de-siècle* development that changes the character of scientific work and inquiry, and it continues to the present day. Whereas modern science had taken inspiration from the ancients to base indubitable knowledge upon the twin sources of demonstrative proof and empirical observation, the modelling attitude adds a third prominent layer involving the construction of figurative, idealised, fictional, or artefactual scenarios within scientific models. In practice, models often mediate between the lofty realms of high explanatory theory, on the one hand, and low-level renditions and records of data and phenomena, on the other (Morgan and Morrison 1999). As such, models continue to take place of pride in scientific work throughout the natural and social sciences – including the physical, chemical, earth, and life sciences, as well as in economics, psychology, or sociology.

Yet, the fortunes of the modelling attitude in the philosophy of science and amongst philosophers have been varied, experiencing ups and downs, and always subject to a measure of controversy. The object of some fierce criticism in the work of Pierre Duhem (1861–1916), the modelling attitude nonetheless experienced much philosophical attention and influence in the early decades of the twentieth century, in the wake of formidable endorsements by the likes of Boltzmann, Norman Campbell (1880–1949), Henri Poincaré (1854–1912), and Hans Vaihinger (1852–1933). However, with the ascent of logical positivism, particularly its North American version from the 1930s onwards, the modelling attitude went into a period of relative philosophical decline. There was for many years scant regard for modelling generally amongst philosophers, and a return to the dismissive cautionary warnings so acutely voiced by Pierre Duhem (Bailer-Jones 1999). A renaissance of philosophical interest began in the 1960s, and the modelling attitude as a philosophical object of inquiry slowly surged back in the wake of pioneering work by authors such as Max Black (1909–1988), Mary Hesse (1924–2016), and Stephen Toulmin (1922–2009). The last years of the twentieth century finally saw the modelling attitude gain centre stage in the philosophy of science once more, with the publication of the celebrated *Models as Mediators* collected volume (Morgan and Morrison 1999), signalling the start of an entire movement that endures to the present day. Philosophical discussions of the nature, role, and practice of modelling are now very prominent and are an absolutely central piece in contemporary philosophy of science, as is shown by even a cursory look at the major philosophy of science journals and publishing houses.

The most striking episode in this remarkable history (gracefully told in Bailer-Jones 1999) is perhaps that unusual, slow, and gradual upsurge in interest in models during the 1960s. Where did authors like Max Black and Mary Hesse gain inspiration from? Not entirely surprisingly, they were mostly inspired by the originators of the modelling attitude, by Hertz and Boltzmann, and, most prominently, by James Clerk Maxwell. Black and Hesse, in particular, both went back to Maxwell to the point of restoring the focus on the sort of analogical thinking practised by Maxwell.[8] 'Analogy' as a form of reasoning thus took the stage again, with 'model' consigned to the secondary role of its main product.

Black focused on analogies that turn fully into metaphors, which he argued required a realist reading distinct from Maxwell's early typically Scottish attitude. As he writes (Black 1962, 228): "One approach uses a detached comparison reminiscent of simile and argument from analogy; the other requires an identification typical of metaphor". The nature of metaphor is debated to this day, and its application to science remains controversial (see Suárez 2024, chap. 3, for an assessment). By contrast, Hesse's nuanced analysis of analogical reasoning caught on quickly and is widely regarded to be central to any understanding of modelling. It informs the sort of philosophy that focuses on scientists' inferential processes and practices at the expense of just analysing their product in ready-made models. In her highly influential *Models and Analogies in Science* (Hesse 1966), Hesse distinguished between three parts in any analogy or model: the positive, negative, and neutral analogies. The first includes those properties and relations shared between the source and the target; the second, those properties denied in the target; the third, those properties about which it is unknown whether they are shared between the source and target. She also helpfully distinguished vertical and horizontal relations in analogical thinking, thus emphasising the fact that model sources are dynamic structured entities endowed with parts and often dynamically evolving in time (see Bartha 2019 for further development). The vertical relations thus capture some causal principles at work. As Hesse puts it (Hesse 1966, 87; quoted in Bartha 2019, 28):

> The vertical relations in the model [source] are causal relations in some acceptable scientific sense, where there are no compelling a priori reasons for denying that causal relations of the same kind may hold between terms of the explanandum [target].

This is exactly in line with Hertz's 'conformity' when suitably extended to capture all kinds of dynamic relations within the model source that may not be ruled out to have correlates in the target. Whether it actually corresponds to existing causal relations in nature is rather a question for the further Hertzian 'correctness' of the model.

7. Lessons for Contemporary Debates

The modelling attitude has Scottish and Cantabrigian origins in Victorian science and is deeply enmeshed in James Clerk Maxwell's work and thought. Yet, it developed most firmly in Berlin, Bonn, and Vienna, as the German-speaking *Bildtheorie* took hold. This key development in the emergence of a modelling attitude characterises much of the twentieth-century science. In the proficient hands of Maxwell, Hertz, and Boltzmann, the modelling attitude gained weight and developed into a formidably precise tool for mathematical and quantitative prediction and understanding. Nevertheless, Maxwell's insights regarding analogy (especially his apt metaphor of a model as a 'conductor' of surrogative reasoning) are deeply embedded in *Bildtheorie*'s conception. The outlines of a twofold conception of scientific representation emerged around two minimal conditions of conformity and informational gain, which every scientific model minimally complies with. These two requirements, as they appear in Hertz's and Boltzmann's work, lead naturally to a deflationary, functionalist, and pragmatist conception of representation.

The dominant accounts of representation in recent literature fall into one of two kinds: substantive and deflationary. The ostensive thought in a substantive account is that every case of representation is the instantiation of a particular type of relation between what we may call the representational source and its target. Thus, there is a substantive relation r of type R, $\{r \in R\}$, such that for any pair of objects or systems {x, y}, x is the source S, and y is the target T of a representation if and only if x and y stand in that relation: r (x, y). It is important to get the order of the quantifiers right in this expression: $\exists r \in R : \forall \{x,y\} : (S(x) \,\&\, T(y)) \leftrightarrow r(x, y)$. That is, the quantifier that determines the domain of the universal substantive relation of representation ranges over all source–target pairs. In other words, a substantive account of representation assumes that a certain type of relation (similarity, isomorphism, or some variety thereof) is invariably instantiated in every case of representation by models in science. Model building is then essentially all about finding out that relation as it applies to each {source, target} pair. The Victorian models of the aether, for example, are attempts to characterise the main properties of the aether through the similarities or isomorphisms that the aether (or its 'structure', whatever that may mean) holds to the mechanical models advanced to represent it, such as the vortex model. If there is no substantive relation to speak of, or none that actually holds, then there is no actual representation. Since the aether is nowadays not a recognised real entity, it seems to follow that Maxwell's model was never a representation in the first place. This seems farfetched to say the least and does poor justice to the historical record, which does not contain any indication that the model worked as anything other than a model and invited the sorts of inferences in practice that any model would. A metaphysical distinction without any practical consequence is arguably, on a pragmatist maxim at least, an idle posit lacking any content.

Substantive accounts of representation suffer from additional problems, canvassed thoroughly in the literature (including Suárez 2003; 2010; 2024). These need not detain us here, though. The historical observation above regarding the representational use of Maxwell's vortex model already ought to prompt a search for an alternative account of representation, one that stays resolutely close to the practice, while avoiding reifying the diversity of representational means and relations into any essential constitutive element in all scientific representations. These accounts are deflationary because they skip any substantive constitutive relation. Thus, another way to characterise the difference is that in a deflationary pragmatic account, the quantifiers appear inverted relative to the statement of a substantive account. Hence, there is in fact no constitutive relation that universally applies to all representations. Rather, for all {x, y} pairs where x is the representational source, S(x), and y is the representational target, T (y), there may be some functional relation $\{r \in R\}$ that applies to that {source, target} pair: $\forall \{x,y\} : (S(x) \& T(y)) \leftrightarrow \exists r \in R : r(x, y)$. Here, the quantifier ranges over all the various relations that instantiate representations; it merely affirms that there is one such relation for every source–target pair. Since the relation is merely a function and the set may contain the null relation, this definition, significantly, does not require all representational sources to have targets. Nor does it require that all those representational sources that do have targets be related to them via the same universally applicable relation. Thus, Maxwell's vortices and idle-wheel model is a representation of the aether, properly speaking, even if it lacks a target. And if we were to insist that Maxwell's model represents not the aether but properties of the electromagnetic field (such as the displacement current), it would not need to be related to it by means of the same type of relation as, say, Maxwell's equations hold to the electromagnetic field. The former ones may be related by similarity, while the latter ones are by convention, or through a statement of some structural morphism in their phase spaces.

There are a number of deflationary accounts in the recent literature, including RIG Hughes' DDI model (Hughes 1998), the artefactual approach (Knuuttila 2011; Carrillo and Knuuttila 2022), and a variety of inferential approaches (Kuorikoski and Ylikoski 2015; De Donato and Zamora-Bonilla 2012; Khalifa et al, 2022). They all have considerable merits and are apt in confronting a large variety of modelling cases. The original inferential conception [inf] (Suárez 2004; 2010; 2024) has the additional virtue to accord with the history of the modelling attitude reviewed in this chapter. The only two necessary conditions on representation, according to [inf], are what I refer to as the 'representational force' of a source, and its 'inferential capacities' with respect to the (real or fictitious) target. Each of these conditions describes, properly speaking, an aspect of the normative practice of reasoning by analogy and is not to be conceived as a relation in any metaphysical sense. Thus [inf] is anticipated by the twofold requirements adumbrated by Hertz and Boltzmann in the wake of Maxwell's innovations: Hertz's conformity requirement anticipates [inf]'s 'representational force', while Boltzmann's information requirement informs [inf]'s 'inferential capacities'.

Acknowledgements

I thank Julia Sánchez-Dorado and the editors of the volume for their helpful feedback. This essay is an elaboration of parts of chapter 2 in my recent book (Suárez 2024), and I thank audiences at meetings of the *Integrated History and Philosophy of Science* and the *Society for the Philosophy of Science in Practice* where different versions of some of the historical material were presented. Many thanks also to reading groups at the Universities of Vienna and Cambridge for their comments and encouragement. Financial support from the Spanish research agency (AEI), research projects PGC2018-099423-BI00 and PID2021-126416NB-I00 is acknowledged.

Notes

1 Warwick (2003) is an unsurpassed account of the Cambridge tripos system in the nineteenth century, while Buchwald (1985) and Darrigol (2000) are key historiographical references.

2 The literature on Helmholtz is large, and the account that follows leans heavily on Darrigol (2000), Eckert (2006), Patton (2010), as well as the superb essays in Cahan (1993). In addition, Hatfield (1991), Patton (2009), and Schiemann (1998) are insightful accounts of Helmholtz's work on perception and his 'sign theory'.
3 This is curiously in contrast with the sorts of pedagogical resistance that field theories encountered initially precisely in Cambridge, where they were first adumbrated – see Warwick (2003, 306–56).
4 There are essentially two kinds of rules, referred to in Suárez (2024) as horizontal and vertical rules of inference, which mirror Mary Hesse's (1966) similar distinctions reviewed later in the chapter.
5 For Hertz's views regarding the underdetermination of ontology, see the essays in Baird et al. (1998). For Boltzmann's epistemology, see Blackmore (1995) and de Regt (1999, 2005).
6 Hertz is uncharacteristically not entirely clear in his presentation of the relation between correctness and conformity. I follow the reconstruction in (Suárez, 2024, pp. 38–41).
7 Boltzmann's entry on 'models' in the Encyclopaedia Britannica in 1902 is also climatic for the *Bildtheorie*, but it had less of an impact on the public and the modelling community in the German-speaking world.
8 A revival of interest in Aristotelian analogy at Cambridge may have been involved too – Lloyd's seminal *Polarity and Analogy* (Lloyd, 1966) was published in the same year as the revised version of Hesse's book.

References

Bailer-Jones, Daniela. 1999. "Tracing the development of models in the philosophy of science". In *Model-based Reasoning in Scientific Discovery,* edited by Lorenzo Magnani, Nancy Nersessian and Paul Thagard, 23–40. New York: Kluwer Academic / Plenum Publishing.

Baird, Davis, RIG Hughes, and Alfred Nordmann, eds. 1998. *Heinrich Hertz: Classical Physicist, Modern Philosopher.* Dordrecht: Kluwer Academic Publishers.

Bartha, Paul. 2019. "Analogy and analogical reasoning." *Stanford Encyclopaedia of Philosophy.* https://plato.stanford.edu/entries/reasoning-analogy/.

Black, Max. 1962. *Models and Metaphors.* Ithaca: Cornell University Press.

Blackmore, John, ed. 1995. *Ludwig Boltzmann: His Later Life and Philosophy, 1900–1906.* Dordrecht: Kluwer Academic Publishers.

Boltzmann, Ludwig. 1902. "Model." *Encyclopaedia* Britannica, 10th Edition, 788–91.

———. 1905/1974. *Theoretical Physics and Philosophical Problems: Selected Writings*, edited by Brian McGuinness. Dordrecht: Reidel.

Buchwald, Jed Z. 1985. *From Maxwell to Microphysics: Aspects of Electromagnetic Theory in the Last Quarter of the Nineteenth Century.* Chicago: University of Chicago Press.

———. 1993. "Electrodynamics in context: Object states, practice and anti-romanticism." In *Hermann von Helmholtz and the Foundations of Nineteenth Century Science*, edited by David Cahan, 334–73. Berkeley: University of California Press.

———. 1994. *The Creation of Scientific Effects: Heinrich Hertz and Electric Waves.* Chicago: University of Chicago Press.

Cahan, David, ed. 1993. *Hermann von Helmholtz and the Foundations of Nineteenth Century Science.* Berkeley: University of California Press.

Carrillo, Natalia and Tarja Knuuttila. 2022. "Holistic idealisation: An artefactual standpoint", *Studies in History and Philosophy of Science* 91: 49–59.

Cat, Jordi. 2001. "On Understanding: Maxwell and the Methods of Illustration and Scientific Metaphor." *Studies in History and Philosophy of Modern Physics* 32(3): 395–441.

Chakravartty, Anjan. 2004. "Stance relativism: Empiricism versus metaphysics." *Studies in History and Philosophy of Science* 35(1): 173–84.

D'Agostino, Salvo 1990. "Boltzmann and Hertz on the Bild-conception of physical theory." *History of Science* 28(4): 380–98.

Darrigol, Olivier. 2000. *Electrodynamics from Ampère to Einstein.* Oxford: Oxford University Press.

Davie, George E. 1961. *The Democratic Intellect: Scotland and her Universities in the Nineteenth Century.* Edinburgh: Edinburgh University Press.

De Donato, Xavier and Jesús Zamora-Bonilla. 2012. "Explanation and modelization in a comprehensive inferentialist approach." In *EPSA Philosophy of Science: Amsterdam 2009*, edited by Henk W. de Regt, Stephan Hartmann and Samir Okasha, 33–42. Springer.

De Regt, Henk. 1999. "Ludwig Boltzmann's 'Bildtheorie' and scientific understanding." *Synthese* 119: 113–34.

———. 2005. "Scientific realism in action: Molecular models and Boltzmann's bildtheorie." *Erkenntnis* 63: 205–30.

Eckert, Michael 2006. *The Dawn of Fluid Dynamics: A Discipline Between Science and Technology*. New York: John Wiley.

Fine, Arthur. 1984/1987. "The natural ontological attitude", in Leplin, ed., *Scientific Realism*. Berkeley: University of California Press, 83–107. Reprinted in Arthur Fine (1987) *The Shaky Game: Einstein, Realism, and the Quantum Theory*, Chicago: University of Chicago Press.

Hatfield, Gary. 1991. *The Natural and the Normative: Theories of Spatial Perception from Kant to Helmholtz*. Cambridge, MA: The MIT Press.

Harman, Peter M, ed. 1990. *The Scientific Letters and Papers of James Clerk Maxwell, vol. 1, 2, and 3*. Cambridge: Cambridge University Press.

Harman, Peter M. 1998. *The Natural Philosophy of James Clerk Maxwell*. Cambridge: Cambridge University Press.

Heidelberger, Michael. 1993. "Force, law and experiment: The evolution of Helmholtz's philosophy of science". In *Hermann von Helmholtz and the Foundations of Nineteenth Century Science*, edited by David Cahan, 461–97. Berkeley: University of California Press.

Helmholtz, Hermann von. 1870. "Über die Bewegungsgleichungen der Elektrizität für ruhende leitende Körper". *Journal für die Reine und Angedwandte Mathematik* 72: 57–129. Reprinted in his *Wissenschaftliche Abhandlungen*, Vol. 1, 545–628.

Hertz, Heinrich. 1894/1956. *The Principles of Mechanics Presented in a New Form*. New York: Dover Publications.

Hesse, Mary. 1966. *Models and Analogies in Science*. Notre Dame: The University of Notre Dame Press. (Revised version of the 1963 1st edition).

Hoffman, Dieter. 1998. "Heinrich Hertz and the Berlin school of physics." In *Heinrich Hertz: Classical Physicist, Modern Philosopher*, edited by Davis Baird, RIG Hughes and Alfred Nordmann, 1–8. Dordrecht: Kluwer Academic Publishers.

Hon, Giora and Goldstein, Bernard. 2020. *Reflections on the Practice of Physics: James Clerk Maxwell's Methodological Odyssey in Electromagnetism*. London: Routledge.

Hughes, RIG. 1997. "Models and Representation." *Philosophy of Science* 64: S325–36.

Hunt, Bruce J. 1991. *The Maxwellians*. Ithaca: Cornell University Press.

Khalifa, Kareem, Jared Millson and Mark Risjord. 2022. "Scientific Representation: An inferentialist-expressivist manifesto." *Philosophical Topics* 50(1): 263–292.

Klein, Martin. 1973. "The Maxwell-Boltzmann relationship." In *Transport Phenomena: Second Annual International Centennial Boltzmann Seminar*, edited by Joseph Kestin, 297–308. American Institute of Physics.

Knuuttila, Tarja. 2011. "Modelling and representing: An artefactual approach to model-based representation." *Studies in History and Philosophy of Science* 42(2): 262–71.

Kuorikoski, Jaakko and Petri Ylikoski. 2015. "External representations and scientific understanding." *Synthese*, 192(12): 3817–37.

Lloyd, G.E.R. 1966. *Polarity and Analogy*. Cambridge: Cambridge University Press.

Longino, Helen. 2001. *The Fate of Knowledge*. Princeton: Princeton University Press.

Magnani, Lorenzo, Nancy Nersessian and Paul Thagard, eds. 1999. *Model-based Reasoning in Scientific Discovery*. New York: Kluwer Academic / Plenum Publishing.

Maxwell, James Clerk. 1856a/1890 "On Faraday's lines of force." In William Davidson Niven, ed. 1890. *The Scientific Papers of James Clerk Maxwell, vol. 1*, 155–229. Cambridge: Cambridge University Press.

———. 1856b/1990. "Analogies in Nature: Essay for the Apostles." In Peter M. Harman, ed. 1990. *The Scientific Letters and Papers of James Clerk Maxwell, vol. 1: 1846–1862*, 376–383.

———.1860/1890. "On physical lines of force." In William Davidson Niven, ed. 1890. *The Scientific Papers of James Clerk Maxwell, vol. 1*, 451–513. Cambridge: Cambridge University Press.

———.1873/2010. *A Treatise on Electricity and Magnetism*. Cambridge: Cambridge University Press.

Morgan, Mary, and Margaret Morrison. 1999. *Models as Mediators: Perspectives on Natural and Social Science*. Cambridge: Cambridge University Press.
Mulligan, Joseph F., ed. 1994. *Heinrich Rudolf Hertz: A Collection of Articles and Addresses*. New York and London: Garland Publishing.
———. 1998. "The reception of Heinrich Hertz's *Principles of Mechanics* by his contemporaries." In *Heinrich Hertz: Classical Physicist, Modern Philosopher*, edited by Davis, Baird RIG Hughes and Alfred Nordmann, 173–81. Dordrecht: Kluwer Academic Publishers.
Nersessian, Nancy. 2008. *Creating Scientific Concepts*. Cambridge MA: MIT Press.
Olson, Richard S. 1975. *Scottish Philosophy and British Physics 1750–1880. A Study in the Foundations of the Victorian Scientific Style*. Princeton: Princeton University Press.
Patton, Lydia. 2009. "Signs, toy-models, and the a priori: From Helmholtz to Wittgenstein." *Studies in History and Philosophy of Science* 40: 281–89.
Patton, Lydia. 2010. "Hemann von Helmholtz." *Stanford Encyclopaedia of Philosophy*. https://plato.stanford.edu/entries/hermann-helmholtz/
Rowbottom, Darrell. 2011. "Stances and paradigms: A reflection." *Synthese* 178(1): 111–9.
Schiemann, Gregor. 1998. "The loss of world in the image: Origin and development of the concept of image in the thought of Hermann von Helmholtz and Heinrich Hertz." In *Heinrich Hertz: Classical Physicist, Modern Philosopher*, edited by Davis Baird, RIG Hughes, and Alfred Nordmann, 25–38. Dordrecht: Kluwer Academic Publishers
Siegel, Daniel. 1991. *Innovation in Maxwell's Electromagnetic Theory: Molecular Vortices, Displacement Current, and Light*. Cambridge: Cambridge University Press.
Simson, Robert. 1756. *Euclid: The Elements*, 1st edition. Glasgow: Robert and Andrew Foulis.
Smith, Crosbie and Norton Wise. 1989. *Energy and Empire: A Biographical Study of Lord Kelvin*. Cambridge: Cambridge University Press.
Suárez, Mauricio. 2003. "Scientific representation: Against similarity and isomorphism." *International Studies in the Philosophy of Science*, 17(3): 225–244.
———. 2004. "An inferential conception of scientific representation." *Philosophy of Science*, 71(5): 767–779.
———. 2010. "Scientific representation." *Philosophy Compass*, 5(1): 91–101.
———. 2014. "Scientific representation." Oxford Bibliographies Online, January 2014.
———. 2024. *Inference and Representation: A Study in Modeling Science*. Chicago: University of Chicago Press.
Turner, R. Steven 1993. "Consensus and controversy: Helmholtz on the visual perception of space." In *Hermann von Helmholtz and the Foundations of Nineteenth Century Science*, edited by David Cahan, 154–204. Berkeley: University of California Press.
Van Fraassen, Bas. 2002. *The Empirical Stance*. New Haven: Yale University Press.
Warwick, Andrew. 2003. *Masters of Theory: Cambridge and the Rise of Mathematical Physics*. Chicago: University of Chicago Press.
Wise, Norton and Crosbie Smith. 1987. "The practical imperative: Kelvin challenges the Maxwellians". In Robert Kargon and Peter Achinstein, eds. 1987. *Kelvin's Baltimore Lectures and Modern Theoretical Physics*: 323–348. Cambridge, MA: The MIT Press.

2
THEORIES AND MODELS

Roman Frigg

1. Introduction

There are models, and there are theories. This invites the question of how the two are related. Traditionally, it was assumed that this question had a simple answer, and attempts have been made to explain the relation between models and theories at a general level. In this chapter, I argue that there is no such thing as "the" relation between models and theories. How models relate to theories depends on the cases at hand, and models can stand in a multiplicity of relations to theories.

The chapter starts with a discussion of the Syntactic View and Semantic View of theories and points out that these views have too narrow a vision of what models are and of how they relate to theories (Section 2). We then discuss different relations between models and theories in descending order of models' independence from theory. We begin by looking at models that are constructed without the aid of a theoretical framework and that therefore end up being independent from theory (Section 3). An interesting class of models serves the purpose of exploring the properties of a theory by providing simplified renderings of a theory's features (Section 4). In some cases, models live in a symbiotic relation with theories, adding specifics about which the theory remains silent (Section 5). In other cases, the reliance of theories on models is even stronger because theories require interpretative and representative models in order to relate to real-world targets (Section 6), which motivates the view that models are mediators between theories and the world (Section 7). Sometimes it is difficult to draw the line between models and theories, and we discuss how, and where, such a line could be drawn (Section 8). Section 9 concludes.[1]

2. Two orthodoxies

Twentieth-century philosophy of science has produced two broad views of what scientific theories are, and both imply a position on how models relate to theories. For better or worse, these two views form the backdrop of most discussions of models and theories today, and so our discussion should begin with them.

DOI: 10.4324/9781003205647-4

The first view, often referred to as the *Syntactic View of Theories* ("Syntactic View", for short), is associated with logical empiricism. Early statements of the Syntactic View include Carnap (1923) and Schlick (1925); full developments can be found in Carnap (1938, sec. 23), Braithwaite (1953, chaps. 1–3; 1954), Nagel (1961, chap. 5), and Hempel (1966, chap. 6; 1970).[2] The Syntactic View regards a theory T as a linguistic entity that satisfies the following three requirements:

(R1) T is formulated in an appropriate system of formal logic.
(R2) T contains axioms, which, when interpreted, are the theory's laws.
(R3) T's extralogical terms are divided into observation terms and theoretical terms, and theoretical terms are connected to observation terms by correspondence rules.

R1 is often said to mean that the theory is formulated in first-order predicate logic, but this restriction is unnecessary and T can be formulated in any system of logic (Lutz 2012). R2 requires there to be general propositions in the logical system which are the theory's laws when the extralogical terms are given an empirical interpretation. As a simple example, consider the sentence $(\forall x)(Fx \rightarrow Gx)$. Taken on its own, this is just a formal sentence (saying that for every object x, if x has property F, then x also has property G). This sentence becomes a statement of a law of nature of a simple theory of electricity if we interpret F as "is a piece of copper" and G as "conducts electricity". Under this interpretation, the sentence says that every object that is a piece of copper also conducts electricity. R3 harbours the view's empiricist commitments. Extralogical terms are terms that relate to objects and properties in the world (in contrast to logical terms like "and" and "or", which concern the structure of sentences). The Syntactic View separates these into observation terms and theoretical terms. The former are terms like "round", "green", "ball", "liquid", "wheel", "hot", "longer than", and "contiguous with", which refer to directly observable objects, properties, and relations. The latter are terms like "electron", "entropy", "orbital", "electromagnetic field", "gene", "quantum jump", "temperature", and "rate of inflation", which (purportedly) refer to objects, properties, and relations beyond direct observation. The view postulates that theoretical terms are related to observation terms by so-called *correspondence rules*. By way of illustration, consider "temperature". The temperature of an object is not directly observable. What is observable are thermometer readings. So the Syntactic View postulates that the term "temperature" be connected to an observation term through a rule like "an object has temperature θ if, and only, a thermometer shows θ when brought in contact with the object".[3]

Let us call the theory's system of formal logic together with its uninterpreted axioms the theory's formalism. The formalism of a theory is a set of formal sentences. Given such a set of sentences, one can always look for a set of objects, along with their properties and relations, which make the sentences true if the sentences' terms are interpreted as referring to those objects, properties, and relations. Such a set of objects constitutes a *logical model*. It is then common to say that the model *satisfies* the formal sentences in the sense that the model makes the sentences true if the terms of the sentences are taken to refer to the objects, properties, and relations in the model. In the context of a discussion of scientific theories, the relevant formal sentences are stated in the language of the formalism of a theory, and hence logical models are sometimes referred to as "models of a theory" or "models for a theory".

If, for the sake of illustration, we assume that the formalism of our theory consists only of the sentence $(\forall x)(Fx \rightarrow Gx)$, then a set of objects is a model for that theory if it is the case that to every object to which the predicate F applies, the predicate G also applies. Earlier we interpreted F as "is a piece of copper" and G as "conducts electricity". But interpretations are not unique, and formalisms can often be interpreted in several different ways. Rather than interpreting F and G in terms of copper and conductivity, we could interpret F as "is a piece of granite" and G as "contains quartz", which also makes the sentence $(\forall x)(Fx \rightarrow Gx)$ true. Hence, a set of objects in which it is the case that every object to which "is a piece of granite" applies is such that also "contains quartz" applies to it is a model of the theory.

In the Syntactic View, *scientific models* are essentially alternative interpretations of a theory's formalism. Braithwaite expresses this clearly when he says that a model is "another interpretation of the theory's calculus" (1962, 225), whereby his "calculus" is synonymous with our "formalism". However, for an alternative interpretation to be useful, it must have an additional feature: the objects of the alternative interpretation must be familiar to us. In Hesse's words, "a model is drawn from a familiar and well-understood process" (1961, 21). Crucially, this requirement applies to *all* terms of the formalism. That is, it applies also to the terms that were considered theoretical terms under the standard interpretation of the theory. In R3, these terms were given an "indirect" interpretation via correspondence rules, which made them difficult to grasp intuitively. In the context of a model, these terms receive a direct interpretation based on something familiar to us. In sum, then, we can say that according to the Syntactic View, a scientific model (often just "model") is a logical model of a theory's *entire* formalism that consists of objects, properties, and relations that are familiar to us.

As an example, consider the kinetic theory of gases. The theory takes a gas to consist of molecules that move freely unless they either collide with each other or the walls of the vessel containing the gas. Since "gas molecule" and "trajectory of a molecule" are theoretical terms, the theory is not easy to comprehend. To get an intuitive grip on the theory, we can reinterpret the theory in terms of billiard balls and their paths. The terms that were formerly interpreted as referring to molecules are now interpreted as referring to billiard balls; the terms that were interpreted as referring to the trajectories of molecules are now interpreted as referring to the paths of billiard balls. A bunch of billiard balls is therefore a model of the kinetic theory of gases. Other well-known examples of models of this kind are water waves as a model of the acoustic theory of sound waves and the solar system as a model of the Bohr theory of the atom.

The second view of theories in 20th-century philosophy of science is the so-called *Semantic View of Theories* ("Semantic View", for short). Historically this view was intended to replace the Syntactic View, which has been reported to suffer from a number of serious problems. It is a matter of controversy whether these problems are as severe as critics have said they were, or whether they are problems at all. However, this is not the place to review this debate and the reader is referred to the relevant literature on the subject.[4] Important statements of the Semantic View include Suppes (2002), van Fraassen (1980), Balzer, Moulines and Sneed (1987), Giere (1988), and Da Costa and French (1990). Different authors develop the view in different ways, but there is a common denominator, the focus on a theory's models. As we have seen previously, a logical model is a set of objects (along with their properties and relations) that make the theory's formalism true. We can then ask what the class of all logical models of a formalism looks like, and this will give us important information about the nature of a theory. Hence, rather than focussing on the formalism itself when characterising a theory, we can focus on its models. The Semantic View submits that this is not just another way of doing the same thing; on the contrary, characterising a theory in terms of its models is superior to characterising it in terms of its formalism. The primary reason for this is that formalisms can change and yet describe the same things. We are familiar with this phenomenon from everyday contexts, where we can say the same thing in different languages. "Copper conducts electricity" and "Kupfer leitet Elektrizität" are different sentences but they have the same truth-maker, namely the fact that copper conducts electricity. In the context of theories, we can choose different formal tools to describe the same models, which, however, would not result in a new theory because such reformulations merely describe the same thing in different ways. This motivates the Semantic View's core posit: a scientific theory is a family of models. For instance, in the Semantic View, Newtonian mechanics is not a set of postulates about motion and force; it is the set of models in which these postulates are true.

Two points deserve note. The first is that different authors have different ontologies of models. Suppes and Balzer, Moulines and Sneed take them to be set-theoretical structures; Da Costa and French take them to be partial structures; van Fraassen takes them to be state spaces; and Giere takes them to be abstract objects. These differences are important in other contexts, but they are immaterial to the discussion in this chapter. The second is the role of a formalism. We introduced the Semantic View by appealing to the notion of a logical model, and indeed, it is that notion that gives the view its name: the view is called the "Semantic" View due to the fact that models provide the formalism's semantics because models are what the formalism is taken to be about. Yet, providing a semantics for a formalism is like Wittgenstein's ladder, which is pushed away after it has been climbed. Proponents of the Semantic View insist that interpreting a formalism is in no way essential, nor is the presence of a formalism to begin with. At bottom, a theory is simply a family of models, no matter how (if at all) they are described by a formalism.

As indicated previously, much can be said about the pros and cons of these two views, but this is not our subject matter. What interests us here is the analysis of the relation between models and theories that the two approaches offer. The core argument of this chapter is that both analyses are too narrow. To see why and how, note that in both conceptions, models play a subsidiary role to theories. In the Syntactic View, they are merely reinterpretations of a formalism in terms of something familiar; in the Semantic View, they are the building blocks of which theories are made up. Both notions capture some cases of modelling. The Syntactic View successfully explicates analogue models, which often connect to their target via a shared formalism.[5] The Semantic View offers a cogent analysis of what happens in certain areas of fundamental physics, most notably in theories of space and time.[6] However, there are many cases, and indeed entire areas of science, where the relation between models and theories fits neither the mould of the Syntactic View nor that of the Semantic View. The plan for the remainder of this chapter is to discuss cases of this kind.

3. Models without theory

There are models that are independent of any theory. An often-discussed example of such a model is the so-called Lotka–Volterra model.[7] Volterra's version of the model is about the fish population in the Adriatic Sea. Volterra conceptualised the problem as a population-level phenomenon with a population of predators interacting with a population of prey. The populations are described solely in terms of their sizes, and no biological facts about the animals that constitute the populations are taken into account (beyond the obvious truism that predators eat prey and not *vice versa*). Let N_1 be the number of prey and N_2 the number of predators. Volterra then asked how these numbers change over time. The change in these numbers is due to intrinsic births and deaths in both populations, as well as to the interaction between the two. The general form of the interaction can therefore be expressed as follows (Kingsland 1985, 109–100):

Change in N_1 per unit of time = Natural increase in N_1 per unit of time minus decrease in N_1 per unit of time due to destruction of prey by predators
Change in N_2 per unit of time = Increase in N_2 per unit of time due to ingestion of prey by predators minus decrease of N_2 due to deaths of predators per unit of time.

These "verbal equalities" can be turned into proper mathematical equations by replacing the natural numbers N_1 and N_2 by the continuous quantities V (for the quantity of prey) and P (for the quantity of predators) and by choosing specific functions for the population growth and the interactions between the populations. The simplest choice is to assume that each population grows linearly and that the interaction between the populations (predators eating prey and growing as result) is proportional to the product of the two densities. Inputting these formal choices into the above equalities leads to the so-called Lotka–Volterra equations (Weisberg and Reisman 2008, 111):

$$\begin{aligned} \dot{V} &= rV - (aV)P \\ \dot{P} &= b(aV)P - mP, \end{aligned} \tag{2.1}$$

where r is the birth rate of the prey population; m is the death rate of the predator population; and a and b are linear response parameters. The dots on V and P indicate the first derivative with respect to time. Intuitively, $\dot{V}$ is the rate of change of V and ditto for $\dot{P}$.

Even though Volterra notes that Darwin had made an observation similar to his own (1926, 559), neither Darwinian evolutionary theory nor any other biological theory is at work in the model. Indeed, the model has been constructed without a theoretical framework, and it does not instantiate theoretical principles. As a result, the model is independent of theory.

The Lotka–Volterra model is not an isolated instance. The Schelling model of social segregation (Schelling 1978), the Fibonacci model of population growth (Bacaër 2011, chap. 1), the logistic model of population growth (May 1976), the Akerlof model of the market for used cars (Akerlof 1970), and complexity models for the behaviour of sand piles (Bak 1997) are "theory-free" in the same way. Models of this kind are sometimes characterised as bottom-up models. A model is *bottom-up* if the process of model construction departs from the basic features of the target and from what we know about the unfolding of events in the domain of interest, while not relying on general theories. Bottom-up models contrast with top-down models. A model is *top-down* if the process of model construction starts with a theoretical framework, and the model is built by working the way down from the theory to the phenomena. The Newtonian model of planetary motion is an example of a top-down model. The process of model construction starts with Newton's general equation of motion and the law of gravity, and then various steps are made to apply these general principles to the phenomenon of interest, namely the movement of planets.

A special case of models that are independent of theories are models that are built with the express aim of aiding the construction of theories. Leplin emphasises the importance of models in the construction of theories and calls models that are constructed with this purpose in mind *developmental models* (1980, 274). A developmental model "opens several lines of research toward the development" of a theory (278). The importance of models in the development of theories has also been emphasised by other authors. Cushing notes that "[a]n important tool in this process of theory construction is the use of models" (1982, 32), and he illustrates this with a detailed case study from high-energy physics. Hartmann observes that "[a]s a major tool for theory construction, scientists use models" (1995, 49), and he illustrates this with how quantum chromodynamics, the fundamental theory of strong interactions, has been constructed "by means of a hierarchy of consecutive *Developmental Models*" (59). Wimsatt, finally, sees "false models as a means to truer theories" and discusses their construction in the context of evolutionary biology (Wimsatt 2007, chap. 6).

4. Models as a means to explore theories

Models can also be used to explore the features of theories. A case in point is the study of non-linear dynamics. For a long time, it was thought that Newtonian mechanics was dynamically stable, meaning that a small variation in the initial condition of the system would result in a small variation in the trajectory of the system. This belief was shattered at the beginning of the 20th century when Poincaré discovered that Newtonian systems can display what is now known as *sensitive dependence on initial conditions*, which is often taken to be the defining feature of chaos.[8] This raises the question of how the dynamic of such systems looks like. Unfortunately, one cannot simply write down the solutions of the equations of motion of such systems and study their properties; and even if one could write down the solutions, they would be objects in high-dimensional mathematical spaces that are hard to trace and impossible to visualise. Thus, other means to understand the behaviour of such systems must be found, and models play a crucial role in this.

Abstract considerations about the qualitative behaviour of solutions in chaotic systems show that there is a mechanism that has been dubbed *stretching and folding*. Nearby initial conditions drift away from each other, which amounts to stretching the area where they lie. The motion of chaotic systems is such that the system's movement is confined to a restricted part of the state space. This means that the stretching cannot continue forever, and the stretched bits must be folded back onto each other. In practice, it is impossible to trace this stretching and folding in the full state space of a system. To obtain an idea of the complexity of the dynamic exhibiting stretching and folding, Smale proposed to study a model of the flow. The model is a simple two-dimensional map, now known as the horseshoe map (Tabor 1989, 200–202), which is illustrated in Figure 2.1.

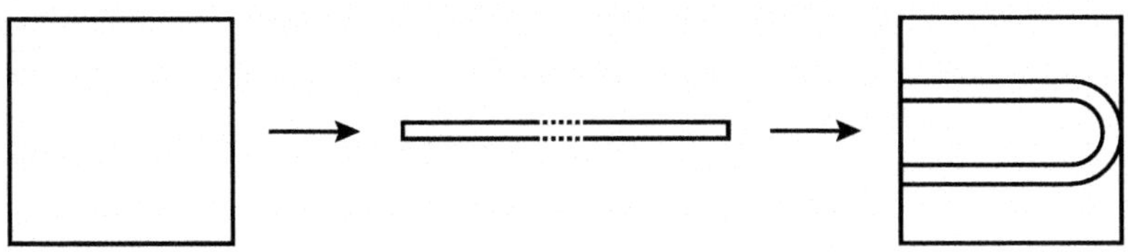

Figure 2.1 The horseshoe map. The dots indicate that the strip is longer than can be shown in the figure.

The map begins by stretching a rectangle horizontally while squeezing it vertically, which turns the rectangle into a strip; it then folds the strip back onto the initial square. The map is designed to "mimic" the stretching and folding motion of the full Newtonian dynamic, but without having any of its mathematical complexities. In this way, the horseshoe map provides a model of an important aspect of the full dynamic of Newtonian theory. The horseshoe map has a number of interesting and important features (Ott 1993, 108–114). An *invariant set* is a set of states that does not change under the dynamic of a model – it is as if the set was not "affected" by the changes that the dynamic brings with it. One can show that the so-called Cantor set is an invariant set of the horseshoe. This is interesting because the Cantor set is a fractal, and so we learn from the model that chaotic dynamical systems can have invariant sets that are fractals. In this way, the simple model of the horseshoe provides a crucial insight into the properties of the theory. The horseshoe is no isolated instance: chaos theory is rife with maps that model certain aspects of the full dynamic and thereby shed light on the nature of the theory itself.[9]

Chaos theory is no exception, and models are used in many contexts to explore the properties of theories. In statistical mechanics, the Kac ring model is used to study the equilibrium properties of the full theory (Jebeile 2020; Lavis 2008). In quantum field theory, the φ^4 model is used to explore theoretical properties like symmetry breaking and renormalisability (Hartmann 1995). The Phillips–Newlyn machine, a material model, is used to explore the properties of Hicks' formalisation of Keynes' theory (Barr 2000; Morgan and Boumans 2004). And the dome model is used to understand causality and determinism in Newtonian mechanics (Norton 2008).

5. Models complementing theories

Theories can be incompletely specified. Models can then step in and add what is missing. The model and the theory thereby enter into a symbiotic relationship in which a model complements the theory. The nature of this "completion" depends on the specifics of the case. Redhead (1980, 147) mentions the case of axiomatic quantum field theory. The theory is an attempt to offer a mathematically rigorous formulation of quantised fields. In its most common formulation, the theory is based on the so-called Wightman axioms. Roughly, the axioms say things like that fields must be invariant under the transformations of Einstein's theory of special relativity and that fields can be expressed as sums of operators acting on the vacuum state.[10] This means that the theory's axioms only impose certain general constraints on fields, and the specifics of particular fields and their interactions are given by models. In doing so, the model provides missing details and enriches the theory. This is not an easy task because it turns out that identifying models that satisfy the axioms of the theory is rather difficult.

Another way in which a theory can be incompletely specified is identified by Apostel when he notes that there are cases where "a qualitative theory is known for a field and the model introduces quantitative precision" (1961, 2). As an example, consider the so-called *quantity theory of money* in monetary economics.[11] The "quantity theory" is purely qualitative and essentially says that the price of goods in an economy is determined by the amount of money in circulation. This law leaves open what the price levels are and how they vary as a function of money supply. To answer these quantitative questions, Fisher constructed a model that is now known as *Fisher's equation of exchange*. The model considers an economy that can be characterised by four quantities: the amount M of money in circulation, the transaction velocity V of money, the level of prices P, and the volume of trade Y. All these are variables with precise numerical values that can, in principle, be measured empirically. The equation of exchange is $MV = PY$. If velocity and volume are constant, the equation says that $P = cM$, where c is a constant. So if the amount of money increases by ΔM, then prices go up by $c\Delta M$. In this way, Fisher's model gives quantitative specificity to the qualitative law of the theory.

Harré (2004) noted that models can complement theories by providing mechanisms for processes left unspecified in the theory but that are nevertheless responsible for bringing about the observed phenomena (2004, chap. 1). In some cases, the model mechanism is known; in other cases, it is hypothesised. The notion of a mechanism is broad, and Harré emphasised that it is not restricted to "anything specifically mechanical": a "[c]lockwork is a mechanism, Faraday's strained space is a mechanism, electron quantum jumps is a mechanism, and so on" (2004, 4).

Models can also step in when theories are too complex to handle. This can happen, for instance, when the equations of the theory are mathematically intractable. In such cases, one can find a model that approximates the theory. As Redhead noted, this can be done in two ways (1980, 150–152): either one finds approximate solutions to exact equations or one finds an approximate equation that one can solve exactly. If one finds either an approximate solution or an approximate equation, these can be seen as approximate models of the theory. However, models can also step in when the relation between the model and the theory is not a clearly defined mathematical approximation. Hartmann (1999) discusses the case of quark confinement in elementary particle physics. The nucleus of atoms is made up of nucleons: protons and neutrons. Nucleons themselves are made up of quarks. How do quarks interact to form a stable nucleon? The general theory covering the behaviour of quarks is quantum chromodynamics. Unfortunately, the theory is too complicated to apply to protons. Computer simulations suggest that at low energies so-called quark confinement occurs, and quarks come together to form nucleons. This, however, leaves the nature of this confinement unexplained and poorly understood, with a number of different kinds of confinement possible and the theory unable to adjudicate between them. To fill in this gap, physicists constructed a phenomenological model, now known as the MIT bag model, which takes the main known features of the theory into account and fills the missing details with postulated configurations. According to the model, nucleons consist of three massive quarks that move freely in a rigid sphere of radius R, where the sphere guarantees that the quarks remain confined within the nucleon. This assumption is motivated by the basic theory, but it does not deductively follow from it. The model then allows for the calculation of the radius R and the total energy of the particle. In this way, the model yields results where the theory is silent, and it fills a gap that the theory leaves open.

6. Applying theories through models

Cartwright argues that models not only aid the application of theories that are somehow incomplete; she submits that models are always involved when a theory with an overarching mathematical structure is applied to a target system. The main theories in physics fall into this category: classical mechanics, quantum mechanics, electrodynamics, and so on. In fact, applying such theories involves two kinds of models: interpretative models and representative models.

Let us begin with *interpretative models*. Overarching mathematical theories like classical mechanics appear to provide general descriptions of a wide range of objects that fall within their scope. However, on closer inspection, it turns out that these theories do not apply to the world directly. The reason for this is that they employ abstract terms, i.e. terms that apply to a target system only if a description couched in more concrete terms also applies to the target. Cartwright offers the following two conditions for a concept to be abstract relative to another concept:

> First, a concept that is abstract relative to another more concrete set of descriptions never applies unless one of the more concrete descriptions also applies. These are the descriptions that can be used to "fit out" the abstract description on any given occasion. Second, satisfying the associated concrete description that applies on a particular occasion is what satisfying the abstract description consists in on that occasion.
> *(1999b, 39)*

She offers the example of *work*. Having responded to an email, having revised a section of a paper, and having attended a meeting is what my having done work this morning consists in. If I tell a friend over lunch what I have done and he responds, "well, you've responded to an email, revised a section, and attended a meeting, but when did you work?", he either does not understand the concept of work or, more likely, is joking with me.

Cartwright submits that important concepts that appear in mathematised theories are abstract in the same way as *work*. The concept of *force*, for instance, is abstract in that it applies only if a more concrete concept also applies. There is no such thing as "nothing but a force" acting on a body. There being a force between two bodies on a particular occasion consists in them gravitationally attracting each other, or electrostatically repelling each other, or … These more concrete claims fit out the abstract claim of there being a force. Force, therefore, is an abstract property and "Newton's law tells that whatever has this property has another, namely having a mass and an acceleration which, when multiplied together, give the […] numerical value, *F*" (1999b, 43). Force, therefore, has no independent existence; it exists only in its more specific forms like gravity, electrostatics, and so on. Specifying what concrete claims fit out abstract claims amounts to specifying an *interpretative model*. An interpretative model then consists of the "actors" that fit out the abstract claims of the theory.

Let us now turn to *representative models*. Cartwright regards representative models as ones that are built to "represent real arrangements and affairs that take place in the world" (1999b, 180). These models have two crucial features. The first is that they are highly idealised. Constructing a representative model involves twisting and distorting the properties of the target in many ways and the result of this process is in no way a mirror image of the target. Indeed, Cartwright notes that "it is not essential that the models accurately describe everything that actually happens; and in general it will not be possible for them to do so" (1983, 140). Second, all these distortions notwithstanding, the model still is a representation of the target, albeit one that is inaccurate in certain respects. The principles of the theory therefore apply to "highly fictionalized objects" (1983, 136) in the representational model. So, one has to distort reality to force it into the corset of the theory: "our prepared descriptions lie" because "in general we will have to distort the true picture of what happens if we want to fit it into the highly constrained structures of our mathematical theories" (139). Without these distortions, the theory would be inapplicable.

We are now in a position to see how the two notions of an interpretative model and a representational model work together in the application of a theory to a real-world target. To apply a theory, scientists must construct a model. This model must be such that it is, at once, an interpretative model of the general theory at hand (which means that it is couched in terms of concepts that fit out the abstract concepts of the theory) and a representative model of the target system (which means that it stands in a certain representational relation to the target).

7. Models as mediators

The relation between models and theories can be even looser than in the cases we have discussed so far. The contributors to a programmatic collection of essays edited by Morgan and Morrison (1999b) rally around the idea of "models as mediators", and so it is apt to call the vision of modelling that emerges from this project the *Models as Mediators View*. This view sees models as instruments that mediate between theories and the world while remaining independent from both. Models are, therefore, as Morgan and Morrison put it, "autonomous agents" (1999a, 10). The autonomy of models has four dimensions: construction, functioning, representing, and learning (10–12). Let us look at each of these in turn.

The first and most important dimension is independence in construction. Morgan and Morrison observe that "model construction is carried out in a way which is to a large extent independent of theory" (1999a, 13), and Morrison locates models as being "between physics and the physical world" (1998, 65). This is because "theory does not provide us with an algorithm from which the model is constructed and by which all modelling decisions are determined" (Morgan and Morrison 1999a, 16). In her contribution to the collection, Cartwright portrays the Semantic View of theories as a "vending machine" view of model construction:

> The theory is a vending machine: you feed it input in certain prescribed forms for the desired output; it gurgitates for a while; then it drops out the sought-for representation, plonk, on the tray, fully formed, as Athena from the brain of Zeus. This image of the relation of theory to the models we use to represent the world is hard to fit with what we know of how science works. Producing a model of a new phenomenon such as superconductivity is an incredibly difficult and creative activity.
>
> *(1999b, 247)*

According to Cartwright, the "vending machine view" of theories is wrong on at least two counts. First, it erroneously assumes that all ingredients that are needed for the construction of a model are already contained in the theory. As we have seen in the previous section, she sees representative models as an essential ingredient for the application of a theory. The construction of such a model requires resources that go beyond what theories can offer. Discussing quantum models of superconductivity, Cartwright notes that theories leave out much of what is needed to produce a model capable of generating an empirical prediction. While theories contain general principles, they contain no information either about the real materials from which a superconductor is built or about the various approximation schemes and the mathematical techniques needed to handle them. Second, the view is wrong in assuming that models embody only one theory. The internal setup of a model is often a complicated conglomerate of elements from different theories. Cartwright illustrates this point with the Ginzburg–Landau model of superconductivity (1999a, 244–245), but the point also holds about other models like the classical London model of superconductivity (Suárez 1999) and models of business cycles (Boumans 1999). The same is also true of contemporary climate models which incorporate elements from different theories, including mechanics, fluid dynamics, electrodynamics, quantum theory, chemistry, and biology (Frigg, Thompson, and Werndl 2015). Models of this kind do not belong to a family of models that form a theory in anything like the way that the Semantic View posits; in fact, they do not belong to any particular theoretical framework at all.

The second dimension of autonomy is functioning: models can perform many functions without relying on theories. One of these functions is to aid theory construction (Morgan and Morrison 1999a, 18). As we have seen previously, models can play a role in theory construction (Section 3) and in exploring theories (Section 4), which they can do only if they are autonomous from theories. Models also serve as a means for policy intervention (Morgan and Morrison 1999a, 24). Central banks use economic models to inform monetary policy decisions, for instance, whether to change the base rate, and models can do this independently from theory.

Representation is the third dimension of autonomy. Morgan and Morrison point out that the "critical difference between a simple tool and a tool of investigation is that the latter involves some form of representation: models typically represent either some aspect of the world, or some aspect of our theories about the world, or both at once" (1999a, 11). They emphasise that representing does not presuppose that there is "a kind of mirroring of a phenomenon, system or theory by a model" because representing is in no way tantamount to producing a copy, or effigy, of the target.[12]

The final dimension of autonomy is learning. Morgan and Morrison point out that we learn from models and argue that this happens in two places: in building the model and in manipulating it (1999a, 11–12). As we have seen earlier in this section, there are no general rules or algorithms for model building and hence insights gained into what fits together and how during the process of construction are invaluable sources for learning about the model (30–31). The second place to learn about the model is when we manipulate it. Morgan (1999) notes that Fisher did not find out about the properties of his monetary models by contemplating them, but by manipulating them to show how the various parts of the model work together to produce certain results.

8. Separating models from theories

So far, we worked under the assumption that models and theories are clearly distinct, and we focussed on the relation between them. In practice, this is not always a realistic assumption. In fact, in some cases it is not clear where the line between them should be drawn, and whether something is a model or a theory. An example is Bohr's account of the atom, which is sometimes referred to as the "Bohr model" and sometimes as the "Bohr theory" of the atom. This problem not only besets philosophical analysis; it also arises in scientific practice. Bailer-Jones interviewed a group of nine physicists about their understanding of models and their relation to theories. She reports that the following views were expressed (2002, 293):

1 There is no real difference between model and theory.
2 Models turn into theories once they are better and better confirmed.

3 Models contain necessary simplifications and deliberate omissions, while theories are the best we can do in terms of accuracy.
4 Theories are more general than models. Modelling becomes a case of applying general theories to specific cases.

The first suggestion is too radical to do justice to many aspects of practice, where a distinction between models and theories is clearly made. The second view is encapsulated in phrases like "it's just a model", which indicate either that scientists take a cautious attitude towards a certain proposition that they regard as speculative or provisional, or that something is known to be false and entertained only for heuristic purposes. But, models and theories are not distinguished by their degree of confirmation. There can be well-confirmed models and unconfirmed theories. The third proposal is up to something, but it ultimately does not hold water. It is true that models involve idealisations and omissions of all kinds, but so do theories. Newtonian mechanics, for instance, deals with point masses that move in a Euclidean space, and it omits most properties of the objects in its target domain (it omits, for instance, colour, temperature, and chemical constitution of its targets) but that does not seem to strip Newtonian mechanics of its status as a theory.

The fourth suggestion is closely aligned with a view that has emerged in the literature on models. In the wake of the debates we have reviewed in this chapter, models have become the focal point of attention and the emphasis has shifted so far away from theories that Morrison detects the need for a "redress of the imbalance" (2007, 195). She asks "where have all the theories gone" and then sets out to articulate how theories are different from models. Morrison points out that models contain a great deal of "excess" structure like approximation methods, mathematical techniques, and highly stylised descriptions of certain parts of the target, and she notes that one would not want to count these as part of a theory (197). This can be avoided if "theory" is reserved for a "theoretical core", which contains the constitutive assumptions of the theory. In the case of Newtonian mechanics, the core consists of the three laws of motion and the law of universal gravitation (197), in the case of classical electrodynamics of Maxwell's equations, in the case of relativistic quantum mechanics of the Dirac equation (205), and in the case of quantum mechanics of the Schrödinger equation (214). The core of a theory constrains the behaviour of objects that fall within the scope of the theory, and it plays a crucial role in the construction of models. Models concretise the abstract laws of the theory and put them to use by adding elements that are specific to the situation. In this way, theories assist the construction of models without determining the way in which they are built. Models are specific in that they are adapted to a particular situation and a particular problem, while the theories on which they are based contain the general principles of wide scope.

The problem with the "theoretical core" view of theories as presented by Morrison is that the notion of a theoretical core is introduced through examples – Newton's laws of motion, Maxwell's equations, and so on – and is then not further analysed. Morrison seems to regard this as an advantage when she observes that "nothing about this way of identifying theories requires that they be formalized or axiomatized" (2007, 205). However, this pragmatism must seem unsatisfactory to those who have contributed to the development of the two grand views of theories and who will feel that we have now come full circle. Neither the Syntactic View nor the Semantic View would disagree that what makes a theory a theory is a theoretical core. The question they are concerned with is how this notion can be analysed and what kind of objects theoretical principles are. This question is left open.

9. Conclusion

We have discussed a number of different relationships between models and theories that can be found in the practice of science. These range from complete independence to total dependence, and many things in between. Many of these cases do not seem to sit well either with the Syntactic View or with the Semantic View, and they show that there is nothing like "the" relation between models and theories.

Notes

1 Sections 3–8 of this chapter are based on Chapter 13 of my (2023).
2 I note that the label "Syntactic View" is a misnomer because it gives the mistaken impression that the view only deals with the syntax of theories. Some readers may object to calling the Syntactic View an orthodoxy because it has been superseded by the Semantic View long ago. This narrative has become untenable in the last decade, when the Syntactic View had a veritable revival. For a discussion, see, for instance, Halvorson (2016).
3 The exact form of correspondence rules has been the subject matter of extensive debates. For a survey, see, for instance, Percival (2000).
4 For a detailed discussion of the problems faced by both the Syntactic View and the Semantic View, see Chapters 1–8 of my (2023) and references therein.
5 The locus classicus for a discussion of analogies is Hesse (1963). For further discussions of analogies and analogical models, see Chapter 10 of my (2023) and references therein.
6 For a discussion, see, for instance, Friedman (1983).
7 The model was formulated by Lotka (1925) and Volterra (1926). Kingsland (1985, chap. 5) gives a historical account of the development of the model. For philosophical discussions, see, for instance, Knuuttila and Loettgers (2017) and Weisberg and Reisman (2008).
8 For basic introductions to chaos and discussions of its philosophical ramifications, see Kellert (1993) and Smith (1998). Argyris, Faust and Haase (1994) and Tabor (1989) offer advanced discussions. Parker (1998) discusses the question of whether it was really Poincaré who discovered chaos.
9 For instance, the dynamics of KAM type systems near a hyperbolic fixed point can be modelled by the baker's transformation. For a discussion, see Berkovitz, Frigg, and Kronz (2006, 680–687).
10 For a discussion of quantum field theory, see, for instance, Ruetsche (2011).
11 Apostel does not provide an example. I am grateful to Julian Reiss for suggesting the quantity theory of money to me. For a discussion of the theory, see Humphrey (1974).
12 For a discussion of how models represent their targets, see Frigg and Nguyen (2020) and Nguyen and Frigg (2022).

References

Akerlof, George A. 1970. "The Market for 'Lemons': Quality Uncertainty and the Market Mechanism." *The Quarterly Journal of Economics* 84(3): 488–500. https://doi.org/10.2307/1879431.

Apostel, Leo. 1961. "Towards the Formal Study of Models in the Non-Formal Sciences." In *The Concept and the Role of the Model in Mathematics and Natural and Social Sciences*, edited by Hans Freudenthal, 1–37. Synthese Library. Dordrecht: Springer Netherlands. https://doi.org/10.1007/978-94-010-3667-2_1.

Argyris, John H., Gunter Faust, and Maria Haase. 1994. *An Exploration of Chaos: An Introduction for Natural Scientists and Engineers*. North-Holland: Elsevier.

Bacaër, Nicolas. 2011. *A Short History of Mathematical Population Dynamics. A Short History of Mathematical Population Dynamics*. London: Springer. https://doi.org/10.1007/978-0-85729-115-8.

Bailer-Jones, Daniela M. 2002. "Scientists' Thoughts on Scientific Models." *Perspectives on Science* 10(3): 275–301. https://doi.org/10.1162/106361402321899069.

Bak, Per. 1997. *How Nature Works : The Science of Self-Organized Criticality*. Oxford: Oxford University Press.

Balzer, Wolfgang, Carlos U. Moulines, and Joseph D. Sneed. 1987. *An Architectonic for Science: The Structuralist Program*. Dordrecht: Reidel.

Barr, Nicholas. 2000. "The History of the Phillips Machine." In *A. W. H. Phillips: Collected Works in Contemporary Perspective*, edited by Robert Leeson, 89–114. Cambridge: Cambridge University Press. https://doi.org/10.1017/CBO9780511521980.013.

Berkovitz, Joseph, Roman Frigg, and Fred Kronz. 2006. "The Ergodic Hierarchy, Randomness and Chaos." *Studies in History and Philosophy of Science Part B: Studies in History and Philosophy of Modern Physics* 37(4): 661–691.

Boumans, Marcel. 1999. "Built-In Justification." In *Models as Mediators: Perspectives on Natural and Social Science*, edited by Mary S. Morgan and Margaret Morrison, 66–96. Cambridge: Cambridge University Press.

Braithwaite, Richard Bevan. 1953. *Scientific Explanation: A Study of the Function of Theory, Probability and Law in Science*. Cambridge: Cambridge University Press.

———. 1954. "The nature of theoretical concepts and the role of models in an advanced science." *Revue Internationale de Philosophie* 8(27/28), 34–40. https://doi.org/10/39899.

———. 1962. "Models in the Empirical Sciences." In *Logic, Methodology and Philosophy of Science*, edited by Ernest Nagel, Patrick Suppes, and Alfred Tarski, 224–231. Stanford: Stanford University Press.

Carnap, Rudolf. 1923. "Über die Aufgabe der Physik: und die Anwendung des Grundsatzes der Einfachstheit." *Kant-Studien* 28(1–2): 90–107. https://doi.org/10.1515/kant-1923-0107.

———. 1938. "Foundations of Logic and Mathematics." In *International Encyclopedia of Unified Science: Foundations of the Unity of Science*, edited by Otto Neurath, Rudolf Carnap, and Charles W. Morris. Chicago: University of Chicago Press.

Cartwright, Nancy. 1983. *How the Laws of Physics Lie*. Oxford: Oxford University Press.

———. 1999a. "Models and the Limits of Theory: Quantum Hamiltonians and the BCS Models of Superconductivity." In *Models as Mediators: Perspectives on Natural and Social Science*, edited by Margaret Morrison and Mary S. Morgan, 241–281. Ideas in Context. Cambridge: Cambridge University Press. https://doi.org/10.1017/CBO9780511660108.010.

———. 1999b. *The Dappled World: A Study of the Boundaries of Science*. Cambridge: Cambridge University Press. https://doi.org/10.1017/CBO9781139167093.

Costa, Newton C. A. da, and Steven French. 1990. "The Model-Theoretic Approach in the Philosophy of Science." *Philosophy of Science* 57: 248–265.

Cushing, James T. 1982. "Models, High-Energy Theoretical Physics and Realism." *PSA: Proceedings of the Biennial Meeting of the Philosophy of Science Association*. Vol. 1982, Volume Two: Symposia and Invited Papers, 31–56. Cambridge University Press.

Fraassen, Bas C. van. 1980. *The Scientific Image*. Clarendon Library of Logic and Philosophy. Oxford: Oxford University Press. https://doi.org/10.1093/0198244274.001.0001.

Friedman, Michael. 1983. *Foundations of Space-Time Theories. Relativistic Physics and Philosophy of Science*. Princeton, NJ: Princeton University Press.

Frigg, Roman. 2023. *Models and Theories. A Philosophical Inquiry*. London: Routledge.

Frigg, Roman, and James Nguyen. 2020. *Modelling Nature: An Opinionated Introduction to Scientific Representation*. Berlin and New York: Springer Nature.

Frigg, Roman, Erica Thompson, and Charlotte Werndl. 2015. Philosophy of climate science part II: modelling climate change. *Philosophy Compass* 10(12): 965-77.

Giere, Ronald N. 1988. *Explaining Science: A Cognitive Approach*. Chicago: University of Chicago Press.

Halvorson, Hans. 2016. "Scientific Theories." In *The Oxford Handbook of Philosophy of Science*, edited by Paul Humphreys, 585–608. Oxford: Oxford University Press.

Hartmann, Stephan. 1995. "Models as a Tool for Theory Construction: Some Strategies of Preliminary Physics." In *Theories and Modelsin Scientific Processes*, edited by William E. Herfel, Władysław Krajewski, Ilkka Niiniluoto, and Ryszard Wójcicki, 49–67. Poznań Studies in the Philosophy of Science and the Humanities 44. Amsterdam: Rodopi.

———. 1999. "Models and Stories in Hadron Physics." In *Models as Mediators: Perspectives on Natural and Social Science*, edited by Margaret Morrison and Mary S. Morgan, 326–346. Ideas in Context. Cambridge: Cambridge University Press. https://doi.org/10.1017/CBO9780511660108.012.

Hempel, Carl G. 1966. *Philosophy of Natural Science*. Princeton, NJ: Princeton University Press.

Hempel, Carl G. 1970. "On the 'Standard Conception' of Scientific Theories." In *Minnesota Studies in the Philosophy of Science Vol 4*, edited by Stephen Winokur, and Michael Radner, Minneapolis: University of Minnesota Press. http://conservancy.umn.edu/handle/11299/184647.

Hesse, Mary B. 1963. *Models and Analogies in Science*. London: Sheed and Ward.

Humphrey, Thomas M. 1974. "The Quantity Theory of Money: Its Historical Evolution and Role in Policy Debates." *Economic Review* 60(May): 2–19.

Jebeile, Julie. 2020. "The Kac Ring or the Art of Making Idealisations." *Foundations of Physics* 50(10): 1152–1170. https://doi.org/10.1007/s10701-020-00373-1.

Kellert, Stephen H. 1993. *In the Wake of Chaos: Unpredictable Order in Dynamical Systems*. Science and Its Conceptual Foundations Series. Chicago, IL: University of Chicago Press. https://press.uchicago.edu/ucp/books/book/chicago/I/bo3645742.html.

Kingsland, S. 1985. *Modeling Nature*. Chicago and London: The University of Chicago Press.

Knuuttila, Tarja, and Andrea Loettgers. 2017. "Modelling as Indirect Representation? The Lotka–Volterra Model Revisited." *The British Journal for the Philosophy of Science* 68(4): 1007–1036. https://doi.org/10.1093/bjps/axv055.

Lavis, David A. 2008. "Boltzmann, Gibbs, and the Concept of Equilibrium." *Philosophy of Science* 75(5): 682–696. https://doi.org/10.1086/594514.

Leplin, Jarrett. 1980. "The Role of Models in Theory Construction." In *Scientific Discovery, Logic, and Rationality*, edited by Thomas Nickles, 267–283. Boston Studies in the Philosophy of Science. Dordrecht: Springer Netherlands. https://doi.org/10.1007/978-94-009-8986-3_12.

Lotka, Alfred J. 1925. *Elements of Physical Biology*. Baltimore, MD: Williams and Wilkins.

Lutz, Sebastian. 2012. "On a straw man in the philosophy of science: a defence of the received view." *HOPOS*, 2(1): 77–120.

May, Robert M. 1976. "Simple Mathematical Models with Very Complicated Dynamics." *Nature* 261(5560): 459–467. https://doi.org/10.1038/261459a0.

Morgan, Mary S. 1999. "Learning from Models." In *Models as Mediators: Perspectives on Natural and Social Science*, edited by Margaret Morrison and Mary S. Morgan, 347–388. Ideas in Context. Cambridge: Cambridge University Press. https://doi.org/10.1017/CBO9780511660108.013.

Morgan, Mary S., and Marcel Boumans. 2004. "Secrets Hidden by Two-Dimensionality: The Economy as a Hydraulic Machine." In *Models: The Third Dimension of Science*, edited by Soraya De Chadarevian and Nick Hopwood, 369–401. Palo Alto: Stanford University Press. http://www.sup.org/book.cgi?isbn=0804739722.

Morgan, Mary S., and Margaret Morrison. 1999a. "Models as Mediating Instruments." In *Models as Mediators: Perspectives on Natural and Social Science*, edited by Margaret Morrison and Mary S. Morgan, 10–37. Ideas in Context. Cambridge: Cambridge University Press. https://doi.org/10.1017/CBO9780511660108.003.

———, eds. 1999b. *Models as Mediators: Perspectives on Natural and Social Science*. 1st ed. Cambridge University Press. https://doi.org/10.1017/CBO9780511660108.

Morrison, Margaret. 1998. "Modelling Nature: Between Physics and the Physical World." *Philosophia Naturalis* 35(1): 65–85.

———. 2007. "Where Have All the Theories Gone?" *Philosophy of Science* 74(2): 195–228. https://doi.org/10.1086/520778.

Nagel, Ernest. 1961. *The Structure of Science*. London: Routledge and Keagan Paul.

Nguyen, James, and Roman Frigg. 2022. "Scientific Representation." *Elements in the Philosophy of Science*, August. https://doi.org/10.1017/9781009003575.

Norton, John D. 2008. "The Dome: An Unexpectedly Simple Failure of Determinism." *Philosophy of Science* 75(5): 786–798. https://doi.org/10.1086/594524.

Ott, Edward. 1993. *Chaos in Dynamical Systems*. Cambridge; New York: Cambridge University Press.

Parker, Matthew W. 1998. "Did Poincare Really Discover Chaos?" *Studies in the History and Philosophy of Modern Physics* 29(4): 575–588.

Percival, Philip. 2000. "Theoretical Terms: Meaning and Reference." In *A Companion to the Philosophy of Science*, 495–514. Oxford: Wiley-Blackwell. https://doi.org/10.1002/9781405164481.ch72.

Redhead, Michael. 1980. "Models in Physics." *British Journal for the Philosophy of Science* 31(2): 145–163. https://doi.org/10.1093/bjps/31.2.145.

Rom, Harré. 2004. *Modeling: Gateway to the Unknown: A Work by Rom Harre*, edited by Daniel Rothbart. 1st edition. Amsterdam: Elsevier Science.

Ruetsche, Laura. 2011. *Interpreting Quantum Theories*. 1st edition. Oxford; New York: Oxford University Press.

Schelling, Thomas C. 1978. *Micromotives and Macrobehavior*. New York: Norton.
Schlick, Martin. 1925. *Allgemeine Erkenntnislehre*. 2nd edition. Berlin: Springer.
Smith, Peter. 1998. *Explaining Chaos*. Cambridge: Cambridge University Press. https://doi.org/10.1017/CBO9780511554544.
Suárez, Mauricio. 1999. "The Role of Models in the Application of Scientific Theories: Epistemological Implications." In *Models as Mediators: Perspectives on Natural and Social Science*, edited by Margaret Morrison and Mary S. Morgan, 168–196. Ideas in Context. Cambridge: Cambridge University Press. https://doi.org/10.1017/CBO9780511660108.008.
Suppes, Patrick. 2002. *Representation and Invariance of Scientific Structures*. Stanford: CSLI Publications (distributed by Chicago University Press).
Tabor, Michael. 1989. *Chaos and Integrability in Nonlinear Dynamics: An Introduction*. 1st edition. New York: Wiley-Interscience.
Volterra, Vito. 1926. "Fluctuations in the Abundance of a Species Considered Mathematically1." *Nature* 118(2972): 558–560. https://doi.org/10.1038/118558a0.
Weisberg, Michael, and Kenneth Reisman. 2008. "The Robust Volterra Principle*." *Philosophy of Science* 75(1): 106–131. https://doi.org/10.1086/588395.
Wimsatt, William C. 2007. *Re-Engineering Philosophy for Limited Beings: Piecewise Approximations to Reality*. Cambridge, MA: Harvard University Press. https://doi.org/10.2307/j.ctv1pncnrh.

3

PRACTICE-ORIENTED APPROACHES TO SCIENTIFIC MODELING

Axel Gelfert

1. Introduction

Whether we are dealing with climate change or the dynamics of an unfolding pandemic, with developing new materials or genetically modifying model organisms: models, and the practice of modeling, are indispensable to our contemporary ways of navigating the world. This is true not only for basic research, but also for applied questions in relation to global social, economic, and ecological challenges, which often require modeling possible scenarios in which events, or the dynamics of global systems, may unfold. For this reason, in conjunction with computer simulations and visualization methods, models are fast becoming the dominant interface between science and the public. Within science, modeling is indispensable whenever theoretical derivation or direct observation of phenomena is beyond our reach. Yet despite this central role, scientific models are still often treated as mere makeshift solutions that we may have to depend on for practical purposes but which we would rather do without. In doing so, models are often treated as isolated *products*, stripped of any contextual information about their origins, the underlying motivating concerns, and the model-building practices that gave rise to them. This chapter attempts to shift the focus to the *practice* of scientific modeling. To this end, it surveys (and adds to) a number of philosophical proposals that conceive of models not as abstract, self-contained entities, but as temporary, dynamically shifting intermediate formations of underlying scientific practices – where the latter, in particular, can serve a multiplicity of epistemic and non-epistemic goals.

The rest of this chapter is organized as follows: Section 1 sketches the historical background to the emergence of scientific modeling as a separate, discernible methodology within scientific practice. Section 2 distinguishes between representational and non-representational uses of scientific models, while arguing that scientific practice casts doubt on the idea that scientific models are primarily representational in character. Section 3 elaborates on artifactualism as an attempt to respond to this practice-based insight and also discusses more radical proposals that seek to eliminate any representational idiom from philosophical discussions of scientific models. Settling for a more pluralistic outlook, Section 4 then argues that the practice of scientific modeling is constituted by a range of component activities, some of which may be illustrated (in an idealized fashion) by a hypothetical

DOI: 10.4324/9781003205647-5

episode of model-building, from the initial articulation of the model all the way to a final assessment of its adequacy relative to the goal in question. Section 5 concludes by arguing that modeling has a stabilizing influence on scientific practice by allowing model users to switch back and forth between different kinds of what has been called "intentionality relations" (Ihde 1990), i.e., ways of relating to the world with, and through, models.

1. Models as products and the practice of modeling

Ludwig Boltzmann (1844–1906), in an entry on "model" he wrote for the 1902 edition of the *Encyclopedia Britannica*, characterizes a model as

> a tangible representation, whether the size be equal, or greater, or smaller, of an object which is either in actual existence, or has to be constructed in fact or in thought. More generally it denotes a thing, whether actually existing or only mentally conceived of, whose properties are to be copied.
>
> *(Boltzmann 1902/1911, 638)*

While this particular definition is no more than a snapshot in the varied and conflicted evolution of the scientific term "model," two aspects of it are noteworthy and cast a light on the term's layered meanings. First, coming from the perspective of a scientific practitioner, this characterization suggests that there is parity between *material* models ("tangible representations") and *abstract* (mental) models ("...mentally conceived of...") – an observation whose full significance to the philosophy of science only came to be appreciated much later. Second, Boltzmann's characterization emphasizes the constructive element of model-building by acknowledging that a model "has to be constructed in fact or in thought." Without reading too much into this passing remark, it is perhaps significant that Boltzmann does not regard models as timeless entities that, as a matter of mere fortune, stand in the right sort of similarity relation to their target system; rather, models need to be constructed – *brought into existence* (as material objects in the world, or in thought) – based on active determinations by their users as to which properties of the target system "are to be copied." Thus understood, a model is not just a mere copy of a segment of reality but rather serves as a tool for attributing corresponding properties to the object it represents: "On this view our thoughts stand to things in the same relation as models to the objects they represent" (1902/1911, 638).[1]

Boltzmann refers to James Clerk Maxwell (1831–1879), whose mechanical ether model prepared the ground for the modern theory of electromagnetism, which he later elaborated. The analysis and explanation of electrical and magnetic phenomena initially faced seemingly insurmountable difficulties, not least since it was unclear what the substrate of these phenomena would have to be in order to explain the variety of newly observed phenomena. According to Boltzmann, Maxwell managed to circumvent these difficulties by combining two lines of thinking. On the one hand, if the "true nature and form" of the "constituents constituting the phenomena" was "absolutely unknown," then one should at least explore how far a genuine attempt at explanation in purely mechanical terms ("a conception of purely mechanical processes") might take us. On the other hand, Maxwell urges us to refrain from attributing any reality to the mechanical processes postulated in this way; they were merely "mechanical analogies" – mere means to the end of reproducing the observed phenomena within a theoretical description. The successful description of observed phenomena by means of such mechanical models was not meant to support the realist claim that the entities and processes posited by the analogies enjoy an independent reality; rather, the goal was to uncover partial similarities, whatever their underlying basis in reality.

Contemporary commentators have no qualms about referring to Maxwell's mechanical analogies as "models" – as, indeed, was done in the previous paragraph. Yet, as Giora Hon and Bernard Goldstein (2012) have argued, one should make an effort to disentangle the terminologies of "analogy," "model," and "hypothesis," lest one engage in the anachronistic projection of contemporary notions onto history. For, at the time Maxwell was first attempting to use mechanical terms to make sense of the newly observed electromagnetic phenomena, he still considered his mechanical analogies to be hypotheses, from which experimentally observed predictions could be derived. Only later, as he moved from physical analogies to what he dubbed "mathematical analogies," did Maxwell realize the artificiality of such representational tools as "lines of force": that is, of "mathematically identical" systems, which were acknowledged from the start to be imaginary (and so not hypotheses *about nature* per se), yet which nonetheless could stand in insightful relations to the real systems under investigation (see Hon and Goldstein 2012, 42). It is this further step of severing the link between models and their hitherto assumed status as actual hypotheses which, Hon and Goldstein argue, creates room for a genuinely new *methodology of modeling*. From our contemporary vantage point, which has a well-developed (albeit not uniformly shared) understanding of the term "model" at its disposal, it is easy to miss that the transition from isolated uses of "physical" or "mechanical" analogies to a methodology of modeling is a significant leap in scientific practice from the 19th century onwards. As Hon and Goldstein (2021) remind us: "That a model (a concept) is invoked in some scientific discussion does not mean that the methodology applied is modeling" (332).

Once a distinction is made between models as finished "products" and the activity of modeling, a space opens up for realizing that *model-building* is not merely some sort of propaedeutic exercise, the details of which become irrelevant once a model has been derived, but is itself an integral – and epistemically significant – part of the *activity of modeling*. This contrasts sharply with traditional views of how models are arrived at. On the simplest, and perhaps most naïve, view, models – specifically, theoretical models – were regarded as approximations or limiting cases of an underlying fundamental theory, serving either as toy examples for didactical purposes or as convenient ways of simplifying a complex situation so as to allow for a more straightforward application of the theory. To this day, this caricature of models as mere simplifications of a theory for particular conditions can be found in introductory science textbooks and popularizations of science. Early philosophical accounts of scientific models tended to maintain this close link between theories and models. Within the syntactic view of scientific theories, the role of models was limited to providing a semantics for a theory by specifying an interpretation on which all of its axioms come out true; models, thus, were primarily treated as a philosophical device for clarifying the nature of theories, not as a way of capturing the complexities of scientific practice. The semantic view replaced this quasi-linguistic approach with a conception of theories as a family of abstract structures – the *models* that constitute them – which stand, at least in part, in a relation of isomorphism to selected aspects of nature. Theories, as the slogan goes, were "collections of models." On the one hand, this moved philosophical accounts of models closer to scientific practice: where the syntactic view demanded "formulating abstract theoretical axioms which remain uninterpreted until observable consequences are derived," the semantic view was able to make sense of how "scientists build in their mind's eye systems of abstract objects whose properties or behavior satisfy certain constraints" (Liu 1997, 154). On the other hand, the austere conception of models as abstract structures still is a far cry from what scientists have in mind when they speak of "models," and it leaves out much of what motivates us to turn to models in inquiry – not least the fact that "they are inherently intended for specific phenomena" (Suárez 1999, 75).

In their seminal work on *models as mediators*, Margaret Morrison and Mary Morgan (1999), made a compelling case that it is the very process of model construction that imbues a model with a certain independence from both theory *and* data. Precisely because "model construction involves a complex activity of integration" (Morrison 1999, 44), it proceeds partly independently of both theory and data, thereby placing models "outside the theory-world axis" (Morrison and Morgan 1999, 18). Even when models aim to represent real-world target systems, or when they are derived from theory, they are not reducible to either, but retain a certain autonomy from both. This way, they acquire characteristics typically associated with tools (such as *multiple utilizability*), rendering them "technologies for investigation": "[W]e make use of these characteristics of partial independence, functional autonomy and representation to learn something from the manipulation" of models (Morrison and Morgan 1999, 32), where such manipulation is enabled through the diversity of formats and media – whether material, mathematical, or diagrammatic – in which a model system is realized. To be sure, earlier philosophical accounts of scientific models had occasionally acknowledged the existence of "surplus content" on the part of models; Ernan McMullin, who coined this phrase, traces this to the fact that "model-structure has some sort of basis in the 'real world'" (McMullin 1968, 395). The "models as mediators" view can be seen as one attempt at making explicit just what, in detail, constitutes this surplus content, and at acknowledging its heterogeneity. A further influence can be traced to Mary Hesse, whose work on models as analogies prepared the ground later for subsequent practice-based accounts of models, and who noted that in addition to *positive* and *negative* analogies (i.e., ways in which a model system is *similar* or *dissimilar* to its target system), there are also *neutral* analogies – viz., additional features contributed by a model, which hold out the promise of novel insights and predictions. When dealing with models, Hesse argues, we are "not dealing with static and formalized theories, corresponding only to the known positive analogy, but with theories in the process of growth" (Hesse 1963, 11–12).

Insisting on the distinction between *models* (understood as specific theoretical, conceptual, or analogical means of representation and inquiry) and an overarching methodology of *modeling* is more than a historical quibble. At the same time, the distinction is easy to miss – and may even seem quaint – from the viewpoint of contemporary science and technology, which are steeped in models and modeling approaches, and which often include explicit, discipline-specific methodologies for generating domain-adequate models. Practitioners of contemporary science are well aware of the fact that their reliance on models is not a matter of the mere ad-hoc use of specific models in isolated instances. Instead, it is widely acknowledged that scientific work across many disciplines is thoroughly infused with modeling approaches. This is evident from such work as Daniela Bailer-Jones' qualitative study, based on a large number of interviews conducted with scientists in early 2001, of how scientists think about scientific models. In the article, one interviewee after another is quoted as acknowledging, variously, that "modeling" as a way of self-describing what their scientific work is all about, "is much more used now amongst theorists, amongst physicists, amongst mathematicians than it used to be" (282); that "they are almost entirely concerned with the process of modeling" (282); and that science, in particular "the whole of physics," "whether you like it or not, is actually building models all the time" (281). While the distinction between modeling (as process) and individual models (as products) is not always strictly kept apart by practitioners, Bailer-Jones' conclusion that "scientific practice [...] is significantly shaped by modeling efforts" (299) certainly stands. Whereas in Maxwell's time, scientists grappled with the emergence of a new, self-reflective methodology of modeling, contemporary scientists take modeling to be a fundamental feature of scientific practice.

2. Representationalism and non-representational uses of scientific models

One of the core functions of modeling is the provision of *representations* of target systems (or target phenomena). This much is uncontroversial. Straightforward representational models are easy to come by: an architectural model may serve as a small-scale replica of a large, existing building; the original stick-and-ball model of the DNA double helix, made famous through the iconic photo of its creators standing next to it, represents the molecule that makes up the genetic material inside our cells; the 3D orrery depicting the planets and their relative position to the sun, and to one another, represents the solar system (or at least one possible configuration of it). Equating models with representations has had a long tradition in philosophical discussions of scientific models – from Boltzmann's definition, quoted earlier, of a model as "a tangible representation [...] which is either in actual existence, or has to be constructed in fact or in thought" to such passing remarks as Paul Teller's statement that "in principle, anything can be a model, and that what makes a thing a model is the fact that it is regarded or used as a representation of something by the model users" (Teller 2001, 397). On such a view, models are to be characterized in terms of the representational relation they stand in with respect to their target system or phenomenon (typically, some observable part of the world around us). The core question, thus, becomes not *whether* models represent, but rather *how* models manage to pull off this remarkable feat of standing in the right sort of relation to a target system that allows us to extract knowledge about the target from the model itself.

Two broad types of approaches to the problem of how models represent can be distinguished. *Two-place accounts* render the problem of model-based representation independent of any third parties (such as model users), reducing it essentially to a two-place relation between the model and the target, where this is typically taken to be a similarity relation or a (partial) isomorphism. Understood in this way, a model represents its target if and only if the two are sufficiently similar to one another or if the elements in one can be mapped onto the elements of the other in a structure-preserving way. *Three-place accounts* include the model user (or, less frequently, the community of model users) in the picture, thereby introducing a wider range of considerations such as relevance *to a particular audience*, and more generally (epistemic and non-epistemic) interests and goals. On this view, model-based representation cannot be a matter solely of the features of models and their targets alone; instead, it is regarded as depending on "the essentially intentional judgments of representation-users," which cannot be reduced "to facts about the source and target objects or systems and their properties" (Suárez 2004, 768).

In recent years, there has been a shift toward acknowledging the pragmatic dimension of models by making the role of the model user more explicit. This also applies to those philosophical accounts that initially highlighted two-place relations such as similarity. As an example, consider Ronald Giere's position on the relationship between models and theories. Giere initially conceived of a scientific theory as "a population of models" with various associated "hypotheses linking those models with systems in the real world," where the links between models and the real world were "again relations of similarity between a whole model and some real system" (Giere 1988, 85–86). In this early version, a link with scientific practice was achieved mainly by likening models to the idealized systems discussed in textbooks (such as the ubiquitous discussions of the harmonic oscillator in physics textbooks). Later, Giere pushes back against the idea "that the model itself represents an aspect of the world because it is similar to that aspect," since, he argues, there exists no "objective measure of similarity between the model and the real system" (Giere 2004, 747). Representation does not spontaneously emerge from any relation between the model and its target according to fixed criteria, but instead requires comparative judgments by the model user; likewise, "judging the fit of a model to the world is a matter of decision, not logical inference" (Giere 1999, 7).

An exclusive focus on representation faces a number of difficulties. For one, there are a number of important non-representational purposes which models frequently serve. Some of these may of course be entirely compatible with a model's *also* serving a (different) representational function: We may employ a model *of X*, not in order to represent *X*, but in order to try out new methods of approximation, develop our skill in modifying or manipulating the model by toying around with it, or even gain an understanding *of the model system* (rather than its target) – all the while acknowledging that someone else may very well use the model in order to represent *X*. In other cases, e.g., the overtly "false models," as discussed by William Wimsatt among others, it is more difficult to see how they could be regarded as anything more than "mere heuristic tools to be used in making predictions or as an aid in the search for explanations" (Wimsatt 2007, 94), let alone as full-fledged representations of any (real or imaginary) system. Even if one widens the scope of targets to include abstract or hypothetical systems, many models do not, in any obvious sense, have such targets which they could be said to represent. The line is not always easy to draw: As an example, consider biological models of sexually reproducing three-sex species. Results from computational biology based on such models demonstrate that any such arrangement would incur a heavy evolutionary cost, which goes some way toward explaining why such systems are not found in nature.[2] In physics, too, models may be constructed (e.g., by restricting or inflating the number of spatial dimensions, or by varying the laws of nature) such that we know that the situation they purport to describe could not possibly be realized in the world. In such a case, there may not be much to choose between saying that a model represents a merely logically possible scenario and describing it as non-representational. Even if one were to insist on the former so as to save one's representational idiom, not much insight for analyzing scientific practice should be expected from such usage. This much seems safe to say, then: The diversity and range of models and their uses – including those that are not easily assimilated to the task of depicting real target systems and processes in simplified terms – put pressure on any narrow version of *representationalism*, where the latter combines an ontological claim – that models *are* representations – with an epistemological assumption (viz., that we learn from models *in virtue of* their being representations).

Indeed, it is the latter assumption – the idea that what makes scientific models *epistemically productive* is solely due to their standing in certain representational relations to their targets – which lies at the heart of the controversy, perhaps more so than definitional issues. On one side are those who think that, in order for us to answer the question of how we can successfully learn from models, we need a unified account of how models represent, since it is *in virtue of* their representational success that we can acquire knowledge from them. Put crudely, we first need to reason our way up to an account of model-based representation before we can legitimately place trust in scientific models as a source of knowledge. On the other side are those who question the centrality of representation to the epistemic utility of models. The epistemic function of models is not reducible to the issue of representation, as if the only way to gain knowledge from a model were by holding it against the target system and assessing its degree of representational fidelity and completeness; rather, we ascribe epistemic value to models because of *how they are being used* – which varies across different domains, research questions, and stages of inquiry. There may be no *general* answer to how models can generate knowledge, and it may simply be misguided to hope for a general theory of scientific representation to hold the key to why models are epistemically valuable.

The various examples discussed thus far point to a great heterogeneity of the kinds of knowledge claims in support of which models are routinely deployed: claims concerning specific aspects of existing target systems, existence claims and impossibility theorems, ways of aggregating data, relations of evidential support, and many others. Which models are deemed most insightful, and which uses they are subsequently put to, depends on the goal and context of the inquiry. Unlike what the representational focus on the abstract two-place relation between model and target might suggest, models are rarely "parachuted in" from the outside, but instead gain their significance from being embedded into theoretical frameworks, experimental practices, and research programs; models need to prove their mettle, and acquire their epistemic merits, through successful uses and applications, typically over an extended period of time. What makes a model epistemically meritorious, and by which standards, depends on the specifics of the case at hand. This is why, in order to give satisfactory answers to these questions, philosophers in recent decades have turned toward case studies, rather than one-size-fits-all proposals, in order to deepen our understanding of model-based scientific practice.

3. Artifactualism and its challenges

What the preceding discussion suggests is that a proper understanding of scientific models need not be premised on, and does not require, a fully developed philosophical theory of scientific representation. Instead, we should acknowledge the variety of uses and functions of models, highlight the active role of model-builders and users – in line with the pragmatic turn and its emphasis on the triad of model/user/world – and begin to characterize scientific models as *instruments of inquiry*.

This, at least, is the recommendation issued by proponents of *artifactualism*. Instead of conceiving of models as abstract entities distinct from any particular (physical, or otherwise cognitively accessible) realization, we should treat models the way we encounter them in scientific practice: as *epistemic tools*, developed for the study of particular scientific questions, which – in virtue of the specific qualities of their concrete realization – afford us opportunities for learning about aspects of reality by interacting with, and actively manipulating, them. Once we shift the focus from the abstract question of how a model relates to, or *represents*, the world, to the question of how a model is constructed and used within a given context of inquiry, the urgency of providing a general account of model-based representation dissipates. As Tarja Knuuttila puts it: "Models are not freely floating objects in need of being linked to the real world: they are already linked to our knowledge of the real world by way of the scientific questions that motivate their construction" (Knuuttila 2011, 267).

By putting the construction and use of models center-stage, artifactualism highlights two important aspects that traditional (representationalist) approaches tend to sideline: first, models are often used for a variety of purposes, not all of which need to be commensurable with one another or need to conform to the goal of providing accurate representations. In this regard, they function like ordinary tools – that is, other artifacts we help ourselves to in daily life – which are to be judged by whether they are fit for the purpose at hand, not by some transcendent ideal. Second, models are *artificial creations*: we have designed them in ways that allow us to manipulate them in order to achieve specific ends. The latter aspect is central to understanding how models function in scientific practice: for, typically we are not passively confronted with models, needing to ascertain which aspects of reality they could possibly apply to, but instead we actively deploy them in order to make sense of, or otherwise engage with, a limited aspect of reality.[3] The scientific questions at hand *constrain* the design of models, while simultaneously their very constructedness allows for the *concrete manipulability* needed to make a model an effective tool of inquiry (see Knuuttila 2011).

The greater amenability of artifactualism to accounts of scientific practice stems not only from its pragmatic orientation, in the sense discussed above as acknowledging the triadic relation between model, target, and user, but also from the way it links models to the social and material dimensions of doing science. Rather than conceiving of models as offering, in the abstract (or at best in the mind of an individual user), a (perhaps distorted) mirror image of its target, models are recognized as essential communicative tools among the community of scientists: "Scientists do not read the minds of each other, and neither are they able to process even modestly complicated relations or interactions between different components without making use of external representational scaffolding" (Knuuttila 2017, 12), and models are often the preferred ways for constructing just such representational scaffolding. Concretely, representational means – be they physical, diagrammatic, or notational in character – provide cognitive access and allow for structured interventions according to shared (or potentially shareable) rules and conventions.

Its reliance on the notion of representational means – by which artifactualists mean such a diverse bunch as "diagrams, pictures, scale models, symbols, natural language, mathematical notations, 3D images on screen" (Knuuttila 2011, 268), each with its specific affordances and limitations – has exposed artifactualism to criticism from a minority of radical anti-representationalists, who fault it for still being "couched in thoroughly representational language" (Sanches de Oliveira 2022, 15). While it may be one thing to pursue the (laudable) project of *shifting the emphasis* from abstract philosophical characterizations of scientific representation to the "actual representational means with which a model is constructed and through which it is manipulated" (Knuuttila 2017), it is quite another, and indeed a more radical step, to give up on the idea of models as ways of representing (some aspect of) a target system, treating them instead as "tools that scaffold the activities of agents as they try to solve problems and make sense of the world" (Sanches de Oliveira 2022, 30).

Yet, upon closer inspection, what such "radical artifactualists" need to reject is not so much the notion of representation *per se*, but what has been called "*targetism*," i.e., the belief that models must be thought of "as the sort of thing that is defined by something else it refers to, something else it is a *source of information about*, because that's what it represents, or is a model *of*" (Sanches de Oliveira 2022, 36). This, however, is not how moderate artifactualists need to think about models. From their perspective, models are constructed for a variety of purposes, including – often enough – goals that explicitly or implicitly demand learning more about the world. We may, of course, learn about the world in the course of pursuing other goals, but if we are to take scientific practice seriously, we must acknowledge that models are often constructed in the search for answers to pertinent scientific questions. As already discussed, such questions need not be about existing targets, but can also be about (non-actual) possibilities, impossibility results, or the reliability of new methods or approximations. When a model represents ("internally," as it were) a counterfactual, hypothetical, or even a physically impossible state (e.g., by tweaking the known laws of physics), it would seem misguided to criticize such activity as unduly representational or as hampered by an excessive focus on an "external" (real, hypothetical, or merely fictional) target.

4. Component activities of modeling practice

Once it is acknowledged that models serve a multiplicity of uses and functions in science, the mired debate about the status of representation may be considered something of a red herring.

Not only are scientific models used for all sorts of goals and purposes, but the scientific practice of modeling is itself constituted by a heterogeneous admixture of component activities that constitute it. Representational uses of models are but one aspect of the complex fabric of modeling activities. Ultimately, both representationalists and anti-representationalists are at risk of overshooting the mark: Those who reduce models to their representational function without attending to the details of how a given model system mobilizes representational resources, tend to abstract away from the process of model construction and instead jump to conclusions about the kind of representational relation in which the finished product – the model – stands (or ought to stand) to reality.[4] By contrast, those who, in a radically anti-representationalist spirit, reject the representational idiom altogether assimilate scientific modeling to the somewhat amorphous cluster of problem-solving activities that human beings have developed as ways of coping with the manifold challenges in their environment. As this juxtaposition already makes clear, most well-developed views on how models work fall somewhere between these two extremes. And for good reason too: For, if we are to understand why models are of special significance to scientific practice, we should aim to be attuned to the variety of recurring patterns and component activities which together constitute the practice of scientific modeling. These patterns and component activities display, if not uniformity, then at least local stability within (and sometimes across) disciplinary boundaries. They are neither necessitated by the abstract demands of representationalism, nor can their local stability – the fact that not anything goes – easily be explained by a view that treats modeling as simply an extension of our regular problem-solving capacities.

A first take on the kinds of component activities that make up the practice of scientific modeling may be gleaned from a somewhat idealized timeline of a hypothetical episode of modeling. In this reconstruction of how modeling proceeds (methodologically, though not necessarily in strict temporal sequence), the first step is typically called "model-building," which may be variously followed by understanding (or "gaining a grasp") of a model, before testing its explanatory and predictive power, and subsequently applying – or, as the case may be, modifying – it in an iterative fashion that accords with the overall goal of inquiry. When model-building targets a specific phenomenon, it requires, first and foremost, settling on a relevant research question and choosing an appropriate medium or format, whether this be a mathematical calculus or a material medium (as in the case of physical scale models). Some properties of the target system will usually be considered negligible, so that no attempt is made to include these in the model system. The practice of deliberately neglecting or ignoring some properties or features of the target, such that only a subset of target properties is included in the model, is typically referred to as *abstraction*. Yet, this is rarely the only type of simplification, and further distortions, e.g., in the form of *approximations* and *idealizations*, are usually required in order to create a workable model.[5] Sometimes, additional variables or parameters will need to be posited, even as modelers are aware of their non-referring character. As is evident from this thumbnail caricature, "model construction involves a complex activity of integration" (Morrison 1999, 44).

Even as a model is being formulated, much work goes into integrating, and calibrating, the various ingredients in such a way that it meets – or at least does not stray too far from – both theoretical background assumptions (where these are available and are sufficiently explicit) and empirical constraints. Which desiderata enjoy priority – e.g., predictive power, explanatory success, generality, or simplicity – will vary across, and even within, disciplines, which specific weights should be attached to them and how the (as Levins (1966) reminds us: inevitable) trade-offs are to be negotiated, depends on standards recognized by other researchers. These are influenced by disciplinary expectations and are context-specific and may well vary across research programs within what is nominally the same discipline. The crucial point is that models do not spontaneously emerge, "fully-formed, as Athena from the brain of Zeus" (as Nancy Cartwright once put it, 1999, 247), but need to be *articulated*. This process of articulation draws on prior commitments, rendering models, as Mieke Boon argues, "embedded in a network consisting of different types of intellectual, epistemic and conceptual aspects" (Boon 2020, 31).

The specific strategies and recurring approaches that make up modeling as a practice, over and above the (representational or non-representational) function of its products and their overall contribution to our generic problem-solving activities, likewise vary across disciplines and research programs. *Which* strategies of abstraction and idealization are appropriate, and *how* the resulting models are to be assessed, is often hotly contested. Two examples illustrate this. First, among philosophers of economics, there has been considerable debate as to whether the core strategy of economic modeling consists in *theoretically isolating* a target phenomenon by means of idealization (akin to what happens in scientific experimentation, where one usually seeks to isolate the system under investigation from external influences) or whether economic models amount to *constructions of credible worlds*.[6] Second, in climate modeling, the tension between what one might call the "models-as-representations" and "models-for-use" views has played out in debates about what constitutes an improvement of existing climate models. In revising our climate models, should we be guided by the representational ideal of completeness, effectively treating climate models as candidates for truth or empirical adequacy, or should we settle for what Wendy Parker (2009) has called "adequacy-for-purpose," for example, by focusing on those (and only those) aspects of the climate system that we have reason to believe are most important to the task of ensuring reliable future predictions? Such questions cannot be answered in the abstract but can only be approached and debated from within an established practice of modeling, taking into account the interlinking activities and recurring strategies of model-building, testing, calibration, and – where necessary – revision.

5. The contribution of modeling to scientific practice

While it is one thing to claim that models "are highly structured entities which are woven into, and give stability to, scientific practice" (as I myself once did; Gelfert 2015, 224), it is quite another to fill such a programmatic statement with meaning and make clear precisely *what* the contribution of modeling to scientific practice consists of and *how* it has become so central to contemporary science. Philosophers have tended to search for a one-size-fits-all answer to this question – especially those who have operated on the assumption that a general account of representation (one that focuses on the dyadic relation between model and target) holds the key to explaining how models function. Others went instrumentalist: Since modeling has proven to be useful across a large number of scientific disciplines, should not its past successes – along with the fact that scientists routinely turn to modeling and profess to find it useful to do so – give us a reason to consider it central to contemporary scientific practice? Yet this amounts to little more than a re-description of the *explanandum*: why is modeling indispensable to contemporary scientific practice? Perhaps, then, no general answer to how models function can be given, precisely *because* models are "technologies for investigation" (Morrison and Morgan 1999) or "epistemic artifacts" (Knuuttila 2005): In order to see what their epistemic role is, one must consider models *in context* – how they were constructed, what they are built for, what potential applications they afford their users, how they are *in fact* used, etc. There simply is no shortcut to answering the question of how models function, and at most one can hope to identify recurring patterns, partially shared characteristics, and preliminary taxonomies of what kinds of uses have proven successful in which domains of inquiry.

Artifactualists acknowledge this when they characterize modeling "as a specific scientific practice in which concrete entities, i.e., models, are constructed with the help of specific representational means and used in various ways, for example, for the purposes of scientific reasoning, theory construction and design of other artifacts and instruments" (Boon and Knuuttila 2008, 689); moreover, "modellers typically proceed by turning the constraints [...] built into the model into affordances" (695). Modeling thus focuses attention on a select number of features of a target system and concretizes them into a model system whose affordances match our cognitive capacities; this way, we can learn about the target system by engaging with the model system, not as an abstract place-holder, but as a concrete, manipulable entity with specific affordances – much like ordinary tools – and with select cognitive "entry points."

Yet there remains a tension between viewing modeling as, basically, a practice of tool use and its epistemic orientation toward generating novel insights and creating knowledge. This tension is not unique to modeling as an epistemic practice: As Karin Knorr-Cetina has argued, while the concept of "practice" is typically associated with habituation and rule-guidedness, contemporary knowledge-based practices thrive on curiosity and innovation and therefore need to be "more differentiated than current conceptions of practice as skill or habitual task performance" (Knorr-Cetina 2001, 184). Put differently, whereas tools tend to blend into the background and, in Heideggerian terminology, are "ready-to-hand" – that is, experienced in a state of immersion into a practice – models are typically being encountered as objects of inquiry: they demand attention and cognitive engagement. Unlike an ordinary tool, the success of a scientific model is not wholly exhausted by how seamlessly and efficiently it allows its user to achieve an intended outcome; nor is this something that artifactualists must assume or endorse. Scientific models, in this regard, are more akin to what Hans-Jörg Rheinberger has called "technical objects": temporarily "defined in a characteristic manner," but able to "gain or regain an epistemic status and [to] be re-transformed into research objects" (Rheinberger 2011, 312). As such, they can impose much-needed (local, temporary) order on the process of inquiry, but at the same time hold out the prospect of novel insight – and, on occasion, are even capable of surprising us.

The philosopher of technology Don Ihde, drawing on Heidegger's distinction between tools being used in a "ready-to-hand" manner and objects being encountered as "present-at-hand," has coined the term *intentionality relations* to refer to such different phenomenological ways of engaging with the world around us, in particular with technological artifacts. Some technologies (e.g., binoculars) blend in with our experience, once we have mastered them; others (e.g., computers) demand constant cognitive engagement. The former give rise to *embodiment relations*, whereby we incorporate them into our experience, by habitually adjusting, in a self-reflexive way, our perceptual and bodily senses, allowing us to perceive "through" such technologies; the latter require a "special interpretive action" (Ihde 1990, 80), akin to deciphering a text, thereby giving rise to *hermeneutic relations*. When it comes to scientific modeling, it seems clear that different types of models cater to both sorts of intentionality relations to varying degrees: material models lend themselves more obviously to embodied engagement, whereas complex mathematical models may require significant hermeneutic input from both the modeler and the user. This does not mean that sustaining a hermeneutic relation is always more strenuous than finding oneself immersed in an embodied state of interaction; after all, the act of reading – the paradigmatic case of a hermeneutic relation – is itself one that, for most of us, has become "second nature." Similarly, the activity of reading a mathematical equation or performing a calculation in physics with the help of a series of Feynman diagrams, for those who use them on a daily basis, may over time become routine. Sometimes, both types of relationships are simultaneously co-present, for example, when modelers use integrated software packages that afford quasi-immersive visualizations while at the same time offering a vast range of options to select from. As Natasha Myers describes the user's phenomenology of engaging in protein modeling with molecular graphics software: "In one window, data will be streaming up the screen, and in another, the crystallographer holds the skeleton-like interactive rendering of a model. She keeps it alive in space and depth, rotating it onscreen and zooming in and out, keeping it visible at multiple angles, constantly shifting her visual and haptic relationship to it" (Myers 2008, 179).

Thus, similar to the way tools afford us different ways of manipulating the world and representations allow us to access information about their targets, scientific models, combining both, enable different forms of encountering the world via models, e.g., (using Ihde's terminology) in an embodied or a hermeneutic fashion.[7] While different types of models may have greater affinities with one or the other, in any real-life cases of scientific models, this is rarely an either/or affair. Not only does working with models often *require* switching between embodied and hermeneutic modes of interaction but also many models are specifically designed to *facilitate* such switching. Perhaps, then, what makes scientific models so valuable to science, and what modeling contributes to scientific practice, is the ability of scientific models to function not only as representations or tools but as mediators between different ways in which we relate with the world.

Notes

1 On the representationalist assumption underlying Boltzmann's account (and, traditionally, most of the other conceptions) of models, see the next section.
2 It is worth pointing out that this only holds for a specific conception of sexual reproduction, so this is at best a *ceteris paribus* conclusion. Since nature has been rather inventive when it comes to matters of reproduction and the evolution of the sexes, it should come as no surprise that scientists have since identified species with three sex phenotypes in a number of taxa such as algae, nematodes, and others.
3 This also applies to contexts of exploratory modeling, where the target may itself be undergoing revision, or where the main concern is with exploring *what is possible*. On this point, see (Gelfert 2019).
4 Whether or not a model "is representational" becomes a moot point, once emphasis is shifted towards its *uses*; for, surely, there are representational and non-representational *uses* of models, as indeed moderate artifactualism concurs.
5 This may later be matched by a process of *de-idealization*, when a model is being applied to a specific case, though it has been doubted whether de-idealization is indeed a frequent occurrence: As Roman Frigg and Stephan Hartmann argue, "it seems that de-idealization is not in accordance with scientific practice because it is unusual that scientists invest work in repeatedly de-idealizing an existing model" (Frigg and Hartmann 2020).
6 See (Mäki 1992) for the former view and (Sugden 2000) for the latter, as well as (Mäki 2009) for a conciliatory review of the debate.
7 This proposal is developed more fully in Section 5.5 ("Models as Enablers of Scientific Knowledge") of (Gelfert 2016).

References

Bailer-Jones, Daniela. 2001. "Scientists' Thoughts on Scientific Models." *Perspectives on Sciences* 10(3): 275–301.
Boltzmann, Ludwig. 1902/1911. "Model." In *Encyclopaedia Britannica*, 10th edition (1902), Vol. 30, 788–791. London: A&C Black. (Reprinted without changes in the 11th edition, 1911).
Boon, Mieke. 2020. "The Role of Disciplinary Perspective in an Epistemology of Scientific Models." *European Journal for Philosophy of Science* 10(31): 1–34.
Boon, Mieke, and Tarja Knuuttila. 2008. "Models as Epistemic Tools in Engineering Sciences: A Pragmatic Approach." In *Handbook of the Philosophy of Science, Vol. 9: Philosophy of Technology and Engineering Sciences*, edited by Anthonie Meijers, 687–720. Amsterdam: Elsevier.
Cartwright, Nancy. 1999. "Models and the Limits of Theory: Quantum Hamiltonians and the BCS Model of Superconductivity." In *Models as Mediators: Perspectives on Natural and Social Science*, edited by Mary S. Morgan and Margaret Morrison, 241–281. Cambridge: Cambridge University Press.
Frigg, Roman and Stephan Hartmann. 2020. "Models in Science." *Stanford Encyclopedia of Philosophy*. Accessed December 23, 2022. https://plato.stanford.edu/entries/models-science/
Gelfert, Axel. 2015. "Between Rigor and Reality: Many-Body Models in Condensed Matter Physics." In *Why More Is Different: Philosophical Issues in Condensed Matter Physics and Complex Systems*, edited by Brigitte Falkenburg and Margaret Morrison, 201–226. Heidelberg: Springer.
———. 2016. *How to Do Science with Models: A Philosophical Primer*. Cham: Springer.
———. 2019. "Probing Possibilities: Toy Models, Minimal Models, and Exploratory Models." In *Model-based Reasoning in Science and Technology (MBR18)*, edited by Francisco Salguero-Lamillar, Cristina Bares-Gomez, and Matthieu Fontaine, 3–19. Cham: Springer.
Giere, Ronald. 1988. *Explaining Science: A Cognitive Approach*. Chicago: The University of Chicago Press.
———. 1999. *Science without Laws*. Chicago: The University of Chicago Press.
———. 2004. "How Models Are Used to Represent Reality." *Philosophy of Science* 71: 742–752.
Hesse, Mary. 1963. *Models and Analogies in Science*. London: Sheed and Ward.
Hon, Giora, and Bernard R. Goldstein. 2012. "Maxwell's Contrived Analogy: An Early Version of the Methodology of Modeling." *Studies in History and Philosophy of Science* 43: 236–257.

———. 2021. "Maxwell's Role in Turning the Concept of Model into the Methodology of Modeling." *Studies in History and Philosophy of Science* 88: 321–333.
Ihde, Don. 1990. *Technology and the Lifeworld*. Bloomington: Indiana University Press.
Knorr-Cetina, Karin. 2001. "Objectual Practice." In *The Practice Turn in Contemporary Theory*, edited by Karin Knorr Cetina, Theodore R. Schatzki, and Eike von Savigny, 175–188. London: Routledge.
Knuuttila, Tarja. 2005. "Models, Representation, and Mediation." *Philosophy of Science* 72(5): 1260–1271.
———. 2011. "Modelling and Representing: An Artefactual Approach to Model-based Representation." *Studies in History and Philosophy of Science* 42(2): 262–271.
———. 2017. "Imagination Extended and Embedded: Artifactual versus Fictional Accounts of Models." *Synthese* 198(3): 1–21.
Levins, Richard. 1966. "The Strategy of Model Building in Population Biology." *American Scientist* 43: 421–431.
Liu, Chang. 1997. "Models and Theories I: The Semantic View Revisited." *International Studies in the Philosophy of Science* 11(2): 147–164.
Mäki, Uskali. 1992. "On the Method of Isolation in Economics." *Poznan Studies in the Philosophy of the Sciences and the Humanities* 26: 319–354.
———. 2009. "MISSing the World. Models as Isolations and Credible Surrogate Systems." *Erkenntnis* 70(1): 29–43.
McMullin, Ernan. 1968. "What Do Physical Models Tell Us?" In *Logic, Methodology, and Philosophy of Science III*, edited by Bob van Rootselaar and J. F. Staal, 385–396. Amsterdam: North-Holland.
Morrison, Margaret. 1999. "Models as Autonomous Agents." In *Models as Mediators: Perspectives on Natural and Social Science*, edited by Mary S. Morgan and Margaret Morrison, 38–65. Cambridge: Cambridge University Press.
Morrison, Margaret and Mary S. Morgan. 1999. "Models as Mediating Instruments." In *Models as Mediators: Perspectives on Natural and Social Science*, edited by Mary S. Morgan and Margaret Morrison, 10–37. Cambridge: Cambridge University Press.
Myers, Natasha. 2008. "Molecular Embodiments and the Bodywork of Modeling in Protein Crystallography." *Social Studies of Science* 38(2): 163–199.
Parker, Wendy. 2009. "Confirmation and Adequacy-for-Purpose in Climate Modelling." *Proceedings of the Aristotelian Society, Supplementary Volumes*, 83: 233–249.
Rheinberger, Hans-Jörg. 2011. "Consistency from the Perspective of an Experimental Systems Approach to the Sciences and Their Epistemic Objects." *Manuscrito* 34(1): 307–321.
Sanches de Oliveira, Gui. 2022. "Radical Artifactualism." *European Journal for Philosophy of Science* 12(36): 1–33.
Suárez, Mauricio, 1999. "Theories, Models, and Representations." In *Model-based Reasoning in Scientific Discovery*, edited by Lorenzo Magnani, Nancy Nersessian, and Paul Thagard, 75–83. New York: Plenum Publishers.
———. 2004. "An Inferential Conception of Scientific Representation." *Philosophy of Science* 71(5): 767–779.
Sugden, Robert. 2000. "Credible Worlds: The Status of Theoretical Models in Economics." *Journal of Economic Methodology* 7(1): 1–31.
Teller, Paul. 2001. "Twilight of the Perfect Model." *Erkenntnis* 55(3): 393–415.
Wimsatt, William. 2007. *Re-Engineering Philosophy for Limited Beings: Piecewise Approximations to Reality*. Cambridge, MA: Harvard University Press.

PART 2

Philosophical accounts of modeling

4
REPRESENTATION

Julia Sánchez-Dorado

1. Introduction

The problem of scientific representation has become a central topic of debate in contemporary philosophy of science. Early considerations can be already found in the nineteenth century: in Hertz's and Boltzmann's accounts of scientific theories as images (*Bildtheorie*), in Maxwell's discussions on analogical reasoning in science, and in Whewell's ideas on colligation and idealization in the inductive sciences. The proposals of these philosophers and scientists are now respectively read as forerunners of structuralist, inferentialist, and pragmatic accounts of scientific representation (see Suárez 2024, chap. 2; van Fraassen 2008; Cristalli and Sánchez-Dorado 2021). Yet, the problem of scientific representation as we normally frame it today only started to be explicitly discussed in the second half of the twentieth century, once the semantic view had gained a central stage in the philosophy of science.

In the semantic view, models were regarded as structures that provide tools for interpreting the axioms of a scientific theory (Suppes 1960). A group of adherents to the semantic view progressively started to emphasize that models should be primarily taken as *representational* structures, that is, structures standing in relation to certain targets in the world (van Fraassen 1980; Giere 1988; see Bailer-Jones 1999, 32–33). By the 1980s, talk on models and talk on representation became closely connected in philosophy of science. That connection was strengthened even more in the early 2000s, when the stance of "models as mediators" acquired popularity (Morgan and Morrison 1999; Cartwright et al. 1995). Only in recent years, have there been explicit attempts to disentangle the by-now familiar conception of models as scientific representations. The observation that models play a diversity of epistemic roles in science has motivated some scholars to contend that such diversity cannot be fairly investigated if models are primarily conceived as representational units (Hacking 1983; Knuuttila 2017, 2021).

In approaching *the* problem of scientific representation, we should be wary not to conflate different questions that are involved in it. At some point in the 2000s, disagreements about scientific representation became so vivid and intricate that some philosophers started to suspect that they were in fact dealing with several problems at the same time (Hughes 1997; Frigg 2006; Suárez 2004). Indeed, when asking "how do models represent natural

DOI: 10.4324/9781003205647-7

phenomena?", one might be interested in identifying the necessary and sufficient conditions for something to be a scientific representation. But it is also possible that what one wants to understand is how a model can become an epistemically adequate representation and be used to make fruitful inferences about the world. It is additionally possible that the problem one wants to address is whether *scientific* representations constitute a genuine way of representing the world, different from representations in other domains. And even one might be interested in determining whether the only, or most genuine function, of scientific models is to represent, or if, on the contrary, it is possible to talk about the epistemic contribution of models independently of their representational capacity.[1] These different questions will be discussed in the following sections:

- Section 2 discusses the question: "*in virtue of what does a model represent a certain target in the world?*". This has been called the "constitutional question" of representation (Suárez 2010; Callender and Cohen 2006), the problem of "mere representation" (van Fraassen and Sigman 1993; Bolinska 2013), and the problem of "representation simpliciter" (Contessa 2007).
- Section 3 discusses the question: "*what makes a model an adequate representation in practice?*". In recent philosophy of science, this has been also understood as the problem of the "faithfulness of representation" (Contessa 2007), the "means of representation" (Suárez 2003; 2010; 2015), or the "standards of accuracy" of representation (Frigg 2006). But this question has been also addressed by scholars in iHPS and STS, with a focus not on model-target accuracy standards, but on the learning process afforded by adequate-for-purpose models (Parker 2020).
- Section 4 addresses the question: "*is there anything distinctive about scientific representations, in contrast to representations in other domains?*". This has been called the "demarcation problem" of representation (Frigg and Nguyen 2021), or the "special problem of scientific representation" (Callender and Cohen 2006; Boesch 2017), and it has encouraged philosophers of science to look at debates on representation in aesthetics, philosophy of language, and philosophy of mind.
- Section 5 deals with the question: "*is representation the epistemic core of scientific modeling?*". This question concerns whether representation is all we need to explain the value of models in epistemological terms. It has been also described as the issue of "representationalism" or "targetism" and has motivated the advancement of artifactual accounts of scientific modeling in response to it (Knuuttila 2017; 2021; Oliveira 2021; 2022).

2. In virtue of what do models represent?

The constitutional question of scientific representation presupposes that representation is a relation between a certain vehicle (model) and a certain target system in the world (natural or social phenomenon), and demands an analysis of such a relation in terms of something else more elementary. One reason why it is important to differentiate the constitutional question from the adequacy question of representation (discussed in Section 3) is that if we want to explain why a certain model is not a good representation of a target, we first ought to accept that the model in question is a representation of the said target. Only then can one discuss the reasons for its inadequacy. In other words, the distinction between these two questions helps to account for the fact that misrepresentation is a species of representation (van Fraassen 2008, 13).

The standard way of answering the constitutional question of scientific representation is to offer a set of necessary and jointly sufficient conditions, which identify in a unique and universal way the existing relation for any vehicle-target pair (Suárez 2010, 93). Probably the most frequent answer to this question has consisted in appealing to a relation of similarity or structural similarity (i.e., isomorphism, partial isomorphism, homomorphism) between the vehicle and the target of the representation. However, similarity-based accounts of representation have faced severe criticisms, so some alternatives have been proposed in response (see Sections 2.2 and 2.3).

2.1 *Similarity-based accounts of representation*

Many philosophers and non-philosophers alike would intuitively agree that, for example, an orrery representing the solar system is similar in some respect to the solar system; a scale model of a river shares important similarities with the behavior of flows in the real river; model organisms (like non-obese diabetic mice) are similar to humans with respect to how a certain condition (type 1 diabetes) develops; and a computer simulation of a tornado is similar to how a tornado progresses in reality.

However, going from these intuitions to explaining the relation of representation in terms of a more fundamental relation of similarity has become a highly contentious project in the debate of scientific representation. A group of philosophers associated with the semantic view attempted to formalize the similarity intuition by appealing to the sharing of structures or "mapping" between models and targets. The central postulate of these accounts is that a model (M) represents a target (T) if and only if M and T instantiate similar structures.[2] When a structure-preserving bijection between M and T is assumed, their structures are isomorphic (van Fraassen 1980); when not all relations defined in the domain of T are mapped into M, their structures are partially isomorphic (French and Ladyman 1999; French and Bueno 2011); and, if some parts of the structure of M do not match any part of the structure of T, and parts of T are not included in the mapping, the structures of M and T are homomorphic (Bartels 2006; Ambrosio 2013).

Similarity-based accounts of representation, including structural versions of it, face at least three fundamental challenges when addressing the problem of the constituents of representation: the logical argument against similarity, the problem of vagueness, and the challenge of misrepresentation. A reference point when spelling out these challenges is Nelson Goodman (1968; 1972). In *Languages of Art*, Goodman argues that "copy theories of representation" endorse the naïve view that a symbol represents an object if and only if it appreciably resembles that object (1968, 3–4). Vestiges of that naïve view were found for Goodman in theories in aesthetics that explained depiction in terms of a similarity relation between a picture and the real scene depicted, producing a visual illusion in the viewer (Gombrich 1960). Philosophers of science have substantially drawn on Goodman's views to also reject what they see as "vestiges of copy theories of representation" in structuralist (and more generally, similarity-based) accounts of scientific representation (Suárez 2003; Hughes 1997; Contessa 2007; Frigg 2006).

The first challenge faced by similarity-based accounts of representation, namely, Goodman's logical argument, says that similarity cannot constitute the relation of representation because while similarity entails symmetrical, reflexive, and transitive relations, representation entails asymmetrical, non-reflexive and non-necessarily transitive relations (1968, 4–5). If A is similar to B, then B is similar to A; but the fact that a painting represents a certain person does not imply that the person represents the painting. Likewise, an object resembles itself to a maximum degree but rarely represents itself. And if object A is similar to object B, and object B is similar to object C, we would say that A and C are similar to one another; in contrast, representation does not necessarily establish that kind of transitive relation (Goodman 1968, 4–5). In short, similarity lacks the logical properties to define representation.

The second challenge of similarity-based accounts of representation is the problem of vagueness. The idea is that the concept of similarity is so poorly defined that it becomes trivial, since anything can be similar in some respect to anything else: "That a given two things are similar will hardly be notable news if there are no two things that are not similar" (Goodman 1972, 443). Thus, it would be pointless to treat similarity as a necessary condition for representation. Some philosophers of science have further maintained, following Goodman, that "an unqualified similarity claim is empty" (Frigg 2006, 61), and that without an objective measure of similarity, similarity-based accounts of representation are relativistic (Chakravartty 2001).

The third challenge to similarity-based accounts of representation concerns misrepresentation. All models simplify, occlude, and distort some aspects of reality, irrespective of whether they are judged highly successful, or plainly inaccurate, models. In either case, we would treat them still as representations of their target system. A satisfactory account of what constitutes the relation of representation ought to be able to accommodate the persistent phenomenon of misrepresentation (Suárez 2003; Frigg 2006). However, this requirement invites one to think that similarity is a poor candidate as a constituent of representation, since misrepresenting involves accepting dissimilarities and distortions of a target, precisely the opposite of what the similarity condition seems to demand.

There have been numerous attempts to respond to these three challenges. For instance, to circumvent Goodman's logical argument, some scholars have resorted to empirical evidence on how epistemic agents actually formulate similarity judgments in everyday situations. The claim is that, in practice, epistemic agents do not always treat similarity relations as symmetrical and transitive, so similarity might actually establish more analogous relationships to representational model-target relations than we may have initially thought (Tversky 1977).[3] However, this move to subjects' judgments in practice can be read more as a refocusing of the problem of similarity than as a solution to Goodman's logical argument, which is directed against analytic attempts to offer universal, *in abstracto*, explanations that reduce the relation of representation to a relation of similarity. A different strategy to respond to this challenge is adopted by Bartels (2006), who endorses homomorphism because this version of structural similarity is explicitly non-symmetrical. That is, Bartels' homomorphism, different from other morphisms, can only occur when a vehicle already refers to a certain target, establishing a unidirectional relation from vehicle to target.[4]

In response to the problem of vagueness, philosophers like Giere (2004) and Teller (2001) have sustained that representing does not require the existence of an objective measure of similarity and that the lack of such a measure does not introduce an undesirable amount of relativity in claims about the similarity between specific models and real systems (Giere 2004, 748). However, if we accept this response and the possibility of adopting different measures of similarity for particular cases, then it appears that the question we start dealing with is what the degree of adequacy between a given representation and a target is (see Section 3), instead of the initial question of what constitutes representation.

Regarding the challenge of misrepresentation, it might seem that appealing to notions like "partial isomorphism" or "homomorphism", which allow for the incomplete matching of properties between vehicle and target, would be enough to account for the phenomenon of misrepresentation. However, these morphisms struggle to accommodate typical ways in which models misrepresent: Bartels' (2006) homomorphism is unable to account for cases of abstraction (when the model neglects some features of the target it refers to) (see Pero and Suárez 2016, 86), while da Costa and French's (2003) partial isomorphism cannot accommodate idealizations unless they are reinterpreted exclusively as approximations (see Pincock 2005, 1257).

In addition to these challenges, structural similarity accounts have given rise to suspicions about how target systems can actually instantiate structures, given that these are typically objects or events (a cell, an earthquake, an economic crisis) and not mathematical entities. We can grant that any target can instantiate a structure if some relationships between its parts (objects, features) are recognized, but then the problem is that each target could instantiate many different structures. The response originally offered by Suppes (1962) was that what models are actually isomorphic to are "models of data", that is, models that do not involve any theoretical assumption. Yet, we might be unsatisfied with an account of representation that only explains how models of data, but not real targets, are represented (in Frigg and Nguyen 2021, § 4). Furthermore, many models are themselves not mathematical entities (e.g., scale models built by civil engineers), so these could also instantiate different structures.

2.2 *Denotation and DEKI*

In response to the shortcomings of similarity-based accounts of representation, some philosophers have advanced alternative proposals that leave similarity aside and bring denotation to the center. Callender and Cohen (2006) defend a version of the denotational account of scientific representation, insofar as they sustain that a vehicle represents a certain target if and only if the user stipulates that the vehicle denotes the target.[5] But the most prominent denotational account of scientific representation in recent years is Frigg and Nguyen's proposal (2018; 2021; 2022).

Inspired by Goodman (1968), where "denotation is the core of representation", and Elgin's (2010) developments of it, Frigg and Nguyen propose the DEKI account. That is, denotation, exemplification, keying up, and imputation are the necessary, and jointly sufficient, conditions for scientific representation. Denotation is defined as a dyadic relation between an existing symbol (e.g., a model) and an existing object (2022, 54). This implies that targetless models do not denote, but it is still possible that they are Z-representations; that is, they can still belong to the class of things that portrays Z (even if they cannot be representations *of* Z)[6] (see Goodman 1968, 30; Frigg and Nguyen 2022, 56–58). The next condition, exemplification, is a special form of symbolization that involves the instantiation of certain properties and the selective reference to some of those properties (Goodman 1968; Elgin 2010; 2017). However, exemplification does not work in a straightforward way, since models frequently do not *literally* instantiate the properties they are meant to refer to (or which are being imputed to a target). To deal with this problem, Goodman and

Elgin appealed to the intricate notion of "metaphorical exemplification", while Frigg and Nguyen refer to the idea of "instantiation under an interpretation". Lastly, the existence of a "key", different in each modeling practice, affords the means to convert model features into target features, some of which are eventually imputed to a specific target.

The DEKI account is a systematic attempt to formalize the compelling idea that things are always represented *as being thus or so*, as much in science as in art. This account has not been free from criticism, however. Salis (2021, 165–168) identifies several problems of the DEKI account when it is used to explain how theoretical models represent (as opposed to physical models, cases it can account for more successfully) (see also Knuuttila 2017), while Millson and Risjord (2022) criticize it for being unable to block unjustified surrogative inferences, that is, appealing to DEKI does not say how the content of a representation (i.e., a map, a model) justifies the inferences drawn from it.

2.3 *Deflationism*

The difficulties in sustaining a general theory of what constitutes the relation of representation have made a group of philosophers wonder if it is actually possible to offer universal conditions for representation. Some have even questioned whether this problem is at all worth addressing epistemologically, and advanced deflationary accounts of scientific representation accordingly (Hughes 1997; van Fraassen 2008; Suárez 2015). A deflationary approach sustains that no attempt should be made to explain representation in terms of something more elementary than itself, such as similarity or denotation. Deflationism is still compatible, however, with studying typical features that representations exhibit in practice.

According to van Fraassen (2008, 23), whose work has noticeably transitioned from the semantic view to a pragmatic conception of models, we should endorse a deflationary account that puts the "use" of representations at the center. That is, he identifies the constituents of representation with its means (or the specific relations that are established by epistemic agents when using particular models to make inferences about particular targets) (Suárez 2015, 45; for a critique of van Fraassen's deflationary approach, see Frisch 2015). Hughes' (1997) DDI account can also be considered an early deflationary account of representation. He proposes taking denotation, demonstration, and interpretation not as necessary conditions for representation, but rather as "three activities" which, if kept in mind when studying a scientific model, could help "achieve some insight into the kind of representation it is" (1997, 329). In recent years, the most notable advocate of the deflationary approach has been Suárez (2015). He suggests reinterpreting "denotation" in deflationist terms as "representational force", a notion that more clearly describes the activity performed by epistemic agents using a vehicle with a denotative function, whether it tracks any real target or not. Together with representational force, the "inferential capacity" of a vehicle relative to a target would be the most typical feature of scientific representations in practice (Suárez 2015, 43–45).

A question that has been raised against deflationary accounts is whether they give up too quickly in the endeavor of giving an answer to the constitutional question of representation (Chakravartty 2009). They would be in a sense conformist, pointing out some "surface features" of representations in practice (Suárez 2004, 771), but potentially disregarding the epistemic virtues that make it possible for scientific representations to in fact be used to learn about the world (Liu 2015, 42; Knuuttila 2021, 3). Despite these observations, something that deflationary views have helped to make manifest is that certain avenues in the research about scientific representation – mainly, the investigation of the constituents of representation – is becoming exhausted, and have encouraged philosophers to shift their attention to the study of pragmatic questions concerning representation.

3. What makes a model an adequate representation in practice?

The conceptual systematicity with which the constitutional question can be and has been addressed is not easily applicable to the study of the adequacy of representation. This is because of the pragmatic, situated, and thus slippery nature of this problem, which demands a practical type of inquiry to study it (Suárez 2010, 91–93). Besides, while the problem of the constituents of representation has been exclusively addressed by a group of philosophers of science in the analytic tradition, the problem of adequate representation has also been discussed from other disciplinary perspectives, such as iHPS (integrated history and philosophy of science) and STS (science and technology studies). To clarify, iHPS and STS accounts of adequate representation are interested in examining how epistemically fruitful models are built and assessed by scientific communities, while analytic accounts have tended to focus on the identification of adequate model–target relations in practice – also referred to as the "standards of accuracy" (Frigg 2006) or "means" of representation (Suárez 2010).

At the risk of encompassing many different perspectives under this heading, it is helpful to recognize two broad methodological approaches to the study of the problem of adequate representation: a generalist approach, which aims to identify general standards for adequate model–target relations in practice; and non-generalist approaches, which draw on specific cases of model construction to advance a pragmatic reading of what it takes for a scientific community to produce adequate-for-purpose models.

3.1 Generalist approach to the adequacy question

There are many different formats and styles of representation in science, including the use of mathematical equations, three-dimensional models, images, computer simulations, and graphs. The generalist approach to the study of the adequacy of representation searches for rules of correctness that go beyond the idiosyncrasies of individual modeling styles.

Examples of the generalist approach are found in the proposals of Giere (2004; 2010) and Weisberg (2013). Surprisingly or not, these are also similarity-based accounts. For Giere (2004; 2010), it is intentional similarity, specified in "respects and degrees", that determines the adequacy of representations. Giere defines similarity as a triadic relation, where scientists are responsible for picking out certain features of models, "claiming them to be similar to features of the real system", and thus building adequate representations for their specific purposes (2004, 747–748). For Weisberg (2013), the *weighted feature-matching account* that he proposes captures the intuition that a model is similar to its target, and therefore an adequate representation, when it shares many features, and does not fail to share too many features that are considered salient, with its target. Which features are considered shared or not shared, as well as their relative importance, is defined by the goals of the scientific community doing the representing.[7]

A question posed to the generalist approach is whether it can in fact account for the variety of models there is in science. Given the huge number of ways in which scientists can potentially employ relations between vehicles and targets to learn about the world, we could expect to find a variety of standards of adequacy or means of representation (Frigg

2006; Suárez 2010). Similarity and structural similarity are two among those means, but they might not exhaust the possibilities to build adequate models: model–target relations could be highly conventional, too (Frigg and Nguyen 2022, 64).

A more fundamental criticism of generalist accounts is that even if they acknowledge that dealing with the problem of adequate representation requires carrying out a practical inquiry, they are largely rational reconstructions of modeling practices. That is, their attention is focused on identifying the correct epistemological criteria that define adequate model–target relations, failing to engage with how scientists actually gain knowledge about the world, as well as about science, throughout their practices of model construction.

3.2 *Non-generalist approaches to the adequacy question*

In contrast to the generalist approach, some approaches to the problem of adequate representation do not aim to find a determinate set of criteria for the adequacy of model–target relations. Instead, they look carefully at concrete practices where scientific communities reach context-dependent agreements about the adequacy (that may be described as accuracy, fruitfulness, or usefulness by scientists) of a certain model depending on their particular goals. The starting point of these approaches is the analysis of the process of designing, constructing, calibrating, and validating specific models in their historical context. The endpoint is usually the advancement of a pragmatic conclusion, more or less broad in scope, about what it takes for scientists to build models that are "adequate-for-a-purpose" (Parker 2020).

Examples of non-generalist analyses of scientific representation are found in Chang's (2004) study of the historical process of representing temperature, going through different iterative stages and uses of instruments; in Schaffer's (2004) work on the production of models of ships on a small scale in late 18th-century London; in Oreskes' (2007) account of the use of compression boxes as material models of orogenesis in the 19th century; and in Knuuttila and Loettgers' (2016) study of the Lotka-Volterra model and the different philosophical readings triggered by it.

Also, collective volumes like Lynch and Woolgar (1990), de Chaderevian and Hopwood (2004), and Coopmans et al. (2014) advance a rich collection of pragmatic accounts of representation sustained on a systematic investigation of case studies across the natural and social sciences. Taken together, these accounts shed light on the understanding of the activity of modeling from both an epistemological and a historical perspective. Furthermore, the case-study perspective sometimes motivates the reevaluation of central assumptions in the philosophy of science, including the pertinence of the notion of representation itself (Daston 2014; Woolgar 2014; Lynch 2014).

Whereas the authors just mentioned focus on singular, historically localized cases, other non-generalist approaches place emphasis on the analysis of a relatively large set of case studies. In so doing, they try to elucidate, also from a practice-based perspective, how certain modeling resources (cognitive and material) end up being entrenched in a scientific field or shared by scientists working across fields. A recent proposal that adopts this mid-level generality approach is Ankeny and Leonelli's (2020) study of model organisms. Working with model organisms proved to be very fruitful in the biological and medical sciences throughout the twentieth century. Ankeny and Leonelli analyze the long processes of standardization that using this type of modeling technique required, and the fundamental epistemic role played by "repertoires" shared by different scientific teams, which served as guidelines to experiment with and extrapolate from living organisms. Sterrett's (2009; 2017) work is another good example of this approach. She examines how practices of scale modeling across the engineering and physical sciences consolidate systematic ways of producing adequate inferences using the principles of physical similarity. Also, Bokulich and Parker (2021) developed a mid-level generality type of account of the entrenched ways in which scientists take data models to represent, differently from other kinds of models.

Looking back at Morgan and Morrison's (1999, 11–12) – by now classic – account of models as mediators helps locate some early motivations for advancing mid-level generality accounts of adequate representation. The understanding of models as autonomous entities requires studying the ways in which scientific communities in specific disciplines (physics and economics in Morgan and Morrison's study) stabilize certain uses of models to learn about the world, about theories, and about models themselves, using the tacit skills and creative strategies characteristic of each field.

4. Is there anything special about *scientific* representations?

Scientific models are not the only vehicles used to represent aspects of the world and learn about them. Paintings, photographs, thought experiments, and narratives are employed across the arts, humanities, and other realms of everyday life to represent objects and states of affairs and gain understanding of them. There is indeed a long tradition in fields like aesthetics, philosophy of language, and philosophy of mind of debating the problem of representation, where questions such as: "How do symbols refer?", "What is the content of mental representations?", or "What is the role of similarity in depiction?" have been thoroughly examined. This seems to be a good reason to think that philosophers of science have a good deal to learn from previous and ongoing debates on representation in other domains. Philosophers who admit the compatibility of debates on representation in different fields have for instance exploited comparisons between scientific models and maps (Winther 2020), caricatures (van Fraassen 2008), and artworks (Suárez 2003; 2004; French 2003; Downes 2009; Ambrosio 2013). Some have made the even stronger claim that there are no significant differences between scientific and artistic representations with regard to their ultimate epistemic aim, namely, understanding (Elgin 2017). Several recent collective volumes show the popularity of this integrative approach to the study of representation across the arts and sciences (Frigg and Hunter 2010; Bueno et al. 2018; Ivanova and French 2020). Yet, the jump from recognizing the strengths of other traditions in dealing with analogous problems to the conclusion that it is possible to bring together various accounts of representation into a unified theory is more problematic than it might first seem (Sánchez-Dorado 2018). Only the careful attention to the specific questions and motivations underlying the debates in each field can grant a fertile integration of perspectives.

There is another line of argumentation regarding the problem of whether *scientific* representation is a unique form of representing. Callender and Cohen (2006, 67) published an influential article where they contended that much of the literature on scientific representation had been "concerned with non-issues". Mental states are, in their view, the fundamental representational objects, from which the rest are derived. Therefore, scientific models, like linguistic utterances and artworks, would be "just one more special case of derivative representation" (Callender and Cohen 2006, 75). It is, thus, only at the fundamental level (of mental representation) at which philosophers need to ask the constitutional question, not at the other levels.

Callender and Cohen's proposal was provocative. There are some who sympathize with it, like Ruyant (2021), who tries to spell out in more detail how scientific representations can be ultimately reduced to mental representations in a non-trivial way. Others, like Liu (2015), however, argue that Callender and Cohen's reductive account of representation – in terms of stipulation – fails to distinguish between mere symbols and epistemic representations (such as scientific models). Also, Boesch (2017) rejects the reduction of scientific representation to mental representation, as the former has a communal nature, while the latter is private.

5. Is representation the epistemic core of scientific modeling?

Philosophers of science have typically avoided making explicit claims about representation being *all that matters* when they discuss the role of scientific models, possibly foreseeing potential criticisms. However, the pervasive identification of models as representational vehicles in the literature suggests a rather strong commitment to "representationalism", that is, the received view that the epistemic role of scientific models is best understood in representational terms (Oliveira 2021). The fact that misrepresentation has been a topic of epistemological concern is also evidence of the assumption that "modeling is an epistemic activity because it is representational" (2021, 212). This assumption has elicited discomfort among a group of commentators in recent times.

It is uncontroversial that models play a diversity of epistemic, as well as non-epistemic, roles in science and beyond. If the epistemic value of models, and our learning from them is, however, explained in terms of their representational capacity, those other roles can hardly be given the prominence they deserve. Design models are, for instance, built with the aim of implementing a new engineering structure or modifying a technological device. They are distinct from representational models either because their target does not exist yet or because the direction of fit is from target to vehicle – instead of from vehicle to target (Poznic 2016). Exploratory models are also targetless or have very roughly defined targets. Without aiming to represent any actual empirical phenomenon, an exploratory model can be used to feature proof-of-principle demonstrations, generate potential explanations, or help scientists gain greater mastery of the repertoire of modeling techniques available in a field (Gelfert 2016, 41, 79). Other epistemic functions that models can play – whether they are representational or not – are testing the compatibility of various concepts, generating hypotheses, constructing other models, and producing new target systems (Luczak 2017; Peschard 2011).

Hacking (1983) was influential in resisting representationalism as the only conceptual framework for the study of scientific practices and emphasized intervention as a fruitful alternative (Cassini and Redmond 2021, 43). In recent years, Knuuttila's artifactual account of models has openly positioned itself against general representationalism (2017; 2021). She proposes a "novel candidate for a unified approach" that neither assumes that models are inherently representational at the outset, nor that the model-target pair should be the fundamental unit of epistemological analysis. Instead, looking at model construction is the key to understanding how a model can achieve its epistemic purposes (2021, 2). Crucially, for Knuuttila the artifactual and the representational approaches do not necessarily clash, as long as one adopts a pragmatist conception of the representational relation (2017; 2021).

While some philosophers of science might still be resistant to endorsing the consequences of the artifactual approach to modeling, others, like Oliveira (2021; 2022), think that artifactualism has not yet gone far enough. The presumed compatibility of the artifactual and the representational approaches – in Knuuttila's (2017) account and also in Morgan and Morrison's (1999) account – is too mild in his view, since it merely shifts the *emphasis* of the debate. A more radical artifactualism would completely avoid the "representationalist quagmire", and focus instead on the skill development and learning transfer afforded by models, which should be literally understood as *tools* and not as referring signs (in analogy to linguistic signs) (Oliveira 2022). It remains to be seen if philosophers of science will accept in the coming years that representationalism is a dead end, or if the debate opens new paths of inquiry on representation in light of the variety of pragmatist and artifactual approaches to scientific modeling that are currently proliferating.

Acknowledgments

This research has benefited from postdoctoral funding from a Talento Doctores PDI fellowship from Junta de Andalucía (Spain) and from an ICI Berlin fellowship.

Notes

1 There are further questions involved in the problem of scientific representation, such as those concerning the ontology of models as representations, as discussed by Frigg and Nguyen (2022).

2 Structural similarity is treated here as a special kind of similarity. There are no explicit attempts in the contemporary literature to advance an account of what constitutes the relation of representation based on a non-structural conception of similarity. Sometimes Giere (2004, 2010) has been treated as an advocate of such an account, but he explicitly clarifies that he is not taking similarity as a necessary condition for representation (2004, 747). Instead, he aims to give a similarity-based response to the problem of adequate representation. Thus, his proposal will be mentioned in Section 5.3.

3 For instance, experimental subjects would normally say that "an ellipse is similar to a circle" and not that "a circle is similar to an ellipse", which questions the symmetry condition (Tversky 1977, 333–336). Other tests showed that similarity is not necessarily treated by experimental subjects as transitive either, while with reflexivity the differences between similarity and representational relations still seem to hold.

4 Bartels (2006) also introduces intentional mechanisms to address the logical argument concerning reflexivity. This move has the problem, however, of making the role of homomorphism in his constitutional account unclear, since it is then the intentional mechanism, and not the structural similarity, that defines the directionality of the representational relation (on reflexivity, see Dipert 1996; for a fully-fledged criticism of Bartels 2006, see Pero and Suárez 2016).

5 Here I am following Contessa's (2011, 124–125) reading. For Frigg and Nguyen (2021, §2), "stipulative fiat", rather than denotation, would be more precisely the condition of representation in Callender and Cohen's (2006) account.

6 Goodman (1968, 22) argues that common expressions such as "a certain painting is a picture *of* an unicorn" are highly ambiguous. The locution "picture *of*" is a two-place predicate, so there cannot be pictures *of* unicorns since there are no such things as unicorns. But we can still produce as many unicorn pictures as we like. Being a Z-picture or a Z-representation is belonging to the set of things that represent Z.

7 For a discussion on whether Weisberg (2013) is actually advancing an account of adequate representation – or, rather, an account of the constituents of representation, or an account of what underlies modelers' judgments in practice – see Parker (2015) and Khosrowi (2020).

References

Ambrosio, Chiara. 2013. "Iconic Representations, Creativity and Discovery in Art and Science." In *Creativity, Innovation, and Complexity in Science*, edited by Wenceslao González, 109–124. A Coruña: Netbiblio.

Ankeny, Rachel, and Sabina Leonelli. 2020. *Model Organisms*. Cambridge: Cambridge University Press.

Bailer-Jones, Daniela. 1999. "Tracing the Development of Models in the Philosophy of Science." In *Model-Based Reasoning in Scientific Discovery*, edited by Lorenzo Magnani, Nancy Nersessian, and Paul Thagard, 23–40. New York: Kluwer/Plenum.

Bartels, Andreas. 2006. "Defending the Structural Concept of Representation." *Theoria* 21: 7–19.

Boesch, Brandon. 2017. "There Is a Special Problem of Scientific Representation." *Philosophy of Science* 84(5): 970–981.

Bokulich, Alisa, and Wendy Parker. 2021. "Data Models, Representation and Adequacy-for-Purpose." *European Journal for Philosophy of Science* 11: 31.

Bolinska, Alisa. 2013. "Epistemic Representation, Informativeness and the Aim of Faithful Representation." *Synthese* 190(2): 219–234.

Bueno, Otávio, George Darby, Steven French, and Dean Rickles. 2018. *Thinking about Science, Reflecting on Art: Bringing Aesthetics and the Philosophy of Science Together*. New York: Routledge.

Callender, Craig, and Jonathan Cohen. 2006. "There Is No Special Problem about Scientific Representation." *Theoria* 21(55): 7–25.

Cartwright, Nancy, Towfic Shomar, and Mauricio Suárez. 1995. "The Tool-Box of Science: Tools for the Building of Models with a Superconductivity Example." *Poznan Studies in the Philosophy of the Sciences and the Humanities* 44: 137–149.

Cassini, Alejandro, and Juan Redmond. 2021. *Models and Idealizations in Science. Artifactual and Fictional Approaches*. Cham: Springer.

Chakravartty, Anjan. 2001. "The Semantic or Model-Theoretic View of Theories and Scientific Realism." *Synthese* 127: 325–345.

———. 2009. "Informational versus Function Theories of Scientific Representation." *Synthese* 72:197–213.

Chang, Hasok. 2004. *Inventing Temperature: Measurement and Scientific Progress*. Oxford: Oxford University Press.

Contessa, Gabriele. 2007. "Scientific Representation, Interpretation, and Surrogative Reasoning." *Philosophy of Science* 74(1): 48–68.

———. 2011. "Scientific Models and Representation." In *The Continuum Companion to the Philosophy of Science*, edited by Steven French and Juha Saatsi, 120–137. London: Continuum Press.

Coopmans, Catelijne, Janet Vertesi, Michael E. Lynch, and Steve Woolgar, eds. 2014. *Representation in Scientific Practice Revisited*. Cambridge: MIT Press.

Cristalli, Claudia and Julia Sánchez-Dorado. 2021. "Colligation in Modelling Practices: From Whewell's Tides to the San Francisco Bay Model." *Studies in History and Philosophy of Science* 85: 1–15.

da Costa, Newton C.A., and Steven French. 2003. *Science and Partial Truth: A Unitary Approach to Models and Scientific Reasoning*. Oxford: Oxford University Press.

Daston, Lorraine. 2014. "Beyond Representation." In *Representation in Scientific Practice Revisited*, edited by Catelijne Coopmans, Janet Vertesi, Michael E. Lynch, and Steve Woolgar, 319–322. Cambridge: MIT Press.

de Chaderevian, Soraya, and Nick Hopwood, eds. 2004. *Models: The Third Dimension of Science*. Palo Alto, CA: Stanford University Press.

Dipert, Randall. 1996. "Reflections on Iconicity, Representation, and Resemblance: Peirce's Theory of Signs, Goodman on Resemblance, and Modern Philosophies of Language and Mind." *Synthese* 106: 373–397.

Downes, Stephen. 2009. "Models, Pictures, and Unified Accounts of Representation: Lessons from Aesthetics for Philosophy of Science." *Perspectives on Science* 17(4): 417–428.

Elgin, Catherine. 2010. "Telling Instances." In *Beyond Mimesis and Convention: Representation in Art and Science*, edited by Roman Frigg and Matthew Hunter, 1–18. Berlin and New York: Springer.

———. 2017. "Nature's Handmaid, Art." In *Thinking about Science, Reflecting on Art: Bringing Aesthetics and the Philosophy of Science Together*, edited by Otávio Bueno, George Darby, Steven French, and Dean Rickles, 27–39. New York: Routledge.

French, Steven. 2003. "A Model Theoretic Account of Representation (or, I Don't Know Much about Art, But I Know It Involves Isomorphism.)" *Philosophy of Science* 70: 1472–1483.

French, Steven, and Otávio Bueno. 2011. "How Theories Represent." *British Journal for the Philosophy of Science* 62: 857–894.

French, Steven, and James Ladyman. 1999. "Reinflating the Semantic Approach." *International Studies in the Philosophy of Science* 13: 103–121.

Frigg, Roman. 2006. "Scientific Representations and the Semantic View of Theories." *Theoria* 55: 49–65.

Frigg, Roman, and Matthew Hunter, eds. 2010. *Beyond Mimesis and Convention. Representation in Art and Science*. Boston Studies in the Philosophy of Science. London: Springer.

Frigg, Roman, and James Nguyen. 2018. "The Turn of the Valve: Representing with Material Models." *European Journal for the Philosophy of Science* 8: 205–224.

———. 2021. "Scientific Representation." *The Stanford Encyclopedia of Philosophy*, Winter 2016 Edition, edited by Edward N. Zalta. https://plato.stanford.edu/archives/win2016/entries/scientific-representation/

Frigg, Roman, and James Nguyen. 2022. *Scientific Representation*. Cambridge: Cambridge University Press.

Frisch, Matthias. 2015. "Users, Structures, and Representation." *British Journal for the Philosophy of Science* 66: 285–306.

Gelfert, Axel. 2016. *How to Do Science with Models*. Springer Cham. https://link.springer.com/book/10.1007/978-3-319-27954-1#bibliographic-information

Giere, Ronald. 1988. *Explaining Science: A Cognitive Approach*. Chicago: University of Chicago Press.

———. 2004. "How Models Are Used to Represent Reality." *Philosophy of Science* 71: 742–752.

———. 2010. "An Agent-based Conception of Models and Scientific Representation." *Synthese* 172(1): 269–281.

Gombrich, Ernst. 1960. *Art and Illusion: A Study in the Psychology of Pictorial Representation*. London: Phaidon.

Goodman, Nelson. 1968. *Languages of Art*. Indianapolis: Hackett.

———. 1972. "Seven Strictures on Similarity." In *Problems and Projects*, edited by Nelson Goodman, 437–447. Indianapolis/New York: Bobbs-Merrill.

Hacking, Ian. 1983. *Representing and Intervening: Introductory Topics in the Philosophy of Natural Science*. Cambridge: Cambridge University Press.

Hughes, R. I. G. 1997. "Models and Representation." *Philosophy of Science* 64: 325–336.

Ivanova, Milena, and Steven French. 2020. *The Aesthetics of Science: Beauty, Imagination and Understanding*. New York: Routledge.

Knuuttila, Tarja. 2017. "Imagination Extended and Embedded: Artifactual versus Fictional Accounts of Models." *Synthese* 198(Suppl 21): 5077–5097.

———. 2021. "Epistemic Artifacts and the Modal Dimension of Modeling." *European Journal for Philosophy of Science* 11(65): 1–18.

Knuuttila, Tarja, and Andrea Loettgers. 2016. "Contrasting Cases: The Lotka-Volterra Model Times Three." *Boston Studies in the Philosophy of Science* 319: 151–178.

Khosrowi, Donal. 2020. "Getting Serious about Shared Features." *The British Journal for the Philosophy of Science* 71: 523–546.

Liu, Chuang. 2015. "Re-inflating the Conception of Scientific Representation." *International Studies in the Philosophy of Science*, 29(1): 51–59.

Luczak, Joshua. 2017. "Talk About Toy Models." *Studies in History and Philosophy of Modern Physics* 57: 1–7.

Lynch, Michael. 2014. "Representation in formation." In *Representation in Scientific Practice Revisited*, edited by Catelijne Coopmans, Janet Vertesi, Michael E. Lynch, and Steve Woolgar, 323–327. Cambridge: MIT Press.

Lynch, Michael, and Steve Woolgar, eds. 1990. *Representation in Scientific Practice*. London: MIT Press.

Millson, Jared, and Mark Risjord. 2022. "DEKI, Denotation, and the Fortuitous Misuse of Maps." In *Scientific Understanding and Representation: Modeling in the Physical Sciences*, edited by Insa Lawler, Kareem Khalifa, and Elay Shech, 301–305. London: Routledge.
Morgan, Mary, and Margaret Morrison, eds. 1999. *Models as Mediators*. Cambridge: Cambridge University Press.
de Oliveira, Guilherme Sanches. 2021. "Representationalism Is a Dead End." *Synthese* 198: 209–235.
———. 2022. "Radical Artifactualism." *European Journal for Philosophy of Science* 12: 36.
Oreskes, Naomi. 2007. "From Scaling to Simulation: Changing Meanings and Ambitions of Models in Geology." In *Science without Laws: Model Systems, Cases, and Exemplary Narratives*, edited by Angela Creager, Elizabeth Lunbeck, and M. Norton Wise, 93–124. Durham: Duke University Press.
Parker, Wendy. 2015. "Getting (even) More Serious about Similarity." *Biology and Philosophy* 30: 267–276.
———. 2020. "Model Evaluation: An Adequacy-for-Purpose View." *Philosophy of Science* 87: 457–477.
Pero, Francesca, and Mauricio Suárez. 2016. "Varieties of Misrepresentation and Homomorphism." *European Journal for the Philosophy of Science* 6:71–90.
Peschard, Isabelle. 2011. "Making Sense of Modeling: Beyond Representation." *European Journal for Philosophy of Science* 1: 335–352.
Pincock, Cristopher. 2005. "Overextending Partial Structures: Idealization and Abstraction." *Philosophy of Science* 72: 1248–1259.
Poznic, Michael. 2016. "Modeling Organs with Organs on Chips: Scientific Representation and Engineering Design as Modeling Relations." *Philosophy & Technology*, 29: 357–371.
Ruyant, Quentin. 2021. "True Griceanism: Filling the Gaps in Callender and Cohen's Account of Scientific Representation." *Philosophy of Science* 88(3): 533–553.
Salis, Fiora. 2021. "Bridging the Gap: The Artifactual View Meets the Fiction View of Models." In *Models and Idealizations in Science. Artifactual and Fictional Approaches*, edited by Alejandro Cassini and Juan Redmond, 159–177. Cham: Springer.
Sánchez-Dorado, Julia. 2018. "Methodological Lessons for the Integration of Philosophy and Aesthetics: The Case of Representation." In *Thinking about Science, Reflecting on Art: Bringing Aesthetics and the Philosophy of Science Together*, edited by Otávio Bueno, George Darby, Steven French, and Dean Rickles, 10–26. New York: Routledge.
Schaffer, Simon. 2004. "Fish and Ships: Models in the Age of Reason." In *Models: The Third Dimension of Science*, edited by Soraya de Chadarevian, and Nick Hopwood, 71–105. Stanford, CA: Stanford University Press.
Sterrett, Susan. 2009. "Similarity and Dimensional Analysis." In *Philosophy of Technology and Engineering Sciences*, edited by Anthonie Meijers, 799–823. Amsterdam: North Holland/Elsevier.
———. 2017. "Experimentation on Analogue Models." In *Springer Handbook of Model-Based Science*, edited by Lorenzo Magnani and Tommaso Bertolotti, 857–878. Dordrecht: Springer.
Suárez, Mauricio. 2003. "Scientific Representation: Against Similarity and Isomorphism." *International Studies in the Philosophy of Science* 17(3): 225–244.
———. 2004. "An Inferential Conception of Scientific Representation." *Philosophy of Science* 71: 767–779.
———. 2010. "Scientific Representation." *Philosophy Compass* 5(1): 91–101.
———. 2015. "Deflationary Representation, Inference and Practice." *Studies in History and Philosophy of Science* 49: 36–47.
———.2024. *Inference and Representation*. Chicago and London: Chicago University Press.
Suppes, Patrick. 1960. "A Comparison of the Meaning and Uses of Models in Mathematics and the Empirical Sciences." *Synthese* 12: 287–301.
———. 1962. "Models of Data." In *Studies in the Methodology and Foundations of Science: Selected Papers from 1951 to 1969*, edited by Patrick Suppes, 24–35. Dordrecht: Synthese Library.
Teller, Paul. 2001. "Twilight of the Perfect Model Model." *Erkenntnis* 55: 393–415.
Tversky, Amos. 1977. "Features of Similarity." *Psychological Review* 84: 327–352.
van Fraassen, Bas C. 1980. *The Scientific Image*. New York: Oxford University Press.
———. 2008. *Scientific Representation: Paradoxes of Perspective*. New York: Oxford University Press.

van Fraassen, Bas C., and Jill Sigman. 1993. "Interpretation in Science and in the Arts." In *Realism and Representation*, edited by George Levine, 73–99. Madison: University of Wisconsin Press.
Weisberg, Michael. 2013. *Simulation and Similarity*. Oxford: Oxford University Press.
Winther, Rasmus G. 2020. *When Maps Become the World*. Chicago, IL: University of Chicago Press.
Woolgar, Steve. 2014. "Struggles with Representation: Could It Be Otherwise?" In *Representation in Scientific Practice Revisited*, edited by Catelijne Coopmans, Janet Vertesi, Michael E. Lynch, and Steve Woolgar, 329–332. Cambridge: MIT Press.

5
IDEALIZATION

Collin Rice

1. Introduction

Scientific models are always idealized to some degree. Indeed, many philosophers have suggested that a model just is an idealized representation of some real or possible target system(s). While perhaps not all models have target systems, certainly the vast majority involve assumptions that are inaccurate with respect to real-world systems. Moreover, idealizations are typically intentionally introduced into scientific models. Scientists' use of myriad idealizations results in most scientific models providing drastically distorted representations of reality. This has led philosophers to investigate the crucial roles that idealizations play within scientific practice. One of the lessons of these investigations has been the discovery of several distinct aims and contexts that motivate the introduction of idealizations into scientific models. These different modeling contexts have, in turn, given rise to a plurality of ways in which scientists justify their use of idealizations. In this chapter, I take idealization to be the *intentional* introduction of distortion into a scientific model or theory *for some purpose*. For example, removing negligible or insignificant factors from a model of a complex ecosystem in order to simplify calculations, or assuming that a system has an infinite number of particles in order to apply various mathematical modeling techniques in physics. The chapter begins by providing a (non-exhaustive) survey of some of the scientific contexts and goals that motivate the introduction and maintenance of idealizations. These different contexts and aims will then be used to discuss various philosophical questions concerning the use of idealization in science.

2. Pluralism about idealization

Rather than a univocal account of idealizations, or of how they are justified, what we find in scientific practice is a plurality of types, motivations, and justifications (Potochnik 2017; Rice 2021; Weisberg 2007; Cassini and Redmond 2021). In this section, I provide a non-exhaustive survey of several "types" of idealization by looking at the reasons that motivate their introduction and how they are justified.

DOI: 10.4324/9781003205647-8

To begin, Michael Weisberg (2007, 2013) has usefully distinguished three kinds of idealization by looking at the reasons for their introduction and what he calls their ultimate "representational ideals" (2007, 639). The three kinds of idealization are Galilean idealization, minimalist idealization, and multiple-model idealization. The two first kinds of idealization are already differentiated in Nowak (1992, 2000), whose work on idealization inspired much of the lively philosophical discussion of idealization in the 1980s and 1990s, resulting in several volumes on idealization in Poznań Studies in the Philosophy of the Sciences and the Humanities. Frigg and Hartmann (2012) introduce a similar kind of distinction between Galilean and Aristotelian idealization.

Galilean idealizations are introduced to address issues of computational tractability and are justified by noting that they make the calculations of the model simpler. Yet, as science advances (e.g., more powerful computers are built), this motivation for idealizing can fade such that "Galilean idealization takes place with the expectation of future deidealization and more accurate representation" (Weisberg 2007, 642). According to Ernan McMullin's account, this means that Galilean idealizations can ultimately "be made more specific by eliminating simplifying assumptions and 'de-idealizing' as it were" (1985, 261). Similarly, William Wimsatt (2007) argues that idealized models can be justified by showing that they eventually lead to truer theories. According to Galilean accounts, idealized models are temporarily justified waystations on the way to the production of more accurate models and theories.

Weisberg also groups a number of views under the category of *minimalist idealization*. Minimalist accounts focus on the aim of providing explanations and suggest that the model that best explains a phenomenon will include only the core causal, or difference-making, factors that gave rise to the explanandum (Weisberg 2007, 643–645). Indeed, several philosophers have suggested that idealized models are able to explain just when they accurately describe the difference-making, contextually salient, or otherwise important explanatory factors. Idealizations are then used to distort other features to emphasize that those features are irrelevant, non-difference-making, not of interest, or negligible. These accounts justify idealizations in scientific models by noting that, while the models distort a variety of irrelevant or non-difference-making features, they still provide accurate descriptions of the system's relevant features used in providing the explanation. Prominent examples of the minimalist approach are provided by Uskali Mäki (1992), Nancy Cartwright (1999), Mehmet Elgin and Elliott Sober (2002), David Kaplan and Carl Craver (2011), and Michael Strevens (2008). As a specific example, both Strevens and Weisberg cite Boyle's law, which assumes that the gas molecules do not collide with each other. This idealization is justified, they argue, because "the no-collision assumption is a way of asserting that collisions are actually irrelevant and make no difference" (Weisberg 2007, 643). However, despite these distortions, the model is still able to explain because "it accurately captures the core causal factors" (Weisberg 2007, 643).

In contrast with minimalist accounts, several philosophers have gone further and argued that idealizations within models that explain, frequently distort difference-making factors as well (Elgin 2017; Potochnik 2017; Rice 2021). For example, several philosophers have noted that an important motivation for introducing idealizing assumptions into models that explain is that they are often *necessary for the use of mathematical frameworks and modeling techniques* (Batterman 2002; Morrison 2015; Rice 2021). As examples, these philosophers have noted that various idealizations are required to enable scientists to apply game-theoretic modeling, statistical modeling, the renormalization group, or homogenization techniques (just to name a few). These idealizations can enable scientists to provide explanations that would otherwise be inaccessible, but their introduction often requires the models to distort features that are known to make a difference and are of interest to the scientists using the model to explain.

Scientific models are also frequently used to produce *understanding* of a phenomenon (Elgin 2017; Potochnik 2017; Rice 2016; Strevens 2013). For example, Yasha Rohwer and Collin Rice have argued that there is a distinctive (fourth) type of idealization used in biology and economics that aims at the production of understanding by investigating hypothetical scenarios (Rohwer and Rice 2013). Rohwer and Rice refer to this as *hypothetical pattern idealization* because the models aim to generate understanding by investigating background assumptions, necessity claims, or how-possibly stories via the construction of hypothetical scenarios that display widely observed patterns. For example, the Hawk-Dove game improves biologists' understanding of restraint in combat by showing that the observed patterns of behavior could *possibly* be produced by individual-level selection in a highly idealized population. What is distinctive about this kind of idealization is that it aims to produce understanding of a general pattern by investigating a hypothetical scenario, i.e., one that is not intended to be actual or even possible. Since this aim is often best achieved by building a highly idealized model of a particular hypothetical scenario, these idealizations typically will not be removed as science progresses. Moreover, these models do not aim at providing an explanation for why the phenomenon *actually* occurred. Consequently, rather than (ultimately) aiming for a model that accurately represents the system or explanatory factors, these models aim to deepen our understanding by exploring a distant counterfactual situation that often is impossible to realize (de Donato Rodríguez and Zamora 2009). Along similar lines, Angela Potochnik (2017) and Catherine Elgin (2017) both analyze a variety of ways that idealized models are used to produce scientific understanding via the embodiment of causal patterns or the exemplification of features of real systems.

Weisberg also discusses "the practice of building multiple related but incompatible models, each of which makes distinct claims about the nature and causal structure giving rise to a phenomenon" (Weisberg 2007, 645). These cases of *multiple-model idealization* are distinguished by "not expecting a single best model to be generated" (Weisberg 2007, 646). Since model builders often have multiple goals that are difficult (if not impossible) to simultaneously achieve with a single model (Levins 1966), scientists often construct multiple models that each make different idealizing assumptions about the phenomenon in order to achieve different modeling goals. As a specific example, Margaret Morrison (2011) discusses the use of over *30* different models of the nucleus that are used to explain, understand, and predict various features of nuclear behavior. As a result, rather than aiming for a single best model that provides an explanation, yields understanding, or can later be deidealized, these contexts typically give rise to the production and maintenance of several conflicting idealized models of the same phenomenon.

Before moving on to some of the philosophical questions surrounding the above kinds of idealization, it is important to note that these reasons, motivations, and justifications are often intertwined in complex ways within scientific practice (Potochnik 2017; Rice 2021). First, a single scientific model might include multiple types of idealization; e.g., a single model might include both Galilean and minimalist idealizations. Second, a single idealizing assumption might have multiple reasons that motivate its introduction. As a result, even if one of those motivations is removed (e.g., through improved computational capacities), there may be several other reasons for maintaining a particular idealizing assumption within the scientific model. Third, because idealizations are often foundational to the application of general modeling frameworks, they often become *deeply embedded* within modeling research programs (Pincock 2012; Potochnik 2017; Rice 2021; Weisberg 2013). These overlapping, intertwined, and embedded reasons for using idealizations give rise to a plethora of philosophical questions that have been the focus of much of the literature on scientific modeling.

3. Can idealization be eliminated?

Accounts of Galilean idealization raise the question of whether, generally, idealizations can be eliminated from scientific models as science progresses. If idealizations can often be removed, then their introduction can be justified as an important first step toward eventually generating more accurate (or truer) models and theories. There are certainly numerous ways in which idealizations contribute to the aims of science by making mathematical or computational models more tractable. However, if most idealizations were Galilean, then we ought to be able to see how they could be removed or replaced by true assumptions without undermining the models' ability to contribute to the aims of science for which they were constructed. Yet, several philosophers have argued that this is not what we find when we look at actual scientific practice.

One reason for this is that, *in practice*, even when deidealization is possible, in many cases, idealizations are not actually removed from scientific models (Knuuttila and Morgan 2019). Indeed, even when more realistic models are available, scientists routinely opt for the more highly idealized model because it is better suited for their purposes (Elgin 2017; Potochnik 2017; Rice 2021). As Catherine Elgin notes, in science, "Elimination of idealizations is not a desideratum" (2017, 62). One example of this is the ideal gas law. Even though more accurate models are available—e.g., models that include van der Waal's equations—the ideal gas law is still widely used. Similar cases can be found throughout biology where idealized models that include only the influence of natural selection are often preferred despite the ability to construct models that would more accurately represent other evolutionary factors like mutation, migration, or drift (Potochnik 2017; Rice 2021). As a result, even if idealizations can sometimes be replaced, in practice they rarely are.

In addition, several philosophers have argued that some idealizations cannot be removed *in principle* without losing the epistemic achievements (e.g., explanation and understanding) enabled by those idealizations. One reason for this is that there are several cases in which scientific explanations require infinite idealizations that are necessary for the mathematical techniques used in providing the explanation. For example, Robert Batterman's (2002) pioneering work on the use of the renormalization group argues that the thermodynamic limit (in which the number of particles goes to infinity) is essential to mathematical modeling techniques that are widely used in physics to explain the universality (i.e., stability) of critical behaviors of various fluids and magnets. In a similar way, Margaret Morrison argues that:

> The occurrence of phase transitions requires a mathematical technique known as taking the 'thermodynamic limit,' $N \rightarrow \infty$; in other words we need to assume that a system contains an infinite number of particles in order to understand the behavior of a real, finite system…[since] the assumption that the system is infinite is *necessary* for the symmetry breaking associated with phase transitions to occur. In other words,

> we have a description of a physically unrealizable situation (an infinite system) that is *required* to explain a physically realizable phenomenon (the occurrence of phase transitions).
>
> *(Morrison 2009, 128)*

Without these limiting idealizations—in which a parameter or variable is taken to infinity or zero—the explanations physicists have provided for various phase transition phenomena would no longer be applicable. In addition, Alisa Bokulich (2008) argues that fictional models, such as Bohr's model of the atom, also play indispensable roles in the explanations provided in other areas of physics. While some philosophers have argued that many of these cases can be subject to reduction or relaxation (Butterfield 2011), it is still debatable whether those less idealized models or theories are able to provide the same explanations or understanding and, if so, why the more idealized models continue to be central to the way physicists investigate these systems.

Other philosophers have argued that several idealizing assumptions are essential to various research programs in biology (Potochnik 2017; Rice 2021). For example, within the adaptationist research program, biological modelers routinely idealize other evolutionary factors (e.g., the processes involved in genetic drift, mutation, migration, or inheritance) of the system in order to focus on the role of natural selection in producing an observed trait. Because these adaptationist models can often provide unique explanations and understandings of a trait, these idealizations often cannot be removed without losing the explanation or understanding provided by the adaptationist model. Another set of cases involves the use of idealizing assumptions to enable the application of statistical modeling techniques within population genetics (Ariew et al. 2015). In these cases, in order to apply various statistical theorems—e.g., the central limit theorem—biological modelers routinely introduce assumptions of infinitely large populations where mating (or other interactions) is completely random. Removing or relaxing these idealizations makes many of these statistical modeling techniques inapplicable.

In fact, across multiple scientific disciplines, we find that many idealizing assumptions cannot be removed in principle because they are *necessary* to apply the modeling techniques that enable scientists to explain and understand complex phenomena (Rice 2021). In short, even if scientists would sometimes prefer to deidealize their models in the way suggested by Galilean accounts of idealization, often the modeling approaches available and the complexity of the phenomenon of interest make it such that eliminating the idealizations from the model or theory would also eliminate the explanations and understanding that motivated their introduction in the first place.

4. Do models accurately represent relevant features or are they holistic distortions?

As I noted above, in order to account for these more permanent contributions of idealizations within models that explain, a number of philosophers have developed views that follow Weisberg's characterization of minimalist idealization. These minimalist accounts require that it is possible (at least in principle) to decompose scientific models into their accurate and inaccurate parts. The idealized parts of the model can then be justified by showing that they only distort features that are irrelevant, non-difference-making, not contextually salient, or otherwise not of interest. Moreover, the models are claimed to be suitable for purposes of explanation because they accurately represent (or describe) the relevant, difference-making, or contextually salient features of interest.

As a first example, many defenders of mechanistic accounts of modeling and explanation have argued that the widespread use of idealization, "should not lead us to dispense with the idea that models can more or less accurately represent features of the mechanism in the case at hand" (Kaplan and Craver 2011, 610). Indeed, despite the use of idealizations, according to most mechanistic accounts, "the goal is to describe correctly enough (to model more or less accurately) the relevant aspects of the mechanism under investigation" (Craver and Darden 2013, 94). Another proponent of this approach is Strevens, who argues that idealized models can provide superior explanations when they accurately represent the causal difference-makers that produced the explanandum and use idealizations to indicate that the distorted features do not make a difference. As Strevens summarizes his view:

> The content of an idealized model, then, can be divided into two parts. The first part contains the difference-makers for the explanatory target...The second part is all idealization; its overt claims are false but its role is to point to parts of the actual world that do not make a difference to the explanatory target. The overlap between an idealized model and reality...is a standalone set of difference-makers for the target.
>
> *(Strevens 2008, 318)*

A similar kind of view is defended by Elgin and Sober (2002) in which they argue that idealizations are "harmless" if correcting them would not make much difference to the predictions of the model (448). The goal of these accounts is to show that the "factors distorted by idealized models are details that do not matter to the explanatory target—they are explanatory irrelevancies. The distortions of the idealized model are thus mitigated" (Strevens 2008, 315).

A related, but importantly different, approach has focused on the features that are of interest within a particular context of inquiry. According to these philosophers, causes or features that make a difference to the phenomenon can be justifiably distorted by idealized models as long as those features are not of interest to the scientists using the model to explain (or understand). For example, Potochnik argues that "significant causal factors that are not central to the research program can still be set aside" (2015, 1178). Rather than appealing solely to difference-making considerations, Potochnik's account uses the research program and context of inquiry to determine which causes are important for providing the desired explanation. However, "posits central to representing the focal causal pattern in some phenomenon must accurately represent the causal factors contributing to this pattern" (Potochnik 2017, 157). That is, in order to explain, the model must accurately represent the features deemed relevant by the context of inquiry. Along similar lines, Elgin's account allows for many difference-making causes to be distorted by models that contribute to scientific understanding (Elgin 2017). Moreover, Elgin also appeals to the interests of scientists in various places by suggesting that idealized models can provide genuine understanding, "because the models are approximately true, or because they diverge from truth in irrelevant respects, or because the range of cases for which they are not true is a range of cases we do not care about" (Elgin 2017, 261). More generally, a wide range of accounts have argued that the best way to justify the use of idealizations to accomplish the epistemic aims of science is to show that the idealizations only distort features that are irrelevant, non-difference-making, or otherwise not of interest to scientists.

In contrast with these views, other philosophers have argued that the idealized models used to explain in scientific practice are far more pervasive distortions. In particular, these philosophers argue that many of the scientific models that are used to explain and understand directly and deliberately distort features that are known to make a difference to the explanandum and that are of interest to the scientists using the models to explain. Many of the foundational ideas of this approach can be found in the pioneering work of Nancy Cartwright (1983), who argues that idealization and abstraction of relevant causes are essential to the ability of models, theories, and laws to explain. One way of developing this approach comes from Bokulich (2008; 2011; 2016), who argues that many of the idealized models used to explain in science are "fictions" that distort difference-making or relevant causes of the phenomenon in a variety of ways. For one thing, Bokulich argues that constructing an accurate representation of the system is not required for scientific modelers to extract explanatory (in her view, modal) information. She contends, "Certainly having an accurate representation is one way to get such modal information, but the success of idealized and fictional models in science teaches us that it is not the only way" (Bokulich 2016, 271). As examples, Bokulich points to Bohr's model of the atom and fictional electron trajectories in quantum dots in which entities that are known not to exist are postulated and play crucial roles within the explanation (Bokulich 2008; 2011).

Another version of this approach argues that the models used to explain in science ought to be construed as *holistic* distortions of their target systems; i.e., they are pervasive misrepresentations of both difference-making and non-difference-making features (Rice 2018; 2021). There are three main arguments for this type of view. First, by looking at a variety of examples from scientific practice, we find that a number of models that are used to explain directly distort known difference-makers that are of interest to the research program in which they are formulated. For example, physicists routinely distort the processes that lead to phase transitions despite the fact that they know that the features distorted by their idealizations make a difference to the real system's critical behaviors and those features are certainly of interest to physicists attempting to explain and understand those behaviors (Rice 2018). Similarly, within evolutionary biology, adaptationists routinely use optimization models to explain although the models distort the very processes of natural selection that are known (or at least assumed) to make a difference to the evolution of the trait and are of interest to biologists studying adaptations (Rice 2021). The second line of argument points to the various cases discussed above in which idealizations are introduced into scientific models because they are necessary for the use of various mathematical modeling frameworks (Rice 2021). In each of these cases, the idealizations are so foundational to the overall mathematical frameworks used in these models that the resulting representations typically distort a wide range of difference-making and non-difference-making features—many of which are of interest to the modelers working within those research programs. This is not to say that models typically distort *all* the features of their target systems, but that very often their distortions are far more pervasive/holistic than is assumed by accounts of minimalist idealization. The final argument for this kind of holistic distortion view is that both scientists and philosophers *are rarely in the required epistemic situation* to be able to identify precisely which features are being accurately represented by a model and which are being distorted. For one thing, this would require us to know what is true of the target system in a way that is often inaccessible when it comes to extremely complex systems. In addition, idealizing assumptions do not make their contributions to models and explanations in isolation, but are rather collaborative members of larger sets of assumptions, inferences, and applications that constitute the model and the explanation it is used to provide (Carrillo and Knuuttila 2022). As a result, the justification offered for using these idealized models to explain and understand ought to be one that can be provided in situations where we are unsure which parts of the system are relevant/irrelevant and which parts of the model are accurate/inaccurate. Characterizing scientific models as holistically distorted representations ensures that the justifications offered for using scientific models to explain and understand take this epistemic limitation seriously. Adopting a holistic distortion view is not the only way to draw attention to those limitations, but it is an effective way of focusing philosophical accounts on the question of how to justify idealizations within that epistemic context.

5. How should we interpret the use of multiple conflicting models?

Weisberg's characterization of multiple-models idealization is related to what has come to be known as *the problem of inconsistent models* (Chakravartty 2010; Massimi 2018; Morrison 2011; 2015; Rice 2021). For example, as Morrison notes, "nuclear spin, size, binding energy, fission and several other properties of stable nuclei are all accounted for using models that describe one and the same entity (the nucleus) in different and contradictory ways" (2011, 349). In addition, Wendy Parker notes that "complex climate models generally are physically incompatible with one another—they represent the physical processes acting in the climate system in mutually incompatible ways and produce different simulations of climate" (2006, 350). These cases raise philosophical questions concerning how multiple conflicting idealized models can contribute to a scientific understanding of the same phenomenon and how models constructed for different scales of the system can be connected to one another. I will briefly discuss these two debates.

The first issue is attempting to clarify how the construction of multiple conflicting models for the same phenomenon could produce genuine scientific explanations or understanding. Specifically, given that most philosophical accounts have required models to provide accurate representations in order to provide scientific explanations or understanding—what Michela Massimi calls the "representationalist assumption" (2018, 335)—it is unclear how constructing multiple conflicting models could produce genuine explanations or understandings of real phenomena.

One way of analyzing these cases is to argue that the models produce understanding when they each target different aspects, features, or patterns within the system (Elgin 2017; Potochnik 2017). For example, Elgin responds to the use of multiple conflicting models of the nucleus by arguing that:

> If what one model highlights is that in some significant respects the nucleus behaves like a liquid drop, and another model highlights that in some other significant respects it behaves as though it has a shell structure, there is in principle no problem. There is no reason why the same thing should not share some significant properties with liquid drops and other significant properties with rigid shells.
>
> *(Elgin 2017, 270)*

Similarly, Potochnik (2017) suggests that many of these cases can be handled by arguing that the models can produce understanding when they target different causal patterns embodied by the real phenomenon. Like Weisberg's characterization of multiple-models idealization, these accounts suggest that we can accommodate instances of multiple conflicting models by showing that the models are each built with different goals in mind and, as a result, they each aim to capture different aspects of their target system(s).

In contrast, Morrison (2011), Rice (2021), and Carrillo and Knuuttila (2022) have argued that this approach fails to capture instances in which each of the models "makes very different assumptions about exactly the same thing" (Morrison 2011, 347). In these cases, we cannot resolve the issue just by arguing that the models are accurate with respect to different aspects of the system because the models aim to capture the same relevant features of their target system(s), but they each do so using contradictory idealizing assumptions. One way to respond to these cases is to separate—or at least put some distance between—the requirements for scientifically understanding a phenomenon and the conditions of accurate representation, exemplification, or truth for the model being used to understand that phenomenon (Massimi 2018; Rice 2021). For example, Massimi (2018) argues that these models should be interpreted as being constructed within different perspectives rather than as models that each aim to accurately describe their target systems in contradictory ways. Alternatively, Rice (2021) argues that multiple conflicting models can produce understanding by providing different sets of modal information about universal patterns that hold across different ranges of real and possible systems. Both of these accounts emphasize the use of multiple conflicting models to explore possibilities and provide information about counterfactual situations rather than attempting to accurately describe real-world systems.

A related set of issues arises in cases of *multiscale modeling* (Batterman 2021; Jhun 2021; Rice 2021; Wilson 2017). In these cases, multiple conflicting models are constructed because the phenomenon of interest depends on features of the system that span across a wide range of spatial and temporal scales, but the available models (or modeling techniques) are restricted to a relatively small range of scales. Batterman and others have referred to this challenge as the *tyranny of scales* (Batterman 2013; 2021; Green and Batterman 2017; Wilson 2017). A key philosophical question here is how multiple conflicting models constructed for each of these scales ought to be combined, coupled, or used to pass information from one scale to another. For example, Eric Winsberg (2006) has analyzed multiscale modeling cases in which "handshakes" between the models are used to combine conflicting idealized models at different scales. This is accomplished by first modeling the boundary regions within one (macroscale) modeling framework, then modeling those same regions with another (more microscale) modeling framework, and then averaging the results for various key parameters. While this is certainly one way to have models communicate across scales, as Julia Bursten (2018) argues, the handshakes will need to be quite different in different modeling contexts. Therefore, we still need to look at the details of particular multiscale modeling cases in order to determine just which multiscale modeling techniques and conceptual strategies ought to be used to integrate the models constructed for different scales of the system (Bursten 2018). As another example, Batterman (2002; 2013; 2021) and Mark Wilson (2017) discuss a number of cases in which renormalization or homogenization techniques are used to bridge between different scales in physics. Rice (2021) discusses similar cases in biological contexts. These multiscale modeling approaches routinely involve a number of idealizing assumptions that enable the application of modeling techniques that identify a set of key parameters across a variety of more macroscale scales that are essential for capturing the general patterns of behavior of the system(s) of interest.

6. Conclusion: how do idealizations relate to the aims of science?

Each of the debates discussed here provides an example of a more general question concerning how the widespread use of idealizations is able to contribute to the aims of science. As we saw above, a number of philosophers have tried to show how idealizations can be made compatible with the widely held assumption that scientific explanations ought to be true descriptions of the reasons why the phenomenon occurred (Kaplan and Craver 2011; Strevens 2008). Others have argued that models can explain even when they distort the explanatorily relevant features of interest (Batterman and Rice 2014; Potochnik 2017; Rice 2021). In a similar way, a number of philosophers have aimed to show how idealized models contribute to scientific understanding. Some of these accounts have suggested that the use of idealizations requires us to adopt non-factive accounts of scientific understanding (Elgin 2017; Potochnik 2017), while others have argued that even pervasively distorted models can give rise to factive understanding of real phenomena (Khalifa and Sullivan 2019; Rice 2021). Finally, several philosophers have argued that idealized models improve our ability to make predictions (Douglas 2009; Odenbaugh 2005). Despite the disagreements about precisely how idealizations contribute to each of these aims, philosophers generally agree that most of the representations provided in scientific practice are "idealized, inaccurate, but successful" (Odenbaugh 2005, 231). As a result, philosophers ought to continue investigating the variety of ways that idealizations contribute to the central aims of science.

References

Ariew, André, Collin Rice, and Yasha Rohwer. 2015. "Autonomous Statistical Explanations and Natural Selection." *British Journal for the Philosophy of Science* 66(3): 635–658.

Batterman, Robert W. 2002. *The Devil in the Details: Asymptotic Reasoning in Explanation, Reduction, and Emergence*. Oxford: Oxford University Press.

———. 2013. "The Tyranny of Scales." In *The Oxford Handbook of Philosophy of Physics*, ed. Robert Batterman, pp. 255–286. Oxford: Oxford University Press.

———. W. 2021. *A Middle Way: A Non-Fundamental Approach to Many-Body Physics*. Oxford: Oxford University Press.

Batterman, Robert W., and Collin Rice. 2014. "Minimal Model Explanations." *Philosophy of Science* 81(3): 349–376.

Bokulich, Alisa. 2008. *Reexamining the Quantum-Classical Relation: Beyond Reductionism and Pluralism*. Cambridge: Cambridge University Press.

———. 2011. "How Scientific Models Can Explain." *Synthese* 180: 33–45.

———. 2016. "Fiction as a Vehicle for Truth: Moving Beyond the Ontic Conception." *The Monist* 99: 260–279.

Bursten, Julia R. 2018. "Conceptual Strategies and Inter-Theory Relations: The Case of Nanoscale Cracks." *Studies in History and Philosophy of Modern Physics* 62: 158–165.

Butterfield, Jeremy. 2011. "Emergence, Reduction, and Supervenience: A Varied Landscape." *Foundations of Physics* 41: 920–959.

Carrillo, Natalia, and Tarja Knuuttila. 2022. "Holistic Idealization: An Artifactual Standpoint." *Studies in the History and Philosophy of Science* 91: 49–59.

Cartwright, Nancy. 1983. *How the Laws of Physics Lie*. Oxford: Oxford University Press.

———. 1999. "The Vanity of Rigour in Economics: Theoretical Models and Galilean Experiments." Discussion paper series 43/99. Centre for Philosophy of Natural and Social Science.

Cassini, Alejandro, and Juan Redmond, eds. 2021. *Models and Idealizations in Science: Artifactual and Fictional Approaches*. Logic, Epistemology, and the Unity of Science 50. Cham, Switzerland: Springer.

Chakravartty, Anjan. 2010. "Perspectivism, Inconsistent Models, and Contrastive Explanation." *Studies in History and Philosophy of Science* 41: 405–412.

Craver, Carl, and Lindley Darden. 2013. *In Search of Mechanisms: Discoveries across the Life Sciences*. Chicago: University of Chicago Press.
de Donato Rodríguez, Xavier, and Jesus Zamora Bonilla. 2009. "Credibility, Idealisation, and Model Building: An Inferential Approach." *Erkenntnis* 70(1): 101–118.
Douglas, Heather. 2009. *Science, Policy, and the Value-Free Ideal*. Pittsburgh: University of Pittsburgh Press.
Elgin, Catherine Z. 2017. True Enough. Cambridge: MIT Press.
Elgin, Mehmet, and Elliott Sober. 2002. "Cartwright on Explanation and Idealization." *Erkenntnis* 57: 441–450.
Frigg, Roman, and Stephan Hartmann. 2012. "Models in Science." *The Stanford Encyclopedia of Philosophy*, ed. Edward N. Zalta. https://plato.stanford.edu/entries/models-science/.
Green, Sara, and Robert W. Batterman. 2017. "Biology Meets Physics: Reductionism and Multi-Scale Modeling of Morphogenesis." *Studies in History and Philosophy of Science Part C: Studies in History and Philosophy of Biological and Biomedical Sciences* 61: 20–34.
Jhun, Jennifer. 2021. "Economics, Equilibrium Methods, and Multi-Scale Modeling." *Erkenntnis* 86: 457–472.
Kaplan, David M., and Carl Craver. 2011. "The Explanatory Force of Dynamical and Mathematical Models in Neuroscience: A Mechanistic Perspective." *Philosophy of Science* 78: 601–627.
Khalifa, Kareem, and Emily Sullivan. 2019. "Idealization and Understanding: Much Ado about Nothing?" *Australasian Journal of Philosophy* 97(4): 673–689.
Knuuttila, Tarja, and Mary S. Morgan. 2019. "De-idealization – No Easy Reversals." *Philosophy of Science* 86(4): 641–661.
Levins, Richard. 1966. "The Strategy of Model Building in Population Biology." *American Scientist* 54: 421–431.
Massimi, Michela. 2018. "Perspectival Modeling." *Philosophy of Science* 85(3): 335–359.
McMullin, Ernan. 1985. "Galilean Idealization." *Studies in History and Philosophy of Science* 16: 247–273.
Morrison, Margaret. 2009. "Understanding in Physics and Biology." In *Scientific Understanding: Philosophical Perspectives*, ed. Henk W. de Regt, Sabina Leonelli, and Kai Eigner. 123–145. Pittsburgh: University of Pittsburgh Press.
———. 2011. "One Phenomenon, Many Models: Inconsistency and Complementarity." *Studies in History and Philosophy of Science Part A* 42(2): 342–351.
———. 2015. *Reconstructing Reality: Models, Mathematics, and Simulations*. New York: Oxford University Press.
Mäki, Uskali. 1992. "On the Method of Isolation in Economics." *Poznań Studies in the Philosophy of the Sciences and the Humanities* 26: 319–354.
Nowak, Leszek. 1992. "The Idealizational Approach to Science: A Survey." In *Idealization III: Approximation and Truth*, ed. Jerzy Brzeziński and Leszek Nowak, 9–66. Amsterdam: Rodopi.
———. 2000. "The Idealizational Approach to Science: A New Survey." In *Idealization X: The Richness of Idealization*, ed. Leszek Nowak and Izabella Nowakova, 109–184. Amsterdam: Rodopi.
Odenbaugh, Jay. 2005. "Idealized, Inaccurate, but Successful: A Pragmatic Approach to Evaluating Models in Theoretical Ecology." *Biology and Philosophy* 20: 231–255.
Parker, Wendy S. 2006. "Understanding Pluralism in Climate Modeling." *Foundations of Science* 11: 349–368.
Potochnik, Angela. 2015. "Causal Patterns and Adequate Explanations." *Philosophical Studies* 172(5): 1163–1182.
———. 2017. *Idealization and the Aims of Science*. Chicago: University of Chicago Press.
Pincock, Christopher. 2012. "Mathematical Models of Biological Patterns: Lessons from Hamilton's Selfish Herd." *Biology and Philosophy* 27: 481–496.
Rice, Collin. 2016. "Factive scientific understanding without accurate representation." *Biology and Philosophy* 31(1): 81–102.
Rice, Collin. 2018. "Idealized Models, Holistic Distortions and Universality." *Synthese* 195(6): 2795–2819.
———. 2021. *Leveraging Distortions: Explanation, Idealization and Universality in Science*. Cambridge: MIT Press.

Rohwer, Yasha, and Collin Rice. 2013. "Hypothetical Pattern Idealization and Explanatory Models." *Philosophy of Science* 80: 334–355.
Strevens, Michael. 2008. *Depth: An Account of Scientific Explanation*. Cambridge, MA: Harvard University Press.
———. 2013. "No Understanding without Explanation." *Studies in History and Philosophy of Science* 44: 510–515.
Weisberg, Michael. 2007. "Three Kinds of Idealization." *Journal of Philosophy* 104(12): 639–659.
———. 2013. *Simulation and Similarity: Using Models to Understand the World*. Oxford: Oxford University Press.
Wilson, Mark. 2017. *Physics Avoidance*. Oxford: Oxford University Press.
Wimsatt, William. 2007. *Re-engineering Philosophy for Limited Beings: Piecewise Approximations of Reality*. Cambridge, MA: Harvard University Press.
Winsberg, Eric. 2006. "Handshaking Your Way to the Top: Inconsistency and Falsification in Intertheoretic Reduction." *Philosophy of Science* 73: 582–594.

6
DEIDEALIZATION

Alejandro Cassini

1. Deidealizing models

All scientific models are idealized to some degree. This presupposes that idealization itself is a matter of degree and, consequently, that we can build more or less idealized models of the same phenomena. In principle, at least, a highly idealized model of a given domain is capable of being deidealized, that is, becoming less idealized. Some idealizations can be removed from the model - or modified, or replaced - in such a way that the resulting model becomes less simple, less abstract, or less distorted than the original model. In positive terms, the deidealized model is more complicated, more concrete, and perhaps more "realistic" than the more idealized model of which it is a deidealization. Some philosophers and scientists would claim that a deidealized model provides a better approximate description of the phenomena, or even that it is more truthlike or verisimilar than its idealized predecessor.

Although many philosophical studies have been devoted to the concept of idealization, the study of deidealization is just getting off the ground. What exactly deidealization is and how it must be carried out are questions whose answers depend essentially on what we understand as idealization. There is no widespread agreement among philosophers of science on how to define the concept, though there exists a body of different approaches to idealization; see Jones (2005), Weisberg (2013), Morrison (2015), Potochnik (2017), Wheeler (2018), Cassini and Redmond (2021), Rice (2021), Frigg (2023), and Shech (2023). By contrast, no monographic book has yet been published on deidealization. For articles specifically devoted to the topic, see those by Knuuttila and Morgan (2019), and Cassini (2021). The recent books by Rice (2021) and Shech (2023) include extensive discussions of deidealization within a broader philosophical context.

Roughly speaking, idealized models are usually described as *simplified*, *abstract*, *distorted*, and *approximate* representations of some domain of phenomena. These concepts, in turn, are in need of elucidation. Sometimes, models are qualified as false representations of the phenomena. However, given that models are not bearers of truth values, at least not primarily, it is convenient to avoid calling them true or false. Some decisions concerning the use of the concepts associated with idealization are unavoidable for starting an analysis

DOI: 10.4324/9781003205647-9

of the notion of deidealization. Without much justification, I will assume here that idealization implies abstracting and distorting procedures, and that the simplified and approximate character of idealized models is the outcome of both abstraction and distortion. Some philosophers, however, have conceived of abstraction as something different from idealization. For a sample of different positions concerning how the concepts of idealization and abstraction are related, see Cartwright (1989, 1999), Jones (2005), Godfrey-Smith (2009), Morrison (2015), Levy (2021), and Portides (2021).

An idealized model contains some abstractions and distortions. Some constants, parameters, or variables that we believe to be relevant to the phenomena to be modeled are not included in the model. Besides, the model contains some constants, parameters, or variables that we regard as non-representational of features of the modeled phenomena, or that are set to values that we do not regard as the correct ones considering our experience (typically, values such as 0, 1, or infinite). Deidealizing a model essentially consists of removing or replacing some of the abstractions and distortions it contains, for instance, adding and/or removing new constants, parameters, or variables, and/or setting some of its parameters to different, more empirically adequate values.

2. A deidealized model

The so-called kinetic theory of gases (the term model was not fashionable when the theory was put forward) provides a good example of how idealizations and deidealizations work. A model of an ideal gas is built based on some general hypotheses concerning the composition of all gases and some idealizations concerning the specific properties of ideal gases. Without intending a complete formulation of the model, we can list four hypotheses and four idealizations.

H_1: All gases are composed of many molecules of different elements (say, hydrogen, helium, oxygen). H_2: All the molecules move spontaneously at random in empty space, colliding frequently with each other and, more often, with the walls of the vessel containing the gas. H_3: The motion of the molecules satisfies all the laws of Newtonian mechanics. H_4: Every macroscopic volume of a gas is composed of a huge number of microscopic molecules (to the order of Avogadro's number, namely 10^{23} or higher).

Given the enormous number of molecules that compose it, a macroscopic volume of any real gas is a very complex physical system, whose dynamical state (the position and momentum of every molecule) we cannot know in practice. Thus, to be able to state some regularities about the behavior of the gases, some simplifying assumptions are required. The ideal gas model is obtained by means of the following idealizations:

I_1: The size of the molecules is negligible compared to the distances between them (provided that the pressure of the gas is not very high). For that reason, the internal structure of the molecules is not taken into account, and they can be regarded as point-like masses. I_2: The collisions of the molecules among themselves and with the walls of the container-from a microscopic point of view, that is just to say they are colliding with other kinds of molecules—are perfectly elastic (that is, the total kinetic energy of the colliding molecules remains constant). I_3: The different components of the velocity of each molecule are statistically independent of each other. I_4: There are no intermolecular forces, meaning no molecule exerts any attractive or repulsive force on other molecules.

As far as we know, the four hypotheses can be regarded as true about real gases, whereas the four idealizations can be regarded as false hypotheses consciously introduced to build the ideal gas model. We do not believe that the molecules that compose a real gas are perfectly elastic, point-like particles that do not exert any force on other molecules. Quite to the contrary, we accept that they have a definite volume, and an internal structure, interact via intermolecular forces, and undergo more or less inelastic collisions (in which a part of their kinetic energy is converted to other forms of energy, such as heat). Nonetheless, we need all the false assumptions contained in the four stated idealizations if we want to build a model from which we can infer some approximately true regularities about the behavior of real gases. We do not believe that ideal gases exist, but the model of the ideal gas allows us to know some general laws that approximate the behavior of real gases in some specified conditions of temperature and pressure.

The equation of state for an ideal gas is $PV = nRT$ (where P is the pressure, V is the volume, n is the amount of substance or number of moles of the gas, T is the thermodynamic temperature, and R is the molar gas constant). An ideal gas, by definition, obeys exactly, among others, Boyle's law (according to which, if a given mass of gas is compressed at a constant temperature, the product PV remains constant) and Joule's law (according to which the internal energy of a gas is independent of its volume). Those laws are only approximately true of real gases at low pressures; they are exactly true in the limit when the pressure tends to zero. The ideal gas equation of state provides a good approximation of the behavior of real gases at relatively low pressures and high temperatures. As pressure increases, however, the approximation worsens. The ideal gas model has, then, a limited dominion of application and does not deliver good approximate predictions for the behavior of gases at high pressures or low temperatures.

The van der Waals gas model can be regarded as a deidealization of the ideal gas model. It keeps all the general hypotheses of the kinetic theory of gases but removes some idealizations of the ideal gas model, specifically I_1 and I_4. The van der Waals equation of state for a gas is $\left(P + a\frac{n^2}{V^2}\right)(V - nb) = nRT$ (where parameters a and b, called the attraction and repulsion parameters, are characteristic of a given substance). When the temperature and volume of a gas are high enough, this equation, in the limit, reduces to the equation of the state of the ideal gases. The van der Waals equation of state takes into account the volume of the molecules and the existence of short-range intermolecular forces, both attractive and repulsive, of electrostatic origin (the van der Waals forces). This equation of state permits a more accurate account of the properties of real gases than the equation of the ideal gas. It can be applied as a good approximation of gases at higher pressures and lower temperatures than the ideal gas equation. It also works for fluids generally, but it delivers worse approximations for the behavior of liquids, where the molecules are tightly packed and move with less freedom than they do in gases.

The van der Waals model also permits explaining why the ideal gas model provides a good approximation of the behavior of gases at low pressures, relatively large volumes, and high temperatures. The explanation hinges on the properties of the van der Waals intermolecular forces. These electrostatic forces are repulsive when the distance r between the molecules is lower than a critical distance d but are attractive when r is higher than d (being d of the order of the size of the molecules). The attractive force between two molecules is inversely proportional to the 7th power of the distance r that separates them $\left(f_{\text{att}} = -\frac{a}{r^7}\right)$, and the repulsive force to the 13th power $\left(f_{\text{rep}} = \frac{b}{r^{13}}\right)$, the net intermolecular force being the addition of the two $\left(f = -\frac{a}{r^7} + \frac{b}{r^{13}}\right)$. This fact explains why intermolecular forces are short-ranged: they quickly tend to zero when the distance *r* increases. Given that in a low-pressure gas all the molecules are very far apart from each other, the attractive and repulsive forces are so low as to be considered negligible. In high-pressure gases, by contrast, the distances between the molecules are much shorter and, consequently, the intermolecular forces become significant and cannot be neglected.

The van der Waals model has been tested by measuring the properties of different gases at a wide range of temperatures and pressures. Its predictions have been confirmed as good approximations to the measured values, except for critical temperatures in which gases approach a change of phase and undergo liquefaction. This indicates the limits of the domain of application of the model, which is, nonetheless, much broader than the domain of application of the ideal gas model. In this sense, it can be said that the deidealized van der Waals model *justifies* the idealizations built into the ideal gas model (compare Shech 2023, 33). The less idealized van der Waals model delimitates the domain of application of the ideal gas model and explains why it works in such a domain. More generally, one is entitled to appeal to deidealized models to justify the use of more idealized models under certain conditions.

3. The realist construal of deidealization

"Deidealization" is not a word that appears very often in the language of science, but the idea is pervasive. Here is how a theoretical physicist characterizes the recipe he calls a "general principle"—a methodological norm—of his discipline:

> Idealize a difficult problem down to a simple one by ignoring as many complications as you can. Get an answer to the simple problem. Then put the complications back in and calculate how they affect the answer to the simple problem.
>
> *(Carroll 2022, 27)*

This advice suggests that the method of physics prescribes first building a simplified and idealized model, and then deidealizing it and comparing the performance of both models in solving the problem you are interested in.

The term *deidealization* (sometimes spelled *de-idealization*) was introduced into the mainstream philosophy of science by Ernan McMullin in a pioneering article devoted to distinguishing different kinds of idealizations (1985). Deidealization is defined therein as "the way in which models can be made more specific by eliminating simplifying assumptions" (1985, 261). According to McMullin, idealizing consists of a "deliberate simplifying of something complicated with the view of achieving at least a partial understanding of that thing" (248). The aim of idealized models is then not just to make some complex phenomena tractable, but rather to understand some features of the real world (248). Idealizations are "false assumptions" and for that reason, idealized models are "departures from truth" (257). This departure from truth may take the form of deliberate neglect of some properties we know the phenomena to possess, or of the deliberate attribution of properties we know the phenomena not to. Those are the strategies of abstraction and distortion by which idealized models are built in the first place.

In McMullin's view, deidealization proceeds by "adding back" details that had been neglected when a model was built. This is the case of the van der Waals "corrections" to the ideal gas model, an example used by McMullin himself (1985, 259). The outcome of this procedure is an "improved" model, that is, a model that delivers a better approximation of the properties or the behavior of a real system. The new model, in turn, can be further deidealized and so on. The first deidealized model then "serves as the basis for a continuing research program" (261), a program that consists of obtaining a sequence of less idealized models of the same domain of phenomena. In McMullin's words, "this technique will work only if the original model idealizes the real structure of the object" (261). Here we find an overtly realist assumption (one that we do not know how to satisfy): How could we possibly verify that a model idealizes "the real structure" of something? From the realist point of view, deidealization basically consists of removing the false assumptions of a very simple model and replacing them with more verisimilar assumptions. Consequently, deidealized models do not only permit better approximations of the data we have collected about a given phenomenon but also provide a more truthlike description of the real world. McMullin, as most present-day realists do, acknowledges that "models are necessarily incomplete" (1985, 262) and, therefore, that no model, no matter how deidealized, could give us a complete true description of any real system in the world. Presumably, the sequence of more realistic models will never come to an end. Nonetheless, from a realist standpoint such as McMullin, it can be said that the more deidealized a model is, the more approximately true the description of the phenomena it provides.

The realist stance toward idealization can be found, sometimes in weaker or implicit ways, in most endorsements of the representationalist conception of models. Models can provide only distorted and incomplete representations of the phenomena precisely because they are idealized. In the best case, idealized models give us partial and inaccurate representations of the modeled phenomena. They always misrepresent the phenomena in one way or another. Nonetheless, some models can provide better (more accurate or more approximate) representations of the structure and behavior of systems in the world. For representationalists, richer and more complex (i.e., less idealized) models provide better representations of the phenomena than simpler and highly idealized models. From this point of view, which has been called the "deficiency account of idealization" (Carrillo and Knuuttila 2022, 50), all idealizations are problematic because they introduce deliberate distortions or false assumptions, whereas the fundamental aim of science is to reach truthlike representations of the world. Specifically, models aim at providing approximately true descriptions of the phenomena in the real world. Among the many representationalist accounts of scientific models within the philosophical literature, those of Laymon (1995), Niiniluoto (1999), Sklar (2000), Giere (2006, 2009), and Teller (2008, 2009, 2012) are overtly realist concerning the question of how idealizations represent the real world, whereas more pragmatically oriented others, such as Jones (2005), Wimsatt (2007), Godfrey-Smith (2009), Morrison (2015), and Strevens (2008, 2017), have dealt with idealizations in terms that show at least some commitment to realist assumptions on ontological and epistemological issues. For more references, see Cassini and Redmond (2021) and Frigg (2023).

A positive assessment of deidealization follows from a realist stance toward modeling and representation. Deidealized models are perceived as epistemically superior to highly idealized models because, to the extent that they eliminate abstractions and remove distortions, they provide more concrete and truthful representations of phenomena. Given that idealizations are understood as "false assumptions", deidealized models are "truer" representations of the phenomena precisely because they dispose of some falsehoods. Although we cannot conceive of a model entirely devoid of idealizations, one that would give us a complete true description of the real world, deidealized models are more verisimilar than the more idealized models from which they have originated. In this view, deidealization is a valuable aim of science. A sequence of deidealized models gives us a way to gradually approach a more accurate representation of phenomena and, in the end, a truer description of the world. In this sense, they constitute progress in our knowledge of the world. To come back to our previous examples, from the realist standpoint, the van der Waals model provides a truer representation of real gases than the ideal gas model, although not a complete or entirely undistorted one. We should notice, however, that all idealizations we regard as false assumptions are relative to a background of accepted knowledge, and the same holds for "truer" deidealizations. Modeling water as a continuous fluid counts as an idealization relative to the accepted atomic theory of matter; in a different historical context—say, Cartesian physics—it would have been regarded as a literally true description.

4. The pragmatic approach to deidealization

A different approach to idealization emphasizes the advantages and benefits of idealized models. The standpoint of the pragmatic approach to idealization consists of acknowledging that idealized models are often very efficient means for exploring, describing, explaining, and predicting some complex domain of phenomena that is inaccessible by other means. This attitude goes beyond the indisputable claim that the world of our experience is extremely complex and human agents have very limited epistemic capacities. It is not that humans must resign themselves to using simplified and distorted models to represent (or rather, misrepresent) an otherwise intractable domain of phenomena. Instead, idealized models are powerful tools for gaining epistemic access to phenomena that cannot be known without employing idealizations. Idealization can be regarded as a virtue in itself, not necessarily as a defect due to the incompleteness of our knowledge or our limited computational powers. For instance, a highly idealized model, no matter how simplified or approximate, can be a flexible tool that may be applied to different domains, including many that were not intended when that model was built. There are many examples of these beneficial side effects of modeling in the history of recent science. The Lotka–Volterra prey–predator model, for instance, has found useful applications, among other domains outside biology, in the field of economic theory. Abstract models are sometimes useful precisely because they are very general; by contrast, deidealized models tend to be more specific and, as a consequence, are hardly applicable to different domains of phenomena.

The pragmatic approach to idealization does not necessarily imply a non-realist construal of the aims of science, and for that reason, it must be distinguished from pragmatism. It is also not necessarily linked to an antirepresentationalist stance toward models. It is an approach focused mainly on the many functions and applications of scientific models. From a pragmatic point of view, models are built primarily to be used to solve a diversity of well-posed problems, rarely to obtain a verisimilar description of real-world systems underlying the phenomena. Sometimes, obtaining predictions of relative accuracy about the values of just one variable, under definite initial conditions, is all that is required from a model. In this respect, a purely predictive model may be regarded as highly successful even though it does not give us any description of the underlying dynamics of a system and, much less, of the causes responsible for the measured values of the variable in which we are interested. Models must be assessed in light of the purposes of the designers and users. They are successful to the extent that they are adequate for those specific purposes. Whether they provide verisimilar representations of the real world or not, assuming we could provide them, might be in some contexts entirely irrelevant to satisfy the purposes of the model users. As Winsberg (2018, 33) concisely puts it, "to be a good model is purpose relative", and this includes relativity to a specific domain of phenomena to which the model is to be applied and to a desired standard of accuracy.

From this point of view, models are primarily useful tools for guiding our actions, regardless of whether they provide a verisimilar representation of the world or not. Maps, for instance, can be regarded as idealized models of a given territory, according to a well-known analogy. A very simplified map, such as the train stops displayed on a straight line, where all the stops are placed at the same distance from each other, can be sufficiently accurate for traveling from one place to another if our purpose is just to get off at the right station. These kinds of idealized maps are usually distorted representations of the real train ride because the real path is not a straight line, and the stops are not located at the same distance from each other. Those maps truthfully represent some topological properties of the territory, such as the order of the stops, but not its metrical properties, such as the distances between them. If our purpose is to calculate the total distance we must travel or the time the trip will take, purely topological maps are not adequate for such purposes. Conversely, a fine-grained representation of a territory may be counterproductive to our purposes; if all that is wanted is to find the shortest route to the main highway, a very detailed representation of the rivers and the mountains of the lands traversed can make the map more difficult to use. That is why road maps are usually simplified and represent just what is useful for the sole purpose of making a car trip.

The pragmatic approach to idealization stresses the benefits of simple idealized models more than their deficiencies. Often what we want is not a detailed representation of a phenomenon, but rather a coarse-grained representation of it. In those cases, we do not need to deidealize a model that works well enough for our purposes, as is the case with road maps. From the pragmatic point of view, deidealizing a model is not always convenient and could even be counterproductive to the purposes for which the model was built. Consequently, deidealization cannot be conceived of as a valuable end in itself, much less as one of the basic aims of science. Deidealizing a model is not always a step forward on the path of science because the primary aim of scientific models is not to provide a sequence of progressively truthlike representations of the world. Deidealized models are valuable—and desirable—to the extent they provide us with better epistemic tools to interact with the phenomena of one's experience and to satisfy the purposes regarding them. Those purposes can be extremely diverse, including all the functions that models may fulfill in scientific contexts, such as exploring, discovering, explaining, and predicting phenomena, to mention just the most relevant ones. A deidealized model is welcome when it contributes to satisfying one's purposes in more efficient or expedient ways. Obviously, sometimes this is not the case. Assume, for instance, that the main purpose of a highly idealized model is to make the dynamic equations that describe the evolution of a physical system mathematically tractable; a deidealized but mathematically intractable model of the same system is far from useful for that purpose, and for that reason, it can hardly be regarded as progress toward the intended aim, even when it provides a finer-grained description of that dynamics.

The pragmatic approach to deidealization is not yet a well-defined stance, although it can be found among the authors, either realists or anti-realists, that have pointed out the benefits of idealization and the possible counterproductive consequences of deidealization (such as Strevens 2008, 2017; Potochnik 2017; Rice 2021). It has been recently developed within the framework of the artifactual conception of models by Carrillo and Knuuttila (2022). One of the main outcomes of this approach has been to alleviate the concerns about the distorted character of idealized models (and the consequent virtues of deidealization) by focusing on the adequacy for purposes of all scientific models. From this standpoint, the costs and benefits of deidealizing models have to be assessed case-by-case in each context, relative to the purposes that such deidealized models intend to fulfill.

Real gases provide a good example of how practical considerations determine which model must be employed in each context. The ideal gas model can be generalized by means of the following equation: $PV = znRT$ (where z is a non-dimensional number called the *compressibility factor*). This law (which reduces to the ideal gas law when $z = 1$) gives a better approximation of the behavior of real gases in conditions of pressure and temperature in which the ideal gas law can be applied. In turn, the van der Waals model does not provide a good approximation of the behavior of real gases in conditions of low temperatures and/or high pressures. For that reason, it is not of very much use to solve many problems in the domains of physics and engineering of low-temperature fluids. Other models (that physicists use to call simply "equations") have replaced the van der Waals model, among them, the Redlich–Kwong model, the Soave model, and the Peng-Robinson model. These models include more complicated equations than the van der Waals model, but they give better approximations of the behavior of real gases in a wide variety of conditions of temperature and pressure, as is required to solve different problems, mainly in the field of engineering.

5. Disputed questions on deidealization

In recent years, there have been several controversies concerning the very possibility of deidealizing models and the costs and benefits of deidealization. There cannot be any doubt that deidealization is possible because we have enough examples of deidealized models. The different models of the physical pendulum have been the standard case study for philosophers of science for years (Morrison 2015; Cassini 2021). The ideal or simple pendulum model is highly idealized, but it can deliver approximate predictions for the period of real pendula when the oscillations of the bob are small enough. In contrast, the compound pendulum model, which is a deidealized model, is built by removing some idealizations of the simple pendulum model, resulting in a model that takes into account mass, moment of inertia, and the distance from the pivot point to the center of mass of the pendulum. The physical pendulum model can also be deidealized in many ways by introducing what physicists call "corrections", a strategy that consists of adding new parameters to the physical pendulum model. These include (i) finite amplitude corrections for different angles of oscillation, (ii) mass distribution corrections, where the finite mass of the bob and the cord are taken into account, (iii) correction for air effects, such as buoyancy and friction, and (iv) elasticity corrections, in which the stretching of the chord and the motion of the support are considered. Those corrections are mathematically complicated (for details see Baker and Blackburn 2005). Besides, the many deidealized models cannot be ordered in a linear sequence of successively less idealized models. In any event, the example suffices to show not only that deidealization is possible but also the many ways in which one model can be deidealized.

The pendulum example, however, does not show that every model can be deidealized or that all models can be deidealized by removing one-by-one the idealizations they contain. Several philosophers of science have argued that certain kinds of models contain ineliminable idealizations. Batterman (2002, 2009, 2010) and Weisberg (2013) have claimed that the so-called minimal models—those that aim to explain the occurrence of some physical regularity by isolating the dominant causal factors responsible for the observed regular behavior—cannot be deidealized. If we were to deidealize a minimal model by introducing new independent parameters, we would lose its explanatory power. We do not get a deeper understanding of a natural regularity by adding more details to a minimal model because those details generally obscure or screen off the dominant causal factors that produce that regularity. In Batterman's words: "adding more details counts as explanatory noise -noise that often obscures or completely hides the features of interest" (Batterman 2010, 17).

Another argument against the possibility of deidealization was put forward by Batterman and Rice (2014) and further elaborated by Rice (2021). According to Batterman and Rice, models represent their targets in a rather holistic way and not through separable components. That is why idealized models must be conceived of as holistic distortions of the phenomena, in which the idealized components cannot be isolated from the non-idealized components. Consequently, a model cannot be gradually deidealized by removing its idealizations one-by-one. Most idealizations are introduced globally into a model to allow for the application of mathematical modeling techniques that would be otherwise inapplicable. Such global idealizations then cannot be removed without impairing the explanatory power of the model, or even without destroying the model as a whole. The argument concludes that at least some idealizations are not eliminable and have to be conceived of as inescapable features of a given model.

A related argument against the possibility of deidealization appeals to the epistemic opacity of highly complex models, such as global climate models. Winsberg (2018) has claimed that those models have a modular architecture that does not permit decomposing them into separately manageable pieces. Climate models are built from many different modules and submodules and involve many parameter options. The interaction between the different modules is itself very complex and the process of coupling some submodules, which include their specific parametrizations, is often a very difficult problem. According to Winsberg (2018, 142), this complex architecture, which he calls "fuzzy modularity", has the consequence that "the overall dynamics of one global climate model is the complex result of the interaction of the modules - not the interaction of the results of the modules". Those complex models are "analytically impenetrable", as Winsberg (2010, 105) has called them. In practice, it is impossible to track the sources of successes and failures of these kinds of models up to single separable modules or submodules, which are epistemically inscrutable. Such complex models cannot be deidealized because one cannot even know precisely which idealizations are embedded into them.

Another holistic argument against the possibility of deidealizing models points out that idealizations come in bundles and, consequently, cannot be separated. In this respect, Knuuttila and Morgan (2019) have argued that the idealizations embedded in several economic models cannot be reversed because they cannot be separated from each other. In many cases, economic models are not decomposable into independent parts, which could eventually be controlled, edited, and corrected. For that reason, it may not be possible to deidealize a definite assumption without collapsing the functionality of the model. Models cannot be deidealized step-by-step because they were not constructed this way, rather all the idealizations were jointly embedded when the model was built.

These arguments, however, do not prove that deidealization is impossible or that no model can be deidealized. They point out that many scientific models holistically represent their targets and, consequently, function as non-decomposable wholes. What the arguments do show is that some models cannot be deidealized step-by-step, as some realist philosophers have thought. The idealizations embedded in a model sometimes cannot be dismantled individually, rather they are subject to an all-or-nothing choice: either we use them as units that represent holistically the intended target, or we have to replace them with entirely different models. While we use a highly idealized holistic model for the purposes for which it was built, its central idealizations cannot be corrected or removed. In any case, this conclusion cannot be extended to all scientific models; as the example of the deidealized models of the physical pendula shows, sometimes one can identify the idealizations a model contains and remove or replace at least some of them in a non-holistic way.

A different controversy deals with the question of whether idealizations provide some understanding of the phenomena in the real world and, if that is the case, what kind of understanding they provide. At first glance, if all idealizations are regarded as false hypotheses, they must lack explanatory power by themselves. On the other hand, if highly idealized models are explanatory and give us a genuine understanding of the phenomena, deidealized models of the same phenomena should give us a better understanding of them. Several philosophers of science have claimed that idealizations are effective means to obtain understanding, either because they explain the phenomena (Bokulich 2016) or because they help identify causal influences by highlighting causally relevant factors (Strevens 2017). From these points of view, deidealized models do not necessarily provide a better understanding of the phenomena. On the contrary, they can make explanations more difficult (for instance, mathematically more complicated) or obscure the causal factors that one wishes to isolate.

The stance according to which idealized models provide understanding of phenomena has been the target of a variety of criticisms. In this respect, Sullivan and Khalifa (2019, 1) have claimed that idealizations have merely an instrumental value: to the extent that they are falsehoods, they are mere "conveniences that aid in easing calculations and making things salient". These authors endorse the idea that deidealized models have more epistemic value because they are more veridical or approximately true than their idealized counterparts. Here, the realist and pragmatic approaches to deidealization show their differences. For the realists, deidealized models are more explanatory than their idealized counterparts, and for that reason, they give us a better understanding of the phenomena. The ideal gas model, again, is a good example. It is understandable why the false assumption according to which there are no intermolecular forces is a good idealization when one grasps the explanation provided by the van der Waals deidealized model. Then, we understand that the ideal gas model is a good approximation for the behavior of gases at low pressures and high temperatures because we know why those forces are negligible in such conditions. By contrast, for the pragmatic approach, some deidealized models actually provide less understanding of the phenomena than the more idealized ones because they are more complex, epistemically opaque, and often mathematically intractable.

The question of whether deidealized models provide a better understanding of the phenomena than their more idealized counterparts is sensitive to several very general issues concerning scientific explanation. There certainly are different kinds of explanations and different types of understanding. Consequently, some explanatory models can provide one or another sort of understanding, depending on which kind of explanation they provide. On the other hand, whether non-explanatory models can provide some understanding of the phenomena will depend on whether we are disposed to accept that it is possible to understand a phenomenon without explaining it in any way. This is a contentious issue on which no consensus has emerged among philosophers of science. There is extensive and growing literature on scientific understanding and its relations to scientific explanations. Strevens (2008) is a classic on the topic. For more references, see Grimm, Baumberger, and Ammon (2017) and Sullivan and Khalifa (2019).

As the ideal gas and ideal pendulum examples have shown, there is no doubt that at least some models can be successfully deidealized. Furthermore, these examples show that deidealized models are sometimes necessary for achieving certain well-defined purposes, such as making precise measurements or obtaining accurate predictions. However, it does not follow from this that deidealization is always possible or desirable. The issue must be resolved in each specific case. When we are faced with a particular model, four different issues should be distinguished. First, there is the question of whether deidealizing is possible at all, or more concretely, whether that model can be deidealized. Second, we can ask whether we know how to deidealize it. If we do not know how to do it, we cannot conclude that deidealization is impossible, because nothing follows from our ignorance. Third, we can try to determine how many ways to deidealize such a model are feasible. Fourth, there is the issue of whether deidealizing that model is convenient or not, considering the purposes for which it was built or subsequently used.

Deidealization is far from a simple or trivial task. Sometimes we simply do not know how to deidealize a model, and sometimes there are several possible ways of deidealizing it. In that case, the different deidealized models must be assessed pragmatically as some of them might be useless or even counterproductive given their intended applications. In any case, from a pragmatic point of view, deidealization is not an end in itself but rather a means of accomplishing more efficiently the purposes for which a given model was designed or used.

References

Baker, Gregory, and James Blackburn. 2005. *The Pendulum: A Case Study in Physics*. Oxford: Oxford University Press.

Batterman, Robert. 2002. “Asymptotics and the Role of Minimal Models.” *The British Journal for the Philosophy of Science* 53(1): 21–38.

———. 2009. “Idealization and Modeling.” *Synthese* 169(3): 427–446.

———. 2010. “On the Explanatory Role of Mathematics in Empirical Science.” *The British Journal for the Philosophy of Science* 61(1): 1–25.

Batterman, Robert, and Collin Rice. 2014. “Minimal Model Explanations.” *Philosophy of Science* 81(3): 349–376.

Bokulich, Alisa. 2016. “Fiction as a Vehicle for Truth: Moving Beyond the Ontic Conception.” *The Monist* 99(3): 260–279.

Carrillo, Natalia, and Tarja Knuuttila. 2022. “Holistic Idealization: An Artifactual Standpoint.” *Studies in History and Philosophy of Science* 91(1): 49–59.

Carroll, Sean. 2022. *The Biggest Ideas in the Universe: Space, Time and Motion*. London: Oneworld Publications.

Cartwright, Nancy. 1989. *Nature’s Capacities and Their Measurement*. Oxford: Clarendon Press.

———. 1999. *The Dappled World: A Study of the Boundaries of Science*. Cambridge: Cambridge University Press.

Cassini, Alejandro. 2021. “Deidealized Models.” In *Models and Idealizations in Science: Artifactual and Fictional Approaches*, edited by Alejandro Cassini and Juan Redmond, 87–113. Cham: Springer.

Cassini, Alejandro, and Juan Redmond, eds. 2021. *Models and Idealizations in Science: Artifactual and Fictional Approaches*. Cham: Springer.

Frigg, Roman. 2023. *Models and Theories: A Philosophical Inquiry*. London: Routledge.

Giere, Ronald. 2006. *Scientific Perspectivism*. Chicago: The University of Chicago Press.

———. 2009. "Why Scientific Models Should Not Be Regarded as Works of Fiction." In *Fictions in Science: Philosophical Essays on Modeling and Idealization*, edited by Mauricio Suárez, 248–258. London: Routledge.

Godfrey-Smith, Peter. 2009. "Abstractions, Idealizations, and Evolutionary Biology." In *Mapping the Future of Biology: Evolving Concepts and Theories*, edited by Anouk Barberousse, Michel Morange, and Thomas Pradeu, 47–55. Dordrecht: Springer.

Grimm, Stephen, Christoph Baumberger, and Sabine Ammon, eds. 2017. *Explaining Understanding: New Perspectives from Epistemology and the Philosophy of Science*. London and New York: Routledge.

Jones, Martin. 2005. "Idealization and Abstraction: A Framework." In *Idealization XII: Correcting the Model. Idealization and Abstraction in the Sciences*, edited by Martin Jones and Nancy Cartwright, 173–217. Amsterdam: Rodopi.

Knuuttila, Tarja, and Mary Morgan. 2019. "Deidealization: No Easy Reversals." *Philosophy of Science* 86(4): 641–661.

Laymon, Ronald. 1995. "Experimentation and the Legitimacy of Idealization." *Philosophical Studies* 77(2–3): 353–375.

Levy, Arnon. 2021. "Idealization and Abstraction: Refining the Distinction." *Synthese* 198(Suppl. 24): 5855–5872.

McMullin, Ernan. 1985. "Galilean Idealization." *Studies in History and Philosophy of Science Part A* 16(3): 247–273.

Morrison, Margaret. 2015. *Reconstructing Reality: Models, Mathematics, and Simulations*. New York: Oxford University Press.

Niiniluoto, Ilkka. 1999. *Critical Scientific Realism*. New York: Oxford University Press.

Portides, Demetris. 2021. "Idealization and Abstraction in Scientific Modeling." *Synthese* 198(Suppl. 24): 5873–5895.

Potochnik, Angela. 2017. *Idealization and the Aims of Science*. Chicago: The University of Chicago Press.

Rice, Collin. 2021. *Leveraging Distortions: Explanation, Idealization, and Universality in Science*. Cambridge, MA: The MIT Press.

Shech, Elay. 2023. *Idealizations in Physics*. Cambridge: Cambridge University Press.

Sklar, Lawrence. 2000. *Theory and Truth: Philosophical Critique within Foundational Science*. New York: Oxford University Press.

Strevens, Michael. 2008. *Depth: An Account of Scientific Explanation*. Cambridge, MA: Harvard University Press.

———. 2017. "How Idealizations Provide Understanding." In *Explaining Understanding: New Perspectives from Epistemology and the Philosophy of Science*, edited by Stephen Grimm, Christoph Baumberger, and Sabine Ammon, 37–49. New York: Routledge.

Sullivan, Emily, and Kareem Khalifa. 2019. "Idealizations and Understanding: Much Ado about Nothing?" *Australasian Journal of Philosophy* 97(4): 673–689.

Teller, Paul. 2008. "Of Course Idealizations Are Incommensurable!" In *Rethinking Scientific Change and Theory Comparison: Stabilities, Ruptures, Incommensurabilities*, edited by Lena Soler, Howard Sankey, and Paul Hoyningen-Huene, 247–264. Dordrecht: Springer.

———. 2009. "Fictions, Fictionalization, and Truth in Science." In *Fictions in Science: Philosophical Essays on Modeling and Idealization*, edited by Mauricio Suárez, 235–247. London: Routledge.

———. 2012. "Modeling, Truth, and Philosophy." *Metaphilosophy* 43(3): 257–274.

Weisberg, Michael. 2013. *Simulation and Similarity: Using Models to Understand the World*. Oxford and New York: Oxford University Press.

Wheeler, Billy. 2018. *Idealization and the Laws of Nature*. Cham: Springer.

Wimsatt, William. 2007. *Re-Engineering Philosophy for Limited Beings: Piecewise Approximations to Reality*. Cambridge, MA: Harvard University Press.

Winsberg, Eric. 2010. *Science in the Age of Computer Simulation*. Chicago: The University of Chicago Press.

———. 2018. *Philosophy and Climate Science*. Cambridge: Cambridge University Press.

7
MODELS, FICTION, AND THE IMAGINATION

Arnon Levy

1. Introduction

Science and fiction seem to lie at opposite ends of the cognitive–epistemic spectrum. The former is typically seen as the study of hard, real-world facts in a rigorous manner. The latter is treated as an instrument of play and recreation, dealing in figments of the imagination. Initial appearances notwithstanding, several central features of scientific modeling suggest a close connection with the imagination, and recent philosophers have developed detailed accounts of models that treat them, in one way or another, as akin to fictions. This chapter will critically discuss the fictions approach. The chapter first motivates the appeal to fiction (section 2); then looks at several ways of developing the basic idea that models are a form of fiction (section 3); and finally considers how models, understood in a fiction-based way, can play the epistemic roles they are typically thought to play, namely as tools of scientific reasoning, representation, and explanation (section 4.) The final section provides a summary and points to some possible directions for further development and expansion.

2. Motivating the models-as-fictions view

A central feature of modeling is idealization: the introduction of false assumptions—infinite populations, point masses, and perfectly rational agents—to facilitate analysis and understanding (Jones 2005; Levy 2021). These entities seem concrete, but they cannot be encountered in any spatiotemporal location, nor can they be studied by empirical methods; they are posits. It is natural to regard such posits as imaginary, and this is often how they are referred to within scientific discourse. Another way to put this, to quote Godfrey-Smith (2006, 734–5), is that "model systems are often treated as 'imagined concrete things'—things that are imaginary or hypothetical, but which would be concrete if they were real." In this, models appear not unlike the persons, places, and events populating novels, films, and other fictions—Holmes, Middle Earth, The War of the Worlds, etc. Thus, a central motivation for treating models as fictions is their shared imagined, concrete-hypothetical character.

DOI: 10.4324/9781003205647-10

Several related features of the practice of modeling strengthen this line of thinking. For one thing, modelers often describe models in a way that is not unlike the introduction and development of a fictional setup. Consider this example, drawn from Chapter 19 of Richard Feynman's celebrated *Lectures on Physics*, which discusses centers of mass. Early on, Feynmann explains how to calculate the center of mass of a compound object:

> Suppose that we imagine an object to be made of two pieces, A and B. Then the center of mass of the whole object can be calculated as follows. First, find the center of mass of piece A, and then of piece B. Also, find the total mass of each piece, M_A and M_B. Then consider a new problem, in which a point mass M_A is at the center of mass of object A, and another point mass M_B is at the center of mass of object B. The center of mass of these two point masses is then the center of mass of the whole object.

This little text has the look and feel of a (very) short (if rather unblemished) story. Moreover, Feynman begins by asking the reader to use their imagination. Such examples can be multiplied,[1] suggesting a kinship between modeling and fictionalizing.

A further aspect of similarity involves the presence of an internal/external distinction. In fiction, it is natural to distinguish what is the case "in" or "according to" the fiction from what is true *simpliciter*. According to Thomas Mann's *The Magic Mountain*, a young shipbuilder named Hans Castrop goes to visit his ailing cousin at a sanitorium near the Alpine resort town of Davos, Switzerland, and ends up staying for seven years. That this is true "in" Magic Mountain is settled by the text of the novel. Whether such a place actually existed is a different question, to be settled "outside" the novel—most straightforwardly by visiting Davos (at least around the time of the novel's writing—i.e., 1924.)[2]

A final, more theoretical, consideration motivating appeals to fiction is the thought that this will allow one to use resources from the philosophy of fiction. There is a fairly rich tradition of philosophical discussion about fiction, including its representational and ontological aspects. As we shall see in the next section, several ideas from this area have been utilized in accounts of modeling.

Before that, two preliminary points to clarify these notions and the use to which they will be put to here should be discussed. First, the term "fiction" is sometimes used to indicate that a statement or narrative is false ("the story he told us was a complete fiction."). But fictions can contain true propositions, and more generally there is no contrast between fictionality and truth. This is clearest in cases of fictions that are extensively grounded in fact, like historical novels, where much of the fiction's content may be factually accurate. Even in many "ordinary" fictions much of what is depicted—including mundane facts such as names of places as well as more subtle aspects of culture and society—may well be factually accurate.[3] Thus, the fictional is not fundamentally opposed to the factual. That said, the fictional, as should be obvious, is not necessarily anchored in the factual. Fictions contain many statements that are not true of any part or aspect of the real world—invented persons, places, and events. More fundamentally, fiction-making is not based on truth in the way factual description normally is, or at least is expected to be. In producing fiction, either in a literary context or in a scientific (modeling) context, one is not attempting to describe reality, at least not in the first instance, and the success of the product—the fiction—isn't to be assessed, at least not in the first instance, in terms of factual accuracy.

The second point concerns the relevant notion of imagination, which can be understood in thinner and thicker ways. In a thin sense, to imagine is to merely entertain, to take under consideration without presuming the truth. In this sense, we imagine whenever we perform hypothetical reasoning. But in a richer sense, to imagine is to "see in the mind's eye." In this sense, imagining involves a special psychological capacity, a kind of offline sensory experience. Philosophers have differed on which of these senses is important in the context of modeling (Salis and Frigg 2020) and on its epistemic significance more broadly (Kinberg and Levy 2022). It could be argued that the significance of appeals to the imagination is diminished if it is confined to the thick sense. Since some authors discussed in this chapter would disagree, this is not presumed in what follows.

One final remark is in order, before moving on to consider fiction-based accounts in more detail. While the idea that models can be understood in terms of fiction has gained some popularity, with more than a few authors arguing for it and developing it, significant dissent has been voiced as well. A moderate criticism, presented by Adrian Currie (2017) is that a fiction-based view is insufficiently general. Currie gives examples from engineering and argues that the fictions approach does not capture them well. Gregory Currie (2016) has made a more far-reaching critique, suggesting that the appeal to fiction is philosophically unilluminating. He is unimpressed, in particular, by the idea that models and fictions are alike in terms of the internal/external distinction. Perhaps the harshest critic of the appeal to fiction is one of the forefathers of the modeling literature, namely Ronald Giere (2009). He holds that models function differently from fictions, in society and culture. Moreover, he worries that the assimilation of models into fiction might provide fodder for anti-scientific forces such as the Intelligent Design movement.

These objections should be taken seriously and may at the very least point to limitations of and problems with the fictions view of models. But they do not seem to nullify the appeal of the view, certainly not to the point where we should avoid looking into it in further detail. That is what the chapter turns to next.

3. Developing the models-as-fictions view

3.1 Models and the imagination

The idea that models are akin to (or perhaps a species of) fiction can be developed along various interrelated dimensions. One such dimension concerns the relationship between a model, construed fictionally, and the imagination. Suppose we accept that modeling, in some of its phases at least, involves the use of the imagination. What exactly is the nature of the imaginative exercise in question and how does it relate to the way we engage our imagination when consuming fiction? This is relevant, not merely to understanding the relationship between models and fiction. It is also central to understanding what determines a fiction's content, and what makes claims regarding it correct or incorrect (note that the notion of "true" is not used here deliberately, as per the remarks about truth and fictionality made above).

Almost everyone in the philosophy of fiction agrees that we engage with novels, films, and other central forms of fiction by means of our imagination. Fleshing out the connection to fiction has been key to most views of the semantics of fiction, but this can be done in several ways. Two will be discussed, as they seem to illustrate not only the similarities but also the potential differences between models and fiction. The first move is to distinguish actual from prescribed uses of the imagination. What makes something fiction could be, on the one hand, that people actually use their imaginations when consuming it. Or it might be that they *ought* to use their imaginations when consuming it. A central insight attributed to Walton (1990) and Currie (1990), now endorsed by most philosophers of fiction, is that fictionality has a normative element to it. It is about what we should imagine when encountering a novel or a painting. While it seems true that people often do in fact use their imaginations when reading a novel or watching a film, that is not what makes it fiction, and not what determines its fictional content. It is the fact that we *should* respond to it a certain way: by employing the imagination (and not, as in the case of non-fictional work, by believing its content).

But what determines *what* we should imagine when consuming fiction? A simple answer is the fictional work at issue, be it a novel, a painting, or a play. However, this simple answer calls for elaboration: how does the work determine what we ought to imagine? At least two sorts of views can be outlined. An "intentionalist" outlook holds that when, say, we read a novel, we are supposed to imagine what its author intends for us to imagine. And to the extent that we have succeeded in doing so, we have correctly grasped the fiction's content, can make correct statements about it, etc. Stated in this simple form, this view seems implausible, but several writers about fiction have developed it in ways that overcome its apparent simplemindedness (Currie 1990; Stock 2017). Without entering into detail, we can note that on such a view, fictions are seen as a form of communication, inasmuch as the reader (or, more broadly, the consumer) is engaged in interpreting the words (or splotches of paint, etc.) of the author. It is for this basic reason that such a view is not well suited to account for models. Modeling is, of course, a human activity, and does involve communications—among modelers, for instance—but a model is not fundamentally a vehicle of interaction between people. Moreover, while the intentions of a modeler (i.e., someone who originates a model) may matter in some contexts, they do not in general determine what the model is about. For this reason, while intentionalism may be a plausible view of fiction it is not suitable for an account of models.

An alternative to intentionalism is the make-believe view from Kendall Walton (1990), an influential account of fiction and artistic representation. In this view, fiction is a regimented, "grown up" version of children's pretend play. A game, in this analysis, involves two key elements: a prop, which is a concrete real-world object; and a set of rules—Walton calls them "principles of generation"—which specify what the game's participants are to imagine given the prop's properties. To use an example from Walton, in a game of "Spot the Bear," the prop might be a tree stump and principles of generation might say, for instance, that when seeing a stump, one is to imagine encountering a bear, that the stump's color should determine how one is to imagine the color of the bear's fur and so on. The extension to fiction is straightforward: a novel or a painting is a prop, which together with principles of generation mandates that certain propositions are "fictional," namely, *to be imagined*. Given the text of Thomas Mann's *The Magic Mountain*, it is fictional, i.e., one is to imagine, that Hans Castrop visits his ailing cousin at a sanitarium near Davos. That is, given the novel's text and the relevant principles of generation, this is a correct (in Walton's terms 'fictional') claim in the game of *The Magic Mountain*. A further distinction made by Walton is significant here: some claims in a game are primary, i.e., they are explicitly specified in the work of fiction. That the hero's name is Hans Castrop, for instance. Other claims are implied, that is, they are inferable from the primary claims, given principles of generation (as well as features of the context). For instance, it is implied, but never stated explicitly, in *The Magic Mountain* that Hans Castrop is killed on the battlefields of WWI. Indeed, for Walton most of a fiction's content is implied, since only a small portion of what is to be imagined when, say, reading a novel, is explicitly stated in it.

Walton's view is much more readily applied to modeling than intentionalism. We can regard a model's equations (or even a verbal description of the model) as a prop that, given suitable (scientific) principles of generation, implies what the model's content is. The primary truths are those propositions that are explicitly specified in the equations or text, and the implied propositions are those that follow from them, given principles of inference from logic, mathematics, and the relevant scientific discipline. Several authors have adopted such a view of modeling (e.g., Frigg 2010; Toon 2010; Levy 2015.) Notice that it suggests that the contents of the model are not dependent on the modeler, or her intentions. They are a matter of accepted principles of generation that (we may suppose) are part of the practice of the relevant scientific community. Moreover, while some such principles may be general, applying across many or all scientific disciplines—perhaps basic principles of logic and mathematics—others may be specific to a given area or modeling tradition.

4. Direct versus indirect

The next point concerns the manner in which models relate to their targets, i.e., the things (systems, phenomena) in the world that we intend to study by means of modeling. Suppose one introduces a model as follows: "imagine an ideal pendulum with length l and period p." We may call this a *model specification*. What is the status of such a specification—what is it about? And how does the specified model relate to its target in the world? Two sorts of answers are possible: a direct and an indirect approach. Let's start with the latter.

It may be easiest and most natural to understand the indirect approach by thinking first of a concrete actual model—such as a stick-and-wire model of a molecule or the San Francisco Bay Model developed by the US Army Corps of Engineers (Weisberg 2013, chap. 1). In these sorts of cases, scientists construct an object—a concrete, actual one, that is—so as to serve as a simple and accessible surrogate for the system they are ultimately interested in. They study this system for a while, figuring out (if successful) how the model behaves under various circumstances. They then apply the lessons to a target, transferring their finding about the model to the system they are ultimately interested in (the chemical, the bay, etc.). This process is indirect in a straightforward manner: to study the actual bay, one first studies a surrogate. The indirect approach to modeling can be seen as a generalization of this, to include models that are not concrete (which is to say, most models). It should be noted that the indirect approach need not be coupled to a fictionalist attitude to models. Indeed, one of its central advocates, Michael Weisberg, has been explicitly skeptical of the connection between models and fiction (2013, chap. 4). But others have developed an indirect fictionalist outlook, and this will be the focus here.

In the indirect approach, fictionally construed, the modeling process involves two "things" corresponding to the two phases of a model-based investigation: *a fictional model system* and a real-world *target system*. While the second of these is fairly straightforward, ontologically speaking, the first is puzzling: what is a fictional model system? Does it genuinely exist and if so, what does its existence consist in? Here too philosophers of science have looked to discussions of fiction for guidance. One option is to view the model as a possible entity. In the context of fiction, this view was developed by David Lewis (1978).[4] It has some initial attraction since it seems that at least many literary fictions describe a possible world, a way the (actual) world might be. It might initially seem that such a view is even more attractive as a view of fictional models. Recall the phrase used by Godfrey-Smith: models appear to be objects which "would be concrete if they were real." Isn't this close enough to saying that models are possibilia? Perhaps (although this is not his view—see Godfrey-Smith 2020 for discussion). Be that as it may, the possibilia view has not garnered much support. One reason for this is that there are well-known cases of models that depict impossible states of affairs (Thomson-Jones 2010).[5] A more basic reason is that many philosophers of science are wary of the metaphysical commitments of such a view (Levy 2015). They do not think acknowledging possibilia is a price worth paying for an account of modeling.

Might models be construed not as concrete hypotheticals but as abstract objects? Some have suggested so. Weisberg identified models with mathematical structures (2013)—although it is not clear that he intends this as an ontological claim. Recently, Thomson-Jones (2020) and Thomasson (2020) have suggested that models be understood as abstract artifacts. Based on Thomasson's previous work on social ontology and the metaphysics of fiction, this approach has it that models are "thin" abstract objects. They are generated—or, more precisely, modelers bring them into being—in the course of scientific modeling but have no more reality to them than is needed to serve as loci of reference and property attribution. They are "hypostasized" objects that serve the purpose of coordinating our talk of models. This kind of view is ontologically economical (unlike modal realism about models). Thomasson and Thomson-Jones argue that there are real similarities between models and fictions inasmuch as both are created systems, and that the artifactual approach captures this. But there are concerns about this approach, too. Perhaps the most serious of these is that abstract artifacts do not play a genuine cognitive role (Godfrey-Smith 2020): they are too thin to constrain the practice of modeling, and are of doubtful explanatory significance (Frigg 2022, chap. 14 provides further discussion.)

So much for the indirect approach. It is fair to say that it sits well with modeling practice but generates an ontological puzzle that is not easy to resolve. To the extent that one is troubled by this puzzle, one might at this point opt for a direct approach to modeling. On such an approach, there is no model system that stands apart from the target and can be explored independently of it. Both Toon (2012) and Levy (2015) developed such an approach. Relying on Walton's make-believe approach to modeling, they suggest that it allows one to view models as ways of thinking about real-world targets, and nothing more.[6] Levy and Toon argue for this approach primarily on the grounds of ontological parsimony: the direct approach does not need to view models as entities in any substantive sense. They merely involve imaginative descriptions of real-world systems.

Toon develops the direct view as a straightforward application of Walton's general ideas about fiction. He thinks that model specifications can serve the role of Waltonian props, with the rest of the account largely parallel to how Walton views fiction in general. Levy's account is a variant of this general strategy, which relies on Walton's notion of a prop-oriented make-believe—the idea, in essence, is that we can play a game of make-believe in which our interest is geared at a real-world system. Thus, (to use an example from Walton 1993) suppose you ask a person where in Italy the town of Crotone is and they reply, "on the arch of the Italian boot." Here, you are enjoined to imagine Italy as a boot as a means for informing you of the location of Crotone. Applying this to models, the idea would be that we imagine certain systems as different than they in fact are so as to highlight certain properties, make evident certain processes, ease certain inferences about them, etc.

As can be seen, the direct approach is ontologically parsimonious. It recognizes only actual target systems and creative descriptions of them. Some have argued that this parsimoniousness is also a source of trouble. Frigg and Nguyen (2016) think that some cases of idealization cannot be accommodated within a direct approach. They also argue that the direct approach cannot handle cases in which models have either generalized targets or no apparent target (see also Salis 2021). If this is so then the direct approach seems doomed, since many models have generalized or non-existent targets, and idealization is central to the practice of modeling. But this topic is still being debated, and the jury is out on whether the direct approach can overcome these difficulties.

Before moving on, two recent accounts should be mentioned. In a sense, these views aim for the best of both worlds—both to neatly capture the practice as the indirect approach does and to remain as ontologically lean as the direct approach. The first view is expounded in Frigg and Nguyen (2016). These authors offer an elaborate account of model-based representation which we cannot fully recap here (although one of its elements is examined in section 4, when discussing keys.) They too employ Walton's make-believe approach but do so while aiming to remain ontologically neutral. As they put it at one point:

> Game-driven make-believe can be seen as a way to refer to, or even create, a Meinongian fictional entity (Priest 2011), as a method to create an abstract artifact of the kind Thomasson (1999) describes, or simply as inducing mental content in those who play the game. (2016, 27)

In the present context, this claim to ontological neutrality cannot be assessed in detail. The main worry about it is that it may cause trouble when we come to account for model–target comparisons (which will be discussed more fully in section 4), inasmuch as such comparisons may pose constraints on the ontology of modeling. Frigg and Nguyen are somewhat terse in their treatment of comparative statements, seeming to suggest that they are less important than some authors hold.

A second best-of-both-worlds attempt is put forth by Fiora Salis. She has recently proposed an approach that incorporates elements of both the direct and indirect views. Salis' suggestion is that models be seen as complex objects: "According to [this] view, model M is a complex object constituted by model description D and content C, so that M = [D, C]." She adds that "From an ontological point of view, the model is analogous to a literary work of fiction; the model description is analogous to the text of a fictional story (the prop that prescribes imagining certain f-truths); and the model content is analogous to the content of a fictional story..." (2021, 729). She goes on to argue that the model's content (the C in the above formula) is no more than the contents of a mental file, having no further, "heavy duty" reality. While this suggestion seems to do a better job with cases of generalized and/or non-existent targets, it arguably faces a version of the criticism leveled by Godfrey-Smith at the abstract artifacts view: can mental files explain the uses to which models are put, in the course of model exploration? Salis does not address this point directly, and she may well have a response. Given the difficulties of both the direct and the indirect approaches, her third option seems promising, and at any rate well worth further development.

5. Knowledge of models

A further set of questions that arises for a view of models as fictions is epistemological. Here we can divide the terrain in two: knowledge of the model itself will be discussed in this section. The next section will discuss issues relating to knowledge of the world outside the model, as it were, under the heading of "exportation."

If a model is a fiction, then investigating the model is akin to figuring out the content of a fictional scenario. But this leads to a concern: why think there is a definite, discoverable content to fictions? Doesn't the fictions approach portray model-based investigation as a much less systematic and objective affair than it is (and should be)?

A basic response to this worry can be obtained by appealing, yet again, to the work of Kendal Walton. Recall the distinction drawn above, in connection with Walton's framework, between primary and implied fictional statements. The former is explicitly stated in the fictional text (or expressed in a non-verbal way, if the fiction isn't literary) whereas the latter is implicit, to be inferred from the explicit ones. If a model is to be construed along these lines, then clearly much of what constitutes modeling is a matter of figuring out what the implied propositions are: what follows from the model's explicitly specified elements. Thus, if we model some system as an ideal pendulum, the bulk of our work would be to solve the pendulum equation for the relevant values, i.e., to figure out what the mathematical expressions (given an interpretation, and given values for variables, boundary conditions, etc.) imply.

Walton's framework supplies a general answer to the question of what governs these implications: it is the relevant principles of generation. Such principles *just are* principles for inferring fictional statements, either from props or (more importantly) from primary propositions. Whether such principles exist for artistic fiction can be disputed: arguably, in literary fiction, there simply is no determinate implied content (Levy 2020). But it is much more plausible that they exist in scientific contexts. Some of the relevant principles are general, including principles of mathematics and logic, while others are domain-specific, i.e., particular to this or that scientific field. But this seems to be as far as we can go within a general discussion of models: the notion of principles of generation supplies an answer to the question of how knowledge of fictional models works in principle, but it also suggests that beyond basic principles of mathematics and logic, there will not be a general account of how model content is determined.

Before moving on to questions about exploration, one final issue pertaining to knowledge of models, to which relatively little attention has been paid in the literature so far, should be mentioned: the role of the imagination. Early on, thinner and thicker senses of the imagination were distinguished. The first is closely related to hypothetical reasoning, while the latter involves a sensory-like component. Which kind of imagination is involved in the exploration of a model? Weisberg (2013, chap. 4) assumes that it is the richer sense, and on this basis voices concerns about the fictions view of models—he worries that some common elements of models (such as probabilistic ensembles) cannot be imagined (in the rich sense of imagining, that is.) Salis and Frigg (2020) argue, to the contrary, that a thin notion of imagination suffices. This allows the view to avert concerns such as Weisberg's. A worry about a view like Salis and Frigg's, however, is that it dilutes the role of the imagination, and consequently of imagination-based views of fiction. These and related issues have not been hashed out in much detail as of yet and remain largely open.

6. From models to targets

The final set of issues to be discussed is perhaps the most important, as it concerns the very purpose of modeling, namely learning about the world. While a modeling project often involves an extensive phase in which the model is explored, that is typically done, ultimately, in service of using the information gathered from the model to predict, understand, and explain some real-world set of targets. Several philosophical issues arise in this context, from relatively general questions pertaining to realism versus anti-realism to more specific questions concerning the manner in which model-based knowledge is exported to worldly targets. These are tackled in order.

First, do models, understood as fictions, generate special problems for a realist standpoint?[7] Here realism is understood as a view both about science's goals—seeking true descriptions of phenomena and their underpinnings; and as a statement about its results—science sometimes succeeds in producing true descriptions of phenomena and their underpinnings. It might at first seem that the fictional perspective on models does indeed generate problems, for isn't the claim that a model is a fiction tantamount to saying that it does not accurately describe reality?

But this is too quick. First, recall the fact that modeling often involves idealization—making false assumptions in order to simplify and facilitate model analysis. Indeed, that was part of *the motivation* for the fictions view. It is not as if treating a model as fiction *adds* further tension with realism. Another point made earlier concerns the relation between fictionality and truth: it is not one of opposition but of independence. Fictions need not, but certainly can, contain true propositions. Indeed, fiction can—and many artistic fictions arguably do—aim at truth. That is, fiction can, by presenting the world in a fictional way, try and sometimes succeed, in telling us a larger (or simply different) truth about the world. This is equally, if not more so, the case in science as it is in art. Thus, the mere fact that a model is seen as fiction does not entail that it cannot also tell us true things about its targets.

That said, the fictions view does remind us that one central argument for realism—the No Miracles Argument (NMA)—may have limited applicability in the context of model-based science. Briefly put, the NMA is an argument that states that since the best explanation for the success of science is that its underlying theories are (at least approximately) correct, we should accept that it is (at least approximately) correct. This argument is seen by many as realism's "master argument" (Musgrave 1988; Psillos 2003). The NMA takes the form of an inference to the best explanation—it suggests that since truth (or approximate truth) is the best explanation of the success of scientific theories, we should believe that at least many of these theories are true (or approximately so). However, like in any case of inference to the best explanation, we cannot use such an inference rule to reach a conclusion that we know, in advance, to be untrue. It may well be that the conspiracy theory according to which the CIA is behind the assassination of JFK, is very attractive in terms of sheer explanatory "loveliness" (Lipton, 2004.) But we have independent information confirming the falsehood of this theory, and so we cannot move from its explanatory prowess to its truth. In the case of modeling the situation is, in a sense, even more extreme: we know that the central elements of the model are idealizations. That is, we know them to be false. So, we cannot use model-based explanations in an IBE, at least not without extreme care. Thus, by focusing our attention on idealizations as the fictions view of models does, it helps us see the limitations of an argument such as the NMA. This is not an in-principle blow to realism, but it does limit its relevance—or at least the relevance of its master argument—in many real-life cases.

But realism, as a broad philosophical question, is not the only or even perhaps the main question on the agenda, when learning from models is at issue. A more immediate set of questions concerns *how* models inform us about their targets. In asking such a question, we presume, at least provisionally, that models do indeed inform us about targets. There is room for further distinctions in this context: we can ask whether and how models allow us to make predictions, how they enter into explanations, and so on. But this direction will not be examined here further. Instead, two broad sorts of answers to the "how models inform" question will be addressed: one based on similarity relations, the other based on the concept of a key.

The notion that we learn from a model about its target by means of, and to the extent that, it is similar to its target, need not be associated with a fictions view of models and is in fact embraced and developed by authors who explicitly oppose the fictions view. But it sits well with a view in which models are concrete hypothetical systems, to return to Godfrey-Smith's locution. In this kind of view, models and targets may share certain properties, or at least have a degree of resemblance in their properties. The ideal pendulum's length may be similar to a real pendulum's; the rate of predation of a model population may resemble the rate at which a real population is preyed upon, etc.

A similarity account of model–target relations merits further elaboration. For one thing, it should spell out an account of similarity and of the kinds of similarities that are relevant in the assessment of model–target relations. The philosopher who has done the most to articulate such an account appears to be Michael Weisberg (2013, chap. 8).[8] Weisberg explicitly distinguishes similarity with respect to the target's *attributes* in contrast to an underlying similarity of *mechanisms.* He then offers an account in terms of feature-matching, inspired by the seminal work of psychologist Amos Tversky (1977). Whether Weisberg's account succeeds in part or in whole is not an issue that will be discussed here (see Parker 2015). But surely some such account is needed if claims about model–target similarities are to be illuminating.

A similarity account of model–target relations should also be seen in light of the discussion of the previous section, concerning model ontology. A simple and straightforward understanding of similarity says, roughly speaking, that two things are similar insofar as they share properties. Obviously, for an object to share a property with some other object, it must have that property. But recall that at least some versions of the fictions view of models contend that the model is a "mere" fiction and not an object at all. It is unclear whether and how such a view is consistent with a similarity-based account of model–target relations.

A second, more abstract approach to model–target relations has been developed by Roman Frigg, partly in collaboration with James Nguyen (2010; 2022; Frigg and Nguyen 2016; 2018). A crucial element in their approach is the notion of a key, namely a mapping from properties of the model to properties of the target. A key tells one how properties of the model translate into properties of the target. In this sense it tells one how to "read" the model inasmuch as one wants to learn from it about the target. A key can utilize relations of similarity—it can map the size of an element in the model to the size of a corresponding element of the target—but similarity need have no role. A key can map size onto, say, a location relative to some point of reference. All that is required is a consistent, one-to-one mapping between relevant elements of the model and elements of interest in the target.

An advantage of the appeal to keys is that it can be applied very widely. As previously indicated, keys can rely on similarity relations but need not. In this sense, the keys approach is a generalization of the similarity approach. This approach is, as noted, rather abstract.

Keys are mappings, and the specific mapping used will vary by context. This means that much of the "action" concerning how models represent targets will depend on the relevant key, a matter which varies as a function of the area of science, and indeed the type of model being used. Frigg and Nguyen probably view this as an advantage of their view. Others may take such generality to deprive the view of some of its explanatory power, relative to a more concrete approach such as Weisberg's. It is possible, perhaps ironically in view of his rejection of it, that Weisberg's view is a better match to the fictions view of modeling, relative to Frigg and Nguyen's more abstract approach.

7. Summary and open questions

The fictions view of modeling is motivated by features of the practice and embodies the thought that a focus on the role of the imagination can illuminate modeling. We have seen that such a view can be fleshed out along several dimensions, with choice points for each of them. Questions arise about the semantics, metaphysics, and epistemology of models, understood as fictions.

Kendal Walton's make-believe account of fiction has been central to the development of the fictions view. It plays a role in accounting for the semantics of models as fictions, inasmuch as modeling differs from artistic practices such as literary fiction (where an intentionalist view is at least a plausible candidate). Walton's account is also central to the metaphysics of fiction, since some philosophers take it to permit an attractive anti-realist stance toward models. This is so especially if the Waltonian framework is combined with an indirect view of modeling that many take to be true to modeling practice.

Finally, we have seen two sorts of accounts of the manner in which the results of model exploration can be exported to the target. One of these, the similarity-based account, is more closely connected with the motivations for the fictions account but raises semantic and ontological concerns. The other, Frigg and Nguyen's keys approach is more abstract and more general, but its fit with the fictions approach may be somewhat less tight.

The fictions approach is still a lively area of research in which several questions remain under active debate, and several avenues for development remain untrodden. The chapter has tried to indicate these throughout. Let us highlight, in closing, two areas outside of the philosophy of science, with which fruitful connections can be made. The first concerns the ontology of modeling—as noted in discussing this, beyond the direct and indirect approach, several recent authors have offered what may be regarded as intermediate stances, and the prospects of these are yet to be fully determined. Here it is notable that there is a large literature in metaphysics concerning fiction, as well as related questions (such as social ontology; see surveys in: Epstein 2021; Kroon and Voltolini 2018; 2019). Contact between the literature on modeling and this larger body of work in metaphysics has, to date, been relatively minimal. Another area with which the fictions view can make contact is the large (and increasing) literature on the imagination (Kind and Kung 2016; Badura and Kind 2021). In particular, much of the recent writing on the imagination, has dealt with its epistemic aspects: whether and how can imagining play a role in justifying belief? How does this relate to other forms of justification and knowledge acquisition? What role do different forms of imagining play in this process? Since modeling is a central epistemic practice within science, and since the fictions approach tightly connects it to the use of the imagination, it seems likely this is an area with which potentially fruitful contact can be made.

Notes

1 The collection edited by Suarez (2009) contains many examples. See especially the chapters by Morrison, Teller and Winsberg.
2 In fact, there was indeed a *Schatzalp Sanatorium* near Davos. So the statement "There exists a sanitorium near Davos" is true (or was, at the time, since it was converted to a hotel in the 1950s.)
3 In this sense, the current discussion departs from ideas familiar from Vaihinger (1925, in which the "philosophy of as if" is explicitly linked to claims that are untrue of, often even in conflict with, reality (on linking Vahinger to modern treatments of fiction, and to its role in science, see Fine 1993).
4 To be precise, Lewis provides a possible worlds semantics for fictional statements. He does not, in his 1978 article, tie this to an ontological stance on possible worlds. But since Lewis is well-known for his modal realism, many understand him as offering, indirectly, a view of the metaphysics of fiction, as well.
5 Lewis, in his 1978 paper, considered the parallel problem for fictions and offered some solutions. But these solutions have been criticized (Byrne, 1993) and in any event it is not clear that they can be transferred as is to the context of modeling.
6 It should be noted, however, that Walton's approach, in and of itself, is neutral as between the direct and indirect views, and more generally is compatible with a variety if takes on the ontology of fictions.
7 These issues are discussed at greater length in Levy (2018).
8 It should be recalled, however, that Weisberg is a critic of the fictions view of models. He regards his account of model-target similarity as independent of such a view.

References

Badura, Christopher and Amy Kind. 2021. *Epistemic Uses of Imagination.* Routledge.
Byrne, Alex. 1993. "Truth in Fiction: The Story Continued." *Australasian Journal of Philosophy* 71: 24–35.
Currie, Adrian. 2017. "From Models-as-Fictions to Models-as-Tools." *Ergo* 4(27). https://doi.org/10.3998/ergo.12405314.0004.027
———. 1990. *The Nature of Fiction.* Cambridge: Cambridge University Press.
Epstein, Brian. 2021 "Social Ontology." The Stanford Encyclopedia of Philosophy (Winter 2021 Edition), edited by Edward N. Zalta https://plato.stanford.edu/archives/win2021/entries/social-ontology/
Fine, A. 1993. Fictionalism. *Midwest Studies in Philosophy* 18: 1–18.
Frigg, Roman. 2010. "Models and Fiction." *Synthese* 172: 251–68.
———. 2022. *Models and Theories, A Philosophical Inquiry.* Routledge.
Frigg, Roman and Nguyen, James. 2016. "The Fiction View of Models Reloaded." *The Monist* 99: 225–242.
———. 2018. The Turn of the Valve: Representing with Materia Models. *European Journal for Philosophy of Science* 8: 205–224.
Giere, Ronald N. 2009. "Why Scientific Models Should Not Be Regarded as Works of Fiction." In *Fictions in Science: Philosophical Essays on Modeling and Idealization,* edited by Mauricio Suárez, 248–58. Routledge.
———. 2006. "The Strategy of Model Based Science." *Biology & Philosophy* 21: 725–740.
Jones, Martin. 2005. "Idealization and Abstraction: A Framework." In *Idealization XII: Correcting the model. Idealization and abstraction in the sciences*, Poznań Studies in the Philosophy of the Sciences and the Humanities, Volume: vol. 86 no. 12, edited by Martin R. Jones, and Nancy Cartwright, 173–217. Amsterdam: Brill.
Kinberg, Ori and Levy, Arnon 2022. "The epistemic imagination revisited. Philosophy and Phenomenological Research." *Philosophy and Phenomenological Research.* https://doi.org/10.1111/phpr.12909
Kind, Amy and Perter Kung, eds. 2016. *Knowledge Through Imagination.* Oxford University Press.

Kroon, Fred and Alberto Voltolini. 2018. "Fictional Entities." In *The Stanford Encyclopedia of Philosophy* (Winter 2018 Edition), edited by Edward N. Zalta. https://plato.stanford.edu/archives/win2018/entries/fictional-entities/

———. 2019. "Fiction.", *The Stanford Encyclopedia of Philosophy* (Winter 2019 Edition), edited by Edward N. Zalta https://plato.stanford.edu/archives/win2019/entries/fiction/

Levy, Arnon. 2015. "Modeling without Models." *Philosophical Studies* 152: 781–798

———. 2020. "Models and Fictions: Not So Similar after All?" *Philosophy of Science* 87(5): 819–28

———. 2021. "Idealization and abstraction: refining the distinction." *Synthese* 198: 5855–5872.

Lewis, David K. 1978. "Truth in Fiction." *American Philosophical Quarterly* 15(1): 37–46.

Lipton, Peter. 2004. *Inference to the Best Explanation*. 2nd Edition. Routledge.

Musgrave, Alan. 1988. "The Ultimate Argument for Scientific Realism". In *Realism and Relativism in Science*, edited by Robert Nola. Springer.

Suárez, Mauricio, ed. 2009. *Fictions in Science: Philosophical Essays on Modelling and Idealization.* London: Routledge

Parker, Wendy. 2015. "Getting (Even More) Serious about Similarity." *Biology & Philosophy.* 30: 267–76.

Priest, Graham, 2011." Creating Non-Existents." In *Truth in Fiction*, edited by Franck Lihoreau, 107–18. Ontos Verlag.

Psillos, Stathis 2003. *Scientific Realism: How Science Tracks Truth*. Routledge.

Salis, Fiora 2021. The New Fiction View of Models. *The British Journal for the Philosophy of Science* 72(3): 717–42.

Salis, Fiora and Frigg, Roman 2020. "Capturing the Scientific Imagination." In *The Scientific Imagination: Philosophical and Psychological Perspectives*, edited by Peter Godfrey-Smith, and Arnon Levy. Oxford: Oxford University Press

Stock, Kathleen. 2017. *Only Imagine*. London: Cambridge University Press.

Suárez, Mauricio, ed. 2009. *Fictions in Science: Philosophical Essays on Modelling and Idealization.* London: Routledge

Thomasson, Amie. 1999. *Fiction and Metaphysics*. New York: Oxford University Press.

———. 2020. "If Models were Fictions Then what would they Be?" In *The Scientific Imagination: Philosophical and Psychological Perspectives*, edited by Arnon Levy, and Peter Godfrey-Smith. Oxford University Press.

Thomson-Jones, Martin. 2010. Missing Systems and Face Value Practice. *Synthese* 172: 283–99.

———. 2020. Realism About Missing Systems. In *The Scientific Imagination: Philosophical and Psychological Perspectives*, edited by Arnon Levy and Peter Godfrey-Smith- New York: Oxford University Press

Toon, Adam. 2010. The Ontology of Theoretical Modelling: Models as Make-Believe. *Synthese* 172: 301–315.

———. 2012. *Models as Make-Believe*. London: Routledge.

Tversky, Amos, 1977. Features of Similarity. *Psychological Review* 84(4):327–52.

Vaihinger, Hans 1925. *The Philosophy of "As If": A System of the Theoretical, Practical and Religious Fictions of Mankind*. New York: Harcourt, Brace & Co.

———. 1990. Mimesis as Make-Believe. Cambridge, MA: Harvard University Press.

Weisberg, Michael. 2013. *Simulation and Similarity: Using Models to Understand the World*. Oxford: Oxford University Press.

8

THE ARTIFACTUAL APPROACH TO MODELING

Tarja Knuuttila

1. Why an artifactual approach to modeling?

The artifactual approach to models is a relatively recent perspective on modeling. Artifactualism serves as a unifying concept for a variety of approaches that regard models as instruments, tools, or artifacts. In and of itself, the fact that scientific models are human-made objects, used in scientific practices for particular purposes, is something that is hardly contested in philosophical discussion. For instance, for van Fraassen "science presents us with representations of the phenomena through artifacts, both abstract, such as theories and mathematical models, and concrete such as graphs, tables, charts, and 'table-top' models" (van Fraassen 2008, 265). The artifactuality of models is implicit in many accounts of surrogate and analogical reasoning (see Nersessian, this volume), and as such, already present in the classic entry on modeling by Boltzmann where he conceives of models as objects, constructed or imagined, that "assist our conceptions of space by figures, by the methods of descriptive geometry, by various thread and object models; our topography by plans, charts and globes; and our mechanical and physical ideas by kinematic models" (Boltzmann 1911, 638). What, then, is specific about the current artifactual perspective to modeling? As human life is surrounded by artifacts of all kinds, ranging from coffee machines to sophisticated technologies, novels, and artworks, considering models as artifacts may not, at first blush, seem too helpful in understanding the place of models in scientific practice. Artifacts are just all too numerous and all too diverse. How could considering models inhabitants of such a multitude possibly enhance our understanding of their epistemic contribution?

What is common to the artifactual approaches discussed below is their pragmatic approach to modeling, *and* the loosening of what could be called the *representational bind*, the idea that models give knowledge because they represent their supposed target systems more or less accurately. From the artifactual perspective, and in agreement with the pragmatic accounts of representation (Suárez 2004; Giere 2010; Hughes 1997), the representational model–target pair is too restrictive a unit of analysis. However, while the pragmatic approaches to representation make room for subjects and communities, the artifactual approaches reach further in not assuming, as Isaac has put it, that "representation [is] conceptually prior to success" (Isaac 2013, 3612). Such a starting point does not rule out

 DOI: 10.4324/9781003205647-11

the possibility of representation playing a role in artifactual approaches, though the artifactualists differ in how they understand the role of representation in modeling (see below). Furthermore, given the artifactual account's purpose of providing an alternative to representational approaches to models, it goes beyond highlighting other uses of models such as prediction, developing testable hypotheses and engineering designs, or providing didactic tools.

This entry will first discuss the main artifactual approaches within philosophy of science, starting from Morrison and Morgan's models as mediators and Knuuttila's models as epistemic artifacts accounts (Sections 2 and 3). Section 4 discusses the relationship between fictional and artifactual accounts, also introducing Currie's models-as-tools account. The question of whether the artifactual accounts can, or should, do without the notion of representation is raised in Section 5, introducing Sanches de Oliveira's radical artifactualism.

2. Models as mediating instruments

The importance of the edited volume *Models as Mediators* (Morgan and Morrison 1999) for the present discussion of modeling is hard to overestimate. Although the philosophical accounts of modeling have traditionally been oriented toward scientific practice (Black 1962; Hesse 1963; Cartwright 1983; Giere 1988), Morrison and Morgan (henceforth MM) put the practice-oriented approaches to modeling right at the center of the philosophical agenda, claiming that "before we even begin to identify criteria for what comprises a model, we need much more information about their place in practice" (Morrison and Morgan 1999, 12).

The four pillars on which the MM account of models as mediating instruments rests are construction, functioning, representing, and learning. MM conceive of mediating mostly in terms of mediation between theory and data, in contrast to science and technology studies, where mediation is understood more widely, e.g., between different social groups and between human and non-human actors (Latour 1994). MM base the ability of models to function "like a tool or instrument" on their *construction*, which is partially independent of theory and data. They claim that "[by] its nature, an instrument or tool is independent of the thing it operates on, but connects with it in some way" (Morrison and Morgan 1999, 11). Apart from this observation, they do not offer any specific argument for why the autonomy of models enables them to function as instruments, but instead invoke an analogy to correlations. They point out that one does not learn much either from perfect or zero correlation, while some correlation between these two extremes provides information of the degree of association, providing a starting point for further investigations. The partial autonomy of models is due to the fact that apart from theory and data, models are also constructed from other elements. In the same volume, Boumans (1999) analyzes the various business cycle theories, showing how they are "baked" with various kinds of ingredients: theoretical notions, mathematical concepts and techniques, stylized facts, empirical data, policy views, analogies, and metaphors. He calls "mathematical molding" the process in which the various ingredients are integrated such that a suitable mathematical form is arrived at. Another side of mathematical molding is calibration, i.e., choosing the parameter values in view of integration.

The functions of models are many in science. MM discuss models in theory construction and exploration, theory application, and the role of models in measurement, design, and intervention. They underline, however, that they do not consider models "simple tools" such as hammers, but as investigative instruments that involve "some form of representation," of either theories or worldly systems, or both. In their view, the investigative function of models ensues from the activities of model building and manipulation. Such experimental and interventive uses of models presuppose that they can be regarded as representations of some systems, however theoretical or hypothetical such systems may be (Morrison and Morgan 1999, 26). Consequently, for MM, the function of models as a means of intervention is intertwined with representation.

MM declare, however, that they do not think about representation in a traditional way. Rather than "mirroring" a natural system or a theory, "representation is seen as a kind of rendering—a partial representation that either abstracts from, or translates into another form, the real nature of the system or theory, or one that is capable of embodying only a portion of a system" (27). Moreover, for them, the legitimacy of representation is "a function of the model's performance in specific contexts" (28). These formulations do not differ much from the present mainstream philosophy of science discussion of representation that stresses the partial, context-dependent, and goal-oriented nature of representation, though MM have surely inspired many of these discussions. The notions of performance and rendering also invoke approaches other than representational ones. For instance, rendering has been used by ethnomethodologists precisely to avoid a commitment to the idea of representation (Lynch 1990).

Morrison and Morgan consider the task of models to represent theories as equally important as that of representing the world. Yet, it is their stress on *learning* from models resting on the combined instrumental and representational role of models that distinguishes their account from the mainstream representational accounts of modeling. Models are not "passive" entities; to be epistemically fruitful, they must be used, built, developed, and manipulated. It is then not merely in virtue of their representational qualities that models give us knowledge, but rather through the activities of building and manipulating a model, understood as "a representative structure" within which learning can take place (Morrison and Morgan 1999, 3). Such learning not only concerns some actual or possible (or impossible) systems, but also the model itself. Clearly then, much of the representation that MM talk about concerns the model world and not just model–world relations. Focusing on the epistemic usefulness of the world in a model (Morgan 2012), the MM account comes close to indirect representation as depicted by Weisberg (Weisberg 2007) and Godfrey-Smith (Godfrey-Smith 2006). However, while Weisberg and Godfrey-Smith approach models as abstract or fictional entities, MM's emphasis on the construction and manipulation of models is more concrete in character, paving the way for artifactual approaches.

3. Epistemic artifacts

Although MM do not explicitly refer to models as artifacts, their discussion of models as investigative instruments, and how scientists learn from building and manipulating them emphasizes the epistemic value of working with purposefully designed artifacts. Building on their account and some more general accounts of artifacts, Knuuttila has argued for considering models as epistemic artifacts (Knuuttila 2005; 2011; 2021). Knuuttila's artifactual account originated in her study of language models within the emerging field of natural language processing (Knuuttila and Voutilainen 2003). These language models process large natural language data sets for useful purposes, yet they cannot be claimed to understand language, nor to represent human linguistic capacity (Bender et al. 2021). To accommodate such tools as models would require another kind of approach to modeling than the representational one that assumes models to be representations of some determinable natural or social systems. Apart from language models and other computational models, the artifactual account can also deal with more traditional models, and model-uses, whose epistemic value is difficult to render in terms of (more or less accurate) representation. For instance, both within economics and biology, there has been a long tradition of criticizing highly simplified and idealized mathematical models that appear far removed from the complexities of natural and social systems (Kingsland 1985; Sugden 2000).

It does seem intuitive to think that for a model to give us knowledge, it would need to represent (more or less) correctly some system that it is constructed to model. However, many models do not have any unique or even actual target systems, and in cases in which they do appear to have determinable targets, they typically grossly misrepresent them. Another problem concerning the supposed representational nature of models is due to the deflationary character of pragmatic accounts of representation. Though pragmatic accounts of representation do not make the problematic assumptions of the structuralist and similarity accounts of representation (Suárez 2003; Frigg 2010; Giere 2004), neither do they have resources to tackle the question of what makes scientific modeling epistemically rewarding (apart from referring to surrogate reasoning). The DEKI account of representation (Demonstration, Exemplification, Keying-Up, and Imputing), which relies on exemplification (Elgin 2004) and pretense theory of fiction (Walton 1990), goes further than other pragmatic accounts of representation in this regard (Nguyen and Frigg 2022; Frigg and Nguyen 2016). The DEKI account nevertheless relies on imputing some features of the model to a target system, though it also admits that a model may not have a target system.

In contrast to the representational accounts, Knuuttila's artifactual approach seeks to explain the epistemic value of models by not building on the model–target relationship, which is the usual unit of analysis of representational approaches. The reasons for dissociating the philosophical account of modeling from the model–target relationship are many. First, given that there is no consensus on how representation should be analyzed, that invoking the representational relationship cannot, in and of itself, account for the epistemic value of modeling. Second, many scientific models are highly idealized and so unrealistic that by structuralist or any other similarity criteria, their continued scientific use appears puzzling. As already mentioned, this has been especially the case in economics and biology, where theoretical models typically do not have the predictive value that many idealized models in physics have. Neither do many such models succeed in isolating some causal difference makers, or even studying some causal difference makers on their own (Strevens 2011; Cartwright 1999; Mäki 1992). Third, as a result of the accumulation of data and the advancement of computational methods, the inventory of different kinds of models is rapidly accumulating, as recent large language models show. Such developments tend to reduce the importance of theoretical modeling, and at any rate, they cannot be properly accommodated by the representational approach (see Knuuttila and Voutilainen 2003; and Knuuttila and Honkela 2005 for early philosophical discussions of language models). One of the benefits of the artifactual approach is precisely its ability to cover many different types of models and modeling practices. Fourth, scientific models typically study generic phenomena and possibilities of various kinds. The representational model–target unit of analysis is not in tune with the modal dimension of modeling (Sjölin Wirling and Grüne-Yanoff 2021; Knuuttila 2021; Knuuttila and Koskinen 2020).

Knuuttila approaches scientific models as epistemic artifacts that are constructed in view of the purposes they aim to accomplish. Such purposes are many: prediction, understanding, data mining, experimental and engineering design, etc. That the artifactual approach is applicable to models, whose primary tasks are instrumental, is without doubt. But what about theoretical models? From the artifactual perspective, the key to the epistemic value of theoretical modeling is the question or problem that a model is constructed to address. Models are epistemic objects that serve as *erotetic devices*, that is, they are purposefully designed artificial systems of dependencies, whose construction enables scientists to tackle pending scientific questions. As such, questions are theoretically and/or empirically motivated; a model will typically incorporate a substantial amount of theoretical and empirical knowledge in its construction. Consequently, despite their artificial nature, models are not accidental things in need of connection to worldly systems via a relation of representation as the representational approach implicitly assumes. The epistemological conundrum of how to analyze the representational relationship between a model and some real-world system shifts to the examination of how the model's design facilitates the investigation of some open scientific problems.

Often the starting point of modeling is a question concerning observed phenomena. For instance, in his design of the Lotka–Volterra model, Volterra explicitly addressed the question of whether "oscillations [...] in the number of the individuals of the various species" could be produced by what he called "internal causes" that are due to the interactions between the populations and "would exist even if [external causes] were withdrawn" (Volterra 1928). To study this question, he constructed a highly idealized model consisting of a pair of nonlinear differential equations that depict two populations, one of which preys on the other. The Lotka–Volterra equations are but one of the models of population dynamics developed by Volterra over the course of more than a decade. These different models depict diverse types of hypothetical situations, taking into account more species, and different kinds of interactions (Knuuttila and Loettgers 2017). As shown also by its popularity in the philosophical discussion, the Lotka–Volterra model is in many ways a paradigmatic model in the sense that it displays several features that are common to many mathematical models. Rather than being a representation of some determinable real-world system, it addresses a particular type of general phenomena, i.e., oscillations in different kinds of populations, and the model is part of an ensemble of related models already in Volterra's work. Moreover, in constructing his version of the Lotka–Volterra model, Volterra made use of mathematical methods and concepts from physics, resulting in a model that itself became a transdisciplinary model template, which would go on to be applied to study the properties of nonlinear dynamics and oscillations in vastly different kinds of material systems from biology and chemistry to social systems, and even technological innovations (Houkes and Zwart 2019; Knuuttila and Loettgers 2017; 2023).

The fact that models characteristically come in multiple versions and families of models and that the same model templates are applied across different disciplinary domains casts doubt on the fertility of the representational model–target unit of analysis. The idea that the epistemic value of a model would primarily derive from its representational relationship to some uniquely identifiable real-world target system does not seem to capture many epistemic enablements of modeling. The simultaneous existence of different versions of the same basic model and the cross-disciplinary dissemination of particular models is a phenomenon familiar to us from the rotation, evolution, and compounding of other cultural artifacts and appears easier to account for from the artifactual viewpoint.

Apart from addressing the purposeful nature of artifacts, the conventional definitions of "artifact" also refer to their production, which involves the modification of materials. Knuuttila's account of models as epistemic artifacts also focuses on the epistemic enablements of the representational tools used in model construction. She emphasizes that irrespective of the representational tools employed, scientific models have a material embodiment that allows for the manipulation of the model and is needed for intersubjective communication between scientists. In order to analyze the epistemic contribution of representational tools, it is useful to make a distinction between their *representational mode* and *representational media* (Kress and Leeuwen 2001). Representational mode refers to the symbolic or semiotic ordering that is rendered by various symbolic, mathematical, diagrammatic, pictorial, and 3-D/geometric devices, while representational media consist of the material means through which the symbolic or semiotic articulation takes place (e.g., ink on paper, electric signals in computers, various materials of physical artifacts or even biological organisms and their parts). The representational mode and media are not necessarily coupled: one can write equations, for example, by using a pen and paper, or chalk and chalkboard.

The focus on representational tools has unifying benefits. First, the artifactual approach does not distinguish between models and "model descriptions" as do the conventional approaches that consider models as either abstract or fictional entities (Giere 1988; Godfrey-Smith 2006; Frigg 2010; Weisberg 2013; Frigg and Nguyen 2016). From the artifactual perspective, the "vehicle" of a model, rendered by representational tools, is an irreducible part of it, as the representational tools employed crucially influence the epistemic affordances of a model. Second, the artifactual approach does not make a sharp distinction between concrete models such as scale models and "nonconcrete" ones such as mathematical models. All models have a material, sensorially perceptible dimension that functions as a scaffold for interpretation, and theoretical or other inferences. However, the representational mode and media play different roles in different kinds of models.

In mathematical modeling, the focus is on the representational mode. For instance, one can model genetic networks using different methods such as coupled ordinary differential equations (ODE), Boolean networks, or stochastic methods. All these different methods make use of different mathematical representational modes. In mathematical modeling, the representational media play a less crucial role than the representational mode, i.e., mathematical methods and notation. The media functions primarily as an external aid for memorizing, reasoning, communication, computing, or demonstration. However, mathematical models are often not analytically solvable and must be turned into simulation models, whose epistemic features are dependent on a physical device. In the case of simulations, several philosophers have pinpointed the important epistemic role of the representational medium: the digital computer (Humphreys 2009; Lenhard 2006).

In contrast to mathematical models, concrete media play a more direct epistemic role in physical 3D models. When working with physical models, scientists typically draw inferences by examining the material features of the model. It is important to note, however, that the material features of the model also embody a symbolic, conceptual dimension—a fact that shows that the distinction between abstract models and concrete models is relative at best. For example, the Phillips–Newlyn hydraulic model is far from being just a physical, three-dimensional object. It embodies and makes visible economic ideas such as the principle of effective demand and the conceptualization of macroeconomy in terms of stocks and flows. Consequently, the way the water pools and flows in the containers and the tubes of the model takes on economic significance—showing how the material and symbolic aspects become coupled in model construction (Morgan and Boumans 2004).

4. Artifacts and fictions

The artifactual approach comes close to (indirect) fictional approaches in that both of them consider models to be objects, separating that which is represented within a model from model–world relations (Godfrey-Smith 2006; Frigg 2010; Frigg and Nguyen 2016). Currie (2018) argues that while the fictional approach suits many kinds of models and model uses, the artifactual account can accommodate fictionalism while thus being broader. By "broadness" Currie refers to the use of models in engineering and design, where models are world-directed, but their success does not derive from their representational success vis-à-vis some actual real-world target systems. Such engineering models often serve as "scaffolds for the construction of real-world systems as well as further models" (Currie 2018, 759). Following Frigg's (2010) discussion of the advantages of fictionalism regarding models, Currie seeks to show that artifactualism can also provide an adequate answer to the semantic, metaphysical, and model–world questions. As for the semantic question concerning the truth of claims about models, the artifactual approach functions just as the fictional account: claims such as "pipe friction pressure is exponentially proportional to flow" are internal to models (Currie 2018, 763).

How should such internal-to-model claims be understood? Instead of referring to, e.g., possible worlds, fictionalist philosophers of science have typically sought to stay metaphysically uncommitted, adopting Kendall Walton's theory of fiction as make-believe (1990). In Walton's games of make-believe, various kinds of props are used according to some rules of generation, prescribing the players to adopt various kinds of imaginings. In scientific modeling, the "model descriptions" function as props. From the artifactual perspective, viewing models as props appears unproblematic, though the artifactual approach does not distinguish between model descriptions and models (more on that below). Where the paths of the artifactual and fictionalist approaches part is that the Waltonian approaches within philosophy of science typically turn on representation, while the artifactualist approaches position themselves as not limited to representational uses of models. In fact, Frigg and Nguyen use the Waltonian approach to develop their DEKI account of representation (Nguyen and Frigg 2022; Frigg and Nguyen 2020). Currie also points out that another advantage of the artifactualist approach is that it does not require acts of imagination on the part of model users. For instance, many modeling processes are increasingly carried out automatically or the goal is to optimize the output, to use the model as a calculating device, to clean data, and so on.

Given the different ways in which models can be useful, Currie emphasizes that the aboutness of models need not be cashed out in representational terms. If the purpose of the model is to scaffold further model-making or to construct various kinds of objects (as kinds of future targets), the success of a model does not depend on model–target comparisons since the target does not yet exist, or its future properties are still in the process of specification. Consequently, there is no access to the possible future object apart from the different models and other renderings that typically are unfolding objects, further developed in the design process.

Currie claims that "understanding models *qua* tools is deeper, more unified and more metaphysically kosher than understanding models *qua* fictions" (Currie 2018, 773).

Following Hilpinen (Hilpinen 1993; 1999), he identifies tools with material objects that are used to manipulate other material objects. Tools as intentional objects have two kinds of features: their material properties and F-properties. The latter are related to the suitability of a tool for some function F. For instance, the size of a sewing needle's eye is an F-property, while its color usually is not an F-property. While Currie distinguishes between the content of a model ("F-properties") and the vehicle of a model ("material properties"), the distinction between them appears contextual; F-properties are a subset of vehicle properties. Consequently, Currie does not separate a model vehicle from the model as the fictional approaches do, arguing instead for equating the model with its vehicle (Currie 2018, 777).

The model vehicle and its material affordances are central for both Knuuttila and Currie. Knuuttila (2021) offers an extended critique of the distinction between model descriptions (i.e., model vehicles) and model systems (see below). Currie (2018) addresses, in turn, the common criticism that proper individuation of models is not possible if model descriptions are allowed to be parts of models (Frigg 2010; Weisberg 2013). The argument is that the *same* model can be realized in different ways, e.g., the Lotka–Volterra model can be expressed by equations on paper or implemented as an algorithm running on a computer. Currie's counterargument is that tools are classifiable objects as well. Indeed, there exists a large discussion on artifact kinds, with different positions regarding whether artifact kinds are similar to or different from natural kinds (Preston 2013). Currie claims that different kinds of vehicles can be classified according to the relevant F-properties that they share. While he appears to invoke the functional features of an artifact, he simultaneously agrees with Hilpinen (1993) and Thomasson (2007) that the intentions of the authors or makers are constitutive of artifact kinds. Currie nevertheless declares not being too moved by the problem of individuation, because the multi-usability of models makes the question of the individuation of models a pragmatic rather than a metaphysical one.

The fictionalists have taken notice of the artifactualist critique of the distinction between model description and model system. Salis (2021a, 2021b) suggests that the fictionalist account should be combined with the artifactual account to amend the shortcomings of both accounts, affirming Knuuttila's criticism concerning the fictionalist separation of the model systems from the model descriptions. Knuuttila (2021) discusses three kinds of problems to which such separation leads (see also Weisberg 2013). First, if the imaginary entities are the locus of representation, this poses the question of how such merely imagined entities are supposed to represent external target systems. Second, there is the problem of how model descriptions are able to provide access to the supposedly more fully-fledged imaginary systems (that are more like concrete systems than what the abstract representations as such entail). Finally, how are the imaginings of different scientists to be coordinated, if not by external representational means? Knuuttila concludes: "Inasmuch as representational tools are merely ascribed the task of describing or generating imagined objects, the imaginary approaches largely ignore the way humans as cognitive agents are able to creatively use different kinds of representational means" (Knuuttila 2017, 5086).

Salis also mentions two other problems of the fictional approaches, claiming that the artifactual approach has in turn problems of its own that require combining the two in a kind of fictionalist synthesis (Salis' "new fiction view" [2021b]). In contrast to Salis, both Knuuttila and Currie think that the artifactual approaches are in fact able to cover the fictional approaches. Fictions, too, can be approached from the artifactual perspective (Thomasson

1999; Thomson-Jones 2020). Salis finds, however, the artifactual account lacking in four different ways, concentrating on Knuuttila's account. She claims, first, that the artifactual approach cannot in fact explain how scientists build and manipulate models, and second, that it does not distinguish between the representational relationship between the model description and the model system, and the representational relationship between the model and a target system. The first claim concerning the inability of scientists to manipulate their models is premised on the idea that the artifactual approach would only be dealing with uninterpreted concrete objects. This is not the case. Knuuttila (2021) distinguishes between internal and external representations and argues that *both* mathematical representations, such as the Lotka–Volterra equations, and concrete things, such as the Phillips–Newlyn machine, need to be interpreted in order to function as scientific models. Consequently, the artifactual approach does not reduce models to mere equations or material objects devoid of any content but rather emphasizes that the representational modes and media used in model construction are important for their cognitive and epistemic functioning—precisely what Salis also recognizes.

The other two criticisms that Salis (2021a) launches against the artifactual approach concern its supposed inability to explain how scientists can attribute concrete properties to (fictional) model systems, as a result of which the artifactual cannot explain model–world comparisons. Both of these things are "difficult to explain without some sort of imagination and pretense," according to Salis (2021a, 171). The question is what qualifies as imagination. Does interpreting the Lotka–Volterra equation in terms of a hypothetical system of two (fictional) species of fish, one of which preys on the other, qualify as imagination? If this is the case, the artifactual approach has the capacity to address both problems raised by Salis. The artifactual approaches do not aim to dispense with imagination. Knuuttila rather focuses on how representational and other artifactual means employed by modelers scaffold their imagination and the construction of fictional or hypothetical systems.

Salis (2021a) concludes that "models are intersubjectively available tools of enquiry and objects of knowledge that crucially rely on the social activity of make-believe for their construction and manipulation in particular scientific communities" (175). An artifactualist could agree, apart from pointing out that modeling does not need to rely on imagination, as Currie (2018) argues in his discussion of engineering practices. Salis needs Walton's (1990) account of fiction to explain how model–world comparisons are possible in the case of theoretical models. While mathematically formulated models are abstract, their target systems are often concrete. Consequently, make-believe and pretense are needed to imagine the model system as concrete, thus making model–world comparisons between pretended concrete systems and the actual concrete systems possible. In contrast, the philosophical gist of the artifactual account is to tackle the epistemic value of modeling without supposing, at the outset, that the model would need to correspond more or less accurately to some determinable actual target system. Such an idea of comparison lies at the heart of the representational accounts of modeling, including the present fictional accounts of modeling. From the artifactual perspective, one does not need to construe fictions representationally, but one can rather approach the *fictional use of models* in terms of them being hypothetical systems constructed to study some pending scientific questions. These questions can concern actual systems, but also possible, or even impossible ones (Knuuttila 2021). One may ask, however, to what extent Salis' criticism of the artifactual approaches applies to radical artifactualism (Sanches de Oliveira 2022).

5. Hybrid vs radical artifactualism

While all the artifactual accounts discussed above seek an alternative way to approach modeling, beyond invoking representation, the question is to what extent their accounts still rely, at least implicitly, on a representational approach to modeling. Recently, Sanches de Oliveira has claimed that the aforementioned artifactual accounts of Morrison and Morgan, Knuuttila and Currie, are in fact hybrid artifactual accounts in being all too wedded to representation, in one way or another.

Sanches de Oliveira criticizes the accounts of Morrison and Morgan, Knuuttila, and Currie on different grounds, respectively. The ties of the models as mediators account to representationalism appear easiest to establish since representing is one of the functions of models according to Morrison and Morgan. In discussing representing, they make a distinction between a "simple tool" and "a tool of investigation" on the basis that the latter "involves some form of representation: models typically represent either some aspect of the world, or some aspect of our theories about the world, or both at once" (Morrison and Morgan 1999, 11).

The arguments that Sanches de Oliveira offers for maintaining that Knuuttila's and Currie's accounts remain within the realm of representationalism are less straightforward. Sanches de Oliveira labels Knuuttila's account representationalist on the basis that she talks about "representational means" in referring to "diagrams, pictures, scale models, symbols, natural language, mathematical notations, 3D images on screen" (Sanches de Oliveira 2022, 5). However, Knuuttila does not claim that models constructed with various representational means necessarily represent any external target system (external, that is, to the model itself). Instead, she takes a departure from representational accounts in not approaching the epistemic value in terms of a model–target relationship, but rather seeking to explain it by viewing models as entities that are constructed to answer some pending theoretical or empirical questions.

In addressing the supposed representational nature of Currie's models-as-tools account, Sanches de Oliveira concentrates on Currie's content/vehicle distinction. Such distinction, according to Sanches de Oliveira, is inherently representational, since assuming that a model as a vehicle "carries content" is another way of saying that a model represents. In choosing to consider any meaning or content in representational terms, Sanches de Oliveira ends up presuming that even such accounts that do not approach the epistemic value of models through the representational model–target relationship are nevertheless representational or "targetist". They are "targetist" since attributing content to a model makes it "defined by something else it refers to, something else it is a *source of information about*" (Sanches de Oliveira 2022, 19).

In his earlier article "Representationalism is a dead end" (2021), Sanches de Oliveira criticized the representationalist approaches mainly for assuming that models stand in for real-world target phenomena (Sanches de Oliveira 2021, 210). However, the notion of targetism introduced in his later article on radical artifactualism also covers fictional models as supposedly being defined by something else they refer to (Sanches de Oliveira 2022). But can a fictional system be separated from the model such that the model would give information about this distinct system? Such an assumption does not agree with how the philosophers of science entertaining fictionalism usually consider fictions, since according to them the (nonconcrete) model *itself* is fictional (Godfrey-Smith 2006; Frigg and Nguyen 2016). Irrespective of this difficulty, the more important question is whether Sanches de Oliveira's own radical alternative succeeds to do without any (representational or other) content.

Sanches de Oliveira's (2022) account of modeling aims to do without invoking the representational idiom entirely, approaching models as "simple tools" rather than as "representational tools." The paradigmatic tool for him is a hammer, through which he approaches scientific modeling. Sanches de Oliveira invokes Heidegger in explaining "the aboutness of a tool," turning to what the model is for, instead of what the model is a model of. Many practice-oriented accounts have indeed approached models also as "models for" as already suggested by Fox Keller (2000). Moreover, most pragmatists of representation would neither contest the claim that "tools (including models) are inherently and objectively meaningful for users engaged in particular practices" (Sanches de Oliveira 2022, 25). However, Sanches de Oliveira uses the Heideggerian approach precisely to help him to approach aboutness non-representationally: tools relate to a totality of equipment such that learning through using a tool amounts to an understanding of how it relates to other tools in some practice (23). Having thus clarified the aboutness of tools, Sanches de Oliveira formulates his radical alternative to the representationalist (and hybrid artifactualist) approaches to modeling in terms of "limited action-relevant similarities that a model bears to some system(s) of interest" (26).

Sanches de Oliveira explicates the action-relevant similarities in the following way: "the action-relevant similarities and dissimilarities between model-artifacts and the systems we usually conceptualize as targets enable scientists to think about interventions in those systems by means of manipulating the model-artifact" (2022, 28). It is unclear whether this notion enables radical artifactualism to shed the remnants of representationalism, however. Sanches de Oliveira claims that the action-relevant similarity between a butter knife and a screwdriver allows users to learn how to employ one by manipulating the other, without involving a representation of any kind. But does this kind of learning apply to scientific models? Sanches de Oliveira thinks that it does. He uses the Phillips–Newlyn hydraulic model of economy as an example, claiming that "actively intervening water flow rates in [this] model supports reasoning about how specific interventions such as changes in tax or investment rates might affect the economy" (26). But it does not seem possible to comprehend the action-oriented similarities between the Phillips–Newlyn machine and the economy of, say Guatemala (Frigg and Nguyen 2018), without assuming that the different parts of this material machine can somehow be related to economic concepts and magnitudes. Morgan and Boumans (2004) discuss the complexity of these linkages, and the theoretical economic thinking and new interpretations that the model gave rise to.

The crucial artifactual point is this: one does not need to relate the Phillips–Newlyn model to some determinable real-world economy, via action-relevant (or any other) similarities to learn from it, as Sanches de Oliveira assumes. The model has economic import by making use of our theoretical and empirical knowledge; it addresses some general features of economic theories and economies without any determinable relationship to some real-world economy. The problem of radical artifactualism is precisely that it does not make a distinction between what can be represented within the model and whether a model is a representation of some real-world target systems, e.g., a representation of some particular economy. This is a distinction that the (indirect) fictional and other artifactual accounts make. As a consequence, Sanches de Oliveira blends together representationalism (i.e., the idea that in order to give us knowledge models would need to correspond to some target systems) with the interpretation of signs and cultural representations. Yet, it is quite a different thing to claim that something is represented within a model than that a model represents or refers to some external target system. It is difficult to analyze fiction or diverse scientific and other displays without invoking the idea that various kinds of semiotic and symbolic devices are able to create meaning and prompt interpretations. Consequently, it seems that hammers and other "simple tools" do not get us all the way to the scientific understanding and explanation that models are able to offer.

6. Conclusion

So far, few philosophers of science have explicitly entertained the artifactual approach, yet the artifactuality of models is implicit in many practice-oriented approaches to modeling (Gelfert 2016; Parker 2020). The artifactual approach also holds promise when it comes to many traditional philosophical topics other than modeling, like idealization (see Carrillo and Knuuttila 2021; 2022). Although within philosophy of science artifactualism is a rather recent approach, in the grand scheme of things, this clearly is not the case. The artifactual approach to models has drawn inspiration from *Science and Technology Studies* (e.g., Lynch and Woolgar 1990), research on mathematical practices (Johansen and Misfeldt 2020), and extended, embodied, enactivist, and distributed approaches to cognition (e.g., Clark 1997; Hutchins 1995). One can expect the cross-pollination between these different fields to further enrich the philosophical discussion of scientific modeling.

Acknowledgments

This chapter and Handbook were made possible thanks to funding from the European Research Council (ERC) under the European Union's Horizon 2020 research and innovation program (Grant agreement No. 818772). They are also funded by the John Templeton Foundation project "Pushing the Boundaries" (Grant ID: 62581).

References

Bender, Emily M., Timnit Gebru, Angelina McMillan-Major, and Shmargaret Shmitchell. 2021. "On the Dangers of Stochastic Parrots: Can Language Models Be Too Big?" In *Proceedings of the 2021 ACM Conference on Fairness, Accountability, and Transparency*, 610–623. FAccT '21. New York, NY, USA: Association for Computing Machinery. https://doi.org/10.1145/3442188.3445922.

Black, Max. 1962. *Models and Metaphors: Studies in Language and Philosophy*. Ithaca, NY: Cornell University Press. https://doi.org/10.7591/9781501741326.

Boltzmann, Ludwig. 1911. "Models." In *Encyclopaedia Britannica*, 11th Edition, edited by Chisholm, Hugh and Walter Alison Phillips, 638–640. Cambridge: Cambridge University Press.

Boumans, Marcel. 1999. "Built-In Justification." In *Models as Mediators: Perspectives on Natural and Social Science*, edited by Mary S. Morgan and Margaret Morrison, 66–96. Cambridge: Cambridge University Press.

Carrillo, Natalia, and Tarja Knuuttila. 2021. "An Artifactual Perspective on Idealization: Constant Capacitance and the Hodgkin and Huxley Model." In *Models and Idealizations in Science: Artifactual and Fictional Approaches*, edited by Alejandro Cassini and Juan Redmond, 51–70. Logic, Epistemology, and the Unity of Science. Cham: Springer International Publishing. https://doi.org/10.1007/978-3-030-65802-1_2.

———. 2022. "Holistic Idealization: An Artifactual Standpoint." *Studies in History and Philosophy of Science* 91(February): 49–59. https://doi.org/10.1016/j.shpsa.2021.10.009.

Cartwright, Nancy. 1983. *How the Laws of Physics Lie*. First Edition. Oxford: New York: Oxford University Press.

———. 1999. *The Vanity of Rigour in Economics Theoretical Models and Galilean Experiments*. London: LSE, Centre for the Philosophy of the Natural and Social Sciences.

Clark, Andy. 1997. *Being There: Putting the Brain, Body, and World Together*. Cambridge: MIT Press.
Currie, Adrian. 2018. "From Models-as-Fictions to Models-as-Tools." *Ergo, an Open Access Journal of Philosophy* 4(27). https://doi.org/10.17863/CAM.17502.
Elgin, Catherine Z. 2004. "True Enough." *Philosophical Issues* 14(1): 113–131. https://doi.org/10.1111/j.1533-6077.2004.00023.x.
Fox Keller, Evelyn. 2000. "Models Of and Models For: Theory and Practice in Contemporary Biology." *Philosophy of Science* 67: S72–S86.
Fraassen, Bas C. van. 2008. *Scientific Representation: Paradoxes of Perspective*. Oxford; New York: Oxford University Press.
Frigg, Roman. 2010. "Models and Fiction." *Synthese* 172(2): 251–268. https://doi.org/10.1007/s11229-009-9505-0.
Frigg, Roman, and James Nguyen. 2016. "The Fiction View of Models Reloaded." *The Monist* 99(3): 225–242. https://doi.org/10.1093/monist/onw002.
———. 2018. "The Turn of the Valve: Representing with Material Models." *European Journal for Philosophy of Science* 8(2): 205–224. https://doi.org/10.1007/s13194-017-0182-4.
———. 2020. *Modelling Nature: An Opinionated Introduction to Scientific Representation*. Vol. 427. Synthese Library. Cham: Springer International Publishing. https://doi.org/10.1007/978-3-030-45153-0.
Gelfert, Axel. 2016. *How to Do Science with Models: A Philosophical Primer*. New York: Springer.
Giere, Ronald N. 1988. *Explaining Science: A Cognitive Approach*. Chicago: University of Chicago Press.
———. 2004. "How Models Are Used to Represent Reality." *Philosophy of Science* 71(5): 742–752. https://doi.org/10.1086/425063.
———. 2010. *Explaining Science: A Cognitive Approach*. Chicago: University of Chicago Press.
Godfrey-Smith, Peter. 2006. "The Strategy of Model-Based Science." *Biology and Philosophy* 21(5): 725–740. https://doi.org/10.1007/s10539-006-9054-6.
Hesse, Mary B. 1963. *Models and Analogies in Science*. London: Sheed and Ward.
Hilpinen, Risto. 1993. "Authors and Artifacts." *Proceedings of the Aristotelian Society* 93: 155–178.
———. 1999. "Artifact." In *The Stanford Encyclopedia of Philosophy*, edited by Zalta Edward and Nodelman Uri. Metaphysics Research Lab, Stanford, CA: Stanford University. https://plato.stanford.edu/archives/win2011/entries/artifact/.
Houkes, Wybo, and Sjoerd D. Zwart. 2019. "Transfer and Templates in Scientific Modelling." *Studies in History and Philosophy of Science Part A* 77(October): 93–100. https://doi.org/10.1016/j.shpsa.2017.11.003.
Hughes, R. I. G. 1997. "Models and Representation." *Philosophy of Science* 64: S325–S336.
Humphreys, Paul. 2009. "The Philosophical Novelty of Computer Simulation Methods." *Synthese* 169(3): 615–626. https://doi.org/10.1007/s11229-008-9435-2.
Hutchins, Edwin. 1995. *Cognition in the Wild*. Cambridge: MIT Press.
Isaac, Alistair M. C. 2013. "Modeling without Representation." *Synthese* 190(16): 3611–3623. https://doi.org/10.1007/s11229-012-0213-9.
Johansen, Mikkel W., and Morten Misfeldt. 2020. "Material Representations in Mathematical Research Practice." *Synthese* 197(9): 3721–3741. https://doi.org/10.1007/s11229-018-02033-4.
Kingsland, S. 1985. *Modeling Nature*. Chicago and London: The University of Chicago Press.
Knuuttila, Tarja. 2005. "Models, Representation, and Mediation." *Philosophy of Science* 72(5): 1260–1271. https://doi.org/10.1086/508124.
———. 2011. "Modelling and Representing: An Artefactual Approach to Model-Based Representation." *Studies in History and Philosophy of Science Part A* 42(2): 262–271. https://doi.org/10.1016/j.shpsa.2010.11.034.
———. 2017. "Imagination Extended and Embedded: Artifactual versus Fictional Accounts of Models." *Synthese*, September. https://doi.org/10.1007/s11229-017-1545-2.
———. 2021. 'Epistemic Artifacts and the Modal Dimension of Modeling'. *European Journal for Philosophy of Science* 11(3): 65. https://doi.org/10.1007/s13194-021-00374-5.
Knuuttila, Tarja, and Timo Honkela. 2005. "Questioning External and Internal Representation: The Case of Scientific Models." In *Computation, Philosophy, and Cognition*, edited by Lorenzo Magnani and Riccardo Dossena, 209–226. London: College Publications.

Knuuttila, Tarja, and Rami Koskinen. 2020. "Synthetic Fictions: Turning Imagined Biological Systems into Concrete Ones." *Synthese*, February. https://doi.org/10.1007/s11229-020-02567-6.

Knuuttila, Tarja, and Andrea Loettgers. 2017. "Modelling as Indirect Representation? The Lotka-Volterra Model Revisited." *The British Journal for the Philosophy of Science* 68 (4): 1007–1036. https://doi.org/10.1093/bjps/axv055.

———. 2023. "Model Templates: Transdisciplinary Application and Entanglement." *Synthese* 201(6): 200. https://doi.org/10.1007/s11229-023-04178-3.

Knuuttila, Tarja, and Atro Voutilainen. 2003. "A Parser as an Epistemic Artefact: A Material View on Models." *Philosophy of Science* 70: 1484–1495. https://doi.org/10.1086/377424.

Kress, Gunther, and Theo Van Leeuwen. 2001. *Kress, G: Multimodal Discourse: The Modes and Media of Contemporary Communication*. First Edition. London : New York: Hodder Arnold.

Latour, Bruno. 1994. "On Technical Mediation-Philosophy, Sociology, Genealogy." *Common Knowledge* 3: 29–64.

Lenhard, Johannes. 2006. "Surprised by a Nanowire: Simulation, Control, and Understanding." *Philosophy of Science* 73(5): 605–616. https://doi.org/10.1086/518330.

Lynch, Michael. 1990. "The Externalized Retina: Selection and Mathematization in the Visual Documentation of Objects in Life Sciences." In *Representation in Scientific Practice*, edited by Michael Lynch and Steven Woolgar, 153–186. Cambridge: MIT Press.

Lynch, Michael, and Steve Woolgar. 1990. "Introduction: Sociological Orientations to Representational Practice in Science." In *Representation in Scientific Practice*, edited by Michael Lynch and Steve Woolgar, 1–18. Cambridge: MIT Press.

Mäki, Uskali. 1992. "On the Method of Isolation in Economics." *Poznan Studies in the Philosophy of the Sciences and the Humanities* 26: 19–54.

Morgan, Mary S. 2012. *The World in the Model: How Economists Work and Think*. Cambridge: Cambridge University Press. https://doi.org/10.1017/CBO9781139026185.

Morgan, Mary S., and Marcel Boumans. 2004. "Secrets Hidden by Two-Dimensionality: The Economy as a Hydraulic Machine." In *Models The Third Dimension of Science*, edited by Soraya De Chadarevian and Nick Hopwood, 369–401. Palo Alto: Stanford University Press. http://www.sup.org/book.cgi?isbn=0804739722.

Morgan, Mary S., and Margaret Morrison, eds. 1999. *Models as Mediators: Perspectives on Natural and Social Science*. First edition. Cambridge, UK: Cambridge University Press. https://doi.org/10.1017/CBO9780511660108.

Morrison, Margaret, and Mary S. Morgan. 1999. "Models as Mediating Instruments." In *Models as Mediators: Perspectives on Natural and Social Science*, edited by Margaret Morrison and Mary S. Morgan, 10–37. Ideas in Context. Cambridge: Cambridge University Press. https://doi.org/10.1017/CBO9780511660108.003.

Nguyen, James, and Roman Frigg. 2022. "Scientific Representation." *Elements in the Philosophy of Science*, August. https://doi.org/10.1017/9781009003575.

Parker, Wendy S. 2020. "Model Evaluation: An Adequacy-for-Purpose View." *Philosophy of Science* 87(3): 457–477. https://doi.org/10.1086/708691.

Preston, Beth. 2013. *A Philosophy of Material Culture*. Routledge Studies in Contemporary Philosophy. https://doi.org/10.4324/9780203069844.

Salis, Fiora. 2021a. "Bridging the Gap: The Artifactual View Meets the Fiction View of Models." In *Models and Idealizations in Science: Artifactual and Fictional Approaches*, edited by Alejandro Cassini and Juan Redmond, 159–177. Logic, Epistemology, and the Unity of Science. Cham: Springer International Publishing. https://doi.org/10.1007/978-3-030-65802-1_7.

———. 2021b. "The New Fiction View of Models." *The British Journal for the Philosophy of Science* 72(3): 717–742. https://doi.org/10.1093/bjps/axz015.

Sanches de Oliveira, Guilherme. 2021. "Representationalism Is a Dead End." *Synthese* 198(1): 209–235. https://doi.org/10.1007/s11229-018-01995-9.

———. 2022. "Radical Artifactualism." *European Journal for Philosophy of Science* 12(2): 1–33. https://doi.org/10.1007/s13194-022-00462-0.

Sjölin Wirling, Ylwa, and Till Grüne-Yanoff. 2021. "The Epistemology of Modal Modeling." *Philosophy Compass* 16(10): e12775. https://doi.org/10.1111/phc3.12775.

Strevens, Michael. 2011. *Depth: An Account of Scientific Explanation*. Cambridge, MA: Harvard University Press.

Suárez, Mauricio. 2003. "Scientific Representation: Against Similarity and Isomorphism." *International Studies in the Philosophy of Science* 17: 225–244. https://doi.org/10.1080/026985903200 0169442.

———. 2004. "An Inferential Conception of Scientific Representation." *Philosophy of Science* 71(5): 767–779. https://doi.org/10.1086/421415.

Sugden, Robert. 2000. "Credible Worlds: The Status of Theoretical Models in Economics." *Journal of Economic Methodology* 7(1): 1–31. https://doi.org/10.1080/135017800362220.

Thomasson, Amie. 2007. "Artifacts and Human Concepts." In *Creations of the Mind: Theories of Artifacts and Their Representation*, edited by Eric Margolis and Stephen Laurence, 52–73. Oxford, UK: Oxford University Press. https://www.academia.edu/20295571/Artifacts_and_Human_Concepts.

Thomasson, Amie L. 1999. *Fiction and Metaphysics*. Cambridge, UK: Cambridge University Press.

Thomson-Jones, Martin. 2020. "Realism about Missing Systems." In *The Scientific Imagination*, edited by Arnon Levy and Peter Godfrey-Smith, 75–101. Oxford, UK: Oxford University Press. https://doi.org/10.1093/oso/9780190212308.003.0004.

Volterra, Vito. 1928. "Variations and Fluctuations of the Number of Individu- Als in Animal Species Living Together." *Journal Du Conseil International Pour L'exploration de La Mer* 3: 3–51.

Walton, Kendall L. 1990. *Mimesis as Make-Believe: On the Foundations of the Representational Arts*. Cambridge, MA: Harvard University Press.

Weisberg, Michael. 2007. "Who Is a Modeler?" *British Journal for the Philosophy of Science* 58(2): 207–233. https://doi.org/10.1093/bjps/axm011.

———. 2013. *Simulation and Similarity: Using Models to Understand the World*. Reprint Edition. Oxford: Oxford University Press.

9
TARGET SYSTEMS

Francesca Pero

1. The rise of the concept of target system

Philosophers conventionally refer to the other end of the representational relation as the model's "target system."[1] The term likely made its first appearance in French and Ladyman (1999), although it is with the representation's "source/target" distinction by Suárez (2003) that its use spread. This term has gradually replaced terms like "phenomena," "empirical system," or "(parts of) the world." The main reason for the wide use of "target system" is philosophers' recognition that modelers tackling a specific research question do not use models to represent a phenomenon in its entirety. The focus is rather on some aspects of the phenomenon that are deemed relevant to address the question.[2] The term "target system" conceptually catches the selective activity exerted on real-world phenomena and denotes the product of these activities. The use of the term suggests a conceptual distinction between what is used to represent (the model), what stands in the representational relation with the model as a reliable, yet simplified version of the targeted phenomenon (the target system), and the phenomenon itself (what the model is used to understand).

To this day, accounts explicitly addressing target systems—what they are and how they are constructed—are only a handful. Such a shortage is surprising given the many philosophical accounts of modeling. Even before the concept of "target system" caught on in the literature, philosophers advocating different approaches to modeling subscribed to the idea that models do not represent the world directly, because some preliminary work is mandatory for fitting phenomena with models. Cartwright (1983) refers to "prepared descriptions" as necessary to ensure that facts can be "fitted to" their (mathematical) treatment. These descriptions belong to an informal stage of modeling that mainly requires "a good deal of practical wisdom" (133) and is not fully standardized by the theory, although it is guided by disciplinary and theoretical goals. After the term "target system" officially entered the philosophical jargon, it has been pointed out that a target must be "determined" (Suárez 2010; Nguyen and Frigg 2021) or "constructed" (Knuuttila and Boon 2011; Tee 2019; Zuchowski 2019). Some accounts even address a process in which the target itself is to be "refined" once it is determined (a refinement that can either lead to amendments in the model or be due to such amendments; see Suárez *forthcoming*). Acts such as determination,

DOI: 10.4324/9781003205647-12

construction, or refinement hint at the fact that target systems are not out there in the world, at least not in the same way phenomena are assumed to be. Consequently, target systems and phenomena are not—conceptually at least—the same. Unfortunately, in most of the aforementioned discussions, the distinction between target systems and phenomena is not further articulated, and often the terms "target system" and "phenomena" are used interchangeably.

Another way to accommodate the relation between models and phenomena has continued to exploit the concept of the "model of data" as presented by Suppes (1962). This concept was developed within the context of the semantic view of scientific theories to spell out the intuition that there exists an intermediate entity between models and phenomena that makes raw phenomena amenable to the model's treatment (French and Ladyman 1999; Bueno, French and Ladyman 2002; French 2020; van Fraassen 2008).

Conceiving target systems as a separate and intermediary item between models and phenomena is a recent perspective in the literature on modeling. So far, there have been few attempts to clarify what target systems are and how they carry out their twofold function, i.e., to be what models are in a representational relation and to stand for the phenomena models are meant to account for.[3] The following sections focus on the philosophical attempts to flesh out the function of target systems. To this end, the chapter first analyses the distinction between target systems and phenomena set forth by authors explicitly dealing with target systems, as well as the concept of *phenomena of interest* they lay out. It then provides an overview of the kinds of target systems philosophers have considered so far. Finally, it considers possible ways the philosophical investigation of scientific modeling could benefit from integrating the analysis of target systems and their construction.

2. Target system construction

Providing a philosophical analysis that covers all instances of target system construction in actual practice is a demanding task. Such a project faces the same practical difficulty as that of analyzing the model concept: given the different kinds of target systems flooding actual scientific practice, is it possible to provide a philosophical account encompassing them? As we are going to see in the following sections, philosophers engaging in this project are quite cautious about this possibility. In fact, they admittedly present their proposals of target system construction as conceptual stretches useful only for the sake of clarification, hardly reflecting any logical or temporal order of phases in actual practice.

The starting point of target system construction is generally identified with the selection of relevant features of a phenomenon that should feature in the target system. Relevance criteria for selecting these features can be cast in terms of *causality*: a feature of the phenomenon is retained in the target system construction if, for different reasons (e.g., scientists' educated guess, indications contained in the background knowledge or provided by experiments, etc.), it is deemed to causally influence the occurrence or the behavior of the phenomenon of interest (Bailer-Jones 2009; Weisberg 2013; Serban and Green 2020; Tee 2019 Elliot-Graves 2020). Independently of whether causality is called upon as a criterion for selecting relevant features, the process of target construction is usually presented in terms of *partitioning*, i.e., as the identification of parts and properties of a phenomenon. Structural approaches to modeling lie on the back of this intuition (Bueno, French, and Ladyman 2002; da Costa and French 2003; Bartels 2006; French 2020). These accounts present the model–target relation as a mathematical morphism of some sort holding between the structures of the model and the target system, respectively. Models successfully represent their targets as they pinpoint their structure, which is identifiable once the target system has been partitioned into a domain of objects and a set of properties and relations among them.

Partitioning plays a role in deflationary approaches as well. Deflationary approaches offer a pragmatic account of modeling practice (Suárez 2003; Giere 2004; Bailer-Jones 2009). Contrary to the structuralist's approach, deflationists deny that the model–target relation can be spelled out in a univocal manner, let alone in mathematical terms. Suárez suggests that the philosophical analysis of modeling should say nothing about the relation *per se* and only account for the capacity of a model to allow its competent user to draw inferences on the target system. The inferential suitability of the model with respect to its target is presented by Suárez as the possibility to employ the model's "internal structure" (2010, 98)—informally intended as its division into parts and relations—as an approximation so that the model's parts and relations can be interpreted in terms of those of the target. Finally, approaches that attempt to retain both structural and pragmatic components in their analysis present partitioning as a crucial moment in representation to make a model's structure ascription meaningful (van Fraassen 2008; Bueno and Colyvan 2011; Suárez and Pero 2019; Nguyen and Frigg 2021).

Structural and inferential accounts mainly focus on spelling out the explanatory and predictive functions of models. Although none of their advocates would likely object that the process of target system construction is part and parcel of the modeling activity, in these accounts such a process is considered a readymade stage and rarely thoroughly addressed. This gap has been noticed by philosophers such as Weisberg (2007, 2013) and Elliot-Graves (2020), who provide thorough analyses of the phase of target system construction.[4] They both label the stage of the modeling activity that corresponds to target system construction as the "specification of the target system" and divide it into phases. The target system specification is generally presented as following the phase of model construction. However, as Weisberg points out, neither this placement nor that of the phases of target system specification actually reflect a logical or empirical order. The first step in specifying a target system is identifying the phenomenon of interest, which involves determining the spatiotemporal regions of the world the scientist wants to study. Also, in this case, partitioning plays a pivotal role. According to Weisberg, after modelers have decided which aspects of the phenomenon should be represented by the model, they carve the target system from the phenomenon of interest. Weisberg presents this process as a form of partitioning as well, guided by the "conceptualization of the target and model into properties" (149).

Elliot-Graves also describes target system specifications in terms of partitioning. Modelers first pinpoint the phenomenon of interest by identifying the boundaries of the spatiotemporal regions of the domain of study. In order to identify the domain, the latter has to be partitioned and arranged into parts and properties. At this stage, there is partitioning only: no criterion of usefulness or relevance is applied. Once the domain has been partitioned, the theorist will decide which parts of a phenomenon she will consider relevant, i.e., objects, properties, or dynamical processes, together with anything exogenous that nonetheless exerts a causal influence on the phenomenon.[5] After the phenomenon of interest is thus identified on the backdrop of a wider domain of study, the modeler will partition the phenomenon of interest into parts and properties as well, select those parts to be retained as relevant for the sake of investigation, and omit the others.

Both Weisberg and Elliot-Graves present the process of target system construction as a gradual thinning-out process that begins by separating the domain of study from the whole world, then the phenomenon of interest from the domain of study, and, finally, the target system from the phenomenon of interest. This process is carried out by partitioning, selecting, and omitting until the final product is carved out.[6]

A few modeling approaches in their analyses include cases where the moments of model and target construction cannot be sharply distinguished and layout target specification as something that happens concurrently with the model development. Such *co-construction* of the target system alongside its model can be either conceptual (Knuuttila and Boon 2011) or physical (Tee 2019). In the act of isolating a phenomenon of interest, scientists rely on scientific concepts that are called on by the very research question. Thus, the targeted phenomenon is endowed with conceptual content without which it "is not even recognized as a phenomenon in need of scientific explanation" (Knuuttila and Boon 2011, 320). The targeted phenomenon is then re-described theoretically by the model (as a target system) to answer the pending research question. The research question, as well as the concepts and principles it calls upon, may undergo amendments or further developments during the construction of the target system.

Bueno and Colyvan (2011) provide a formal reconceptualization of this process of mutual construction and adjustments appealing to "composite mappings" that allow scientists to go back and forth between the structures of model and target. In particular, Bueno and Colyvan aim at accommodating cases where the model structure needs to be adjusted, as it turned out to be empirically inadequate after it has been applied to the target system. Conversely, the possibility to amend the "initial" structure of the target system on the grounds of refinements and revisions informed by the model is also accommodated.

With the exception of Weisberg and Elliot-Graves, most of these accounts focus on *where* to place the target system construction throughout the modeling activity and rarely acknowledge or articulate (or give us reasons not to develop) a distinction between target systems and phenomena. The following section focuses on those accounts that lay out this distinction.

3. The distinction between target systems and phenomena

Philosophers dealing explicitly with the issue of target systems present the latter as conceptually different from phenomena. The concepts of target system and phenomena that emerge from these analyses are pragmatic in the sense of being strictly determined by the discipline at stake (and the background knowledge it is built upon), its research question, and the focus it determines.

According to Weisberg (2013), the identification of the phenomenon of interest is preliminary to the construction of the target system. A phenomenon is identified by circumscribing a spatiotemporal region of the world with the main objects and properties and whatever may have a *causal* influence on them. Modelers are not interested in all the properties of the phenomenon (the *total state* of the phenomenon), but in a subset of these properties. The identification of objects, properties, and potential causal factors by a theorist is constrained by her background knowledge (and the background theories it brings into play) and the procedural rules followed by her scientific community. Target systems are "abstractions performed over these phenomena" (90): anything featuring the total state of the phenomenon but lying outside the intended scope determined by the research question is abstracted away.[7]

Analogously, Elliot-Graves (2020) presents the identification of a phenomenon of interest as preliminary to target construction. In order to identify the phenomenon, we first have to determine the spatiotemporal location at which the phenomenon takes place (the *domain of* study). Target systems are generated by partitioning the domain of study into elements, their properties, and relations: those deemed relevant to answer the research question are retained, and the others are omitted. Elliot-Graves also casts relevance in causal terms: the retained factors are those that *causally* influence the phenomenon.[8] The identification of relevant causal factors is based on the available background knowledge (or, in case no background knowledge is available, on educated guesses).[9]

What is particularly relevant about Weisberg's and Elliot-Grave's conception of *phenomenon of interest,* and their rendition of how a target system is carved out of it, is that it does not rely on any metaphysical stance about phenomena. Phenomena are not conceived as entities out there to be observed (or detected, if not observable). They are constructed during the modeling process, which is strictly guided by the discipline at stake. Massimi (2011) insightfully articulates the idea of "constituted phenomena", which fits nicely with the notion of *phenomenon of interest* presented by Weisberg and Elliot-Graves: "Phenomena are not ready-made in nature; instead we have somehow to make them. And we make them by first ascribing certain spatiotemporal properties to appearances [objects given in sensibility], and then by subsuming them under a causal concept" (2011, 110). The construction of target systems can be conceived as the further step one needs to take to make these "conceptual constructions" amenable to model treatment by partitioning and arranging them in an ordered ensemble of properties and relations. The selection (and omission) of features used to build up the target system is a function of the modelers' goals and, according to which desideratum the modeler wants to comply with, different models and target systems are considered to fit better the same phenomenon of interest—provisionally, at least.

4. What is a target system: ontology and taxonomy

This section examines what kind of objects target systems are. This issue can be tackled in a twofold, complementary manner. First, if we subscribe to the distinction drawn in the previous section, according to which target systems are the product of abstraction over phenomena, we may wonder whether target systems are by default abstract entities or if they can be concrete. Second, we may wonder what kinds of target systems there are in modeling practice, a question that may be tackled by providing a sort of taxonomy. In the following, full-blown attempts to provide a classification of possible target systems are considered, as well as other insights into the subject from contributions not directly dealing with this issue.

The previous section points out that target systems are mainly obtained by selecting relevant features of otherwise too complex phenomena. The supposition that target systems are somehow carved out from presumably actual phenomena and that selective activity is guided by abstract theoretical guidelines, concepts, or (mental) conjectures might lead to conceiving target systems themselves as abstract. However, philosophers engaged with this issue do not take this answer for granted. Peschard (2010) argues that target systems are both abstract and concrete. Target systems are abstract in two senses. First, they can be conceived of as *types* instantiated in different contexts or experimental settings. For example, a spring can be chosen by the scientific community as a target system and used in those laboratories in which the dynamical properties of a mechanical system are investigated. Therefore, two springs, each used in a different laboratory, are *tokens* of the same type, i.e., of the spring that is preliminary (or conventionally) chosen as a fit target system for that specific purpose. Second, materiality is not sufficient to individuate a target system.

The same material system can be analyzed or manipulated by two scientists of different disciplines in different ways and for different purposes, thus being conceivable as two different target systems on the grounds of functional criteria. On the other hand, target systems are also concrete. In fact, whether they are tokens of a type of target system (as in the case of the two springs) or two different target systems *tout court*, they are particular, spatiotemporally identifiable objects.

Elliot-Graves (2002) argues that target systems gain their ontological status from their domain of study. The fact that target systems are generally construed via partitioning and omitting characterizing features of a phenomenon does not necessarily make them abstract: if the domain of study is concrete, so will the target system carved out of it.[10] The core of the argument is that the parts of the phenomenon that have been selected to compose the target system are not ontologically modified in the process of partitioning the domain: "All we do when we partition and identify relevant parts and properties is group a part of the world in a particular way. But this does not change the parts themselves. [...] If we think that my laptop is concrete and real, then the 'R' key on the keyboard is also concrete and real" (10).

Weisberg does not seem to take a stance on what determines, if anything, the abstractness or concreteness of target systems. Surely, he does not consider models as determining whether their target systems are abstract or concrete. He stresses that there might be cases where a target system is even more abstract than its model. This can happen when a concrete model is constructed out of model organisms, or in cases of "individual-based modeling" where populations of organisms, generally treated as aggregates, are represented by focusing on individuals and their properties.

While there is not a univocal view on what determines target systems' abstractness or concreteness, different authors seem to agree that the modeling practice will determine the kind of target system employed. In the following, a taxonomy of target systems is considered, expanding on that provided by Weisberg (2013) to incorporate other authors' insights:

> *Specific targets*: The target is a specific entity (or group of entities), phenomenon, or process. Instances of specific targets are those represented by scale models (e.g., the San Francisco Bay modeled by a hydraulic scale model), a particular species used as an exemplar to study a class of species (e.g., the Australian rabbits investigated to study invasive species).
>
> *Generalized targets*: The target is a *class* of phenomena, not a specific instance (e.g., the target of the model of sexual reproduction is sex in general rather than reproduction as performed by a specific species).

This level of generality of the target system is achieved by identifying the relevant features shared by all the specific targets and finding out those generalizable properties, which Weisberg considers to be at "the intersection of the total states for each target" (116). In the case of the model of sexual reproduction of two-sex species, the generalized target is meant to have the set of properties shared by all sexually reproducing species. This level of abstraction in a target system is useful when the model is supposed to provide a *how-possibly explanation* for some kind of phenomenon (e.g., "What is a possible reason for sexual reproduction when asexual reproduction is less costly?").

In the cases considered by Weisberg, the generalization is performed over classes of phenomena made of similar elements (species, autonomous agents, etc.) that exhibit the same behavior (sexual reproduction, segregation, etc.). There are cases where the generalization is performed over target systems of unrelated phenomena, also pertaining to different disciplines, whose behavior nonetheless exhibits similar patterns. These different phenomena can be modeled by a *model template*, i.e., a "mathematical structure that is coupled with a general conceptual idea that is capable of taking on various kinds of interpretations in view of empirically observed patterns in materially different systems" (Knuuttila and Loettgers 2016, 379). In this case, the construction of the generalized target system is suggested by the conceptual idea the template is endowed with. There are also cases where the generalized target system is constructed not by the intersection of the total states of specific and similar target systems, nor by the intersection of specific yet different target systems whose similarity pattern is suggested by a template. These are cases, as considered by Godfrey-Smith (2009), where the generalization is suggested by the description of a specific target system that "acts as a 'hub' that anchors a large number of other cases" (Godfrey-Smith 2009, 107). Once the modeler determines the target's hub role, she will combine the knowledge of the hub-target system with *ad hoc* tools (concepts and methods external or internal to her discipline) relevant to apply the "exact knowledge" of the hub-target system to other target systems she is interested in.

Non-existent targets: In this case, targets stand for non-existent phenomena. Modeling non-existent targets is dubbed *hypothetical modeling* by Weisberg. The non-existence of the target can be either contingent or nomically necessary. In the case of contingently non-existent target systems, the target does not exist although the laws of nature would not prevent its existence (e.g., the xDNA whose physical model led scientists to conclude that DNA is likely *not* the only possible genetic system, that is, its existence is contingent).[11]

The non-existence of the target system can be also *nomically necessary*, i.e., it is physically impossible for the target to exist. This is, for example, the case of perpetual motion machines described by models such as the *ratchet and pawl machines*. The existence of these models' target systems is impossible as it would violate the second law of thermodynamics.

Both contingently and nomically non-existent target systems are to be conceived as mere *possibilities*, and models that provide information about such targets are *hypothetical models*, i.e., models that tell us something about real-world phenomena by telling us something about a target that exists *ex hypothesis* only. Both kinds of hypothetical models are useful as they provide counterfactual knowledge, i.e., "what the world would be like if the model's structure and behavior were instantiated in our own world" (128), which deepens our understanding of actual phenomena, showing how they could have been different and even why they are not so.

No target system at all: In this case, the object of the representation is the model itself "without regard to what it tells us about any specific real-world system" (Weisberg 2013, 129).[12] This is often the case in mathematical and computer simulation modeling (see Parker 2009). An example is the *Game of Life* by Conway, a two-dimensional cellular automaton made of a grid of square cells whose possible states are determined by some rules of interaction inspired by real-life behaviors (e.g., being in a neighborhood, living, dying, surviving, etc.). The simulation is employed to study the behavior of the cellular automaton, hence of a mathematical model.[13] Levy (2015) refers to the *Game of Life* to reach the opposite conclusion, i.e., there are no targetless models: the *Game of Life* was originally presented as recreational mathematics, i.e., as a piece of mathematics that taken by itself has no target system yet; once the model has been put to use to predict real-life human behaviors, it has instantly gained a target system. A similar claim appears in Cassini's work (2018), with attention to the target construction processes.

Bokulich (2003) and Zuchowski (2019) argue that modeling with no target systems at all is typical of the "horizontal model construction." While vertical models are constructed either top-down from theory or bottom-up from empirical data, horizontal models are constructed by mathematically manipulating, e.g., a set of equations for investigative reasons. For example, in complexity science, it is common to modify the dynamics of an "ancestor model" (used to model a target system). The dynamics thus generated lead to a "lineage of models" none of which, contrary to their ancestor, has a target system for its dynamics as it is automatically generated by artificial intelligence (Zuchowski 2019). Analogously, in physics, quantum maps are models generated by discretizing the equations of classical models and for the sole purpose of clarifying the relation between quantum and classical mechanics (Bokulich 2003).

5. Placing target systems between models and phenomena: possible consequences for philosophical analysis

The conflation of targets with phenomena and data models has proven to be quite problematic when one of the major conundrums of scientific representation via models is at stake, i.e., how can something abstract—such as a mathematical model—represent something concrete, e.g., the behavior of a physical system (van Fraassen 2008 calls this the "link to reality objection"; see also Nguyen and Frigg 2021). On the other hand, what is missing in accounts that distinguish target systems from phenomena and data models is a fully-fledged analysis of the possible consequences of rethinking the relation between (abstract) models and (concrete) targeted phenomena in light of this distinction.[14] The scope of this section is to highlight these consequences and to briefly consider how the philosophical analysis of scientific modeling could benefit from integrating the issue of target system construction.

Conflating target systems with phenomena is particularly problematic for those accounts that make representation depend on some intrinsic property of the model and the target system, such as that of sharing (some) structure (Bueno, French and Ladyman 2002; da Costa and French 2003; French 2020.) These accounts provide a formal rendition of this relation in terms of different kinds of morphisms between the structures of the model and the target (see Pero and Suárez 2016 for a comparative analysis). The issue at stake is that the only justification for the fact that models (structures) are applicable is that phenomena naturally exhibit some kind of structure and that the model correctly pinpoints such structure. However, metaphysical justifications of the form "models successfully represent (are morphic to their target systems) as they correctly identify that the structure phenomena are actually equipped with" (Ladyman 1998; French 2000) would jibe the issue of scientific representation with philosophical problems it is proclaimed to be neutral about. As stressed in a recent and milder formulation of these accounts, such as Bueno and Colyvan's inferential conception, the way we carve up phenomena and arrange them into a set of objects and relations (a structure) is something we obtain under the guidance of our theories and not something the world comes equipped with (2011, 347). Notwithstanding, as pointed out by Nguyen and Frigg (2021), the target system's structure in Bueno and Colyvan's account is only "assumed" as necessary for articulating their mapping account, and no story is given on where the assumed structure of the target system comes from. Nguyen and Frigg (2021) claim to fill this lacuna by offering an account of how structures for target systems are "actually" generated via *structure generating descriptions*. These are descriptions of phenomena that "strip" away the physical nature of their elements and relations and replace them respectively with featureless dummy objects and pure extensions of the relations (the latter specify between which object the relation holds, but not what the relation itself is).

A different strategy to fill in the gap between (abstract) models and (concrete) targeted phenomena has been developed by van Fraassen, who identifies data models as intermediate elements between models and phenomena. Placing data models between models and phenomena is presented by van Fraassen (2008) as a possible solution to the "link to reality objection" as he claims that "construction of data models is precisely the selective relevant depiction of the phenomena *by the user of the theory* required for the possibility of representation of the phenomenon" (253). Such identification is nonetheless problematic as data models are conceived as (mathematical) structures. The same question arises regarding how to justify the use of *their* (the data models') structure to represent phenomena (Brading and Landry 2006; Le Bihan 2012; Nguyen 2016).

It has been previously highlighted that the few accounts explicitly dealing with target systems, their ontology and construction, present target systems as the final recipient of the representational relation having models on the other side. Target systems are carved out from phenomena and, consequently, rephrasing van Fraassen's pragmatic tautology, claiming that the model represents a target system is the same as claiming that it represents the phenomenon the target "stands for." However, it has been pointed up that the notion of *phenomenon of interest* is not metaphysically loaded, nor does it imply any ontological commitment: a phenomenon is a conceptual construction identified as the content of a spatiotemporal region in the world that a competent research isolates according to the focus posed by the research question raised by the discipline at stake. Target systems are possible ways of carving out the phenomenon of interest into elements and relations that are conjectured to have a causal influence in determining the phenomenon or its behavior.

This conception of target systems and phenomena could help to spell out the representation relation between models and targeted parts of the world more precisely and neutrally:

i the representation relation holds between a model and its target system;
ii we are entitled to draw inferences from the model to the phenomenon the target system was carved out from;
iii the "reality" that is being addressed via (i) and (ii) need not be that of the world but of the content of the spatiotemporal region of the domain of study the phenomenon pertains to, which includes all the objects, properties and relations characterizing the phenomenon, together with the exogenous causes that affect its behavior.

The relation between target systems and phenomena, as pointed out by Elliot-Graves (2020), could be analyzed in terms of *aptness:* a target system is apt for explaining a phenomenon if the partition displayed in the target is useful for understanding the domain of study and the selected parts contain *all* the factors that are deemed relevant in order to understand the domain of study.[15]

A final remark concerns the relevance of the issue of target system construction for a philosophical analysis of modeling that aims at being closer to scientific practice. Which features should be included in the construction of a target system is itself subject to scientific research. Moreover, assessing the adequacy of target systems as refined versions of phenomena of interest requires some normative constraints that govern the construction of target systems. As these processes may be performed differently, there may be more than one target system for a single phenomenon. Hence some standards of adequacy are to be set.[16]

As pointed out in Section 1, the idea that phenomena are to be "prepared" for model application floated from the beginning of the philosophical debate on modeling as representing. However, it has taken some lexical creativity and conceptual efforts to identify target systems and their construction as a distinct moment in the reconstruction of the modeling practice with respect to model construction or phenomena preparation, as well as to identify target systems as a proper item with respect to the other components that modeling is traditionally reduced to. This entry has focused on such recent efforts and tried to underline the insights they could bring to the philosophical analyses of modeling and representation.

Acknowledgments

Francesca Pero acknowledges financial support from MIUR through the PRIN 2017 project "The Manifest Image and the Scientific Image" (Prot. 2017-ZNWW7F-004).

Notes

1 Representation is not the only way the model-target relation can be cast (see Peschard 2009; Knuuttila 2011, Tee 2019). However philosophical the accounts considered here mostly stick to the representational conception of the relationship.

2 Another reason for the replacement is that these terms are ontologically loaded, and philosophers rarely engaged with the issue of modeling are concerned with taking an ontological stance.

3 Cases of models with no target systems at all are also considered in the literature (see Section 4).

4 Elliot-Graves does not take side in the debate between structuralists and inferentialists. On the other hand, Weisberg does subscribe to the structuralist approach as he considers models interpreted structure that stand in a mapping relation to their target systems (2013, chap. 2), and he shares with champions of structural approaches the view that "comparing structures to structures is at the core of modeling" (ibid., 15, fn. 3). In spite of that, his account of target-system construction and partitioning, as Elliot-Grave's, is spelled out in neutral terms. Weisberg's structural setting resurfaces when, once the target has been construed, it is put in relation to the model. This is the *coordination* phase, when there are "specifications of how parts of real or imagined target systems are to be mapped onto parts of the model" (39).

5 Elliot-Graves acknowledges that presenting the *partitioning* of the domain into parts and properties and the *omission of parts and properties* as irrelevant, as two separate steps may be superfluous since they may be indistinguishable steps in actual practice. However, she claims there is a conceptual difference between partitioning and omitting. Partitioning can be performed in different manners thus leading to different arrangements of parts and their properties, yet "all those partitions will contain the same amount of 'stuff'" (2020, 28). On the other hand, omission leads to thinning out elements of the partitioned domain as they are deemed irrelevant.

6 The following quote from Weisberg nicely takes stock of the process of target-construction: "When a scientist is interested in studying some phenomenon in the *world*, she begins by identifying a *spatio-temporal region of interest*. [...] Call the entire set of these properties the *total state* of the *phenomenon*. In almost every instance, modelers are not interested in studying the total states of phenomena, but rather some scientifically important *subset* of these properties. These restricted subsets are *target systems*. In other words, when scientists choose a focus, or an intended scope ([...]), they focus on some set of properties and abstract away the others. This yields a target system, a subset of the total state of the system." (2013, 90).

7 Winsberg (2009) distinguishes as well "target systems" from "objects," yet he provides a different interpretation of both the terms. Target systems are the class of systems of scientists' interest, while the "object" of the investigation is the *artifact* that they observe and intervene on during investigation. In this context, the target system is what Weisberg identifies as the phenomenon of interest, not the outcome of the abstraction performed over the latter.

8 Elliot-Graves prefers to lay out the process of target system construction in terms of omission rather than abstraction as the latter has not a univocal meaning in the philosophical literature on models (see Frigg and Hartmann 2020).

9 Peschard (2010) questions Weisberg's point that the individuation of the phenomenon of interest is a prerequisite for the modeling activity to get started (and her concern can be applied to Elliot-Grave's analysis as well): if the phenomenon of interest is the item upon which the research question is built, how could the modeler know what to include in the preparation of the target system as relevant to answer the question? In particular, Peschard does not subscribe the claim that modeling and targeting phenomena amounts to an investigation of the causes for a given effect since there might not be a clear-cut effect in the first place.
10 In her contribution, Elliot-Graves is not only defending the equivalence of statuses for the target system and its domain of reference. She is actually claiming that target systems are *always* "real parts of the world" and concrete. What is missing in Elliot-Graves' argument in favor of the concrete status of target systems is the explicit assumption, or premise, that domains of study are always concrete.
11 For a thorough analysis of alternative genetic systems see Koskinen (2017), Knuuttila and Koskinen (2020).
12 Weisberg seems to partially contravene this understanding of models with no target at all when he claims few lines later that "the development of such models has often been motivated by ideas about the way the world might work [...] even if they are not intended to be models of such phenomena" (130).
13 Only later, because of the possibility the *Game of Life* offered to simulate real-life processes, the model has been ascribed to target systems, thus becoming target directed.
14 Weisberg presents the model-target relationship in terms of similarity, in line with Giere's (2004) understanding of the concept as more a theoretical hypothesis by scientists (that the model is similar to the phenomenon of interest) than substantive properties of the model and the target. As such, similarity comes "in respects and degrees", according to scientists' goals.
15 Elliot-Graves does acknowledge that both the selection of some parts as relevant and of other parts as irrelevant can be incorrectly performed. A re-examination of the target system thus constructed may reveal whether this is the case (2020, 11).
16 The possibility of multiple target systems at stake here is not the one due to the difference among research questions. As exemplified by Weisberg (2007): two scientists might be studying the Adriatic Sea after World War I with two different research questions in mind. One might be interested in predator-prey relations after the conflict, and another in the effect of surface temperature on algae blooms. Different target systems correspond to each of these two questions. The issue is rather if, once the research question is determined, a target system could in principle be arbitrarily generated.

References

Bailer-Jones, Daniela M. 2009. *Scientific Models in Philosophy of Science*. Pittsburgh: University of Pittsburgh Press.
Bartels, Andreas. 2006 "Defending the Structural Concept of Representation." *Theoria* 21(1): 7–19.
Bokulich, Alisa. 2003. "Horizontal Models: From Bakers to Cats." *Philosophy of Science* 70: 609–627.
Brading, Katherine and Elaine Landry. 2006. "Scientific Structuralism: Presentation and Representation." *Philosophy of Science* 73: 571–581.
Bueno, Otavio, Steven French, and James Ladyman. 2002. "On Representing the Relationship between the Mathematical and the Empirical." *Philosophy of Science* 69: 497–518.
Bueno, Otavio and Mark Colyvan. 2011. "An Inferential Conception of the Application of Mathematics." *Nous* 45(2): 345–374.
Cartwright, Nancy. 1983. *How the Laws of Physics Lie*. Oxford: Oxford University Press.
Cassini, Alejandro. 2018. "Models without a Target." *ArtefaCToS. Revista de estudios de la ciencia y la tecnología* 7(2): 185–209.
da Costa, Newton and Steven French. 2003. *Science and Partial Truth: A Unitary Approach to Models and Scientific Reasoning*. Oxford: Oxford University Press.
Elliot-Graves, Alkistis. 2020. "What Is a Target System?" *Biology and Philosophy* 35: 1–28.
French, Steven. 2000. "The Reasonable Effectiveness of Mathematics: Partial Structures and the Application of Group Theory to Physics." *Synthese* 125: 103–120.
———. 2020. *There Are No Such Things as Theories*. Oxford: Oxford University Press.

French, Steven and James Ladyman. 1999. "Reinflating the Semantic Approach." *International Studies in the Philosophy of Science* 13: 103–121.
Frigg, Roman and Stephan Hartmann. 2020. "Models in Science." In *The Stanford Encyclopedia of Philosophy*, edited by Edward N. Zalta. https://plato.stanford.edu/archives/spr2020/entries/models-science/.
Giere, Ronald N. 2004. "How Models Are Used to Represent Reality." *Philosophy of Science* 71: 742–752.
Knuuttila, Tarja. 2011. "Modelling and Representing: An Artefactual Approach to Model-based Representation." *Studies in History and Philosophy of Science* 42(2): 262–271.
Knuuttila, Tarja and Mieke Boon. 2011. "How Do Models Give Us Knowledge? The Case of Carnot's Ideal Heat Engine." *European Journal for Philosophy of Science* 3: 309–334.
Knuuttila, Tarja and Andrea Loettgers. 2016. "Model Templates Within and between Disciplines: From Magnets to Gases—and Socio-Economic Systems." *European Journal for Philosophy of Science* 6: 377–400.
Knuuttila, Tarja and Rami Koskinen, 2020. "Synthetic Fictions: Turning Imagined Biological Systems into Concrete Ones." *Synthese* 198(9): 82338250.
Koskinen, Rami. 2017. "Synthetic Biology and the Search for Alternative Genetic Systems: Taking How-Possibly Models Seriously." *European Journal for Philosophy of Science* 7(3):493–506.
Ladyman, James. 1998. "What Is Structural Realism?" *Studies in History and Philosophy of Science* 29: 409–424.
Le Bihan, Soazig. 2012. "Defending the Semantic View: What It Takes." *European Journal for Philosophy of Science* 2: 249–274.
Levy, Arnon. 2015. "Modeling without Models." *Philosophical Studies* 172(3): 781–798.
Massimi, Michela. 2011. "From Data to Phenomena: a Kantian Stance." *Synthese* 182: 101–116.
Nguyen, James. 2016. "On the Pragmatic Equivalence between Representing Data and Phenomena." *Philosophy of Science*, 83(2): 171–191.
Nguyen, James and Roman Frigg. 2021. "Mathematics Is Not the Only Language in the Book of Nature." *Synthese* 198: S5941–S5962.
Parker, Wendy. 2009. "Does Matter Really Matter? Computer Simulations, Experiments, and Materiality." *Synthese* 169: 483–496.
Pero, Francesca and Mauricio Suárez. 2016. "Varieties of Misrepresentation Homomorphism." *European Journal for Philosophy of Science* 6(1): 71–90.
Peschard, Isabelle. 2009. "Making Sense of Modeling: Beyond Representation." *Society for Philosophy of Science SPSP 2009 conference 2009*. http://philsci-archive.pitt.edu/5110/.
———. 2010. "Target Systems, Phenomena and the Problem of Relevance." *The Modern Schoolman* 87: 267–284.
Serban, Maria and Sara Green. 2020. "Biological Robustness: Design, Organization and Mechanisms." In *Philosophical Perspectives on the Engineering Approach in Biology: Living Machines?*, edited by Sune Holm and Maria Serban, 141–164. New York: Taylor and Francis.
Suárez, Mauricio. 2003. "Scientific Representation: Against Similarity and Isomorphism." *International Studies in the Philosophy of Science* 17: 226–244.
———. 2010. "Scientific Representation." *Philosophy Compass* 5: 91–101.
———. Forthcoming. *Inference and Representation: A Study of Modelling In Science*. Chicago: University of Chicago Press.
Suárez, Mauricio and Francesca Pero. 2019. "The Representational Semantic Conception." *Philosophy of Science* 86: 344–365.
Suppes, Patrick. 1962. "Models of Data." In *Logic, Methodology, and Philosophy of Science: Proceedings of the 1960 International Congress*, edited by Ernest Nagel and Patrick Suppes, 252–261. Stanford: Stanford University Press.
Tee, Sim-Hui. 2019. "Constructing Reality with Models." *Synthese* 196: 4605–4622.
———. 2008. *Scientific Representation*. Oxford: Oxford University Press.
———. 2007. "Who Is a Modeler?" *British Journal for the Philosophy of Science* 58: 207–233.
———. 2013. *Simulation and Similarity. Using Models to Understand the World*. Oxford: Oxford University Press.
Winsberg, Eric. 2009. "A Tale of Two Methods." *Synthese* 169: 575–592.
Zuchowski, Lena. 2019. "Modeling and Knowledge Transfer in Complexity Science." *Studies in the History and Philosophy of Science* 77: 120–129.

10
MINIMAL MODELS

Christopher Pincock

1. Introduction

This chapter considers some defenses of the use of minimal models or toy models, treating them interchangeably (cf. Gelfert 2019). As these terms suggest, minimal models are especially simple, and so they seem to be too unrealistic to be used for ordinary modeling purposes such as accurate description, prediction, or explanation. Section 2 sketches three examples of minimal models and uses these examples to motivate a provisional definition. Section 3 discusses some modal strategies for making sense of the use of minimal models. Section 4 engages with some alternative reinterpretation approaches to minimal models. Section 5 considers the argument developed by Batterman and Rice that there is a special kind of "minimal model explanation" where a minimal model plays a crucial part. This short survey suggests that there are many questions about minimal models and their viable uses that remain open to debate. While there may be no single answer to the question of how minimal models are useful in science, a variety of strategies can be fruitfully combined to cover many of the initially puzzling cases.

2. A puzzle for the use of minimal models

In many cases of model-based science, a model is used to describe, predict, or explain some aspect of a target system only after it has been extensively tested. Users of the model then often maintain that the model can afford new, justified beliefs. While there is no consensus on how testing a model leads to new, justified beliefs, one popular proposal is that said testing often involves establishing a representational relation between a model and its intended target. This representational relation provides some assurance that a generic feature of the model will also be present in the target. In general, a minimal model or a toy model will fail these sorts of tests, in part because it is too simple to stand in such a representational relation. This suggests the following provisional definition: a scientific model is a minimal model just in case the users of the model believe that it lacks a representational relation to its target that would license a user to infer that a generic feature of the model is present in the target (cf. Grüne-Yanoff 2009, 83). The puzzle for the use of minimal models is then

DOI: 10.4324/9781003205647-13

immediate: when is it appropriate for a user of a minimal model to infer that some specific feature of the model is present in the target, given that they believe that there is no license for this inference in general?

This provisional definition of minimal model fits well with two of the most commonly mentioned minimal models: the Ising model of phase transitions and the Schelling model of racial segregation. The Ising model of phase transitions is typically identified with a special sort of system that undergoes a phase transition: "The most important and simplest system that exhibits a phase transition is the Ising model" (Gould and Tobochnik 2010, 248). The two-dimensional Ising model is composed of elements placed on a square lattice, where each element is assigned spin-up or spin-down. In the simplest case, interactions are allowed only between nearest neighbors. These interactions then generate blocks of aligned elements, which produce some net magnetic field. Below some critical temperature, the system will generate a magnetic field via the contribution of blocks of similarly aligned elements. However, when the critical temperature for the system is reached, the model's magnetic field goes to 0. This is a "continuous" phase transition because the magnetic field, the salient "order parameter," "vanishes continuously rather than discontinuously" (Gould and Tobochnik 2010, 266). More generally, a phase transition occurs when a system's order parameter changes from non-zero to zero. The same basic approach can be used to formulate a three-dimensional Ising model where elements are arranged in a three-dimensional cubic lattice. Other phase transitions can be modeled in terms of the vanishing of other kinds of order parameters. For example, for materials that may be in either a liquid state or a gas state, the order parameter is the difference between the density of the liquid and the density of the gas. Phase transitions occur in these materials when the difference between these densities goes to 0. This occurs when a critical temperature and critical pressure are reached.

The Schelling model of racial segregation is often introduced as a 64-square checkerboard, where some of the squares are occupied by nickels and some of the other squares are occupied by dimes (Schelling 1978, 147–155). We suppose that the coins are initially randomly distributed across the checkerboard, with many squares left blank (e.g., 19). Each coin is then given an opportunity to move. A coin's movement is determined by the occupants of its neighboring squares: if 33% or fewer of its neighbors are the same coin as it, then it moves to an unoccupied square where more than 33% of its neighbors will match its coin type. Otherwise, the coin stays where it is. Schelling found that for nearly all starting configurations, the initially random arrangement would be transformed into a highly segregated pattern of coins over several rounds of movement. Just as the Ising model of phase transitions exhibits a phase transition, the Schelling model of racial segregation exhibits a process of "racial" segregation, where coins of different types wind up in homogenous groups on the checkerboard.

Hamilton's selfish herd model of gregarious behavior considers an infinitely large field with cows randomly distributed across it (Hamilton 1971; Pincock 2012a). Every so often, a lion emerges from a random location in the field and consumes the cow that is spatially closest to it. Hamilton used this setup to consider various rules that cows could follow to avoid being eaten. He argued that the best movement rule for an individual cow would be to move toward its nearest neighbor. This results in "gregarious behavior," that is, animals of the same species staying in close spatial proximity to one another. A cow following this movement rule would reduce its chance of being eaten by lions as this movement rule would be the best way for the cow to reduce its so-called "domain of danger," i.e., the region made up of points closer to that cow than any other cow.

Each of these models is a model of some phenomenon, i.e., phase transitions, racial segregation, and gregarious behavior, and yet the simplicity of the model creates doubts about the use of the model for generating new, justified beliefs about these phenomena. If taken at face value, hardly any of the features present in the model system are also present in the target phenomenon. The same point holds for other popular examples of minimal models: hardly any features of the model systems can be found in the target phenomena. Other common examples are the Hawk–Dove model of restraint in combat (Rohwer and Rice 2013; Fumagalli 2016), the Lotka–Volterra model of predator–prey interaction (Weisberg 2013; Knuuttila and Loettgers 2017; Reutlinger, Hangleiter, and Hartmann 2018), and the Hotelling model of market competition (Aydinonat and Köksal 2019). The puzzle for these minimal models is that such models are widely used throughout the sciences, and yet their simple character stands in the way of understanding their use.

3. Modal approaches to minimal models

One proposed solution to the puzzle of the use of minimal models is that model users are careful to restrict the sort of properties that they transfer from what they find in the model to the target phenomenon. Grüne-Yanoff argues that only modal properties are apt to be transferred from a minimal model to a target. A model user who respects this restriction can then learn something about the phenomenon, even though the model lacks the usual representational relation to the target. For Grüne-Yanoff this "learning" from a minimal model involves a rational revision of the credence of some hypothesis. In summary form, he proposes that "If we are to learn from a model ... it must (1) present a relevant possibility that (2) contradicts an impossibility hypothesis that is held with sufficiently high confidence by the potential learners" (Grüne-Yanoff 2009, 97). He applies this proposal to the Schelling model. Suppose that some scientists believed that it was impossible for racial segregation to arise in the absence of strong racial preferences. The model presents a "relevant possibility" where racial segregation arises on the checkerboard, even though the racial preferences are very weak (Grüne-Yanoff 2009, 96). A similar point can be made for Hamilton's selfish herd model: biologists had claimed that gregarious behavior required group selection, but the model indicates how gregarious behavior could arise through ordinary, individual selection. This would then count as another example of learning from a minimal model. (See also Knuuttila (2021) for another proposal along these lines.)

Grüne-Yanoff's proposal seems adequate for cases where scientists are interested in all the possible instances of some phenomenon. Suppose, though, that scientists are interested only in the actual causes of actual instances of the phenomenon. For this sort of investigation, the modal strategy is not well suited to make sense of the use of a minimal model. As Fumagalli puts the worry, even though Schelling's model "may prompt a justified change in confidence in hypotheses about the segregation processes figuring in the *possible* worlds envisioned by his [Schelling's] model ... [this] does not imply a justified change in modelers' confidence in hypotheses about any *real-world* segregation process" (Fumagalli 2016, 445, emphasis in original). Similarly, one could complain that Hamilton's model does not establish that any actual instances of gregarious behavior arose (or even could have arisen) as the model depicts, i.e., entirely through individual selection (cf. Sjölin Wirling 2021).

Another modal strategy is pursued by Reutlinger, Hangleiter, and Hartmann. Their proposal is based on a distinction between models that are "embedded" within a theory and models that are "autonomous" from a theory. An embedded model can draw on the theory it is tied to in order to provide "an interpretation and justification of the idealizations" of the model (Reutlinger, Hangleiter, and Hartmann 2018, 1086). Reutlinger et al. admit that many minimal models fail to satisfy these conditions: even if they are models of some theory, no theory is able to legitimate or interpret the idealizations so that they correspond to true claims about genuine causes. This is how they characterize the Schelling model. In such cases, Reutlinger et al. maintain that a minimal model can still provide a how-possibly explanation. This means that the model does not explain how racial segregation actually arose, but instead how racial segregation could have arisen (Reutlinger, Hangleiter, and Hartmann 2018, 1094). This is quite similar to Grüne-Yanoff's proposal, except Reutlinger et al. add that legitimating this sort of modal claim also explains what they call a "modal phenomenon" (Reutlinger, Hangleiter, and Hartmann 2018, 1094). A modal phenomenon in their sense involves the necessity or possibility of something. Two examples of modal phenomena would be the possibility of extraterrestrial life and the impossibility of a perpetual motion machine. (See Verreault-Julien (2019) for a general discussion of this kind of explanation.)

As with Grüne-Yanoff's point about learning about possibilities, it is important to be clear on what these how-possibly explanations target. Neither Schelling's model nor Hamilton's model show, for any actual instance of racial segregation or gregarious behavior, that this instance could have arisen in the way the model depicts. Instead, the models show that some non-actual instances of the phenomenon arose in this way. This is a very weak explanatory claim. Analogously, one could explain how life could have arisen by invoking a minimal model that includes special creation: this does not explain how actual life on earth could have arisen, but only how life arose in some remote possible scenario. It is not clear that the importance of minimal models can be clarified if these claims were all that minimal models could offer.

4. Reinterpretation approaches to minimal models

Reutlinger et al. allow that some autonomous minimal models can be used to provide explanations of the actual features of some phenomenon. However, they seem to assume that the most common way that this occurs is by developing a new, more complicated model (Reutlinger, Hangleiter, and Hartmann 2018, 1092). This section considers three attempts to legitimate the explanatory use of the minimal model without developing another, more realistic model. These attempts all involve reinterpreting the minimal model so that its features are changed. This then licenses relating these new features to the features of the target phenomenon.

Perhaps the most well-known reinterpretation strategy for minimal models has been developed by Weisberg as part of his account of what he calls "minimalist idealization." Minimalist idealization is a process of model construction that aims at what Weisberg calls a "minimalist model": "a minimalist model contains only those factors that *make a difference* to the occurrence and essential character of the phenomenon in question" (Weisberg 2007, 642). Weisberg allows for a variety of ways that this sort of model could be obtained. The goal is to obtain an explanation using the model that relies on "a special set of explanatorily privileged causal factors" (Weisberg 2007, 645) found in the model, which is taken to make a difference to the target phenomenon. There is a considerable gap, though, between the usual presentation of minimal models like the three introduced in Section 2 and an interpretation of the model of the special sort that Weisberg describes. For example, Weisberg says of the Ising model: "What it seems to capture are the interactions and structures that really make a difference, or the core causal factors giving rise to the target phenomenon" (Weisberg 2007, 642–643). However, it is not clear what notion of difference-making or core causal factors we should use to interpret the Ising model so that it presents "only those factors." When compared to the typical instance of the target phenomenon, many features of the Ising model seem more like idealizations or deliberate distortions than genuine difference makers. For example, as was noted in Section 2, the Ising model only allows interactions between an element and its nearest neighbors. Should we dismiss this feature of the model as a distortion, or does it somehow reflect a difference-making factor for phase transitions quite generally? Reutlinger et al. are also unsure how to implement Weisberg's proposal for the Schelling model. How should we reinterpret the assumption that agents know the color of their neighbors so that it reflects a genuine difference maker for racial segregation quite generally (Reutlinger, Hangleiter, and Hartmann 2018, 1090)? Or, if we are to dismiss this assumption as a distortion that is not really part of the intended reinterpretation of the model, then how is this reading to be identified? Until this procedure is clarified, the widespread use of minimal models remains unmotivated.

Nguyen considers a more open-ended process of reinterpretation in "It's Not a Game: Accurate Representation with Toy Models" (Nguyen 2020). He criticizes approaches to the representational relationship between models and targets that rely on similarity. For Nguyen, a better way to think about this relationship is in terms of more flexible interpretation functions that lead a feature X of the model to stand for a distinct feature Y of the target: once these functions are applied "the model can generate true claims about a target system, and thereby accurately represent said system, despite failing to share any relevant feature with its target ..." (Nguyen 2020, 1024). One way to capture this process of reinterpretation is to suppose that the initial presentation of the model offers only a superficial or naïve interpretation. For example, in the Schelling model, we have a model system made up of differently colored coins, and in the Hamilton model, we have an infinitely large grassy plain occupied by randomly placed cows. However, users of the model reinterpret the features of the model system so that the model generates claims that are apt to be exported from the model and applied to real-world targets, such as residential patterns in Chicago or some actual school of fish. Nguyen argues that this sort of reinterpretation can lead to explanations of the actual features of real-world instances of these phenomena.

Two sorts of reinterpretation are central to Nguyen's analysis of how minimal models can afford explanations of real-world phenomena. First, a reinterpretation may take a specific, inevitable process found in the model and translate it into a claim about a less specific tendency that is present in the target. For the Schelling model, this less specific claim is that "A city whose residents have weak preferences regarding the skin colour of their neighbors has a susceptibility towards global segregation" (Nguyen 2020, 1030). So, what is more or less guaranteed to result in the model is now taken to represent only a tendency or susceptibility of the target phenomenon. The second sort of reinterpretation that Nguyen emphasizes concerns the idealizations of the minimal model. For example, it is not initially clear how to interpret the assumption that agents in the model (i.e., the coins) know the makeup of their neighbors. Nguyen argues that "As long as a model user understands the idealizations in question, then they shouldn't interpret those features in a way that entails exporting them, incorrectly, to the model's target" (Nguyen 2020, 1035). The enlightened user should appreciate that a claim like this assumption of the Schelling model involves "precisely the sorts of features that get altered by the interpretation function" (1035). Nguyen does not say much about how idealized assumptions are altered, but he may think that we will typically weaken those claims so they are true of the target. For example, in this case, we will interpret the model claim that agents in the model know everything about the makeup of their neighbors to the exportable claim that humans know a lot about this makeup. In both reinterpretations, then, the minimal model is used to generate less specific claims about processes or features that are arguably present in the target phenomenon.

Nguyen's flexible reinterpretation strategy is quite promising and similar to Pincock's account of how to deal with these sorts of highly idealized mathematical models. For example, "an idealization transforms a representation that only obscurely represents a feature of interest into one that represents that same feature with more prominence and clarity. This involves, among other things, decoupling some parts of the representation from their original interpretation" (Pincock 2012b, 97). In Pincock's discussion of Hamilton's model, this approach was generalized to allow for what he called gambit idealizations: "we sacrifice truth with respect to one feature with the aim of accurately representing some other features" (Pincock 2012a, 493). On this understanding of Hamilton's model, there are aspects of the model that are idealized and also essential to the modeling purpose. So we do not reinterpret these aspects of the model. For example, recall that Hamilton's model uses the size of the domain of danger of a cow to estimate the chance that the cow will be eaten. This is an idealization of any actual biological population, but it is required in order to evaluate the fitness of various movement rules. These fitnesses need to be well-defined so that Hamilton can argue that his preferred movement rule has evolved through ordinary processes of individual selection. The idealization may be necessary to use the model to explain how gregarious behavior evolved, even though users of the model are aware that the idealization is false. However, the model can still be used to explain if there is good reason to think that this falsity is consistent with accurately representing genuine causes of the evolution of that trait. The general suggestion, then, is to allow for even more options for reinterpretation than Nguyen explicitly considers. Some uses of minimal models will require only the sorts of weakenings that he mentions, while other uses will involve a more involved reinterpretation or selective appeal to interpreted aspects of the model.

The key point to keep in mind is that even when users of a model lack assurances that a generic feature of the model will be found in the target, they can still have good reason to think that some special features associated with the model will be found in the target. These special features may not be immediately apparent and so investigations of the model may motivate novel or creative reinterpretations of its representational content. If we consider the Schelling model or Hamilton's model in this light, then there is no puzzle about how these minimal models can be used to generate accurate descriptions, predictions, or even explanations of real-world phenomena.

5. Minimal model explanations

In their paper, "Minimal Model Explanations," Batterman and Rice argue that minimal models may be used in a special sort of explanation. This involves "a fundamentally different kind of story about how these minimal models 'latch onto the world' ..." (Batterman and Rice 2014, 350). To illustrate their proposal, I will consider their account of how the Ising model may function in an explanation of the universality of critical phenomena (Batterman 2019). Recall from Section 2 that the Ising model exhibits how the magnetic field of a system vanishes at some critical temperature T_c. Other systems exhibit a second sort of phase transition tied to the vanishing of the difference between the densities of the liquid and gas present at some critical temperature T_c and pressure P_c. Somewhat remarkably, it turns out that the order parameters involved in both transitions change in the same way as the temperature is raised to T_c: in both cases, the order parameter is proportional to ε^β. While the interpretation of ε varies from the magnetic case to the liquid/gas case, β has the same value: for two-dimensional systems, β is 1/8, while for three-dimensional systems, β is approximately 0.324. This striking correspondence was discovered decades before it was explained in the 1970s.

The apparently unified character of phase transitions, despite their many physical differences, motivates what Kadanoff has called a "hypothesis of universality": "All phase transition problems can be divided up into a small number of different classes depending upon the dimensionality of the system and the symmetries of the order state" (given at Batterman 2019, 33). For Batterman and Rice, this is the sort of target that requires a minimal model explanation: why are these critical phenomena divided up into a small number of "universality" classes?

To reconstruct Batterman and Rice's argument, suppose that an explanation of the universality of some phenomenon has a special sort of target that forces a special sort of explanation. For a phenomenon to be universal, that phenomenon must arise in the same way across a wide variety of systems *despite* their differences. So to explain the universality of some phenomenon, one must indicate how the differences between these systems fail to matter for the outcome, and also indicate what common aspects of these systems do matter for the outcome. In our case, the universality of the phenomenon in question partly consists in the fact that two-dimensional systems of this kind have a critical exponent of 1/8, while three-dimensional systems of this kind have a critical exponent of 0.324.

To explain the identity of these critical exponents despite the differences between these systems, Batterman and Rice first invoke "a space of possible systems." The next step in the explanation is to group these systems together based on how a special sort of transformation maps one system to another. This sort of transformation is identified through "renormalization group" methods. If the transformation is appropriately chosen, it "in effect eliminates details or degrees of freedom that are irrelevant." Systems S1 and S2 can then be grouped into a universality class when this transformation takes both S1 and S2 to the same fixed point S*, i.e., applying the transformation to S* yields S*. According to Batterman and Rice, "A derivative, or *by-product*, of this analysis is the identification of the shared features of the class of systems" (Batterman and Rice 2014, 362–363). In our case, the transformation takes systems with very different characters to the same fixed point. The transformation is chosen so that the critical exponent is the same for all systems in a given universality class. But the only other common features of significance of the systems in a class are the dimensionality of the system and the structure of the order parameter. All remaining differences between the systems have thereby been shown to be irrelevant to the value of the critical exponent. The explanation should thus be clear: these systems share a critical exponent because they have the same dimension and the salient order parameters have a common structure, and not because of any additional features that differentiate those systems, e.g., their microphysical constitution.

Notice that minimal model explanations involve three different elements: (i) a minimal model like the Ising model, (ii) models for ordinary real-world systems, and (iii) a transformation that appropriately connects them all together. The combined role of all these models and the choice of transformation highlight the limitations of both the modal and reinterpretation strategies. Neither approach can make sense of the role of minimal models in explanations of universality. As we have seen, both the modal and reinterpretation strategies deal with the inaccuracies of the minimal models by limiting the features of the model that a scientist should use to characterize the target. Modal approaches supposed that these special features would be modal in character, while reinterpretation approaches allowed for more flexible shifts in how the model depicts the target. However, not even the most liberal reinterpretation can transform the Ising model into an explanation of the universality of critical phenomena. What is needed instead is a way of relating or connecting the Ising model to other models in a way that fits this special sort of explanatory target.

Batterman and Rice emphasize this point when they insist that "What makes such models explanatory has nothing to do with representational accuracy to any degree" (Batterman and Rice 2014, 356). As they elaborate in a footnote, "in the case of minimal models the features that correspond are inadequate to explain why so many diverse systems, including the model system, will display the same macroscale behavior" (Batterman and Rice 2014, 356, fn. 7). This rejection of a central role for accuracy in these model-based explanations has prompted a number of objections (Povich 2018; Franklin 2018; Sullivan 2019; Rodriguez 2021). One concern emphasized by Lange is that if we give up focus on the common features between some model and its target, then we will lose the asymmetry that is central to the contrast between genuine explanation and mere description. As Lange puts it, "The target system and the minimal model are simply two systems in the universality class. Why does the behavior of one of these systems help to explain the behavior of the other?" (Lange 2015, 295). Here Lange is thinking of a target system that exhibits a phase transition with the very same critical exponent as we find in the Ising model. This is arguably a misunderstanding of the explanatory target that Batterman and Rice emphasize: as was noted above, they aim to explain the universality of the phenomenon, which is that many systems exhibit this feature despite their differences. In a reply to Lange, McKenna makes the same point: "the explanatory target of minimal model explanations is in the first place the ubiquity of the macrobehavior" (McKenna 2021, 737).

Another concern raised by Lange is that Batterman and Rice's explanation relies on some common features between the minimal model and the other systems that exhibit the phase transition. As we have seen, at the core of the explanation is the way that a transformation unites the Ising model with the other systems that exhibit that phase transition. One option that Lange considers is that "the given fluid's macroscale behavior is explained by its possessing the property of being such that it is brought to a certain fixed point in the state space (the same point for every member of the universality class) when it repeatedly undergoes a certain transformation ... Since this property is common to all members of the universality class, it constitutes a 'common feature' of the kind that B&R deny explains the system's macrobehavior" (Lange 2015, 299–300). That is, some real-world systems and the Ising model both have the same critical exponent because they get mapped to the same fixed point by this transformation. So, there is a common feature present after all.

It seems that Batterman and Rice should concede this point, but argue that their claims about common features and accuracy were more limited in scope: the explanation does not consist of simply pointing to this common feature between the Ising model and these real-world systems. Instead, the core of the explanation is the way the transformation works to connect the Ising model to these real-world systems. There is a common feature, but, as McKenna says, "this common feature does not furnish us with the accuracy conditions that are required for the model to explain" (2021, 740). The Ising model does not explain because it accurately represents real-world systems to be in the same universality class as the Ising model. Instead, the Ising model may figure into the explanation, as a proper part, because it is in the same universality class as these real-world systems.

A residual worry could be raised on Lange's behalf, though: is a minimal model like the Ising model essential to a minimal model explanation of the universality of some phenomenon? Batterman and Rice's label of "minimal model explanation" certainly suggests that some minimal model is essential for these explanations to work. There are two different explanatory targets that are easy to confuse. The first target is the division of systems exhibiting critical phenomena into a small number of classes. The second target is the very same division with the additional stipulation that the two-dimensional Ising model is in one class and the three-dimensional Ising model is in another class. The first target thus makes no mention of the Ising model, and for this reason, there is no need to mention the Ising model or any other minimal model in the explanation. All one needs to do is show how the real-world systems that exhibit this phenomenon are mapped to distinct fixed points, and how this transformation accounts for the critical exponents that are shared. By contrast, a scientist who considers the second target has already included the features of the Ising models in their explanatory target. Thus, for this target, it is essential that one mention the Ising models and illustrate how they are affected by the transformation in question. Historically, it seems clear that the Ising models played a central role in the investigation of critical phenomena, and so it is plausible to suppose that most scientists were interested in this second target. But for others who cared only about the first target, there is an explanation of the universality of real-world critical phenomena that does not rely essentially on a minimal model.

In more recent work, Batterman and Rice have emphasized how a minimal model can contribute to scientific goals like prediction and explanation by being appropriately linked to a larger ensemble of models. In *A Middle Way: A Non-Fundamental Approach to Many-Body Physics,* Batterman (2021) emphasizes the importance of a result in statistical mechanics known as the fluctuation–dissipation theorem. This theorem considers a many-body system such as a gas or fluid. If such a system starts in an equilibrium state, it may transition to a non-equilibrium state through either a spontaneous internal fluctuation or a small external disturbance, followed by a transition back to an equilibrium state. The theorem claims that, in Batterman's words, "That evolution … is the same regardless of the origin of the non-equilibrium" (Batterman 2021, 21). In addition, Batterman argues, minimal models like the Ising model prove to be the right models to use to appreciate the mesoscale structures of these systems that mediate between the various microscale differences between such systems and the macroscale commonalities that they exhibit. The very same mesoscale structures that govern the processes of returning to equilibrium are prominent in minimal models. As a result, minimal models are "so apt … because they do not model 'fundamental' properties of systems, but they do model the *natural* properties of many-body systems" (Batterman 2021, 131). These non-fundamental, natural properties are best modeled by minimal models. Of course, Batterman is clear that the features of the minimal model must be carefully chosen if they are to allow for the identification of these natural properties. A key result of the book is that these minimal models arise in a more general setting than cases where renormalization group methods are available.

Another sort of generalization is developed by Rice by considering various ways that universality classes can be identified through minimal models. Rice considers several instances of complex phenomena where a minimal model is used to explain (Rice 2022).

Each explanation involves two steps. First, "show how the observed macroscale pattern (the explanandum) depends on (changes to) the features that characterize/distinguish the universality class," such as the dimensions of these systems. Second, "demonstrate that the remaining heterogeneous features of the systems within the universality class (e.g., the features ignored or idealized by the minimal model) are irrelevant to displaying the universal patterns of behavior" (Rice 2022, 28). This combination of information about what is relevant and irrelevant is often achieved through the use of a minimal model. However, the explanation consists in relating this minimal model to other models in the right way. As any number of modeling techniques can furnish these relations, the scope of these explanations is much wider than it might initially seem to be.

6. Conclusion

This survey of debates about minimal models has focused on the simplicity of minimal models and the barriers that this places on the use of minimal models for description, prediction, and explanation. While it is clear that this simplicity stands in the way of any indiscriminate extension of the features of the model to the model's target, a number of more sophisticated uses are defensible. First, one could focus on the modal properties found in the model and consider the appropriate ways to apply these modal properties to the target. Second, one could allow for various reinterpretations of the model so that some non-modal properties could be ascribed to the target. Third, one could embed the minimal model in a larger class of systems through various mathematical transformations. This last embedding seems to permit a special sort of explanation where the minimal model plays a central role. One point to emphasize in conclusion is that a combination of strategies may be needed to clarify the scientific value of models as different as the Ising model, Schelling model, and Hamilton's model. For this reason, future work on minimal models can be expected to develop all of these approaches further as part of a broader attempt to make sense of the central role of minimal models in many scientific investigations.

References

Aydinonat, N. Emrah, and Emin Köksal. 2019. "Explanatory Value in Context: The Curious Case of Hotelling's Location Model." *The European Journal of the History of Economic Thought* 26(5): 879–910.

Batterman, Robert. 2019. "Universality and RG Explanations." *Perspectives on Science* 27: 26–47.

———. 2021. *A Middle Way: A Non-Fundamental Approach to Many-Body Physics*. New York: Oxford University Press.

Batterman, Robert, and Collin Rice. 2014. "Minimal Model Explanations." *Philosophy of Science* 81: 349–376.

Franklin, Alexander. 2018. "On the Renormalization Group Explanation of Universality." *Philosophy of Science* 85: 225–248.

Fumagalli, Roberto. 2016. "Why We Cannot Learn from Minimal Models." *Erkenntnis* 81(3): 433–455.

Gelfert, Axel. 2019. "Probing Possibilities: Toy Models, Minimal Models, and Exploratory Models." In *Model-Based Reasoning in Science and Technology*, edited by Angel Nepomeceno-Fernandez and Lorenzo Magnani, 3–19. Cham: Springer.

Gould, Harvey, and Jan Tobochnik. 2010. *Statistical and Thermal Physics*. Princeton: Princeton University Press.

Grüne-Yanoff, Till. 2009. "Learning from Minimal Economic Models." *Erkenntnis* 70(1): 81–99. https://doi.org/10.1007/s10670-008-9138-6.

Hamilton, William D. 1971. "Geometry for the Selfish Herd." *Journal of Theoretical Biology* 31: 295–311.
Knuuttila, Tarja. 2021. "Epistemic Artifacts and the Modal Dimension of Modeling." *European Journal for Philosophy of Science* 11(3): 65.
Knuuttila, Tarja, and Andrea Loettgers. 2017. "Modelling as Indirect Representation? The Lotka-Volterra Model Revisited." *The British Journal for the Philosophy of Science* 68: 1007–1036.
Lange, Marc. 2015. "On 'Minimal Model Explanations': A Reply to Batterman and Rice." *Philosophy of Science* 82: 292–305.
McKenna, Travis. 2021. "Lange on Minimal Model Explanations: A Defense of Batterman and Rice." *Philosophy of Science* 88(4): 731–741. https://doi.org/10.1086/713890.
Nguyen, James. 2020. "It's Not a Game: Accurate Representation with Toy Models." *The British Journal for the Philosophy of Science* 71: 1013–1041.
Pincock, Christopher. 2012a. "Mathematical Models of Biological Patterns: Lessons from Hamilton's Selfish Herd." *Biology & Philosophy* 27(4): 481–496.
———. 2012b. *Mathematics and Scientific Representation*. New York: Oxford University Press.
Povich, Mark. 2018. "Minimal Models and the Generalized Ontic Conception of Scientific Explanation." *The British Journal for the Philosophy of Science* 69(1): 117–137.
Reutlinger, Alexander, Dominik Hangleiter, and Stephan Hartmann. 2018. "Understanding (with) Toy Models." *The British Journal for the Philosophy of Science* 69(4): 1069–1099. https://doi.org/10.1093/bjps/axx005.
Rice, Collin. 2022. "Modeling Multiscale Patterns: Active Matter, Minimal Models, and Explanatory Autonomy." *Synthese* 200(6): 432. https://doi.org/10.1007/s11229-022-03885-7.
Rodriguez, Quentin. 2021. "Idealizations and Analogies: Explaining Critical Phenomena." *Studies in History and Philosophy of Science Part A* 89: 235–247.
Rohwer, Yasha, and Collin Rice. 2013. "Hypothetical Pattern Idealization and Explanatory Models." *Philosophy of Science* 80(3): 334–355.
Schelling, Thomas C. 1978. *Micromotives and Macrobehavior*. W. W. Norton & Co.
Sjölin Wirling, Ylwa. 2021. "Is Credibility a Guide to Possibility? A Challenge for Toy Models in Science." *Analysis* 81(3): 470–478.
Sullivan, Emily. 2019. "Universality Caused: The Case of Renormalization Group Explanation." *European Journal for the Philosophy of Science* 9: 36.
Verreault-Julien, Philippe. 2019. "How Could Models Possibly Provide How-Possibly Explanations?" *Studies in History and Philosophy of Science Part A* 73(February): 22–33. https://doi.org/10.1016/j.shpsa.2018.06.008.
Weisberg, Michael. 2007. "Three Kinds of Idealization." *Journal of Philosophy* 104: 639–659.
———. 2013. *Simulation and Similarity: Using Models to Understand the World*. New York: Oxford University Press.

11
COMPUTER SIMULATIONS

Juan M. Durán

1. Introduction

Computer simulations are found in a myriad of scientific fields and practices. In some cases, they constitute whole lines of research (e.g., climate modeling and molecular simulations in chemistry (Goldman 2014). The debate over their philosophical merits involves a wide range of topics, including, but not restricted to, their function as experiments (e.g., Beisbart 2017; Boge 2019; El Skaf and Imbert 2013); their value as sources of scientific evidence (e.g., Morgan 2004; Parker 2020); their role as measuring devices (e.g., Morrison 2009; Tal 2011); their place in the scientific methodological map (e.g., Rohrlich 1990); and their scientific and philosophical novelty (e.g., Humphreys 2009; Frigg and Reiss 2009).

A key issue common to many of these debates is how philosophers have conceived—and even defined—computer simulations and the models they implement. This chapter presents and discusses three chief views found in the literature. The first one takes computer simulations to implement mathematical models *simpliciter*. A second one takes computer simulations to be a richer and more complex unit of analysis than mathematical models, yet still related to mathematics. A third viewpoint is sketched, where computer simulations depart even further from implementing mathematical models, gaining the status of modeling in its own right. To simplify the analysis, the focus will primarily be on equation-based simulations and their application to medicine and the natural sciences. Since significant philosophical issues also emerge in relation to diverse fields such as biology, sociology, and psychology, and in relation to a variety of other kinds of computer simulations such as cellular automata, agent-based simulations, and Monte Carlo simulations, let us first look briefly at these. The chapter ends with a discussion on *epistemic opacity*, arguably a chief philosophical issue pertaining to all computer simulations.

2. Kinds of computer simulations

Cellular automata are the first of our examples of computer simulations. They were devised in the 1940s by Stanislaw Ulam and John von Neumann while Ulam was studying the growth of crystals using a simple lattice network as a model and von Neumann was

 DOI: 10.4324/9781003205647-14

working on the problem of self-replicating systems. It is said that Ulam suggested to von Neumann that the latter use the same kind of lattice network to create a two-dimensional, self-replicator algorithm.

Cellular automata are simple forms of computer simulations. Their simplicity inheres in both their programming and underlying conceptualization. A standard cellular automaton is an abstract mathematical system in which space and time are considered to be discrete; it consists of a regular grid of cells, each of which can be in any state at a given time. Typically, all the cells are governed by the same rule, which describes how the state of a cell at a given time is determined by the states of itself and its neighbors at the preceding moment. Wolfram defines cellular automata as:

> [...] mathematical models for complex natural systems containing large numbers of simple identical components with local interactions. They consist of a lattice of sites, each with a finite set of possible values. The value of the sites evolves synchronously in discrete time steps according to identical rules. The value of a particular site is determined by the previous values of a neighborhood of sites around it.
>
> *(Wolfram 1984, 1)*

Although a rather general characterization of this class of simulation, the definition already provides the first ideas as to their domain of applicability. Cellular automata have been successfully used for modeling many areas in social dynamics (e.g., Thomas Schelling's social segregation model), biology (e.g., patterns of some seashells), and chemical types (e.g., the Belousov–Zhabotinsky reaction). But perhaps the most canonical example is Conway's *Game of Life*. This simulation is remarkable because it constitutes a key example of self-organization dynamics and the emergence of patterns seen in some real-world systems. In this simulation, a cell can survive only if there are either two or three other living cells in its immediate neighborhood. Without these companions, the rule indicates that the cell dies either from overcrowding if it has too many living neighbors or from loneliness if it has too few.

Cellular automata embody a unique set of methodological and epistemological virtues. To name a few, they deal better with errors because they render exact results of the model they implement. Since there is rarely any attempt to approximate the detailed setup of the target system, any disagreement between the model and the empirical data can be ascribed directly to the model that realized the set of rules. Another epistemologically interesting characteristic of cellular automata pointed out by Fox-Keller is that they lack theoretical underpinning in the familiar sense of the term: "what is to be simulated is neither a well-established set of differential equations [...] nor the fundamental physical constituents (or particles) of the system [...] but rather the phenomenon itself" (Fox-Keller 2003, 208). Consequently, approximations, idealizations, abstractions, and the like are concepts that worry the practitioner of cellular automata very little.

Having said that, cellular automata have been criticized on several grounds. One of these criticisms touches on the metaphysical assumptions behind this class of simulation. It is not clear, for instance, that the natural world is characterized by discrete rather than continuous phenomena, as assumed by the cellular automata. Much contemporary work in science and engineering work assumes that phenomena are, in fact, continuous. On less speculative grounds, it is a fact that cellular automata lack presence in many scientific and engineering fields. The reasons for this might be partially cultural. The physical sciences are still the accepted viewpoint for describing the natural world, which largely takes form in the language of partial differential equations (PDEs) and ordinary differential equations (ODEs).

Advocates of cellular automata have made efforts to demonstrate their relevance. It has been argued that cellular automata are more adaptable and structurally similar to empirical phenomena than are PDEs or ODEs. Lesne (2007) points out that discrete and continuous behaviors coexist in many natural phenomena (with their proportions depending on the scale of observation) and suggests that this is an indicator not only of the metaphysical basis of natural phenomena, but also of the need to deploy cellular automata to understand them. In a similar vein, Gérard Vichniac believes that cellular automata not only seek numerical agreement with a physical system, but also attempt to match the simulated system's own structure, its topology, its symmetries, and its "deep" properties (Vichniac 1984, 113). Despite these and many other authors' efforts to show that the world might be more adequately described by cellular automata, the majority of scientific and engineering disciplines have not made a significant shift in that direction as of yet. Most of the work done in these disciplines is predominantly based on agent-based and equation-based simulations. As mentioned before, in the natural sciences and engineering, most physical and chemical theories used in astrophysics, geology, climate change, and the like implement PDEs and ODEs, the primary forms of equation-based simulations. Social and economic systems, on the other hand, are better described and understood by means of agent-based simulations.

While there is no general agreement on what precisely an "agent" is, the term typically refers to self-contained programs that control their own actions based on perceptions of their overall operating environment: agent-based simulations "intelligently" interact with their peers as well as their environment.

A key characteristic of these simulations is that they can show how the total behavior of a system emerges from the collective interaction of their parts. Deconstructing these simulations into their constituent elements would remove the added value provided in the first place by the computation of the agents. It is a fundamental characteristic of these simulations, then, that the interplay of the various agents and their environment generates unique behavior in the entire system.

Good examples of agent-based simulations come from the social and behavioral sciences, where they are heavily represented. Perhaps the most well-known example of an agent-based simulation is Schelling's Model of Social Segregation.[1] A very simple description of Schelling's model consists of two groups of agents living in a 2-D,[2] n by m matrix "checkerboard" where agents are placed randomly. Each individual agent has a 3 by 3 neighborhood, which is evaluated by a utility function that indicates the migration criteria. That is, the set of rules that indicates how to relocate—if possible—in case of discontent by an agent.

Schelling's model is a canonical example, but other, more complex agent-based simulations can also be found in the literature. It is now standard for researchers to model a range of different attributes, preferences, and overall behavior in agents. Gilbert and Troitzsch list the attributes that are typically modeled by agent-based simulations, including knowledge and beliefs of the agents, inferences from beliefs, goals, overall planning, and language (Gilbert and Troitzsch 2005).[3]

Monte Carlo methods are the second of our examples of computer simulations. Their basic operation is to use stochastic techniques to compute the properties of a model. A key feature of these methods is that they use random sampling for target systems that could in principle be deterministic. Monte Carlo is a very powerful technique that is typically applied to systems with many coupled degrees of freedom, such as fluids, gases, crystallizable polymers, and strongly coupled solids, among others. Within the philosophical literature, there has been some debate over its status as a method for discovery and experimentation. Grüne-Yanoff and Weirich, for instance, indicate that "the Monte Carlo approach does not have a mimetic purpose: It imitates the deterministic system not in order to serve as a surrogate that is investigated in its stead but only in order to offer an alternative computation of the deterministic system's properties. In other words, the probabilistic analogy does not serve as a representation of the deterministic system" (Grüne-Yanoff and Weirich 2010, 30). To these authors, then, Monte Carlo experiments are merely methods of calculation and not simulations in a proper sense, for the latter are "used to learn something about the world, and they are used as stand-ins or surrogates for whatever is of interest for the simulationist" (Grüne-Yanoff and Weirich 2010, 30). Beisbart and Norton seem to agree with this idea when they claim that "Monte Carlo simulations are like experiments that discover novel results. We will argue, however, that these sorts of similarities are superficial. They do not and cannot make them function like real experiments epistemically" (Beisbart and Norton 2012, 404).

In what follows, the focus is on the use of computers to find solutions to a set of equations. Equation-based simulations are most commonly used in scientific domains in which the governing theories and models are based on differential equations.

3. Equation-based computer simulations

Suppose we are interested in a simulation of a satellite orbiting around a planet under tidal stress such that it stretches along the direction of the radius vector. Suppose further that this model represents the orbit as non-circular with variable stress, making the satellite expand and contract periodically along the radius vector. Since the satellite is not perfectly elastic, the mechanical energy is converted into heat and radiated away. Despite this, the system as a whole is capable of conserving angular momentum (see, for details, Woolfson and Pert 1999, 18–19). In this context, we have equations of total energy (e.g., Eq. (1) below), angular momentum, and others. We also have other relevant components of the system and their interactions represented in the model. The planet has mass M; the satellite mass m ($<< M$); the orbit is of semi-major axis a; and the gravitational constant is represented by G; and so forth. The masses are represented by connected springs, each of unstressed length l, and the same spring constant, k. Thus, a spring constantly stretched to a length l' will exert an inward force (e.g., Eq. (2)—see also Woolfson and Pert 1999, 19, fig. 1.8).

$$E = -\frac{GMm}{2a} \tag{1}$$

$$F = k(i' + l) \tag{2}$$

For simplicity, the above set of equations will be referred to as a *mathematical model*[4] that describes the behavior of and interaction between any planet and any satellite under the specified conditions. Now, to have a simulation, this mathematical model needs to be implemented in the form of an *algorithmic structure*. That is, the sets of variables, procedures, data, functions, and other structures that are tractable in a digital computer (e.g., algorithms (3) and (4) partially implementing the mathematical equations). Let us call this algorithmic structure a *simulation model*.

```
TOTM = CM(1) + CM(2) + CH(3) + CM(4);

EN = -G * TOTM + 0.5 * V2                                    (3)

R = SQRT(POS(1)**2 + POS(2)**2 + POS(3)**2)                  (4)
```

The above algorithms suggest that mathematical equations can be implemented as a simulation model rather straightforwardly. These algorithms effectively do so. Algorithm (3) partially implements equation (1) *simpliciter*, and algorithm (4) does something similar with equation (2). Naturally, the simulation model will require some *discretizations* for tractability reasons (i.e., continuous equations cannot be implemented on physical computers), aggregation of procedures for the treatment of *errors*, and a handful of *ad hoc modifications* for smooth numerical integration (e.g., computers cannot represent infinite orbiting).

A critical issue that divides philosophers is how to interpret the simulation model that is at the basis of computer simulations, as well as the computer simulations themselves. To some, computer simulations are numerical methods for finding sets of solutions to mathematical models. To some others, computer simulations are more than numerical methods destined to have merely instrumental value. Instead, they are part of—or stand for—a novel and more comprehensive form of scientific methodology. Thus understood, simulation models are conceived as a new type of model, related to but not entirely obtained from mathematical models and modeling. Key observations favoring this latter view are that any given simulation model will, in fact, involve several layers of models, each potentially requiring differing modeling practices; it will represent structures that are not necessarily present in mathematical models nor secured by mathematical modeling; and it will not necessarily derive from a chain of inferences and varying adjustments and aggregations that started with one or more mathematical models. This second view revolves around the idea that a proper methodology of simulations requires a distinctive ontology leading to specific epistemic and methodological issues.

The remainder of this chapter discusses some of these interpretations and their resulting characterization of simulation models and computer simulations.

3.1 Simulations for analytically intractable mathematics

Let us start with an often-quoted working definition of computer simulation:

> A computer simulation is any computer-implemented method for exploring the properties of mathematical models where analytic methods are unavailable.
>
> *(Humphreys 1990, 501)*

According to this working definition, computer simulations are instrumental in finding the set of solutions to an analytically intractable mathematical model. Understood as numerical methods, they explore the mathematical properties of the simulation models. Hartmann presents a similar definition. According to him, (a) a simulation is the result of solving the equations of a dynamic model, and (b) a computer simulation is the result of having a simulation run on a physical computer. Taken together, (a) and (b) entail that a computer simulation results when a dynamic mathematical model is solved by a physical computer (Hartmann 1996). Let us note that Hartmann is also claiming that the physical dimension of the computer plays a relevant role in imitating the dynamics of a real-world system. Interestingly, some philosophers have developed this idea (e.g., Parker 2009, and Boge 2020), arguing for meaningful morphisms between the (physical) computer processes and the target system.[5] Others, opposing this claim (e.g., Beisbart 2014; Durán 2018), argue that the multirealizability of physical processes means that the resulting analogy is thin and contrived.

These definitions come with varying methodological and epistemological assumptions. For starters, the adjustments required for implementing the mathematical model onto the computer must be minimal. That is, the discretizations and ad hoc modeling must go only as far as is required for the tractability of the mathematical model. By themselves, simulations do not possess—nor should they possess—any representational value other than that inherited from the mathematical models they deploy. No aggregates to the simulation model could suggest a deviation from the implemented mathematical models.

Humphreys' and Hartmann's definitions loom large in the philosophical and technical literature. Parker, for instance, adopts Hartmann's definition in her analysis of the experimental value of simulations. In her 2009 paper, she makes explicit reference to it by characterizing a computer simulation as a time-ordered sequence of states that represents another time-ordered sequence of states. In her latest publication, however, she seems to have distanced herself from this commitment. She states that "a *computer simulation model* is a computer program that is designed to iteratively solve a set of dynamical modeling equations, either exactly or approximately, following a particular algorithm" (Parker 2020, sec. 2). Moreover, Parker also calls attention to the plurality of models in simulation practice and their role in computer simulations in climate models (see the next section). It would require some argumentative acrobatics to make a convincing case that climate simulations hold nontrivial morphisms at the physical level.

Guala has also made explicit reference to Hartmann's definition in discussing the time evolution of systems, the use of simulations to provide numerical solutions to sets of mathematical equations, and in distinguishing between static and dynamic models (Guala 2002). Krohs (2008) adopts Humphreys' and Hartmann's definitions to account for the role and merits of computer simulations in scientific explanation (Durán 2017). Frigg and Reiss largely base their disapproval of the philosophical novelty of computer simulations on a *narrow* sense of simulations, assuming that they are, ultimately, about mathematical models (Frigg and Reiss 2009, 596).

Recently, Boge has claimed that a simulation model "will usually (if not always) be based on some previously existing numerical, i.e., discrete mathematical model of a system of interest (the 'target system'), which in many cases is an approximation to another model based on continuous mathematics, and hence not suited for a translation into algorithms" (Boge 2019, 3). Boge goes on to discuss simulations in terms of mathematical language and derivations, as well as the physical characteristics of the target system mimicked by, and emerging from, the execution of such simulations.

3.2 Simulations as a "new type" of mathematical model

The alternative viewpoint takes that simulation models are related to, but not entirely obtained from, mathematical models and modeling. Weisberg, in his analysis of the anatomy of models, considers simulation models as "a subset of mathematical models" (Weisberg 2013, 30) but holds that they constitute an especially important subset. Morrison has also urged that more philosophical attention must be given to computer simulations, in light of their being a special kind of experimental practice related to modeling (Morrison 2015).[6] In his recent book, Lenhard explicitly refers to simulations as a "*new* type" of mathematical model. There are two sides to this interpretation. Whereas simulation models must be "counted into the established classical and modern class of mathematical modeling," one must also take stock on how they "contribute to a novel explorative and iterative mode of modeling characterized by the ways in which simulation models are constructed and fitted" (Lenhard 2019, 7). Lenhard cements this view by saying: "[o]ne direction seems self-evident: the (further) development of computers is based primarily on mathematical models. However, the other direction is at least just as important: the computer as an instrument channels mathematical modeling" (Lenhard 2019, 8). Simulations are a "new type" of model primarily because of the *plasticity* of their modeling, which "draws on the effects that arise from the ways in which the (artificial) parameters are set. The more flexible a model is, the more significant is the phase of modeling during which the parameters are adjusted."[7] (Lenhard 2019, 11).

What does the methodology of simulations as a "new type" of mathematical model look like? Winsberg provides an answer to this question. This author advances a hierarchy of models that begins, at the top, with a given theory (i.e., general physical and modeling assumptions) and terminates, after a series of specifications, alterations, and inferences at each level of modeling with a model of the phenomena, which represents the outcome of the simulation research in question (Winsberg 1999, 277). In Winsberg's view, this inferential hierarchy suggests a distinct epistemology—and, it could be added, a distinct methodology—for simulations whose chief features are being *downwards*, *autonomous*,[8] and *motley* (Winsberg 2001, S447). It follows that "simulations often do not bear a simple, straightforward relation to the theories from which they stem" (Winsberg 1999, 276).

Humphreys also offers an elaborated, multi-level methodology and epistemology for simulation models. He presents it in the following way: "System S provides a core simulation of an object or process B just in case S is a concrete computational device that produces, via a temporal process, solutions to a *computational model* [...] that correctly represents B, either dynamically or statically. If in addition the computational model used by S correctly represents the structure of the real system R, then S provides a core simulation of system R with respect to B" (Humphreys 2004, 110, emphasis added). The computational model comprises six different elements, each performing a specific function. These are the computational template, the construction assumptions of that model, the correction set, an interpretation, an initial justification, and the output representation (see Humphreys 2004, 102). The first element of this sextuple, i.e., the computational template, is the heart of the computational model and can essentially be understood as a set of computationally tractable equations (61).

Taking stock of these interpretations, simulation models are still obtained from mathematical models in varying degrees and fashions. With Winsberg, this comes through the hierarchical-inferential process that ultimately results in a model of the phenomena. For Humphreys, the unit of analysis for computational science is the *computational template*. Following his example, a simulation utilizing Newton's Second Law consists of a theoretical template that "describes a very general constraint on the relationship between any force, mass, and acceleration, but to use it in any given case, we need to specify a particular force function, such as a gravitational force, an electrostatic force, a magnetic force, or some other variety of force" (Humphreys 2004, 60). A computational template emerges when "the resulting, more specific, equation form is computationally tractable" (60). Finally, Lenhard intends to balance the transformations of mathematical models introduced by the computer with the role of simulations as instruments that channel mathematical modeling.

One must then ask, to what extent are these interpretations aligned or misaligned with the notion of simulations as a way of approaching analytically intractable mathematics? While there is some evident overlap, there are also a handful of reasons to separate these two notions. For starters, simulation models are conceived as a richer structure than mathematical models by philosophers arguing for the novelty of simulation modeling (e.g., they use external databases, involve multiple layers of models). This also means that the goal of simulations has substantially shifted from finding solutions to a set of equations representing a complex target system. Finally, scientific research involving computer simulations does not necessarily reflect the same epistemic and methodological principles, social organization, and research questions as those involving mathematical models.

Climate simulations have made visible the rich and complex structure of simulations, primarily through the implementation of a plurality of models. In fact, many philosophers agree that model pluralism is an inherent and inevitable feature of simulation models. As Lenhard and Winsberg (2010, 261) put it, "pluralism is not a temporary failure that eventually will be overcome, but will remain for principled reasons of simulation modeling methodology." Parker has argued that "complex climate models generally are physically incompatible with one another—they represent the physical processes acting in the climate system in mutually incompatible ways and produce different simulations of climate" (Parker 2006, 350). Durán (2020) has reflected on the plurality of models in regard to the architecture of simulation models. There, simulation models *recast* a host of models pertaining to different kinds of representational values, methodological principles, and epistemic goals. The resulting architecture includes *kernel simulations*, understood as the implementation of each individual model in the formalism of a programming language, and *integration modules*—modules "which play two fundamental roles, namely, they integrate external databases, protocols, libraries and the like with [each kernel simulation], and ensure the synchronization and compatibility among [the kernel simulations]" (Durán 2020, 307). Computer simulations are therefore conceived as non-hierarchical, non-inferential, and non-homogeneous units of analysis.

3.3 Can simulations be autonomous from mathematical models?

The view that simulations are a "new type" of mathematical models tends to obscure the tension between acknowledging that simulation models both provide an unprecedented form of modeling and a forceful attempt to stay rooted in mathematical modeling. For instance, Winsberg introduced the idea of ad hoc modeling, understood as "relatively simple mathematical relationships designed to approximately capture some physical effect in nature. When 'coupled' to the more theoretical equations of a simulation, they allow the simulation to produce outputs that are more realistic than they could have been without some consideration of that physical effect" (Winsberg 1999, 282). Another distinctive methodological practice in simulation is "kludging," roughly understood as adding bits of code to simulation that are not principled in their design and whose purpose is to optimize the performance and improve the simulation in a "quick and dirty" way (Lenhard 2019).

But kludging is not the only distinctive methodological trick implemented in simulations. Fuzzy modularity (i.e., the piecemeal adjustment of models for their use in multiple simulations) and generative entrenchment (i.e., the multiple sources on which the model depends because they played a role in generating it) also cement claims about confirmatory holism and explain the failure of analytic understanding in climate models, for instance (Lenhard and Winsberg 2010, 256–257). Yet another interesting example is the so-called Arakawa operator, also discussed by Lenhard and Winsberg, which can be used to overcome the nonlinear instability of the mathematics in meteorological models. In this respect, Lenhard says: "[i]n my opinion, this was a decisive point: the discreteness of the model required artificial and also nonrepresentative elements in the simulation model whose dynamic effects could be determined only in a (computer) experiment" (Lenhard 2019, 36). Finally, parametrizations further engross the list as they are "pragmatic decisions that balance fidelity to what we know about the target system with the need for effective implementation" (Lenhard and Winsberg 2010, 256).

What does this alleged distinctive form of modeling mean for the representational merits of simulations? In principle, not much. Ad hoc modeling takes it that "more" modeling is added to the simulation for reasons of tractability, but there is no claim of added representational value. Kludging, fuzzy modularity, Arakawa-like operators, and parametrization are genuine simulation-inspired practices, but they are also "nonrepresentative" of the target system (Lenhard 2019, 36). Again, they are solely dedicated to making the simulation model tractable.

Interestingly, it is increasingly the case that mathematical and logical formalism is omitted in favor of readymade algorithmic structures. Researchers prefer to dispense with the trouble of first developing a mathematical model and then figuring out how to implement it as (part of) a simulation model by representing target systems directly into their codes. For instance, DeAngelis and Grimm (2014) and Peck (2012) show how a (total or partial) representation by the simulation model might take place directly at the level of algorithmic structures and without the mediation of any formal mathematical modeling. The representation is built from hypothesized relational structures abstracted from the target system and directly coded as the simulation model.

One could object at this point that readymade algorithmic structures are conducive to other forms of modeling. That the practice of dispensing with the writing of mathematical equations before coding the algorithm does not necessarily imply that there is no mathematical model underpinning the algorithm.[9] But the critical point here is that, on occasion, researchers encode forms of behavior of the target system that do not correlate with mathematical modeling. To put this idea somewhat differently: if we want to recreate the algorithm as a mathematical model, we would face the problem that specific structures and patterns of behavior relevant to the representation of the target system and encoded in the algorithm do not correspond to mathematical machinery. Durán (2020) explores this idea, arguing that programming languages allow researchers to encode into their simulation-specific structures and patterns of behavior of the target system. The key intuition here is that a given simulation might represent two non-trivially different target systems depending on the chosen programming language, code execution, and the like. Constraints on behavior and behavioral decisions are, on many occasions, conditional on circumstances. For example, *if-then statements* and other forms of programming conditionals might constrain the behavior of the simulation and, as such, configure non-trivially different target systems. Durán (2022) illustrates this with a simulation of spatiotemporal patterns of respiratory anthrax infection in a population (see Cooper et al. 2004). In this simulation, the network of nodes and subnodes can be directly coded into the simulation through nested conditionals (i.e., no mathematical formalism is required). As such, and depending on the conditional executed, the simulation would represent different valid paths in the proliferation and spread of the infection, distinctive states of the infection at any given time, and the like.

Can it be assumed that programming languages and code execution constitute legitimate forms of representation that are not necessarily reliant on mathematical models? Some researchers seem to think so (Aronis et al. 2020). Simulation models also seem to allow this kind of philosophical speculation. Clearly, more research is needed in this direction. It remains an open question, whether kludging, Arakawa-like operators, and other computational-inspired practices have representational value or are solely instrumental to the tractability of the simulation model.

4. A new scientific methodology

Where can computer simulations be located in the methodological map? Famously, Rohrlich placed them somewhere intermediate between theoretical physical science and its empirical methods of experimentation and observation (Rohrlich 1990, 507). This view strikes now as too narrow, even for equation-based simulations. The prevailing view is that computer-based methodologies rather extend the class of tractable mathematics and representation and thereby broaden the ranges of modeling (Morgan 2003), observations (Beisbart 2017), predictions (Parker 2014), measurements (Morrison 2009; Tal 2011), and explanation of phenomena (Durán 2017), among several other scientific endeavors. That is to say, computer simulation is not just an intermediate between two familiar ends, but rather a scientific methodology in its own right. Furthermore, there are good reasons to believe that computer simulations raise new epistemological issues, arguably without a precedent in the philosophy of science. This point has forcefully been made by Humphreys and constitutes a central element of his understanding of computer-based methodologies. To be precise, Humphreys distinguishes between anthropocentric epistemologies, which "involve representational intermediaries that are tailored to human cognitive capacities" (Humphreys 2009, 617), and non-anthropocentric epistemologies, where "there now exist superior, non-human, epistemic authorities" (Humphreys 2009, 617). Computer simulations belong to the latter class.

In this context, the claim arises that computer simulations are *epistemically opaque* in that "no human can examine and justify every computational step performed by the computer, because those steps are too numerous" (Parker 2014).[10] What, more precisely, does *epistemic opacity* amount to? Humphreys discusses two related but distinct definitions. The first definition—sometimes referred to as *general epistemic opacity* (GEO) (Alvarado 2021; Beisbart 2021)—says that a given process is opaque to an agent to the extent that the said agent does not know (that is, cannot check, trace, or survey) all of the epistemically relevant elements of the process. Here, a process is broadly understood as the different methods, devices, systems, or instruments of interest. What constitutes an epistemically relevant element of the process will depend on the kind of process involved (Humphreys 2009, 618). For instance, a mathematical proof can be considered the process, and a given lemma is a relevant element in that process. The second definition specifies that a process is *essentially epistemically opaque* (EEO) to an agent if it is *impossible*, given the nature of the agent, to know all the epistemically relevant elements of the process. For instance, the weather forecast for the next two years is impossible to predict by climatologists given their cognitive limitations to handle all the variables involved in such complex systems.

Philosophically speaking, there are a few distinctions of interest between GEO and EEO. For instance, the former is tailored to diverse contingencies, such as context, efforts, goals, and the current state of knowledge of the agent(s). In other words, GEO comes in degrees.[11] Consider Humphreys' own example: "for a mathematical proof, one agent may consider a particular step in the proof to be an epistemically relevant part of the justification of the theorem, whereas to another, the step is sufficiently trivial to be eliminable" (Humphreys 2009, 618). The first agent's knowledge of the proof might change over time, say, in light of a new piece of information. This agent then decides to join the second agent in that that particular step in the mathematical proof is utterly irrelevant. Context, goals, efforts, and the current state of an agent's (or agents') knowledge vary over time, as does practice, and the agents themselves. In contrast, EEO takes it that it is the very nature of agents that prevents knowing all the relevant elements of the process: "[m]any, perhaps all, of the features that are special to simulations are a result of this inability of human cognitive abilities to know and understand the details of the computational process" (Humphreys 2009, 618–619). In other words, a process is essentially epistemically opaque, not because the agent does not know a given relevant epistemic element in the process, but because the agent will never know, given their nature, any of the relevant epistemic elements in the process. EEO is not contingent upon the agent's epistemic context, goals, or efforts, but rather it is an absolute matter about the nature of the agent.

Here we should note that both GEO and EEO are understood from the agent-relative perspective. Whereas in GEO there might be a point in the future where a process ceases to be opaque (e.g., because the mathematician decides that the step is irrelevant for the proof), in EEO agents are by their constitutional nature unable to access the relevant elements of the process. This might either be because they are cognitively limited (e.g., a computer algorithm involves too many steps) or time-restricted (e.g., the algorithm would take long to compute). Agent-relative epistemic opacity is very much the way in which the literature has discussed this issue so far (Beisbart 2021; Durán and Formanek 2018), including the most recent and, sadly, last article on computer simulations by Humphreys (Humphreys, 2022). Interestingly, in this article, Humphreys extends the interpretation of "agent" to also include computer algorithms, with the result that, if we ask questions about ameliorating opacity, one could always think of a third-party algorithm fulfilling this role. This idea is extensively exploited in the literature on *transparency*, especially in the context of machine learning. This said, while trading human agents for algorithms does have some appeal, it does not come cheap. A particularly pressing issue is the *algorithmic regress* that transparency presupposes. To illustrate this, consider an algorithm A that is epistemically opaque. Suppose we make use of A_1, a third-party algorithm that can, presumably, provide knowledge on the relevant elements e in A. Given that A_1 is by definition also epistemically opaque, we are not yet in a position to claim knowledge of e. For this, we need to turn to a second algorithm, A_2 for dealing with the opacity of A_1. The regress continues until either we reach a simple algorithm A_n of which we know all the relevant elements or we abruptly decide to stop the regress.

In a later work, Alvarado challenges the agent-based view on opacity on the basis that "there are instances of epistemic opacity that are either neutral to and/or independent from the limitations of agents. That is, they arise in virtue of factors that are not responsive to or are not related to agential resources" (Alvarado 2021, 9). Whereas Alvarado admits that this description of agent neutrality remains close to agent-based viewpoints (e.g., "as far as accounts of epistemic opacity go, agent-neutral instances of opacity can still be formulated in relation to agential limitations" (10)), agent independency poses an interesting departure from both standard views. According to Alvarado, "an account of agent-independent opacity must include both the fact that the opacity does not arise in virtue of anything related to an agent *and* the fact that it is not responsive to agential resources and/or efforts" (13). In other words, a process is EEO to an agent if it is impossible, given the nature of the *process*, to know all its epistemically relevant elements.[12] Borrowing Alvarado's example, we can say that a stochastic process is agent-independent opaque in virtue of "the combination of its stochasticity (the randomness of paths chosen) and the vast overdetermination (the fact that many—too many—different paths lead to the same outcome) [which makes] inquiry into the actual paths taken (the relevant epistemic elements of the process) inaccessible" (Alvarado 2021, 14).

This more nuanced, process-centered approach to EEO proposed by Alvarado is a welcome addition to the literature, particularly because it offers a way to account for cases where opacity cannot be explained by the cognitive limitations of agents. However, more needs to be said. For instance, it remains unexplained on what grounds a process is to be considered inherently opaque. Without this, it is difficult to distinguish between processes that permanently remain opaque from those that might cease to be opaque at some point in the future. Furthermore, an argument must be provided such that it excludes non-human agents (e.g., algorithms) from accessing inherently opaque processes. Indeed, Alvarado's argument doesn't seem to work if the agent is non-human. Let us recall that Humphreys accepts that algorithms can channel insight into the epistemically relevant elements of a process (Humphreys, 2004, p. 150).

Complementary to these debates are attempts to deal with opacity. Above, I mentioned *transparency*, nowadays gaining significant traction in philosophical debates over machine learning. The core idea of transparency is to make algorithms accessible by showing the inner workings and properties of the algorithm (e.g., Creel 2020). The opposing view is *computational reliabilism*, understood as a set of methods and practices that credit reliability to an algorithm under conditions of opacity (Durán and Formanek 2018; Humphreys 2022; Durán, forthcoming). In other words, whereas transparency makes efforts to grant (human) access to algorithms, computational reliabilism accepts their opacity and focuses instead on the conditions for epistemically trusting them.

There is still plenty of room for further philosophical debate on epistemic opacity and the different specific conceptions of it that figure in debates over computer simulations. But perhaps the greatest contribution of these debates to our understanding of computer simulations (and machine learning) is to bring to the fore their merits as units of philosophical analysis in their own right.

Acknowledgments

Many thanks to Edoardo Datteri, Florian Boge, Ramón Alvarado, and Nico Formanek for comments on earlier versions of this chapter. Their acute insights have made their way into this chapter. This chapter is in great debt to the editors of this Handbook, who read and carefully commented on the several versions of the chapter. Thank you also for your encouragement and patience. All errors are my sole responsibility.

This work is partially supported by the EU program under the scheme "INFRAIA 2020-2024-SoBigData++: European Integrated Infrastructure for Social Mining and Big

Data Analytics," grant agreement 871042. It is also partially supported by the EU program under the scheme "ICT48 Humane AI Net," grant agreement 952026. Their support is gratefully acknowledged.

Notes

1 Although nowadays Schelling's model is implemented using computers, Schelling himself warned against their use for understanding the model. Instead, he used coins or other elements to show how segregation occurred. In this respect, Schelling says: "I cannot too strongly urge you to get the nickels and pennies and do it yourself. I can show you an outcome or two. A computer can do it for you a hundred times, testing variations in neighborhood demands, overall ratios, sizes of neighborhoods, and so forth. But there is nothing like tracing it through for yourself and seeing the thing work itself out. In an hour you can do it several times and experiment with different rules of behavior, sizes and shapes of boards, and ... subgroups of dimes and pennies that make different demands on the color compositions of their neighborhoods" (Schelling 1971, 85). Schelling's warning against the use of computers is an amusing anecdote that illustrates how scientists could sometimes fail in predicting the role of computers in their own respective fields.

2 Schelling also introduced a 1-D version, with a population of 70 agents, with the four nearest neighbors on either side, the preference consists of not being minority, and the migration rule is that whoever is discontented moves to the nearest point that meets her demands (Schelling 1971, 149).

3 For a more thorough review of kinds of computer simulations, see (Durán chap. 1).

4 Here, a *mathematical model* is a generic term covering any scientific, non-physical model, such as theoretical models, data models, phenomenological models, and the like (Frigg and Hartmann 2020).

5 Thanks to Florian Boge for pressing on this point.

6 In her view, computer simulations are the "result of applying a particular kind of discretization to the theoretical/mathematical model [...] There are several reasons for characterizing this type of investigation as an experiment, or more properly, a computer experiment" (Morrison 2015, 219). Thanks to Ramón Alvarado for this reminder.

7 The flexibility of a model is measured as the capacity to implement "generic structures" and the associated possibility of reusing the model in different contexts.

8 Autonomy is attributable to the scarcity of data rather than being a methodological principle of models and modeling.

9 Thanks to Edoardo Datteri for pressing on this point.

10 There is a burgeoning literature that discusses other forms of opacity, such as social opacity (Longino 1990), methodological opacity (Beisbart 2021), corporate opacity (Burrell 2016), and representational opacity (Humphreys 2022), just to mention a few.

11 In Humphrey's words, "[i]t is obviously possible to construct definitions of 'partially epistemically opaque' and 'fully epistemically opaque'" (Humphreys 2009, n. 5).

12 Alvarado provides his own working definition; see (Alvarado 2021, 13).

References

Alvarado, Ramón. 2021. "Explaining Epistemic Opacity." Forthcoming in *The Science and Art of Simulation II*, edited by Andreas Kaminski, Michael Resch, and Petra Gehring. Berlin: Springer Verlag. (References in the text are to the 2021 online preprint version.)

Aronis, John M., Jeffrey P. Ferraro, Per H. Gesteland, Fuchiang Tsui, Ye Ye, Michael M. Wagner, and Gregory F. Cooper. 2020. "A Bayesian Approach for Detecting a Disease that is not Being Modeled." *PLoS One* 15(2):1–15. https://doi.org/10.1371/journal.pone.0229658.

Beisbart, Claus. 2014. "Are We Sims? How Computer Simulations Represent and What This Means for the Simulation Argument." *The Monist* 97(3): 399–417.

———. 2017. "Are Computer Simulations Experiments? And If Not, How Are They Related To Each Other?" *European Journal for Philosophy of Science* 8(2): 171–204.

———. 2021. "Opacity Thought Through: On the Intransparency of Computer Simulations." *Synthese* 199: 11643–11666.
Beisbart, Claus and John D. Norton. 2012. "Why Monte Carlo Simulations are Inferences and Not Experiments." *International Studies in the Philosophy of Science* 26(4): 403–422.
Boge, Florian J. 2019. "Why Computer Simulations Are Not Inferences, and in What Sense They Are Experiments." *European Journal for Philosophy of Science* 9(13). https://doi.org/10.1007/s13194-018-0239-z.
———. 2020. "How to Infer Explanations from Computer Simulations." *Studies in History and Philosophy of Science Part A* 82: 25–33.
Burrell, Jenna 2016. "How the Machine 'Thinks': Understanding Opacity in Machine Learning Algorithms." *Big Data & Society* 3(1): 1–12.
Cooper, Gregory F., Denver Dash, John Levander, Weng-Keen Wong, William R. Hogan, and Michael M. Wagner. 2004. "Bayesian Biosurveillance of Disease Outbreaks." In *Proceedings of the 20th Conference on Uncertainty in Artificial Intelligence*, ed. Christopher Meek, Max Chickering and Joseph Halpern, 94–103. Arlington, VA: AUAI Press.
Creel, Kathleen A. 2020. "Transparency in Complex Computational Systems." *Philosophy of Science* 87(4): 568–589.
DeAngelis, Donald L. and Volker Grimm. 2014. "Individual-based Models in Ecology after Four Decades." *F1000Prime Reports* 6(39): 1–6.
Durán, Juan M. 2017. "Varying the Explanatory Span: Scientific Explanation for Computer Simulations." *International Studies in the Philosophy of Science* 31(1): 27–45.
———. 2018 *Computer Simulations in Science and Engineering. Concept, Practices, Perspectives.* Switzerland: Springer.
———. 2020. "What is a Simulation Model?" *Minds and Machines* 30: 301–323.
———. 2022. "Models, Explanation, Representation, and the Philosophy of Computer Simulations." In *Philosophy of Computing* (*Philosophical Studies* 143), ed. Björn Lundgren and Nancy A. Nuñez Hernández. Berlin: Springer pages 221–249.
Durán, Juan M. and Nico Formanek. 2018. "Grounds for Trust: Essential Epistemic Opacity and Computational Reliabilism." *Minds and Machines* 28(4): 645–666.
Durán, Juan M (forthcoming) "Beyond transparency: computational reliabilism as an externalist epistemology for algorithms" ed Juan M. Durán and Giorgia Pozzi. Synthese Library. Springer Cham.
El Skaf, Rawad and Cyrille Imbert. 2013. "Unfolding in the Empirical Sciences: Experiments, Thought Experiments and Computer Simulations." *Synthese* 190: 3451–3474.
Fox-Keller, Evelyn. 2003. "Models, Simulations, and 'Computer Experiments'." In *The Philosophy of Scientific Experimentation*, ed. Hans Radder, 198–215. Pittsburgh: University of Pittsburgh Press.
Frigg, Roman and Stephan Hartmann. 2020. "Models in Science." In *The Stanford Encyclopedia of Philosophy*, edited by Edward N. Zalta, https://plato.stanford.edu/archives/spr2020/entries/models-science/.
Frigg, Roman and Julian Reiss. 2009. "The Philosophy of Simulation: Hot New Issues or Same Old Stew?" *Synthese* 169(3): 593–613.
Gilbert, Nigel and Klaus G. Troitzsch. 2005. *Simulation for the Social Scientist*, 2nd edn. Maidenhead; New York: Open University Press.
Goldman, Nir. 2014. "A Virtual Squeeze on Chemistry." *Nature Chemistry* 6(November): 1033–1034.
Grüne-Yanoff, Till and Paul Weirich. 2010. "The Philosophy and Epistemology of Simulation: A Review." *Simulation & Gaming* 41(1): 20–50.
Guala, Francesco. 2002. "Models, Simulations, and Experiments." In *Model-based Reasoning: Science, Technology, Values*, edited by Lorenzo Magnani and Nancy J. Nersessian, 59–74. Boston, MA: Springer US.
Hartmann, S. 1996. "Modelling and Simulation in the Social Sciences from the Philosophy of Science Point of View." In *Modelling and Simulation in the Social Sciences from the Philosophy of Science Point of View*, ed. Rainer Hegselmann, Ulrich O. Mueller, and Klaus G. Troitzsch, 77–100. Berlin: Springer.
Humphreys, Paul W. 1990. "Computer Simulations." *PSA: Proceedings of the Biennial Meeting of the Philosophy of Science Association* 2: 497–506.
———. 2004. *Extending Ourselves: Computational Science, Empiricism, and Scientific Method.* Oxford: Oxford University Press.

———. 2009. "The Philosophical Novelty of Computer Simulation Methods." *Synthese* 169(3): 615–626.
———. 2022 "Epistemic Opacity and Epistemic Inaccessibility." Accessed 21 Sep 2022 https://wordpress.its.virginia.edu/Paul_Humphreys_Home_Page/files/2016/02/epistemic-opacity-and-epistemic-inaccessibility.pdf
Krohs, Ulrich. 2008. "How Digital Computer Simulations Explain Real-World Processes." *International Studies in the Philosophy of Science* 22(3): 277–292.
———. 2019. *Calculated Surprises*. Oxford: Oxford University Press.
Lenhard, Johannes and Eric Winsberg. 2010. "Holism, Entrenchment, and the Future of Climate Model Pluralism." *Studies in History and Philosophy of Science Part B: Studies in History and Philosophy of Modern Physics* 41(3): 253–262.
Lesne, Annick. 2007. "The Discrete versus Continuous Controversy in Physics." *Mathematical Structures in Computer Science* 17(2): 185–223.
Longino, Helen E. 1990. *Science as Social Knowledge: Values and Objectivity in Scientific Inquiry*. Princeton, NJ: Princeton University Press.
Morgan, Mary S. 2003. "Experiments without Material Intervention: Model Experiments, Virtual Experiments and Virtually Experiments." In *The Philosophy of Scientific Experimentation*, ed. Hans Radder, 216–235. Pittsburgh, PA: University of Pittsburgh Press.
———. 2004. "Simulation: The Birth of a Technology to Create 'Evidence' in Economics." *Revue d'histoire des sciences* 57(2): 339–375.
Morrison, Margaret. 2009. "Models, Measurement and Computer Simulation: The Changing Face of Experimentation." *Philosophical Studies* 143(1): 33–57.
———. 2015. *Reconstructing Reality: Models, Mathematics, and Simulations*. Oxford: Oxford University Press.
Parker, Wendy S. 2006. "Understanding Pluralism in Climate Modeling." *Foundations of Science* 11: 349–368.
———. 2009. "Does Matter Really Matter? Computer Simulations, Experiments, and Materiality." *Synthese* 169(3): 483–496.
———. 2014. "Computer Simulation." In *The Routledge Companion to Philosophy of Science*, ed. Martin Curd and Stathis Psillos 135–145. New York: Routledge.
———. 2020. "Evidence and Knowledge from Computer Simulation." *Erkenntnis* 87: 1521–1538.
Peck, Steven L. 2012. "Agent-based Models as Fictive Instantiations of Ecological Processes." *Philosophy and Theory in Biology* 4: 1–12.
Rohrlich, Fritz. 1990. "Computer Simulation in the Physical Sciences." *PSA: Proceedings of the Biennial Meeting of the Philosophy of Science Association* 2: 507–518.
Schelling, Thomas C. 1971. "On the Ecology of Micromotives." *National Affairs* 25: 61–98.
Tal, Eran. 2011. "From Data to Phenomena and Back Again: Computer-simulated Signatures." *Synthese* 182(1): 117–129.
Vichniac, Gérard Y. 1984. "Simulating Physics with Cellular Automata." *Physica D: Nonlinear Phenomena* 10(1–2): 96–116.
Weisberg, Michael. 2013. *Simulation and Similarity: Using Models to Understand the World*. Oxford: Oxford University Press.
Winsberg, Eric. 1999. "Sanctioning Models: The Epistemology of Simulation." *Science in Context* 12(2): 275–292.
———. 2001. "Simulations, Models, and Theories: Complex Physical Systems and Their Representations." *Philosophy of Science* 68(3): S442–S454.
Wolfram, Stephen. 1984. "Universality and Complexity in Cellular Automata." *Physica* 10D: 1–35.
Woolfson, Michael M. and Geoffrey J. Pert. 1999. *An Introduction to Computer Simulation*. Oxford: Oxford University Press.

12
SCIENTIFIC LAWS AND THEORETICAL MODELS

Jarosław Boruszewski and Krzysztof Nowak-Posadzy

1. Introduction

In contemporary philosophy and methodology of science, discussions have focused on models and modeling with special attention put to the polysemy of the very term "model." At the same time, the polysemy of the term "law" has been somewhat omitted. However, if one focuses on the problem of law-model relation it turns out that these semantic ambiguities are related in specific ways. A certain distinction between laws of science and laws of nature is therefore needed – a distinction found already in 19th-century methodological treatises (i.e., Mill 1843). Traditionally, discussions on laws concerned either the logical and cognitive status of law statements or ontological and metaphysical properties of the objective dependency expressed by law statements. However, in modern discussions on law-model relation, the term "law of nature" is understood differently, namely as a law of science describing or representing a certain natural regularity (Cartwright 1983, 54–55; Giere 1999, 86). This semantic shift is only seemingly insignificant as it determines the status of laws of nature understood this way. These laws, thus, adopt the status of universal statements which refer to the real world and are applied to empirical objects. Such an understanding of laws was the basis for the deductive-nomological account of explanation. Some contemporary authors believe this account provides an answer to the question about the relations between laws and theoretical models. When it comes to the law-model relation, traditional approaches to theoretical models contain at least one law of nature as well as initial and boundary conditions (Carrier 2004; Carrier, Gölzhäuser, Kohse-Höinghaus 2018). This understanding of models, as an auxiliary to theoretical laws, has been subject to numerous critical analyses most extremely expressed as "science without laws."[1] The latter expression is elliptic because it is about science without laws *of nature* or, to put it differently, about a transition from laws of nature to *laws-of-models* (van Fraassen 1989, 188). What played an important role in this turn was the tradition of the semantic view of theories and the approach of Nancy Cartwright, who pointed out that "situations that fall under the fundamental laws are generally the fictional situations of a model" (Cartwright 1983, 160). However, not only was there no consensus on the precise understanding

DOI: 10.4324/9781003205647-15

of law-model relation, but there also emerged a new incarnation of the problem of the logical and cognitive status of laws of science.

An instructive starting point for such a new understanding of laws would be to identify the various ways of using law sentences, as proposed by Norwood Russell Hanson. Using broad material from the history of science, Hanson established a view that theoretical law sentences or formulae "can be used to express – definitions, *a priori* statements, heuristic principles, empirical hypotheses, rules of inference, etc." (Hanson 1958, 112). Contemporary philosophy of science reconstructs the multitude of usages of law sentences in scientific modeling, although this reconstruction differs depending on the philosophical approaches adopted. This analysis presents different options for those usages, including usages not mentioned by Hanson. A review of various ways of using laws as laws-of-models is included in the second section of the chapter. The third section deals with yet another turn in the discussion on the law-model relation, which can be described as a transition from laws of models to *laws-for-modeling*.[2] The fourth section demonstrates different usages of law sentences based on the example of the Copernicus–Gresham Law, as this law still raises interest among philosophers of science and economists because it is seen both as "an easily understandable" and as a "complex issue" (Bernholz and Gersbach 1992, 288).[3]

At this point, several reservations need to be mentioned. Firstly, while in the context of law-model relation the distinction between fundamental and phenomenological laws plays an important role (for instance, in Nancy Cartwright's approach), such a strict dichotomy is questionable (i.e., Laymon 1989; Weinert 1995). Therefore, following Ronald Giere, there is no reason to exclusively consider fundamental laws as laws-of-models or as laws-for-modeling. Secondly, when it comes to the division into quantitative and qualitative laws, it can be questioned whether only the former is treated as important in scientific modeling, as a review of usages of the Copernicus–Gresham Law demonstrates that qualitative laws are also important in model-building. Thirdly, it seems justified to speak about laws in special sciences – the Copernicus–Gresham Law is actually treated as an "archetypal special-science law" (Shahvisi 2019). Last but not least, the ways of presenting and formulating laws do matter, as demonstrated in the terminology chosen by the scholars referred to in this chapter. On the one hand, some authors often use expressions in the form of alternatives, for instance, "principles, equations or laws" (Lorenzano and Díaz 2020, 164). On the other, Giere, for instance, avoids speaking about laws, because "[i]nterpreting the equations as laws assumes that [...] there is an implicit universal quantifier out front" (Giere 1999, 92). Giere, therefore, suggests that the way of speaking about laws presupposes the way of their logical reconstruction. However, one needs to differentiate between explicit formulations of laws functioning in the research practice of a given science from their reconstructed forms, which anyway have to include what is implicitly embedded in laws according to a given scholar. For instance, Cartwright claims that laws include implicit *ceteris paribus* clauses. In what follows, no implicit content is attributed to or imputed to laws and they are treated at face value. As far as the logical status of a given law is concerned, laws are then propositional schemata or simply open formulae (Mejbaum 1977).

2. Laws-*of*-models

This section discusses different approaches to separating models from laws. Laws are characterizations of models, thus resulting in *laws-of-models*. What is specific here is that modeling is understood as an indirect representation – therefore a triad emerges:

specification–model–target. The notion of specification is the least ambiguous, as underlined by Cartwright: "laws of the theory *are true of the objects* in the model, *and they are used to derive* a specific account of how these objects behave" (Cartwright 1983, 17; emphasis added). This duality of laws is manifested in Cartwright's work (1983) by explicit references to a definitional understanding of laws and allusions to understanding them as rules of inference.

The approach to laws as definitions in the context of theoretical modeling has been developed since the 1970s by Giere (1979), although he most frequently wrote about principles[4] or equations that define model systems recognized as non-linguistic abstract objects. To be more precise, principles or equations are stipulative definitions of abstract model systems, in which they are perfectly satisfied. Linguistic formulations of principles incorporated into the specification of a given model are always true of the model system, although in a trivial sense. On the other hand, the function of principles in modeling is far from trivial: "principles thus help both to shape and also constrain the structure of these more specific models" (Giere 2006, 62). Giere also deals with the problem of low-level generalizations (phenomenological laws); however, he declines to treat them as implicit *ceteris paribus* laws because it could lead to eventually reducing them to trivial statements such as a "law holds except where it does not." He believes it is much better "to keep the single law statements, but understand them as part of characterization of an abstract model and thus being true of the model" (Giere 2004, 749). Contrary to the "science-without-laws" thesis, all laws, not only high-level theoretical principles, play an important role in Giere's approach to modeling: "laws are to be interpreted as providing definitions of various models" (Giere 1988, 84). The law-model relation can then be briefly referred to as *stipulation-and-satisfaction*.

The trial set by Giere is followed by Michael Weisberg and Peter Godfrey-Smith, who in the first place differentiate between model system and model description. Equations, diagrams, or language expressions are included in the model description while a model executes its description, although there is no unequivocal assignment between model and model description. A given model can have many descriptions and a given description can specify many models. Therefore, a many-to-many relation is obtained. A relation between description and model is understood as specification, which is a weaker relation than definition. A description is only a partial characteristic of models. Therefore, it can be assumed that principles or equations incorporated in the model description are partial definitions of the model system. What is extremely important is that in modeling, the model description does not have to precede the model: "In some cases, the model is constructed before or without description. In others, the description comes first. And perhaps most commonly, the two are produced in tandem" (Weisberg 2013, 38). A model description plays a crucial role in the case of mathematical models because such models can be explored and manipulated only via their description. And although a mathematical model in this approach is not a system of equations, it can only be used by proxies in the form of equations. This can explain the propensity of some scholars to call equation models. It makes the strict dualism description system difficult to maintain in practice – "It would be a mistake to insist that one of these is 'the model' and the other is not. Each kind of talk can constrain the other" (Godfrey-Smith 2006, 736). This can lead to stating that a model need not be considered distinct from its description.

Roman Frigg and James Nguyen present a more expanded version of model description on the grounds of indirect fictionalism, namely the DEKI (denotation, exemplification, keying up, and imputation) account. They make a distinction between a description of the model's carrier and the principles of generation. The former functions as a prop that prescribes scientists to imagine something (i.e., two-body system), which is a basic assumption of a given model and is presupposed to be true within this model. The principles of generation, on the other hand, play a crucial role, especially in the context of the law-model relation:

> The 'working out' of the details of a model consist in deriving conclusions from the basic assumptions of the model and some general principles or laws that are taken to be in operation in the context in which the model is used. […] The laws and principles that are used in these derivations play the role of principles of generation.
>
> *(Frigg and Nguyen 2020, 122)*

Models are incomplete without principles of generation as these principles provide models with certain internal dynamics and properties not specified in their basic assumptions. Getting acquainted with those principles means in fact learning from the model and deriving implicit truth from it: "implicit fictional truths can be *inferred according to certain principles of generation*" (Salis and Frigg 2020, 45; emphasis added). The status of principles of generation is difficult to determine in a general way because it is always relative to specific domains of knowledge. Sometimes those principles are ad hoc, but more significant are those that have the status of intersubjective, though often implicit, rules of inference (Frigg 2010, 258). It can therefore be assumed that on the grounds of indirect fictionalism laws and principles of science are used as rules of inference in such a way that secondary truths, not directly specified in model assumption, are inferred from the model.

A different approach to the problem of the status of laws and their functions in modeling can be found in some proposals in the field of philosophy and methodology of biology, namely proposals treating biological laws as *a priori* laws (Sober 1997; Elgin 2003). A widely discussed example here is the Hardy–Weinberg law of population genetics. The discussions reject both the empiricist view of laws and the definitional approach; this law is understood neither as an empirical statement nor as a stipulative definition of a model, and becomes an *a priori* conditional tying contingent statement. The conditional is *a priori* because important conceptual relations occur between its antecedent and consequent. Proponents of a priori laws as key elements of mathematical models pay attention to the fact that those laws are important guides in understanding the living world and enable grasping it precisely, thus contradicting the view that a priori equals uninformative:

> For those who find the idea of the synthetic a priori unattractive, the a priori tends to suggest examples like 'Bachelors are unmarried men'; such statements merely provide definitional abbreviations and furnish zero insight into the nature of reality. […] If a priori generalizations figure in explanations and predictions in the same way that empirical laws do, we should regard these a priori generalizations as laws.
>
> *(Sober 2011, 588)*

Laws, therefore, play a role in modeling: "laws do some work in the models they are part of" (Elgin 2010, 442). It is worth noticing that this finding, in a way, restores the utility of laws. Laws are thus an integral part of models, they are laws *of models*, although this comes at the expense of changing their logical status.

3. Laws-*for*-modeling

Cartwright's investigation provides another valuable insight into the law-model relation problem as she uses the concept of theoretical instrumentalism, namely a toolbox of science. While in her earlier works Cartwright considered theoretical laws tell the truth when it comes to objects in models and "lie" when it comes to the real world, she subsequently understood theoretical laws as purely instrumental. In the toolbox approach, the answer to the question about what a law represents is categorical – nothing (Cartwright, Shomar, and Suárez 1995, 139–140). Laws "do not model anything, but are rather useful tools to build models" (Suárez and Cartwright 2008, 75). Laws are an important starting point in model-building, because they provide a wider theoretical context. This function, however, is instrumental – laws are only theoretical context providers that are evident when scientists improve the model, construct a more accurate model, or customize it to special needs. In Giere's approach, a more accurate model still meets model-defining principles, but according to Cartwright, such a statement is overly optimistic. Corrections made to a model rarely, if at all, result in a model consistent with the principles that served as a starting point in its construction. Generally, model customization often results in a situation where the model fails to fulfill the initial law (Cartwright 1999, 250–252). This is why it is imprecise to call the toolbox approach one consistent with the "laws-of-models." It is rather consistent with "laws-for-modeling" as laws are only one of many tools used to build a model, usually used at the early stages of model construction and can be even dropped.

Yet another understanding of laws-for-modeling comes with direct fictional approaches. They are a reaction to indirect fictionalism which keeps the model separated from its description. The main difference is that direct fictional approaches reject the very existence of model systems. According to Adam Toon, in this approach, a model is what in indirect fictionalism is viewed as model description. A law being a part of a model loses its descriptive status at the expense of the prescriptive one; a law functions as a prop that prescribes imaginings about target systems. Toon's solution to the law-model relation question is categorical:

> [it] is simply to deny that we need to regard theoretical principles formulated in modelling as genuine statements. Instead, they are prescriptions to imagine. If theoretical principles are understood in this way then there is no reason to think that there needs to be any object which they describe.
>
> *(2012, 44)*

The antirealistic approach to model systems results in a situation where there is no room for a satisfaction relation – there are no abstract, fictional, or any other objects satisfying the laws or equations of models. A rejection of the existence of mediating model systems does not imply that modeling becomes a purely subjective issue. Toon's and Arnon Levy's criticism of indirect modeling and abstract or fictional model systems can even go further: "on a direct account, there is no model system, not even imaginary one" (Levy 2015, 792). Moreover, Levy considers that the biggest weakness of indirect approaches is that they discredit models formulated in natural language; they treat models merely as model descriptions. Levy thus shares the view of some representatives of deflationism (Downes 1992) – another influential side in the discussion of the law-model relation.

Deflationary approaches are characterized by a very liberal attitude – in fact, everything can be a model, and it is not possible to discern any intrinsic property of a given object which makes it a model or to point to a constitutive property of the representational relation (Teller 2001; Callender and Cohen 2006). One of the most important proposals here is the inferential approach to representation as a variant of use-based deflationism (Suárez 2016). While analyzing the notion of representation, Maurizio Suárez distinguishes its vehicle – source S and its object – target T. In modeling practice, sources of modeling are multiple, "from concrete physical objects and diagrams to abstract mathematical structures or laws" (Suárez 2015, 41). Therefore, a law can be a model; for instance, the second law of thermodynamics represents entropy as an abstract property and asserts entropy's increase in closed systems.

Deflationism is therefore liberal when it comes to vehicles of representation. Moreover, representation is not considered a conceptual relation, but an activity making the source S useful as a representation of target T. The usefulness of S means that some users of the model "draw inferences about T from S" (Suárez 2010, 93). If a law is a model, then its inferential function becomes principal. Such inferences are not based on properties of S which are necessary and sufficient conditions of representation of T. Deflationary "flattening" of representation relation implies that the complex issue of representational vehicles becomes significant, thus exposing key semiotic aspects of modeling: "representational vehicles and the content they express *are* the models. We might say models are nothing over and above their mode of presentation" (Odenbaugh 2021,11; emphasis in original).

Therefore, the question of whether the models' content is derivative from the individual mental states of modeling agents or whether it goes beyond them becomes important. If the former is the case, then, similarly to direct fictionalism, this account can be accused of lacking the guarantee of intersubjectivity, which for scientific models is a non-negligible issue. What emerges from this account is a rather naïve and highly dubitable image of modeling, in which agents' intentions attribute content to models, regardless of models' history, reception, and usages. The inherent intersubjective aspect of models then goes missing: "Scientists do not merely start using a model however they would like, without recourse to the history of the use of the model. There are autonomous elements of the model which are carried with it" (Boesch 2017, 978). These autonomous elements constitute salient yet long-neglected dimensions of models, namely materiality and, more importantly, *semioticity*. A model's semioticity is not an intrinsic feature of models; it is formed and transformed by model builders, users, and recipients in specific socio-cultural functionings of models. Generally, models are then ascribed to the status of culturally established artifacts (Knuuttila 2017). Bringing out issues of materiality and semioticity makes it possible to move to the artifactual approach to scientific modeling, which is nowadays gaining recognition. Theoretical models are built by making use of various "tools and other resources" (Knuuttila and Loettgers 2017). When it comes to the question of law-model relation, laws can be resources for models or ingredients of models as built-in dependencies (Knuuttila 2021a, 2021b). In the artifactual approach, questions of linguistic formulations, omitted in indirect approaches, regain importance. Representational means are an ineliminable part of the model itself, not secondary to abstract or imagined entities. Therefore, attention is put on the cultural significance of models and the question of style (Boruszewski, Nowak-Posadzy 2021).

4. One law – many uses

The variety of uses of laws in modeling will now be demonstrated with the Copernicus–Gresham Law as an example. When addressing the questions concerning the uses of this law, it is hardly possible not to refer to the work of Renaissance polymath Nicolaus Copernicus, "Monetae cudendae ratio," the final version of which was released in 1528:

> While it is quite inappropriate to introduce new and good money at a time when the old, cheaper money remains in circulation, how much greater is the fault of introducing new and cheaper money while the old and better remains in circulation: it not only corrupted the old but, so to speak, conquered it entirely.
>
> *(Copernicus 1979, 307)*

This excerpt shows that Copernicus captured the idea behind the law which became explicitly formulated as a scientific law by Henry Macleod as late as in the 19th century when scholars quested after theoretical economic laws analogous to the principles of classical mechanics: "good and bad coin cannot circulate together, but the bad coin will drive out the good" (Macleod 1872, 375).

However, with time, the status of the Copernicus–Gresham Law was systematically weakened: it moved from being treated as a great fundamental law (Macleod) to a principle of economics (Jevons 1875) and a universal law (Fetter 1932), until finally ending up being treated by some as a trivial law (Schumpeter 1954). Almost parallelly, proposals started to appear of treating the Copernicus–Gresham Law as an empirical generalization, set relatively independently of theory and having its own historical exemplifications. It was François R. Velde who explicitly pointed out that the disputes over the nature of the Copernicus–Gresham Law were carried out by those who viewed it as a theoretical proposition and by those who read it as an empirical regularity (2008, 769). What was little discussed then, was the nature of relations between the Copernicus–Gresham Law and the explanatory models offered. Two options can be distinguished here: either the law can be located on the explanandum side or on the explanans side. In the first option, models can provide a theoretical explanation of the Copernicus–Gresham Law operation – the Copernicus–Gresham Law is then *explanandum* and the theoretical model is the *explanans*. Currently, such models include mainly, but not exclusively, theoretical models of commodity money. They differ in terms of the theoretical apparatus used from, for instance, asymmetric information theory (Akerlof 1970), search theory (Velde, Weber, Wright 1999), or game theory with prisoner's dilemma (Selgin 2020).

In the second option, the Copernicus–Gresham Law (located on the explanans side) is a useful tool for building economic models. Although this law is far from being new, it is by no means irrelevant or redundant. Greenfield and Rockoff's model built on the quantitative theory of money is a good case in point here: the authors conclude that "Gresham's law still belongs in the monetary economist's tool kit" (1992, 1). This law can be used in different ways as will be demonstrated below.

Let us start with the usage of the Copernicus–Gresham Law as an empirical generalization, which is the starting point for Charles P. Kindleberger, who, however, generalizes the scope of the law from two kinds of money to two different kinds of assets, both financial and non-financial. Secondly, Kindleberger extends the law to a model in which the question of the quantity of money is separated from market instability and asset convertibility, which are linked to the operation of the Copernicus–Gresham Law:

> Gresham's law thus extended is a highly useful analytical model for the economic historian to keep in his toolbox [...]. Convertibility of one money into another, of money into assets, and of normally marketable assets into money is the touchstone. When such convertibility is maintained, Gresham's law is held at bay.
>
> *(Kindleberger 1989, 44–53)*

The generalization of the Copernicus–Gresham Law range and extending it to a model of market instability is an example of how the Copernicus–Gresham Law can be used in modeling. After extension, the monies appearing in the law no longer represent only coins:

> Foreign-exchange crises can be assimilated to Gresham's law, with *the two monies representing* one national money on the one hand, and all other currencies into which it is convertible on the other.
>
> *(Kindleberger 1989, 57; emphasis added)*

In research concerning monetary history and the application of the method of ideal types, one contends with a completely different use of the Copernicus–Gresham Law in model-building. Although it is still subject to methodological controversies, ideal types tend to be accepted as theoretical models in the social sciences (i.e., Weinert 1996; Aronovitch 2012). The Copernicus–Gresham Law then "provides historians with concrete (although qualitative) comparative counterfactual ideal types" (Elliott 2020, 165). Models formulated as ideal types determine boundaries of market conditions within which certain values, for instance, the exchange rates, are set. As researchers do not have credible data concerning these values, they commonly adopt certain extreme values and use them for comparative purposes. In such modeling strategies, the Copernicus–Gresham Law is not used as an empirical law: "It is better to deploy Gresham's law as a complex and interconnected set of conditions and premises" (Elliott 2020, 171). The conditions concern the structure of the monetary system, while the premises concern the motives of money users' behaviors. The ideal type determined by the Copernicus–Gresham Law provides insights into those conditions and improves understanding of the premises. This use of the Copernicus–Gresham Law in economic modeling is in line with Giere's approach – a law defines a model system, in this case, as an ideal type.

George Akerlof offers yet another use of the Copernicus–Gresham Law, which can be understood here as a heuristic principle. In his seminal paper, "Market for 'Lemons'," Akerlof points out that a modified form of the Copernicus–Gresham Law appears in his model of the market of bad quality commodities and considers such a *reappearance* as "instructive" (Akerlof 1970, 480). The use of the Copernicus–Gresham Law by Akerlof is a heuristic device enabling a better understanding of the target system ("market of lemons").

The Copernicus–Gresham Law can also be used not as an empirical generalization but as an intended theoretical statement, although this use is quite specific and not entirely clear. For instance, Arthur Burns pointed out that when in historical research a theoretical model is constructed, a historian's investigation starts to resemble the investigation of a theoretical economist:

> the historian to be making a calculation in which theoretical concepts were being used as they would have been had he been proving Gresham's Law *qua* economist, and not just illustrating it *qua* economic historian. [...] Preeminently this will be so whenever the historian has need to construct a model.
>
> *(Burns 1960, 66)*

Friedrich von Hayek, by continuing Burns' considerations, advocated for such usage. Hayek concluded that using the Copernicus–Gresham Law as a theoretical statement is a useful research tool because it orientates the research process toward the search for the causes of the driving-out dependency, as long as the condition of having at least two types of money is met: "which are of equivalent value for some purposes and of different value for others" (Hayek 1962, 101). This is how researchers can acquire additional valuable information of a general kind, to which they have no direct access. Hayek's approach has been taken up by Richard Mundell, who introduces two significant qualifications. The first qualification is that using the Copernicus–Gresham Law in historical research allows one "to draw inferences about the monetary policies at the time the coins disappeared" (Mundell 1998). Secondly, he points out that on top of the condition set by Hayek, an additional condition has to be introduced, namely whether the two types of money exchange for the same price. Only then, according to Mundell, the Copernicus–Gresham Law becomes a powerful research tool. However, as Alex Rosenberg noted when discussing the status of the Copernicus–Gresham Law, "this qualification comes dangerously close to making the law a necessary truth" (2018, 29). Mundell suggested the usage of the Copernicus–Gresham Law points to the possibility of converting the law into an inference rule. Rosenberg's friendly warning seems to apply more to instances of using the Copernicus–Gresham Law as an *a priori* statement.

Last but not least, the Copernicus–Gresham Law can also function as a *commentary* on a model, which was for instance the case with Schelling's segregation model. The author directly refers to the Copernicus–Gresham Law, while introducing the question of dependencies obtained in the model: "small incentives can lead to striking results; Gresham's Law is a good example" (Schelling 1969, 488). This function of the law is fairly modest, but its discernable character undoubtedly enables achieving the intended rhetorical effect. Such a reference while discussing the model can be called an illustrative function of the law.

5. Conclusions

This chapter explored the many ways in which scientific laws and theoretical models intersect. Investigating this relation seems particularly important as philosophers of science have gradually reoriented their inquiries from the analysis of laws, through the analysis of models, to the analysis of modeling practices. A clear illustration of this tendency is the science-without-laws thesis or thought-provoking questions about what science without laws looks like (Morgan 2007, 271). The answer depends on the interpretation of the science-without-laws thesis with at least two interpretations possible. The first interpretation offers what can be called an unqualified version of the thesis and states that scientists' interest in formulating, using, and applying laws will continue to decrease. Accepting this version seems premature, although laws did lose their privileged status in the realm of science. The second interpretation offers what can be called a qualified version of the thesis and says science without laws is possible only if laws are understood as laws of nature. This version can be accepted provided it includes various cognitive and extra-cognitive functions performed by scientific laws and theoretical models.

It is certainly possible to speak about a transition from laws to models. This also implies an interchange of their respective *functions* – functions traditionally attributed to laws are currently performed by models and the other way around. For instance, explanatory and predictive cognitive functions, once exclusive to laws, are now attributed to models. At the same time, educational and heuristic functions once reserved to models are now assigned to laws. The difficulties stemming from explanation by means of laws have been widely discussed in the literature and currently, it is the explanation by means of model-building that seems most promising, although it also raises some controversies. The issue of the explanatory function of science will likely remain subject to animated discussions. As far as predicting is concerned, it was noticed a long ago that laws are not needed for making predictions. Already Rudolf Carnap stated that "the use of laws is not indispensable for making predictions. Nevertheless, it is expedient, of course, to state universal laws in books on physics, biology, psychology, etc." (Carnap 1950, 574). This leads to the communication-educational function of scientific laws. The dominant belief nowadays is that theoretical models are able to perform the indicated cognitive functions of science much better than laws. However, one can question whether models are able to better execute the communication-educational function. Scientific handbooks, reliable popular science works, and even good science-fiction literature cannot do without scientific laws. While models can provide more efficient knowledge, scientific laws seem to be irreplaceable when it comes to the communication-educational function.

The general message of this chapter is that the law-model relation can be understood in a simple or complex way. The simple understanding suggests substitutability or rivalry of laws and models, as exemplified by the use of such words as "without," "or," "versus." For instance, it is not the quest for laws but model constructing that dominates contemporary research activity. However, if the diversity of laws' usages is taken into account, the issue becomes complex in the sense that while model-building remains at the forefront, the role of laws in modeling is far from marginal. Another complex issue is that the discussion concerns differently understood laws-of-models and laws-for-modeling. One cannot forget that laws and their various formulations belong not only to the history of science but also to up-to-date resources of scientific thought. Sometimes scientific laws are close at hand, sometimes they need to be dusted off, but their use is always a matter of scientific invention.

Notes

1 Giere (1999), Creager, Lunbeck, Wise (2007), Hardt (2017).

2 Differences between laws-of-models and laws-for-modeling take the form of differentiating "model view" (canonical work on the matter by Giere 1988) from "hybrid view" (canonical work on the matter by Morrison and Morgan 1999) (Teller 2001; Contessa 2014).

3 The Copernicus–Gresham Law continues to raise numerous controversies; for instance, it is referred to in debates concerning the issue of multi-realizability (Fodor 1974) and the conditional form of laws (Friend 2016); philosophers of science continue to argue whether this law is causal (Loewer 2009) or functional (Rosenberg 2018).

4 Apart from principles of physics, Giere also refers to the principle of natural selection and "economics boasts of various equilibrium principles" (2006, 61).

References

Akerlof, George A. 1970. "The Market for 'Lemons': Quality Uncertainty and the Market Mechanism.", *Quarterly Journal of Economics* 84: 488–500.

Aronovitch, Hilliard. 2012. "Interpreting Weber's Ideal-Types." *Philosophy of the Social Sciences* 42: 356–369.

Bernholz, Peter, and Hans Gersbach. 1992. "Gresham's Law: Theory." In *The New Palgrave Dictionary of Money and Finance*, Vol. 2, edited by Peter Newman, Murray Milgate and John Eatwell, 286–288. London: Macmillan.

Boesch, Brandon. 2017. "There *Is* a Special Problem of Scientific Representation." *Philosophy of Science* 87: 970–981.

Boruszewski, Jarosław, and Krzysztof Nowak-Posadzy. 2021. "Economic Models as Cultural Artifacts: A Philosophical Primer." *Filozofia Nauki – The Philosophy of Science* 29: 63–87.

Burns, Arthur Lee. 1960. "International Theory and Historical Explanation." *History and Theory* 1: 55–74.

Callender, Craig, and Jonathan Cohen. 2006. "There Is No Special Problem About Scientific Representation." *Theoria* 21: 67–85.

Carnap, Rudolf. 1950. *Logical Foundations of Probability*. Chicago: The University of Chicago Press.

Carrier, Martin. 2004. "Knowledge Gain and Practical Use: Models in Pure and Applied Research." In *Laws and Models in Science*, edited by Donald Gillies, 1–17. London: King's College Publications.

Carrier, Martin, Armin Gölzhäuser, and Katharina Kohse-Höinghaus. 2018. "Understanding Phenomena by Building Models. Methodological Studies on Physical Chemistry." In *Progress in Science, Progress in Society*, edited by Alain Tressaud, 19–36. Cham: Springer International Publishing AG.

Cartwright, Nancy. 1983. *How the Laws of Physics Lie*. Oxford: Clarendon Press.

———. 1999. "Models and Limits of Theory: Quantum Hamiltonians and the BCS Models of Superconductivity." In *Models as Mediators*, edited by Mary S. Morgan, Margaret Morrison, 241–281. Cambridge: Cambridge University Press.

Cartwright, Nancy, Towfic Shomar, and Mauricio Suárez. 1995. "The Tool Box of Science." In *Theories and Models in Scientific Processes*, edited by William E. Herfel, Władysław Krajewski, Ilkka Niiniluoto, Ryszard Wójcicki, 137–149. Amsterdam-Atlanta: Rodopi.

Contessa, Gabriele. 2014. "Scientific Models and Representation." In *The Bloomsbury Companion to the Philosophy of Science*, edited by Steven French, Juha Saatsi, 120–137. London-New York: Bloomsbury Academic.

Copernicus, Nicholas. 1979. "A Theory Concerning the Minting of Money." *Journal of the History of Ideas* 40: 304–313.

Creager, Angela N.H., Elisabeth Lunbeck, and M. Norton Wise, eds. 2007. *Science Without Laws. Model Systems, Cases, Exemplary Narratives*. Durham-London: Duke University Press.

Downes, Stephen M. 1992. "The Importance of Models in Theorizing: A Deflationary Semantic View." In *PSA: Proceedings of the Biennial Meeting of the Philosophy of Science Association*, edited by Arthur Fine, Micky Forbes, Linda Wessels, 1992 vol. I, 142–153. Chicago: The University of Chicago Press.

Elgin, Mehmet. 2003. "Biology and a Priori Laws." *Philosophy of Science* 70: 1380–1389.

———. 2010. "Mathematical Models, Explanation, Laws, and Evolutionary Biology." *History and Philosophy of The Life Sciences* 32: 433–452.

Elliott, Colin P. 2020. *Economic Theory and the Roman Monetary Economy*. Cambridge: Cambridge University Press.

Fetter, Frank W. 1932. "Some Neglected Aspects of Gresham's Law." *The Quarterly Journal of Economics* 46: 480–495.

Fodor, Jerry. 1974. "Special Sciences (or: The Disunity of Science as a Working Hypothesis)." *Synthese* 28: 97–115.

Friend, Toby. 2016. "Laws are Conditionals." *European Journal for the Philosophy of Science* 6: 123–144.

Frigg, Roman. 2010. "Models and Fiction." *Synthese* 172: 251–268.

Frigg, Roman, and James Nguyen. 2020. *Modelling Nature: An Opinionated Introduction to Scientific Representation*. Cham: Springer.

Giere, Ronald. 1979. *Understanding Scientific Reasoning*. New York: Holt, Rinehart, and Winston.

———. 1988. *Explaining Science: A Cognitive Approach*. Chicago-London: The University of Chicago Press.

———. 1999. *Science without Laws*. Chicago-London: The University of Chicago Press.

———. 2004. "How Models Are Used to Represent Reality." *Philosophy of Science* 71: 742–752.

———. 2006. *Scientific Perspectivism*. Chicago-London: The University of Chicago Press.

Godfrey-Smith, Peter. 2006. "The Strategy of Model-based Science." *Biology & Philosophy* 21: 725–740.

Greenfield, Robert L., and Hugh Rockoff. 1992. "Gresham's Law Regained.", Accessed January 21, 2022. https://www.nber.org/papers/h0035, DOI: 10.3386/h0035

Hayek, Friedrich A. 1962. "The Uses of 'Gresham's Law' as an Illustration in Historical Theory." *History and Theory* 2: 101–102.
Jevons, William S. 1875. *Money and the Mechanism of Exchange*. New York: D. Appleton & Co.
Kindleberger, Charles. 1989. *Economics Laws and Economic History*. Cambridge: Cambridge University Press.
Knuuttila, Tarja. 2021a. "Imagination Extended and Embedded: Artifactual versus Fictional Accounts of Models." *Synthese* 198: 5077–5097.
———. 2021b. "Epistemic Artifacts and The Modal Dimension of Modeling." *European Journal for Philosophy of Science* 11. https://doi.org/10.1007/s13194-021-00374-5
Knuuttila, Tarja, and Andrea Loettgers. 2017. "Modelling as Indirect Representation? The Lotka-Volterra Model Revisited." *The British Journal for the Philosophy of Science* 68: 1007–1036.
Hanson, Norwood R. 1958. *Patterns of Discovery*. Cambridge: Cambridge University Press.
Hardt, Łukasz. 2017. *Economics without Laws*. Cham: Palgrave Macmillan.
Laymon, Ronald. 1989. "Cartwright and the Lying Laws of Physics." *The Journal of Philosophy* 86: 353–372.
Levy, Arnon. 2015. "Modeling without models." *Philosophical Studies: An International Journal for Philosophy in the Analytic Tradition* 172: 781–798.
Loewer, Barry. 2009. "Why Is There Anything Except Physics?" *Synthese* 170: 217–233.
Lorenzano, Pablo, and Martín A. Díaz. 2020. "Laws, Models, and Theories in Biology: A Unifying Interpretation." In *Life and Evolution*, edited by Lorenzo Baravalle and Luciana Zateka, 163–207. Cham: Springer.
Macleod, Henry D. 1872. *The Principles of Economical Philosophy*. London: Longmans, Brown, Green, Reader and Dyer.
Mejbaum, Wacław. 1977. "A Law of Science as an Open Formula." *Reports on Philosophy* 1: 43–49.
Mill, John Stuart. 1843. *System of Logic*. London: John W. Parker.
Morgan, Mary S. 2007. "Afterword: Reflections on Exemplary Narratives, Cases, and Model Organisms." In *Science without Laws. Model Systems, Cases, Exemplary Narratives*, edited by Angela N.H. Creager, Elisabeth Lunbeck and M. Norton Wise, 264–274. Durham-London: Duke University Press.
Morgan, Marry S., and Margaret Morrison, eds. 1999. *Models as Mediators*. Cambridge: Cambridge University Press.
Mundell, Robert. 1998. "Uses and Abuses of Gresham's Law in the History of Money." *Zagreb Journal of Economics* 2: 3–38.
Odenbaugh, Jay. 2021. "Models, Models, Models: A Deflationary View." *Synthese* 198: 1–16.
Rosenberg, Alex. 2018. "Making Mechanism Interesting." *Synthese* 195: 11–33.
Salis, Fiora, and Roman Frigg. 2020. "Capturing the Scientific Imagination." In *The Scientific Imagination*, edited by Arnon Levy, Peter Godfrey-Smith, 17–50. Oxford: Oxford University Press.
Schelling, Thomas C. 1969. "Models of Segregation." *The American Economic Review* 59: 488–493.
Schumpeter, Joseph. 1954. *History of Economic Analysis*. London: Allen & Unwin Publishers Ltd.
Selgin, George. 2020. "Gresham's Law." In *Handbook of the History of Money and Currency*, edited by Stefano Battilossi, Youssef Cassis and Kazuhiko Yago, 199–219. Singapore: Springer.
Shahvisi, Arianne. 2019. "Particles Do Not Conspire." *Journal for General Philosophy of Science* 50: 521–543.
Sober, Elliott. 1997. "Two Outbreaks of Lawlessness in Recent Philosophy of Biology." *Philosophy of Science* 64, S458–S467.
———. 2011. "A Priori Causal Models of Natural Selection." *Australasian Journal of Philosophy* 89: 571–589.
Suárez, Mauricio. 2010. "Scientific Representation." *Philosophy Compass* 5: 91–101.
———. 2015. "Deflationary Representation, Inference, and Practice." *Studies in History and Philosophy of Science* 49: 36–47.
———. 2016. "Representation in Science." In *The Oxford Handbook of Philosophy of Science*, edited by Paul Humphreys, Oxford: Oxford University Press.
Suárez, Mauricio, and Nancy Cartwright. 2008. "Theories: Tools versus Models." *Studies in History and Philosophy of Modern Physics* 39: 62–81.
Teller, Paul. 2001. "Twilight of the Perfect Model Model." *Erkenntnis* 55: 393–415.

Toon, Adam. 2012. *Models as Make-Believe. Imagination, Fiction and Scientific Representation*. New York: Palgrave Macmillan.

Van Fraassen, Bas. 1989. *Laws and Symmetry*. Oxford: Clarendon Press.

Velde, François R. 2008. "Gresham's Law." In *The New Palgrave Dictionary of Economics*, Second Edition, edited by Steven N. Durlauf, Lawrence E. Blume, 768–771. Basingstoke: Macmillan Publisher's Ltd.

Velde François R., Warren E. Weber, and Randall Wright. 1999. "A Model of Commodity Money, with Applications to Gresham's Law and the Debasement Puzzle." *Review of Economic Dynamics* 2: 291–323.

Weinert, Friedel. 1995. "Laws of Nature – Laws of Science." In *Laws of Nature*, edited by Friedel Weinert, 3–64. Berlin: De Gruyter.

———. 1996. "Weber's Ideal Types as Models in Social Sciences." *Royal Institute of Philosophy Supplements* 41: 73–93.

Weisberg, Michael. 2013. *Simulation and Similarity: Using Models to Understand the World*. Oxford: Oxford University Press.

13

THE PUZZLE OF MODEL-BASED EXPLANATION

N. Emrah Aydinonat

1. Introduction

Among the many functions of models, explanation is central to the aims and functions of science; models explain in various ways. However, the discussions surrounding modeling and explanation in philosophy have remained largely separate from each other. Accounts of models have mainly focused on questions of representation, idealization, and fiction, mostly paying attention to the relation between models and their targets (e.g., Weisberg 2013; Frigg and Nguyen 2020). Accounts of explanation, on the other hand, have predominantly concentrated on the nature and types of explanation, developing alternative accounts of explanation (e.g., Woodward 2003; Strevens 2008). As philosophers generally agree that idealizations play indispensable roles both in modeling and explanation, one possible way to bring these two lines of inquiry together is to focus on the role of idealized models in explanation. In both literatures, idealizations are commonly conceived of as distortions (however, see Carrillo and Knuuttila 2022): like fictions, they introduce falsehoods into models. There is also a common presumption that explanations must be true. The question is, if idealizations and fictions are "false," how can idealized models provide true explanations? This is the puzzle of model-based explanation (henceforth, the puzzle). To solve it, one would need to resolve many debates in the philosophy of science and, ideally, provide compatible accounts of models, truth, fiction, idealization, representation, understanding, and explanation. This chapter has the more modest aim of giving a selective and critical overview of the available strategies to solve the puzzle, mainly considering idealized models—although the discussion naturally extends to the case of fictional models. The chapter does not explicitly address applied models (i.e., those fine-tuned to a specific particular real-world target) or statistical models (including econometric models, machine learning models, and the like), although some of the strategies for solving the puzzle may apply to them as well.

2. What is the puzzle?

The puzzle has been discussed in a variety of ways. Let us look at *some* examples—reformulated here as dilemmas or trilemmas.

 DOI: 10.4324/9781003205647-16

Strevens (2008, 297) discusses the puzzle in terms of the difficulty of explaining the widespread use of idealizations for causal accounts of explanation.

(S_i) Nonveridical models cannot explain.
(S_{ii}) Idealized causal models misrepresent their targets.
(S_{iii}) Idealized causal models are commonly used to provide explanations.

Bokulich (2008, 140, fn. 9) focuses on the tension between the requirement of truth for explanation, and the practice of providing model-based explanations that are "not entirely true" (Bokulich 2009, 105).

(B_i) "Widely received philosophical accounts of scientific explanation" have a "strict requirement of truth." (2009, 104)
(B_{ii}) Scientists nevertheless explain using idealized or fictional models and provide explanations that are "not entirely true."

In the philosophy of economics, the puzzle is dubbed an explanation paradox:

(R_i) "Economic models are false.
(R_{ii}) Economic models are nevertheless explanatory.
(R_{iii}) Only true accounts can explain." (Reiss 2012, 49)

Love and Nathan (2015, 768) underscore the conflict between the goal of accurate representation in explanation and the "deliberate misrepresentation" of mechanisms in models:

(LN_i) Accurate representation is necessary for mechanistic explanations.
(LN_{ii}) Idealized models of mechanisms that are cited in mechanistic explanations misrepresent those mechanisms.

Potochnik (2017) highlights the contradiction between the beliefs that explanations must be true and that idealizations are untrue:

(P_i) Explanations must be true.
(P_{ii}) Idealizations are patently untrue.
(P_{iii}) Idealized models explain.

Examples can be multiplied. Formulations of the puzzle assume that (i) a good explanation is a true explanation, (ii) idealized models explain, and (iii) idealizations are falsehoods or distortions. Proposed solutions to the puzzle often involve the rebuttal of one or more of these assumptions.

To solve the puzzle, philosophers of science have employed multiple strategies (cf. Reiss 2012): (A) abandoning the requirement of truth for explanation (*Explanations need not be true*), (B) arguing that models cannot explain (*Models cannot explain ... but they might help)*, (C) arguing that models can contain truths, enable correct inferences, or provide true explanations despite (or thanks to) idealizations (*Models explain)*, and (D) arguing that *Models are not explanations, but tools*. Without trying to be exhaustive, let us look at examples from each strategy.

3. Explanations need not be true

Catherine Elgin (2004; 2017) famously argued that "laws, models, idealizations, and approximations which are acknowledged not to be true [...] figure *ineliminably* in the success of science" (2004, 113–114, emphasis added). Thus, she said, if we were to stick to the requirement of truth strictly, we would have to conclude that "much of our best science" is "epistemologically unacceptable" (2004, 114). Thinking of the puzzle, one way to follow Elgin is to argue that explanations need not be true. This would be a straightforward solution since nothing is puzzling about "false" models providing false explanations. Even so, philosophers rarely follow this strategy explicitly, most likely because they commonly subscribe to the factivity of explanation.[1] One notable exception is Potochnik (2017), who argues that "because idealizations are patently untrue," (93) model-based explanations cannot be true either (134). Because Potochnik accepts that models are "false" and that models can explain, she sacrifices the factivity of explanation. However, on closer inspection, she does not give up on the truth completely. She argues that "idealized representations can *truly depict* causal patterns" and that "scientific representations generate understanding of phenomena *in virtue of being true of* causal patterns" (2017, 119, emphasis added). She also substitutes the truth requirement with the following: explanations must depict real causal patterns. That is, according to Potochnik, a good explanation "must capture what is responsible for the explanandum" and "depict dependence relations" (135). Therefore, Potochnik transforms the puzzle into a new one: how can patently "untrue" models depict what is truly responsible for the explanandum? Consequently, we are no closer to the solution of the original puzzle than we started. Before moving on, note that if we were to brush Potochnik's points about explanation aside, her account would find a better home under *Models explain ... thanks to representational failure.*

4. Models cannot explain ... but they might help

The second strategy is to reject the premise that "models explain," saying that most idealized models cannot provide true explanations *by themselves,* but are nevertheless explanatorily useful. There are variations on this theme.

Consider McMullin's (1978) *hypothetico-structural* (HS) account of explanation. McMullin conceives of structural explanations as causal explanations that explain the "properties or behavior of a complex entity [...] by alluding to the structure of that entity" (139). He argues that HS explanations, where a structure is *postulated with a theoretical model* (HS model) to explain a phenomenon, are common in science. They are *hypothetical* because "a different structure might also account for the features to be explained" (139). They are provisional and tentative because they do not satisfy the truth requirement and cannot be considered complete definitive explanations. In Hempel's terms, HS explanations are *potential explanations,* i.e., explanations where the truth or falsity of the propositions constituting the explanans are not known yet (Hempel 1965, 338). They can be turned into true explanations if their explanans can be justified by de-idealization.

Craver's (2006) account of mechanistic models also acknowledges the usefulness of models for explanation, while introducing strong requirements for explanations. According to Craver, models have many explanatory functions, including tools for demonstration, sketching explanations, and conjecturing how-possibly explanations (355). However, to be an explanation or to explain, a model needs to "characterize the phenomenon" and "describe the behavior of an underlying mechanism," and the components it describes "should correspond to components in the mechanism" in its real-world target (361). Accordingly, Craver sees models on a continuum based on how well they satisfy these requirements: (i) "phenomenal" models, which are mere descriptions that do not explain (2006, 358); (ii) how-possibly models, which are "loosely constrained conjectures" (2006, 361); (iii) how-plausibly models, which are how-possibly models that fit better into what we already know; and (iv) how-actually models, which give complete descriptions of the actual mechanism "that in fact produces the phenomenon" and "show how a mechanism works, not merely how it might work" (2006, 361).

Craver's account does not accept anything less than a complete description of a mechanism for a true explanation. Note, however, that this statement concerns the descriptions of explanatory mechanisms in an explanation, not models. It does not assume that more detailed models are better (Craver and Kaplan 2020). In this account, most idealized models cannot be considered explanations, but they can be helpful in providing explanatorily relevant information that can be used in explanations. On the other hand, if a model explains, it must be because it captures the truths about actual mechanisms, and idealizations must have been harmless in this sense. Either way, the puzzle is resolved.

Kaplan's (2011) 3M account, which is related, introduces "a model-mechanism-mapping (3M) constraint on explanatory mechanistic models" (347): components of the model should map onto and match the actual mechanisms producing the phenomenon. Models that do not satisfy this requirement can only provide how-possibly explanations, not true explanations. The 3M account does not necessarily ask for de-idealization for explanatory usefulness. If there is some "model-mechanism correspondence [...] the model will be endowed with explanatory force," Kaplan argues (348). Nevertheless, according to Kaplan, anything short of a complete description of the actual mechanism(s) will be an incomplete explanation (348).

McMullin, Craver, and Kaplan agree that even though most idealized models cannot be considered explanations, they are still explanatorily useful. Many philosophers agree, and some openly propose a weaker reading of models. For example, Alexandrova (2008) suggests that we should conceive of models as *open-formulae* that help in formulating explanatory hypotheses. In this account, models are not explanations in and of themselves, but just recipes, schemata, or templates for explanatory causal claims (397). Using models in explanations requires further steps like identifying the relevant causal hypothesis and ensuring that it holds for the case at hand.

As should be clear by now, the philosophers who argue that most models cannot explain do not deny that models can be useful in the process of producing true explanations. Models have many functions, most of which can help in producing explanations: they can generate explanatory hypotheses, help explore possible explanations, provide conceptual frameworks, assist in sketching explanations, aid in devising potential explanations, etc. (e.g., see Pielou 1981; Wimsatt 1987; Odenbaugh 2005). There is considerable literature on the exploratory role of models (Aydinonat 2007; 2008; Gelfert 2015; Shech and Gelfert 2019; Massimi 2019), their modal functions (e.g., Rappaport 1989; Massimi 2019; Sjölin Wirling and Grüne-Yanoff 2021), and the relation between idealized models and how-possibly explanations (e.g., Craver 2006; Ylikoski and Aydinonat 2014; Bokulich 2014; Verreault-Julien 2019; Nguyen 2022). Most of this literature agrees with Craver, Kaplan, Alexandrova, and others that idealized models can help us discover true explanations. Interestingly, as we will see shortly, philosophers who argue that models can and do explain are also happy to accept this claim, arguing that some models are useful in developing how-possibly explanations, potential explanations, sketches, or comparison cases. All this suggests that perhaps the solution to the puzzle is to be sought by analyzing how models are used as *tools* for explanatory purposes rather than conceiving models *as* explanations (more on this below).

5. Models explain

Another way to solve the puzzle is to argue that models can provide true explanations thanks to their (i) representational adequacy, (ii) capacity to be used to make correct inferences, or (iii) falsities.

5.1 … thanks to representational adequacy

Showing that idealized models can be true or contain truths would make their ability to explain less puzzling. Many philosophers take this route. Consider Mäki's *functional decomposition account.* Mäki argues that idealized models represent selective aspects of their targets and isolate explanatorily relevant factors, and with respect to these aspects and factors, they can be true (e.g., Mäki 1992; 2010). Similarly, Strevens (2008) thinks that the function of idealizations is to remove explanatorily irrelevant aspects of the explanandum phenomenon from the model. He argues, *if "done right"* (300), an idealized model contains two parts: idealizations and "difference-makers for the explanatory target" (318). In both accounts, idealizations do not distort or misrepresent explanatory factors; they help in isolating them. If this were true, the puzzle would be resolved.

Both accounts presume that models have modular components and can be decomposed into idealized and difference-making parts. But can we decompose models in this way? Rice (2019) argues that most models do not decompose this way for two main reasons. First, idealizations are indispensable for many mathematical techniques employed in model building without which explanation would not be possible (193). Second, the assumption that idealizations will not distort a model's representation of explanatorily relevant (e.g., difference-making) relations is often not true. Hence, it is often not possible to "map the accurate parts of the model onto what is relevant and its inaccurate parts onto what is irrelevant" (194). This would at least require further steps, such as some *interpretation* of and *commentary* on the model, by the *model user*.

If Rice is right, and if some idealizations are ineliminable (Batterman 2009; see also Elgin 2004), then it becomes difficult to solve the puzzle with a naïve decompositional strategy. However, a closer look reveals that Mäki and Strevens' strategies are not so naïve after all. For example, Strevens agrees that some interpretation might be required to determine explanatory (ir)relevance and even gives a role to the *explanatory framework*, which could include the "nature and goals of a particular conversation" (2008, 151); hence the explanatory practices, conventions, and norms within a field. Similarly, Mäki (2010, 180) emphasizes the importance of the intention and *purpose* of the model user, and *model commentary* that connects a model's elements with the real world. Both Mäki and Strevens are aware that determining whether a model explains requires some interpretation and information about the context, but they do not provide enough guidance about concepts such as explanatory framework and model commentary. Moreover, both accounts allow for incomplete model-based explanations with varying degrees of explanatory power and how-possible explanations.

To overcome the difficulties that these accounts face with regard to ineliminable idealizations, Pincock (2020; 2021) recommends abandoning the commitment to the truth of model parts that perform the explanatory task and accepting that generalizations generated by models are often only *partially true*. But how can partially false generalizations provide wholly true explanations? According to Pincock, the presence of falsehoods in models is consistent with true model explanations if "there is an appropriate truth underlying each falsehood" (2021, 18). The problem with this is that we do not know how to determine the truths underlying falsehoods any better than we know the answer to the original puzzle. While Pincock talks about underlying truths, Niiniluoto (2018, 57) argues that although each idealization might not be partially or approximately true, "together with other claims, an idealized theory or model as a whole may be truthlike or sufficiently similar to the real system." Either way, the basis on which the model user infers the true claims that will constitute the explanans remains unclear.

An alternative route is to argue that model-based explanations are *partial* in the Hempelian sense. In a partial explanation, "the explanans does not account for the explanandum-phenomenon in the specificity with which it is characterized by the explanandum-sentence" (Hempel 1965, 416). Elgin and Sober (2002) think that models can provide partial explanations without necessarily being false. They argue that idealized models can explain if their idealizations are *harmless* in the sense that removing these idealizations would not "make much difference in the predicted value of the effect variable"; that is, the explanandum (448). In this account, the explanandum, *E*, need not be entailed by the explanans or be derivable from it: it is enough if the explanans implies *E′*, provided that it is *close enough* to *E* (448). The difficulty is that this approach presumes not only that successful idealizations ("done right") will be harmless in the sense that they will distort the model results only slightly, but also that the idealizations do not influence the truth of the explanans. However, if idealizations are ineliminable, how can we know that they are harmless in both senses? The similarity between *E* and *E′* will not do. Robustness analysis might help (e.g., Levins 1966), but it has limited use without empirical evidence (Orzack and Sober 1993). So, after all, it appears that idealized models can explain only if we can make sure that their idealizations play no role whatsoever in explanations, other than removing disturbing factors. Hence, given the ineliminability of idealizations, the puzzle remains (see also Bokulich 2011, 36).

5.2 … *thanks to correct inferences*

The preceding accounts in this section agree that explanatory inferences are made possible if a model (M) successfully represents a real-world target (T). An alternative approach is to reconsider what "M represents T" means and to reverse the relation between explanatory inferences and representation. The inferential conception of representation does just this, saying that if one can draw inferences about T by using M, then M represents T (e.g., Suárez 2004). Can this approach solve the puzzle?

Recall that the puzzle is a puzzle because it starts with the premise that idealized models are "false" and explanations are true. The inferentialist approach does not impose truth conditions for inferences, only requiring that the model user can make inferences about T using M. That M represents T does not imply that M provides a true explanation. Hence, conceived this way, the inferentialist approach does not even address the puzzle, let alone solve it. However, there is a version of inferentialism that explicitly addresses the puzzle.

Kuorikoski and Ylikoski (2015) amend the inferentialist approach to argue that "model-based (explanatory) reasoning" is "a matter of drawing conclusions from given assumptions using external inferential aids" (i.e., models) and this basically explains the "epistemic role of models" (3827). In this account, models help answer what-if questions and in making what-if inferences. It is argued that if M can be used to make *correct* inferences about T, then M represents T (3827).

The puzzle is then transformed into a new one: how can "false" models help in making correct inferences about their targets, and what ensures the reliability of these inferences and the truth of their conclusions? In answering these questions, Kuorikoski and Ylikoski drift away from the basic inferentialist view and draw close to Mäki and Strevens. First, they argue that some assumptions of a model help *isolate* real-world dependency relations and as such, they are not the source of falsities in a model (2015, 3829). These substantial assumptions allow model users to use what they learn about models as guides to inferences about real-world phenomena: an explanatory model, despite the falsities introduced by idealizations, "get[s] the target explanatory dependence right" (3831) thanks to its substantial assumptions. Second, they argue that derivational robustness analysis (Woodward 2006; Kuorikoski, Lehtinen, and Marchionni 2010) increases the reliability of model inferences.

In brief, in this account, substantial assumptions and robustness analysis are doing the heavy lifting with respect to the solution of the puzzle. There is a concern, however. The ineliminability of idealizations also undermines robustness analysis since altering ineliminable idealizations will change the nature of the model, and this would make model comparisons, which are required for robustness analysis, problematic (Lisciandra 2017). Thus, the advertised epistemic benefits of robustness analysis might not be realized, and the puzzle would remain (see also Verreault-Julien 2021).

On the positive side, Kuorikoski and Ylikoski avoid overemphasizing representation and settle for the modest claim concerning model explanation that models "capture a small set of explanatory dependencies that are assumed to be central" (2015, 3830), and when they are used to explain particular empirical phenomena, they do not necessarily provide complete or actual explanations: a model can sometimes bemerely "a part of a how-possibly explanation" (3831). By both emphasizing the role of robustness in enabling model-based inferences and acknowledging the selectiveness and partiality of representation, Kuorikoski and Ylikoski establish that model-based explanations cannot be fully understood by examining an isolated model, a family-of-models perspective often being needed (Ylikoski and Aydinonat 2014; see also Love and Nathan 2015).

5.3 ... *thanks to representational failure*

We have seen that accounts that focus on representational adequacy encounter difficulties with the ineliminability of idealizations. Batterman (2009, 45) argues that some idealizations are necessary for explanation, and de-idealization might even reduce the explanatory power of some models. Batterman and Rice (2014) take this argument one step further, arguing that "highly idealized models can play explanatory roles despite *near complete representational failure*" (2014, 355, emphasis added). They argue that accounts that focus merely on representational adequacy *fail to explain* why idealizations are explanatory (365). To make their point, Batterman and Rice focus on a class of explanations of macro-level patterns across systems using highly idealized models. They show that as a representation of any *particular* system, these models are inadequate because they leave out the important particular details of individual systems. Nevertheless, they argue that these models are explanatory exactly because they leave these details out. If one asks why a set of different systems are strikingly similar in a certain aspect (e.g., a macro-level pattern or feature), this might make the details of individual systems unnecessary from an explanatory point of view: the reason why these systems are similar might have nothing to do with their particular details but with some general features that are shared by all of them. If this is the case, adding detail—to increase the representational adequacy of the model from the perspective of one given individual system—would hinder the explanatory focus and power of the model. Thus, in such a case, idealization would in fact be necessary for explanation.

This point is well taken, but does it really go against the representational adequacy point of view? Representational adequacy depends on the explanatory task at hand. If the task is to explain common macro features of heterogenous systems, a model that focuses only on a small number of common features among these systems would be representationally adequate, even according to a hardheaded representationalist. When Batterman and Rice talk about "complete representational failure," they are talking about the representational adequacy of the model with respect to a particular system, which is not relevant given the explanatory task. Thus, contrary to appearances, the disagreement is not that severe (see also Lange 2015; Reutlinger 2017). Whereas representationalists argue that falsities introduced by idealizations are irrelevant, Batterman and Rice ask for *an explanation* of why the details left out are irrelevant. They argue that at least for the class of models they discuss, "the real explanatory work is done by showing why the various heterogeneous details of these systems are irrelevant and, along the way, by demonstrating the relevance of the common features" (2014, 365). Using examples from fluid dynamics and biology, they argue that these models are explanatory because they have a *backstory* showing that the model and the heterogenous systems it is supposed to explain belong to the same universality class. Note that merely providing a model that is in the same universality class as the phenomena it is supposed to explain does not provide much information. Batterman and Rice are asking for more: a demonstration, a story that explains the explanatoriness of the model. "The models are explanatory in virtue of *there being a story* about why large classes of features are irrelevant," they say (2014, 356, emphasis added).[2] For the class of models that Batterman and Rice are analyzing, this appears to solve the puzzle, in principle. In practice, however, explaining explanatory irrelevance involves considering the context of modeling and explanation. This is perhaps the larger lesson to extract from Batterman and Rice: answering why the relevant isolations are in place, why they were introduced, what modelers discovered by employing certain idealizations, etc. is crucial to an understating of explanatory value. In this regard, studying the broader context of modeling is often superior to just studying an isolated model-target pair (Aydinonat and Köksal 2019). As we will see, philosophers who see models as tools take this suggestion one step further.

Although many philosophers offer potential solutions to the puzzle, only very few address it directly. Bokulich is one of these exceptions and sets her task to show that "idealizations themselves are capable of doing some real explanatory work" (2011, 36). She first defines model-based explanation or *model explanation* as an explanation whose explanans "makes essential reference to" (38) an idealized or fictional model. Next, she defines what it means for a model to explain: a *model explains* when it shows how its elements "correctly capture the patterns of counterfactual dependence in the target system" (2017, 106) or can "'reproduce' the relevant features of the explanandum phenomenon" (2011, 39), enabling model users to answer a wide range of what-if questions. How does this solve the puzzle? How can a "false" model get the counterfactual structure right (i.e., provide a true explanation)? To answer this, Bokulich introduces another step, a *justificatory step* that specifies the model's domain of applicability, shows that the explanandum "falls within that domain" and ensures that it "adequately capture[s] the relevant features of the world" (39).

According to Bokulich, justification might come from theory, showing that the "model can be trusted as *an adequate representation* of the world" or "through various empirical investigations" (39, emphasis added). Moreover, a justificatory step is "to be understood as playing a role analogous to Hempel's condition of truth [...]." It is "intended to rule out as explanatory those models that we know to be merely phenomenological" (39, fn. 11). So, in this account, the justificatory step does "the heavy lifting" (2012, 736).

Where are we at concerning the puzzle? Bokulich's account is not very different from representationalist accounts insofar as the justificatory step is intended to ensure that falsities or fictionalizations in the model are *harmless* with respect to the model's ability to capture the truths about the counterfactual structure of the explanandum phenomenon given the explanatory task. A model might be idealized or refer to fictional entities, but what matters for explanation is whether it gets the explanatory relations, connections, structures, etc. right. The important point is that without the justificatory step, which is often contextual and dependent on the current state of knowledge (Bokulich 2012), we cannot know whether the explanatory hypotheses generated using the model are true or not. Without it, we only have sketches, templates, and potential explanations.

Nguyen (2021) argues that to get the counterfactual dependence right, a model must represent the dependence relation in its target, say, between A and B, correctly. However, in contrast to Bokulich, he contends that since the explanation concerns the relation between A and B, it cannot be said that the falsities in the model play any role in the explanation even though they "play an essential role in *generating* the explanation" (2021, 3232, emphasis added). More generally, according to Frigg and Nguyen's (2020) DEKI (Denotation, Exemplification, Keying-up, and Imputation) account of representation, idealized and fictional models *can* explain provided that they represent the target appropriately. This, however, requires (i) an appropriate *interpretation* of the model given the goals of modeling and explanation, and (ii) a *key* that translates the model's properties to the properties that will be imputed to the target. Although Frigg and Nguyen's solution to the puzzle is like Bokulich's solution in that it argues that models can explain thanks to representational failure, it does not assume that models explain by themselves. Without interpretation and keying-up there would be no model explanation according to the DEKI account. Frigg and Nguyen argue that idealizations and fictions could play an essential role in *producing* the explanation; they do not argue that they are necessarily a part of the explanation. In this sense, their account would perhaps be more at home next to those who argue that models explain thanks to their representational adequacy.

The importance of context and goals of modeling and explanation appears to be a point agreed upon by most philosophers, despite their differences. Another point of agreement, without explicit acknowledgment, seems to be that merely focusing on the model-target relation is not entirely helpful in understanding or solving the puzzle since such things as interpretation, model commentary, model use, explanatory goals, model justification, and exploration have been repeatedly invoked in dealing with the puzzle.

6. Models are not explanations, but tools

6.1 Models are not explanations

If one assumes that *explanations must be true* and *idealized models are false*, then considering false models as explanatory seems paradoxical. However, the paradox arises if we also assume either that (i) models are explanations, or (ii) that models are featured in the set of explanans directly, without any interpretation. If models are not explanations and are not commonly used in the explanans without modification, the puzzle would dissolve because the fact that models contain idealizations would not necessarily mean that the explanantia of model-based explanations are false.

Consider the first assumption. Can an idealized or fictional model be an explanation? One difficulty with equating a model with an explanation is that models and explanations might be different sorts of things. If this is true, conceiving of models as explanations would be misguided. However, even if we assume that models and explanations are the same sort of things, it is hard to conceive of idealized or fictional models as explanations. For the sake of argument, Rohwer and Rice (2016) assume that both models and explanations can be "characterized or reinterpreted as sets of propositions" (2016, 1130) and explore where this assumption leads us. They show that if this assumption were true, a model and an explanation would be identical only for some simple cases that do not involve idealizations or fictions. For a model to be identical to an explanation, its assumptions (or a subset of these assumptions) must constitute the explanans, and the model result they imply must be identical to the explanandum. If a model were to employ idealizing assumptions, this would mean that the explanans of the model explanation cannot be true—unless the model's idealizing assumptions are reinterpreted in some way. In short, in the case of idealized and fictional models, it is hard to say that there would be an identity-preserving matching between the elements of a model and an explanation if we cling to the truth requirement for an explanation. In fact, Rohwer and Rice (2016) show that in most cases, some interpretation of a model is required for an explanation. Relatedly, Marchionni (2017) argues that seeing models as explanations is too limiting and leaves out many explanatory models, particularly explanatory idealized ones. In most cases, models *help* explain rather than being explanations in themselves.

If most idealized models are not explanations, perhaps the second assumption is true, and models are featured in the set of explanans directly, without any interpretation. Recall that Bokulich argues that the explanans of a model explanation "makes an essential reference to" (2011, 38) a model. Thus, Bokulich does not equate models with explanations but argues that models are featured in explanations. In her other work, she uses alternative formulations: "makes central use of" (2018, 144) and "appeal[s] to certain properties or behaviors observed in" (2017, 104) a model. But what do these mean? Essential in what sense? What kind of reference, use, or appeal? Bokulich does not answer these questions. Moreover, her justificatory step requirement, which is external to the model, implies that there must be some interpretation of the model involved in a model explanation. In conclusion, there does not appear to be good reasons to believe in either of the two presumptions of the puzzle. This constitutes yet another solution: it is perhaps a pseudo puzzle after all.

Even though clarifying the relation between a model and an explanation is a promising strategy to resolve the puzzle, there are only a few explicit attempts at doing this. We have seen that Bokulich tells us that a model explanation makes an essential reference to a model. In contrast, Marchionni (2017) argues that we should not consider any explanation that cites a model as explanatory. She recommends asking whether the model provides *explanatorily relevant information* independently of whether the model or some of its parts are cited in the explanans. Lawler and Sullivan (2020), on the other hand, advise us against seeing model-based explanations as a special kind of explanation. The sheer diversity of models and their explanatory uses suggest that they might have a point. They argue that in most cases "model explanations" are just *model-induced explanations*, rather than models being explanations.

The statements of the puzzle appear to make the implicit assumption that idealized models, their premises, or results are or could be somehow added to the set of explanantia without modification and that the falsity of idealizations is preserved in the explanatory context. However, throughout the chapter, we have seen that when challenged, philosophers repeatedly invoked concepts such as justification, interpretation, commentary, and context to defend their versions of how models explain. In most cases, they have argued that models contribute to explanations in several ways.

6.2 *Models are tools*

Taking seriously the arguments concerning various explanatory functions of models, the importance of context, exploration, and justification suggests that we should not ignore what scientists do with their models and how they use them to explain. Looking at how models are used and manipulated for explanatory purposes can provide a key to the puzzle. There are several arguments to this effect. For example, Kennedy (2012), and Jebeile and Kennedy (2015) argue that false idealizations enable model-based explanation by allowing scientists to produce *comparison cases*. Idealizations then allow "scientists to determine what *is* causally relevant" (Kennedy 2012, 327) by comparing the model to the real-world case at hand. Jebeile and Kennedy suggest that merely focusing on representational adequacy is a mistake: explanatory functions of models can be better understood if we consider *models as "epistemic tools* that are designed by and for scientists to make inferences, and explanations" and *explanation as* "a process or *an activity*, rather than simply a product" (2015, 384, emphases added). In other words, model-based explanation cannot be fully understood without studying how model users use models to explain.

Another example is an argument by Boesch (2021) who says that dissimilarities found in models enable "novel forms of manipulation" (504) and thereby facilitate the attainment of epistemic aims, such as explanation. Many representationalists would agree on the point about dissimilarity or function of false idealizations: "It is thanks to the dissimilarities we are able to focus on what matters," they would say (see, e.g., Mäki 2011). However, Boesch, Kennedy, and Jebeile are right in arguing that representationalists put too little emphasis on how model use and manipulation make explanatory inferences possible, crippling their ability to solve the puzzle.

This point is closely related to and follows from the view that sees models as tools that scientists build and manipulate to learn about the world (Morgan and Morrison 1999; Morgan 2012). On this view, models have been characterized in a variety of related ways: as mediators (Morgan and Morrison 1999), epistemic artifacts (Knuuttila 2005), and erotetic devices (Carrillo and Knuuttila 2022; Knuuttila 2021). In contrast to the representationalist accounts of models, which start from questions concerning representation and model-target relations, this view focuses on how models are built, used, and manipulated to allow epistemic access to the world. It is argued that the widely held view that idealizations are distortions is misleading since it moves the focus away from the process and context of modeling to mere comparisons between models and their targets (e.g., Carrillo and Knuuttila 2022). This approach emphasizes that understanding models as tools capable of performing useful epistemic functions such as explanation requires moving beyond the model-target dyad and taking the purposes of model building and manipulations into account, as well as the context of modeling and its place in scientific practice (Knuuttila 2010; 2011; see also Morgan 2012).

How does this so-called artifactual approach view the puzzle? First, it sees the puzzle as pointless, since its proponents assume that there is no independent way of accessing the world without representation. Nevertheless, one lesson we can extract is that, faced with the ineliminability of idealizations, solving the puzzle appears to require more than a focus on the model-target dyad (Knuuttila 2010; Carrillo and Knuuttila 2022). Following up on this point requires getting rid of the straightjacket of representationalist and inferentialist accounts, and more detailed case studies on actual model-based explanations. Second, more recent work that characterizes models as erotetic devices provides a more explicit link between models and explanations. Recall that several philosophers argued that models provide how-possibly explanations. Knuuttila (2021) argues that by seeing models as erotetic devices that are constructed to answer theoretical and explanatory questions, we can understand the modal functions of models and hence how they can provide how-possibly explanations better. This appears to be a fruitful line of research that could help in resolving the puzzle conceived as an *inference gap*; i.e., one between what we know about the model and our model-based inferences concerning the real world.

7. Concluding remarks

This chapter started by saying that to solve the puzzle, one needs to resolve many debates in the philosophy of science and ideally provide compatible accounts of models, truth, fiction, idealization, representation, understanding, and explanation. This is because the puzzle is about all of these things. Philosophical accounts of models and explanations, on the other hand, are like scientific models in that they employ many abstractions and idealizations. They set out to answer very specific questions concerning a limited set of philosophical problems, but not about the full set of questions relating to how models help us explain. For this reason, although each account provided insights into how model-based explanations work and what they might be, they were also vulnerable to criticism, being limited by their assumptions. This short discussion suggests that we still have a long way to go in explicating how model-based explanations explain.

What should the next steps be?

Firstly, it should be obvious that preconceptions concerning what model explanations are can only take us so far. Given that there are several ways in which models can contribute to explanations, more detailed studies of how explanations are produced using models are needed (Rice, Rohwer, and Ariew 2019). Moreover, the roles of interpretation, model commentary, and explanatory context (and all other escape routes we encountered) in model-based explanation need to be investigated further, and with more case studies. Doing this might require a more historical approach (Aydinonat and Köksal 2019). It will also be useful if such studies explicitly and clearly state the explananda and explanantia of the model-based explanations that they discuss.

Secondly, and relatedly, we should pay more attention to the diversity of types of models and model-based explanations. Both Aydinonat (2008) and Marchionni (2017) suggest that in discussing model-based explanations one needs to make further elementary distinctions. Model-based explanations have different types of explananda. Some explain singular events, some explain generic events, and some explain laws and law-like generalizations. Accordingly, we have singular and generic model-based explanations, as well as model-based explanations of laws. Some model-based explanations are complete, others are incomplete, and incomplete ones are such in different ways. Then we have potential explanations, possible explanations, actual explanations, causal explanations, structural explanations, non-causal explanations, equilibrium explanations, etc. Moreover, in practice, explanations are never perfect, being far from the ideals set by philosophers. Consequently, as Marchionni (2017) suggests, if we would like to study model-based explanations, we should also be willing to incorporate varying degrees of explanatory power into our frameworks.

Thirdly, it appears that seeing models as tools or epistemic artifacts will serve the useful purpose of settling many debates, if proponents of this view can show how model use and manipulation contribute to explanation, understanding, or learning—i.e., providing an account of how the inference gap is closed.

Fourthly, recognizing that in practice many explanations make use of multiple models (e.g., Aydinonat 2018) will help in seeing the actual explanatory contribution of individual models.

And finally, more attention needs to be paid to models that fail to explain—to avoid the positive results bias in the philosophy of science.

Notes

1 It is possible for a pragmatist to argue that an explanation need not be true, but as Achinstein (1984, 290) notes, "a pragmatic theory of explanation does not commit one to anti-realism" (or realism). Even versions of a pragmatic theory of explanation employ some conditions concerning the truth or correctness of the explanation.

2 In later work, Rice (2019, 201) loosens this requirement: "scientists can justifiably use idealized models within a universality class to explain the behaviours of real-world systems in that class even when they fail to have a complete explanation of why that universality class occurs." Also, see Woodward (2018) on the sufficiency of information about irrelevance for explanation.

References

Achinstein, Pete. 1984. "The Pragmatic Character of Explanation." *PSA: Proceedings of the Biennial Meeting of the Philosophy of Science Association* 1984(2): 274–292. https://doi.org/10.1086/psaprocbienmeetp.1984.2.192509.

Alexandrova, Anna. 2008. "Making Models Count." *Philosophy of Science* 75(3): 383–404. https://doi.org/10.1086/592952.

Aydinonat, N. Emrah. 2007. "Models, Conjectures and Exploration: An Analysis of Schelling's Checkerboard Model of Residential Segregation." *Journal of Economic Methodology* 14(4): 429–454. https://doi.org/10.1080/13501780701718680.

———. 2008. *The Invisible Hand in Economics: How Economists Explain Unintended Social Consequences*. London: Routledge.

———. 2018. "The Diversity of Models as a Means to Better Explanations in Economics." *Journal of Economic Methodology* 25(3): 237–251. https://doi.org/10.1080/1350178X.2018.1488478.

Aydinonat, N. Emrah, and Emin Köksal. 2019. "Explanatory Value in Context: The Curious Case of Hotelling's Location Model." *The European Journal of the History of Economic Thought* 26(5): 879–910. https://doi.org/10.1080/09672567.2019.1626460.

Batterman, Robert W. 2009. "Idealization and Modeling." *Synthese* 169(3): 427–446. https://doi.org/10.1007/s11229-008-9436-1.
Batterman, Robert W., and Collin C. Rice. 2014. "Minimal Model Explanations." *Philosophy of Science* 81(3): 349–376. https://doi.org/10.1086/676677.
Boesch, Brandon. 2021. "Scientific Representation and Dissimilarity." *Synthese* 198(6): 5495–5513. https://doi.org/10.1007/s11229-019-02417-0.
Bokulich, Alisa. 2008. *Reexamining the Quantum-Classical Relation: Beyond Reductionism and Pluralism*. Cambridge; New York: Cambridge University Press.
———. 2009. "Explanatory Fictions." In *Fictions in Science: Philosophical Essays on Modeling and Idealization*, edited by Mauricio Suárez, 91–109. London: Routledge.
———. 2011. "How Scientific Models Can Explain." *Synthese* 180(1): 33–45. https://doi.org/10.1007/s11229-009-9565-1.
———. 2012. "Distinguishing Explanatory from Nonexplanatory Fictions." *Philosophy of Science* 79(5): 725–737. https://doi.org/10.1086/667991.
———. 2014. "How the Tiger Bush Got Its Stripes: 'How Possibly' vs. 'How Actually' Model Explanations." *The Monist* 97(3): 321–338.
———. 2017. "Models and Explanation." In *Springer Handbook of Model-based Science*, edited by Lorenzo Magnani and Tommaso Bertolotti, 103–118. Dordrecht: Springer.
———. 2018. "Searching for Non-Causal Explanations in a Sea of Causes." In *Explanation beyond Causation: Philosophical Perspectives on Non-Causal Explanations*, edited by Alexander Reutlinger and Juha Saatsi, 141–163. Oxford University Press. https://doi.org/10.1093/oso/9780198777946.003.0008.
Carrillo, Natalia, and Tarja Knuuttila. 2022. "Holistic Idealization: An Artifactual Standpoint." *Studies in History and Philosophy of Science* 91(February): 49–59. https://doi.org/10.1016/j.shpsa.2021.10.009.
Craver, Carl F. 2006. "When Mechanistic Models Explain." *Synthese* 153(3): 355–376. https://doi.org/10.1007/s11229-006-9097-x.
Craver, Carl F., and David M. Kaplan. 2020. "Are More Details Better? On the Norms of Completeness for Mechanistic Explanations." *The British Journal for the Philosophy of Science* 71(1): 287–319. https://doi.org/10.1093/bjps/axy015.
Elgin, Catherine Z. 2004. "True Enough." *Philosophical Issues* 14: 113–131. https://doi.org/10.1111/j.1533-6077.2004.00023.x.
———. 2017. *True Enough*. Cambridge: MIT Press.
Elgin, Mehmet, and Elliott Sober. 2002. "Cartwright on Explanation and Idealization." *Erkenntnis* 57(3): 441–450. https://doi.org/10.1023/A:1021502932490.
Frigg, Roman, and James Nguyen. 2020. *Modelling Nature: An Opinionated Introduction to Scientific Representation*. Cham, Switzerland: Springer.
Gelfert, Axel. 2015. *How to Do Science with Models*. New York: Springer Berlin Heidelberg.
Hempel, Carl G. 1965. "Aspects of Scientific Explanation." In Aspects of *Scientific Explanation and Other Essays in Philosophy of Science*, 331–496. New York: Free Press.
Jebeile, Julie, and Ashley Graham Kennedy. 2015. "Explaining with Models: The Role of Idealizations." *International Studies in the Philosophy of Science* 29(4): 383–392. https://doi.org/10.1080/02698595.2015.1195143.
Kaplan, David Michael. 2011. "Explanation and Description in Computational Neuroscience." *Synthese* 183(3): 339. https://doi.org/10.1007/s11229-011-9970-0.
Kennedy, Ashley Graham. 2012. "A Non Representationalist View of Model Explanation." *Studies in History and Philosophy of Science Part A*, Structures and Strategies in Ancient Greek and Roman Technical Writing, 43(2): 326–332. https://doi.org/10.1016/j.shpsa.2011.12.029.
Knuuttila, Tarja. 2005. *Models as Epistemic Artefacts: Toward a Non-Representationalist Account of Scientific Representation*. Philosophical Studies from the University of Helsinki 8. Vantaa: Department of Philosophy, University of Helsinki.
———. 2010. "Some Consequences of the Pragmatist Approach to Representation: Decoupling the Model-Target Dyad and Indirect Reasoning." In *EPSA Epistemology and Methodology of Science: Launch of the European Philosophy of Science Association*, edited by Mauricio Suárez, Mauro Dorato, and Miklós Rédei, 139–148. Dordrecht: Springer. https://doi.org/10.1007/978-90-481-3263-8.

———. 2011. "Modelling and Representing: An Artefactual Approach to Model-based Representation." *Studies in History and Philosophy of Science Part A* 42(2): 262–271. https://doi.org/10.1016/j.shpsa.2010.11.034.
———. 2021. "Epistemic Artifacts and the Modal Dimension of Modeling." *European Journal for Philosophy of Science* 11(3): 65. https://doi.org/10.1007/s13194-021-00374-5.
Kuorikoski, Jaakko, Aki Lehtinen, and Caterina Marchionni. 2010. "Economic Modelling as Robustness Analysis." *The British Journal for the Philosophy of Science* 61(3): 541–567. https://doi.org/10.1093/bjps/axp049.
Kuorikoski, Jaakko, and Petri Ylikoski. 2015. "External Representations and Scientific Understanding." *Synthese* 192(12): 3817–3837. https://doi.org/10.1007/s11229-014-0591-2.
Lange, Marc. 2015. "On 'Minimal Model Explanations': A Reply to Batterman and Rice." *Philosophy of Science* 82(2): 292–305.
Lawler, Insa, and Emily Sullivan. 2020. "Model Explanation versus Model-Induced Explanation." *Foundations of Science*, February. https://doi.org/10.1007/s10699-020-09649-1.
Levins, Richard. 1966. "The Strategy of Model Building in Population Biology." In *Conceptual Issues in Evolutionary Biology*, edited by Eliot Sober, 18–27. Cambridge: MIT Press.
Lisciandra, Chiara. 2017. "Robustness Analysis and Tractability in Modeling." *European Journal for Philosophy of Science* 7(1): 79–95. https://doi.org/10.1007/s13194-016-0146-0.
Love, Alan C., and Marco J. Nathan. 2015. "The Idealization of Causation in Mechanistic Explanation." *Philosophy of Science* 82(5): 761–774. https://doi.org/10.1086/683263.
Mäki, Uskali. 1992. "On the Method of Isolation in Economics." In *Idealization IV: Intelligibility in Science (Poznan Studies in the Philosophy of the Sciences and the Humanities 26)*, edited by C. Dilworth, 317–351. Amsterdam: Rodopi.
———. 2010. "Models and Truth: The Functional Decomposition Approach." In *EPSA Epistemology and Methodology of Science: Launch of the European Philosophy of Science Association*, edited by Mauricio Suárez, Mauro Dorato, and Miklós Rédei, 177–187. Dordrecht: Springer.
———. 2011. "Models and the Locus of Their Truth." *Synthese* 180(June 2009): 47–63. https://doi.org/10.1007/s11229-009-9566-0.
Marchionni, Caterina. 2017. "What Is the Problem with Model-based Explanation in Economics?" *Disputatio* 9(47): 603–630. https://doi.org/10.1515/disp-2017-0020.
Massimi, Michela. 2019. "Two Kinds of Exploratory Models." *Philosophy of Science* 86(December): 869–881. https://doi.org/10.1086/705494.
McMullin, Ernan. 1978. "Structural Explanation." *American Philosophical Quarterly* 15(2): 139–147.
Morgan, Mary S. 2012. *The World in the Model: How Economists Work and Think*. Cambridge; New York: Cambridge University Press.
Morgan, Mary S., and Margaret Morrison. 1999. *Models as Mediators: Perspectives on Natural and Social Science (Ideas in Context)*. Vol. 1. Cambridge: Cambridge University Press.
Nguyen, James. 2021. "Do Fictions Explain?" *Synthese* 199(1–2): 3219–3244. https://doi.org/10.1007/s11229-020-02931-6.
———. 2022. "Scientific Modeling." In *The Palgrave Encyclopedia of the Possible*, edited by Vlad Petre Glăveanu, 1–10. Cham: Springer International Publishing. https://doi.org/10.1007/978-3-319-98390-5_183-1.
Niiniluoto, Ilkka. 2018. "Explanation by Idealized Theories." *Kairos. Journal of Philosophy & Science* 20(1): 43–63. https://doi.org/10.2478/kjps-2018-0003.
Odenbaugh, Jay. 2005. "Idealized, Inaccurate but Successful: A Pragmatic Approach to Evaluating Models in Theoretical Ecology." *Biology and Philosophy* 20(2–3): 231–255. https://doi.org/10.1007/s10539-004-0478-6.
Orzack, Steven Hecht, and Elliott Sober. 1993. "A Critical Assessment of Levins's The Strategy of Model Building in Population Biology (1966)." *The Quarterly Review of Biology* 68(4): 533–546.
Pielou, Evelyn C. 1981. "The Usefulness of Ecological Models: A Stock-Taking." *The Quarterly Review of Biology* 56(1): 17–31. https://doi.org/10.1086/412081.
Pincock, Christopher. 2020. "Concrete Scale Models, Essential Idealization, and Causal Explanation." *The British Journal for the Philosophy of Science*, December. https://doi.org/10.1093/bjps/axz019.

———. 2021. "A Defense of Truth as a Necessary Condition on Scientific Explanation." *Erkenntnis*, January. https://doi.org/10.1007/s10670-020-00371-9.

Potochnik, Angela. 2017. *Idealization and the Aims of Science*. Chicago: The University of Chicago Press.

Rappaport, Steven. 1989. "The Modal View of Economic Models." *Philosophica* 44: 61–80.

Reiss, Julian. 2012. "The Explanation Paradox." *Journal of Economic Methodology* 19(1): 43–62. https://doi.org/10.1080/1350178X.2012.661069.

Reutlinger, Alexander. 2017. "Do Renormalization Group Explanations Conform to the Commonality Strategy?" *Journal for General Philosophy of Science* 48(1): 143–150. https://doi.org/10.1007/s10838-016-9339-7.

Rice, Collin. 2019. "Models Don't Decompose That Way: A Holistic View of Idealized Models." *The British Journal for the Philosophy of Science* 70(1): 179–208. https://doi.org/10.1093/bjps/axx045.

Rice, Collin, Yasha Rohwer, and André Ariew. 2019. "Explanatory Schema and the Process of Model Building." *Synthese* 196(11): 4735–4757. https://doi.org/10.1007/s11229-018-1686-y.

Rohwer, Yasha, and Collin C. Rice. 2016. "How Are Models and Explanations Related?" *Erkenntnis* 81: 1127–1148. https://doi.org/10.1007/s10670-015-9788-0.

Shech, Elay, and Axel Gelfert. 2019. "The Exploratory Role of Idealizations and Limiting Cases in Models." *Studia Metodologiczne* 39(195–232): 38. https://doi.org/10.14746/sm.2019.39.8.

Sjölin Wirling, Ylwa, and Till Grüne-Yanoff. 2021. "The Epistemology of Modal Modeling." *Philosophy Compass* n/a (n/a): e12775. https://doi.org/10.1111/phc3.12775.

Strevens, Michael. 2008. *Depth: An Account of Scientific Explanation*. Cambridge, MA: Harvard University Press.

Suárez, Mauricio. 2004. "An Inferential Conception of Scientific Representation." *Philosophy of Science* 71(5): 767–779. https://doi.org/10.1086/421415.

Verreault-Julien, Philippe. 2019. "How Could Models Possibly Provide How-Possibly Explanations?" *Studies in History and Philosophy of Science Part A* 73(February): 22–33. https://doi.org/10.1016/j.shpsa.2018.06.008.

———. 2021. "Factive Inferentialism and the Puzzle of Model-Based Explanation." *Synthese* 199(3–4): 10039–10057. https://doi.org/10.1007/s11229-021-03235-z.

Weisberg, Michael. 2013. *Simulation and Similarity: Using Models to Understand the World*. Oxford: Oxford University Press.

Wimsatt, William C. 1987. "False Models as Means to Truer Theories." In *Neutral Models in Biology*, edited by Matthew H. Nitecki and Antoni Hoffman, 23–55. Oxford: Oxford University Press.

Woodward, James. 2003. *Making Things Happen: A Theory of Causal Explanation*. Oxford: Oxford University Press.

———. 2006. "Some Varieties of Robustness." *Journal of Economic Methodology* 13(2): 219–240. https://doi.org/10.1080/13501780600733376.

———. 2018. *Some Varieties of Non-Causal Explanation*. Vol. 1. Oxford: Oxford University Press. https://doi.org/10.1093/oso/9780198777946.003.0007.

Ylikoski, Petri, and N. Emrah Aydinonat. 2014. "Understanding with Theoretical Models." *Journal of Economic Methodology* 21(1): 19–36. https://doi.org/10.1080/1350178X.2014.886470.

PART 3

Methodological aspects

Model construction, evaluation and calibration

14
ROBUSTNESS ANALYSIS

Wybo Houkes, Dunja Šešelja and Krist Vaesen

1. Introduction: what is robustness analysis?

In most modeling practices, researchers do more than *construct* and *manipulate* models. In order to draw conclusions on the phenomena that these models are taken to address, they also *vary* features of the model and study the impact of these changes on the model's behavior. These practices are found across disciplines and contexts of application and, in many of these, are known as *robustness analysis.*[1] Under this heading, we may find, e.g., ecologists examining how changes in parameter settings affect the behavior of Lotka–Volterra equations, taken to represent interacting populations of organisms, physicists studying the impact of perturbation terms on Navier–Stokes equations that represent turbulence, and social scientists checking how Schelling models of segregation depend on particular relocation rules.

For philosophers of science, the main interest has been to understand why modelers engage in this practice, i.e., what is epistemically valuable in robustness analysis (henceforth: RA).

As James Woodward put it in the context of economic modeling, the aim is to understand whether and, if so, why 'robustness (of inferences, measurements, models, phenomena and relationships discovered in empirical investigations etc.) is a Good Thing' (2006: 219). Robustness here stands for the stability of these inferences / measurements / models / phenomena under perturbations affecting the broader context or the system they belong to.

While robustness in a broader sense has been used to capture different notions of stability, we focus on the robustness of results obtained by means of scientific models and RA as a method of examining this property.[2]

The most prominent explanation, which arguably started the current discussion, is found in Richard Levins' work. Levins describes RA as a powerful strategy available to modelers like him:

> [...] we attempt to treat the same problem with several alternative models each with different simplifications but with a common biological assumption. Then, if these models, despite their different assumptions, lead to similar results we have what we

 DOI: 10.4324/9781003205647-18

can call a robust theorem which is relatively free of the details of the model. Hence *our truth is the intersection of independent lies.*

(Levins 1966: 423, emphasis added)

Levins' description makes evident the potential value of RA: it would allow modelers to derive true claims from models that are in important respects inaccurate or (over)simplified.

In the extreme case, genuine insights into complex real-world systems could be gained by studying only a variety of highly unrealistic, minimal, or 'toy' models. Although this would clearly be 'a Good Thing', philosophers have understandably suspected that it is too good to be true.

In this chapter, we review the ensuing debate. In the philosophy of science, a key role was played by William Wimsatt (1981), who identified the three central elements of RA that philosophers are still mainly concerned with: its *core definition* and *varieties*; its epistemic *value*; and the *conditions* under which it realizes this value. We briefly review each, also to set the stage for this paper.

Regarding the *central definition*, Wimsatt notes that a broad variety of practices can be gathered under the heading of 'robustness analysis'. This includes checking which implications of models remain the same under change to those models, but also practices such as triangulation, which check whether observational results remain the same under change of method. In all of these, the aim is to determine whether something is 'robust', where:

> 'X is robust = X remains invariant under a multiplicity of (at least partially) independent derivations'
>
> *(Soler et al. 2012: 3, paraphrasing Wimsatt 1981)*

Wimsatt's reasons for discussing the practices under the same heading refer directly to RA's most contentious features: its overriding purpose or epistemic value, as well as the conditions for realizing this purpose or value – the reason for engaging in these practices, and their proper implementation. For both, Wimsatt extends and partly specifies Levins' characterizations. Regarding purpose, '[a]ll the variants and use of robustness have a common theme in ... distinguishing ... which is regarded as ontologically and epistemologically trustworthy and valuable from that which is unreliable, ungeneralizable, worthless, and fleeting' (Wimsatt 1981/2012, 63). More extensively than Levins, Wimsatt identifies necessary conditions for realizing this, as well as a risk of engaging in RA:

> [a]ll these procedures require at least some partial *independence* of the various processes across which invariance is shown. And each of them is subject to a kind of systematic error leading to a kind of *illusory robustness* when we are led, on less than definitive evidence, to presume independence
>
> *(1981/2012, 64; emphasis in original)*

As the latter part of the quote makes clear – more so than Levins' much-quoted claim – there is a risk to engaging in RA. Because of this systematic error, which Wimsatt claims is intrinsic to the practice, it makes sense to investigate which, if any, of the varieties of RA meet which conditions for successfully realizing the envisaged purpose.

In this chapter, we review this debate and its results so far. We do so by focusing, like most philosophers of science, on the role of RA in the testing of model-derived theorems for an epistemic (rather than ontological) purpose. Some authors in the debate defend that RA can realize the purpose envisaged by Levins and Wimsatt – albeit only in some forms and under strict conditions and qualifications. Others reject this, mainly by problematizing Wimsatt's condition of independence; they submit that any robustness will, on closer inspection, turn out to be illusory for evidential purposes. However, critical authors have identified alternative epistemic purposes of robustness analysis. Interestingly, in some cases *negative* results (i.e., the 'fragility' of an implication) can be equally or even more valuable than positive results. So, where Woodward's framing suggests that lack of robustness is a 'Bad Thing', modeling practice does not always conform, and modelers might have many options to manage the risk of 'illusory robustness' mentioned by Wimsatt and emphasized by many philosophers.

We start by introducing some terminology and reviewing the three most prominent types of RA that have been distinguished by philosophers of science (Section 2). In Section 3, we turn from types of RA to the various roles or epistemic functions of RA, focusing on the contested issue of its evidential import. Section 4 concludes the chapter.

2. Different types of robustness analysis

Before presenting the most prominent types of RA discussed in the philosophical literature, we define some key terms. In the literature, 'robustness analysis' refers to any practice of varying aspects of the model and studying which implications remain invariant; and 'robustness' refers to any invariance revealed. In RA, relevant aspects of a model are changed, and it is established whether particular implications of this model are invariant under those changes. Implications that are invariant to a relevant degree are called 'robust'; and we refer to the models that share the implication as the 'robustness set' for the implication. Some authors, following Levins (1966), take the result of (successful) robustness analysis to be a *robust theorem* rather than an implication. This requires an additional analytical step, to identify the minimal features shared by members of the robustness set that entail the invariant implication (Weisberg 2006; Weisberg and Reisman 2008).[3]

Robustness analysis is a systematic way or strategy of identifying a robustness set: it starts from a model M, varies it in some respect, and checks whether some relevant implication p is conserved. Here, M and p may be called the 'targets'. RA is thus a generative method, rather than merely a comparative one, in which one would search for some arbitrary alternative model that has a sufficiently similar implication. Finding out, for instance, which (if any) implications are shared by magnetohydrodynamic models of fusion plasmas and Schelling's checkerboard models of segregation would not be called 'robustness analysis', if it is a meaningful scientific practice at all.

Following the relevant literature in the philosophy of science, one can distinguish three prominent types of RA. Each concerns a different way of generating the robustness set, i.e., each type primarily indicates in which respect a target model is changed to determine the effects on a target implication. In the literature, different typologies as well as nomenclatures can be found.[4] We follow Weisberg and Reisman (2008) both in the nomenclature and in distinguishing these three types of RA.

In *parameter RA*, it is checked whether some implication of a model and its auxiliary assumptions is robust to the extent that the implication holds over different parameter settings. Thus, the robustness set is generated by varying the parameters of a target model over some interval. Take, for instance, Schelling's model of social segregation. The model was designed to examine factors concerning individual preferences that lead two groups within a society to segregate. Schelling approached this question in terms of an abstract model: by randomly placing members of two groups of an equal size on a checkerboard, he examined how the population changes if we assume that individuals have a specific preference about the composition of their neighborhood. One striking result of this model is that, even when agents prefer as little as one-third of their neighborhood to consist of members of their own group, the society ends up clustered in homogenous neighborhoods: there is '*de facto* segregation with mild in-group preferences'. To examine the parameter robustness of this implication of the Schelling model, we can test whether similar *de facto* segregation obtains once we change the size of the population, the size of the checkerboard, and so forth.

Some authors have called the parameter RA 'sensitivity analysis' (e.g., Raerinne 2013; Gräbner 2018). In some disciplines, such as many forms of economic modeling, practices under the latter name indeed largely match what we described immediately above (i.e., checking to what extent implications are conserved under varying parameter settings). However, in some contexts and disciplines, 'sensitivity analysis' refers to a broader set of practices, in which one investigates how the output of a model changes under variations in input parameters (see, e.g., Saltelli et al. 2008 for an overview of techniques). Here, modelers are not specifically interested in output invariance, i.e., robustness; rather, they seek a more general understanding of the relations between a model's input and output, e.g., to identify which input variables most strongly affect output ('importance assessment'; Saltelli 2002).

Structural RA pertains to structural features of the target model, in particular its central assumptions.[5] In this case the modeler aims to find out which parts of the model's structure govern an implication. Such an analysis can take two forms. First, the modeler might remove or relax certain existing assumptions. Second, the modeler might add assumptions or replace existing ones. In either of these ways, modelers may find out which assumptions are genuine difference-makers with regard to the implication. In particular, structural robustness may test the implication's dependence on what Kuorikoski et al. (2010) distinguish as 'tractability assumptions' and 'substantial assumptions'. The former are mathematical formulations allowing for an easier or more efficient solution to the represented problem.

Such assumptions usually have no clear causal interpretation and/or are highly unrealistic. They are a 'necessary evil', intended to facilitate derivations or even to make them feasible at all. Substantial assumptions, on the other hand, are empirically informed and they serve to identify the causal structure of the target phenomenon.[6] While tractability assumptions may impact the formal representation of substantial assumptions, substantial assumptions may impact the tractability of the model. Such dependencies may restrict the scope of structural RA for some implications and assumptions: for lack of tractable results, it may be impossible to determine the effects of target implications for some relevant changes.[7]

For instance, network epistemology models, which study the impact of social networks on the production of knowledge, usually represent the structure of information flow in terms of directed graphs, with nodes standing for agents and edges between them for communication channels. This allows for the representation of communities that have varying degrees of connectivity, that is, a varying degree of information flow. Structural RA can, on the one hand, be used to examine whether changing such a tractable representation of information flow impacts the result of the model. For instance, Borg et al. (2017) examine whether the results of their model remain stable once a network in which the probability that an agent shares information with others is a parameter of the model, is replaced with networks that have stable links between agents. On the other hand, structural RA can be used to study the impact of different substantial assumptions, such as those that underpin the representation of learning. For example, if agents stand for scientists who are trying to identify the better of two available theories, we can represent their research in different ways. We could, for instance, assume that scientists have 'inertia' toward their preferred theory in the sense that they do not immediately abandon it even if they learn from others that an alternative theory appears to be better. Because such behavior of scientists may be more characteristic of some contexts of inquiry over others, the assumption is an empirical issue. Frey and Šešelja (2020) use structural RA to examine the impact of adding such inertia to the process of scientific research in Zollman's (2010) network epistemology model to specify the context of learning to which the results of the model apply.

Representational RA goes beyond structural RA in varying the representational framework, modeling technique, or modeling medium. The aim here is to determine the extent to which the target model's specific representational framework or implementation makes a difference with respect to its implications. For instance, the Volterra principle was originally derived from a set of differential equations, which describe predation at the population level. Using representational RA, one may study whether the principle also holds if the predatory system is represented in terms of individuals and their individual-level properties. Indeed, Weisberg and Reisman (2008) present a set of such agent-based models and find that they too produce the Volterra principle. From this, the authors conclude that the principle is robust across at least two representational frameworks. Another example is evolutionary game-theoretic modeling, which is based either on mathematical analytical frameworks or on computational frameworks such as agent-based models (ABMs). As de Marchi and Page (2009) argue, ABMs allow for the representation of features that may be impossible to represent in analytical models due to tractability constraints. Again, implications that are shared by ABMs and analytical models may be called (representationally) robust; here, one may conclude specifically that these implications are not artifacts of the constraints inherent to analytical frameworks. Accordingly, representational RA can, like structural RA, serve to study the impact of certain tractability assumptions in the models. Finally, modelers may vary the medium in which models are realized or implemented: Knuuttila and Loettgers (2021) discuss how, in synthetic biology, a particular network design (the repressilator model) was implemented in multiple media to test whether it produced robust oscillations in genetic networks.

Intuitively, the change made in representational RA is 'larger' than the one in structural RA: it concerns the very formal modeling technique rather than a particular tractability assumption made in implementing a technique. The robustness set in representational RA thus also consists of models that hold a stronger (intuitive) claim to being independent, since they are not constructed with the same technique or, more broadly, epistemic means. In light of Levins' claim, this would seem to make *positive results* of representational RA more valuable than those of structural or parameter RA. Admittedly, examples of such positive results are also difficult to find, whereas variations of parameter settings and structural features are part and parcel of modeling practice. This, however, may only underline how valuable representationally robust implications are if they can be obtained (cf. Houkes and Vaesen 2012; Lisciandra 2017).

The main purpose of representational RA is perhaps in negative findings: failing to replicate a result with a different framework may help to identify a set of difference-making assumptions in the original model, which may otherwise remain overlooked. For instance, in the above-mentioned field of network epistemology, Borg et al. (2018) use an agent-based model (ABM) based on argumentation dynamics to examine the robustness of results previously obtained with an ABM employing a Bayesian framework based on bandit models (Zollman 2010). While Zollman's results are representationally robust with respect to a number of ABMs employing the epistemic landscape framework (e.g., Lazer and Friedman 2007; Grim et al. 2013), Borg et al. fail to reproduce the same findings. In light of this, Borg et al. identify assumptions in their model, absent from the previous ones, which are responsible for this outcome. This in turn helps to specify the context of learning to which previous results apply.

3. Epistemic roles

Philosophers of science have discussed various epistemic roles that robustness analysis can play. Most of the discussion has focused on the question under which conditions (if any) this role can be evidential – roughly, when modelers have indeed found a 'truth at the intersection of independent lies'; and slightly less roughly, whether positive results of RA should increase one's credence in the truth of some hypothesis. Insofar as other epistemic roles have been discussed, this was mainly to identify an alternative, which would make sense of modelers' engaging in RA even when it cannot play an evidential role. In this section, we first outline the main arguments regarding the evidential role of RA and then review some of the alternative roles that have been identified.

3.1 Does robustness analysis have evidential value?

Levins' original claim can be read in a strong way: showing that an implication is robust provides evidence for regarding this claim as true, i.e., by studying whether a set of models behaves similarly, one can learn something about the world. Furthermore, Levins suggests that RA could play this strong evidential role regardless of any observational evidence for this implication or a robust theorem. This would make RA especially valuable if it is difficult or impossible to validate a model or its implications in another way, e.g., by successful prediction. Such an epistemic situation obtains in many modeling contexts across research fields, e.g., in economics, evolutionary biology, climate science, and computational philosophy. Consequently, many contributions to the debate draw on one or more of these contexts to illustrate their general claims – positive or negative – about the role of RA.

It is broadly acknowledged (e.g., Cartwright 1991; Orzack and Sober 1993; Sugden 2000) that RA does not have the strong, complementary evidential role suggested by Levins' dictum – or at least that the conditions for RA playing this role are so strict that this cannot reasonably explain the widespread use of the practice. To see why, take an extended Schelling model in which agents' behavior is governed by their 'range of vision' $R \in \mathbb{N}$ over the grid, rather than only their immediate neighbors (corresponding to $R = 1$) (Laurie and Jaggi 2003). Suppose for the sake of our argument that some interesting implication p holds for all ranges R, i.e., that p is parameter-robust with respect to R. Then, we may conclude that p is true for actual urban areas – or other target systems to which Schelling models are applied – only if a modeler has reason to believe that the correct model of the target system was to be found in this robustness set, consisting of models in which $\mathrm{R} \in [1, m]$, where m is the measure of the grid length. If the modeler does not know whether this is the case, let alone if she has reason to think that all members of the robustness set are unrealistic in some relevant respect, R-robustness alone does not have sufficient evidential impact to warrant accepting the target implication. In Levins' terms, something has been found at the intersection of lies, but it cannot be said to be a truth.

In response, it could be pointed out that this analysis ignores one important aspect of Levins' statement: the models in the robustness set need to be *independent*. Recall that according to Wimsatt, failure of independence produces illusory robustness and that the models in the set need to have 'at least some *partial* independence' (see Section 1). Only if the models are mutually independent can RA play a role similar to triangulation, making it less likely that the implication is false.

A well-established line of argumentation shows the difficulties in spelling out a suitable notion of independence. As Orzack and Sober (1993) point out, competing models of the same phenomenon cannot be *logically* independent, since the truth of one implies the falsity of all the others. Models in robustness sets tend to be competing. Take, for instance, our case from above: at most one value of *R* can be descriptively adequate for a given urban area. The models in a robustness set are not *statistically* or *probabilistically* independent, in the sense that a certain result following from one model has no bearing on the probability that the same result will be detected by the other model (cf. Schupbach 2018, who also discusses other notions of independence in this context). However, when doing RA, modelers do not review models that are independent in this way. Reviewing whether target implications still hold under changes in parameter settings requires holding fixed a model's structural assumptions. While the latter assumptions may be relaxed or changed (in structural RA), deriving implications typically requires holding fixed the model's tractability assumptions. Finally, checking whether implications hold under changes in tractability assumptions requires holding fixed substantial assumptions (including structural assumptions and those concerning parameter values). Even if this is done via representational RA, the chosen representational frameworks need to have the core substantial assumptions in common. Therefore, in a crucial sense, the models in a robustness set must share some of their assumptions. As a result, robustness might still only reflect commonalities of the models and/or the representational frameworks (cf. Odenbaugh and Alexandrova 2011, 763). In Orzack and Sober's words, there is always the possibility that 'robustness simply reflects something common among the frameworks and not something about the world those frameworks seek to describe' (1993, 539). Phrased more negatively, using Wimsatt's terms, no notion of 'partial' independence seems available that would dispel the suspicion that robustness might be illusory and confer evidential value on RA.

A recent, powerful defense of the evidential role of RA grants the validity of this critical argument, but submits that it largely misses the point of how RA can be and is used in modeling practice. According to Kuorikoski et al. (2010; 2012), epistemically impactful RA does not feature just any change to a model (let alone every possible change); rather, it focuses on specific assumptions to show that a target implication does not crucially depend on them. While this does not amount to empirical confirmation of the implication, it should also not be dismissed as epistemically futile. According to the authors, the primary value of RA lies in making our inferences *more reliable* and *increasing our confidence* in them by showing that they do not depend on problematic modeling assumptions. Since RA serves to identify assumptions that the result of the model depends on, if such assumptions are problematic, this will lower our confidence in the given inference. However, if the result appears to depend mainly on plausible substantial assumptions, we should have more confidence in its validity than prior to conducting the RA. Importantly, for RA to play such an evidential role, the substantial modeling assumptions need to be 'reasonably realistic'. In other words, RA can increase our confidence in the given inference only in combination with empirical evidence supporting the assumptions of the model.[8] Moreover, for RA to have this effect, there should be no reason, prior to RA, to think that differences in tractability assumptions of the studied models 'have a similar mathematical and empirically interpretable impact on the modelling result' (Kuorikoski et al. 2012, 898). In Levins' terms, RA requires independence of the specific *lies* inherent to each model in the set; then, a robust result might still not be true, but it is at least not an artifact of one specific lie.

This debate on the evidential role of RA has revealed that this role is tightly connected to the empirical underpinnings of the studied models. For models with realistic substantial assumptions, RA can serve to insulate (some) implications from (some) specific lies, such as particular parameter settings, auxiliary assumptions, idealizations, or even tractability assumptions. It might also provide indirect confirmation if the robustness set of the implications consists of models that have other confirmed results (Lehtinen 2018).[9] Defenders of this evidential value admit, however, that robustness could always prove to be illusory, because implications could be the result of shared and unquestioned assumptions within or even across modeling frameworks. The use of a large number of such frameworks may alleviate this worry to some extent, since they are unlikely to all share such assumptions. Whether or not they do, however, remains an empirical question; there is no strength in numbers here per se.

3.2 *Which other epistemic roles can robustness analysis play?*

An interesting side effect of the debate on the evidential role of RA has been the identification of various alternative purposes that RA can and does serve in modeling practices. The reason is, of course, that if RA cannot or hardly ever increases our credence in hypotheses, it becomes all the more puzzling 'what modelers get out of it': why is the practice so widespread if positive robustness checks do not give (additional) reasons to believe that particular modeling results are true? Even if one would assign an evidential role to RA, alternative roles could be used as supplementary reasons to engage in the practice. Here, we briefly describe several alternatives that have been identified.

3.2.1 *Discovery of causal structure*

Even those who are not convinced that RA might have evidential value often subscribe to its usefulness in generating causal hypotheses. Specifically, RA allows exploration of the implications of substantial assumptions, together with varying parameter settings, tractability assumptions, auxiliary assumptions, etc. If such substantial assumptions identify the causal structure of a phenomenon, these explorations allow statements about the conditions in the model world under which the causal mechanism holds. In this way RA allows for the formulation of more precise causal hypotheses,[10] or to identify the common causal mechanism in a family of models, rather than providing evidence for any implications. Thus, Knuuttila and Loettgers (2011) distinguish 'causal isolation' RA from the 'independent determination' RA on which most of the philosophical literature has focused. In this epistemic role, RA can also help to formulate pursuit-worthy hypotheses. It does so by providing 'inquisitive reasons' (Fleisher 2022), which are reasons that concern promoting successful inquiry (such as showing that a hypothesis is testable, that it is based on a heuristic analogy, etc.).

By identifying specific conditions under which the given causal mechanism holds in the model world, RA helps to delineate the application domain in which the causal hypothesis should be further pursued in terms of empirical studies.

3.2.2 Deepened causal understanding

Relatedly, and perhaps a bit more distinctively, RA might help to develop and deepen our causal understanding of real-world systems and phenomena. It may do so by presenting a way in which to vary systematically the factors that could be causally responsible for certain system behavior – albeit through their representation in substantial assumptions, and heavily mediated by tractability assumptions and other auxiliaries. All forms of RA would appear to be useful in this respect. Parameter RA helps to study the range under and extent to which factors cause behavior (e.g., how the 'range of vision' influences segregation in Schelling models; Laurie and Jaggi 2003). Structural RA contributes to developing more sophisticated causal understanding, because it allows studying the effects of adding or removing factors as well as possible confounders and mediators. Finally, representational RA allows studying alternative or supplementary causal mechanisms, perhaps at different levels of organization (e.g., population-level versus individual).[11]

3.2.3 Elimination of (alternative) potential explanations

As a complement to the previous role, RA might serve an eliminative role in explanatory reasoning, as argued by Schupbach (2018). Suppose that we have a model that has some empirically validated implications and we are trying to explain why the model gives this result. Then, studying how these implications of the model vary under changes to the model may serve to rule out competing possible explanations of this kind. For instance, in the case of the Volterra principle, this means ruling out various explanations which stipulate that the result is due to idealizing and simplifying assumptions in the model. Specifically, if such competing alternatives entail that implications fail to hold under particular changes, this provides a way of discriminating between them and the target explanation. In the case of the above example, this means that RA can help to discriminate between two explanations: that the model accurately represents the given predator-prey dynamics and therefore continues to behave in accordance with the Volterra principle if we relax certain unrealistic assumptions; or that the result is due to the given unrealistic assumption (so that, once this assumption is removed, we should fail to observe the same output). RA could thus amount to a strategy of systematically and incrementally generating such explanatorily discriminating means.[12]

3.2.4 Calibration of alternative modeling techniques

RA may have a role in *constructing* models rather than in studying and evaluating their implications. This is most straightforwardly illustrated with representational RA. When developing a modeling technique as an alternative to existing approaches, some implications may be used to calibrate or even test the alternative: only if those implications can be replicated, the alternative will be considered. Houkes and Vaesen (2012, 361) argue that this applies to Weisberg and Reisman's agent-based alternative to Lotka–Volterra models: an alternative that does not display the Volterra property (i.e., the desired implication) is discarded in favor of another, more sophisticated agent-based model.

Structural RA might play out similarly, for instance if changing structural features of a model only reproduces desired results under specific parameter settings or with additional auxiliary assumptions. This calibrative role of RA is, in many ways, complementary to the eliminative role discussed above. Clearly, it has no bearing on one's credence in any hypothesis, since there is not even the semblance of independence; thus, if one adopts Levin's and Wimsatt's characterization of RA, this practice may be taken as a degenerate case of the practice.

4. Conclusion

Robustness analysis is commonly used in modeling practices as the method of examining the stability of results under various perturbations of features of the model. In light of this, philosophers of science have inquired which kinds of RA there are, and what exactly their epistemic function is. In this chapter, we have reviewed this debate. We started by defining key terms and distinguishing between parameter RA, structural RA, and representational RA. While each kind of RA can increase our understanding of the studied models, philosophers have debated whether any of them can have an evidential, confirmatory value in the sense that a robust modeling result can be considered true of real-world phenomena. Even though there is general consensus in the literature that RA on its own does not provide an evidential import of that kind, different proposals of its alternative epistemic functions have been put forward. As our discussion shows, RA can help to improve not only our understanding of the inner functioning of models, but also our causal and explanatory insights obtained by them. Yet, for RA to play such a role, it has to be combined with empirical methods, on the basis of which the model and its results can be empirically embedded in the first place. Whether and to which extent this is possible remains a challenge for each domain of modeling, especially for those researchers that employ either highly idealized, theoretical models or highly complex but difficult-to-validate models. Moreover, which types of RA are most epistemically useful in such cases – and whether *negative* results of RA can be as much of a Good Thing as positive results – is another question that may vary from one modeling context to another.

Acknowledgements

The research by Dunja Šešelja is supported by the Deutsche Forschungsgemeinschaft (DFG, German Research Foundation) – project number 426833574.

Notes

1 See, e.g., Soler et al. (2012) for discussions of robustness analysis in various contexts of application.
2 This means that we leave out other forms of robustness analysis, which would fit under Wimsatt's more encompassing 'multiple-determination' heading. For instance, scholars have written about evidence robustly corroborating theories (Eronen 2015; Calcott 2011), about phenomena being robustly present in different contexts (Calcott 2011), or about robustness of scientific knowledge in a given domain (Šešelja and Straßer 2014).
3 The same goes for understanding robustness analysis in terms of *robustness arguments*, e.g., Stegenga and Menon (2017), in which the set of statements in our scheme are the premises for the conclusion that p is more likely to be true.

4 For instance: many authors follow Woodward (2006) in referring to parameter and structural robustness as 'derivational robustness'; Kuhlmann (2021) calls representational robustness 'multiple-model robustness'; etc.
5 We prefer the term 'structural' to 'derivational' RA since, similar to 'parameter' RA, it indicates the aspect of a model that is varied during the generative process of analysis.
6 Kuorikoski et al. (2010) also distinguish 'Galilean assumptions' which are idealizations used to isolate the purported causal mechanism from all other interfering factors (see also, e.g., Mäki 1994).
7 While Kuorikoski et al. (2010) consider derivational RA as an RA with respect to tractability assumptions, Raerinne (2013) introduces RA with respect to substantial assumptions as 'sufficient parameter RA' since different parameter values could be based on different substantial assumptions in the model.
8 In a similar defense of RA, Michael Weisberg (2006) refers to the 'low-level confirmation' of central modeling assumptions. Houkes and Vaesen (2012) identify some complications in this account. See Lloyd (2010) for an application of evidential RA to climate models based on Weisberg's account, and Parker (2011) and Justus (2012) for a discussion of complications.
9 Schupbach (2018; Section 2) provides an in-depth review of other attempts to coin out the evidential value of RA. Also see Fuller and Schulz (2021) and Casini and Landes (2022).
10 One way to develop this idea is in terms of open formulae – templates for formulating hypotheses that should then be empirically examined (Odenbaugh and Alexandrova 2011, 769).
11 Paternotte and Grose (2017) discuss this and other explanatory roles of RA, focusing on evolutionary biology.
12 Schupbach (2018; Section 3.2) reconstructs this role of RA so that it can have evidential value (e.g., with regard to mutually exclusive competing explanations). We discuss it as an alternative role here since identifying this eliminative role does not seem to depend strictly on this reconstruction; Forber (2010), for instance, identifies a similar role for RA prior to empirical testing.

References

Borg, AnneMarie, Daniel Frey, Dunja Šešelja, and Christian Straßer. 2017. "Examining Network Effects in an Argumentative Agent-Based Model of Scientific Inquiry." In *International Workshop on Logic, Rationality and Interaction*, ed. Alexandru Baltag, Jery Seligman and Tomoyuki Yamada, 391–406. Lecture Notes in Computer Science 10455. New York: Springer.

———. 2018. "Epistemic Effects of Scientific Interaction: Approaching the Question with an Argumentative Agent-Based Model." *Historical Social Research* 43(1): 285–309.

Calcott, Brett. 2011. "Wimsatt and the Robustness Family: Review of Wimsatt's *Re-engineering Philosophy for Limited Beings*." *Biology & Philosophy* 26: 281–293.

Cartwright, Nancy. 1991. "Replicability, Reproducibility, and Robustness." *History of Political Economy* 23: 143–155.

Casini, Lorenzo, and Jürgen Landes. 2022. "Confirmation by Robustness Analysis: A Bayesian Account." *Erkenntnis*. DOI: 10.1007/s10670-022-00537-7

De Marchi, Scott, and Scott E. Page. 2009. "Agent-Based Modeling." In *The Oxford Handbook of Political Methodology*, ed. Janet M. Box-Steffensmeier, Henry E. Brady, and David Collier, 71–94. Oxford: Oxford University Press.

Eronen, Markus I. 2015. "Robustness and Reality." *Synthese* 192(12): 3961–3977.

Fleisher, Will. 2022. "Pursuit and Inquisitive Reasons." *Studies in History and Philosophy of Science* 94: 17–30.

Forber, Patrick. 2010. "Confirmation and Explaining How Possible." *Studies in History and Philosophy of Science Part C* 41: 32–40.

Frey, Daniel, and Dunja Šešelja. 2020. "Robustness and Idealization in Agent-Based Models of Scientific Interaction." *British Journal for the Philosophy of Science* 71(4): 1411–1437.

Fuller, Gareth P., and Armin W. Schulz. 2021. "Idealizations and Partitions: A Defense of Robustness Analysis." *European Journal for Philosophy of Science* 11: 1–15.

Gräbner, Claudius. 2018. "How to Relate Models to Reality? An Epistemological Framework for the Validation and Verification of Computational Models." *Journal of Artificial Societies and Social Simulation* 21(3): 8.

Grim, Patrick, Daniel J. Singer, Steven Fisher, Aaron Bramson, William J. Berger, Christopher Reade, Carissa Flocken, and Adam Sales. 2013. "Scientific Networks on Data Landscapes: Question Difficulty, Epistemic Success, and Convergence." *Episteme* 10(4): 441–464.

Houkes, Wybo, and Krist Vaesen. 2012. "Robust! Handle with care." *Philosophy of Science* 79(3): 345–364.

Justus, James. 2012. "The Elusive Basis of Inferential Robustness." *Philosophy of Science* 79(5): 795-807.

Knuuttila, Tarja, and Andrea Loettgers. 2011. "Causal Isolation Robustness Analysis: The Combinatorial Strategy of Circadian Clock Research." *Biology & Philosophy* 26: 773–791.

———. 2021. "Biological Control Variously Materialized: Modeling, Experimentation and Exploration in Multiple Media." *Perspectives on Science* 29(4): 468–492.

Kuhlmann, Meinard. 2021. "On the Exploratory Function of Agent-Based Modeling." *Perspectives on Science* 29(4): 510–536.

Kuorikoski, Jaakko, Aki Lehtinen, and Caterina Marchionni. 2010. "Economic Modelling as Robustness Analysis." *British Journal for the Philosophy of Science* 61(3): 541–567.

———. 2012. "Robustness Analysis Disclaimer: Please Read the Manual before Use!" *Biology & Philosophy* 27(6): 891–902.

Laurie, Alexander J., and Narendra K. Jaggi. 2003 "Role of 'Vision' in Neighbourhood Racial Segregation: A variant of the Schelling Checkerboard Model." *Urban Studies* 40(13): 2687–2704.

Lazer, David, and Allan Friedman. 2007. "The Network Structure of Exploration and Exploitation." *Administrative Science Quarterly* 52(4): 667–694.

Lehtinen, Aki. 2018. "Derivational Robustness and Indirect Confirmation." *Erkenntnis* 83(3): 539–576.

Levins, Richard. 1966. "The Strategy of Model Building in Population Biology." *American Scientist* 54(4): 421–431.

Lisciandra, Chiara. 2017. "Robustness Analysis and Tractability in Modeling." *European Journal for Philosophy of Science* 7(1): 79–95.

Lloyd, Elisabeth A. 2010. "Confirmation and Robustness of Climate Models." *Philosophy of Science* 77(5): 971–984.

Mäki, Uskali. 1994. "'Isolation, Idealization, and Truth in Economics." In *Idealization VI: Idealization in Economics*, ed. Bert Hamminga and Neil B. De Marchi, 147–168. Amsterdam: Rodopi.

Odenbaugh, Jay, and Anna Alexandrova. 2011. "Buyer Beware: Robustness Analyses in Economics and Biology." *Biology & Philosophy* 26(5): 757–771.

Orzack, Steven H., and Elliott Sober. 1993. "A Critical Assessment of Levins's 'The Strategy of Model Building in Population Biology' (1966)." *Quarterly Review of Biology* 68(4): 533–546.

Parker, Wendy S. 2011. "When Climate Models Agree: The Significance of Robust Model Predictions." *Philosophy of Science* 78(4): 579–600.

Paternotte, Cédric, and Jonathan Grose. 2017. "Robustness in Evolutionary Explanations: A Positive Account." *Biology & Philosophy* 32(1): 73–96.

Raerinne, Jani. 2013. "Robustness and Sensitivity of Biological Models." *Philosophical Studies* 166(2): 285–303.

Saltelli, Andrea. 2002. "Sensitivity Analysis for Importance Assessment." *Risk Analysis* 22(3): 579–590.

Saltelli, Andrea, Marco Ratto, Terry Andres, Francesca Campolongo, Jessica Cariboni, Debora Gatelli, Michaela Saisana, and Stefano Tarantola. 2008. *Global Sensitivity Analysis: The Primer*. Hoboken, NJ: John Wiley & Sons.

Schupbach, Jonah N. 2018. "Robustness Analysis as Explanatory Reasoning." *The British Journal for the Philosophy of Science* 69(1): 275–300.

Šešelja, Dunja. 2021. "Exploring Scientific Inquiry via Agent-Based Modelling." *Perspectives on Science* 29(4): 537–557.

Šešelja, Dunja, and Christian Straßer. 2014. "Epistemic Justification in the Context of Pursuit: A Coherentist Approach." *Synthese* 191(13): 3111–3141.

Soler, Léna, Emiliano Trizio, Thomas Nickles, and William C. Wimsatt, eds. 2012. *Characterizing the Robustness of Science: After the Practice Turn in Philosophy of Science*. Boston Studies in the Philosophy of Science 292. Dordrecht: Springer.

Stegenga, Jacob, and Tarun Menon. 2017. "Robustness and Independent Evidence."*Philosophy of Science* 84(3): 414–435.
Sugden, Robert. 2000. "Credible Worlds: The Status of Theoretical Models in Economics." *Journal of Economic Methodology* 7: 169–201.
Weisberg, Michael. 2006. "Robustness Analysis." *Philosophy of Science* 73: 730–742.
Weisberg, Michael, and Kenneth Reisman. 2008. "The Robust Volterra Principle." *Philosophy of Science* 75: 106–131.
Wimsatt, William C. 1981. "Robustness, Reliability, and Overdetermination." In *Scientific Inquiry in the Social Sciences*, ed. Marilynn B. Brewer and Barry E. Collins, 123–162. San Francisco: Jossey-Bass. Reprinted in: (Soler et al. 2012), 61–87.
Woodward, James. 2006. "Some Varieties of Robustness." *Journal of Economic Methodology* 13(2): 219–240.
Zollman, Kevin J.S. 2010. "The Epistemic Benefits of Transient Diversity." *Erkenntnis* 72(1): 17–35.

15
MODEL EVALUATION

Wendy S. Parker

1. Introduction

Assessment of model quality occurs informally throughout the model development process. For instance, when constructing a model, the aim is not to produce just any model of the target system but to produce a good model, and this informs the choices made. Model evaluation, however, is also frequently identified as a distinct step in model development, occurring after a model has been fully constructed. It is this formal evaluative step in model development that will be the focus of the present chapter.

In several scientific and engineering domains, there has been extensive discussion of appropriate terminology and methods to employ in model evaluation, in some cases resulting in official guides for evaluation under the auspices of professional societies (e.g., AIAA 1998). In many other modeling contexts, however, conceptual frameworks and standards of practice for model evaluation are not articulated explicitly, and evaluation activities are only selectively reported. This can make it difficult for individuals not directly involved in the evaluation process to interpret evaluative claims (e.g., that a model is "credible") or to have a sense of the strength of evidence that underlies those claims.

The topic of model evaluation has received relatively little attention from philosophers of science. An influential contribution by Oreskes et al. (1994) called attention to the limits of what can be learned in model evaluation. Teller (2001) emphasized the purpose-relativity of model quality, understood as relevant similarity (see also Cartwright 1983; Giere 1988). More recently, Weisberg (2013) has offered an account of model-target similarity intended to facilitate the evaluation of scientific models, and Parker (2020) has advocated for an adequacy-for-purpose approach to model evaluation. A number of other contributions have emerged as a byproduct of work on the epistemology of computational modeling (e.g., Winsberg 1999; 2010; 2018; Lloyd 2010; Lenhard and Winsberg 2010; Jacquart 2016). Very recently, a massive volume edited by Beisbart and Saam (2019), *Computer Simulation Validation*, brings together both philosophical and scientific perspectives on the evaluation of computational models and constitutes a major addition to the literature.

The present chapter situates existing work within a general philosophical discussion of model evaluation.[1] Section 2 addresses a foundational question: what does it mean for

DOI: 10.4324/9781003205647-19

a model to be a good model? Three common answers are presented: quality as accurate and comprehensive representation, quality as relevant similarity, and quality as fitness-for-purpose. Section 3 considers the task of model evaluation from the perspective of each of these three conceptions of model quality and discusses allied approaches to evaluation that have been advocated by scientists and philosophers. Section 4 outlines several obstacles and challenges that can arise when performing model evaluation, which can prevent evaluators from reaching confident conclusions about model quality. Finally, Section 5 summarizes key points and identifies some directions for future research.[2]

2. Models and model quality

Assessment of the quality of a scientific model depends, at least implicitly, on some conception of model quality, i.e., of what constitutes a good model. This section presents three common conceptions of model quality, which are associated with different views of what scientific models are: quality as accurate and comprehensive representation, associated with a view of models as representations; quality as relevant similarity, associated with a view of models as representational tools; and quality as fitness-for-purpose, associated with a view of models as tools or artifacts, not necessarily representational.

What will here be called the *mirror view* of model quality is only sometimes explicitly espoused, but it seems implicit in much modeling practice (see also Saltelli et al. 2020). In this view, a model is a representation, and it is of higher quality *the more accurately and comprehensively it represents its target system*. The hypothetical limit is a model that mirrors the target system, in the sense that every element (part, property, relationship) of the target system is represented by a corresponding element in the model, and with perfect accuracy.[3] Increasing the comprehensiveness of a model by adding a representation of a target system process that was previously unrepresented, or increasing the fidelity with which some feature of the target system is represented, will count as improving the model on the mirror view, regardless of the purposes for which the model will be used. Conversely, idealizations, distortions, and omissions in representation necessarily detract from model quality on this view, regardless of the purposes for which the model will be used.

On many other views of model quality, however, the intended use of the model *is* relevant to the assessment of model quality. In the philosophy of science, a prominent view is that model quality is a matter of *relevant similarity*: a good model is *similar enough to its target in the relevant respects*, where the relevant respects are determined by the model user's purpose (Giere 1988; 2004; Teller 2001; Weisberg 2013). A closely related view is expressed in terms of representation: a good model represents its target system with sufficient fidelity in the relevant respects, given the modeler's purpose. This way of thinking of model quality is associated with a view of models as representational tools: they are representations, intended to be useful for particular purposes (e.g., predicting X with specified accuracy, explaining Y).

If model quality is a matter of relevant similarity (or relevant representational fidelity), then idealizations, distortions, and omissions in modeling do not necessarily detract from model quality; it depends on whether they render the model dissimilar to its target in ways that impede achieving the purposes of interest. Indeed, idealizations, distortions, and omissions can even enhance the quality of a model in many cases, insofar as the resulting model represents the target system in a way that better serves the purpose of interest (see also Bokulich 2013; Potochnik 2018). For example, "artificial viscosity" in fluid simulations is a distortion that allows for a more accurate prediction of the evolution of shock waves (see Winsberg and Mizra 2017 for more examples). Likewise, if the aim is to learn whether a particular causal process plays an important role in producing a phenomenon, it might be advantageous for a computer simulation model to omit that process (while representing other contributing processes with sufficient fidelity) in order to reveal how the phenomenon changes, if at all, when the process is absent.

A third perspective on model quality is closely associated with an understanding of scientific models as tools or artifacts (Caswell 1976; Beck 2002; Knuuttila 2005; 2011; NRC 2007; Boon and Knuuttila 2009; Currie 2017). On this *fitness-for-purpose* view, a model is a good model to the extent that it *has properties that make it a suitable tool for the task at hand*. These properties will often include more than representational properties – properties like manipulability, computational tractability, cognitive accessibility, and so on, can contribute to a model's quality. Moreover, whether a model has such properties can vary with the context of the use, i.e., with the model user, with the methodology employed, and with the background conditions in which the use of the model will occur. For example, a model might be computationally tractable for a user who has access to a supercomputer, but not for a user who has only an ordinary desktop computer. The fitness-for-purpose of a model thus can vary with the context of use (Parker 2020).[4]

As with the relevant similarity view, idealizations, simplifications, and omissions need not detract from the model's quality on a fitness-for-purpose view and are sometimes advantageous. Here, however, they can be advantageous not only for reasons having to do with how the model relates to a target system but also for reasons having to do with how the model relates to model users and other features of the context of use. For example, compared to a complex, hyper-realistic model, a simpler model, which omits many processes at work in the target system and represents others in an idealized way, might better facilitate understanding of a particular phenomenon, given humans' (i.e., users') cognitive limitations (see also Isaac 2013; Potochnik 2018). Indeed, such a view regarding the value of simple models for purposes of understanding is frequently expressed in the study of complex systems.

3. Model evaluation

The aim of model evaluation is to learn about model quality, whether quality is conceptualized as accurate and comprehensive representation, relevant similarity, fitness-for-purpose, or in some other way.[5] Put differently, model evaluation activities are directed at obtaining evidence regarding hypotheses of interest about model quality, such as the hypothesis that the model is similar enough to the target in the relevant respects, given the modeling purpose of interest. This section considers the task of model evaluation from the perspective of each of the views of model quality introduced in Section 2 and discusses allied approaches to model evaluation that have been advocated by scientists and philosophers. Throughout, the analysis attends to two complementary sources of evidence regarding model quality: evidence related to the model's *composition*, i.e., its ingredients and how they are put together, and evidence related to the model's *performance*, i.e., its behavior or output.[6] Although it will not be emphasized below, it is important to recognize that evaluation is typically an iterative process: what is learned when evaluating a model often leads to further adjustments to the model, after which the new version of the model is evaluated, and so on.[7]

Mirror view. From the perspective of the mirror view, model evaluation is an activity that seeks to learn to what extent a model accurately and comprehensively represents a target system. When examining a model's composition, the mirror-view evaluator will be interested in whether any elements of the target system are omitted from (i.e., not represented at all in) the model as well as how closely, from the perspective of theoretical and other background knowledge, the elements of the model come to perfectly representing the corresponding elements of the target system. For example, the evaluator of a mathematical model of an ecosystem might note that some species in the ecosystem have not been represented at all in the model and that interactions among other species have been represented in a quite simplistic way relative to what is known about those species' interactions; this will be judged to detract from the model's quality.

When examining model performance, the mirror-view evaluator will be interested in how closely the behaviors of the model resemble those observed for the target system in corresponding circumstances. For mathematical models and computer simulation models, this typically will involve comparing the values of model variables to observational data. Assessing the fit between model results and observational data is considered a crucial part of the evaluation of such models regardless of the conception of model quality adopted. For the mirror-view evaluator, output for *every* model variable (and combinations/aggregations of such variables) for which observations are available will in principle be of interest, since any such model-data comparison can provide some (indirect) evidence regarding the extent to which the model accurately and comprehensively represents the target system. Performance scores for individual variables might even be averaged or otherwise aggregated to produce an indication of "overall" performance.

In scientific publications, evaluative discussions of computational models sometimes are strongly suggestive of a mirror view of model quality. Reasons are given for thinking that a model is a "credible" representation of the target system in some general or overall sense. For instance, it might be reported that a model "includes" (i.e., includes some representation of) many target system processes, that the model's core equations are grounded in established theory, and that the model achieves a relatively good fit with available observational data across a range of output variables. In some cases, this approach to model evaluation might reflect a simple commitment to a mirror view of model quality. In other cases, however, it may be intended as a kind of "purpose-neutral" evaluation, motivated by the expectation that the model will be used for a wide range of (perhaps yet-to-be-fully-specified) purposes.[8] Either way, from the fact that a model is "credible" in this general or overall sense, it does not follow that any particular results from the model will be accurate, since a model that represents a target system reasonably well in some overall sense might represent relatively poorly the aspects that matter for a specific question or task.

Relevant similarity view. Unlike the mirror-view evaluator, the evaluator of relevant similarity will be interested in only some aspects of a model's composition and performance, namely, those for which sufficient similarity to the target system is needed in order for the model to serve the purpose of interest. For example, when evaluating an animal model to be used in investigating the toxicity of a chemical, the relevant-similarity evaluator might check whether a particular set of biochemical pathways operative in humans – and expected to mediate any toxic effects of the chemical – are also operative in the animal; the evaluator will not be concerned with aspects of the animal's composition that are expected to make no difference to whether it will be informative about the toxicity of the chemical in humans. Continuing with the example, when focusing on model performance, an evaluator might investigate how the response of the animal to other toxic chemicals compares to the known effects of those chemicals in humans, but many other aspects of the animal's behavior – such as whether it is quieter than humans when eating, whether it wakes up earlier in the morning than humans – are unlikely to be of interest.

The selectiveness of relevant similarity evaluation is formalized in Weisberg's (2013) weighted-feature-matching account of model-target similarity. On his account, models can be assigned a similarity score, depending on the extent to which they have specific features (attributes and/or mechanisms) that match those of the target system, where the features of interest and their relative importance are determined by the modeling purpose (e.g., predicting X, explaining how-possibly Y). What counts as a "match" between features of a model and target system on this account merits further attention, however; in general, relevant features of a model do not need to be identical to those of the target system in order for a model to serve a purpose of interest, yet what it means for features to be "sufficiently similar" is not so clear either (Parker 2015; Khosrowi 2020). A further question is whether Weisberg's account can be usefully applied in practice (Jacquart 2016).

Fitness-for-purpose view. Fitness-for-purpose evaluation seeks to determine whether a model is a suitable tool for the task at hand. In contrast to mirroring and relevant similarity evaluation, fitness-for-purpose evaluation will often need to consider more than just how a model relates to a target; it will need to consider how the model relates to the model user and other aspects of the context of use (Parker 2020). The evidence cited regarding the model's fitness-for-purpose can likewise be broader. Consider, for example, an evaluation of the fitness of a computer model for the purpose of ranking the effectiveness of various possible interventions to curb algae blooms in a given lake. Evidence that the model is fit for purpose could include not only facts about how the model represents certain biological and chemical processes in the lake but also the fact that the model has an interface that allows its users to easily adapt the model to represent the different possible interventions and the fact that the model takes only a short period of time to run on available computers.[9]

A fitness-for-purpose approach to the evaluation of models has been advocated by a number of practitioners in the earth and environmental sciences. An important early contribution comes from Caswell (1976) in the context of ecological modeling. He argues that, since models are artificial systems designed to serve particular purposes, they should be evaluated relative to their intended task environment; for some purposes, such as gaining insight or understanding, whether a model produces output that closely fits observations may be relatively unimportant. Building on this, Beck (2002) notes that environmental models are used not only for "scientific" purposes, such as making predictions or gaining understanding, but also for various "pragmatic" purposes, such as supporting decision-making, formulating public policy, or communicating scientific information to lay audiences, and he raises the question of how to evaluate the fitness of models for such pragmatic purposes. Some progress in this regard is made in a report from the U.S. National Academies of Science, *Models in Environmental Regulatory Decision Making* (2007). It develops an extensive list of considerations relevant to evaluating the fitness-for-purpose of environmental models in regulatory contexts, including considerations like model transparency to stakeholders.

Many other discussions of fitness-for-purpose evaluation, however, largely ignore the context of use of a model, focusing attention instead on how to probe whether a model represents its target system accurately enough in relevant respects to provide sought-after information. Here, the language of fitness-for-purpose (or adequacy-for-purpose) is adopted, but the evaluation is essentially concerned with relevant similarity or relevant representational fidelity. For instance, Baumberger et al. (2017) develop a framework for evaluating the fitness-for-purpose of climate models for projecting long-term changes in climate, but the potential lines of evidence that they identify – coherence with background knowledge, sufficient fit with relevant observational data, and robustness of projections across models – are of interest because they bear on whether models represent sufficiently well the causal processes that will shape the long-term evolution of climate characteristics (see also Knutti 2018 on process understanding and Kawamleh 2022 on process-based evaluation). Another example can be found in the context of hydrological modeling. Beven (2018) argues for the benefits of a falsificationist approach to fitness-for-purpose evaluation, whereby hydrological models – understood as hypotheses about how water catchments function – are tested against relevant observational data and rejected if they fail to meet pre-specified performance criteria identified in light of the modeling purpose.

Pre-specified performance criteria are also an important part of the evaluation of the fitness-for-purpose of computational models in engineering contexts. Here, it is well recognized that the fitness-for-purpose of a model can depend on more than how it represents a target system: computational demands, adaptability, ease of use for model users of a given experience level, etc., can all be relevant (Oberkampf and Roy 2010, 37). Nevertheless, the core of model evaluation is often conceptualized as consisting of two activities: *verification* investigates whether the model's computational algorithm delivers results that approximate closely enough the solutions of the modeling equations that have been selected; *validation* investigates whether those modeling equations represent the target system with sufficient fidelity in relevant respects for the application of interest, primarily by comparing results obtained from the computational model with observational data (see contributions in Beisbart and Saam 2019 for further discussion of these concepts and related practices). Ideally, this comparison is pursued in a systematic way such that individual model components (representing a particular process or part of the target system), and then various combinations of those components, are tested against high-quality observational data obtained from specialized validation experiments, in order to see if pre-specified levels of accuracy are met, where those levels of accuracy are determined by the model application (Oberkampf and Roy 2010). Though verification and validation are often conceptualized as distinct activities, Winsberg (2010; 2019) argues that in practice they are not so neatly separable (see also Lenhard 2018; 2019; and further discussion by Beisbart 2019a).

Evidence synthesis. Regardless of the conception of model quality that is adopted, evaluators may also wish (or be expected) to provide some summary judgment or conclusion about model quality. Doing so in effect involves a kind of evidence synthesis, where the evidence consists of what has been learned about model composition and/or performance. How to perform this synthesis, and when evidence is sufficient to warrant various conclusions about model quality, are complicated matters. Not infrequently, practitioners seem to adopt a kind of informal Bayesian perspective (Schmidt and Sherwood 2015), where particular findings about model composition or performance – such as the finding that the model's results for a given variable closely track observations – are taken to confirm or disconfirm (and thus build or reduce confidence in) a hypothesis about model quality, e.g., the hypothesis that the model is fit for a particular purpose or is a credible representation of the target system (see also Baumberger et al. 2017; Beisbart 2019b; Gelfert 2019).

A quite different sort of approach involves specifying criteria in advance of model evaluation which, if met, will be considered sufficient to warrant a conclusion of interest about model quality. For example, Haasnoot et al. (2014, 112), evaluating a model for screening and ranking different water policy pathways, conclude that their model is fit for purpose after reaching affirmative answers to a series of questions about the model's composition and performance. Similarly, in engineering contexts, evaluators sometimes specify accuracy requirements (with respect to high-quality data from experiments) for a series of model variables, such that meeting those requirements will be sufficient to consider the model (or its results) accurate enough for its intended use. In many modeling contexts, however, it is difficult to confidently specify such a set of sufficient criteria, much less to demonstrate that they are met by a given model, in part for reasons discussed in the next section.

4. Obstacles and challenges in model evaluation

Ideally, the activity of model evaluation will deliver strong evidence regarding model quality, such that confident conclusions – e.g., that a model is fit for purpose P – will be warranted. For a number of reasons, however, confident conclusions can remain out of reach. This section surveys some of these reasons.

Limited observations of the target system. First, scientific models are often employed when, for practical or ethical reasons, target systems are inaccessible to observation and experiment under conditions of interest. As a consequence, there are limited relevant observations of the target system, which can significantly hinder model assessment. For example, assessment of the fitness of today's climate models (for the purpose of projecting future temperature change in response to rising greenhouse gas concentrations) is hindered by the fact that, during the past periods for which reliable observations of the climate system are available, greenhouse gas concentrations were lower than in the scenarios for which projections are being made. In such situations – when available data were collected under conditions quite different from those that are ultimately of interest – it can be difficult to tell what a model's performance on the data indicates about its fitness-for-purpose (Parker 2009). This is especially so when models could have been constructed in awareness of, or even partially tuned to reproduce, the available data (Baumberger et al. 2017).

Model opacity. Another obstacle is model opacity, i.e., the inscrutability or incomprehensibility of aspects of a model, including its behavior, to an evaluator (see also Humphreys 2004 on epistemic opacity). Especially when models are complex and nonlinear, they are somewhat opaque even to individuals intimately involved in their development. A relevant-similarity evaluator, for instance, may find it difficult to understand – just by observing the behavior of a complex computational model – why it behaves in a particular way and may thus be unsure what that behavior indicates about the fidelity with which the model represents relevant target system processes (Baumberger et al. 2017; see also Lenhard and Winsberg 2010 on analytic impenetrability). Opacity can be just that much greater for evaluators who were not involved in the development of a model, especially when that development involved ad hoc elements (e.g., kludging) and when the model is poorly documented, i.e., when little accompanying information is provided and/or the model code is undocumented. Such an evaluator may have a difficult time deciding where to focus their evaluation efforts and determining whether the results of model tests provide strong evidence regarding model quality. They may also be left unaware of how non-epistemic (social, political, ethical) values shaped choices in the model's development, which in some cases might be relevant to their evaluation (see, e.g., Parker and Winsberg 2018; Hirsch Hadorn and Baumberger 2019; Lusk and Elliott 2022).

Holism in assessment. Holism is a challenge that arises primarily when assessing relevant similarity or fitness-for-purpose: in many cases, what is learned about the composition or performance of a model component in isolation cannot on its own serve as evidence regarding model quality (Parker 2020; see also Lenhard and Winsberg 2010; Lenhard 2018; 2019). Suppose that the purpose of a modeling study is to predict whether applications to a university will increase or decrease in number over the next several years. Finding that a model grossly underestimates a factor that is an important determinant of application numbers might or might not be evidence that the model is not fit for purpose, depending on whether that error is sufficiently compensated for by errors elsewhere in the model or by the broader methodology in which the model is embedded (e.g., a bias correction step). Likewise, whether the degree of similarity between a component of a model and a part of a target system counts as evidence for or against a relevant similarity hypothesis (for the model overall) can depend on how similar other components are and in what ways. The fact that components of models sometimes cannot be assessed in isolation makes evaluation a more complicated task, both practically and cognitively, especially when models are complex.[10]

Quantifying quality. Further challenges arise when evaluators seek to quantify model quality, i.e., to assign each of several models a quantitative score indicative of its quality. Such scores might be used, for instance, to differentially weight results from different models or to select from a set of models the ones that are best for a given purpose. A fundamental challenge here is quantifying the contribution of model composition to model quality (see also Baumberger et al. 2017). Weisberg's (2013) weighted feature-matching approach, mentioned in Section 3, might be one way forward for relevant-similarity evaluators, insofar as its scoring procedure takes account of both mechanisms (pertaining to composition) and attributes (covering performance aspects). Yet relevant-similarity evaluators will still need to determine how to assign weights indicating the relative importance of various mechanisms, how to avoid double-counting when both mechanisms and attributes they help to bring about are among the relevant features, and more.[11]

A different approach that is sometimes employed in practice is for evaluators to limit their attention to models that, based on expert judgment, seem of at least roughly equal quality from the perspective of composition, and then assign quality scores based on performance metrics.[12] Challenges here include determining which performance metrics should be employed and how they should be combined to produce an overall quality score. Mirror-view evaluators will need to choose from a host of measures of model-data fit (root mean square error, max absolute error, etc.) for each model variable for which observational data are available and will need some method for aggregating findings across variables into an overall score. Relevant-similarity and fitness-for-purpose evaluators will, in addition to choosing among measures of model-data fit, need to identify which model variables to focus on and how to weight performance on these variables to produce an overall quality score (Knutti 2018). Typically, there will be many reasonable ways to proceed for all three types of evaluators, with different choices resulting in somewhat different assessments of the relative quality of different models. In other words, there will be uncertainty about the models' relative (and absolute) quality.

5. Concluding remarks

Model evaluation is an important part of the model development process, occurring informally even during the building of models, and more formally once they are fully constructed.

The aim of model evaluation is to learn about the quality of one or more models, whether the quality is conceptualized as accurate and comprehensive representation, relevant similarity, fitness-for-purpose, or in some other way. The conception of model quality that is adopted carries implications for the practice of model evaluation, including whether evaluation must attend to the purposes for which models are being used and whether factors other than how the model relates to a target system, such as aspects of the context of use, are relevant.

Whatever the operative conception of model quality, evidence regarding model quality can come via two complementary routes: by examining the model's composition, i.e., its ingredients and how they are put together, and by examining the model's performance, especially its performance against observations of the target system. A number of obstacles and challenges can arise in the course of gathering such evidence and attempting to reach conclusions about model quality, including limited observations of the target system, model opacity, holism in assessment, and uncertainty about how to quantify model quality. Because of obstacles and challenges like these, it is sometimes difficult to reach confident conclusions about model quality.

Many questions about model evaluation merit further attention from philosophers of science. To name just a few: How do practices of model evaluation vary across different types of scientific models and in different scientific fields? How should evidence regarding model quality be synthesized to reach conclusions about model quality? How and to what extent should non-epistemic values figure in the evaluation of scientific models? A topic that can be expected to attract attention in the near future is the evaluation of "models" produced via machine learning methods; they present an especially interesting case for philosophical analysis, given their opacity, their questionable representational status, and their increasing use in high-stakes practical applications.

Notes

1 The discussion of existing work will – of necessity – be far from comprehensive, especially when it comes to scientific work on model evaluation. The author apologizes for omissions of important works.

2 This chapter is concerned with the evaluation of scientific models whose targets are real systems or phenomena, such as earth's atmosphere or the spread of flu virus through a population. The evaluation of models that have only imagined/imaginary target systems will not be addressed, e.g., a model of the population dynamics of a hypothetical species with four sexes and particular mating strategies. Likewise, the evaluation of statistical/data models, which are intended to capture relationships among variables in datasets, may merit separate treatment.

3 All elements of the target system might be represented in a model if the target system is specified such that it encompasses only a finite set of elements, e.g., particular relationships in a set of chemical reactions.

4 A "fitness-for-purpose" view of model quality is often adopted in scientific practice today, though exactly what practitioners mean by "fitness-for-purpose", and whether they understand it to be relative to a context of use, is sometimes unclear.

5 This is not to suggest that practitioners always have a clear and explicit conception of model quality; in some cases, for instance, evaluation proceeds in a way that simply follows what is usually done in a particular lab, community, or field.

6 Jacquart (2016) understands relevant similarity to be a matter of a model's composition and adequacy-for-purpose to be a matter of a model's performance. This differs from the present discussion, which allows that a model's performance might make it relevantly similar to a target, a model's composition might be essential to its fitness-for-purpose, etc.

7 Likewise, even after a model is fully constructed and put to use, it may subsequently undergo further development and evaluation. This is common, for instance, in weather and climate modeling,

and is reflected in the labels given to successive versions of a model, e.g., CESM1.0, CESM1.1, CESM1.2.

8 Thanks to Donal Khosrowi for prompting me to consider this possible motivation and for supplying the language of "purpose-neutral" evaluation.

9 For a similar, real example of fitness-for-purpose evaluation, see Haasnoot et al. (2014).

10 Rice (2019) argues that many highly idealized models are "holistically distorted representations" that are "greater than the sum of their accurate and inaccurate parts." If so, then even when a mirror view of model quality is adopted, it might be misguided in some cases to assess models by examining the representational fidelity of each component in isolation and aggregating the findings. (Note, however, that Rice's analysis is not concerned with the assessment of model quality; it is intended to challenge the view that, when models are used successfully for explanation and understanding, it is because their idealized/inaccurate parts do not "get in the way" of the accurately representing parts that do the real work.) Taking an artifactual perspective, Carrillo and Knuuttila (2022) offer a view of "holistic idealizations" that downplays the idea that they are distortions and emphasizes that they "result from more systematic research programs that integrate different concepts, analogies, measuring apparatus and mathematical approaches" (50).

11 In the context of statistical model selection, scoring criteria like the Akaike information criterion (AIC) take account of model composition by penalizing models for having more adjustable parameters; models receive a higher quality score to the extent that they can fit some set of data well with a smaller number of adjustable parameters. When it comes to models of real-world phenomena, the quality of a model's composition is usually understood to be a matter of much more than the number of adjustable parameters it contains.

12 Note that, for fitness-for-purpose evaluators, composition will need to be evaluated taking account of the model user, methodology, and background circumstances, not just the model's target system.

References

AIAA. 1998. "Guide for the Verification and Validation of Computational Fluid Dynamics Simulations." *American Institute of Aeronautics and Astronautics*, AIAA-G-077-1998. Reston, VA.

Baumberger, Christoph, Reto Knutti, and Gertrude Hirsh Hadorn. 2017. "Building Confidence in Climate Model Projections: An Analysis of Inferences from Fit." *WIREs Climate Change* 8(3): e454.

Beck, Bruce. 2002. "Model Evaluation and Performance." In *Encyclopedia of Environmetrics*, Volume 3, edited by Abdel H. El-Shaarawi and Walter W. Piegorsch, 1275–1279. Chichester: Wiley and Sons.

Beisbart, Claus. 2019a. "Should Validation and Verification Be Separated Strictly?" In *Computer Simulation Validation*, edited by Claus Beisbart and Nicole J. Saam, 1005–1028. Switzerland: Springer.

———. 2019b. "Simulation Validation from a Bayesian Perspective." In *Computer Simulation Validation*, edited by Claus Beisbart and Nicole J. Saam, 173–202. Switzerland: Springer.

Beisbart, Claus, and Nicole J. Saam, eds. 2019. *Computer Simulation Validation*. Switzerland: Springer.

Beven, Keith J. 2018. "On Hypothesis Testing in Hydrology: Why Falsification of Models Is Still a Really Good Idea." *WIRES Water* 5(3): e1278.

Bokulich, Alisa. 2013. "Explanatory Models versus Predictive Models: Reduced Complexity Modeling in Geomorphology." In *EPSA11 Perspectives and Foundational Problems in Philosophy of Science*, edited by Vassilios Karakostas and Dennis Dieks, 115–128. Switzerland: Springer.

Boon, Mieke, and Tarja Knuuttila. 2009. "Models as Epistemic Tools in Engineering Sciences: A Pragmatic Approach." In *Handbook of the Philosophy of Science, vol. 9, Philosophy of Technology and Engineering Sciences*, edited by Antoni Meijers, 687–720. Amsterdam: Elsevier.

Carrillo, Natalia, and Tarja Knuuttila. 2022. "Holistic Idealization: An Artifactual Standpoint." *Studies in History and Philosophy of Science* 91: 49–59.

Cartwright, Nancy. 1983. *How the Laws of Physics Lie*. Oxford: Oxford University Press.

Caswell, Hal. 1976. "The Validation Problem." In *Systems Analysis and Simulation in Ecology*, vol. 4, edited by Bernard C. Patten, 313–325. Cambridge, MA: Academic Press.

Currie, Adrian. 2017 "From Models-as-Fictions to Models-as-Tools." *Ergo* 4(27): 759–781.

Gelfert, Axel. 2019. "Assessing the Credibility of Conceptual Models." In *Computer Simulation Validation*, edited by Claus Beisbart and Nicole J. Saam, 249–270. Switzerland: Springer.

Giere, Ronald N. 1988. *Explaining Science: A Cognitive Approach*. Chicago: University of Chicago Press.

———. 2004. "How Models Are Used to Represent Reality." *Philosophy of Science* 71(5): 742–752.

Haasnoot, Marjolin, W. P. A. van Deursen, Joseph H. A. Guillaume, Jan H. Kwakkel, Ermond van Beek, and Hans Middelkoop. 2014. "Fit for Purpose? Building and Evaluating a Fast, Integrated Model for Exploring Water Policy Pathways." *Environmental Modelling and Software* 60: 99–120.

Hirsch Hadorn, Gertrude, and Christoph Baumberger. 2019. "What Types of Values Enter Simulation Validation and What Are Their Roles?" In *Computer Simulation Validation*, edited by Claus Beisbart and Nicole J. Saam, 961–980. Switzerland: Springer.

Humphreys, Paul W. 2004. *Extending Ourselves: Computational Science, Empiricism, and Scientific Method*. Oxford: Oxford University Press.

Isaac, Alastair M. 2013. "Modeling without Representation." *Synthese* 190: 3611–3623.

Jacquart, Melissa. 2016. *Similarity, Adequacy, and Purpose: Understanding the Success of Scientific Models*. PhD diss, University of Western Ontario. https://ir.lib.uwo.ca/etd/4129.

Kawamleh, Suzanne. 2022. "Confirming (Climate) Change: A Dynamical Account of Model Evaluation." *Synthese* 200: 122.

Knutti, Reto. 2018. "Climate Model Confirmation: From Philosophy to Predicting Climate in the Real World." In *Climate Modelling: Philosophical and Conceptual Issues*, edited by. Elisabeth A. Lloyd and Eric Winsberg, 325–359. Palgrave MacMillan.

Khosrowi, Donal. 2020. Getting Serious about Shared Features. *British Journal for the Philosophy of Science* 71(2): 523–546.

Knuuttila, Tarja. 2005. "Models as Epistemic Artefacts: Toward a Non-Representationalist Account of Scientific Representation." *Philosophical Studies from the University of Helsinki* 8: 12–78.

———. 2011. "Modeling and Representing: An Artifactual Approach." *Studies in History and Philosophy of Science A* 42(2): 262–271.

Lenhard, Johannes. 2018. "Holism, or the Erosion of Modularity: A Methodological Challenge for Validation." *Philosophy of Science* 85(5): 832–844.

———. 2019. "How Does Holism Challenge the Validation of Computer Simulation?" In *Computer Simulation Validation*, edited by Claus Beisbart and Nicole J. Saam, 943–960. Switzerland: Springer.

Lenhard, Johannes, and Eric Winsberg. 2010. "Holism, Entrenchment, and the Future of Climate Model Pluralism." *Studies in History and Philosophy of Science A* 41(3): 253–262.

Lloyd, Elisabeth A. 2010. "Confirmation and Robustness of Climate Models." *Philosophy of Science* 77(4): 971–984.

Lusk, Gregory, and Kevin C. Elliott. 2022. "Non-Epistemic Values and Scientific Assessment: An Adequacy-for-Purpose View." *European Journal for Philosophy of Science* 12: 35.

NRC (National Research Council). 2007. *Models in Environmental Regulatory Decision Making*. Washington, DC: National Academies.

Oberkampf, William L., and Christopher J. Roy. 2010. *Verification and Validation in Scientific Computing*. Cambridge: Cambridge University Press.

Oreskes, Naomi, Kristin Shrader-Frechette, and Kenneth Belitz. 1994. "Verification, Validation and Confirmation of Numerical Models in the Earth Sciences." *Science* 263(5147): 641–646.

Parker, Wendy S. 2009. "Confirmation and Adequacy-for-Purpose in Climate Modeling." *Aristotelian Society Supplementary Volume* 83: 233–249.

———. 2015. "Getting (Even More) Serious about Similarity." *Biology and Philosophy* 30(2): 267–276.

———. 2020. "Model Evaluation: An Adequacy-for-Purpose View." *Philosophy of Science* 87(3): 457–477.

Parker, Wendy S., and Eric Winsberg. 2018. "Values and Evidence: How Models Make a Difference." *European Journal for Philosophy of Science* 8(1): 125–142.

Potochnik, Angela. 2018. *Idealization and the Aims of Science*. Chicago: Chicago University Press.

Rice, Colin. 2019. "Models Don't Decompose That Way: A Holistic View of Idealized Models." *British Journal for the Philosophy of Science* 70(1): 179–208.

Saltelli, Andrea, Gabriele Bammer, Isabelle Bruno, Erica Charters, Monica Di Fiore, Emmanuel Didier, Wendy Nelson Espeland, John Kay, Samuele Lo Piano, Deborah Mayo, et al. 2020. "Five Ways to Ensure that Models Serve Society: A Manifesto." *Nature* 582: 482–484.
Schmidt, Gavin A., and Steven Sherwood. 2015. "A Practical Philosophy of Complex Climate Modelling." *European Journal for the Philosophy of Science* 5(2): 149–169.
Teller, Paul. 2001. "Twilight of the Perfect Model Model." *Erkenntnis* 55: 393–415.
Weisberg, Michael. 2013. *Simulation and Similarity*. New York: Oxford University Press.
Winsberg, Eric. 1999. "Sanctioning Models: The Epistemology of Simulation." *Science in Context* 12(2): 275–292.
———. 2010. *Science in the Age of Computer Simulation*. Chicago: Chicago University Press.
———. 2018. *Philosophy and Climate Science*. Cambridge: Cambridge University Press
———. 2019. "Computer Simulations in Science." In *The Stanford Encyclopedia of Philosophy* (Winter 2022 Edition), edited by Edward N. Zalta and Uri Nodelman. https://plato.stanford.edu/archives/win2022/entries/simulations-science/
Winsberg, Eric, and Ali Mizra. 2017. "Success and Scientific Realism: Considerations from the Philosophy of Simulation." In *The Routledge Handbook of Scientific Realism*, edited by Juha Saatsi, 250–260. London: Routledge.

16
MATHEMATIZATION

Marcel Boumans

1. Introduction

In most disciplines, models, as embodiments of knowledge, are mathematical objects, where the mathematics can range from algebra to algorithms to geometry. In other words, most models are built with mathematical material. The building of models as mathematical expressions of knowledge is often the way the mathematization in a discipline has taken place.[1] This chapter discusses mathematization in terms of this kind of model building.

This chapter starts from the viewpoint that sees models as instruments of investigation (Morgan and Morrison 1999), modelmaking as the integration of several "ingredients" in such a way that the resulting model meets certain *a priori* criteria of quality (Boumans 1999), and the process of model building as being epistemologically compared with the process of instrument making (Boumans 2005). This particular starting point has been chosen because it allows for a model-based mathematization account in which the role of mathematics is not that of translator but one in which mathematics functions as material and, as such, plays a critical role in the model construction process. The ingredients mentioned by Boumans (1999) are metaphors, analogies, mathematical concepts and techniques, stylized facts, data, and policy views. As the focus of this latter account is the integration process, it does not detail the considerations that play a role in the selection of the ingredients. However, when designing a new instrument, the choice of the materials from which the instrument will be made is a critical aspect of its design. This chapter shows that for designing a mathematical model, the selection process of the appropriate mathematical ingredients is equally critical.

The more general approach that includes the above models-as-instruments accounts is the artifactual account, which sees models as epistemic artifacts (Knuuttila 2021). The artifactual account views models as purposefully designed objects that are used in view of particular questions or aims in the context of specific scientific practices; in other words, they function as erotetic devices. The advantage of this approach is that, due to its view of models as epistemic artifacts, it directs attention to questions like how the model construction facilitates the answering of pending scientific questions or to the materials that are used and modified as constituents for its construction.

DOI: 10.4324/9781003205647-20

Relevant for understanding mathematization is that the type of question confines the type of model that will function as an appropriate erotetic device: it defines the criteria that a model should meet and thereby conditions how the model should be constructed and what kind of materials are needed. For example, the answer to a "why" question is an explanation, and the answer to a "how much" question is a measurement. Boumans (2006) and (2009) have shown that for "why" questions, the model should be a white-box model, that is, a model that includes a representation of the structure of the target system, while for "how much" questions, the model can be a black-box model, for which any representation of the target's structure is no longer required. The assessment of the appropriateness of an answer depends on the kind of question investigated, e.g., in the case of measurement, the answer's accuracy is established by calibration. Moreover, the choice of the materials from which the model is made is contingent on the kind of question the model should investigate. Since this chapter discusses mathematization, it focuses on the choice of the mathematical forms that are needed to make the model as satisfactory as possible. A model is satisfactory when it satisfies criteria that are defined by the kind of question to be addressed (Morgan 1988).

In brief, to answer a model-based question satisfactorily, the model should meet specific criteria that are closely intertwined: The structure of the model, its validation as well as the chosen mathematical materials should meet specific requirements, which in close interdependence with each other, determine whether the model provides satisfactory answers.

The next section shows how model structure and validation criteria are mutually dependent. Section 3 discusses how rigorous the structure and validation requirements have to be met by the model to be satisfactory. The mathematical materials are selected in such a way as to allow the model to perform its purpose as satisfactorily as possible. Section 4 examines how this selection is done. Section 5 presents the process of modeling as the integration of all these requirements and materials. Section 6 presents the tradition in which this model-based mathematization is embedded.

2. Structure and validation

> Remember that all models are wrong; the practical question is how wrong do they have to be not to be useful.
>
> *(Box and Draper 1987, 78)*

The relevant question about models as erotetic devices is not, "How true are they?" but rather, "How useful are these instruments to answer specific questions?" The *validity* of a model is therefore defined as its usefulness with respect to some purpose. Barlas (1996) notes that for the exploration of the validation of models, it is crucial to make a distinction between white-box models and black-box models. In black-box models, what matters is the output behavior of the model. The model is assessed to be valid if its output matches the behavior of the target system within some specified range of accuracy, without any question of the accuracy of the individual model equations. White-box models, in contrast, are statements on how the target system actually operates in some aspects. Generating accurate output behavior is not sufficient for model validity; the accuracy of the model's internal structure is also critical.

Barlas (1996) discusses three stages of model validation: "direct structure tests," "structure-oriented behavior tests," and "behavior pattern tests." Direct structure tests assess the validity of the model structure by direct comparison with knowledge about the target system structure. The model structure here is a system of mathematical equations. The direct structure test then involves taking each model equation individually and comparing it with available knowledge about the target system. The list of direct structure tests includes tests such as chi-square tests. The structure-oriented behavior tests assess the validity of the structure indirectly by applying certain behavior tests to model-generated behavior patterns. These tests include the extreme-condition test, the behavior sensitivity test, and the Turing test. Pattern and point prediction tests are examples of behavior pattern tests. For the validation of white-box models, all three stages are equally important. For black-box models, only the last stage is required.

While the second-stage tests—the structure-oriented behavior tests—do not give direct access to the model structure, they nevertheless can provide information on potential structural flaws. To see this, we first need to further qualify what is meant by model structure. The notion of model structure is not limited to a system of equations that is assumed to represent the systems of relationships of the target system, as in the case of white-box models. It can also include other arrangements, like modular organizations, in which these modules are lower-level models or sub-models.

In systems engineering, a module is defined as a self-contained component with a standard interface to the other components within a system (White 1999). Each module can be validated prior to assembly, and new systems can be realized by new combinations of existing and improved modules. The notion of structure, then, refers not only to relationships between causal factors, but also to relationships between modules. These modular-designed models – in line with the labeling of the other two types of models – are called gray-box models. The modules can then themselves be a white-, gray-, or black-box model. For the validation of these gray-box models, they should pass structure-oriented behavior tests and behavior pattern tests.

To answer "why" questions, we need white-box models, and for "how much" questions, we can do with black-box models. Boumans (2006) shows that for "what-is-the-effect of" questions, gray-box models are most appropriate. In other words, there is a close connection between the kind of question one is investigating, the required model structure, and the way the model should be validated.[2]

In the case of black-box models, the choice of the mathematical forms is only constrained by the applied behavior pattern tests. The objects are chosen such that a specific combination of them produces the required pattern. For example, if the required output pattern is cyclical, the input-output relationship could well be a differential equation, without supposing that this differential equation is an accurate representation of the target system.

For the construction of white-box models, the composition of the mathematical objects must comply with both specific behavior pattern tests and structure-oriented behavior tests, as well as specific direct structure tests. This does not mean that each selected mathematical component needs to satisfy every test. Some components are selected to make the model meet the behavior pattern test and some to meet the direct structural test. Because the structure of the white box is considered to be a representation of the target system, only the mathematical components selected to meet the direct structure tests are chosen to represent parts of the target system directly. For example, if the target system is a cyclical mechanism, the chosen mathematical objects could again be differential equations, but now with the claim that they represent the mechanism of the target system.

The structure of gray-box models is a specific combination of modules. This combination of modules can represent the structure of the target system (when the structure of the target system is also modular), but it may also just be an arrangement of the modules such that the overall behavior meets the behavior pattern tests and structure-oriented behavior tests. The paper in which this modeling methodology was proposed for the first time is von Neumann's paper, "The general and logical theory of automata," first published in 1951. In this paper, this methodology was called the 'Axiomatic Procedure' and was explained as follows:

> The natural systems are of enormous complexity, and it is clearly necessary to subdivide the problem that they represent into several parts. One method of subdivision [...] is this: The organism can be viewed as made up of parts which to a certain extent are independent, elementary units. We may, therefore, to this extent, view as the first part of the problem the structure and functioning of such elementary units individually. The second part of the problem consists of understanding how these elements are organized into a whole, and how the functioning of the whole is expressed in terms of these elements.
>
> *(von Neumann 1963, 289)*

Instead of the more familiar mathematical equations, the interaction between the modules can also be formulated as algorithms. This is usually the case when the model is used for simulation purposes, that is, to answer "what-would-happen-if" questions.

3. Formalization and rigor

> Studying the methods of solving problems, we perceive another face of mathematics. Yes, mathematics has two faces; it is the rigorous science of Euclid but it is also something else. Mathematics presented in the Euclidean way appears as a systematic, deductive science; but mathematics in the making appears as an experimental, inductive science.
>
> *(Polya 1957, vii)*

The process of model-making has often been labeled as "formalization." In her account of how models are made, Morgan (2012, 19–20) makes a useful distinction between two meanings of formalization in order to understand what model-making entails. If we think about its active form, 'to formalize' implies to give form to, to shape, or to provide an outline of something. The second meaning can be clarified if we take its passive form 'formal.' Formal implies something rule-bound, following prescribed forms. According to Morgan, making models involves both meanings: "models give form to, in the sense of providing a more explicit or exact representation of our ideas about the world, and in creating those forms we make them subject to rules of conduct or manipulation" (20).

These rules of conduct or manipulation, which are the rules for reasoning with a model, come, according to Morgan (2012), from two distinct aspects of the model: First, these rules should be in accordance with "the kind of the *stuff* that the model is made from, or language it is written in, or the format it has," or in other words, "they are given and fixed by the *substance* of the model" (26, italics added). Second, these rules are also determined and constrained by the subject matter represented in the model. This chapter focuses on the first aspect of rules, namely the constraining features of the model's substance on the kind of reasoning one can do with the model. This implies that in the selection of the mathematical ingredients, one also has to take into account the kind of reasoning one wishes to perform with the model.[3]

In answering the crucial question about modeling, "How can we get knowledge from models?" Morrison (2015) also emphasizes the role of constraints. They are induced not only by what we already know about the phenomenon to be modeled, but also by the materials from which the model will be built: "Once we decide what needs to be modelled [i.e. what the target is], these constraints determine, to some extent, how to do it. They function like rules in a game" (153).

The rule-bound aspect of formalization is usually referred to as *rigor*. What is taken to be rigorous depends on the underlying assumption of what a model is: whether it is seen as an epistemic instrument or as a formal object.

The model accounts that see models as formal objects are the axiomatic approaches; for them, rigor means consistency of the rules, famously expressed by Hilbert (1902, 448): "If contradictory attributes be assigned to a concept, I say, that mathematically the concept does not exist." But rigor has not always been identified with axiomatics. Israel (1981) shows that a shift from rigor in its older (19th-century) meaning of meeting empirical requirements to the current meaning of logical consistency came with a loss of the applicability of mathematics in empirical science: "What appears to be missing, is a codification of the rules which should define and guide the use of mathematics as an instrument for the description, interpretation and control of phenomena" (Israel 1981, 219). This means that modeling for dealing with practical issues requires a different codification of rules than models that aim at solving axiomatic problems.

To fulfill its purpose, a model has to meet a set of requirements. For practical problems, it is often the case that these requirements are not consistent with each other. According to the axiomatic view, it would mean that in these cases, such a model cannot be built. If one nevertheless wishes to keep to this kind of axiomatic rigor, it means that one has to decide which of the requirements has to be abandoned based on some theoretical considerations, such that the remaining set of requirements is consistent. An instrumental approach to this problem is that one seeks an appropriate balance or compromise between these requirements, in the sense that one decides to what extent each requirement should be met.

This instrumental approach towards rigor can be nicely illustrated by the problem of designing a world map, which is a two-dimensional projection of the world globe. To flatten out a globe, one must stretch and/or shrink it in certain directions and tear it at several places. In mathematical terms, the world map and the world globe are not topological equivalents. In particular, there is a trade-off between interruption and distortion: only by increasing the interruptions of the map can we lessen distortion. In a book on the design of world maps (Fisher and Miller 1944, 27–28), the requirements a world map should meet are stated as the following objectives:

1 to have distances correctly represented;
2 to have shapes correctly represented;
3 to have areas correctly represented;
4 to have great circles represented by straight lines.

It is a geometric impossibility to have all four objectives met on a flat surface and to have them in every part. So, Fisher and Miller concluded that "projections are confessedly compromises, being perfect in none of the four ways but balancing the different kinds of errors against one another" (34).

4. The choice of the mathematical ingredients

> It is imperative to notice that whenever we apply a definition to nature we must wait to see if it will correspond to it. With the exception of pure mathematics we can create our concepts at will, even in geometry and still more in physics, but we must always investigate whether and how reality corresponds to these concepts.
>
> *(Mach quoted in Ellis 1966, 185)*

Mathematical models are compositions of mathematical objects. The selection of them is determined by the question the model needs to address, the related model structure, as well as to what extent the validation requirements should be met. With respect to these three aspects of model building, mathematical objects have different functions in model construction. In line with Morgan's conceptualization of formalization, on the one hand, the mathematical objects are stuff that the model is made from, and on the other hand, they are also determined and constrained by the nature of the target system. But this does not mean each mathematical object has to be constrained in both ways. Some objects are selected to enable the model to fulfill its purpose, and other – not necessarily the same – objects are chosen to enable a representational relationship to the target system.

To see the difference between these two roles of mathematical objects in models, it is helpful to compare a mathematical model with a physical model, for example, with the Newlyn-Phillips machine. This hydraulic machine is a physical 3-D model made of Perspex, water, springs, wire, etc., built to represent a Keynesian economy, in which the circulating water represents money (Phillips 1950).[4] One of the most important characteristics of 3-D physical objects is that they are subject to gravity. The circulation in this hydraulic machine worked because of this force and an electronic motor to pump the water up. Both gravity and the electronic motor do not have economic equivalents. Because the machine was meant as a representation of an economy, the motor was hidden. Besides the motor, the machine consisted of many other parts, hidden or not, which had no economic equivalents but were critical to the working of the machine. For such a model, it is not expected that every part of the model represents something of the Keynesian economy. There are always some things, which are likely to be untranslatable or just plain wrong. But these elements do not necessarily cause difficulties in the functioning of the model. On the contrary, they are installed to enable its functioning.

This physical model also makes us better aware of the *material* aspects of model building. Morgan and Boumans' (2004) study of the model building process of this 3-D hydraulic machine showed that model building involves dealing with both a great many constraints imposed from the physical side and a whole lot of commitments about how the economics are physically represented.

Working with mathematics means taking into account the same kind of constraints. Just as one has to choose which material is both strong and transparent enough to carry the colored water and keep it visible, the different kinds of mathematical objects need to be chosen to make the model carry out its purpose. This constraining aspect is typical of materiality. The substance aspect of materiality constrains the kinds of things one can do with any given material. Wood does not conduct electricity, but iron does. According to

Fleischhacker (1992), this is because substance has structure and because mathematical objects also have structure, he characterizes mathematical objects as "quasi-substantial." This structural aspect of mathematical objects conditions their functioning.

This structural aspect of mathematical objects means that one has to consider the kind of mathematics that allows the kind of functioning one is aiming for. Because each mathematical object has its own structure with its own structural properties, one has to take these properties into account when deciding which of them may be useful for the model in question.

To better understand how mathematical objects are selected based on their structural properties, it is useful to draw on material selection accounts in mechanical design.[5] Each material can be thought of as having a set of properties, such as density, modulus, strength, toughness, and thermal conduction. But it is not a material, per se, that the designer seeks; it is a specific combination of these properties, a specific *property-profile*. The material name can then be seen as the identifier for a particular property-profile. Knowing the property-profile is relevant because these material properties constrain performance.

The selection process works as follows: A material has properties, such as its density and strength. A design demands a certain profile of these, for example, a low density and a high strength. The problem is that of identifying the desired property-profile and then comparing it with those of real engineering materials to find the best match. The immensely wide choice is narrowed, first, by applying *property limits* that screen out the materials which cannot meet the design requirements. Further narrowing is achieved by ranking the candidates by their ability to maximize performance. Performance is generally limited not by a single property, but by a combination of them. For example, the best materials for a light, stiff tie-rod are those with the greatest value of stiffness, which is a specific ratio of modulus and density. Combinations such as these are called *material indices*: they are groupings of material properties which, when maximized, maximize some aspect of performance. There are many such indices. They are derived from the design requirements for a device through an analysis of *function*, *objectives*, and *constraints*. Property limits isolate candidates that are capable of doing the job; material indices identify those among them that can do the job well.

To show that the selection of mathematical objects in model construction is similar to the choice of materials in instrument design, the case of business cycle modeling in the 1930s by the founders of mathematical model building in economics, Frisch and Tinbergen, will be briefly presented.[6] As there were no mathematical theories available at that time which could instruct them on how to build these models, they had to start almost from scratch. They were looking for mathematical equations that could represent the business cycle mechanism. First of all, such an equation would have to be dynamic. This meant, that the equation must at least have a term that denotes a rate of change with respect to time. They considered the following terms: $x(t-\theta)$, $\dot{x}(t)$, $\ddot{x}(t)$, and $\sum_t x(t)$ or $\int x(t)dt$. Second, the dynamic equation should describe a specific kind of cyclical behavior. These latter conditions were called "wave conditions" by Tinbergen. This meant that the values of the coefficients of the dynamic equation must be chosen in such a way that the resulting cyclical behavior meets some specified characteristics, such as the periodicity and amplitude of a real business cycle.

This case shows that mathematical objects, like physical materials, have properties that need to be accounted for when building a model for a specific purpose. The property profile one was looking for in business cycle modeling is a particular equation that consists of a variable, say $x(t)$, to which are added specific dynamic terms, such as $x(t-\theta)$ or $\dot{x}(t)$, in such a way that the equation represents cyclical behavior. Any such dynamic equation can be considered a material index, that is to say, a combination of dynamic properties. The values of the equation's coefficients determine its specific property profile. The builder of a business cycle model then seeks a property-profile that meets some specified wave conditions.

5. The process of model building

> I think of a modeler as starting with some disparate pieces – some wood, a few bricks, some nails, and so forth – and attempting to build an object for which he (or she) has only a very inadequate plan, or theory. The modeler can look at related constructs and can use institutional information and will eventually arrive at an approximation of the object that they are trying to represent, perhaps after several attempts.
>
> *(Granger 1999, 6–7)*

Knowledge of materials is necessary, but it is not the only epistemic requirement of model building. Model building is an attempt at a successful integration of various ingredients so that they meet the validation criteria (Boumans 1999). The ingredients include, besides the mathematical objects, theoretical notions, analogies, and metaphors, as well as empirical data and facts. Because of the integration of the latter ingredients, the positivist distinction between "discovery" and "justification" cannot be sustained.

To clarify this integration process, Tinbergen's attempts to arrive at a model of the business cycle mechanism which culminated in his (1931) ship-building model will be taken as an exemplary case. This ship-building model consists of one equation:

$$\dot{x}(t) = -ax(t-\theta)$$

where x represents available world tonnage, t time, and θ production time of a new ship, thus new tonnage.

This model was, in Tinbergen's view, the successful result of a long search for a representation of the business cycle mechanism that had to integrate the following two ingredients: Aftalion's crisis theory and the empirical fact that the business cycle period is about eight years.

Aftalion's theory was, according to Tinbergen (1927, 715; my translation), the only economic theory that could explain "most clearly ... that every cycle already contains the seed for the next cycle and thus real periodicity occurs." Aftalion's thesis was "that the chief responsibility for cyclical fluctuations should be assigned to one of the characteristics of modern industrial technique, namely, the long period required for the production of fixed capital" (Aftalion 1927, 165). For producers, the value of a product depends on the price it is expected to fetch; that is to say, their values depend on the forecast of future prices. Aftalion assumed that the expectations of those directing production are, alternately, either too optimistic or too pessimistic. The cycle is a consequence of the long delay, which often separates the moment at which the production is decided upon and a forecast is made from the moment when the manufacture is terminated, because the forecast of future prices is based on the present prices and the present state of demand.

It took Tinbergen about five years before he could find a satisfactory cycle profile, that is, the right combination of the mathematical dynamic terms, that would integrate both ingredients. His starting point was harmonic oscillation, whose dynamics can be mathematically described by a second-order differential equation. However, the differential terms, $\dot{x}(t)$ and $\ddot{x}(t)$, had to be combined with a lag term $x(t-\theta)$ to integrate Aftalion's theory. He tried out several combinations of dynamic terms, of which each combination had to include the lag term. Each of them implied either an unrealistic production time or a periodicity that was too short or too long. Only the ship-building equation led to satisfactory results. With a production time of two years, $\theta = 2$, and the equation's parameter a having a value that confirms the data he had about the ship-building market, the resulting cycle has a period equal to eight years.

Tinbergen's ship-building model is a nice example of the model-building process as the satisfactory (to the model builder) integration of several ingredients, such as theoretical ideas (Aftalion's crisis theory), analogies (harmonic oscillation), mathematical concepts (dynamic time functions), stylized facts (the cycle's period of eight years), and empirical data (data of the ship-building market). It was the result of a long trial-and-error process to get all the ingredients integrated. Because this set of ingredients also contained the facts the model was supposed to explain, justification was built in.

6. The artifactual view of mathematization

> But scientific accuracy requires of us that we should in no wise confuse the simple and homely figure, as it is presented to us by nature, with the gay garment which we use to clothe it. Of our own free will we can make no change whatever in the form of the one, but cut and colour of the other we can choose as we please.
>
> *(Hertz 1962, 28)*

This model-based mathematization finds its roots in Hertz's Kantian account presented in his *The Principles of Mechanics Presented in a New Form* (1956):[7]

> It is impossible to carry our knowledge of the connections of the natural systems further than is involved in specifying models of the actual systems. We can then, in fact, have no knowledge as to whether the system which we consider in mechanics agree in any other respect with the actual systems of nature which we intend to consider, than this alone, – that the one set of systems are models of the other.
>
> *(Hertz 1956, 177)*

Hertz formulated three requirements a model should fulfill: *logical permissibility*, *correctness,* and *appropriateness*.[8] Hertz considered correctness as the "fundamental requirement": models are incorrect "if their essential relations contradict the relations of external things" (2). Hertz was thinking about this requirement in terms of the model's predictive performance, but one could state more generally that a model must be empirically validated. It should, however, be noted that the requirement of correctness applies only to the model as a whole and not to the individual equations or terms of the model, so it was not a direct-structure-test requirement, or in other words, a white-box requirement.

The second criterion, logical permissibility, is analytic: a model is not permissible if it "contradicts the laws of thought" (2). In other words, the mathematics or logic used to formulate the model should not consist of any contradictions. This refers to the above rigor requirement of the axiomatic approaches. The approach that emphasized this logical requirement evolved into the semantic account of models, according to which a model is an interpretation of a theory in which all the axioms of that theory are true. But such a model can only exist if the axioms are logically consistent. According to Hertz, we can decide "without ambiguity" whether a model meets these two criteria.

In the model literature, these two requirements, or variations of them, are usually mentioned, while the appropriateness criterion is often ignored. But, according to Nagel (1961), it is "important to remember" that a model is a human artifact, and therefore "likely to contain some elements that are simply expressions of the special objectives and idiosyncrasies of their human inventors, rather than symbols with a primary referential or representative function" (103). This point was also stressed by Hertz's criterion of appropriateness.

A model will unavoidably contain what Hertz called "superfluous or empty relations"—mathematical objects that are not representative of anything in the subject matter for which the model is devised. According to Hertz, these "empty relations cannot be altogether avoided: they enter into the images because they are simply images, – images produced by our mind and necessarily affected by the characteristics of its mode of portrayal" (Hertz 1956, 2).

According to the criterion of appropriateness, of two models equally permissible and correct, the better model is the simpler one, that is, the one which contains "the smaller number of superfluous or empty relations" (2) and that is more "distinct" if it "pictures more essential relations of the object" (2). In modern terms, a more distinct model has a larger scope. According to Lützen (2005), the issue of simplicity is related to the avoidance of "conceptual and mathematical complication" (92) and involves "such properties as intuitive clarity, elegance, and beauty" (93). In other words, as meeting the permissibility and correctness criteria still allows for several different models, the final choice for a model was determined by balancing between the scope of analysis and tractability. Nevertheless, the relations that are empirically "empty" were needed to enable the model to be correct.[9]

A 20th-century version of the artifactual view is Simon's (1969) artifact account. Simon defines an artifact as an "interface":

> between an "inner" environment, the substance and organization of the artifact itself, and an "outer" environment, the surroundings in which it operates. If the inner environment is appropriate to the outer environment, or vice versa, the artifact will serve its intended purpose.
>
> *(Simon 1969, 7)*

The advantage of factoring an artificial system into goals, outer environment, and inner environment is "that we can often predict behavior from knowledge of the system's goals and its outer environment, with only minimal assumptions about the inner environment" (8), or in Hertz's terminology, a model can meet the fundamental correctness requirement with only minimal assumptions about its structure, and therefore it also complies with the appropriateness requirement. Different materials and organizations can accomplish identical goals in similar outer environments. For example, both weight-driven clocks and spring-driven clocks measure the same time.

The choice of the inner environment of the model, its material, and its organization, is thus determined by the kind of question one is aiming to address and the characteristics of the outer environment. Whether a clock will, in fact, tell the time accurately is also dependent on its location. A sundial performs very well in sunny climates, but to devise a clock that would tell the time on a rolling and pitching ship, it has to be endowed with many delicate properties, some of them largely or totally irrelevant to the performance of a chimney clock. The design of the model must be such that there is an invariant relation between the inner system and goal across some specified range in most of the parameters that characterize the outer environment (see also Simon 1969, 9). According to Simon, therefore, the model needs to be assessed for its validity, at least by the structure-oriented behavior tests and behavior pattern tests. The direct structure tests are only needed for a rather restricted set of questions, such as 'why' questions, for which a white-box structure is needed.

7. Conclusion

> Its modest aim is to elaborate the point that informal, quasi-empirical, mathematics does not grow through a monotonous increase of the number of indubitably established theorems but through the incessant improvement of guesses by speculation and criticism.
>
> *(Lakatos 1976, 5)*

Mathematization, in the sense of finding a mathematical expression of what we would like to know about a certain phenomenon, is a modeling operation. In the process of building a mathematical model, we hope to find an answer to a specific question we have about this phenomenon. These questions can be of various kinds, such as "why" questions, "how much" questions, or "what would happen if" questions. Each answer has to meet specific requirements to be satisfactory. These different requirements can come from different directions; they can come from specific theoretical frameworks, from methodological demands about validation, and from what is already known about the phenomenon. The kind of mathematics that must be used can also be defined in advance, for example, the mathematical expression has to be in terms of calculus. But even if a mathematical framework is set in advance, it still does not tell the modeler which mathematical forms of that framework are the most appropriate. This selection of the most appropriate mathematical objects is similar to the selection of materials in mechanical design: one is to take into account the properties of the considered materials and what the (optimal) performance is of combinations of them. These materials are not only selected for enabling a representational relationship with the target system, that is, meeting Hertz's requirement of correctness. Some of the mathematical objects are chosen only in order to enable the model to achieve its goal.

In this chapter, mathematization is thus seen as mathematical modeling, where modeling is the attempt to integrate various kinds of ingredients, such as specific theoretical notions, specific facts, and data about the phenomenon in question, (mathematical) analogies, and metaphors. Finding the appropriate mathematical forms is crucial for the success of this integration. Although material knowledge and knowledge of the phenomenon to be investigated, as well as further background knowledge and training are essential, finding the right combinations of the materials remains an explorative process, largely comparable to empirical research.[10] Modeling is a trial-and-error process, "not driven by a logical process but rather involves the scientist's intuitive, imaginative, and creative qualities" (Morgan 2012, 25). The design of epistemic artifacts is an experimental, inductive process.

Notes

1 Although the standard view holds that mathematization takes place through translation of verbal expressions of knowledge into mathematical language, there is little to no (historical) evidence for this view.
2 See Tieleman (2021) for a more recent discussion of the validation of grey-box models.
3 This chapter focuses on mathematical functions. Chao (2018) provides nice cases of reasoning in which geometrical shapes, like hexagons, triangles, and circles, are used.
4 A Keynesian economy is not a real economy but a theoretical model, designed to account some macroeconomic features of actual economies. So, what we have here is actually a material model of a theoretical model, a "nesting of models" (Hughes 1997). I thank Tarja Knuuttila for reminding me about Hughes's DDI account which nicely fits with the model account presented here, see also Section 5.
5 This discussion is based on Ashby (1999).
6 This highly condensed presentation will only discuss the main choices that have been made. See Boumans (2005) and Morgan (2012) for more detailed accounts of this kind of model building.
7 See Nagel (1961, 103) and Hughes (1997, 333) for similar accounts.
8 See Lützen (2005) for a detailed discussion of these three criteria.
9 A similar view can be found in Cartwright's (1983) simulacrum account of explanation. According to this account, some properties of the objects in the model are "properties of convenience," "to bring the objects modelled into the range of the mathematical theory", of which some are "not even approached in reality. They are pure fictions" (153).
10 In this sense, the "logic" of mathematical modeling is similar to Lakatos's "logic of mathematical discovery."

References

Aftalion, Albert. 1927. "The theory of economic cycles based on the capitalistic technique of production." *Review of Economic Statistics* 9: 165–170.

Ashby, Michael F. 1999. *Materials Selection in Mechanical Design*, 2nd edition. Oxford: Butterworth-Heinemann.

Barlas, Yaman. 1996. "Formal aspects of models validity and validation in system dynamics." *System Dynamics Review* 12(3): 183–210.

Boumans, Marcel 1999. "Built-in justification." In *Models as Mediators: Perspectives on Natural and Social Science,* edited by Margaret Morrison and Mary S. Morgan, 66–96. Cambridge: Cambridge University Press.

———. 2005. *How Economists Model the World into Numbers*. London and New York: Routledge.

———. 2006. "The difference between answering a 'why' question and answering a 'how much' question." In *Simulation. Pragmatic Construction of Reality*, edited by Johannes Lenhard, Günter Küppers, and Terry Shinn, 107–124. Dordrecht: Springer.

———. 2009. "Understanding in economics: Gray-box models." In *Scientific Understanding: Philosophical Perspectives*, edited by Henk W. de Regt, Sabina Leonelli, and Kai Eigner, 210–229. Pittsburgh: University of Pittsburgh Press.

Box, George E. P., and Norman R. Draper. 1987. *Empirical Model-Building and Response Surfaces*. New York: Wiley.

Cartwright, Nancy. 1983. *How the Laws of Physics Lie*. Oxford: Clarendon Press.

Chao, Hsiang-Ke. 2018. "Shaping space through diagrams: The case of the history of location theory." *Research in the History of Economic Thought and Methodology* 36B: 59–72.

Ellis, Brian D. 1966. *Basic Concepts of Measurement*. Cambridge: Cambridge University Press.

Fisher, Irving, and Osborn Maitland Miller. 1944. *World Maps and Globes*. New York: Essential Books.

Fleischhacker, Louk. 1992. "Mathematical abstraction, idealization and intelligibility in science." In *Idealization IV: Intelligibility in Science*, edited by Craig Dilworth, 243–263. Amsterdam, Atlanta: Rodopi.

Granger, Clive W.J. 1999. *Empirical Modeling in Economics: Specification and Evaluation*. Cambridge University Press.

Hertz, Heinrich.1956. *The Principles of Mechanics Presented in a New Form*. New York: Dover.
———. 1962. *Electric Waves*. New York: Dover.
Hilbert, David. 1902. "Mathematical problems, lecture delivered before the international congress of mathematicians at Paris in 1900", translated by M. Winston Newson. *Bulletin of the American Mathematical Society* 8: 437–479.
Hughes, R.I.G. 1997. "Models and representation." *Philosophy of Science* 64, Supplement: S325–S336.
Israel, Giorgio. 1981. ""Rigor" and "axiomatics" in modern mathematics." *Fundamenta Scientiae* 2, 205–219.
Knuuttila, Tarja. 2021. "Epistemic artifacts and the model dimension of modeling." *European Journal for Philosophy of Science* 11: 65.
Lakatos, Imre. 1976. *Proofs and Refutations. The Logic of Mathematical Discovery*, edited by John Worrall and Elie Zahar. Cambridge: Cambridge University Press.
Lützen, Jesper. 2005. *Mechanistic Images in Geometric Form. Heinrich Hertz's Principles of Mechanics*. Oxford: Oxford University Press.
Morgan, Mary S. 1988. "Finding a satisfactory empirical model." In *The Popperian Legacy in Economics*, edited by Neil De Marchi, 199–211. Cambridge: Cambridge University Press.
———. 2012. *The World in the Model*. Cambridge: Cambridge University Press.
Morgan, Mary S., and Marcel Boumans. 2004. "Secrets hidden by two-dimensionality: The economy as a hydraulic machine." In *Models. The Third Dimension of Science*, edited by Soraya de Chadarevian, and Nick Hopwood, 369–401. Stanford University Press.
Morgan, Mary S., and Margaret Morrison 1999, eds. *Models as Mediators*. Cambridge: Cambridge University Press.
Morrison, Margaret. 2015. *Reconstructing Reality. Models, Mathematics, and Simulations*. Oxford and New York: Oxford University Press.
Nagel, Ernest. 1961. *The Structure of Science*. London: Routledge and Kegan Paul.
Phillips, Alban William. 1950. "Mechanical models in economic dynamics." *Economica, New Series*, 17(67): 283–305.
Polya, George. 1957. *How to Solve It: A New Aspect of Mathematical Method*. Princeton, NJ: Princeton University Press.
Simon, Herbert A. 1969. *The Sciences of the Artificial*. Cambridge: MIT Press.
Tieleman, Sebastiaan. 2021. "Towards a validation methodology for macroeconomic agent-based Models." *Computational Economics* 60: 1507–1527. https://doi.org/10.1007/s10614-021-10191-w.
Tinbergen, Jan. 1927. "Over de mathematies-statistiese methoden voor konjunktuuronderzoek." *De Economist* 11: 711–723.
———. 1931. "Ein Schiffbauzyklus?" *Weltwirtschaftliches Archiv* 34: 152–164.
von Neumann, John. 1963. "The general and logical theory of automata." In *John von Neumann Collected Works*, vol. 5, edited by Abraham H. Taub, 288–318. Oxford: Pergamon Press.
White, K.P. 1999. "System design." In *Handbook of System Engineering and Management*, edited by Andrew P. Sage and William B. Rouse, 455–481. New York: Wiley.

17

EPISTEMOLOGY AND PRAGMATISM

The debated role of models in statistics

Johannes Lenhard

1. Introduction

Statistics occupies a special place both in the sciences and in philosophy. In the sciences, statistical methods are at work whenever empirical data are of concern. Students of many scientific professions will likely have to go through a mandatory statistics course. Even if many of such courses are infamous for teaching standard recipes rather than critical thinking, working with statistical tools is a widely accepted indicator of being scientific. From a philosophical perspective, statistics deals with the interface between the world and scientific apparatuses. For example, when do data falsify a hypothesis? When does inconclusive evidence change into conclusive evidence? Neither the data nor the theory or hypothesis alone can tell. What is needed to get an answer from statistics is a statistical model. In short, statistical modeling occupies a special place because it is involved in mediating between (almost all kinds of) data and (almost all kinds of) theory. On the one hand, statistical modeling is part of everyday scientific practice, on the other hand, operating with data is a fundamental condition of scientific epistemology. This chapter acknowledges this tension between pragmatism and epistemology.

Furthermore, modeling has not yet received proper attention from the philosophical side. The philosophy of statistics is infamous for the longstanding and deeply entrenched opposition between Bayesian and classical standpoints regarding probability.[1] Although the concept of statistical model has an important function in both classical and Bayesian accounts, the role of modeling in statistics is seriously under-examined.[2]

This chapter presents an uncommon cut through the philosophy of statistics, namely a cut that follows the concept of modeling. The hope is to invite philosophical and historical research into hitherto under-explored terrain. The following text has three parts that entertain three different—though related—perspectives on statistical modeling. The first part (Section 2) is devoted to the classical standpoint and the origins of the concept of a statistical model. Ronald A. Fisher introduced this concept in (1922) to mathematize the logic of inference. A model mediates between mathematics, data, judgment, and economy of computation. The philosophical significance of this mediating role elucidates a controversy about modeling between the main proponents of the classical camp (Fisher, Neyman

 DOI: 10.4324/9781003205647-21

and Pearson). Section 3 discusses the counter-movement of "Exploratory Data Analysis" (EDA) led by John W. Tukey in the 1960s and 1970s who pleaded to abandon models and let the data speak for themselves. EDA makes use of computer software and visualization. Based on recent computer methods, in connection with big data and machine learning, the prospect of letting the data speak for themselves has attracted a range of new followers. Finally, Section 4 turns to the career of Bayesian models in statistical practice, told as a tale about the impact of computer use on epistemology. A remarkable upswing of Bayesian methods in the 1990s is tied to a modeling practice that challenges Bayesian epistemology. The section closes with a brief look at recent accounts of practicing statisticians (of varying camps) who discovered the notion of modeling as a new focus and as a common ground.

2. Models mathematize the logic of inference

The mathematical theory of statistical inference—the classical account—was developed during the 1920s and 1930s mainly by three scholars: Ronald A. Fisher (1890–1962), Jerzy Neyman (1894–1981), and Egon S. Pearson (1895–1980). While Neyman and Pearson argued their account would provide a mathematical foundation to Fisher's older approach, Fisher disagreed fiercely and an embittered controversy set in that was never settled (compare Hacking 1965, 89). This section argues that the controversy rests, aside from any personal aspects, on a profound conceptual basis, while both sides held conflicting views about statistical modeling.[3]

2.1 *Fisher's account of modeling*

Fisher elaborated a comprehensive logic of inductive inference, as he called it. His presumably philosophically fundamental innovation consists of precisely describing what is to be understood by a model, and how models are to be embedded in the logic of inference. In 1922, Fisher published his seminal contribution, "On the Mathematical Foundations of Theoretical Statistics," where we find a number of influential new concepts, among them, the level of significance (for rejecting a null hypothesis) and the parametric model, whose systematic role within statistical inference was elaborated for the first time. Fisher describes the general goal of statistics as follows:

> In order to arrive at a distinct formulation of statistical problems, it is necessary to define the task which the statistician sets himself: briefly, and in its most concrete form, the object of statistical methods is the reduction of data. A quantity of data, which usually by its mere bulk is incapable of entering the mind, is to be replaced by relatively few quantities which shall adequately represent the whole, or which, in other words, shall contain as much as possible, ideally the whole, of the relevant information contained in the original data.
>
> *(1922, 311)*

At first glance, it may seem that Fisher's concern is merely a technical question of the reduction of data. This, however, is not the case, for the problem of whether certain standard quantities "adequately represent" the entirety of data cannot be solved based on the data alone. The same holds for "relevant information"—whether it is still contained in a data-reducing statistic will have to be evaluated according to further criteria. In other words, the mathematical part first requires modeling. Fisher continues:

> This object is accomplished by constructing a hypothetical infinite population, of which the actual data are regarded as constituting a random sample. The law of distribution of this hypothetical population is specified by relatively few parameters, which are sufficient to describe it exhaustively in respect of all qualities under discussion. *(311)*

Fisher explicitly mentions the constructive character of this undertaking, which conceives of the data observed as an instance of the underlying model-type population. The merit of this is that such a population, i.e., its distribution law, is exhaustively ("in respect of all qualities under discussion," i.e., with regard to a concrete question of application) described by a small number of parameters. It is this law, in combination with specified parameters, that transfers the testing problem into a mathematical problem.

Fisher subdivided the general task of statistics into three types of problems:

1 Problems of Specification. "These arise in the choice of the mathematical form of the population" (1922, 366). This step thus is part of the modeling activity; and it cannot be derived, but requires deliberations, like those on the basis of practical experience gained in similar situations.
2 Problems of Estimation. They are formulated on the basis of a mathematical-statistical model. Fisher saw his own work as a solution to these problems.
3 Problems of Distribution. The matter here is mathematical tractability. The most beautiful model is good for nothing if it yields no distribution curves (with available mathematical means).

For Fisher, the main task of modeling consists in balancing judgment and experience with tractability. A model may assume a certain family of distributions whose parameters have to be specified by estimation from the data. A simple, admittedly very simplified, example may explain the terminology. During his work, Fisher was intensely engaged in agro-science experiments such as estimating the effect of a certain fertilizer. A model could look as follows: the yield of the various acreages is equally distributed, that is, normally distributed to the two parameters (m, σ^2). This establishes essential assumptions of the model. The effect of the fertilizer, it is further assumed, will only change the parameter m. In other words, the yield of a fertilized acreage is normally distributed to a mean m'. A typical question regarding the statistical inference to be drawn from the data, i.e., the yields of all acreages, would then be: Which effect is produced by treating with the fertilizer? The null hypothesis, which is part of Fisher's logic, H_0 would be that the fertilizer has no effect at all, that is, that the means are equal, and all differences observed are random: H_0: $m = m'$.

Based on the modeling assumptions, all information contained in the data not concerning the two parameters is irrelevant. Given the model, the specification is achieved by assigning the values of these parameters: It is a mathematical fact that the normal distribution is characterized by mean and variance. In Fisher's terms, the normal distribution is part of the model while assigning concrete values to the parameters specifies a hypothesis. In this way, only the assumption of a model makes it possible to speak of the "relevant information" contained in the data and to assess the hypothesis mathematically.

2.2 The Neyman–Pearson theory: the fundamental lemma

During the following decade, Jerzy Neyman and Egon Pearson elaborated the theory of statistical inference that bears their names. Their seminal essay "On the Problem of the Most Efficient Test of Statistical Hypothesis" of 1933 can be considered the founding document—an essay referred to by the authors as "the big paper." The theoretical backbone of the Neyman–Pearson theory is expressed by their "fundamental lemma." Only further specification of what modeling should consist of allowed them to prove this lemma.

Neyman and Pearson criticized the asymmetrical treatment of the null-hypothesis as a deficit of Fisher's logic of testing. Fisher started with the null hypothesis that no effect could be observed, and a test might lead to accepting another hypothesis, thereby rejecting the null hypothesis. This name alone already testifies to the asymmetrical conception. Neyman and Pearson insisted that a model should produce a symmetrical situation where two hypotheses compete with each other ("hypothesis" versus "alternative"); observing the data should lead to the decision on which hypothesis was to be preferred. For guiding this decision, Neyman and Pearson introduced the errors of the first and second kinds. Choosing one of two competing hypotheses can be wrong. One can commit errors of the first kind (accepting a false hypothesis) and errors of the second kind (rejecting a true hypothesis), and one should therefore make the relative assessment between the two types of error an object of the method as well.

From their analysis of the two types of statistical error, Neyman and Pearson derived two further concepts, namely, the concept of the size of a test that corresponds to the level of significance and the concept of the power of a test that corresponds to the analogous quantity for the error of the second kind. According to the Neyman–Pearson account, modeling must create a situation in which two hypotheses confront one another, and then, one has to fix a test's size before optimizing its power. The Fundamental Lemma states that, in the case of a simple dichotomy of hypotheses, there exists, for any possible size, a uniquely most powerful test of that size.

Consequently, modeling is not concerned with individual cases, but rather with what happens if one proceeds in accordance with such a rule. Framed by a model in this way, the (remaining) possible courses of action have mathematical properties, namely, they form convex risk sets. Technically speaking, there is a unique element in this set with minimal distance (maximal power) to any point specified by size. Neyman–Pearson realized that the proof of their lemma required a strict delineation of modeling: at stake is an iterated procedure with two alternatives: one first determines size and then maximizes power.

2.3 Controversy about modeling

Although Neyman and Pearson see their work as a mathematical rounding off and improvement of Fisher's approaches, Fisher responded with a polemical attack. In the literature, this controversy has repeatedly been treated both mathematically and philosophically.[4] Cutting through the controversy from the perspective of modeling offers a view of why the controversy has not been resolved: models should fulfill incompatible tasks.

In the frame of the Neyman–Pearson theory, the reiterated application of a procedure forms the basis for statistical inferences. The paradigmatic example is a procedure for accepting or rejecting shipments of some product based on a random sample taken from the shipment. The Neyman–Pearson theory then suggests an optimal rule by considering the statistical properties when the procedure is applied over and over again. This particular conceptualization was the only way that Neyman and Pearson could provide an objective basis for the logic of inference, thereby dispensing with Fisher's hypothetical infinite populations. Therefore, Neyman and Pearson rely on a model concept that includes many more preconditions, according to which much of the statistician's method is already fixed. According to Fisher, a statistician uses mathematical reasoning within the logic of inference, e.g., building and adjusting a model according to the data at hand and the questions under discussion. In the Neyman–Pearson theory, the reasoning of the statistician (e.g., finding an appropriate acceptance procedure) has become *subject* to modeling.

With this, however, they place themselves in strict opposition to Fisher. For him, modeling creates the objects one can argue about mathematically, whereas Neyman and Pearson shape the basic situation in which modeling takes place, requiring reiterated procedures and competing hypotheses. Fisher considered the applied mathematician's situation fraught in principle with many subjective components—working on an applied problem requires a high degree of "judgment" and is also sensitive to the concrete case at hand. According to Fisher, reflecting this application situation and its non-mathematical components is an integral part of applied mathematics or statistics. Modeling thus has the task of mediating between real-world problems and mathematics. Hence, Neyman and Pearson intended to get rid of precisely the constructive act of modeling at the center of Fisher's inductive inference logic. This somewhat ironic point teaches a cautionary lesson about modeling that is relevant far beyond statistics. In modeling, mathematization is not neutral but can impose critical conditions that change the concept of modeling.

3. Abandon models and let the data speak for themselves

This section takes a look at the anti-modeling standpoint. It is not at all misplaced in a chapter on modeling because modeling is about mediation and the data-centric standpoint holds that much of the modeling task can be replaced by the data themselves. There are many variants of this standpoint. This section focuses on an early example, Tukey's work on "Exploratory Data Analysis" (EDA), and at the end takes a brief look at recent computer methods that have brought new prominence to the data-centric view.

EDA was initiated and propagated by John Wilder Tukey in the 1960s, and Tukey's programmatic book, "Exploratory Data Analysis," appeared in 1977. In contrast to its influence on the practice of statistics, EDA is often neglected in philosophically oriented considerations. In the context of models, EDA is of great interest because Tukey combined his programmatic design with a strong critique of the concept and use of models. What is data analysis about? The *Encyclopedia of Statistics* summarizes:

> Exploratory data analysis is the manipulation, summarization, and display of data to make them more comprehensible to human minds, thus uncovering underlying structure in the data and detecting important departures from that structure
>
> *(Kruskal 1978, 3)*

This statement expresses a fine, but decisive difference to Fisher's account of statistics in which, "reducing the data to relevant information," was key, which requires reference to an underlying model. EDA, in contrast, concerns a process preceding the construction of a model. Tukey conceived of EDA very consciously as a counter-model and as a necessary complement to what he called hypothesis testing-oriented confirmatory data analysis (CDA). Not working with a model should liberate the skilled judgment of the statistician. In a certain sense, Tukey considered mathematical models in statistics to be a dangerous gift, as they suggested the applicability of rigorous mathematical arguments. Often, Tukey says, the complex difficulties arising from amorphous data are passed over too quickly. In other words, Tukey was convinced that application-oriented statistics must begin methodologically even before the data are inserted into the Procrustes bed of a model. For Tukey, mathematical, model-dependent arguments should enter at a late stage of the application process that would have to begin with exploring the data without a potential bias by modeling assumptions. For instance, the judgment of what part of the data are outliers and may therefore be ignored is often decided too quickly by reference to a model. For him, the very process of model building has to be guided by EDA—a position quite contrary to Neyman and Pearson's effort to integrate model building into a mathematical framework.

Tukey illustrated the relationship between exploratory and confirmatory data analysis with the metaphor of the detective and the judge:

> Unless the detective finds the clues, judge or jury has nothing to consider. Unless exploratory data analysis uncovers indications, usually quantitative ones, there is likely to be nothing for confirmatory data analysis to consider.
>
> *(1977, 3)*

Was that not the initial motivation for modeling as well? Modeling was indeed also one of the prerequisites for applying mathematical propositions to reality, by having models bring a practical situation into a sufficiently exact form. While Tukey does not challenge this, he insists on the systematic importance of the first preparatory steps in the process of modeling. His main issue is to clarify how the judgment necessary to construct an adequate mathematical-statistical model can depend on an investigation by means of mathematical tools. This extended frame of mathematical tools (far from deductive reasoning) then encompasses decidedly less precise concepts. In this context, Tukey pleads in favor of vague concepts, a rather uncommon recommendation, at least in a mathematical context:

> Effective data analysis requires us to consider vague concepts, concepts that can be made definite in many ways. To help understand many definite concepts, we need to go back to more primitive and less definite concepts and then work our way forward
>
> *(Mosteller and Tukey 1977, 17)*

At the very outset of a problem of application, Tukey says, there is mostly quite a number of possible ways to attain a more abstract, more rigorous, or more precise formulation of the problem. This view recalls Fisher's position that there is a multitude of possible infinite populations which come under consideration during the first steps of modeling. Confirmatory data analysis assumes a class of models and then makes the data decide which the best model is in said class, while explorative data analysis aims to let the data speak for themselves. Fisher's and Tukey's conceptions do not contradict one another; rather, what becomes evident if one integrates the two is that the process of modeling is based on an interplay of data and models in the course of which both have to be considered variable. When Tukey and Wilks (1970) underline that using models to evaluate data is different from using data to evaluate models, they do not intend to play down the use of models, but rather assign some autonomy to both approaches that then require mediation.

Tukey introduced a set of new tools like stem-and-leaf diagrams that are intended to make the explorative analysis of the data possible. These tools are fundamentally based on the capacities of modern computers, in particular, visualization. EDA may well be seen as the herald of instrument-driven and ongoing multifaceted changes in modern statistics that have been triggered by the computer.

The advent of cheaply available and networked computers enlarged these possibilities. Some of them address elements of the modeling process,[5] but some even claim to replace modeling altogether. One example that created particularly big waves in philosophy is Bayesian networks. Formal epistemologists (Spohn 2001; Pearl 2000) claimed that causal reasoning can be completely expressed in the technique of Bayesian networks (technical details do not matter here). When Spirtes, Glymour, and Scheines (1993) claimed that they had coded an algorithm that would automatically construct the causal network for given data, a heated controversy set in. Can data processing replace (causal) modeling? Cartwright or Humphreys and Freedman (1996) insisted on a negative answer—as Cartwright put it: no causes in, no causes out (1989, Ch. 2). According to them, causal inference requires statistical (and causal) modeling that cannot be substituted by computational statistics.[6]

Another instance is the purported impact of data-driven science and machine learning on epistemology. Most variations of this claim (see Kitchin 2014 for a sample) hold that deep learning, combined with a sufficient amount of data, will be able to detect all kinds of patterns, independent of any foregoing theory. In other words, statistical modeling allegedly becomes obsolete because there will be one general, powerful model (a deep neural network, much like a human brain) that is able to handle all tasks. I am skeptical whether such a broad claim is warranted. My point here is that the vision of getting rid of statistical modeling, and all the related problems of mediating between the world and our conceptions of it, is getting fueled by computer methods, especially AI.

4. The career of Bayesian models in statistical practice

Philosophers have discussed Bayesian statistics vigorously and elaborated Bayesianism as a *philosophical* position.[7] Bayesian epistemology lays claim to capturing knowledge acquisition in a fairly general manner. The central piece is Bayes' rule which prescribes how one should update prior beliefs in light of new evidence. This rule captures how to calculate conditional probabilities. Let $\pi(H)$ stand for the probability of a statement or hypothesis H, and $\pi(H \mid D)$ for the conditional probability of H given D. Now, both H and D happen if (for the moment, think of temporal order) D happens and then H happens given D, or equivalently, H happens and then D happens given H. In other signs: $\pi(D) \times \pi(H \mid D) = \pi(H) \times \pi(D \mid H)$. Separating $\pi(H \mid D)$ on the left side provides Bayes' rule:

$$(*) \quad \pi(H \mid D) = \pi(H) \times \pi(D \mid H) / \pi(D)$$

It is named after Reverend Thomas Bayes (c. 1701–1761), a Presbyterian minister, philosopher, and statistician. Bayesianism starts with a special interpretation of this rule. Consider you have some hypothesis H—for example, that it will rain tomorrow. You do not know for sure, so (in a Bayesian mood) the degree of your belief can be expressed as a probability, $\pi(H)$. Now there arrives new evidence D—say, you wake up the next morning and have a look at the sky. This should give you additional evidence and will change your (subjective) probability of rain on this day. Therefore, $\pi(H)$ is also called the "prior" that will be updated. The updated probability, written $\pi_{D(H)}$, of your hypothesis given the data is also called the "posterior." Which numerical value does it have? Bayesians take the position that updating needs to happen by conditionalization. The posterior is the conditional probability: $\pi_{D(H)} = \pi(H \mid D)$. In other words, equation (*) answers the question: The posterior is proportional to the (subjective) prior $\pi(H)$ and to $\pi(D \mid H)$, the so-called likelihood—that is, the probability of the data given your hypothesis (how likely it is that the sky looks like it does in the morning if it were to rain). The term $\pi(D)$ plays the role of a (normalizing) constant.

Although Bayes' rule works with basic concepts, actually calculating with it, i.e., determining the conditional probability on the left side of (*) from the terms on the right side—the probability of a hypothesis $\pi(H)$, the probability of the data $\pi(D)$ (often expressed via conditioning on different possibilities), and the conditional probability $\pi(D \mid H)$—requires a detailed model. Moreover, even if a model is given that determines these values, computing them was restricted to the most simple cases, which made Bayesian statistics impractical. The use of Bayesian approaches in scientific *practice* has an illustrious history. Despite their philosophical prominence, they remained a small minority group in science with a consistent share of only 2–4% among papers in leading traditional statistical journals. However, the 1990s saw an increase in interest, and Bayesian methods quickly acquired a high level of popularity (about 20% of papers).

4.1 *Exploration and flexibility*

A common viewpoint holds that Bayesian modeling was initially impractical because of computational difficulties, and later became practical thanks to computational methods, all without changing its rationale. This section looks at the matter from a different perspective. Working with computational methods might change the concept of modeling and, consequently, change the rationale of Bayesian epistemology. Namely, these methods undercut the interpretation of priors, turning them from an expression of beliefs held prior to new evidence into an adjustable parameter that can be manipulated flexibly by computational machinery.

At this point, the argument rests on an analysis of the computational methods of which this section can only provide a glimpse (see Lenhard 2022 for details). By the 1980s, it had become a widely shared view that computational methods were the key to making Bayesian statistics practical. The statistician A. F. M. Smith, a leading voice, argued in a sort of manifesto that efficient numerical integration procedures were needed for the success of Bayesian methods (Smith 1984). There is wide agreement that Markov chain Monte Carlo (MCMC) methods provided these procedures.

MCMC methods *simulate* relevant properties of mathematical objects (such as integrals or distributions) in numerous iterated trials to gain a picture or approximation of these properties. One can compare MCMC with sounding out unknown territory by taking simulated random walks.[8] This modeling approach thus explores the behavior of a (complex) mathematical object, like a posterior distribution, with the help of the MCMC machinery. When proponents such as Smith and Roberts (1993) state that MCMC methods are for "exploring and summarizing posterior distributions in Bayesian statistics" (p. 3), the point about exploration is important. In a way, MCMC explores mathematical properties with the help of probabilistic and iterative means. One can see a *frequentist* element sneaking in here.[9]

However, there is another point about exploration to be made. The speed of MCMC is also an invitation to engage in an exploratory mode of modeling in the following sense. Modelers can work with incompletely specified models that contain parameters that get adjusted only in a feedback loop where model behavior is observed and modified. Researchers do not need to determine parameters from the beginning; rather, they can adapt them during the process to obtain a better match. For Bayesian modeling, MCMC made exploration on this level feasible. With the help of adjustable parameters, a model can be specified in flexible ways. The MCMC trick brings this flexibility to Bayesian modeling.

However, the exploratory-iterative mode affects the Bayesian rationale. The core of Bayesian epistemology, indeed the defining feature for many philosophers, is the subjective stance. The modeling process starts with one's degree of belief. We have seen, however, that this characteristic of Bayesian epistemology fades away over the course of the development of MCMC approaches. Priors now appear as part of the adaptation machinery. Importantly, seen as adjustable parameters, priors lose their interpretation as prior knowledge. To the extent that they are treated like adjustable parameters, the resulting values no longer express (degrees of) *prior* belief, but rather correspond to an overall fit of a model to data, *resulting* from the exploratory-iterative process of modeling. In a nutshell, *the priors cease to be prior*.

4.2 Modeling and pragmatism

Bayesian approaches are a success story in statistics that began in the 1990s. This story pivots on the co-development of computational methods and a concept of modeling that utilizes flexibility, much like a pragmatic tool that comes with more philosophical laissez-faire. The situation looks different from the seasoned positions in the philosophy of statistics. This pragmatic turn has the potential to fundamentally affect the philosophy of statistics. How the new situation should be captured conceptually is not yet clear. However, leading statisticians have engaged in a philosophical debate.

According to Bradley Efron, classical frequentist and Bayesian approaches work together and mutually *complement* each other in computer modeling. Especially when analyzing large amounts of ("big") data—according to Efron (2005)—it is often hopeless to construe priors subjectively. Sander Greenland (2010) argues that Efron's stance on mutually complementing virtues is not correct and that it would be better to use the term "ecumenism" to describe how statistical methods come together. He traces this back to G. E. P. Box's (1983) plea for ecumenism. Despite its prominent advocates—according to Greenland—ecumenism has not yet had a large impact on the teaching or practice of statistics.[10] Robert Kass is another prominent statistician who reflects on the ongoing changes in a conceptual way. He advocates what he calls "statistical pragmatism," a position that sees modeling as the core activity of statistics (Kass 2011). He makes a careful attempt to sketch the common ground between Bayesian and frequentist positions regarding how statistical models are connected with data. Thus, the dynamics of computational modeling seem to be a uniting feature of formerly separated camps of philosophy of statistics: "The loyalists of the 1960s and 1970s failed to realize that Bayes would ultimately be accepted, not because of its superior logic, but because probability models are so marvelously adept at mimicking the variation in real-world data" (Kass, cited according to McGrayne 2011, 234).[11] Steven Goodman (2011) disagrees because Kass' pragmatism looks like a mere truce rather than a new foundation. Also commenting on Kass, Hal Stern (2011, 17) worries "more broadly that pragmatism might appear to reinforce the notion of statistics as a set of techniques that we 'pull off the shelf' when confronted with a data set of a particular type." Finally, Andrew Gelman (2011, 10) observes that this pragmatism, though thriving on the flexibility of methods to obtain calibration between model and data, is still objective.

In sum, notions such as complement, truce, ecumenism, or pragmatism signal how statisticians capture conceptually what is going on in recent practices of modeling. All philosophically minded practitioners as well as practice-oriented philosophers should welcome the debate around the conception of modeling. It breathes fresh air into the philosophy of statistics. Furthermore, following the practices of modeling provides a lens, both to practitioners and philosophers, on how new instrumentation, i.e., the computer and computational methods, reconfigures the relationship between scientific knowledge and scientific data—the primary reason why the philosophy of statistics is so intriguing.

Acknowledgment

Johannes Lenhard's research was funded by Deutsche Forschungsgemeinschaft, DFG LE 1401/9-1.

Notes

1 The entry on philosophy of statistics by Romeijn (2017) in the Stanford Encyclopedia provides a good overview with many references. I would like to highlight accounts of classical statistics, written by the pioneers (Neyman 1957; Fisher 1955), recent philosophical work on the classical account (Mayo 1996; Spanos 2011), and also overviews of the Bayesian standpoint (Press 2002; Howson and Urbach 2006; Gelman et al. 2013; Earman 1992).
2 An exception is the literature on model selection, i.e., finding the optimal model, including the philosophical discussion on what criteria are adequate (see Romeijn 2017 for references). However, this literature takes modeling for granted and starts from there.
3 Lenhard (2006) provides much more historical and mathematical detail to this argument.
4 For a sample, see Hacking (1965), Gigerenzer et al. (1989), or Lehmann (1993).
5 Examples are *principal component analysis* for data reduction, see Jolliffe (2002), or *support vector machines*, see Vapnik (2006).
6 Pearl (2000) has elaborated the machinery of causal inference based on (Bayesian) networks. However, he has dropped the claim of doing without modeling, but assumes a causal model and then shows how to refine and modify it based on the data.
7 The *Stanford Encyclopedia of Philosophy* has entries on the philosophy of statistics (Romeijn 2017) and a separate one on Bayesian epistemology (Talbott 2016). Part of "formal epistemology," too. Taken together, these provide a guide to the large body of philosophical literature on Bayesianism.
8 At the heart of MCMC is how fast Markov chains converge to their equilibrium distribution. Obviously, simulating random walks fits exactly to the iterative capacity of the computer.
9 Quite different from a Bayesian, a frequentist considers the probability of an event as the fraction of occurrences in repeated trials.
10 Greenland further acknowledges that this theme is not new, but also has been brought up repeatedly by Good (1983), Diaconis and Freedman (1986), or Samaniego and Reneau (1994).
11 This capability is based heavily on adaptable parameters, especially on priors that can be changed to increase the ability of a model to mimic the data—quite in line with our prior analysis of MCMC.

References

Box, George E. P. 1983. "An Apology for Ecumenism in Statistics." In *Scientific Inference, Data Analysis, and Robustness*, edited by George Box, Tom Leonard, and Chien-Fu Wu, 51–84. New York: Academic Press.

Cartwright, Nancy. 1989. *Nature's Capacities and Their Measurement*. Oxford: Oxford University Press.

Diaconis, Persi, and David Freedman. 1986. "On the Consistency of Bayes Estimates (with discussion)." *Annals of Statistics* 14: 1–67.
Earman, John. 1992. *Bayes or Bust? A Critical Examination of Bayesian Confirmation Theory*. Cambridge: The MIT Press.
Efron, Bradley. 2005. "Bayesians, Frequentists, and Scientists." *Journal of the American Statistical Association* 100(469): 1–5.
Fisher, Ronald A. 1955. "Statistical Methods and Scientific Induction." *Journal of the Royal Statistical Society B*, 17: 69–78.
———. 1922. "On the Mathematical Foundations of Theoretical Statistics." *Philosophical Transactions of the Royal Society A* 222: 309–368.
Gelman, Andrew. 2011. "Bayesian Statistical Pragmatism." *Statistical Science* 26(1): 10–11.
Gelman, Andrew, and Cosma Shalizi. 2013. "Philosophy and the Practice of Bayesian Statistics (with Discussion)." *British Journal of Mathematical and Statistical Psychology* 66: 8–18.
Gigerenzer, Gerd, Zeno Swijtink, and Theodore Porter. 1989. *The Empire of Chance*. Cambridge: Cambridge University Press.
Good, Irving J. 1983. *Good Thinking*. Minneapolis: University of Minnesota Press.
Goodman, Steven N. 2011. "Discussion of "Statistical Inference: The Big Picture" by Robert E. Kass." *Statistical Science* 26(1): 12–14.
Greenland, Sander. 2010. "Comment: The Need for Syncretism in Applied Statistics." *Statistical Science* 25(2): 158–161.
Hacking, Ian 1965. *Logic of Statistical Inference*. Cambridge: Cambridge University Press.
Howson, Colin, and Peter Urbach. 2006. *Scientific Reasoning: The Bayesian Approach*. La Salle: Open Court, 3rd edition.
Humphreys, Paul, and David Freedman. 1996. "The Grand Leap." *British Journal of the Philosophy of Science* 47: 113–123.
Jolliffe, Ian T. 2002. *Principal Component Analysis*. New York: Springer, 2nd edition.
Kass, Robert E. 2011. "Statistical Inference: The Big Picture." *Statistical Science* 26(1): 1–9.
Kitchin, Rob 2014. "Big Data, New Epistemologies and Paradigm Shifts." *Big Data & Society* 1–12. DOI: 10.1177/2053951714528481.
Kruskal, William H. 1978. *International Encyclopedia of Statistics*. New York: Free Press.
Lehmann, Erich L. 1993. "The Fisher, Neyman-Pearson Theories of Testing Hypotheses: One Theory or Two?" *Journal of the American Statistical Association* 88(424): 1242–1249.
Lenhard, Johannes. 2022. "A Transformation of Bayesian Statistics: Computation, Prediction, and Rationality." *Studies in History and Philosophy of Science* 92: 144–151.
———. 2006. "Models and Statistical Inference: The Controversy between Fisher and Neyman-Pearson." *British Journal for the Philosophy of Science* 57: 69–91.
Mayo, Deborah G. 1996. *Error and the Growth of Experimental Knowledge*. Chicago: The University of Chicago Press.
McGrayne, Sharon B. 2011. *The Theory That Would Not Die. How Bayes' Rule Cracked the Enigma Code, Hunted Down Russian Submarines, and Emerged Triumphant from Two Centuries of Controversy*. Yale: Yale University Press.
Mosteller, Frederick, and John W. Tukey. 1977. *Data Analysis and Regression: A Second Course in Statistics*. Reading (MA): Addison-Wesley.
Neyman, Jerzy. 1957. "Inductive Behavior as a Basic Concept of Philosophy of Science", *Revue Institute Internationale De Statistique* 25: 7–22.
Neyman, Jerzy, and Egon S. Pearson. 1933. "On the Problem of the Most Efficient Tests of Statistical Hypotheses." *Philosophical Transactions of the Royal Society A* 231: 289–337.
Pearl, Judea. 2000. *Causality: Models, Reasoning and Inference*. New York: Cambridge University Press. Second edition, 2009.
Press, S. James. 2002. *Bayesian Statistics: Principles, Models, and Applications*. New York: Wiley.
Romeijn, Jan-Willem. 2017. "Philosophy of Statistics." In *The Stanford Encyclopedia of Philosophy*, ed. Edward N. Zalta. https://plato.stanford.edu/archives/spr2017/entries/statistics.
Samaniego, Francisco J., and Dana Reneau. 1994. "Toward a Reconciliation of the Bayesian and Frequentist Approaches to Point Estimation." *Journal of the American Statistical Association* 89, 947–957.
Smith, Adrian F. M. 1984. "Present Position and Potential Developments: Some Personal Views. Bayesian Statistics." *Journal of the Royal Statistical Society A*, 147(2): 245–259.

Smith, Adrian F. M., and Gareth O. Roberts. 1993. "Bayesian Computation via the Gibbs Sampler and Related Markov Chain Monte Carlo Methods." *Journal of the Royal Statistical Society B* 55(1): 3–23.

Spanos, Aris. 2011. "Curve Fitting, the Reliability of Inductive Inference, and the Error-Statistical Approach." In *Philosophy of Science* 74(5): 1046–1066.

Spohn, Wolfgang. 2001. "Bayesian Nets Are All There Is To Causal Dependence." In *Stochastic Causality*, ed. Maria Carla Galavotti et al., 157–172. Stanford: CSLI Publications.

Spirtes, Peter, Clark Glymour, and Richard Scheines. 1993. *Causation, Prediction and Search*. Springer Lecture Notes in Statistics 81. New York: Springer.

Stern, Hal. 2011. "Discussion of "Statistical Inference: The Big Picture" by R. E. Kass." *Statistical Science* 26(1): 17–18.

Talbott, William. 2016. "Bayesian Epistemology." In *The Stanford Encyclopedia of Philosophy*, ed. Edward N. Zalta. https://plato.stanford.edu/archives/win2016/entries/epistemology-bayesian.

Tukey, John W. 1977. *Exploratory Data Analysis*. Reading (MA): Addison-Wesley.

Tukey, John W., and Maurice B. Wilks. 1970. "Data Analysis and Statistics: Techniques and Approaches." In *The Quantitative Analysis of Social Problems*, ed. Edward R. Tufte, 370–390. Reading (MA): Addison-Wesley.

Vapnik, Vladimir N. 2006. *Estimation of Dependences Based on Empirical Data*. New York: Springer.

18
MODELS, DATA MODELS, AND BIG DATA

Leticia Castillo Brache and Alisa Bokulich

1. Introduction

Although the distinction between models and data may be intuitively clear, adequate definitions can be surprisingly subtle and elusive, and the relationships between models and data can turn out to be quite complex. Data can be defined as records of a process of inquiry, involving causal interactions with features of the world (e.g., Bokulich and Parker 2021; Leonelli 2016). Data are typically the results of experiments, measurements, or observations and are usually (though not necessarily) represented numerically. Data models, by contrast, are usually thought of as an organized or processed version of a data set designed to help the data serve as evidence for various purposes. The line between models and data can be blurred due to what Paul Edwards (2001, 2010) calls *model–data symbiosis*, according to which not only are models data-laden, but data are also model-filtered.

This chapter provides an introduction to these issues as well as other issues that arise out of the complex relationship between data and models. It starts off by exposing different views about data. The chapter moves then, in the section titled "Data Processing and Model–Data Symbiosis," to explain the complexities that arise out of the relationships between data and modeling. In the subsequent section, the processes of data reuse, data repurposing, and data rescue are explored and the differences between them and how they are useful under different circumstances are explained. In the penultimate section, the importance of metadata and data empathy and issues in big data are discussed. The chapter concludes by highlighting central debates in data ethics, including the problem of "dirty data."

2. Measurements, raw data, and data models

Although there are different conceptual views of data, including the relational account of data (Leonelli 2015) and the pragmatic-representational account of data (Bokulich and Parker 2021), these views agree that data are made, not given, and that while data may be causally tied to the world, they are not perfect in capturing it.

Typically, in science, data are the outcomes of various measurements or observations. Thus, further philosophical insight into the nature of data can be gained by relating it

 DOI: 10.4324/9781003205647-22

to work in the philosophy of measurement. Following official guidelines on measurement from metrological organizations such as the International Bureau of Weights and Measures (BIPM), it has become standard to distinguish between a measurement "indication" and a measurement "outcome." Eran Tal (2017), for example, explains that an *indication* is a preliminary property of the measuring instrument, whose information is to be used as a basis from which to infer a measurement *outcome*, which by contrast is a claim about the object or system being measured. A measurement outcome requires taking the instrument indication—often along with other measurement indications, background knowledge, or other resources—and using it as a basis from which to calculate or infer that a particular property or value can be ascribed to the object or system being measured. Because this process requires having an abstract and idealized model of the measurement process, Tal (2012, 2017) refers to it as a *model-based account of measurement*. The output of this process can then be collected as data.

Even after the data are collected, the resulting data set often needs to be further processed, converted, or corrected before it can be used as reliable evidence. Take, for example, a mercury thermometer. In addition to the implicit conversion of data about the height of a mercury column (measured in millimeters) into data about temperature (measured in degrees Fahrenheit or Celsius) that is automatically performed by a well-calibrated thermometer, a doctor may need to further adjust the temperature data based on how the thermometer reading was taken (e.g., orally) and perhaps involving a child who did not keep the instrument properly under their tongue for the full time (an imperfect measurement procedure was followed). In other cases, one might make multiple temperature measurements, taking the average before ascribing a final temperature to the system being measured. In all these cases, one is taking what might be described as "raw data" and converting it into a data model that can more reliably be used as evidence about some claim, such as the health of the patient. The notion of raw data is a slippery one and is often used in a relative rather than absolute sense to mean any given data set before some further data processing is applied (Bokulich 2018/2021; Bokulich and Parker 2021). To further complicate the distinction between raw data and data models, many instruments (such as the thermometer described above) have some form of data processing built in, so that even the seemingly raw data coming out of the instrument already contain a significant level of theory-based data processing.

In sum, data models are data sets that have been processed in some way in order to make salient some features that the data are intended to capture, hence enabling them to better serve as evidence in some context of inquiry. The next section describes the various ways data can be processed in order to construct a data model and the central role that traditional theoretical models can play in that process.

3. Data processing and model–data symbiosis

Paul Edwards (1999, 2010) has argued that data and theoretical models are part of an interdependent and mutually beneficial relationship he calls *model–data symbiosis*. Model–data symbiosis involves two components. On the one hand, models are data-laden, in that large quantities of data can go into the construction, calibration, and evaluation of scientific models. On the other hand, and more controversially, data are also model-filtered—theoretical models can play a central role in data processing. There are many different ways that data can be processed into a data model in order for it to be used as evidence, many of which can make use of substantive theoretical models. Bokulich (2020b) provides a taxonomy of seven different ways that data can be model-filtered. Each of these is explained below.

The first processing technique, which was already discussed, is *data conversion*, where a measurement of one quantity is converted into a measurement of a different quantity. Data conversion can involve conversions about the same quantity or different quantities. For example, one may use data conversion to figure out what the temperature is in Celsius if one has the value in Fahrenheit. In this case, the conversion is about the same quantity, namely temperature. Alternatively, when one uses data conversion to figure out, say, the momentum of an object by combining its mass and velocity, one can say a conversion between different quantities is being done. Data conversion is one of the most common processing techniques used in everyday life.

Second, one can model unwanted influences on a measurement process and then remove them in a process of *data correction*. This involves knowing the magnitude of an error in order to correct for it. For example, one can use data correction when a kitchen or bathroom scale is not calibrated in order to correct for the incorrect value. In instances in which it is observed that a scale does not start off with a zero quantity, it must be corrected for by adding or subtracting the right value after getting the measurement indication in order to get an accurate measurement result. Certainly, no measurement is perfect, so modeling sources of error is of key importance in order to get accurate results. How precise one needs to be with the measurements depends on how accurate of a result a research project needs. The uncertainty budget for any project will depend on how fine a resolution is needed to achieve the purpose of the research project. Knowing what level of precision is needed aids in making the process of data correction successful.

Third, models of how a field quantity might vary spatially can be used to fill in gaps in sparse data measurements through a process of *data interpolation*. For example, in the medical field, doctors may be able to use the process of data interpolation to fill in missing medical records from a patient, such as heart rate values or body temperature. In a case where a doctor has recorded values every two hours, they can interpolate using the known data to find the missing values of the hours they did not record. Data interpolation can also be used to find patients' missing individual variables such as body mass index (BMI), systolic and diastolic blood pressure, and arterial oxygen saturation ($Sa0_2$). The use of data interpolation allows scientists to have more complete data samples for their studies.

Fourth, theoretical models can also be used to upscale data, which involves going from a small to a large scale, or to downscale data, which involves going from a coarse to a fine scale—these processes are known as *data scaling*. Examples of upscaling can be seen in biological research, where scientists must upscale their laboratory findings in order to understand how their results would affect complex ecosystems. This sort of upscaling is necessary for the results of laboratory research to be usable in a broader sense. Examples of downscaling, on the other hand, can often be seen in climate science with the development of Regional Climate Models (RCM), which are used to understand local meteorological parameters. RCMs have a much finer scale than their counterparts Global Climate Models (GCM), which exist at a much larger scale.

Fifth, models can be used to assist in integrating diverse data sets in what is called *data fusion*, also called data integration. The main function of data fusion is to combine heterogeneous data sources into a coherent product. For example, data fusion is often used in neuroscience in order to get more complete images of the brain. A doctor might order different types of scans, such as MRIs, fMRIs, and EEGs, to better understand how a patient's neural activity might be currently affected by a disease, which in turn aids them in recommending a more adequate treatment. Given that different kinds of data can be used for data fusion, researchers must make sure that all the data used is commensurable and meaningfully integrated.

Sixth, models can be used to address the uncertainty in both data and theoretical models through an iterative, dynamic process known as *data assimilation.* Wendy Parker (2016) defines data assimilation as "a process that relies on both observations and model-based forecasts to estimate conditions" (2016, 1565). This processing technique happens when models adjust their initial conditions to be more consistent with observed data. The adjustment does not happen only once, but it is rather an iterative back and forth between model prediction and empirical data. There are many examples of data assimilation in weather forecast models. Parker (2017) explores a specific way in which atmospheric data assimilation can play a role in computer simulations and highlights how the use of data assimilation complicates the picture of what counts as a measurement given the entanglements between observed measurements and model-based outputs.

Lastly, artificial or *synthetic data* can be generated as the output of computer simulation models. Synthetic data can be used to test algorithms in such a way that private information can be fixed and exchanged for synthetic identifiers, which in turn helps protect an individual's privacy. For example, instead of using private individual information, one can run the information through an algorithm to get the values of interest and replace the individual's private information with synthetic identifiers. The synthetic data produced has the information that needs to be recorded without it being attached to someone specific.

These complex and interdependent relations between models and data illustrate the idea of model–data symbiosis and are essential practices across the sciences and, indeed, most areas of data-intensive inquiry. Processes that allow data to be more accurate for research purposes are, e.g., data conversion, data correction, and data assimilation. Other processes that allow for more complete sets of data for model evaluation are, e.g., data interpolation, data fusion, and data assimilation. Lastly, synthetic data allow models to explore possible worlds and test various data processing methods. All in all, model–data symbiosis highlights the beneficial reciprocal relationship between models and data.

4. Data reuse, data repurpose, and legacy data

Although data are often collected for specific purposes, data sets can also be reused and reprocessed for different ends. Although data reuse and data repurpose are sometimes used interchangeably, Bokulich and Parker (2021) argue that a key difference should be drawn between them. *Data reuse* is best understood as using a given data set to reinterrogate the *same* question multiple times, typically refining the analysis and improving the study's reliability. On the other hand, *data repurpos*ing involves using the same data set to answer a *different* question. Data repurpose highlights the ability of a data set to answer a wide variety of different questions, not just the initial purpose it may have been collected for. Both data reuse and data repurpose require that different methods of data processing (e.g., data correction or data conversion) be applied and are often undertaken when new discoveries, methods, or technologies come to light that allow further information to be extracted from a given data set. Indeed, it is precisely the ability of a data set to be reused and reprocessed that drives the open data movement to preserve data and make it permanently findable and accessible in public databases (more on what are known as the FAIR data principles below).

Two further key concepts are legacy data and data rescue. *Legacy data* (also sometimes referred to as *dark data*) are data that are no longer accessible or usable in their current form. This can occur for a variety of reasons: the data may be stored in a substrate (e.g., handwritten in a ship's log or scientist's lab notebook) that either has not been digitized yet or perhaps even if digitized, might not be properly processed or stored to be usable today. It is important to exercise proper care to update data and data storage systems, backup systems, and error-checking procedures. Data require different kinds of maintenance in order to continue to be usable. Moreover, it is also important to update data standards as they evolve with time. Legacy data can also arise when the data have been collected or processed using instruments or information that is out of date. For example, the values of the fundamental physical constants are periodically remeasured and updated to more accurate values (e.g., Bokulich and Bocchi 2024). In order for these data sets that make use of those constants to be integrated (e.g., through data fusion) or meaningfully compared with other more recent data sets, they need to first be reprocessed in light of the new community-accepted constant values, standards, and protocols (Bokulich 2020a).

When researchers set out to find legacy data and make it accessible and usable again, this is known as *data rescue*. Why rescue legacy data? Why not just perform new measurements with the latest instruments and protocols? There are a number of reasons: Many data sources are ephemeral (e.g., historical weather events) and so cannot be remeasured because they no longer exist. Further, data can be extremely difficult and expensive to collect. These are in fact key drivers of the open data movement, which emphasizes the importance of ensuring that data remains accessible and usable for future projects (i.e., reuse and repurpose). Whether it is changing the substrate of the data set as part of a data rescue or reprocessing the data in light of new information or purposes, these transformations illustrate what is more broadly called *data journeys* (Leonelli and Tempini 2020).

5. Metadata and data empathy

Proper interpretation and use of data typically require what is known as metadata. *Metadata* (i.e., data about the data) is information about how, when, why, and by whom the data were initially collected. It can involve a detailed specification of what above was called a model of the measurement process: What types of instruments or measurement protocols were used to collect that data? When were they collected, and under what circumstances? What data correction or processing has already been applied? If fundamental constants were used in the production of the data set, what values for those constants were used (e.g., Bokulich and Bocchi 2024)? Metadata is essential, because as new sources of error are identified in the measurement or data-collection process, or more generally as new theoretical insights come to light, metadata allows researchers to assess the impact of new information on the data set and correct it appropriately, thereby extending the life of the data. Not only this, but also having the necessary metadata along with different lines of evidence allows for what Nora Boyd (2018) calls "enriched evidence," which she argues allows the results of scientific research to be repurposed across different contexts.

In many scientific contexts, there has been a call for some standardization of the metadata collected as a way to advance the project of open science. However, it is important to recognize that different pieces of metadata are of different significance to each field. Therefore, one would expect that the standardization of metadata will look different for different fields if the purpose is to make the data more widely useful. This raises several questions: How can scientists create open databases that include all the different kinds of metadata needed for different scientific fields? How do scientists decide what information should be included and what information can be ignored? These are some of the difficult questions that must be confronted in efforts to standardize metadata and figure out best practices.

While the notion of metadata is relatively straightforward, some have gone a step further, arguing that researchers should also consider what has been called "data empathy." James Faghmous and Vipin Kumar argue, "Every dataset has a story, and understanding it can guide the choice of suitable analyses; some have labeled this data understanding as *data empathy*" (2014, 157). Similarly, Anissa Tanweer and colleagues write, "data empathy refers to developing this ability for sharing and understanding different data valences, or the values, intentions, and expectations around data. Data empathy is an ethical and epistemological approach" (2016, 2). In the context of climate data, Stefan Brönnimann and Jeannine Wintzer emphasize that knowledge about the broader context in which the data were collected or produced is an essential part of data empathy. They write, "atmospheric data sets also embed political, economic, technological, and cultural histories. The context, however, is often overlooked, and not provided along with the data. We term awareness of and sensitivity to context-dependence *climate data empathy*" (Brönnimann and Wintzer 2019, 1). The history, philosophy, and sociology of science, broadly construed, have an important role to play in recontextualizing data, identifying their valences, and drawing out their epistemic, social, and moral implications (some of these moral dimensions of data are further discussed in the Data Ethics section below).

6. Big data

Technological advancements have allowed for much faster collection, storage, and processing of data sets from many different sources, including people's digital footprints. It has become common to characterize big data in terms of a number of various "Vs" (e.g., Leonelli and Beaulieu 2022). These can include volume, velocity, variety, validity, volatility, and vulnerability. *Volume* obviously refers to the large quantity or "bigness" of big data. What counts as big data has certainly evolved over time. For example, William Whewell's "Great Tide Experiment" of 1835 (e.g., Reidy 2008), which collected half a million data points on simultaneous tides around the globe, every 15 minutes over two weeks, measured by a hodge-podge of deputized "scientists," ranging from members of the British Navy and officers in their colonial outposts to various missionaries and ordinary citizens around the globe, was certainly a "big data" project for its time, though not one that would be considered a large quantity of data today. Some, such as Leonelli (2020), have argued that size is not big data's most salient feature. Equally important, if not more so, is big data's *velocity*, which refers to the speed at which large quantities of data can now be gathered, processed, searched, and analyzed—aiding in the rapid identification of patterns and correlations, transforming the traditional research process.

Big data's *variety* refers to the great diversity of data sources that are being amalgamated into a single database. In big data contexts, this often takes place in a more forced, haphazard way than in traditional data fusion or data integration contexts, where the commensurability of the data sets being combined is given more careful consideration. Related to variety, Japec et al. (2015) discuss how one of the important characteristics of big data that often goes unrecognized is its derived, secondary nature—i.e., how it is "found" or borrowed from a variety of primary data collections, rather than the data being "made" or produced specifically for some intended purpose. Big data users often do not carry out any observations, measurements, or experiments of their own; rather they compile different kinds of data collected by others. This at times indiscriminate amalgamation of various sources of differing quality—and perhaps even incommensurabilities—raises a number of epistemological issues about *validity* (or *veracity*) and whether statistical methods can overcome, or see through, the noisy data. For example, recently, social media data have been used to examine things ranging from social media usage to national political sentiment. As Japec et al. (2015) point out, however, social media data are not evaluated for accuracy and can lead to erroneous results. Social media data also raise a number of issues, such as those related to data ownership and privacy, which are discussed more below.

Another characteristic of big data is *volatility,* which refers to whether data remain available and usable despite changes in storage technologies or hosts. The continuous availability of big data depends on substantial investments in infrastructure and maintenance, which are required to host, back up, maintain, and update databases regularly and in perpetuity. Finally, *vulnerability* raises familiar concerns, such as privacy and whether all data should be open access. Information is power, and big data can be used to reveal sensitive information even when it has supposedly been anonymized or redacted. Furthermore, making big data available to anyone increases the potential for misuse of this data. Such concerns arise, for example, in the context of climate data, where climate deniers may cherry-pick data or analyze it using inappropriate methods to advance misinformation, disinformation, and political agendas. Benchmarking tools attached to open databases might be one way to address unintentional misuse of open data, though they are unlikely to address many problems. These last characteristics of big data point to the urgent need to develop adequate data ethics theories, frameworks, and guidelines, as well as appropriate legislation.

The availability and prominence of big data are bringing about many transformations in how data are collected, used, and stored, and in how research is conducted. Some such as Foster et al. (2017) and Japec et al. (2015) have argued that big data poses a paradigm shift in the social sciences, which traditionally have relied on survey data, given the new ways in which human behavior is now being measured. Social scientists must adapt their methodologies in order to successfully harness big data. Moreover, social scientists must take precautions in using big data so as to prevent injustices that may arise due to the problematic nature of the data and algorithms being used.

A number of challenges also arise when big data is used for the development of contemporary generative AI systems such as Large Language Models (LLMs), including high-profile examples such as ChatGPT. Current evidence suggests that these challenges are not being adequately considered or addressed. Birhane et al. (2022) examined 100 highly cited machine learning papers, only to find that the researchers rarely justify how their project helps society (15%) and barely ever discuss potential negative effects (1%). This study gives evidence for how the values currently used in machine learning further centralize power and therefore continue to disproportionally benefit the already advantaged and harm the disadvantaged. Additionally, Birhane (2021) recognizes the possible ethical downsides to the use of big data for machine learning due to the potential recurrence of unjust and discriminatory patterns (such as the encoded values examined in Birhane et al. 2022) and calls for critical work to be done on AI ethics, fairness, and justice. The next section elaborates further on issues related to data ethics, which are now more prominent due to the spread of big data.

7. Data ethics

Data ethics is a topic of critical importance, though one that has only recently begun to attract attention, and there remains an enormous amount of philosophical work to be done. Given the tremendous harm that may arise through big data, it is imperative that data be produced, gathered, analyzed, and disseminated in ethical ways that take into account the various stakeholders and the significant risks and harms that might arise for different groups in various contexts. While some of these harms can be easily anticipated, others may require investing in sustained interdisciplinary inquiry to identify (e.g., Creel and Hellman 2022).

Traditionally, ethical considerations about data have been very limited, with a focus on thin principles such as FAIR, which stands for Findability, Accessibility, Interoperability, and Reusability (e.g., Wilkinson et al. 2016). Despite the acronym "fair", these principles are more concerned with maximizing the instrumental value or exploitability of data, rather than any deeper issues of fairness or justice. For example, de Lima et al. (2022) point out how rainforest data that satisfy the FAIR principles can still be extremely unfair for the people who are actually on the ground making the forest measurements, i.e., gathering the data, and adhering to the FAIR principles can even endanger the very natural resources that the data were intended to protect.

Indigenous leaders have been at the forefront of developing deeper ethical frameworks, such as the CARE principles of indigenous data governance (Carroll et al. 2020, Jennings et al. 2023). CARE stands for Collective benefit, Authority to control, Responsibility, and Ethics. *Collective benefit* calls attention to the importance of developing data inclusively and for equitable outcomes. *Authority to control* recognizes the rights of self-determination, especially when it comes to whom the data is about. For example, genetic data has long been collected from indigenous communities and used by researchers in ways that have harmed—rather than benefited—those communities (Fleskes et al. 2022). During the COVID-19 pandemic, data about rates of infection, hospitalization, and death among Native Americans were aggregated by the U.S. government into a generic racial and ethnic category of "Other" obscuring the impact that this disease was having on their communities. *Responsibility* is understood as an obligation to nurture respectful relationships, in this context with indigenous peoples, lands, and worldviews. *Ethics* requires paying attention to one's moral obligations, minimizing harms, maximizing benefits, and advancing justice.

In addition to the FAIR and CARE principles, another central topic in the ethics of data is *privacy*. Data are routinely gathered, aggregated, interrogated, and sold to other parties about almost every aspect of our lives, from our online searches to our grocery shopping habits. Even more troublingly, those data are used to manipulate everything from what we buy to whom we vote for, with little regard to privacy and why it matters (e.g., Solove 2015). Currently, there are few data protections, and click-the-box informed consent approaches have proven woefully inadequate (Nissenbaum 2011). Helen Nissenbaum (2019), for example, argues that we need a new, more complex approach to data privacy that she calls the contextual integrity approach, which better governs data flow.

A final class of issues in data ethics concerns the ways in which data can encode and reinforce cultural biases, such as sexism and racism. In their landmark article, Rashida Richardson and colleagues introduce an expanded notion of *dirty data* to mean "data that is derived from or influenced by corrupt, biased, and unlawful practices, including data that has been intentionally manipulated or 'juked,' as well as data that is distorted by individual and societal biases" (Richardson et al. 2019, 195). Although their primary focus is on data from corrupt police reports, racially motivated arrests, tampered evidence, and over-policing of minority neighborhoods (data which then get fed back into policing algorithms, sentencing algorithms, algorithms for risks of recidivism, in an ever-reinforcing and self-fulfilling loop), the term "dirty data" as they note can be used to describe data tainted by any sort of societal biases.

Big data that is indiscriminately collected from online and social media sources brings with it the sexist and racist biases of that culture. In her paper "How Our Data Encodes Systematic Racism," Deborah Raji (2020) notes that Google image searches for "Black girls" return primarily pornography; searches for "healthy skin" return only images of White skin, despite the fact that Black/Brown/Colored skin is the norm worldwide. This biased or "dirty" data then infects any machine learning or AI algorithms that are trained on it, from education algorithms to the use of AI in medicine. In their *AI Now Report 2018,* Meredith Whittaker and colleagues discuss a high-profile case from Amazon corporation, whose hiring algorithm "learned" that men are more frequently CEOs and so down-graded women applicant's CVs from being considered for more prestigious and higher paying jobs at the company (Whittaker et al. 2018, 38). These biases inherent in the data then become entrenched in opaque and automated systems that are difficult to interrogate, challenge, or change. These big data problems have profound and pernicious social consequences and are just some of the issues that data ethics will have to confront.

8. Conclusion

This chapter has provided a philosophical introduction to the concept of a data model, discussing the complex multi-layered relationship between data and theoretical models, as well as the various processes by which scientists transform the "raw" data of measurement indications into data models that can begin to serve as evidence for various claims. Data can be model-filtered in many different ways, through processes like data correction, conversion, and interpolation—leading to what is more generally known as model–data symbiosis. These data processing techniques are critical for projects like data reuse, data repurposing, and data rescue. The chapter emphasized the importance of metadata—that is, data about data—and even more subtly, what has been termed data empathy—a sensitivity to the values and valences inherent in data sets. These are often the aspects of data that are overlooked in the era of big data and can lead to various epistemic and ethical problems. After reviewing some of the key characteristics of big data, the chapter discussed what are known as the FAIR data principles and their limitations and concluded with a discussion of the many ethical issues that arise in big data, ranging from "dirty data" to privacy. Finally, a complementary set of data principles, arising from the work of indigenous scholars, known as the CARE data principles, which reorient the traditional discussions about data to broader ethical considerations, was outlined. Although any one of the topics in this chapter could be its own volume, hopefully, the overview given here provides a foundation for further critical philosophical work to be done on the philosophy of models, data models, and big data.

References

Birhane, Abeba. 2021. "Algorithmic Injustice: A Relational Ethics Approach." *Patterns* 2: 1–9. https://doi.org/10.1016/j.patter.2021.100205

Birhane, Abeba, Pratyusha Kalluri, Dallas Card, William Agnew, Ravit Dotan, and Michelle Bao. 2022. "The Values Encoded in Machine Learning Research." In *2022 ACM Conference on Fairness, Accountability, and Transparency* (FAccT '22), June 21–24, 2022, Seoul, Republic of Korea 173–184. https://doi.org/10.1145/3531146.3533083

Bokulich, Alisa. 2020a. "Calibration, Coherence, and Consilience in Radiometric Measures of Geologic Time." *Philosophy of Science* 87(3): 425–456. https://doi.org/10.1086/708690.

———. 2020b. "Towards a Taxonomy of the Model-Ladenness of Data." *Philosophy of Science* 87 (5): 793–806. https://doi.org/10.1086/710516.

———. 2018/2021. "Using Models to Correct Data: Paleodiversity and the Fossil Record." *Synthese* 198(S24): 5919–5940. https://doi.org/10.1007/s11229-018-1820-x.

Bokulich, Alisa, and Wendy Parker. 2021. "Data Models, Representation and Adequacy-for-Purpose." *European Journal for Philosophy of Science* 11(1): 31. https://doi.org/10.1007/s13194-020-00345-2.

Bokulich, Alisa, and Federica Bocchi. 2024. "Kuhn's '5th Law of Thermodynamics': Measurement, Data, and Anomalies." In *Kuhn's The Structure of Scientific Revolutions at 60*, edited by K. Brad Wray, 55–78. Cambridge: Cambridge University Press.

Boyd, Nora M. 2018. "Evidence Enriched." *Philosophy of Science* 85(3): 403–421. https://doi.org/10.1086/697747.

Brönnimann, Stefan, and Jeannine Wintzer. 2019. "Climate Data Empathy." *WIREs Climate Change* 10(2). https://doi.org/10.1002/wcc.559.

Carroll, Stephanie Russo, Ibrahim Garba, Oscar L. Figueroa-Rodríguez, Jarita Holbrook, Raymond Lovett, Simeon Materechera, Mark Parsons, et al. 2020. "The CARE Principles for Indigenous Data Governance." *Data Science Journal* 19 (November): 43. https://doi.org/10.5334/dsj-2020-043.

Creel, Katlheen, and Deborah Hellman. 2022. "The Algorithmic Leviathan: Arbitrariness, Fairness, and Opportunity in Algorithmic Decision-Making Systems." *Canadian Journal of Philosophy* 52(1): 26–43. https://doi.org/10.1017/can.2022.3

Edwards, Paul N. 1999. "Global Climate Science, Uncertainty and Politics: Data-laden Models, Model-filtered Data." *Science as Culture* 8(4): 437–472. https://doi.org/10.1080/09505439909526558.

———. 2001. "Representing the Global Atmosphere: Computer Models, Data, and Knowledge about Climate Change." In *Changing the Atmosphere: Expert Knowledge and Environmental Governance*, edited by Clark A. Miller and Paul N. Edwards, 31–65. Politics, Science, and the Environment. Cambridge: MIT Press.

———. 2010. *A Vast Machine: Computer Models, Climate Data, and the Politics of Global Warming*. Cambridge: MIT Press.

Faghmous, James H., and Vipin Kumar. 2014. "A Big Data Guide to Understanding Climate Change: The Case for Theory-Guided Data Science." *Big Data* 2(3): 155–163. https://doi.org/10.1089/big.2014.0026.

Fleskes, Raquel E., Alyssa C. Bader, Krystal S. Tsosie, Jennifer K. Wagner, Katrina G. Claw, and Nanibaa' A. Garrison. 2022. "Ethical Guidance in Human Paleogenomics: New and Ongoing Perspectives." *Annual Review of Genomics and Human Genetics* 23(1): 627–652. https://doi.org/10.1146/annurev-genom-120621-090239.

Foster, Ian, Rayid Ghani, Ron S. Jarmin, Frauke Kreuter, and Julia Lane, eds. 2017. *Big Data and Social Science: A Practical Guide to Methods and Tools*. Boca Raton, FL: CRC Press.

Japec, Lilli, Frauke Kreuter, Marcus Berg, Paul Biemer, Paul Decker, Cliff Lampe, Julia Lane, Cathy O'Neil and Abe Usher. 2015. "Big Data in Survey Research." *Public Opinion Quarterly* 79(4): 839–880. https://doi.org/10.1093/poq/nfv039

Jennings, Lydia, Talia Anderson, Andrew Martinez, Rogena Sterling, Dominique David Chavez, Ibrahim Garba, Maui Hudson, Nanibaa' A. Garrison and Stephanie Russo Carroll. 2023. "Applying the 'CARE Principles for Indigenous Data Governance' to ecology and biodiversity research." *Nature, Ecology & Evolution*. https://doi.org/10.1038/s41559-023-02161-2

Leonelli, Sabina. 2015. "What Counts as Scientific Data? A Relational Framework." *Philosophy of Science* 82(5): 810–821. https://doi.org/10.1086/684083.

———. 2016. *Data-Centric Biology: A Philosophical Study*. Chicago; London: The University of Chicago Press.

———. 2020. "Scientific Research and Big Data", *The Stanford Encyclopedia of Philosophy* (Summer 2020 edition), edited by Edward N. Zalta. https://plato.stanford.edu/archives/sum2020/entries/science-big-data/

Leonelli, Sabina, and Anne Beaulieu. 2022. *A Critical Introduction to Data and Society*. 1st edition. Thousand Oaks: Sage Publications Ltd.

Leonelli, Sabina, and Niccolò Tempini. 2020. *Data Journeys in the Sciences*. Cham, Switzerland: Springer Open.

Lima, Renato A. F. de, Oliver L. Phillips, Alvaro Duque, J. Sebastian Tello, Stuart J. Davies, Alexandre Adalardo de Oliveira, Sandra Muller, et al. 2022. "Making Forest Data Fair and Open." *Nature Ecology & Evolution* 6(6): 656–658. https://doi.org/10.1038/s41559-022-01738-7.

Nissenbaum, Helen. 2011. "A Contextual Approach to Privacy Online." *Daedalus* 140(4): 32–48. https://doi.org/10.1162/DAED_a_00113.

———. 2019. "Contextual Integrity Up and Down the Data Food Chain." *Theoretical Inquiries in Law* 20(1): 221–256. https://doi.org/10.1515/til-2019-0008.

Parker, Wendy. 2016. "Reanalyses and Observations: What's the Difference?" Bulletin of the *American Meteorological Society* 97, 1565–1572. https://doi.org/10.1175/BAMS-D-14-00226.1.

———. 2017. "Computer Simulation, Measurement, and Data Assimilation." *The British Journal for the Philosophy of Science* 68(1): 273–304. https://doi.org/10.1093/bjps/axv037

Raji, Deborah. 2020. "How Our Data Encodes Systematic Racism." *MIT Technology Review*. https://www.technologyreview.com/2020/12/10/1013617/racism-data-science-artificial-intelligence-ai-opinion/.

Reidy, Michael S. 2008. *Tides of History: Ocean Science and Her Majesty's Navy*. Chicago: University of Chicago Press.

Richardson, Rashida, Jason M Schultz, and Kate Crawford. 2019. "Dirty Data, Bad Predictions: How Civil Rights Violations Impact Police Data, Predictive Policing Systems, and Justice." *New York University Law Review* 94: 42.

Solove, Daniel J. 2015. "The Meaning and Value of Privacy." In *Social Dimensions of Privacy*, edited by Beate Roessler and Dorota Mokrosinska, 1st edition, 71–82. Cambridge University Press. https://doi.org/10.1017/CBO9781107280557.005.

Tal, Eran. 2012. *The Epistemology of Measurement: A Model-based Approach*. Ph.D. Dissertation, University of Toronto.

———. 2017. "A Model-based Epistemology of Measurement." In *Reasoning in Measurement*, edited by Nicola Mößner and Alfred Nordmann, 233–253. London and New York: Routledge.

Tanweer, Anissa, Brittany Fiore-Gartland, Gina Neff, and Cecilia Aragon. 2016. "Data Empathy: A Call for Human Subjectivity in Data Science." In *19th ACM Conference on Computer-Supported Cooperative Work and Social Computing*. San Francisco, CA: ACM.

Wilkinson, Mark D., Michel Dumontier, IJsbrand Jan Aalbersberg, Gabrielle Appleton, Myles Axton, Arie Baak, Niklas Blomberg, et al. 2016. "The FAIR Guiding Principles for Scientific Data Management and Stewardship." *Scientific Data* 3(1): 160018. https://doi.org/10.1038/sdata.2016.18.

Whittaker, Meredith, Kate Crawford, Roel Dobbe, Genevieve Fied, Elizabeth Kaziunas, Varoon Mathur, Sarah Myers West, Rashida Richardson, Jason Schultz, and Oscar Schwartz. 2018. *AI Now Report 2018*. New York: AI Now Institute at New York University.

19
MODELS AND MEASUREMENT

Eran Tal

1. Introduction

Modeling and idealization play central roles in measurement. This may not be immediately apparent. Measuring weight with a kitchen scale, for example, seems to be as simple as placing an object on the weighing platform and reading the result off the display. Yet the reliability of the result is established by a long chain of inferences, which direct-reading instruments like kitchen scales are designed to conceal. The complex epistemic "work" involved in measuring is revealed when one investigates the design, construction, and calibration of measuring instruments and the measurement standards and unit systems that guarantee their comparability. Such investigations reveal that models and idealizations of various kinds are necessary for establishing what, and how well, instruments measure. This holds true for physical measuring instruments like clocks and thermometers as it does for non-material instruments like psychological tests and questionnaires.

This chapter will discuss three kinds of models involved in measurement. Section 2 will focus on how mathematical logic and model theory are used to elucidate the concept of measurement scale. Section 3 will discuss two other kinds of models: statistical models of data and theoretical models of the measurement process. The role of statistical and theoretical models in measurement has received increasing attention from philosophers over the past two decades. Section 4 will elaborate on a specific view of measurement, known as the model-based account, that has emerged from these discussions. Section 5 will offer concluding remarks.

2. Models, homomorphisms, and measurement scales

The term "model" has multiple meanings in scientific discourse. Accordingly, there are different senses in which measurement can be said to involve models and modeling. One important sense of "model" comes from mathematical logic. Here, a model is understood as a set of entities that satisfy a theory. A "theory" in mathematical logic is a linguistic entity, namely a set of sentences in a formal language. A model of the theory is a non-linguistic

DOI: 10.4324/9781003205647-23

entity of which those sentences are true (Suppes 1960, 290). For example, consider a theory that contains only the following sentence:

For all a, b, and c: $a \circ (b \circ c) \sim (a \circ b) \circ c$,

where $\circ$ is a binary operation and ~ a binary relation. One of the models of this theory is the real numbers with the binary operation of addition (+) and binary relation of equality (=) among them. This is because for any three real numbers, x, y, and z, it is true that $x+(y+z)=(x+y)+z$. The sentences of the theory are thus satisfied by the model. The model is also called a "structure," because it is composed of a set of entities along with relations and operations among them.

Another model of the same theory is a set of physical, rigid rods, when ~ is interpreted as a relation of equivalence among the lengths of two rods, and $\circ$ is interpreted as the operation of end-to-end concatenation (combination) of two rods.[1] The sentence above is then interpreted as the claim that the combined length of three rigid rods is indifferent to the order in which they are combined.

This example shows that the same theory can sometimes be satisfied by both a mathematical model and an empirical model. This insight turns out to be useful for clarifying the mathematical foundations of measurement. Numbers and other mathematical objects[2] are commonly used to express the results of measurements. Such mathematical expressions are meant to represent something empirical. For example, the outcome of measuring the length and width of a desk with a measuring tape is intended as a representation of aspects of that desk. Such representations are often expressed numerically, e.g., the desk's length is measured as 120 cm and its width as 60 cm. These measurement outcomes are mathematical representations of aspects of the desk on a particular scale, namely the centimeter scale.

The mathematical representation of empirical objects gives rise to a central question in measurement theory: when is it justified to represent empirical objects and events mathematically? Philosophers of science, as well as scientists, have written extensively on the nature and types of measurement scales, and on the conditions under which objects and events may be represented on measurement scales (Helmholtz 1887; Campbell 1920; Stevens 1946; Ellis 1966). Starting in the 1950s, Patrick Suppes and his colleagues showed that an axiomatic, set-theoretical approach is useful for such investigation (Suppes 1951). In the decades that followed, this approach was developed into the Representational Theory of Measurement (RTM) (Krantz et al. 1971).

In RTM, one begins with a set of formal assumptions ("axioms") about the relations among empirical objects or events. Suppose that one is interested in measuring the lengths of solid rods in a given set. The first step is to list axioms, namely, sentences that are assumed to be true for the solid rods in the set. One such sentence may be the one mentioned above ("For all a, b, and c: $a \circ (b \circ c) \sim (a \circ b) \circ c$ "). This axiom is called "weak associativity." Additional axioms may be listed, together forming a theory. For example, the theory called "positive closed extensive structures" lists weak associativity alongside four other axioms (Krantz et al. 1971, 73).[3] If the set of solid rods and their relations satisfy all five axioms, then the solid rods and their relations constitute a model of the theory of positive closed extensive structures. This model is called an "empirical relational structure" because it is composed of empirical objects and relations among them.

The crucial move in justifying the assignment of numbers to the lengths of solid rods is to show that the same five axioms are also satisfied by another structure, namely a *numerical* relational structure. RTM proves that the "positive real numbers with the usual ordering

≥ and addition + provide a model for the axioms" of positive closed extensive structures (Krantz et al. 1971, 77). The two models – the empirical structure of solid rods along with ordering and combination relations among them, and the numerical structure of positive real numbers along with ordering and addition relations among them – satisfy the same axioms and therefore have a shared structure. The statement of this shared structure is called a "representational theorem" because it guarantees the possibility of representing empirical entities with mathematical ones. Due to the shared structure of the two models, it is possible to construct a mapping function (a "homomorphism") that matches specific rods with specific numbers and specific operations among rods (such as concatenation) with specific operations among numbers (such as numerical addition).

Measurement scales are such mapping functions. For example, the meter scale of length can be understood as a homomorphic function, from physical objects (along with specific relations among them) into the positive real numbers (along with specific relations among them), which assigns the number 1 to the standard meter. The analysis of measurement scales as structure-preserving functions has resulted in a systematic typology of measurement scales and a clear understanding of the invariance and meaningfulness of quantitative representations (Narens 2002). It has also led to unexpected new results. For example, RTM shows that under specific conditions, a quantitative representation of an empirical attribute is justified even without assuming the existence of a concatenation operation (Luce and Tukey 1964). This result is often cited as vindicating the quantification of mental attributes, for which concatenation operations are not available. An in-depth introduction to representational measurement theory is by Luce and Suppes (2002).

3. Statistical and theoretical models

In addition to their role in elucidating the concept of measurement scale, models are also involved in the process of designing measurement procedures and analyzing their results. The two main kinds of models used at these stages are statistical models of data and theoretical models of the measurement process. These are models in a different sense of "model" than in mathematical logic. Contemporary philosophers of science use the term "model" with a variety of meanings, several of which are covered in other chapters of this Handbook. In what follows, the term "model" will be used to denote an abstract entity that is used to approximately represent a system or a type of system. Models are constructed from assumptions that may be theoretical, statistical, pragmatic, or of some other kind. Models are idealized, that is, they involve deliberate distortions of the target system, such as point particles and massless springs. While models often borrow assumptions from a theory, models function autonomously from theories and are more detailed and narrower in scope than theories (Giere 1988; Cartwright, Shomar, and Suárez 1995; Morgan and Morrison 1999).[4]

3.1 Statistical models of data

The first kind of model in the above sense that this chapter will discuss is statistical models of data. The concept of data is itself multifaceted, and different definitions have been offered. This chapter will follow Bokulich and Parker in understanding data as "records of the results of a process of inquiry that involves interacting with the world" and as "taken to be about one or more aspects of the world, namely, those thought to be involved in a particular process of inquiry" (Bokulich and Parker 2021, 6–7).

Measurement involves the production of at least two kinds of data. First, measuring procedures produce records of instrument indications. Instrument indications are the final states of the measuring system once the measurement process is complete. Examples of instrument indications are the displacement angle of an ammeter needle, the color of a pH test strip after being dipped in a solution, and a subject's responses to a questionnaire. Instrument indications are usually recorded in some form, such as handwritten marks or symbols, photographs, graphs, audio recordings, or bits in digital computer memory. Records of instrument indications are an important kind of data and serve as evidence for knowledge claims about the values of the attribute intended to be measured. To continue the same examples, records of the displacement angle of the ammeter needle may be used to infer the intensity of electric current; the color of the pH test strip may be used to infer the acidity of the solution; and responses to the questionnaire may be used to infer the subject's degree of happiness. Knowledge claims about these attributes are known as "measurement outcomes" (or "measurement results"). Measurement outcomes are claims about the object or event being measured, rather than about the final state of the measurement process. They are often expressed in numerical form on a specific scale and involve uncertainty, such as the claim that the current in the wire is 0.5 ± 0.02 ampere.

A second kind of data produced in measurement, then, are records of measurement outcomes. These often take the form of numerals, graphs, or maps, and may appear on paper or be stored digitally. In some cases, records of measurement outcomes seem deceptively similar to records of instrument indications. This is especially the case for direct-reading instruments such as household measuring tapes and kitchen scales, which are pre-calibrated to indicate numerals corresponding to an estimate of the value of the quantity of interest. The design of such instruments provides the illusion that the value of the quantity is read directly off their displays. Yet the road from instrument indications to measurement outcomes turns out to involve non-trivial and often complex model-based inferences. Instrument users typically "outsource" these inferences to the scientists and engineers who design measuring instruments, and to metrologists, i.e., scientists who specialize in accurate calibration and maintain measurement standards.[5]

One source of inferential complexity in measurement is that data of the first kind – records of instrument indications – tend to be idiosyncratic and high-dimensional. Instrument indications are idiosyncratic insofar as they are the product of many local factors besides the attribute of interest. Indications are affected by the way instruments are designed and operated, by the way the object of interest is isolated and prepared, and by elements in the environment. Many of these factors are difficult to predict or control, such as small temperature fluctuations in a physics lab or day-to-day fluctuations in the mood of participants in a survey. Some data artifacts, such as the effects of eye blinking on EEG recordings or the geometric distortion of fMRI images, can be predicted and corrected, but often only imperfectly. Instrument indications are also often of much higher dimensionality than the variable of interest. For example, the verbal comprehension index of the Wechsler Adult Intelligence Scale (WAIS-IV) is calculated from responses to three or four subsets of questions. Each subject generates up to 92 distinct data points – answers to individual questions – that are then used to calculate a single number representing the subject's level of verbal comprehension. This sort of steep reduction of dimensionality from instrument indications to measurement outcomes is commonplace.

Statistical models of data are abstract and approximate representations of data that are used to reduce the complexity and dimensionality of data and to identify patterns of interest in data. They do so by employing a wide array of statistical techniques, from simple linear regression to sophisticated Monte Carlo methods.

A common application of statistical models is the analysis of indications from repeated measurements of the same attribute. When a measurement procedure is repeatedly applied to the same (or relevantly similar) object or event, the resulting instrument indications often vary. This is because some extrinsic factors vary as the measurement is repeated. Modeling the distribution of repeated indications is helpful for evaluating the influence of extrinsic factors on indications, and hence for evaluating measurement precision. For example, when a stopwatch is repeatedly used to measure the period of a pendulum, it usually yields a somewhat different reading with each use. One cause of variability is that human users are somewhat inconsistent in identifying the beginning and end of pendulum periods. The numerals displayed by the stopwatch are recorded, thus producing data of the first kind, i.e., records of instrument indications. A common method of inferring the period of the pendulum (the measurement outcome) from stopwatch indications is to model the distribution of indications as a Gaussian ("normal") distribution. This is an example of a statistical model of indication data.

The Gaussian model is an abstract and approximate representation of the data. The concrete data – records of individual stopwatch readings – are discrete and have a finite range. By contrast, the ideal Gaussian distribution is defined over a continuous variable of infinite range. Nonetheless, the Gaussian model is a highly useful simplification that allows scientists to infer a value range of the quantity of interest from the data. In this case, the period of the pendulum can be estimated as the distribution mean of stopwatch indications. Similarly, the uncertainty concerning the period of the pendulum due to varying extrinsic factors can be evaluated as the standard deviation of the mean.[6]

Extracting the mean and standard deviation from a Gaussian model of repeated instrument indications is often useful but is neither necessary nor sufficient for arriving at a reliable estimate of the quantity value of interest (e.g., the period of the pendulum). It is not necessary because repeated instrument indications do not always approximate a Gaussian distribution. Depending on the kind of measuring system and object being measured, other statistical models may be a better fit. For example, electrical engineers use a variety of statistical models to characterize the random fluctuations of an oscillator, such as a quartz crystal oscillator used in many clocks. These models represent different patterns of noise – such as white noise, flicker noise, and random walk noise – as different power functions of the oscillator's Fourier frequency. This in turn allows engineers to calculate the contribution of random fluctuations to the uncertainty of clocks at different run times. To return to the same example, the noise associated with stopwatch indications is an additional source of uncertainty about the pendulum period, in addition to the variability of the operation of the stopwatch by humans. Hence several different statistical models of instrument indications may be combined to evaluate measurement uncertainty.

Despite their usefulness, statistical models of repeated indications are generally insufficient to arrive at a reliable measurement outcome. This is because other sources of uncertainty may be present that cannot be identified by such models. Measurement is usually affected by systematic biases, that is, biases that do not behave randomly. The person operating the stopwatch may have a delayed response, resulting in systematically biased time readings. The stopwatch may have been imperfectly calibrated, such that its "second" is somewhat longer or shorter than the standard, SI second. This would lead to a clock indication error that increases linearly with time. The stopwatch may also suffer from a systematic frequency drift, such that its "second" becomes shorter or longer over time. This would lead to a non-linear clock indication error. Additional sources of uncertainty depend on the precise definition of the quantity of interest. For example, one may be interested in measuring the pendulum period at sea level. If the pendulum itself is located somewhat above or below sea level, a correction to the indicated period would be required to infer the period at sea level. This correction would involve a secondary measurement of the difference in gravitational potential between the location of the pendulum and sea level. This secondary measurement itself would involve some uncertainty, which would affect the total uncertainty associated with the outcome of the pendulum measurement.

In all the examples in the previous paragraph, the extent of uncertainty cannot be calculated as a statistical property of repeated instrument indications. The biases have a non-zero expectation value – they do not "average out" – across repeated applications of the measurement procedure and are therefore not estimable from the variation of indications alone. Rather, the uncertainty in these examples depends on substantive features of the instruments used, the object being measured, the persons performing the measurement, and the environment, as well as on the quality of background knowledge, measurement standards, and calibration procedures. The evaluation of such uncertainties requires theoretical models of the measurement process, which will be discussed below.

Besides statistical models of instrument indications, measurement may involve statistical models of measurement outcomes. Such models are often useful for comparing different measurement outcomes for mutual compatibility. In the physical sciences, measurement outcomes are commonly reported alongside an uncertainty margin. Such uncertainty margins may be over- or under-estimated, and this can be discovered when different measurements of the same quantity are compared to each other. For example, the velocity of light in a vacuum has been measured in different ways since the late 19th century. Some of the reported values, especially during the 1920s and 1930s, were significantly lower than the currently accepted value, even after taking into account their reported uncertainty margins (Henrion and Fischhoff 1986). This suggests that the uncertainties of those measurements were under-estimated. A common method of determining whether different measured values agree within their respective reported uncertainties is to calculate their Birge ratio (Birge 1932). This ratio is based on a statistical model that views each measurement as an independent sample from a larger set of potential measurements. The ratio is equal to 1 (agreement) when the reported uncertainties match the variability among measured values. Large deviations from 1 indicate that uncertainties have been over- or under-estimated. Such tests for agreement among measurement outcomes are especially important for adjusting the accepted values of fundamental physical constants (Grégis 2019).[7]

3.2 *Theoretical models of the measurement process*

The previous section showed that statistical models of data are highly useful for measurement. At the same time, inferring measurement outcomes from instrument indications requires more than a statistical analysis of indications. Patterns of distribution and correlation among the indications of instruments cannot by themselves establish which – if any – attribute the instrument is measuring, nor how well it is measuring that attribute. Substantive assumptions about the measurand – i.e., the attribute intended to be measured – and the measurement process are also needed. Examples of substantive assumptions already encountered above are the assumption that the stopwatch suffers from a constant frequency drift and that the pendulum's period is affected in a specific way by its vertical distance from sea level. These assumptions are theoretical, that is, they concern the constitution, internal dynamics, and mutual interactions of elements of the measurement process, as well as elements of the calibration process.[8]

Taken together, such assumptions are often used to construct a theoretical model of the measurement process. As in the previous section, the term "model" is meant to denote an abstract and approximate representation of a system. Like statistical models of data, theoretical models of measurement processes are idealized, and describe the components and dynamics of the measurement process in a somewhat simplified way. The frequency drift of a real stopwatch is not exactly constant and the formula that corrects the period of a pendulum for elevation differences is not exact. Corrections can be introduced into the model to make it more realistic, but no model captures the full complexity of the measurement process, and some degree of idealization is inescapable. As will be discussed below, idealization is not a weakness, but an essential feature of measurement. For example, idealizations are necessary for justifying claims about measurement accuracy. They are also necessary for establishing which quantity an instrument measures, and for deciding whether different instruments measure the same quantity (Tal 2019).

There are different ways to model a measurement process theoretically. If a full-fledged theory of the measurand is available, it will usually contribute to the construction of a theoretical model of the measurement process. For example, contemporary acoustic gas thermometry exploits known relations between the temperature of a monatomic gas such as helium and the speed of sound in that gas. These relations are predicted by thermodynamics, and used as key assumptions in the theoretical model of an acoustic gas thermometer (Moldover et al. 2014). Nonetheless, a mature theory of the measurand is not necessary for the construction of a theoretical model of the measurement process. During the 1830s and 1840s, the study of temperature lacked an agreed-upon theory, and thermometry was developed mainly empirically, by comparing the behaviors of different putative thermometers (Chang 2004, chap. 2). Still, some substantive assumptions had to be made to make such comparisons possible. For example, temperature was assumed to be a single-valued (i.e., one-dimensional) property, to be roughly correlated with human sensations of heat and cold, and to cause the monotonic expansion of thermometric fluids such as mercury and air. These assumptions formed the basis for an elementary and crude theoretical model of early thermometers that was later refined.

The last example shows that a theoretical model of the measurement process need not be quantitative. In many cases, substantive assumptions about the measurand and the instrument are qualitative. For example, a widely used method in educational assessment is to specify a construct map, which describes the skills and content a student is expected to command at each level of their study of a given topic (Wilson 2009). This map is used to design tests for assessing student achievement, and is iteratively improved with feedback from educators, test designers, and test scores themselves. Such construct maps can be viewed as sets of qualitative theoretical assumptions about what the test is measuring and how specific questions on the test assess different levels of achievement.

In other cases, a theoretical model of the measurement process is specified either partially or completely in quantitative terms. When designing a new measuring instrument, contemporary metrologists typically express each of their theoretical assumptions as an equation that relates two or more physical quantities to each other and then use these equations to derive the expected relationship between the indications of the instrument and measurement outcomes. For example, a Kibble balance (also known as watt balance) is a sensitive instrument for realizing the definition of the kilogram. It works by linking the mass of an object placed on the pan of the balance with the Planck constant, a fundamental physical constant that since 2019 has served to define the kilogram. The linking of mass to the Planck constant is achieved by specifying a set of theoretical equations that describe how different quantities are related to each other through the operation of the balance. According to these equations, the balance relates (i) mass to electric current and magnetic flux density, (ii) magnetic flux density to voltage, and (iii) voltage and electric current to the Planck constant.[9] The theoretical model of the balance, therefore, establishes an inferential link that allows metrologists to use the indications of the balance to measure the mass of the object in terms of the Planck constant. It is impossible to understand the design and function of a Kibble balance without being familiar with the quantitative theoretical model of the instrument and with background physical theories, such as quantum mechanics and electromagnetism.

This example illustrates that in contemporary physical sciences and engineering, a theoretical model of the measurement process is an essential part of measurement itself. The theoretical model specifies the quantity intended to be measured, provides the rationale for the design and operation of the instrument, provides the justification for inferring values of the measurand from instrument indications, and underlies (together with statistical models of data) the evaluation of accuracy and error. The centrality of theoretical models to measurement is closely linked to the centrality of theory itself. As Bas van Fraassen notes, scientific theories provide the logical space in which measurement locates objects and events and specify which kinds of objects or events can be located in that space (2008, 164).

The distinction between theoretical and statistical models is useful as an analytical tool, but in practice, it is often blurry. Scientists who design, test, and calibrate measuring instruments frequently use a combination of statistical and theoretical assumptions to construct their models. For example, when a caliper is calibrated against gauge blocks (metal objects of known length), scientists are interested in learning the functional relation between the indications of the caliper and the lengths of gauge blocks. The response function of the caliper is typically calculated via simple linear regression, that is, by finding a linear function that best fits the data. Among other assumptions, the model assumes that the caliper's response function is linear and that the variance of errors does not depend on the value of the independent variable (i.e., the length of the gauge blocks). The first assumption would usually be considered theoretical and the second statistical. However, this classification matters little for the practical conduct of measurement, and the resulting model can rightly be called "theoretical-statistical." The distinction between the two types of models is an abstraction that helps philosophers trace different traditions and bodies of knowledge that are involved in model building, rather than a substantive demarcation.

4. A model-based account of measurement

The understanding that theoretical and statistical models are central to measurement has led to a novel understanding of measurement itself and the ways measurement produces knowledge. Starting in the early 2000s, the centrality of theoretical models to measurement became increasingly recognized by philosophers including Marcel Boumans (1999; 2005), Mary Morgan (2001; 2007), and Margaret Morrison (2009). Boumans and

Morgan showed that theoretical models in economics are used to generate measurements of economic variables, such as price levels, and that such models are calibrated in a similar way to physical measuring instruments. Morrison argued that theoretical models of physical measuring instruments, such as a pendulum for measuring gravitational acceleration, are necessary for justifying the approximation techniques that guarantee the accuracy of the measurement outcome (2009, 35).

These insights, along with lessons from the empirical sciences, gave rise to the model-based account of measurement (Tal 2011, 2016b, 2019; Parker 2017; McClimans, Browne, and Cano 2017; Basso 2017, 2021). According to the model-based account, the aim of measurement is to evaluate one or more parameters in an abstract and idealized model of a process, based on the final states of that process and additional information. Measurement is considered successful to the extent that the evaluation meets certain desiderata, including coherence, objectivity, and accuracy. This section will briefly clarify how the aims and quality of measurement are conceptualized under the model-based account.

Under the model-based account, measurement consists of two levels, one concrete, and one abstract. The concrete level is a process, such as the process of a triple point cell interacting with a platinum resistance thermometer and generating an indication, or the process of a person responding to questions on the WAIS-IV. The second, abstract level, is a model (or sometimes several models) representing the processes mentioned above and the elements that compose them. The model is constructed from theoretical and statistical assumptions about the nature, structure, composition, and dynamics of different elements of the measurement process and the interactions among them.

Viewing measurement in this way provides new solutions to long-standing epistemological problems. One such problem is the possibility of evaluating measurement accuracy. A naive realist may think of measurement accuracy, as the closeness of the measured value to the true value of the measurand, where a value true is taken to be independent of human beliefs and practices.[10] The main difficulty with this view is that scientists have no reliable cognitive access to measurand values other than through measurement, which involves human beliefs and practices. Consequently, scientists have no access to true measurand values and no way to evaluate measurement accuracy in accordance with its naive realist conception. Indeed, for a naive realist it is possible for all the measurements ever taken of a given quantity – say, the melting point of copper – to be arbitrarily distant from its true value, even if the measured values are mutually consistent and cohere with accepted theories.

Alternatively, under an extreme form of operationalism, the quantity to be measured is defined by the operation of its measurement (Bridgman 1927). Temperature, for example, is defined by the operation of a given thermometer. By definition, each measurement operation produces a perfectly accurate evaluation of its own *sui generis* quantity. If two thermometers seem to disagree, it is only because each of them measures a different type of quantity, which may be labeled "temperature-A" and "temperature-B." A slightly more moderate version of operationalism would maintain that measurement outcomes are accurate relative to the outcomes of a standard measurement procedure. In this case, the difficulty for an operationalist is justifying the claim that the standard procedure measures the same type of quantity as the procedure being evaluated for accuracy (Tal 2019, 865–866). As with naive realism, accuracy evaluation turns out to be impossible under strong versions of operationalism, although for very different reasons.

Under the model-based account, theoretical and statistical models are necessary for justifying claims about measurement accuracy. The model-based account takes measurement accuracy to be a multifaceted concept, which can be defined metaphysically, epistemically, or operationally, among other ways (Tal 2011, 1084). Regardless of how it is defined, measurement accuracy is evaluated relative to a model of the measurement process. Specifically, accuracy is evaluated by how tightly the indications produced by the measurement process (along with other available information) constrain the values that may be reasonably attributed to the measurand under a given model of the measurement process. The same stopwatch, for example, may be justifiably deemed more or less accurate depending on how it is modeled. Suppose that the stopwatch is represented using a detailed model that corrects for the stopwatch's time offset, frequency offset, and frequency drift. The accuracy of measurements of time duration under such a model is higher than if the stopwatch were represented with a simpler model that does not account for such errors. Under a simpler model of the stopwatch, the extent of errors would be less precisely known, and thus the range of values of duration that can be reasonably attributed to events based on the indications of the stopwatch would be larger than under the more detailed model. This model-relativity of accuracy claims is consistent with metrological practice, which emphasizes models as preconditions for evaluating accuracy (Joint Committee for Guides in Metrology [JCGM] 2008).

The idealized nature of models is of central importance to the possibility of evaluating measurement accuracy. Rather than evaluating accuracy against a true value (as a naive realist would maintain)[11] or against an arbitrarily chosen standard measurement procedure, the model-based account takes accuracy and error to be evaluated relative to an idealized model of the measurement process. Error is evaluated by comparing the predictions of an idealized model to the actual indications of the instrument. For example, an idealized model of a cesium atomic clock assumes that the cesium atoms are at absolute zero temperature and are completely unperturbed by magnetic or gravitational fields. These conditions cannot be completely fulfilled by a real clock. The extent of error associated with the frequency produced by a real atomic clock is calculated by theoretically predicting the extent to which it deviates from the ideal (Jefferts et al. 2002; Heavner et al. 2005). Accordingly, the accuracy of the clock depends on the uncertainty of these model-based theoretical predictions. The more accurately the model can be used to predict the deviation of the real clock from its idealized representation, the more accurate the clock is under that model. This is again consistent with modern metrological practice, which takes measurement accuracy to be the predictability of error (i.e., uncertainty) rather than the absence of error (Giordani and Mari 2014).

Another aspect of measurement that models shed light on is the nature of calibration. According to the model-based account, the calibration of a measuring instrument is a modeling activity, namely, the activity of constructing, testing, and improving a theoretical-statistical model of the measurement process (Tal 2017b). During calibration, scientists assess the degree of fit between their model and the measurement process and attempt to improve this fit by modifying the model, the measurement process, or both. When assessing model fit, scientists often make use of known and stable objects or phenomena, such as standard weights or the triple point of water (Franklin 1997; Boumans 2007, 236; JCGM 2012, sec. 2.39).[12] These stable objects are helpful for determining parameters in the model and for testing whether the measurement process behaves as the model predicts. Nonetheless, the ultimate goal of calibration is not simply to replicate the known values associated with such objects, but to construct an accurate model of the measurement process.

Once the model is deemed sufficiently accurate, it is used to predict the indications that the instrument will produce when it interacts with objects of various quantity values. For example, the model of the caliper is used to predict the indications the caliper will produce when it interacts with objects of various diameters. This "calibration function" is then inverted and used to infer the quantity values associated with objects based on the indications that the instrument will produce (JCGM 2012, sec. 2.39). For example, the inverted calibration function of the caliper is used to infer which diameters will produce a given indication of the caliper. When the caliper is used to measure some concrete object, the measurement outcome is taken to be the diameter value range that best predicts the observed indications of the caliper under the model. Putting things more generally, measurement outcomes are the *best predictors* of patterns of instrument indications under a specific model of the measurement process. The model-based account, therefore, reveals the centrality of prediction to measurement.

The model-based nature of inferences from instrument indications to measurement outcomes has important implications for the objectivity of measurement. A measurement outcome constitutes an objective knowledge claim when one is justified in attributing the quantity values to the object (or event) being measured, rather than to an artifact of the measurement procedure or to one or more background assumptions. The inevitable reliance on models means that measurement outcomes cannot be assessed for truth or accuracy independently of any model. At best, measurement outcomes can be deemed objective to the extent that they are robust under a wide variety of measurement procedures and assumptions. Robustness does not mean model independence, but a coherent fit between different model-based predictors (Basso 2017). Hence, an important lesson of the model-based account is that the standard of objectivity in measurement is context-invariance, rather than context-independence (Tal 2017a). This conclusion has important implications for understanding how measurement can serve as a source of scientific evidence, and how measurement differs from other model-based, data-driven procedures, such as computer simulation (Morrison 2009; Tal 2016a; Parker 2017).

5. Conclusions

This chapter briefly surveyed three ways in which models and modeling are central to measurement. First, a measurement scale can be helpfully understood as a mapping function between models, i.e., structures that satisfy a common set of axioms. This provides insight into the possibility of representing empirical objects mathematically. Second, statistical models play a central role in analyzing measurement data, evaluating some types of measurement uncertainty, and detecting inconsistencies among measurement outcomes. Third, theoretical models of the measurement process are crucial for specifying what an instrument is measuring, for the design and calibration of the instrument, and for evaluating uncertainties that are not accessible through the application of statistical tools alone.

Recent scholarship on the philosophy of measurement has benefited from close attention to the roles of models in measurement, especially statistical and theoretical models. This literature is fast evolving, and this chapter is meant to provide an entry point into the discussion rather than a comprehensive introduction. Interested readers are encouraged to follow the references provided in this chapter for more detailed treatments of the topics covered above. Being a relatively new subdiscipline, many open problems and research areas in the philosophy of measurement remain to be explored. Among these are: the way measurement produces scientific evidence, the role of causality in measurement, differences and commonalities in conceptions of measurement across scientific disciplines, the role of ethical and social values in measurement, the relationship between measurement, prediction, and information, and the conditions for detecting quantitative structure in empirical data. Progress on many of these topics will likely involve an appeal to models.

Acknowledgments

This chapter was written with support from the Canada Research Chairs Program and the Templeton Foundation (Grant 62424).

Notes

1 Equivalence is sometimes interpreted as empirical indistinguishability, e.g., the rods appearing to have the same length. This interpretation leads to complications when the sensitivity of empirical comparisons is low. See Krantz et al. (1971, 2–3) for discussion.
2 Such as vectors and geometric line segments.
3 The other axioms are weak order, monotonicity, positivity, and the Archimedean axiom.
4 This chapter does not presuppose any specific view about the ontology of models or their representational capacity.
5 Other kinds of data are commonly produced in the course of measurement besides the two discussed here. For example, measurement typically involves the production of data about the properties and performance of various components of the measuring system, about the properties of measurement standards, and about properties of the environment in which the measurement takes place. Measurement may also involve the collection of data about the individual people who take measurements, e.g., to determine their individual biases in reading and recording instrument indications.
6 The standard deviation of the mean is the standard deviation of the sample divided by the square root of the sample size: $\sigma_X = \frac{s}{\sqrt{N}}$.
7 This brief survey of statistical models of data is not meant to be comprehensive. Measurement involves many other uses of statistical models of data not discussed here, such as regression, factor analysis, data smoothing, signal-noise separation, uncertainty propagation, significance testing, and the generation of simulated data as a tool for accuracy evaluation, to mention a few.
8 The distinction between theoretical and statistical assumptions is not strict. The assumption that a stopwatch's frequency drift follows a random-walk pattern, for example, could plausibly be categorized as both theoretical and statistical.
9 This description is a vast oversimplification of the measurement procedure. Detailed descriptions of the design of Kibble balances and how they use quantum effects to link voltage and current to the Planck constant can be found in (Robinson 2011; Sanchez et al. 2014).
10 More precisely, a naive realist takes the true value *on a given measurement scale* to be independent of human beliefs and practices. This leaves room for the arbitrary choice of, e.g., measurement unit and zero point, depending on the type of scale. For a discussion and critique of realism about measured values, see (Teller 2018).
11 Being an epistemological rather than metaphysical position, the model-based account is compatible with many realist and anti-realist views about measurement. For example, it is consistent with the (non-naive) realist view that instrument indications are caused by mind-independent magnitudes that are themselves unknown (Trout 1998, chap. 2). At the same time, the model-based account is compatible with anti-realism about quantity values of the sort defended by (Teller 2018).
12 These known signals may be, but need not be, metrologically certified standards. While measurement standards are often helpful for testing the predictions of models of a measurement process, calibration often proceeds by comparing measurement procedures directly to each other in a round robin. For examples see (Philipona et al. 1998; Cabibbo et al. 2012; Dennison et al. 2016).

References

Basso, Alessandra. 2017. "The Appeal to Robustness in Measurement Practice." *Studies in History and Philosophy of Science Part A* 65: 57–66.

———. 2021. "From Measurement to Classificatory Practice: Improving Psychiatric Classification Independently of the Opposition between Symptom-Based and Causal Approaches." *European Journal for Philosophy of Science* 11(4): 104. https://doi.org/10.1007/s13194-021-00424-y.

Birge, Raymond T. 1932. "The Calculation of Errors by the Method of Least Squares." *Physical Review* 40(2): 207.

Bokulich, Alisa, and Wendy Parker. 2021. "Data Models, Representation and Adequacy-for-Purpose." *European Journal for Philosophy of Science* 11(1): 1–26.

Boumans, Marcel. 1999. "Representation and Stability in Testing and Measuring Rational Expectations." *Journal of Economic Methodology* 6(3): 381–402.

———. 2005. *How Economists Model the World into Numbers.* London; New York: Routledge. http://www.taylorfrancis.com/books/9780203324073.

———, ed. 2007. *Measurement in Economics: A Handbook.* Amsterdam: Elsevier.

Bridgman, Percy Williams. 1927. *The Logic of Modern Physics.* New York: Macmillan.

Cabibbo, M., P. Ricci, R. Cecchini, Z. Rymuza, J. Sullivan, S. Dub, and S. Cohen. 2012. "An International Round-Robin Calibration Protocol for Nanoindentation Measurements." *Micron* 43(2–3): 215–222.

Campbell, Norman Robert. 1920. *Physics: The Elements.* London: Cambridge University Press.

Cartwright, Nancy, Towfic Shomar, and Mauricio Suárez. 1995. "The Tool Box of Science: Tools for the Building of Models with a Superconductivity Example." *Poznan Studies in the Philosophy of the Sciences and the Humanities* 44: 137–149.

Chang, Hasok. 2004. *Inventing Temperature: Measurement and Scientific Progress.* New York: Oxford University Press.

Dennison, J. R., Justin Christensen, Justin Dekany, Clint Thomson, Neal Nickles, Robert E. Davies, Mohamed E. Belhaj, Kazuhiro Toyoda, Kazutaka Kawasaki, and Isabel Montero. 2016. "Absolute Electron Emission Calibration: Round Robin Tests of Au and Graphite." In *14th Spacecraft Charging Technology Conference, ESA/ESTEC*, 1–7.

Ellis, Brian. 1966. *Basic Concepts of Measurement.* Cambridge: Cambridge University Press.

Fraassen, Bas C. van. 2008. *Scientific Representation: Paradoxes of Perspective.* Oxford: Oxford University Press.

Franklin, Allan. 1997. "Calibration." *Perspectives on Science* 5(1): 31–80.

Giere, Ronald N. 1988. *Explaining Science: A Cognitive Approach.* Chicago and London: University of Chicago Press.

Giordani, Alessandro, and Luca Mari. 2014. "Modeling Measurement: Error and Uncertainty." In *Error and Uncertainty in Scientific Practice*, edited by Marcel Boumans, G. Hon, and A.C. Petersen, 79–96. Pickering and Chatto.

Grégis, Fabien. 2019. "Assessing Accuracy in Measurement: The Dilemma of Safety versus Precision in the Adjustment of the Fundamental Physical Constants." *Studies in History and Philosophy of Science Part A* 74: 42–55.

Heavner, Thomas P., S. R. Jefferts, E. A. Donley, J. H. Shirley, and T. E. Parker. 2005. "NIST-F1: Recent Improvements and Accuracy Evaluations." *Metrologia* 42(5): 411.

Helmholtz, Hermann von. 1887. *Counting and Measuring.* Translated by Charlotte Lowe Bryan. New York: D. Van Nostrand co.

Henrion, Max, and Baruch Fischhoff. 1986. "Assessing Uncertainty in Physical Constants." *American Journal of Physics* 54(9): 791–798.

JCGM. 2012. *International Vocabulary of Metrology—Basic and General Concepts and Associated Terms.* https://www.bipm.org/utils/common/documents/jcgm/JCGM_200_2012.pdf.

Jefferts, Steven R., J. Shirley, T. E. Parker, T. P. Heavner, D. M. Meekhof, C. Nelson, Filippo Levi, G. Costanzo, A. De Marchi, and R. Drullinger. 2002. "Accuracy Evaluation of NIST-F1." *Metrologia* 39(4): 321.

Joint Committee for Guides in Metrology (JCGM). 2008. "Evaluation of Measurement Data — Guide to the Expression of Uncertainty in Measurement." https://www.bipm.org/en/committees/jc/jcgm/publications.

Krantz, David H., Patrick Suppes, R. Duncan Luce, and Amos Tversky. 1971. *Foundations of Measurement Volume 1: Additive and Polynomial Representations.* New York and London: Academic Press.

Luce, R. Duncan, and Patrick Suppes. 2002. "Representational Measurement Theory." In *Stevens' Handbook of Experimental Psychology*, edited by Hal Pashler and John Wixted, 3rd ed. John Wiley & Sons, Ltd. https://doi.org/10.1002/0471214426.pas0401.

Luce, R. Duncan, and John W. Tukey. 1964. "Simultaneous Conjoint Measurement: A New Type of Fundamental Measurement." *Journal of Mathematical Psychology* 1(1): 1–27.

McClimans, Leah, John Browne, and Stefan Cano. 2017. "Clinical Outcome Measurement: Models, Theory, Psychometrics and Practice." *Studies in History and Philosophy of Science Part A* 65–66(October): 67–73. https://doi.org/10.1016/j.shpsa.2017.06.004.

Moldover, Michael R., Roberto M. Gavioso, James B. Mehl, Laurent Pitre, Michael de Podesta, and J. T. Zhang. 2014. "Acoustic Gas Thermometry." *Metrologia* 51(1): R1.

Morgan, Mary S. 2001. "Making Measuring Instruments." *History of Political Economy* 33(5): 235–251.

———. 2007. "An Analytical History of Measuring Practices: The Case of Velocities of Money." *Measurement in Economics: A Handbook*, edited by Marcel Boumans (ed.), 105–132. Amsterdam: Elsevier.

Morgan, Mary S, and Margaret Morrison. 1999. *Models as Mediators: Perspectives on Natural and Social Sciences.* Cambridge; New York: Cambridge University Press.

Morrison, Margaret. 2009. "Models, Measurement and Computer Simulation: The Changing Face of Experimentation." *Philosophical Studies* 143(1): 33–57.

Narens, Louis. 2002. *Theories of Meaningfulness.* Theories of Meaningfulness. Mahwah, NJ: Lawrence Erlbaum Associates Publishers.

Parker, Wendy S. 2017. "Computer Simulation, Measurement, and Data Assimilation." *The British Journal for the Philosophy of Science* 68(1): 273–304. https://doi.org/10.1093/bjps/axv037.

Philipona, Rolf, Claus Fröhlich, Klaus Dehne, John DeLuisi, John Augustine, Ellsworth Dutton, Don Nelson, Bruce Forgan, Peter Novotny, and John Hickey. 1998. "The Baseline Surface Radiation Network Pyrgeometer Round-Robin Calibration Experiment." *Journal of Atmospheric and Oceanic Technology* 15(3): 687–696.

Robinson, Ian A. 2011. "Towards the Redefinition of the Kilogram: A Measurement of the Planck Constant Using the NPL Mark II Watt Balance." *Metrologia* 49(1): 113.

Sanchez, C. A., B. M. Wood, R. G. Green, J. O. Liard, and D. Inglis. 2014. "A Determination of Planck's Constant Using the NRC Watt Balance." *Metrologia* 51(2): S5. https://doi.org/10.1088/0026-1394/51/2/S5.

Stevens, Stanley Smith 1946. "On the Theory of Scales of Measurement." *Science* 103(2684): 677–680.

Suppes, Patrick. 1951. "A Set of Independent Axioms for Extensive Quantities." *Portugaliae Mathematica* 10(4): 163–172.

———. 1960. "A Comparison of the Meaning and Uses of Models in Mathematics and the Empirical Sciences." *Synthese* 12(2): 287–301.

Tal, Eran. 2011. "How Accurate Is the Standard Second?" *Philosophy of Science* 78(5): 1082–1096.

———. 2016a. "How Does Measuring Generate Evidence? The Problem of Observational Grounding." *Journal of Physics Conference Series* 772: 012001.

———. 2016b. "Making Time: A Study in the Epistemology of Measurement." *The British Journal for the Philosophy of Science* 67(1): 297–335.

———. 2017a. "A Model-Based Epistemology of Measurement." In *Reasoning in Measurement*, edited by Nicola Mößner and Alfred Nordmann, 233–253. Routledge.

———. 2017b. "Calibration: Modelling the Measurement Process." *Studies in History and Philosophy of Science Part A* 65: 33–45.

———. 2019. "Individuating Quantities." *Philosophical Studies* 176(4): 853–878.

Teller, Paul. 2018. "Measurement Accuracy Realism." In *The Experimental Side of Modeling*, edited by Isabelle F. Peschard and Bas C. van Fraassen, 273–298. University of Minnesota Press.

Trout, J. D. 1998. *Measuring the Intentional World: Realism, Naturalism, and Quantitative Methods in the Behavioral Sciences.* New York: Oxford University Press.

Wilson, Mark. 2009. "Measuring Progressions: Assessment Structures Underlying a Learning Progression." *Journal of Research in Science Teaching: The Official Journal of the National Association for Research in Science Teaching* 46(6): 716–730.

20
MODEL TRANSFER IN SCIENCE

Catherine Herfeld

1. Introduction

Scientific research is characterized by strong disciplinary specialization that often manifests in highly abstract models tailored to particular target systems. At the same time, a pertinent feature of contemporary science is that there is increasing interaction across different fields or even disciplines. One way this interaction manifests is in an intensified transfer of models across domains. For instance, the well-known Lotka–Volterra model has not only been used to explain predator–prey interactions in population biology but has also been transferred into medicine and economics to study phenomena as disparate as the growth of cancer cells and the business cycle (for a historical account, see Knuuttila and Loettgers 2017). Modern science is full of examples of such model transfer. This transfer can be successful but might also confront severe challenges and can even sometimes fail. Beyond its pertinence in scientific practice, model transfer can also have critical functions, such as potentially serving as a catalyst for scientific progress and a driver of innovation (e.g., Boumans and Herfeld 2023; Price 2020).

Although knowledge transfer generally and model transfer in particular have recently gained more attention in philosophy of science (e.g., Bokulich 2015; Donhauser and Shaw 2019; Du Crest et al. 2023, Grüne-Yanoff 2011; Herfeld and Lisciandra 2019; Houkes and Zwart 2019; Humphreys 2019; Knuuttila and Loettgers 2014; 2016; 2023; Knuuttila and García-Deister 2019; Lin 2022; Marchionni 2013; Price 2019; 2020; Tan 2023; Zuchowski 2019), the phenomenon has not yet been extensively studied by philosophers of scientific modeling. There is a vast amount of literature that has studied the cross-domain transfer of a variety of epistemic objects and what could be considered to belong to the category of knowledge generally. It concerns the circulation of knowledge (e.g., Ash 2006; Herfeld and Lisciandra 2019; Howlett and Morgan 2011; Lipphardt and Ludwig 2011; Kaiser 1998; Nersessian 2002), the nature of interdisciplinarity, the transfer of facts, and the travel journeys of data (e.g., Andersen 2016; Howlett and Morgan 2011; Leonelli and Tempini 2020; Mäki et al. 2019). While this literature proves instructive in locating the phenomenon of model transfer in the scholarly landscape, it has not straightforwardly been concerned with model transfer itself. Rather, the debate on model transfer is still in its early stages.

DOI: 10.4324/9781003205647-24

The main goal of this chapter is, therefore, to systematically survey the existing literature on model transfer, thereby pointing to possible routes for future research. The review is structured around three kinds of closely connected questions that have been addressed so far. The first question is how we can conceptually think of those models that are transferred, asking for a proper account of the unit of analysis for such transfer. The second question is why some models are transferable across domains to address often fundamentally distinct problems, asking for an explanation of model transfer. The third set of questions is whether and in what way the practice of model transfer can contribute to scientific progress.

2. Approaches to model transfer

Most of the existing literature has focused on the question of why models are transferrable across domains. The question is, when we take specialized models as representations of some target system, how can the same model provide insights about fundamentally distinct systems from different domains? Most philosophers have sought the answer by asking whether the object of transfer has specific characteristics that allow for its transfer; in other words, they have provided an answer to the question of what exactly the object of transfer is. However, there is neither agreement about the exact object of transfer nor, more importantly, about the core characteristics that enable or prevent cross-domain model transfer. Some philosophers argue that the generality, tractability, and flexibility of models explain their transfer (Humphreys 2002; 2004). Others argue that rather general conceptual features and their justification – in addition – make the model attractive for using them in other domains (Knuuttila and Loettgers 2014; 2016; 2023).

Many philosophers either defend an analogy-based or a template-based approach to thinking about the unit of transfer. An analogy-based approach refers to analogical reasoning as a cognitive or research strategy that allows for using concepts, models, or methods that are familiar in one domain in one in which they are less familiar. They do so by positing shared features of the respective phenomena (or some theoretical descriptions of them) in both domains and/or similarities of the models used in both domains to study phenomena (e.g., Hesse 1963; 2017). What explains the transfer of concepts, formal structures, and methods across domains is this similarity relation either between two target systems or between the models of both target domains. On this view, positing such material or formal relations licenses inferences from a system in the source domain to a system in the target domain, or inferences from one model used to theorize about a target system in the source domain to a model used to theorize about a target system in a new target domain (e.g., Hesse 2017; Knuuttila and Loettgers 2016; Bokulich 2015; Zuchowski 2019).

Jhun et al. (2018) provide an analysis of the Johansen–Ledoit–Sornette (JLS) model of critical market crashes from econophysics in light of proclaimed analogies between critical phase transitions in statistical physics and stock market crashes. While they show the limitations of this analogy in enabling the unconstrained use of common explanatory strategies from physics in economics, they argue that the JLS model is useful in that it can offer different kinds of explanations of and theoretically describe stock market crashes as critical phenomena analogous to critical phase transitions of physical phenomena. Generally, relying on analogical inference in explaining model transfer implies a commitment to the view that models represent their target system not by correspondence or isomorphism but by analogy (e.g., Hesse 2017, 305). On this view, the justification of predictions from transferred models about new target systems – potentially in different domains – becomes a matter of the strength of analogous argument (see the entry on analogies and metaphors in this Handbook).

The template-based accounts originate in Paul Humphreys' suggestion that contemporary computational science is organized around a limited number of computational templates for the use of which explicit assumptions can often be formulated (Humphreys 2002; 2004). As such, the decision of whether such templates can be transferred across domains does not have to rely on vague or implicit similarities between phenomena. Templates are general, representational devices that ground the construction of computational models. On this view, what is thus transferred across domains is not the model itself, but the template that a model can be built on. Such templates are more easily transferred than most models because they are based on mathematical or computational forms and methods that are flexible enough to study a variety of different problems in distinct domains (Humphreys 2002; 2004; 2019). Different kinds of templates differ on various levels with respect to their degree of abstractness, their relation to an existing scientific theory, and the degree to which they were originally developed for a specific target system, etc. Examples are the Poisson distribution from probability theory, mathematical models from game theory, the Ising and the Lotka–Volterra models, Newton's second law, or the Barabási-Albert preferential attachment model of network formation. Albeit to different degrees, those templates have in common that they – apart from being abstract – are general and as such subject-independent, which is why they are highly flexible and applicable to fundamentally different target systems. Besides their generality, a second distinguishing feature is their tractability. While some templates are mathematically tractable, most templates become computationally tractable when turned into computational models. Both features explain why templates are transferrable across domains.

In addressing model and template transfer, Humphreys (2019) introduces the notion of a formal template contrasting it to theoretical templates. He defines a theoretical template as:

> a general representational device occurring within a theory, containing at least one schematic, second order, property variable (where a second order variable is one that has n-ary predicates as substitution instances) and is such that, when all of the schematic variables have been substituted for, can be successfully used to represent a variety of different phenomena within the domain of that theory.
>
> *(Humphreys 2019, 3)*

So defined, theoretical templates are to some degree domain-specific, if they are grounded in a specific theory which in turn determines the scope of the domain. To turn them into computational templates, they go through a construction process in which they are complemented by construction assumptions and a correction set. The resulting computational template becomes complemented by an interpretation and an output representation to turn it into a computational model that is ready to be applied to some target system. While a concrete ontology is only specified in the construction process, theoretical templates usually originate in some interpreted scientific theory and are, as such, accompanied by a physical interpretation that constrains them. A theoretical template is, as such, often considered to be part of the fundamental principles of this theory (Humphreys 2019, 3). This is why the domain of application of theoretical templates – before and after its transfer – can be constrained in part by the scientific theory the template is derived from.

This last characteristic is what mainly differentiates theoretical from formal templates. Although formal templates are applied to a variety of different systems, they constitute at first instance purely mathematical objects that are either independent or that have been fully separated from any scientific theory (Humphreys 2019, 3) – they only come with a mathematical or computational interpretation. Also, the assumption is that formal templates only have mathematical content. Humphreys takes a representative example to be the Barabási-Albert preferential attachment model. It provides a formal derivation of the result that networks with a scale-free topology (i.e., those whose distribution follows a power law) emerge by way of a two-step procedure: First, there is a continuously growing network whose number of nodes steadily increases (*growth*); second, new nodes tend to connect to those nodes that are already highly connected within the network (*preferential attachment*). This result is simulated and relies only on a few mathematical assumptions about, for example, how the initial connections of the nodes look (Barabási and Albert 1999). The construction process of formal templates, such as the power law distribution template, differs from theoretical ones in that the former does not require a correction set that specifies the need for adjustment of the computational template to match the empirical data in light of its empirical falsity (Humphreys 2019, 3, fn. 10). At the same time, Humphreys argues that the proper empirical justification of the transfer of formal representational devices – e.g., of the power law distribution template – is given by empirically checking whether the construction assumptions – e.g., of the preferential attachment template – are "representationally correct" when applied to a particular system. It is only then that we are justified and thus can acquire knowledge about the causal process bringing about the phenomenon that the template is meant to represent. This is one reason why analogical reasoning alone is not a good guide to re-apply those formal templates to seemingly fitting systems. In the case of the power law distribution template, for example, the template might seem to be a representationally correct system. However, given that this and (more generally) other templates could in principle be derived from different kinds of construction assumptions, its application is properly justified only by making sure that the construction assumptions are representationally accurate in that application (see also Knuuttila and Loettgers 2012).

Because of their high degree of abstractness, formal templates lend themselves to interdisciplinary transfer. According to Humphreys, psychological aspects, analogical reasoning, and thus, anticipation and identification of vague resemblances between different target systems can only be heuristic devices in re-applying a template. Also, analogical reasoning stemming from the previous success of a template in some other domain could be used for justificatory purposes; both might even explain to some extent cases of template transfer. However, they cannot provide a proper justification for a template's transfer or re-application: "[T]he empirical justification for transferring a formal template ultimately rests on the satisfaction of the construction assumptions in the new domain" according to Humphreys (2019, 4). Because those assumptions are explicit, analogical reasoning is not necessary for template transfer. It is their abstractness, their independence from any physical interpretation, and the fact that assumptions are specified only according to the need given by the system in the target domain that is necessary for their transfer. In such a view, all empirical content of a formal template is only located in its empirical mapping, which implies that only knowledge of the target domain is required for its re-application (Humphreys 2019, 6). This is a view that appreciates the analogy-based approach as capturing the psychological and heuristic function of analogies in the context of discovery but rejects them as necessary to think about the object of transfer and thus understands model transfer within the context of justification.

Humphreys' template account offers important insights into model transfer in science. It is applicable to cases of model transfer within and across both the natural and the social sciences, especially those areas of the social sciences that use mathematical and computational models. Besides the criterion of generality, Humphreys' tractability criterion explains, for example, why models from the natural sciences have been adopted by highly mathematized social scientific domains such as economics (e.g., Hindriks 2006; Lisciandra 2019). However, the concept of a template has limitations in understanding model transfer. While Humphreys has refined it toward a more fine-grained distinction between different kinds of templates (Humphreys 2019), the template concept is often still too general to be properly applied. For instance, when the model transfer occurs from the physical to the social sciences, the template-based account neglects a number of elements. There might be potential preconditions for enabling the transfer, such as the commitment of scientists in the target domain to a specific modeling methodology, a set of concepts, or specific theories long accepted in that domain. Many analytical sociologists, for example, hold strong methodological commitments that profoundly shape their modeling choices, such as that human agents should be modeled as rational choice makers. Such factors conducive to transferring a specific kind of model are not acknowledged in the template-based account.

Indeed, epistemic and methodological features such as the structural similarity of phenomena in the source and target domain, a shared methodology in both domains, shared validation criteria for models depending on their purpose, or the goal of theoretical unification have been shown to play a role in enabling model transfer (e.g., Grüne-Yanoff 2011; Marchionni 2013; Tieleman 2022). Or, there might be specific methodological, epistemological, or conceptual features originating in the source domain that play a crucial role in preventing the transfer (e.g., Anzola 2019). For example, economists are strongly committed to epistemic values such as the predictive power of economic models rather than their ability to give causal explanations, or to conceptual commitments such as that their models should be conceptually compatible with the equilibrium concept. Economists have therefore been generally more open to model transfer if the transferred unit can accommodate their main commitments; they have been hesitant towards the transfer leading either to a more fundamental conceptual and/or theoretical change of the neoclassical paradigm (Basso et al. 2017; Sent 2004; Thébault et al. 2018; Bradley and Thébault 2019) or to questioning their explanatory desiderata of using micro-founded models and providing general explanations (e.g., Marchionni 2013; Lisciandra 2019). In such cases, "when disciplinary conventions about ... modelling play a larger role in dictating modifications of common templates, the tendency toward the kind of interdisciplinary organisation Humphreys identifies may not take place after all; disciplinary rather than interdisciplinary unity remains stronger" and thus, prevents model transfer (Marchionni 2013, 348). A further factor that could be relevant for explaining model transfer, especially from the physical to the social sciences and vice versa, is that social scientists might also implicitly hold on to non-epistemic values that play a role in their modeling choices. Such commitments might not only partially explain why a template is transferred, but also why some templates, although also general and tractable, might not be.

Humphreys' template-based account implicitly allows for building epistemological, conceptual, and methodological commitments via the construction assumptions. However, the specification process is not part of the explanation of the template's transfer. His account also leaves open questions regarding whether such factors are transferred together with the template and whether that would be relevant for explaining serious challenges and even failures of such transfers. More needs to be said about the nature and role of such commitments in enabling or preventing model transfer. In the next section, a few attempts by philosophers to take up those issues will be presented.

3. Open issues in the literature on model transfer

Most of the recent literature starts from a template-based account to think about model transfer. Taking a template as the unit of transfer seems to best capture essential features of model-based disciplines of contemporary science on the one hand, and an increasing cross-disciplinary engagement via the transfer of abstract mathematical tools and computational models on the other. However, the previously mentioned factors have been neglected. This section will present some of the efforts to extend the template-based account by focusing on three sets of issues.

One set of issues relates to the disagreement about the object of transfer and its features, which could explain why some models are transferred while others are not. This disagreement goes beyond the nature of models as analogies or templates and also relates to the level at which model transfer should be studied. Some philosophers analyze modeling frameworks (Lin 2022) as the object of transfer. Others focus on case studies of particular models, e.g., the Fisher model or the Ising model, to study how such models can give rise to transferrable templates (Morrison 1997; Knuuttila and Loettgers 2014; 2016; 2020; Price 2020). Yet others, most prominently Humphreys, consider highly generic mathematical forms, namely those that Humphreys calls formal templates, e.g., coupled harmonic oscillators, network models, or even probability theory, to be the objects transferred (Humphreys 2004; 2019). Such disagreement might partly originate in the fact that in scientific practice, all those objects could be or have been transferred; depending not only on what the object of transfer is, but also what can explain it. Consequently, sometimes the relevant object of philosophical analysis is a specific model, e.g., an interpreted or otherwise contextualized formal structure, while at other times it is the formal template alone. To capture this potential diversity, the existing conceptual proposals of formal, theoretical, computational, and model templates provide a useful starting point for thinking about the nature of the study object, about the justification of its transfers, and about the level of abstraction at which model transfer should be studied in different cases to understand how it can be explained.

Recent philosophical research on model transfer has further elaborated on Humphreys' account to address the relevant unit of analysis for model transfer. Houkes and Zwart (2019) point to a tension arriving from the functions of a template as a representational device on the one hand, and allowing for quantitative manipulation on the other. According to them, this tension arises because computational performance can compromise the representational function in the template's reduction to computation-enabling formal structures (2019, 93). By studying the case of the Lotka–Volterra model as applied to the diffusion of technological innovations, they do not define the notion of a template in terms of a purely formal structure from which its interpretation can be detached before its transfer, proposing instead to reconceptualize Humphreys' notion of a "template.". They show that in some cases, such a formal structure comes with an inseparable and intended "thin" intentional interpretation reflected in the construction assumptions that, for example, specify the variables of a differential equation or the generic mechanism the equations describe. This interpretation is different from Humphreys' interpretation, which they call "analytic interpretation," which is added when turning a template into a computational model. Acknowledging the difference between both kinds of interpretations allows for distinguishing between transferring a mere formalism (a formal template in Humphreys' proposal) and a template, in which case the formalism together with the intentional interpretation is transferred across domains. The distinction between analytic and intentional interpretation allows for templates to fulfill a dual function. It allows retaining the usefulness of the notion of a template to study cross-domain transfer while acknowledging that what is transferred can have proper representational functions despite being different from the application-specific computational model that is grounded in a template.

An account that aims to revise Humphreys' proposal more fundamentally in light of scientific practice has been proposed by Knuuttila and Loettgers (2014; 2016, 2023). They introduce the notion of what they call a "model template" to not only account for what is transferred, but also to explain why it is transferred. By studying a variety of cases such as scale free network, the Sherrington-Kirkpatrick model, the Kuramoto model, the Ising and spin glass models, Knuuttila and Loettgers show that generality and tractability are not the only characteristics that explain their inter- and intradisciplinary transfer. Rather, it is also the general conceptual idea associated with the mathematical form together with a set of computational methods that makes them attractive for model transfer. According to their definition, a model template consists of the mathematical structure – the template – that is complemented by a general conceptual vision associated with it, "that is capable of taking on various kinds of interpretations in view of empirically observed patterns in materially different systems" and that explains its transfer via its mediating capabilities between different target systems (Knuuttila and Loettgers 2016, 396; see also 2023). Such a conceptual idea is equally independent from a specific target system, but at the same time, allows for the application of computational methods and equations associated with a specific template. It is thus the conceptual framework coupled with a formal template that renders the model template applicable to a specific set of phenomena in different domains and thereby explains the model transfer. Importantly, this application can be achieved by relying on analogical reasoning.

The notion of a model template is in many respects a significant advancement in further clarifying Humphreys' different templates. In its implication to exclude the transfer of a piece of pure mathematics as an instance of template transfer, the idea of a conceptual vision is also similar to the thin interpretation by Houkes and Zwart. Whether conceptual (as well as methodological) features play a role in enabling the transfer is also particularly important to consider when those features might not align across domains that concern substantially different subject matters. The conceptual features of a model in physics to predict magnetic moments, for example, might *prima facie* not be shared and easily justifiable in the social sciences that model the behavior of human agents.[1] If they indeed played a role in enabling the transfer, the underlying ontology and conceptualization might constrain the kinds of domains the template can be transferred into in each case. To what extent the general conceptual idea constrains a template and its application also seems to depend on the level of abstraction at which model transfer is studied. If the formal template is the unit of transfer, for example, in cases of transferring specific distributions or purely mathematical equations, the conceptual vision attached to the template can help in identifying specific patterns that those equations could describe.

However, at a high level of abstraction, while a conceptual vision often seems to be a central ingredient of a template, it does not have to be part of all our philosophical template concepts. If we consider a theoretical template, such as Newton's second law, a conceptual vision of some person being described as behaving as the sum of individual forces that can be added up by vector addition indeed constrains the application of this template to only entities whose behavior can be described in this way (Humphreys 2019, 3). In this case, Humphreys' concept of a theoretical template might be satisfactory to capture the conceptual vision through the larger theoretical framework the template is a part of. That is not to say that the concept of a theoretical template is always sufficient. In other examples, it becomes clear that the concept of a theoretical template might fit specific cases but not others. One clear difference between the concepts of theoretical and model templates is that the latter is not bound to a specific theoretical context while the former is. Dynamical systems theory and network models are thus examples of templates that do not seem to be part of a specific theoretical framework but nevertheless come with conceptual content attached to them, for instance, different systems behaviors, kinds of interactions, or network structures that the mathematics of those templates could capture. Consequently, given that both concepts are useful supports the view that a pluralism of templates is required if the diverse set of transferred objects should be captured by our philosophical concepts. Thus, even though Knuuttila and Loettgers argue for the unificatory power of the notion of a model template, a one-case-fits-all template concept might neither be desirable nor possible. Provided the varieties in which models occur (for an overview, see Frigg and Hartmann 2020) and the existing disagreement about the exact object of transfer, the question of what concept is best used to capture the object of transfer might have to be answered on a case-by-case basis.

A second set of issues relates to the question of what the characteristics of the transfer process are. For some time philosophical analyses implicitly assumed that model transfer occurs without substantially changing the model throughout the process; and indeed, this can be the case. Rational choice models in economics – subsumed under the label of "economics imperialism" – have been transferred into fields such as sociology, political science, or anthropology, often without any conceptual or theoretical change. However, that the object transferred does not undergo any change seems unrealistic, as illustrated by a number of recent case studies. Rather, significant changes in the model are often crucial for its transfer (Herfeld and Lisciandra 2019; Knuuttila and Loettgers 2013; 2014). For example, models from engineering have been used in synthetic biology only after extensive modification and rational choice models also had to be adapted to the various target systems in the domains into which they were transferred (e.g., Knuuttila and García-Deister 2019; Grüne-Yanoff 2011; Herfeld and Doehne 2019). To capture such modifications, template-based approaches focus on the transfer process as a model construction process (or "template-to-target mapping" as Kaznatcheev and Lin (2022) have labeled it) by adding construction assumptions and an interpretation that allow for a derivation of an output representation (Humphreys 2004; 2019; Tieleman 2022). In this view, the template itself remains unchanged and its modification is dependent on the target system.

Price (2019; 2020) has studied in detail how the target domain shapes the modification of the model template being transferred and how the target domain might itself be changed to enable the model transfer. Employing Knuuttila and Loettgers' notion of a model template, Price notes that the general conceptual vision and thus some basic ontological commitments that come with the template have to be compatible with the target system in the domain the template is transferred into. Price thinks of this as a preparation process for which he introduces the notion of a "landing zone," basically referring to a model's envisaged target system providing an ontology that enables the template transfer. Price discusses the case of the topological atom as a landing zone for transferring a set of model templates from physics into chemistry to construct and apply the so-called quantum theory of atoms to molecules. Broadly defined as a mathematical model's target system, a landing zone enables the transfer and use of the model's mathematics – of a model template – in a new target domain by shaping the way in which the model becomes designed to ensure its applicability in the new domain (Price 2019, 22). Because the landing zone identifies the ontological features of a target system that enable the use of a template in that domain, it can also suggest possible modifications of the template in light of changes in ontological commitments needed to apply the template.

Other philosophers who acknowledge that templates are not static entities have proposed different notions to describe this modification process. Bradley and Thébault (2019), for example, introduce the distinction between "model imperialism," an extension of the scope of problems addressed with the same, unmodified interpreted model, and "model migration" which describes model modification in terms of a radical reinterpretation of the original model requiring what they call a "re-sanctioning" of the fundamental idealizing assumptions to enable the model's application in the new target domain. Others have proposed the view of this process as one of translation (Herfeld and Doehne 2019) and have discussed the role of informal features as complementing features of the formal model, such as model narratives in this translation (Quack and Herfeld 2023). Given that such discussions are highly case-dependent, more systematic and conceptual work is needed to work out what such "translation" exactly entails. Moreover, the relations between a model template, the source domain, and the target domain can be very complicated. Enabling the transfer of a model might entail rethinking basic principles and methodological commitments, or revising accepted theoretical frameworks in the target domain (Knuuttila and Loettgers 2014). Transferring a model from mathematical game theory into political science, for example, required not only the specification of construction assumptions and an interpretation, but also a substantial reconsideration of the methodologies accepted to study political phenomena (Quack and Herfeld 2023).

The transfer might also lead to such substantial modifications of the model in that its original identity as an epistemic object is affected. Kaznatcheev and Lin (2022) show how model transfer can imply that the template switches from a theoretical modeling mode into an experimental measurement mode. This implies, in turn, that the process of template-to-target mapping can be quite complex. To appreciate this complexity, they introduce the distinction between conceptual and concrete mapping. The former maps the formal template and the theoretical concepts in the target domain, which they understand to be similar to the intentional interpretation introduced by Houkes and Zwart (2019). While after the conceptual mapping, the template still lacks empirical content, the concrete mapping from concepts to concrete objects in the target domain allows empirical content to enter the template, which Kaznatcheev and Lin (2022) understand to be similar to Houkes and Zwart's analytic interpretation. They also show that in their case of the transfer of game theory from mathematical oncology into experimental cancer biology, it is already in the first step that the conceptual mapping could be separated from the template, which suggests that not all templates come with a conceptual vision attached to it.

How a model's identity is affected by the transfer also raises questions about the role of the modeler in enabling model transfer and the kind of knowledge that is required on the side of the researcher to do so. While most template-based accounts keep the modeler out of the picture and limit the expert knowledge needed to that of the target domain (e.g., Humphreys 2019), some cases of model transfer require knowledge of the source domain, for example, about its theoretical and technical languages as well as its modeling practices, to engage with the template as a formal framework, interpret the template, repurpose the template, and anticipate its epistemic potentials (Bradley and Thébault 2019; Kaznatcheev and Lin 2022; Lin 2022). Lin (2022) has furthermore argued that sometimes so-called "spillovers" – defined as knowledge-claims that are "indispensable to the justification of another knowledge-claim" (Lin 2022, 6) – are essential for the justification of a model's re-application. An important question is why scientists engage in model transfer in the first place. Aside from the general importance of having tractable representations in science, the structural similarity of the target system, a shared methodology, or the goal of theoretical unification, different social and psychological factors might be involved: opportunism, attempts to imitate success, the lack of a comparable alternative, and finally "imperialist" tendencies have certainly initiated model transfer processes in the past (e.g., Mäki et al. 2018).

Some philosophers have pointed out the importance of considering the relation between the researchers involved in the transfer and those in the source and target domains to understand the degree of model modification in the transfer. Grüne-Yanoff (2011) discusses the degree of modification in the case of transferring game theoretic models from economics to biology and back. He argues that a modeler's knowledge of, the degree of modification of, and reference to, the original model is inversely proportional to the influence of the modeler in the transfer process and their distance to the source domain. Such relations can tell us a lot about the degree of model modification in transfer processes. In the case of imperialistic transfers, for example, a model from some source domain is applied to a set of problems traditionally tackled in some target domain that is distant from the modeler that applies the model to those problems. Economists applied rational choice models to problems – be that crime, addiction, discrimination, marriage decisions, or breastfeeding – traditionally studied in fields that were distant to them and did so without substantially changing the models (e.g., Becker 1976). In contrast, when biologists transferred game theoretic models into their own discipline, core concepts and formal results of game theoretical models – such as players, strategies, and payoff matrices – were re-interpreted and successively replaced by biologists' own theoretical constructions (e.g., Grüne-Yanoff 2011, 389). For instance, while core concepts and formal results of game theoretical models – such as players, strategies, and payoff matrices – were initially imported into, and re-interpreted in biology, biologists would successively replace them with their own theoretical constructions (e.g., Grüne-Yanoff 2011, 389). Considering this distance between the modeler and the respective domains to which a model is transferred can thus be informative in that it tells about the nature of the transfer and the degree of modification it brings with it.

A final set of issues that have not yet been extensively addressed in the literature on model transfer concerns the relationship between model transfer and scientific progress. In part, this gap in the literature originates in the lack of an explicit discussion of the challenges that hamper model transfers or prevent them from being successful. The existence of such challenges most likely depends upon the factors that need to be in place to enable a model's transfer in the first place (e.g., Price 2019). The aforementioned factors that might hamper model transfer, such as structural dissimilarities between templates and target systems or differences in accepted methodologies in both domains, might certainly play a role (Grüne-Yanoff 2011; Knuuttila and Loettgers 2016). However, given that the philosophical literature has focused mostly on cases of successful model transfer, there is not yet a systematic discussion about how model transfer might generally lead to empirical, theoretical, or conceptual progress – or prevent it. A philosophical analysis of the relationship between model transfer and progress would be important. Progress may only be an apparent result of model transfer. Particularly when models are substantially modified in the transfer process or when they imply profound theoretical and methodological changes in the target domain, their epistemic contribution to a better empirical understanding of phenomena in the target domain, and more generally, might not be straightforward.

For template-based accounts, model transfer and scientific progress seem to be closely connected, in that progress is frequently achieved by applying tractable mathematics. For instance, Humphreys observes that "whenever you have a sudden increase in usable mathematics, there will be a sudden, concomitant increase in scientific progress in the area affected" (Humphreys 2004, 55). To discuss conceptual progress, Price (2020) focuses on the relation between the unit of transfer and the target system to which it is applied (i.e., the landing zone). In his view, the reconceptualization of the phenomenon in the target domain required for model transfer can lead to conceptual progress. Template transfer can provoke discussions about the appropriate ontology for applying the model and about the appropriate assumptions, that can motivate theorizing in the new domain. Insofar as the resulting conceptual pressure leads to changes in, or replacements of, concepts of the target domain, it can lead to the emergence of new concepts and thus to conceptual progress in that domain. Similarly, such conceptual pressure can be perceived as a threat to an existing conceptual framework that needs to be avoided. The existence of such pressures can thus challenge or even prevent model transfers.

Boumans and Herfeld (2023) offer another proposal to appreciate the different ways in which model transfer can lead to epistemic benefits in the target domain. By studying a historical case from econometrics, they explore the way in which a functional account to progress can be used to analyze ways in which model transfer can lead to progress. This so-called new functional approach defines progress in terms of usefulness for defining and solving problems (Shan 2019). Applied to model transfer, epistemic benefit is then translated into the usefulness of a model not only for solving concrete problems but also for proposing, refining, and specifying new problems and thereby guiding future research in some domains. Templates are part of a "common recipe" consisting of a set of concepts; a set of practical guides specifying the procedures and methodologies as a means to solve a problem; a set of hypotheses, and a set of patterns of reasoning indicating how to use other components to solve a problem (Shan 2019, 745). As such, this account already provides indicators to think concretely about success conditions for model transfer as well as reference points to identify some of the major challenges to model transfer in science. By adopting this account and the concept of a model template, Boumans and Herfeld show that the conceptual vision of the business cycle in a core econometric model template was essential to its construction and transfer, but that the resulting progress was also disrupted when the conceptual vision of the phenomenon changed in such a way that the template transferred is no longer considered to be sufficiently representative of the phenomenon in question.

4. Conclusion

Given that the philosophical analysis of model transfer as a prominent phenomenon in modern science is only in its beginning stages, this survey has pointed out open issues that should be addressed to advance the debate further. Surveying the literature not only reveals the relevance of the phenomenon but also shows its philosophical importance for multiple areas within the philosophy of science. Reaching a deeper understanding of model transfer and its challenges in science is therefore highly desirable. Results promise to have profound implications for the way in which we think about scientific models, the practice of modeling, and model integration in the philosophy of scientific modeling.

Acknowledgments

This article is part of a project that has received funding from the European Research Council (ERC) under the European Union's Horizon 2021 research and innovation program (Grant Agreement No. 101043071) with a project on Model Transfer and Its Challenges in Science: The Case of Economics. I thank the editors for inviting me to contribute to this handbook and for useful feedback on some earlier version of this chapter. A special thanks to the late Paul Humphreys for motivating me to dive more into the literature on model transfer.

Note

1 For a case from econophysics suggesting that the conceptual vision underlying physical models of the behavior atoms might hinder applying them to freely choosing agents, see Bradley and Thébault (2019).

References

Andersen, Hanne. 2016. "Collaboration, Interdisciplinarity, and the Epistemology of Contemporary Science." *Studies in History and Philosophy of Science Part A* 56: 1–10.

Anzola, David. 2019. "Knowledge Transfer in Agent-based Computational Social Science." *Studies in History and Philosophy of Science Part A* 77: 29–38.

Ash, Mitchell G. 2006. "Wissens- und Wissenschaftstransfer – Einführende Bemerkungen." *Berichte zur Wissenschaftsgeschichte* 29(3): 181–189.

Barabási, Albert-László, and Réka Albert. 1999. "Emergence of Scaling in Random Networks." *Science* 286(5439): 509–512.

Basso, Alessandra, Chiara Lisciandra, and Caterina Marchionni. 2017. "Hypothetical Models in Social Science." In *Springer Handbook of Model-Based Science*, edited by Lorenzo Magnani and Tommaso Bertolotti, 413–433. Cham: Springer.

Becker, Gary S. 1976. *An Economic Approach to Human Behavior*. Chicago: University of Chicago Press.

Bokulich, Alisa. 2015. "Maxwell, Helmholtz, and the Unreasonable Effectiveness of the Method of Physical Analogy." *Studies in History and Philosophy of Science* 50: 28–37.

Boumans, Marcel, and Catherine Herfeld. 2023. "Progress in Economics." In *New Philosophical Perspectives on Scientific Progress*, edited by Yafeng Shan, 224. Routledge Studies in the Philosophy of Science. New York and London: Routledge.

Bradley, Seamus, and Karim P.Y. Thébault. 2019. "Models on the Move: Migration and Imperialism." *Studies in the History and Philosophy of Science Part A* 77: 81–92.

Donhauser, Justin, and Jamie Shaw. 2019. "Knowledge Transfer in Theoretical Ecology: Implications for Incommensurability, Voluntarism, and Pluralism." *Studies in History and Philosophy of Science Part A* 77: 11–20.

Du Crest, Agathe and Martina Valković, André Ariew, Hugh Desmond, Philippe Huneman, Thomas A. C. Reydon. 2023. *Evolutionary Thinking Across Disciplines: Problems and Perspectives in Generalized Darwinism*. Springer.

Frigg, Roman, and Stephan Hartmann. 2020. "Models in Science." *The Stanford Encyclopedia of Philosophy*, edited by Edward N. Zalta. https://plato.stanford.edu/entries/models-science/

Grüne-Yanoff, Till. 2011. "Models as Products of Interdisciplinary Exchange: Evidence from Evolutionary Game Theory." *Studies in History and Philosophy of Science Part A* 42(2): 386–397.
Herfeld, Catherine, and Malte Doehne. 2019. "The Diffusion of Scientific Innovations: A Role Typology." *Studies in History and Philosophy of Science Part A* 77: 64–80.
Herfeld, Catherine, and Chiara Lisciandra. 2019. "Knowledge Transfer and Its Contexts: Editorial." *Studies in History and Philosophy of Science Part A* 77: 1–10.
Hesse, Mary B. 1963. *Models and Analogies in Science*. London and New York: Sheed and Ward.
———. 2017. "Models and Analogies." In *A Companion to the Philosophy of Science*, edited by William H. Newton-Smith, 299–307. Hoboken, NJ: John Wiley and Sons.
Hindriks, Frank A. 2006. "Tractability Assumptions and the Musgrave-Mäki-Typology." *Journal of Economic Methodology* 13(4): 401–423.
Houkes, Wybo, and Sjoerd D. Zwart. 2019. "Transfer and Templates in Scientific Modelling." *Studies in History and Philosophy of Science Part A* 77: 93–100.
Howlett, Peter, and Mary S. Morgan. 2011. *How Well Do Facts Travel? The Dissemination of Reliable Knowledge*. Cambridge: Cambridge University Press.
Humphreys, Paul. 2002. "Computational Models." *Philosophy of Science* 69(3): 1–11.
———. 2004. *Extending Ourselves: Computational Science, Empiricism, and Scientific Method*. Oxford: Oxford University Press.
———. 2019. "Knowledge Transfer across Scientific Disciplines." *Studies in History and Philosophy of Science Part A* 77: 112–119.
Jhun, Jennifer, Patricia Palacios, and James O. Weatherall. 2018. "Market Crashes as Critical Phenomena? Explanation, Idealization, and Universality in Econophysics." *Synthese* 195(10): 4477–4505.
Kaiser, David. 1998. "A ψ is just a ψ? Pedagogy, Practice, and the Reconstitution of General Relativity, 1942–1975." *Studies in the History and Philosophy of Modern Physics* 29(3): 321–338.
Kaznatcheev, Artem, and Chia-Hua Lin. 2022. "Measuring as a New Mode of Inquiry That Bridges Evolutionary Game Theory and Cancer Biology." *Philosophy of Science* 1–21.
Knuuttila, Tarja, and Vivette García-Deister. 2019. "Modelling Gene Regulation: (De)Compositional and Template-based Strategies." *Studies in History and Philosophy of Science* 77: 101–111.
Knuuttila, Tarja, and Andrea Loettgers. 2012. "The Productive Tension: Mechanisms vs. Templates in Modeling the Phenomena." In *Models, Simulations, and Representations*, edited by Paul Humphreys and Cyrille Imbert, 163–177. New York: Routledge.
———. 2013. "Synthetic Biology as an Engineering Science? Analogical Reasoning, Synthetic Modeling, and Integration." In *New Challenges to Philosophy of Science*, edited by Hanne Andersen, Dennis Dieks, Wenceslao J. Gonzalez, Thomas Uebel and Gregory Wheeler, 163–177. The Philosophy of Science in a European Perspective 4. Dordrecht: Springer.
———. 2014. "Magnets, Spins, and Neurons: The Dissemination of Model Templates across Disciplines." *The Monist* 97(3): 280–300.
———. 2016. "Model Templates Within and Between Disciplines from Magnets to Gases – and Socio-Economic Systems." *European Journal for the Philosophy of Science* 6(3): 377–400.
———. 2017. "Modelling as Indirect Representation? The Lotka–Volterra Model Revisited." *British Journal for the Philosophy of Science* 68(4): 1007–1036.
———. 2020. "Magnetized Memories: Analogies and Templates in Model Transfer." In *Philosophical Perspectives on the Engineering Approach in Biology: Living Machines?*, edited by Sune Holm and Maria Serban, 123–140. London: Routledge.
———. 2020. "Model Templates: Transdisciplinary Application and Entanglement." *Synthese* 201: 200.
Leonelli, Sabina, and Niccolò Tempini. 2020. *Data Journeys in the Sciences*. Cham: Springer International Publishing.
Lin, Chia-Hua. 2022. "Knowledge Transfer, Templates, and the Spillovers." *European Journal for Philosophy of Science* 12(1): 1–30.
Lipphardt, Veronika, and David Ludwig. 2011. "Knowledge Transfer and Science Transfer." *European History Online* ed. Heinz Duchhardt. http://ieg-ego.eu/en/threads/theories-and-methods/knowledge-transfer/veronika-lipphardt-david-ludwig-knowledge-transfer-and-science-transfer
Lisciandra, Chiara. 2019. "The Role of Psychology in Behavioral Economics: The Case of Social Preferences." *Studies in History and Philosophy of Science Part A* 72: 11–21.

Mäki, Uskali, Miles C. MacLeod, Martina Merz, and Michiru Nagatsu. 2019. "Investigating Interdisciplinary Practice: Methodological Challenges (Introduction)." *Perspectives on Science: Philosophical, Historical, Sociological* 27(4): 545–552.

Mäki, Uskali, Adrian Walsh, and Manuela Férnandez-Pinto. 2018. *Scientific Imperialism*. Abingdon and New York: Routledge.

Marchionni, Caterina. 2013. "Playing with Networks: How Economists Explain." *European Journal for Philosophy of Science* 3(3): 331–352.

Morrison, Margaret. 1997. "Physical Models and Biological Contexts." *Philosophy of Science* 64: 315–324.

Nersessian, Nancy J. 2002. "Maxwell and 'the Method of Physical Analogy': Model-based Reasoning, Generic Abstraction, and Conceptual Change." In *Reading Natural Philosophy: Essays in the History and Philosophy of Science and Mathematics*, edited by David Malament, 129–166. Lasalle, IL: Open Court.

Price, Justin. 2019. "The Landing Zone – Ground for Model Transfer in Chemistry." *Studies in History and Philosophy of Science Part A* 77: 21–28.

———. 2020. "Model Transfer and Conceptual Progress: Tales from Chemistry and Biology." *Foundations of Chemistry* 22(1): 43–57.

Sent, Esther-Mirjam. 2004. "Behavioral Economics: How Psychology Made Its (Limited) Way Back into Economics." *History of Political Economy* 36(4): 735–760.

Shan, Yafeng. 2019. "A New Functional Approach to Scientific Progress." *Philosophy of Science* 86(4): 739–758.

Tan, Peter. 2023. "Interdisciplinary Model Transfer and Realism about Physical Analogy." *Synthese* 201: 65

Quack, Alexandra, and Catherine Herfeld. 2023. "The Role of Narratives in Transferring Rational Choice Models into Political Science." *History of Political Economy* 55(3): 549–576.

Thébault, Karim P.Y., Seamus Bradley, and Alexander Reutlinger. 2018. "Modelling Inequality." *British Journal for the Philosophy of Science* 69(3): 691–718.

Tieleman, Sebastiaan. 2022. "Model Transfer and Universal Patterns: Lessons from the Yule Process." *Synthese* 200(4): 267.

Zuchowski, Lena. 2019. "Modelling and Knowledge Transfer in Complexity Science." *Studies in History and Philosophy of Science Part A* 77: 120–129.

PART 4

Related topics

21
MODELS AS SYMBOLS

Catherine Z. Elgin

1. Introduction

Scientific models are a motley crew: some are concrete, others abstract; some are static, others dynamic; some represent states of affairs, others simulate processes; some have targets, others do not; some closely resemble their targets, others drastically distort. Nevertheless, scientific models of all sorts function epistemically. They embody and advance understanding. A critical question is how they do so.

The answer might seem obvious: models are similar to their targets; by investigating a model, we learn about its target. Sadly, this is too simple. First, it cannot accommodate models without targets. Phlogiston models, ether models, and caloric models turned out to lack targets. Nevertheless, their status as models was not rescinded. Nor are all targetless models the results of mistakes. Biologists invoke a model species with four sexes in order to investigate population dynamics.[1] Physicists devise models of perpetual motion machines to deepen our understanding of their impossibility (Weisberg 2013, 126–134). Second, when a model has a target, grounding modeling in similarity makes success too easy. Since any two items are similar in some respect, and each is maximally similar to itself, every item qualifies as a model of every item. Such ubiquity renders similarity epistemically inert. Moreover, if similarity suffices, the ubiquity of similarity makes it hard to see how a model can mislead. Accounts that ground modeling in isomorphism, homomorphism, and the like restrict the range of similarity to structural similarity (see Bartels 2006). Still, the same failings apply. They cannot accommodate models without targets, and too easily succeed if targets are available.

Giere (1988) attempts to evade the problem of easy success by maintaining that a successful model is similar to its target in relevant respects; irrelevant similarities are idle. However, problems remain. First, what we might call "accidental matching" is possible. A model designed to resemble its target in a specified, relevant respect fails to do so but happens to resemble it in unspecified, perhaps undiscerned, but nevertheless relevant respects. Since similarity is ubiquitous, this is a likely scenario. Second, a rococo model might include so much irrelevant information that it occludes relevant similarities. In that case, a

 DOI: 10.4324/9781003205647-26

relevant similarity obtains but is swamped by irrelevancies. This is a problem of too much information. Models streamline and simplify. They advance understanding by omitting what should be ignored. Moreover, models distort. Rice (2021) argues that some models are effective not despite but because of their pervasive, drastic distortions. If so, then however circumscribed, similarity seems the wrong metric.

Similarity approaches, whether selective or not, apparently assume that once we establish that a model stands in the proper relation to its target, the way the model affords understanding of its target will be evident. Models are construed as mirrors—as intentional replicas that reflect a portion of reality. But to a large extent, the value of models lies in their being vehicles for surrogative reasoning. Reasoning with and about a model enables scientists to better understand its target. An effective model fosters and facilitates epistemically fruitful surrogative reasoning. Rococo models include obtrusive, irrelevant information that impedes effective reasoning. Excessively simple models display relevant similarities but fail to facilitate reasoning. Both, however, mirror the phenomena they concern.

Advocates of relevant similarity might try to accommodate this by incorporating considerations pertaining to effective surrogative reasoning into the criteria on relevant similarity. Still, there is a problem. We resort to surrogative reasoning because reasoning directly about the target is too difficult, cumbersome, or time-consuming. The target is too obscure, too complex, entwined with confounding factors, mathematically intractable, or whatever. For a model to be an effective vehicle for surrogative reasoning, it must be suitably and often substantially dissimilar to the target.

A separate issue concerns the selection of reasoning strategies. Enabling the same reasoning we use when we directly confront the target is ill-advised. There is no basis for thinking that reasoning appropriate to the full complexity of the phenomena is equally appropriate when things are pared down. What sorts of reasoning are to be permitted? There need not be a one-size-fits-all answer to this question. But in any particular case, it should be clear what inferences are permissible. Is abductive inference allowed? Is analogical reasoning? Nothing about the similarity of a model to its target, or the structural relations between a model and its target settles, or even addresses, this issue.

We have uncovered a number of features that an adequate account of the epistemic function of models should accommodate. (1) Some models have no targets. Still, they seem to function epistemically. (2) Models can be ineffective because they provide too much information, even if that information is accurate. (3) Effective models distort in ways that are illuminating, not misleading. (4) Some models mislead. An adequate account of models should explain how such models impede understanding or foster misunderstanding. (5) Models are used for surrogative reasoning. Hence an adequate account should explain how the reasoning they promote figures in or advances understanding.

Models are not mirrors; nor are they transparent windows to the world. They are complex symbols whose epistemic contributions derive from multiple interacting symbolic functions. As symbols, they are subject to syntactic, semantic, and pragmatic constraints. They are artifacts—epistemic tools that equip us to understand the world in ways that otherwise we could not (see Knuuttila 2011). Drawing on Goodman (1968), the following sections begin by explicating a number of symbolic devices that figure in Hughes's DDI account (1997). This account will then be presented and extended. It will also be shown how the extended account satisfies the requirements listed above, and how models so construed embody and advance understanding.

2. Symbolic resources

Denotation is the relation of a name to its bearer, a predicate to the items in its extension, a portrait to its subject, a map to the terrain it maps, and a general picture (such as the picture of a sparrow in a field guide) to each of the things it depicts. If a model has a target, it denotes that target. If a symbol has no object, it does not denote. Thus, fictional names, such as 'Huck Finn', and fictional maps, such as the map of the route to Mordor, fail to denote. So do terms like 'phlogiston', 'the ether', and 'the Northwest Passage', which were once thought to denote but turned out to have no objects. Nevertheless, such symbols are not gibberish; nor are they parts of speech like prepositions or adverbs that play distinct, non-denotative grammatical roles. Despite having no referent, they are denoting symbols. 'Huck Finn' remains a name, even though it names no one. It is a denoting term. A model of the ether remains a model, even though it models nothing. It too is a denoting symbol. A critical question is how such symbols function.

Goodman distinguishes two dimensions along which a denoting symbol functions. One is the relation of the symbol to what it is a symbol *of* 'Feynman' denotes a particular physicist. The name is a representation of a particular person. The other dimension concerns what sort of representation it is. To mark the difference, Goodman introduces the concept of a *p-representation* (1968, 127–131). The term 'Feynman' is a symbol of the sort that is capable of denoting Feynman. It is a Feynman-representation, a physicist-representation, and so on. The formula 'O_2' is the sort of symbol capable of denoting oxygen. It is an oxygen-representation, a gas-representation, and so on. 'Representation of' is a two-place relation linking a representation with its object. Where its denotation is null, the symbol is not a symbol *of* anything. Still, such a symbol is of the same syntactic sort as symbols that successfully denote. Its grammar makes it capable of denoting. '*p-representation*' is a schema for a one-place predicate whose members all have the same putative object. It is a classification of denoting symbols themselves, without regard to what, if anything, they denote. In contextually relevant circumstances, 'Huck Finn' is a Huck-Finn-representation and a runaway-boy-description, just as 'Richard Feynman' is a Feynman-representation and an-expert-in-quantum-mechanics-representation. 'Phlogiston' is a phlogiston-representation and a-source-of-combustion-representation, just as 'oxygen' is an oxygen-representation and a-sustainer-of-combustion-representation. What qualifies various symbols to be members of the same class of *p*-representations is their relations to one another, not their relation, if any, to a denoted object.

Through the device of *p*-representation, we see how multiple representations qualify as being of the same putative item. A variety of terms in a novel coalesce to constitute a fictional character like Huck Finn, a variety of symbols in biology papers coalesce to characterize a fictional species with four sexes, a variety of nouns and pronouns, descriptions, and names in a factual report coalesce to characterize an actual avalanche. By being instances of the same *p*-representation, distinct terms and distinct uses of the same term count as being about the same real or ostensible item. The various instances of the same *p*-representation constitute a small genre—the genre of Hobbit-representations, phlogiston-representations, four-sex-species-representations, avalanche-representations (see Elgin 2010, 3). Over time, the genre evolves, as increasingly numerous and varied representations become recognized as members of a given class of *p*-representations. Thus, there were increasingly detailed phlogiston-models even though they turned out not to be models of anything. *p*-representations enable us to understand both why targetless models are representational, and how hypothetical representations function. At the outset, we may not know whether anything answers to a given symbol—that is, whether the symbol denotes anything. The putative item begins its career as a posit. To figure out whether anything answers to the posit requires elaboration—endowing it with more robust characteristics, incorporating it into models, and seeing what happens. The posit acquires a distinctive profile as it is elaborated, and increasingly detailed commitments are incorporated into it. The genre evolves over time, homing in on what it would take for something to constitute an answer to the posit—that is, what would be required for the symbol to denote. Elaborating a model that involves a posit then is a matter of extending, refining, and emending the *p*-representations that collectively come to constitute the identity conditions on the posited object.

Denotation and *p*-representation underwrite *representation-as*. For Winston Churchill to be represented as a bulldog is for a bulldog-representation to denote Churchill. For a nucleus to be represented as a liquid drop is for a liquid-drop-representation to denote the nucleus. Denotation can be affected by stipulation. A user can simply stipulate that *m* shall denote *n*, and *m* thereby comes to denote *n*. So any *p*-representation can, by stipulation, be used to denote any object. A bulldog-representation could represent the nucleus as a bulldog, and a liquid-drop-representation could represent Churchill as a liquid drop. If representation-as is to serve as a vehicle for modeling, further restrictions are required to exclude unwanted cases. This is where exemplification enters the picture.

Exemplification is a mode of reference by which an item refers to some of its own features, a feature being a property or relation at any level of abstraction. Exemplification involves both denotation and instantiation. For a symbol *s* to exemplify feature *t*, *s* must instantiate *t* and must refer to *t* via that instantiation (see Goodman 1968, 50–68, Vermeulen, Brun and Baumberger 2009). Commercial paint companies provide sample cards that exemplify the colors of the paints they sell. Problems worked out in textbooks exemplify the reasoning strategies students are expected to learn. Exemplars are not mere instances of features; they are telling instances. They highlight select features, making them manifest. Some exemplars, such as textbook cases and displays on paint cards, are highly regimented. Others are ad hoc. Anything can serve as a sample of any of its features, simply by being used as such. An ornithologist might point to a bird as an example of a goldfinch; if it is in fact a goldfinch, that bird comes to exemplify its species. It was, of course, a goldfinch anyway. What the ornithologist's gesture did was make it an example of its kind. Nor is it the case that exemplification is simply a vehicle for conveying what is already known. The chef samples the soup to see whether it needs more sage. Until he tastes it, no one knows. He is not especially interested in whether that particular spoonful of soup needs more sage. He treats the spoonful as a representative sample of soup in the pot it was drawn from. He draws inferences about the rest of the soup from what is exemplified by the spoonful he tastes.

Exemplars may require processing to bring the features they exemplify to the fore. Merely looking or tasting is not always enough to ascertain what an exemplar exemplifies. Like the chef, the mining inspector takes samples to exemplify something no one yet knows—in this case, the proportion of different gases at different levels of the mine. But unlike the chef who can trust his senses, the inspector needs to run his samples through a gas chromatograph to determine what the samples exemplify.

Exemplification is selective. To highlight some of an item's features requires bracketing, downplaying, or marginalizing others. In its standard use, a paint card exemplifies the colors on its face. It does not exemplify its position. In a non-standard use—for example, when used as a bookmark—the card exemplifies a place in a book, disregarding color completely. In the sciences, processing often requires more than a reorientation of emphasis. It often involves removing confounding factors. Then scientists work with pure samples rather than relying on what is found in nature. Processing may involve adding reagents to bring a particular feature to the fore or subjecting an item to extreme conditions, in order to highlight features that are not manifest in standard conditions. Experimentation is in large measure a matter of enabling items to exemplify features that are not ordinarily epistemically accessible.

Exemplification is not a matter of conspicuousness. To exemplify subtle factors, conspicuous features often need to be bracketed. A risk assessor may find that a manufacturing process exemplifies a subtle vulnerability to sabotage. To do so, he ignores the deafening din in the factory and the firm's annual production figures. Figuring out how to extract epistemically valuable information requires determining which aspects of the phenomena are significant, and which are irrelevant. Clearly, this is a contextual matter. Depending on the issue under investigation, and the conceptual, instrumental, and methodological resources available, the same phenomena can be interpreted as exemplifying any of a variety of features. What is a signal in one investigation may be noise in another.

In principle, an item can exemplify any of its features. But not all features are easily exemplified. Some are semantically unmarked; we have no readily available labels for them. When this is so, it may be far from obvious how far the exemplified feature extends. Even when a feature is semantically marked, the way it is represented may be unintelligible to those who seek to access it. Innovation is needed to bring it to the fore. In January 1986, the Challenger space shuttle exploded because its O-rings failed to seal due to the low temperature at the launch site. Hearings were held during which scientists presented myriads of relevant information. The Congressmen conducting the hearings did not understand the scientists' charts, graphs, equations, and explanations. Then Richard Feynman dropped an O-ring into a glass of ice water and showed that it became brittle in the cold (Feynman 2001, 146–153). His demonstration exemplified to scientific novices what the more learned explanations could not effectively convey. It displayed the connection between the temperature, the resulting brittleness of O-rings, and their inability to expand to form a seal. In this case, the epistemic limitation was only on the side of the lay audience. In other cases, the limitation is general. A situation may be so complicated that no one knows how to handle it in its full complexity. The task then is to exclude irrelevant details in order to focus on telling features. This is one reason we resort to models.

3. Models as symbols

Scientific models are schematic representations that systematically and rigorously omit irrelevancies. They make no pretense of being accurate. I have characterized epistemically effective models as felicitous falsehoods (see Elgin 2017). Some distort. A model representing planets as point masses ignores the breath of each planet and the fact that its mass is not evenly distributed. For certain purposes, such factors are irrelevant. Only the center of gravity and overall mass need to be exemplified. Other models augment. Maxwell's 'idle wheels' are fictional devices that forge an analogy between electromagnetic and mechanical systems, thereby exemplifying an abstract structure that electromagnetic and mechanical systems share (see Nersessian 2008, 19–61). Still others exaggerate. According to Kepler's first law, the Earth travels around the sun in an elliptical orbit. Diagrammatic models typically represent the major axis as considerably longer than the minor axis. In fact, the two axes are almost equal in length. But the models are effective because they exemplify only the property of *being elliptical*, not the precise shape of the ellipse. Statistical models may be true or true enough in the aggregate, but nowhere near true of any particular. Although there are no rational economic agents, irrational idiosyncrasies cancel out. Models that represent populations as infinite elide the effects of chance that finite populations are subject to, exemplifying the role non-random factors play in the behavior of the phenomena.

Patterns emerge when details are excluded. The Lotka–Volterra model is a pair of differential equations that characterize the interdependent dynamics of predator and prey population sizes. It is a simplified model that represents predators as insatiable and prey as immortal unless eaten. By bracketing the question of how the populations modulate their sizes, it reveals a pattern that holds of rabbits and foxes, mollusks and starfish, fish in the Adriatic, and even loan sharks and their victims. The bracketed mechanisms make no difference (see Strevens 2008). The model thus exemplifies a widespread pattern. To be sure, there are limits. The pattern plainly breaks down if the predators drive their prey to extinction. It is considerably more complicated if the predators are themselves prey, if multiple species target the same prey, and so forth. The model thus operates against background assumptions.

Qualms about its epistemic status may persist. The Lotka–Volterra model involves assumptions that are inaccurate. No members of any species are insatiable. No members of any species are immortal unless eaten. So how does a model that describes the population dynamics of such fictional species tell us anything about the dynamics of real populations? The contention that a distortion, simplification, or amplification is not a difference maker at best assures that we make no mistake in resorting to it; this does not yet explain how it advances understanding. To answer that, we need to look in more detail at how models function.

Effective models foster understanding by facilitating fruitful reasoning that illuminates the phenomena they concern. The liberties they take, the divergences from overall accuracy, are justified by their epistemic payoffs. A number of philosophers of science have emphasized that models are things we think with; they are neither windows nor mirrors, but vehicles for surrogative reasoning (see Suárez 2009). Hughes (1997) connects the referential and inferential roles. Drawing on Goodman (1968), he characterizes a model as a complex symbol that performs three interanimating functions: denotation, demonstration, and interpretation. His discussion is schematic. Here it has been elaborated to bring out features that he sketched.

Denotation, as we have seen, is the relation of the model to whatever it is a model *of*. The harmonic oscillator, being a model *of* a spring, denotes a spring; the Phillips–Newlyn machine, being a model *of* an economy, denotes an economy. *Demonstration* consists of reasoning with the model according, as Hughes says, to 'its own internal dynamic'. *Interpretation* consists in identifying the fruits of that reasoning and imputing them to the target. Denotation has already been discussed. Demonstration and interpretation require explication.

The *demonstration* phase of the modeling process is the locus of surrogative reasoning. A model's internal dynamic sets limits on permissible modes of inference, facilitating informative, fruitful, relevant, non-trivial inferences pertaining to its target while impeding misleading, irrelevant, and idle inferences. Just how the fruits of that reasoning pertain to the target depends on how they are interpreted. Before turning to that, more needs to be said about demonstration.

A model's internal dynamic specifies the resources that can permissibly be deployed and the uses to which they can permissibly be put. These resources both facilitate and rein in reasoning. They include background assumptions, auxiliary hypotheses, forms of inference, categories, standards of relevance and precision, and so forth. The recognition that the model is designed to afford epistemic access to a particular target and answer specific questions about that target guides the choice of constraints. Descriptions, inferences, and actions that take reasoning too far afield are sidelined.

'Inference' is construed broadly. In addition to familiar, rigorous modes of inference, a model's internal dynamic may (but need not) license analogical reasoning, associative reasoning, and/or abductive reasoning. It issues more focused licenses as well. It may license simplifications or distortions, such as treating a discrete function as continuous, ignoring or focusing on what happens in the limit, representing finite populations as infinite, or treating huge objects as point masses. It determines the choice of scale and grain. Reasoning according to an internal dynamic involves action as well as deliberation. Using a Phillips–Newlyn machine to figure out the effects of tweaking economic policy requires physically manipulating a flow of water, for it is by seeing how the water flows through the apparatus that one draws conclusions about the flow of money through an economy. Nor are practical inferences solely the province of material models. The internal dynamic of a purely abstract model or of a computer simulation licenses certain actions when particular results are reached. One important action is terminating demonstration—ceasing to draw further inferences. The internal dynamic determines when to stop. A model's internal dynamic thus specifies the range of permissions and prohibitions for reasoning with it.

Chains of inference are, in principle, endless. Further conclusions could always be drawn. Opportunities for inference radiate out in all directions. To properly use a model, we need to know what direction to take in drawing inferences and when to stop. Unrestricted inference licenses would generate a plethora of disparate conclusions, with no obvious way to tell which ones could be legitimately imputed to the target. It follows from $pV = nRT$ that $1 \neq 0$, that either $pV = nRT$ or Shanghai is in Spain, that if $pV = nRT$ then (q or $\sim q$), and so forth. Such inferences, although sound, are idle. The proper use of the model brackets them; it takes them offline. If a model's demonstration phase promoted drawing valid inferences indiscriminately, irrelevant inferences would swamp and likely deflect our thinking. To function as an effective device for surrogative reasoning, a model must block irrelevant and unproductive inferences.

Objects can be described in indefinitely many ways. Most are irrelevant to the purposes for which the model is to be used. So the internal dynamic also constrains representation. It dictates that model-representations are to take a particular form, grain, semantic character, and orientation.

The internal dynamic channels both inference and representation via exemplification. Models are exemplars. Like paint samples, they are designed to make some of their features salient. The features may be monadic or polyadic, static or dynamic, abstract or concrete. By representing a population as infinite, the Hardy–Weinberg model exemplifies the extent to which allele redistribution is insensitive to random fluctuations. By ignoring reproductive mechanisms, the Lotka–Voltera model exemplifies a widespread pattern in predator–prey dynamics. Exemplification, as we have seen, is selective. To highlight some features, an exemplar marginalizes or occludes others.

The inferences that a model's internal dynamic licenses are vehicles of exemplification. They show how changes in one parameter affect changes in others, how a system evolves over time, and how robust or fragile linkages are. They disclose patterns and discrepancies that might otherwise be hard to discern. A model does not exemplify the results of irrelevant inferences; its internal dynamic does not license them. So even when they are logically impeccable, they are idle. By functioning as an exemplar, the model constrains and directs reasoning toward features that can responsibly be imputed to the target. It facilitates relevant, informative inferences while blocking or bracketing irrelevant ones.

In the demonstration phase, features are exemplified only in the model. The molecule-representations in the model-gas-representation are represented as spherical, as perfectly elastic, and more generally as exemplary of the pattern displayed by '$pV = nRT$'. What remains is to link the results to the target.

Interpretation involves identifying the features exemplified in the model's demonstration phase and imputing them and only them to the target. Hughesian interpretation is not literal denotation. We know perfectly well that gas molecules are not spherical. So, in imputing sphericality to the molecules in the target gas—in representing actual gas molecules as spherical—we do not maintain that they really *are* spherical. Rather, we construe actual gas molecules in effect as spheres with distortions. In general, in imputing features of a model to a target, we represent the target as having the features exemplified by the model, distended, distorted, or overlaid by confounding factors. We then ignore the confounds as circumstantially irrelevant.

Frigg and Nguyen are sympathetic to this approach but consider it incomplete (2020, 159–204). Their reservations concern the lack of explicit rules of interpretation. Following Hughes, context and established practice may be allowed to determine how the fruits of demonstration are to be interpreted so as to illuminate the target. Because Frigg and Nguyen favor further regimentation, they have added a key. This yields the DEKI model (DEKI = Denotation, Exemplification, Key, and Interpretation). The key specifies the correlation between the features exemplified in the model and the features of the target. The question is whether the key needs to be separately articulable and specifiable independently of its use. It is doubtful that this is the case. An articulable key may be heuristically valuable for a novice learning to use a certain sort of model, but once a scientist has mastered a particular modeling strategy, it is obvious to her what, and with what precision, results of the demonstration are to be read onto the phenomena. Still, the addition of a key highlights the fact that interpretation is subject to public standards.

A model is designed to make particular features of its target salient. Its effectiveness depends on whether the features it exemplifies illuminate the target, enabling model users to understand the phenomena it bears on. By exemplifying a feature, a model affords epistemic access to it. The model equips users to discern factors that may have been overlooked and to appreciate their significance. $pV = nRT$ exemplifies the relation between temperature, pressure, and volume, omitting any mention of attractive force. If the results of the inferences drawn in the demonstration phase hold up when imputed to the target, we have reason to think that intermolecular forces play no significant role in the thermodynamics of the system we are investigating at the grain at which we are investigating it. We know, of course, that every material object attracts every other. So, we do not conclude from the effectiveness of the model that there is no attraction. Rather, we conclude that for the sort of understanding we seek, at the level of precision that concerns us, for the phenomena that concern us, intermolecular attraction is negligible. It is not a difference-maker. Similarly, representing gas molecules as spherical does no harm. Indeed, it helps. By representing the molecules as spheres, we omit the delicate, dynamic differences in the molecules' actual shapes, which would make calculations intractable and impede our understanding of the interdependence of pressure, temperature, and volume in a gas. The effectiveness of the model lies in its being fruitful to think of the target in terms of the features it exemplifies. A model invites us to think of actual gases as ideal gases with distortions, of springs as harmonic oscillators with friction as a confounding factor, of investors as rational economic agents with (perhaps irrational but anyway irrelevant) quirks, and so forth.

Because models omit, distend, distort, and amend, they are context- and purpose-relative. An inaccuracy that is illuminating in one context or for one purpose may be misleading in or for another. A psychologist interested in why people are drawn to conspiracy theories would not represent her subjects as rational agents. Such a model would obscure the very features that she sought to investigate. Devising an appropriate model requires recognizing what factors are and what factors are not likely to be difference-makers for the question one is investigating. Figuring this out may be an iterative process where models with a variety of internal dynamics are tested against one another. To use a model correctly requires understanding how it functions—what phenomena it denotes, what range of features it can exemplify, what modes of inference it licenses, what sorts of features it imputes, what assumptions it makes, and what scaffolding it relies on.

Models distort (see Rice 2021). When they are effective, the distortions illuminate. The fact that, for a given range of purposes, it makes no difference that gas molecules are not spherical reveals something significant about gases. Illumination may be indirect. An effective species-with-four-sexes model exemplifies allele distributions that differ in specific, significant ways from the allele distributions found in otherwise-similar two-sex species. Scientists can discover something important about an actual case by investigating a suitably constructed counterfactual.

The very same phenomenon can be modeled in mutually inconsistent ways, each of which is appropriate for a different range of problems. One model represents the nucleus as a rigid shell; another as a liquid drop. A shell model exemplifies features that depend on the stability of nuclides. A liquid drop model exemplifies features that bear on binding energy (see Massimi 2022, 94–110). The selectivity of exemplification explains why the features that the liquid drop model highlights are appropriately absent from the rigid shell model (see Elgin 2017, 249–272). Each facilitates some inferences and blocks others. The question for the user is which, if either, suits her current epistemic purposes. An effective model is a felicitous falsehood. It is false in that it misrepresents features that are non-difference-makers. Its doing so enables it to exemplify features that make a difference. This is what makes it felicitous.

Streamlining is epistemically valuable. The omission of irrelevancies figures in a model's capacity to advance understanding of its target. Strevens argues that it is permissible to omit these (irrelevant) factors since they are not difference-makers (2008). However, in omitting these factors, models exemplify something about the phenomena that we otherwise would not, or not easily, appreciate.

Models figure in the understanding of a range of phenomena when it is epistemically fruitful to represent the phenomena as if they had the features the model imputes to them, whereas something is epistemically fruitful only if it either fosters or challenges the integration of the behavior of the phenomena into our evolving understanding of the world. For example, because it is as if the traffic on the highway was a continuous fluid, we can use fluid flow models to understand the movements of traffic. The model explains why the traffic flows more smoothly in the center lanes than at the edges of the road.

It makes no difference that, rather than actually being a continuous fluid, the traffic consists of discrete cars.

Every object has indefinitely many properties and stands in indefinitely many relations to other things. The vast majority of these are of no interest. Some of the interesting and important ones are neatly labeled by our literal vocabulary. They can be directly and literally represented. Others are semantically unmarked. There is, for example, no term capable of accurately describing the exact shape of a carbon dioxide molecule. If properties and relations that lack literal labels are to be recognized, they need to be indicated indirectly. One way to do so is by characterizing the objects that display them *as-if-ishly* (see Vaihinger 2009). It is as if gas molecules were spheres, or as if predators were insatiable, or as if the moon were falling toward the earth. Such as-if-ish representations can capture something epistemically important. The reason is not just that it won't be wrong in a particular context to think of gas molecules as spherical or predators as insatiable or the moon as a falling body; the important point is that *the fact that it won't be wrong* discloses something significant about the phenomenon. The effectiveness of the model discloses that a particular aspect of things—for example, the molecule's shape being somewhat spherical—is significant. The model then provides emphasis and focus. It affords insight not only into what properties the object has but also into which of its properties are worth registering.

4. Conclusion

The account of models presented satisfies the requirements set out above. Models without targets are bereft of denotation. Ether-models are not models of the ether because 'ether' turns out to fail to denote. 'Four-sex-species-models' are not a model of a species with four sexes because 'four-sex-species' fails to denote. Scientists once thought 'ether' denoted; they were wrong. They never thought 'four-sex-species' had a non-null denotation; there was no mistake. In both cases, however, reasoning in the demonstration phase can be carried out. The models have their own internal dynamics, which constrain and channel reasoning, enabling scientists to explore the implications of the items they posit. They investigate what would happen if items of the sort posited behaved in the ways the dynamic mandates. Since 'what would happen if...?' is often a good question, models without targets are often epistemically valuable.

Because exemplification is selective, it enables us to evade the problem of too much information. An enormously complicated phenomenon can be idealized, bracketing the information that makes no difference to the question being examined. So an effective model excludes irrelevancies and focuses on what, in a given context, is significant.

Although models simplify, amplify, streamline, and distort, they illuminate their targets when the features they exemplify can be imputed to their targets in such a way that the problems at issue can be fruitfully addressed. When the effects of intermolecular attraction are negligible, a model that sets them aside enables scientists to appreciate the interrelation of pressure, temperature, and volume in an actual gas. When, however, they are non-negligible, $pV = nRT$ misleads. Misleadingness can take different forms. If intermolecular forces are significant, $pV = nRT$ can be imputed to the target, but its imputation does not supply enough relevant information to be useful. The result is an interpretation that is unacceptably sparse. It incorrectly suggests that no additional information is required. If a model is just irrelevant, imputation simply fails. A population of mice cannot plausibly be represented as an ideal gas. There is no non-arbitrary way to impute the pattern exemplified in $pV = nRT$ to the mice. A misleading use of a model exemplifies features that cannot be fruitfully imputed to its target. A misleading model of a given target exemplifies features that cannot plausibly be imputed to the target at all. Still, such a model, construed as targetless or imputed to a different target, would not necessarily mislead. Whether a model is misleading then depends on how it is used.

This chapter began by saying that modeling is a powerful epistemic tool. The power lies in its ability to simultaneously generate representations that afford focus and show why that focus (even when provided as-if-ishly) is valuable. In effect, models not only say, 'This is what you should be looking at', they also say, 'This is why you should be looking at it this way and ignoring factors that interfere with doing so.' They thereby extend our epistemic range.

Acknowledgment

This chapter is part of a project that has received funding from the European Research Council (ERC) under the European Union's Horizon 2021 research and innovation program (Grant Agreement No. 101043071) with a project on Model Transfer and Its Challenges in Science: The Case of Economics.

Note

1 Weisberg discusses three-sex models. As it turns out, there are species that have three sexes. Since the point concerns the epistemic value of targetless models, I changed the number to four. Regardless of the number (n) of sexes actual species have, it is fruitful to be able to consider how alleles would redistribute if there were (n + 1) sexes. Such a targetless model can be informative.

References

Bartels, Andreas. 2006. "Defending the Structural Concept of Representation." *Theoria* 21(1): 7–19.

Elgin, Catherine. 2010. "Telling Instances." In *Beyond Mimesis and Convention: Representation in Art and Science,* edited by Roman Frigg and Matthew Hunter. Berlin: Springer, 1–18.

———. 2017. *True Enough*. Cambridge: MIT Press.

Feynman, Richard. 2001. *What Do You Care What Other People Think?* As told to Ralph Leighton, New York: W.W. Norton.

Frigg, Roman, and James Nguyen. 2020. *Modelling Nature: An Opinionated Introduction to Scientific Representation*. Cham: Springer.

Giere, Ronald. 1988. *Explaining Science: A Cognitivist Approach*. Chicago: University of Chicago Press.

Goodman, Nelson. 1968. *Languages of Art*. Indianapolis: Hackett Publishing Co.

Hughes, R.I.G. 1997. "Models and Representation." *Philosophy of Science* 64: S325–S336.

Knuuttila, Tarja. 2011. "Modelling and Representing: An Artefactual-Based Approach to Model Based Representation." *Studies in the History and Philosophy of Science* 42(2): 262–271.

Massimi, Michaela. 2022. *Perspectival Realism*. Oxford: Oxford University Press.

Nersessian, Nancy. 2008. *Creating Scientific Concepts*. Cambridge: MIT Press.

Rice, Collin. 2021. *Leveraging Distortion*. Cambridge: MIT Press.

Strevens, Michael. 2008. *Depth*. Cambridge: Harvard University Press.

Suárez, Mauricio. 2009. *Fictions in Science. Philosophical Essays on Modelling and Idealization*. London: Routledge.

Vermeulen, Inga, Georg Brun, and Christoph Baumberger. 2009. "Five Ways of (Not) Defining Exemplification." In *Nelson Goodman. From Logic to Art*, edited by Gerhard Ernst, Jakob Steinbrenner and Oliver R. Scholz. Frankfurt a.M.: Ontos. 219–250.

Vaihinger, Hans. 2009. *The Philosophy of As-If*. Mansfield: Martino Publishing Co.

Weisberg, Michael. 2013. *Simulation and Similarity*. Oxford: Oxford University Press.

22
SCIENTIFIC UNDERSTANDING

Insa Lawler

1. Introduction

What is the epistemic goal of scientific inquiry? A widespread assumption—also reflected in laypeople's concept of science—is that science aims to accumulate *knowledge* (see, e.g., Bird 2010). For example, science provides us with empirically tested hypotheses and theories. It also increases our *understanding* of the phenomena studied.

Understanding as an epistemic good has been a central subject of epistemological inquiry over the past three decades. Among other things, it has been systematically questioned that (explanatory) understanding is a form of knowledge. An increasingly widespread position is that understanding is distinct from knowledge. Several epistemologists have argued that understanding possesses epistemic features that knowledge does not have, such as being an epistemic achievement or not transmittable via testimony (Section 3.1). Section 3.2 focuses on the claim that understanding—but not knowledge—is not factive. One key reason is that scientists increase their understanding of phenomena with the help of heavily idealized scientific models, which are not intended to be even approximately accurate.

Models also challenge the common idea that understanding a phenomenon involves grasping a correct *explanation* of it. Section 4.1 is concerned with the position that models can increase our understanding of the phenomenon studied without providing us with an explanation, for example, by exploring possible causes. Section 4.2 shows how models call into question the position that understanding requires *propositional* explanations. It seems that non-propositional models can afford understanding by virtue of being the desired explanation that provides an empirically adequate account of the phenomenon.

These challenges support an analysis of understanding in terms of *abilities*. Understanding a phenomenon seems to involve the acquisition of relevant abilities, such as being able to manipulate the phenomenon (or its representation), answer what-if questions, or generalize to other cases. Models qua non-static scientific devices shed light on the dynamic aspects of scientific understanding. Section 5 highlights key takeaways concerning this connection between scientific understanding and models, but also explains how the view that understanding can be defined in terms of abilities has been questioned.

DOI: 10.4324/9781003205647-27

Regardless of which analysis of scientific understanding is ultimately correct, an upshot of the debate is that scientific models play a key role in settling this question.

2. Scientific models

Before analyzing the relationship between models and understanding, let me briefly characterize what I take as a scientific model. Models are devices that scientists employ for examining past, current, future, or even fictional objects or phenomena. They are used in natural sciences, like physics, chemistry, or biology, but also in "softer" sciences, like psychology, linguistics, economics, and the social sciences. For example, geological models are used to study the earth's past geomorphology, biological models support the analysis of enzyme-substrate interaction, potential developments of climate change are examined with the help of climate models, and economists use models to explore the economic behavior of fictional, ideally rational agents.

Models are usually accounts of the examined phenomena, i.e., the target phenomena, but they are not exact replicas or representations of them. As Hughes (1990, 71) puts it, "[t]o have a model [...] is not to have a literally true account of the process or entity in question." Typically, scientists create models to examine particular aspects of the target phenomenon. Take the lock-and-key model for enzyme-substrate interaction as a paradigmatic example. It is used to explore the structural relationship between enzymes and their substrates and does not aim at correctly capturing other aspects of enzymes and substrates, like their weight.

Scientists construct models based on stipulations about the target phenomena. Often, these stipulations are *idealizations*. For instance, the lock-and-key model assumes that enzymes and their matching substrate have complementary geometric shapes that fit exactly into each other (like a lock and its key). Only when bound to their "key" can enzymes catalyze a chemical reaction. Such idealizations play a prominent role in the debate about scientific understanding, as detailed in Section 3.2.

Models come in various forms. Some models are sets of mathematical equations, others are graphs, diagrams, simplified material replications of the target phenomenon, and so forth. Whether (and how) models represent their targets is a controversial topic, which is orthogonal to most of the considerations treated in what follows.

3. Understanding vs. knowledge

Until the early 2000s, a widespread view—especially among philosophers of science— was that understanding is a subjective matter. For example, understanding was considered to be purely "psychological," or "pragmatic" (e.g., Bunge 1973), or a subjective feeling of confidence (e.g., Trout 2002). Consequently, as Hempel (1965) emphasized, "understanding" was classified as a relative notion; what counts as understanding depends on individual attitudes and cannot be objectively analyzed.

This view was challenged by de Regt (2004) and de Regt and Dieks (2005), among others, who make the case that there is an epistemically important, albeit pragmatic, sense of scientific understanding that does not require a subjective feeling of understanding. Views like this sparked an ongoing examination of understanding in general, and scientific understanding in particular. However, endorsing its epistemic relevance does not imply classifying understanding as a special kind of epistemic good. A still popular position is *reductionism about understanding*, i.e., the position that understanding reduces to some kind of knowledge.

There are different forms of understanding. The most prominent ones are *explanatory* understanding and *objectual* understanding. We can understand *how* things work, *why* something occurred, or is the way it is (explanatory understanding). We can also understand larger subject matters, like understanding language acquisition in early infancy (objectual understanding). On the reductionist view of understanding, for each kind of understanding, we can identify corresponding knowledge forms. As Sliwa (2015, 58) puts it, "instances of understanding reduce to the corresponding instances of knowledge." For example, explanatory understanding is analyzed as follows: S understands why p if and only if S knows why p. Knowing why p is often defined as knowledge of causes or dependencies.[1] Among others, Lipton (2004, 30) claims that "[u]nderstanding is not some sort of super-knowledge, but simply more knowledge: knowledge of causes." Similarly, Bird (2007, 84) remarks that "[t]o understand why something occurred is to know what causes, processes, or laws brought it about."

3.1 *Epistemic features of understanding*

Although reductionism about understanding is still defended (e.g., Grimm 2014; Kelp 2015; Sliwa 2015; Khalifa 2017), non-reductive views have become popular, according to which understanding is distinct from knowledge. For various reasons, understanding does not seem to be a form of justified and not "accidentally true" belief. Some epistemologists, like Kvanvig (2003), suggest that understanding has a different *epistemic value* than knowledge or any of its parts (e.g., truth, justification, or belief). It has been argued that understanding—unlike knowledge—can involve so-called *epistemic luck* to some degree, such as the luck of getting true information from a reliable source among unreliable sources (e.g., Kvanvig 2003; Pritchard 2010; Morris 2012). Relatedly, some epistemologists question that understanding requires *justification* (see, e.g., Hills 2016; Dellsén 2017). It is also argued that understanding cannot be *transmitted via testimony* (e.g., Pritchard 2008; Zagzebski 2008; Hills 2016). While we can acquire knowledge by processing what a reliable and knowledgeable source tells us, we cannot gain understanding in this way. Understanding requires a "first-hand" grasping of the relation between, say, cause and effect, or the ability to utilize the information in question (see, e.g., Elgin 2007; Zagzebski 2008; Hills 2016). Similarly, understanding is considered to be an *epistemic achievement* (e.g., Pritchard 2008) or an *ability* (e.g., Elgin 2007; 2017; Grimm 2014; Hills 2016). Another line of thought is that knowledge, but not understanding, can be atomistic or isolated. You can know various unrelated things. For example, you can know that a tree fell without knowing why, when, or how it fell, or what kind of tree it is, or you can know a single random piece of information about a topic. By contrast, understanding requires some *systematicity*, *interconnectedness*, or *coherence*. To understand something, we need to comprehend how things are structured, depend on each other, or coherently come together (see, e.g., Elgin 2007; Zagzebski 2019; Dellsén 2020). Objectual understanding clearly requires grasping complex subject matters, but explanatory understanding also involves grasping how things are connected. For example, to understand why the tree fell requires grasping how the tree's internal constitution or an external event caused it to occur. Relatedly, understanding is considered to be *gradable*, whereas knowledge is not (see, e.g., Kvanvig 2003; Elgin 2004; Grimm 2014; Hills 2016; Dellsén 2020). We can understand in more detail *why* the tree fell, but we cannot know in more detail *that* it fell.

All these views question that understanding is a form of knowledge. It goes without saying that all of them have been met with objections, but I do not explore the controversies in this chapter (for a comprehensive overview of the debate, see Grimm 2021). In Section 4, I return to some of the views to explore the connection between scientific understanding, models, and abilities.

3.2 The factivity of understanding

Whereas epistemologists have focused on exploring various epistemic features of understanding, philosophers of science have focused on the *truth requirement* for knowledge. Only truths (or at least approximations of the truth) can be known. By contrast, it appears that understanding is not necessarily factive for two reasons. One focuses on the past. It seems that we gained understanding based on empirically successful theories or models that turned out to be inaccurate (e.g., Elgin 2007; de Regt 2017; de Regt and Gijsbers 2017).[2] For example, de Regt (2022) argues that Newton's theory of gravitation provided us with some understanding—and can still achieve that—although it has been contradicted by Einstein's theory of general relativity, according to which the local curvature of space-time is decisive. Despite its inaccuracy, de Regt maintains that Newton's theory gave and still gives us some understanding of how some gravitational phenomena work, such as realizing why acceleration is independent of mass, among other things, because his theory can correctly predict them.

The other key reason against a truth requirement for understanding is that scientists increase their understanding of phenomena with the help of substantial idealizations and (heavily) idealized scientific models, such as the lock-and-key model or the Lotka-Volterra model, according to which prey populations reproduce exponentially when not preyed upon. A key difference from the historic case is that scientific idealizations are typically *known* to be false and are thus what Strevens (2017, 37) calls "deliberate falsehoods." The majority of scientific models involve such falsehoods.

The use of scientific idealizations and idealized models is mostly considered legitimate because they serve critical purposes. They are used, for example, to achieve better mathematical tractability of the phenomenon (e.g., Weisberg 2007), to exemplify a critical property (e.g., Elgin 2017), or to afford "epistemic access to different aspects of the target [phenomenon]" (Elgin 2017, 267). Potochnik (2017) stresses that because the phenomena studied by scientists tend to be highly complex causal networks, it is inevitable that scientists focus on some of the causal patterns. This focus involves idealizations and simplifications of the whole network. Strevens (2008; 2017) suggests that successful idealizations highlight factors that do not make a causal difference to the target phenomenon. For example, an idealization that assumes that a given population is arbitrarily large communicates that the population's size (after a certain threshold) is insignificant for the phenomenon studied. (For more reasons to use idealizations, see, e.g., Potochnik 2017, 48.)

The nature of scientific idealization is controversial. Among other things, there are discussions about whether good models need to feature idealizations that can be *de-idealized* in the long run, how models can be de-idealized, whether some models involve *indispensable* idealizations, and whether idealizations can be isolated from accurate parts of the models (for an overview of the debate, see, e.g., Weisberg 2007; Elliott-Graves and Weisberg 2014; Knuuttila and Morgan 2019; Shech 2023).

Several scholars, like Elgin, argue that scientific understanding is not factive (i.e., does not require only truths) because it can be gained with the help of idealizations or idealized models. The contributions that such idealized models or idealizations make cannot be reduced to their accurate counterparts; "[...] their divergence from truth or representational accuracy fosters their epistemic functioning" (2017, 1; see also, e.g., de Regt 2017; Potochnik 2017). Understanding still "[...] somehow answers to facts" (Elgin 2007, 37) because an idealized model only provides understanding if it can correctly capture key empirical facts of the phenomenon studied. Instead of truth, it is required that the scientific models or accounts be "true enough" (Elgin 2017), intelligible (de Regt 2017), or correctly capture causal patterns (Potochnik 2017).

Others have tried to rebut the argument from historical cases and the argument from idealizations against a factive account of understanding. For example, it has been argued that in historic cases like Newton's theory, the truths grasped account for what is understood (e.g., Ross 2023) or that only "proto-understanding" is gained due to the lack of an accurate explanation (Khalifa 2017). To handle the second case, Khalifa (2017, chapter 6.3) offers several strategies, among other things, the proposal that understanding why p only requires that the subject *accepts* that q explains why p and that q explains why p is (empirically) *effective*. Since acceptance—unlike belief—does not require truth, this suggestion supports a (quasi-)factive view of understanding. Lawler (2021) describes and defends what she calls an *extraction account* of idealization, which builds on works by Alexandrova (2008), Pincock (2014), Bokulich (2016), and Rice (2016; 2018; 2019b). According to this account, idealizations and idealized models merely *enable* scientists to gain explanations, theories, or understanding, but these falsehoods are not an element of the explanations, theories, or understanding. Scientists can extract truths when they work with empirically successful idealized models. Pincock (2021) promotes a similar view, according to which idealizations and idealized models are only explanatory when truths underlie each falsehood relevant to the explanation in question (see also Pincock 2014).[3]

Regardless of which of the many views on the relationship between understanding and knowledge is ultimately correct, idealized models play a key role in settling this debate. A proper account of understanding needs to account for their perhaps unique epistemology.

4. Understanding vs. explanation

Understanding—especially explanatory understanding—is closely related to explanations. Understanding a phenomenon seems to involve a correct explanation of it, and an explanation seems to fail if it does not provide any understanding. Scientific models challenge traditional conceptions of understanding and explanation. Models challenge the idea that (explanatory) understanding necessarily involves explanations. We seem to have model-based understanding without explanations. Models also challenge the idea that only theory-like scientific products can be explanations. It seems that there are model explanations where the non-propositional model itself is the explanation. I consider both challenges in what follows.

4.1 Understanding without explanation

That there is no understanding without explanation is a popular thesis. Understanding requires "[...] grasp[ing][4] a correct scientific explanation of that phenomenon," as Strevens

(2013, 510) puts it.[5] Lipton (2009) prominently questioned this thesis in a posthumous paper by arguing that we can get the same epistemic benefits that explanations offer (such as knowledge of causes or unification) without a proper explanation. Since these epistemic benefits are what matters for understanding, we can have understanding without explanation (see also Dellsén 2020).

Among other things, Lipton argues that non-explanatory deductive inferences can provide a subject with knowledge of what is necessary for a given phenomenon. Similarly, non-explanatory analogies can provide tacit knowledge of unification. Knowledge of necessity or unification suffices to gain some understanding of the phenomenon in question. The focus of this chapter is not these challenges (for discussions, see, e.g., Grimm 2006; Strevens 2013; Khalifa 2017). Instead, the focus is on how scientific models can challenge the link between understanding and explanation.

Lipton uses scientific models to support his view that understanding does not require explanation. For example, a subject can gain a tacit understanding of retrograde motion by studying a visual model of the solar system:

> These visual devices convey causal information without recourse to an explanation. And people who gain understanding in this way may not be left in a position to formulate an explanation that captures the same information. Yet their understanding is real.
>
> *(Lipton 2009, 45)*

Tacit knowledge of causes suffices for understanding, in Lipton's view, but we cannot have a "tacit explanation," as explanation requires an explicit representation of the explanans (45; see Khalifa 2017 for a rebuttal). Similarly, de Regt (2014) stresses how visualization can be an effective tool for achieving scientific understanding. (For more on the epistemic role of scientific visualization, see, e.g., Mössner 2018.)

Regardless of whether there can be tacit explanations or understanding based on tacit knowledge, models challenge the connection between understanding and explanation in other ways. It has been argued that so-called *how-possibly models* can afford some scientific understanding. Numerous scientific models cannot explain a given phenomenon and are not even designed to do that. For example, models can be used for exploratory purposes. Models are used to calculate possible climate scenarios (see, e.g., Parker 2006; Werndl and Steele 2016) or to explore the behavior of ideal rational agents (see, e.g., Mäki 2005; Alexandrova 2008; Alexandrova and Northcott 2013; Grüne-Yanoff 2013; Marchionni 2017). (For more examples, see, e.g., Kennedy 2012; Rohwer and Rice 2013; Gelfert 2016, chap.4.)

It has been argued—both by scientists and philosophers—that such models can afford some understanding, although they do not offer explanations of actual phenomena. They can provide us with so-called *how-possibly explanations*, which, for example, specify a possible cause of a phenomenon and allow for causal "what-if-things-had-been-different" inferences about counterfactual scenarios for that phenomenon (see, e.g., Grüne-Yanoff 2009; Rohwer and Rice 2013; Ylikoski and Aydinonat 2014). Such information allows us to better understand the nature of the phenomenon in question without explaining it. Verreault-Julien (2017) argues that how-possibly models can also provide understanding by virtue of offering how-possibly *mathematical* explanations that highlight how a potential phenomenon mathematically depends on the assumptions made about the phenomenon.

Such information can afford some understanding by allowing for "what-if-things-had-been-different" inferences concerned with mathematical dependencies. Koskinen (2017) and Knuuttila and Koskinen (2020) stress that how-possibly models need not even be concerned with providing explanatory information about actual phenomena. For instance, in synthetic biology, how-possibly models are an indispensable tool for building novel biological systems. Such models can afford a crucial understanding of these systems.

Last but not least, *machine-learning models* pose a new challenge to the traditional relationship between understanding and explanation. As Sullivan (2022b) stresses, most machine-learning models are explanatorily opaque "black boxes" to us. We typically do not understand how exactly they work or how they arrive at their conclusions. Sullivan (2022b) proposes that they might still afford some understanding if "link uncertainty" can be reduced by providing external empirical support that connects the model to the phenomenon studied (see also Sullivan 2022a). Meskhidze (2023) argues that some usages of machine-learning models in cosmology can provide us with understanding if their mechanisms are deemed to be insignificant to the object of investigation.

To sum up: scientific models question in various ways the popular thesis that understanding requires explanation. All of them would need to be rebutted in order to save the thesis.[6]

4.2 *Understanding and model explanations*

We have considered how scientific models challenge the relevance of explanations for understanding. They also challenge traditional explanation accounts in a different way. Scientists use models when they explain phenomena, and some of these models seem to be the desired explanation. For example, Bokulich (2011, 44) states: "[...] [o]n my view, Bohr's model [of atoms] does genuinely explain the Balmer series [...]." Strevens (2017, 38) wonders "[...] how to interpret the ideal gas model, when it is proffered as an explanation of gases' Boylean behavior," and so forth (see, e.g., Craver 2006; Kaplan 2011). Such "model explanations" or "model-based explanations" question traditional explanation accounts, according to which explanations are correct answers to why- or how-questions.[7] A common view is that answers to questions are sets of propositions.[8] Accordingly, explanations are propositional and veridical, i.e., they must contain only true propositions.

Models are rarely propositional, and, as we have seen in Section 2, often involve idealizations, i.e., falsehoods. Model explanations thus seem to involve idealizations. As Wayne (2011, 831) puts it in the case of physics:

> Explanation in physics relies essentially on idealizations (idealized models) of physical systems, and the explanations themselves contain false statements about both the explanatorily relevant features of the physical system and the phenomenon to be explained.

Bokulich (2017) argues that heavily *de-idealized* models typically cannot provide explanations due to their complexity. It thus seems that we often need idealizations to get model explanations and the understanding they afford. (For similar arguments, see, e.g., Batterman 2009; Batterman and Rice 2014; Bokulich 2011; 2012; Kennedy 2012.)

This challenge to a factive account of explanation is closely related to the idealization challenge discussed in Section 3.2. The proposals that are used to defend the truth requirement for understanding can typically be used to address the factivity challenge for explanation. Here, I want to focus on the concept of a model explanation itself. What are the precise conditions for a model explanation? Are they necessarily non-propositional? Are they a distinct kind of explanation? Based on critically examining definitions by Bokulich (2011; 2017) and Rohwer and Rice (2016), Lawler and Sullivan (2021, 1056) propose the following definition:

> **Model explanation (Core):** An explanation is a model explanation if the model's core content is identical to the core of the explanation.

This definition is intended to cover all kinds of model explanations, including noncausal ones.[9] It is inspired by Bokulich's proposal that the explanantia of model explanations make an essential and justified reference to a model (such as having a partially isomorphic counterfactual structure), as well as by Rohwer and Rice's suggestion that the model's content is identical to the explanans (see also van Riel 2017). Lawler and Sullivan (2021) argue that the first proposal gives us too weak a connection between the model and the explanation and that the second one is too demanding. A model's content might contain less than what is necessary for explaining the target phenomenon but still constitute the core of the explanation. Models are selective and do not capture all aspects of a phenomenon of interest. What constitutes an explanation's core depends, among other things, on the kind of explanation used, what drives the explanatory power of the explanation, what uniquely discriminates the model in question, and on the "robust" or substantial elements of the model (on the latter, see, e.g., Weisberg 2006; Woodward 2006; Kuorikoski, Lehtinen, and Marchionni 2010). On this account, model explanations need not be a distinct kind of explanation. The explanations that are based on models could be instances of familiar kinds of explanation, such as covering-law explanations, mechanistic explanations, etc. (cf. the taxonomy of model explanation by Bokulich (2011, sec. 2, sec. 3)).

Defining what a model explanation is helps to state more precisely how idealized models challenge the factivity of an explanation or understanding. A model explanation contains the model's idealizations only if these "play any real role in the explanation itself," as Bokulich (2011, 36) puts it. If the model's core contains the idealizations, then the model explanation is non-factive. If only the accurate parts of the model constitute the explanation's core, then the model explanation is factive. It is ultimately an empirical question of whether existing idealized models belong to the first or second category.

To summarize: scientific models cast doubt on traditional concepts of the relationship between understanding and explanation. They seem to show that we can gain scientific understanding without explanation and that there might be non-propositional, non-veridical explanations that afford understanding.

5. Understanding vs. abilities

As we have seen, several considerations give rise to the thesis that understanding is an *ability* or that abilities are *constitutive* of understanding. This section surveys these considerations, as well as reasons that speak against the thesis.

As mentioned in Section 3.1, several epistemologists argue against the view that understanding can be reduced to knowledge. Many of the reasons given support the view that understanding is an ability or that abilities are constitutive of it, such as the position that understanding is not transmittable via testimony because it demands a "first-hand" grasping of the relevant explanatory relations (such as cause-and-effect relations or other explanatorily relevant dependence relations). Grimm (2014) and de Regt (2017), among others, insist that such a grasp is only possible when a subject figures out how these relations work or are structured, which involves abilities.

Relatedly, it is stressed that understanding involves being able to utilize the information in question to explain the phenomenon, to answer questions about it, to draw inferences about it, or to analyze or explain analogous cases (see, e.g., Elgin 2007; 2017). For example, Kvanvig (2003, 198) claims that "an ability [to answer questions about something] is surely constitutive of understanding." The grasping required for understanding is connected to these abilities, as Hills (2016, 663) explains:

> Understanding why p [...] requires a grasp [...] of the relationship between p and q. [...] if you understand why p (and q is why p) then you have cognitive control over p and q and thus you can (in the right circumstances) manipulate the relationship between p and q.

Such cognitive control is claimed to be non-constitutive of ordinary knowledge.

The view that abilities are decisive for understanding is part of several analyses of scientific understanding. For instance, de Regt (2004; 2017; 2022) claims that it is essential to scientific understanding that the subject is at least able to infer a prediction or explanation of the target phenomenon. Kuorikoski and Ylikoski (e.g., 2015) *identify* (explanatory) understanding with the ability to draw correct counterfactual what-if inferences about the target phenomenon, i.e., the ability to infer what would happen to the phenomenon if things were different (see also, e.g., Woodward 2003; Grimm 2014). Wilkenfeld (2013) defends the view that scientific understanding is a form of representational capacity that involves a representation of the target phenomenon that can be manipulated.

Prominent accounts of scientific models also lend support to the view that understanding is an ability. For example, Suárez (2002) makes the case for an *inferential* account of scientific models, according to which a model must enable its users to draw inferences regarding the phenomenon. These inferences drawn are crucial for obtaining the desired understanding. Knuuttila and Boon (2011; Knuuttila 2011) defend what they call an *artifactual* account of models. They argue that scientific models are *epistemic artifacts* or *epistemic tools*. The knowledge and understanding we can gain from them is closely intertwined with the activity of modeling. Among other things, a model's target phenomenon is often (co-)constructed when the model is developed, and the construction of models can involve conceptual innovations, which can be crucial for understanding the target phenomenon. The view that understanding is an ability goes well with these dynamic aspects of scientific understanding and scientific practice. de Regt (2022) similarly stresses that modeling is an "art," which cannot be captured in terms of strict rules or algorithms, and requires the skill to use the right idealizations, among other things.

Not everyone agrees that understanding should be defined in terms of abilities. Khalifa (2017) argues that the abilities involved in scientific understanding are not special and do not exceed the ones needed for scientific knowledge. For example, evaluating and testing hypotheses and their alternatives is part of acquiring scientific knowledge and involves modal abilities, such as drawing what-if inferences. Similarly, Sullivan (2018) argues that, upon closer examination, ordinary knowledge involves the same kind of cognitive abilities as understanding, such as the ability to answer what-if questions. For example, you only know a proposition if you can track when it is no longer true.

Hazlett (forthcoming, Section 2.3) challenges the view that abilities are *constitutive* of understanding. He argues that the desired abilities, such as cognitive control, are the *result* or *consequence* of obtaining the relevant understanding. Because a subject understands a phenomenon, they have the cognitive control that is characteristic of understanding. The understanding generates this control, so to speak, and not the other way around.

Resolving the question of whether abilities result from understanding or are constitutive of it will advance our analysis of the connection between understanding and abilities. Either way, it is undisputed that abilities are crucial for understanding.

6. Concluding remarks

That science aims at increasing our understanding of phenomena is a commonplace assumption. What exactly understanding is, whether it can be reduced to knowledge, whether understanding requires (propositional) explanations, and whether abilities are constitutive of understanding remain controversial. It is abundantly clear that scientific models and their characteristics play a decisive role in advancing and settling these debates.

Notes

1 It has been questioned that knowledge of causes or dependencies is sufficient for understanding. To understand why something happened, it does not suffice to merely know the decisive explanatory factor. You should also know *how* that factor caused the phenomenon. For details, see, e.g., Pritchard 2008; 2014; Skow 2017; de Regt 2022. For a critical discussion see, e.g., Khalifa 2017; Lawler 2018. Either way, this controversy does not necessarily undermine reductionism about understanding. One could argue that understanding reduces to knowledge of dependencies plus further knowledge.

2 The theories or models were successful insofar as that they correctly predicted observable phenomena.

3 While such attempts to defend a factive account of understanding are still discussed, so-called "quasifactivism" has been widely rejected. According to this view, it only matters that the elements central to understanding are true, but peripheral elements may be false (e.g., Kvanvig 2003; Mizrahi 2012). It has been shown that in many cases of idealized models, the idealizations matter for the model's core and cannot be treated as peripheral (see, e.g., Rice 2019a; Lawler 2021; de Regt 2022; Knuuttila and Carrillo 2022).

4 What it means to grasp an explanation is still debated, but it is widely assumed that grasping is the required epistemic attitude for understanding (see, e.g., Kvanvig 2003; Trout 2007; Wilkenfeld 2013; Grimm 2014; Hills 2016; Khalifa 2017).

5 The thesis that there is no understanding without explanation should not be confused with the claim that explanations must produce understanding. This thesis is compatible with the view that explanations can be proper explanations even when they do not generate understanding. For discussions on this connection between explanation and understanding, see, e.g., Hempel 1965; Scriven 1962; Friedman 1974; Trout 2002; Woodward 2003; Khalifa 2012; Skow 2017.

6 If there is understanding without explanation, we would need new conditions on understanding that are sharp enough to differentiate between genuine understanding and misunderstanding (see, e.g., Verreault-Julien 2019 for an attempt).

7 Not every answer to a why- or how-question is an explanation. But the challenge that models pose is sufficiently independent of how to exactly define what an explanation is. That is why I am not discussing additional constraints on explanations here.

8 Non-propositional answers to a question are thought to be describable in terms of propositions. For example, Strevens (2013, 510) claims that the content of explanations that use visual information can be expressed propositionally.

9 Lawler and Sullivan (2021) distinguish model explanations from what they call "model-induced explanations,"—explanations in which constructing or using the model constitutes a decisive part of arriving at the explanation; the model enables the explanation, so to speak. Such explanations are not a distinct kind of explanation, but rather highlight a specific epistemological relation between a modeler, a model, and an explanation. (For more on this epistemic role, see also Lawler 2021.)

References

Alexandrova, Anna. 2008. "Making Models Count." *Philosophy of Science* 75(3): 383–404.

Alexandrova, Anna, and Robert Northcott. 2013. "It's Just a Feeling: Why Economic Models Do Not Explain." *Journal of Economic* Methodology 20(3): 262–267.

Batterman, Robert. 2009. "Idealization and Modeling." *Synthese* 169: 427–446.

Batterman, Robert and Collin Rice. 2014. "Minimal Model Explanations." *Philosophy of Science* 81(3): 349–376.

Bird, Alexander. 2007. "What Is Scientific Progress?" *Noûs* 41(1): 64–89.

———. 2010. "The Epistemology of Science-A Bird's-eye View." *Synthese* 175(S1): 5–16.

Bokulich, Alisa. 2011. "How Scientific Models Can Explain." *Synthese* 180: 33–45.

———. 2012. "Distinguishing Explanatory from Nonexplanatory Fictions." *Philosophy of Science* 79(5): 33–45.

———. 2016. "Fiction as a Vehicle for Truth: Moving Beyond the Ontic Conception." *The Monist* 99(3): 260–279.

———. 2017. "Models and Explanation." In *Springer Handbook of Model-Based Science*, edited by Lorenzo Magnani and Tommaso Bertolotti, 103–118. Heidelberg/London/New York: Springer.

Bunge, Mario. 1973. "Philosophy of Physics." *Journal for General Philosophy of Science / Zeitschrift für Allgemeine Wissenschaftstheorie* 4(2): 407–409.

Craver, Carl. 2006. "When Mechanistic Models Explain." *Synthese* 153: 355–376.

de Regt, Henk. 2004. "Discussion Note: Making Sense of Understanding." *Philosophy of Science* 71(1): 98–109.

———. 2014. "Visualization as a Tool for Understanding." *Perspectives on Science* 22(3): 377–396.

———. 2017. *Understanding Scientific Understanding*. New York: Oxford University Press.

———. 2022. "Can Scientific Understanding Be Reduced to Knowledge?" In *Scientific Understanding and Representation: Modeling in the Physical Sciences*, edited by Insa Lawler, Kareem Khalifa, and Elay Shech, 17–32. London: Routledge.

de Regt, Henk and Dennis Dieks. 2005. "A Contextual Approach to Scientific Understanding." *Synthese* 144: 137–170.

de Regt, Henk and Victor Gijsbers. 2017. "How False Theories Can Yield Genuine Understanding." In *Explaining Understanding: New Perspectives from Epistemology and Philosophy of Science*, edited by Stephen R. Grimm, Christoph Baumberger, and Sabine Ammon, 50–74. London: Routledge.

Dellsén, Finnur. 2017. "Understanding without Justification or Belief." *Ratio* 30(3): 239–254.

———. 2020. "Beyond Explanation: Understanding as Dependency Modelling." *The British Journal for the Philosophy of Science* 71(4): 1261–1286.

Elgin, Catherine Z. 2004. "True Enough." *Philosophical Issues* 14(1): 113–131.

———. 2007. "Understanding and the Facts." *Philosophical Studies* 132(1): 33–42.

———. 2017. *True Enough*. Cambridge: MIT Press.

Elliott-Graves, Alkistis, and Michael Weisberg. 2014. "Idealization." *Philosophy Compass* 9(3): 176–185.

Friedman, Michael. 1974. "Explanation and Scientific Understanding." *The Journal of Philosophy* 71(1): 5–19.

Gelfert, Axel. 2016. *How to Do Science with Models. A Philosophical Primer*. Cham: Springer.

Grimm, Stephen R. 2006. "Is Understanding a Species of Knowledge?" *British Journal for the Philosophy of Science* 57(3): 515–535.

———. 2014. "Understanding as Knowledge of Causes." In *Virtue Epistemology Naturalized*, edited by Abrol Fairweather, 329–345. Synthese Library 366. Cham: Springer.

———. 2021. "Understanding." In *The Stanford Encyclopedia of Philosophy*, edited by Edward N. Zalta. Metaphysics Research Lab, Stanford University, Summer 2021 edition. https://plato.stanford.edu/archives/sum2021/entries/understanding/

Grüne-Yanoff, Till. 2009. "Learning from Minimal Economic Models." *Erkenntnis* 70(1): 81–99.

———. 2013. "Genuineness Resolved: A Reply to Reiss' Purported Paradox." *Journal of Economic Methodology* 20(3): 255–261.

Hazlett, Allan. Forthcoming. "Understanding and Testimony." In *The Oxford Handbook of Social Epistemology*, edited by Jennifer Lackey and Aidan McGlynn. New York: Oxford University Press.

Hempel, Carl. 1965. *Aspects of Scientific Explanation, and Other Essays in the Philosophy of Science*. New York: Free Press.

Hills, Alison. 2016. "Understanding Why." *Noûs* 50(4): 661–688.

Hughes, R.I.G. 1990. "The Bohr Atom, Models, and Realism." *Philosophical Topics* 18: 71–84.

Kaplan, David. 2011. "Explanation and Description in Computational Neuroscience." *Synthese* 183: 339–373.

Kelp, Christoph. 2015. "Understanding Phenomena." *Synthese* 192(12): 3799–3816.

Kennedy, Ashley Graham. 2012. "A Non Representationalist View of Model Explanation." *Studies in History and Philosophy of Science, Part A* 43(2): 326–332.

Khalifa, Kareem. 2012. Inaugurating Understanding or Repackaging Explanation? *Philosophy of Science* 79(1): 15–37.

———. 2017. *Understanding, Explanation, and Scientific Knowledge*. Cambridge: Cambridge University Press.

Knuuttila, Tarja. 2011. "Modelling and Representing: An Artefactual Approach to Model-based Representation." *Studies in History and Philosophy of Science, Part A* 42(2): 262–271.

Knuuttila, Tarja, and Mieke Boon. 2011. "How Do Models Give Us Knowledge? The Case of Carnot's Ideal Heat Engine." *European Journal for Philosophy of Science* 1: 309–334.

Knuuttila, Tarja, and Natalia Carrillo. 2022. "Holistic Idealization: An Artifactual Standpoint." *Studies in History and Philosophy of Science Part A* 91(C): 49–59.

Knuuttila, Tarja, and Rami Koskinen. 2020. "Synthetic Fictions: Turning Imagined Biological Systems into Concrete Ones." *Synthese* 198(9): 8233–8250.

Knuuttila, Tarja, and Mary S. Morgan. 2019. "Deidealization: No Easy Reversals." *Philosophy of Science* 86(4): 641–661.

Koskinen, Rami. 2017. "Synthetic Biology and the Search for Alternative Genetic Systems: Taking How-Possibly Models Seriously." *European Journal for Philosophy of Science* 7(3): 493–506.

Kuorikoski, Jaakko, Aki Lehtinen, and Caterina Marchionni. 2010. "Economic Modelling as Robustness Analysis." *British Journal for the Philosophy of Science* 61(3): 541–567.

Kuorikoski, Jaakko, and Petri Ylikoski. 2015. "External Representations and Scientific Understanding." *Synthese* 192(12): 3817–3837.

Kvanvig, Jonathan L. 2003. *The Value of Knowledge and the Pursuit of Understanding*. Cambridge: Cambridge University Press.

Lawler, Insa. 2018. "Understanding Why, Knowing Why, and Cognitive Achievements." *Synthese* 196(11): 4583–4603.

———. 2021. "Scientific Understanding and Felicitous Legitimate Falsehoods." *Synthese* 198(7): 6859–6887.

Lawler, Insa, and Emily Sullivan. 2021. "Model Explanation versus Model-induced Explanation." *Foundations of Science* 26(4): 1049–1074.

Lipton, Peter. 2004. *Inference to the Best Explanation*. 2nd edition. New York: Routledge.

———. 2009. "Understanding without Explanation." In *Scientific Understanding: Philosophical Perspectives*, edited by Henk de Regt, Leonelli Sabine, and Kai Eigner, 43–63. Pittsburgh: University of Pittsburgh,

Mäki, Uskali. 2005. "Models Are Experiments, Experiments Are Models." *Journal of Economic Methodology* 12(2): 303–315.

Marchionni, Caterina. 2017. "What Is the Problem with Model-based Explanation in Economics?" *Disputatio* 9(47): 603–630.

Meskhidze, Helen. 2023. "Can Machine Learning Provide Understanding? How Cosmologists Use Machine Learning to Understand Observations of the Universe." *Erkenntnis* 88(5): 1895–1909.
Mizrahi, Moti. 2012. "Idealizations and Scientific Understanding." *Philosophical Studies* 160(2): 237–252.
Morris, Kevin. 2012. "A Defense of Lucky Understanding." *The British Journal for the Philosophy of Science* 63(2): 357–371.
Mössner, Nicola. 2018. *Visual Representations in Science - Concept and Epistemology*. London and New York: Routledge.
Parker, Wendy. 2006. "Understanding Pluralism in Climate Modeling." *Foundations of Science* 11(4): 349–368.
Pincock, Christopher. 2014. "How to Avoid Inconsistent Idealizations." *Synthese* 191: 2957–2972.
———. 2021. "A Defense of Truth as a Necessary Condition on Scientific Explanation." *Erkenntnis* 88 (2):621–640.
Potochnik, Angela. 2017. *Idealization and the Aims of Science*. Chicago: University of Chicago Press.
Pritchard, Duncan. 2008. "Knowing the Answer, Understanding and Epistemic Value." *Grazer Philosophische Studien* 77(1): 325–339.
———. 2010. "Achievements, Luck, and Value." *Think* 9(25): 19–30.
———. 2014. "Knowledge and Understanding." In *Virtue Epistemology Naturalized*, edited by Abrol Fairweather, 315–327. Synthese Library 366. Cham: Springer.
Rice, Collin. 2016. "Factive Scientific Understanding without Accurate Representation." *Biology & Philosophy* 31(1): 81–102.
———. 2018. "Idealized Models, Holistic Distortions, and Universality." *Synthese* 195(6): 2795–2819.
———. 2019a. "Models Don't Decompose That Way: A Holistic View of Idealized Models." *The British Journal for the Philosophy of Science* 70(1): 179–208.
———. 2019b. "Understanding Realism." *Synthese* 198(5): 4097–4121.
Rohwer, Yasha, and Collin Rice. 2013. "Hypothetical Pattern Idealization and Explanatory Models." *Philosophy of Science* 80(3): 334–355.
———. 2016. "How Are Models and Explanations Related?" *Erkenntnis* 81: 1127–1148.
Ross, Lewis. 2023. "The Truth about Better Understanding?" *Erkenntnis* 88(2): 747–770.
Scriven, Michael. 1962. "Explanations, Predictions, and Laws." In *Scientific Explanation, Space, and Time. Minnesota Studies in the Philosophy of Science: Vol. 3*, edited by Herbert Feigl, and Grover Maxwell, 170–230. Mineapolis, MN: University of Minnesota Press.
Shech, Elay. 2023. *Idealizations in Physics*. Cambridge: Cambridge University Press.
Skow, Bradford. 2017. "Against Understanding (As a Condition on Explanation)." In *Making Sense of the World: New Essays on the Philosophy of Understanding*, edited by Stephen R. Grimm. New York: Oxford University Press.
Sliwa, Paulina. 2015. "Understanding and Knowing." *Proceedings of the Aristotelian Society* 115(1): 57–74.
Strevens, Michael. 2008. *Depth. An Account of Scientific Explanation*. Cambridge, MA: Harvard University Press.
———. 2013. "No Understanding without Explanation." *Studies in History and Philosophy of Science* 44(3): 510–515.
———. 2017. "How Idealizations Provide Understanding." In *Explaining Understanding: New Perspectives from Epistemology and Philosophy of Science*, edited by Stephen R. Grimm, Christoph Baumberger, and Sabine Ammon, 37–48. London: Routledge.
Suárez, Mauricio. 2002. "An Inferential Conception of Scientific Representation." *Philosophy of Science* 71(5): 767–779.
Sullivan, Emily. 2018. "Understanding: Not Know-How." *Philosophical Studies* 175(1): 221–240.
———. 2022a. "Inductive Risk, Understanding, and Opaque Machine Learning Models." *Philosophy of Science* 89(5): 1065–1074.
———. 2022b. "Understanding from Machine Learning Models." *British Journal for the Philosophy of Science* 73(1): 109–133.
Trout, J.D. 2002. "Scientific Explanation and the Sense of Understanding." *Philosophy of Science* 69: 212–233.
———. 2007. "The Psychology of Scientific Explanation." *Philosophy Compass* 2(3): 564–591.

van Riel, Raphael. 2017. "What Is the Problem of Explanation and Modeling?" *Acta Analytica* 32(3): 263–275.
Verreault-Julien, Philippe. 2017. "Non-causal Understanding with Economic Models: The Case of General Equilibrium." *Journal of Economic Methodology* 24(3): 297–317.
———. 2019. "Understanding Does Not Depend on (Causal) Explanation." *European Journal for Philosophy of Science* 9(2): 18.
Wayne, Andrew. 2011. "Expanding the Scope of Explanatory Idealization." *Philosophy of Science* 78(5): 830–841.
Weisberg, Michael. 2006. "Robustness Analysis." *Philosophy of Science* 73(5): 730–742.
———. 2007. "Three Kinds of Idealization." *The Journal of Philosophy* 104(12): 639–659.
Werndl, Charlotte, and Katie Steele. 2016. "The Diversity of Model Tuning Practices in Climate Science." *Philosophy of Science* 83(5): 1133–1144.
Wilkenfeld, Daniel. 2013. "Understanding as Representation Manipulability." *Synthese* 190: 997–1016.
Woodward, James. 2003. *Making Things Happen: A Theory of Causal Explanation*. New York: Oxford University Press.
———. 2006. "Some Varieties of Robustness." *Journal of Economic Methodology* 13(2): 219–240.
Ylikoski, Petri, and N. Emrah Aydinonat. 2014. "Understanding with Theoretical Models." *Journal of Economic Methodology* 21(1): 19–36.
Zagzebski, Linda. 2008. *On Epistemology*. Belmont, CA: Wadsworth.
———. 2019. "Toward a Theory of Understanding." In *Varieties of Understanding*, edited by Stephen R. Grimm, 123–136. New York: Oxford University Press.

23
MODALITIES IN MODELING

Ylwa Sjölin Wirling and Till Grüne-Yanoff

1. Introduction

Several scientific modeling practices have an important *modal* aspect. In the most straightforward of cases, scientists explicitly state either the aim or the results of certain modeling practices in modal terms, involving reference to e.g. possible causes, dispositional properties, or counterfactual histories. However, there is also a wide range of other cases, where philosophers of science have interpreted the results of modeling practices in modal terms, sometimes because it is difficult to make sense of the epistemic contribution of these practices while understanding the models as accurate representations of actual target systems.

This chapter concerns such *modal modeling* practices. We first give some illustrative examples of modal modeling and indicate the extant philosophy of science literature that has identified some of these practices. We then draw on the modal epistemology literature to distinguish different kinds of modality and show how these are relevant for modal modeling. This is followed by our discussion of three distinct but related sets of issues that modal modeling raises: first, what constitutes methodologically sound modal modeling; second, *under what conditions* and *in virtue of what* models are reliable tools for making justified modal claims; and third, what specific roles models can or should play in such justifications. We conclude by highlighting some lacunae in the literature where further work is needed.

2. Modal modeling practices in the sciences

Scientists in a wide range of disciplines employ models to explore and justify modal claims, often as part of types of modeling that are discussed in other chapters of this book, including how-possibly models, minimal and toy models, and exploratory modeling.[1] We call such practices "modal modeling." The modal claims thus explored and supported come in various forms. They can, for instance, concern possible developments or possible causes, what would have happened under counterfactual (i.e. non-actual) circumstances, or the potential properties of certain systems.[2] This goal of obtaining modal information might be explicitly articulated by the researchers who use the model, or it may be that the model

DOI: 10.4324/9781003205647-28

in question, in the opinion of philosophers of science, is best reconstructed as providing modal information.

Modal modeling contrasts with modeling that provides information about what is, was, or will actually be the case. To be clear, some possibilities explored in modal modeling might be realized in the future or are realized already but unbeknownst to the modeler. The point is that modal models are not intended to provide information about what is actually the case. Now, that is not to say that researchers studying modal modeling are not interested in the real world. On the contrary, they often look for modal information as a means to further investigate, explain or better understand aspects of the real world. Modal modeling is characterized by providing modal information as an immediate result, regardless of whether or not these outcomes are then used as a springboard to another epistemic target.

Modal modeling occurs in several scientific contexts, for a number of different purposes. The clearest examples of modal modeling are modeling methods related to *how-possibly explanations*. Philosophers disagree on how to best characterize how-possibly practices, but most parties admit that (i) models play a crucial role in supporting how-possibly explanations, and (ii) proposing a how-possibly explanation involves making some kind of modal statement. In short, a how-possibly explanation makes a possibility claim based on a model result (see e.g. Bokulich 2014; Grüne-Yanoff 2009; 2013; Reutlinger, Hangleiter, and Hartmann 2018; Verreault-Julien 2019; Weisberg 2013, chap. 7). For example, the Hawk-Dove model supports a how-possibly explanation of the restraint phenomenon in fights between members of the same species. The model is used "to test whether it is possible even in theory for individual selection to account for 'limited war' behavior" (Maynard Smith and Price 1973, 15). Such how-possibly explanations might serve a number of different purposes, for example, providing a menu of possible explanations (Ylikoski and Aydinonat 2014) or refuting claims of necessity or impossibility (Grüne-Yanoff 2013). Furthermore, there is a growing and unresolved controversy about the contribution of how-possibly explanations to understanding (for a skeptical view, see Khalifa 2017; for a defense, see Reutlinger, Hangleiter, and Hartmann 2018; Verreault-Julien 2019).

Exploratory models can also be modal. Gelfert (2016) lists several purposes for exploratory modeling, some of which arguably involve modal claims, including providing how-possibly explanations, but also *proof-of-principle* demonstrations. Massimi (2019, footnote 1) adds two further items to this list of functions for exploratory models, namely that modal models can "provide knowledge of causal possibilities and provide knowledge of objective possibilities for hypothetical entities". Such exploratory modeling can also provide "observation-seeking explanations" (Sugden 2011), i.e. representations of possible properties or possible explanations for understanding such phenomena when they become real.

Many examples of modal modeling relate explicitly to possibility or counterfactual scenarios, but not all. Consider Nguyen's (2020) claim that the highly simplified toy models common in biology and economics can at least sometimes support claims that attribute properties—e.g. *capacities* or *susceptibilities*—to actual targets. For instance, in Akerlof's (1970) "market for lemons" model, asymmetric information distribution between traders prevents car transactions from taking place even if, at a given price, there are sellers ready to sell their car and buyers ready to buy it. According to Nguyen, the model supports claims to the effect that for specific, actual markets, asymmetric information *increases their susceptibility to fail* to reach effective Pareto equilibrium. On this view, Akerlof's model is an example of modal modeling, as it assigns "responsiveness"—a dispositional and hence modal property—to a particular system.

Most of the models mentioned above are abstract and mathematical in nature, but there are also examples of concrete modal models. For example, in synthetic biology, non-real possibilities are explored using concrete models to represent minimal cells, alternative genetic systems, and genetic networks, although in practice such goals may only be partially (or even not at all) achievable (Knuuttila and Koskinen 2021; Knuuttila and Loettgers 2022; Koskinen 2017).

3. Kinds of modalities

For anyone interested in modal modeling, it is vital to keep in mind that philosophers distinguish between different kinds of modality. Of particular note is the distinction between *epistemic* and *objective* modality. An epistemic modal claim is relative to a body of epistemically privileged (e.g. known, justified, evidenced) propositions. For instance, on one standard definition, to say that *p* is epistemically possible for us is roughly to say that we cannot, given *what we know*, rule out that *p* is true (see e.g. Edgington 2004, 6; Chalmers 2011, 60–61; Vetter 2015, 216; Weatherson and Egan 2011, 1).[3] Epistemic modal claims express something about one's epistemic situation. In contrast, a natural way to think about objective (sometimes 'circumstantial') modality is as expressing something about the world, independently of our epistemic situation. For instance, the notion of objective possibility perhaps makes the best sense in light of the assumption that many things are only *contingently* the way they are. That is, the world could have been different from how it is, and there is more than one way the world can be in the future, even if there is just one way it *will* be.

Many philosophers hold that objective modality comes in several different *flavors*, to borrow a term from linguistics (though some are skeptical of this, see e.g. Norton, 2022). That is, we can distinguish between e.g. metaphysical, physical, biological, economic, practical, and technological objective possibilities (e.g. Kment 2021; Mallozzi, Vaidya, and Wallner 2021; Williamson 2016). What distinguishes these are the facts that restrict or determine what is possible. These might be laws—for instance, some think that *p* is physically possible if *p* is compatible with the laws of physics. In contrast, being biologically or technologically possible requires compatibility with quite different, and arguably more demanding, sets of facts. But exactly what makes an objective modal claim true, even within the various subcategories, is subject to extensive debate among metaphysicians, and we will not get into that issue here. The important thing, for current purposes, is that the truth of an objective modal claim is independent of humans' epistemic situation. Note that although the specific constraints imposed by notions like 'technologically possible' or 'practically possible' depend at least partly on human interest and knowledge, these are nonetheless notions of objective possibility: whether *p* is possible in the relevant sense—once the sense is fixed—depends not on human interests or knowledge, but on whether *p* is a way the world could be *given* the facts that restrict the relevant modal space.

In modal modeling practices, scientists relate both epistemic modality and a variety of objective modality notions. These notions are therefore relevant to a philosophical understanding of scientific modeling practices and their modal dimension. In particular, philosophical analyses of modal modeling practices need to consider just what kind of modal notion is at issue in any given case. Disregarding this can lead to disputes in which philosophers talk past one another. For example, Sjölin Wirling and Grüne-Yanoff (forthcoming) argue that this is what underwrites the disagreement between philosophers of science on whether how-possibly explanations are just steps on the way to a how-actually explanation, and thus should be subsumed under one's preferred account of explanation, or whether they are their own kind of explanation, the understanding of which requires additional conceptual resources. Arguably, both sides describe practices that have a legitimate claim to being characterized as explaining how-possibly, but they are focusing on different types of cases. In particular, the types of cases that drive the *sui generis* camp plausibly target objective possibilities, whereas the other camp tends to put forward cases where the relevant possibility is epistemic.

If a how-possibly explanation of a phenomenon *X* is supposed to provide an *epistemically* possible cause of *X*, it *is* reasonable to think that how-possibly explanations are just stages toward a how-actually explanation and that its epistemic contribution can be subsumed under whatever one's favorite account of explanation is. Alisa Bokulich's description of how-possibly explanations of the tiger bush is an excellent example of this. Scientists do not know what actually explains this phenomenon, where vegetation in semi-arid areas grows in strips separated by barren land, thus creating a pattern reminiscent of a tiger's fur. Nevertheless, they construct models that are supposed to provide possible explanations. These "how-possibly explanations are explanations that, though not known to be the case, do not conflict with known facts" (Bokulich 2014, 334). That is, they are *epistemically* possible. As more empirical evidence is gathered, however, some of these how-possibly explanations will be culled, that is to say, scientists will rule out this or that mechanism as *not* in fact responsible for producing the phenomenon. Therefore, there is movement on a spectrum toward a how-actually explanation. The cases put forward by the *sui generis* camp tend to be different. For instance, biochemists have famously synthesized so-called xDNA—a new, size-expanded geometry that seemingly retains the functions that natural DNA has in nature's genetic system (Knuuttila and Koskinen 2021). After having explored and studied such xDNA, some researchers have concluded that such alternative systems *could have* existed in nature and that the evolution of life could have been based on them, either instead of or in addition to DNA. However, these scientists know full well that, *in fact* the evolution of life was based on RNA/DNA. That is, such a how-possibly explanation of life, supported by this research, *does* "conflict with known facts," so the relevant possibility here cannot be epistemic. Rather, it provides a non-actual but allegedly *objectively* possible explanation of how life could have developed. The epistemic contribution of such a how-possibly explanation cannot easily be explained as being steps toward the how-actually explanation, and so it indeed seems *sui generis* and in need of separate methodological evaluation. In short, the distinction between epistemic and objective modality shows that the two camps are not necessarily in conflict after all.

Distinguishing between different modalities is also important for the *evaluation* of modal modeling practices. This is because different modalities are subject to different epistemologies. Differently put, what is required for the justification of a particular modal claim depends on what kind of modal claim it is. First, inquiry into objective possibilities—at least insofar as one wants to investigate a *range* of possibilities rather than settle the truth of a single preconceived possibility claim—will require bracketing some of one's knowledge. In particular, since many objective possibilities will be counterfactual, some contingent but actually obtaining facts need to be bracketed. In contrast, inquiry into what is epistemically possible for an agent might in principle involve considering *all* that agent's knowledge (relevant to the matter at hand). Second, with respect to objective possibility, what knowledge one *should* take into account depends on the flavor of objective possibility at issue.

Claims of e.g. economic and physical possibility presumably need very different kinds of justification. Third, claims of objective possibility require *positive* support—reasons that really speak to whether or not *p* is a way the world could objectively be. That current knowledge does not indicate that not-*p* is not sufficient for that purpose, because knowledge may be too scarce. In contrast, one can be justified in taking *p* to be epistemically possible even if one knows very little if anything relevant to whether *p*. At least, this is so on standard definitions of epistemic possibility according to which *p* is possible when *p* is *compatible with* (Vetter 2015, 215) or *not ruled out by* (Chalmers 2011, 61) the relevant body of knowledge, or when *not-p is not part of that body of knowledge* (Weatherson and Egan 2011). E.g. one can be perfectly justified in claiming that *p* is epistemically possible even though scientific knowledge relevant to *p* is very scarce—it is just a matter of judging the relation between *p* and a body of knowledge, whatever it contains. Thus, epistemic possibility claims, on many existing definitions of epistemic possibility, are subject to what one might call *justification from ignorance*.

4. Methodological problems of modal modeling

Modal modeling practices are widespread in the sciences and enjoy a number of distinct uses. However, it isn't obvious that all of these uses are legitimate or well-justified. Our discussion in the previous section suggests that modal modeling aimed at both epistemic and objective possibilities might be methodologically problematic, as it might either lack any substantial justification or might not be appropriately constrained for the purpose at hand.

The first problem arises when modelers exploit modal modeling in order to give their otherwise vacuous modeling results the sheen of a justified exercise. Let's call this the *apologetic function* of modal modeling. In other words, if a modeler fails to justify their model results with reference to more demanding model functions, e.g. as accurate predictions or genuine explanations, they might almost always revert to the claim that their model at least represents a possibility. The apologetic function might partly arise out of confusion about semantics and evidential standards for the relevant possibility claims. As long as those remain unspecified, anything might identify *some* possibility—such purported justifications would be pointless, amounting to little more than an apology for spurious modeling exercises. However, such methodological problems can be encountered by stressing that all possibility claims have a truth value and that at least in principle evidential standards can be specified for them (Grüne-Yanoff and Verreault-Julien 2021).

The scenario of the apologetic function of modal modeling can, however, become complicated due to *justification from ignorance*. Under standard definitions of epistemic possibility, a claim *p* is epistemically possible for agent A if *p* is not excluded by A's knowledge *K*. From this, it directly follows that the set of epistemically possible claims increases as *K* decreases. In effect, A's ignorance offers them additional opportunities to justify modal claims. While such a justification in itself need not be problematic, it offers an opportunity for those working with highly speculative models (where *K* is small or empty) to always claim that their otherwise vacuous modeling exercise actually performs a justifying function. In particular, such a methodological flaw cannot be rectified by simply clarifying semantics and evidential standards: justification from ignorance proceeds with unambiguous semantics and a clear evidential standard according to standard definitions of epistemic possibility. Instead, this version of the apologetic function can only be properly addressed by revising the definition of epistemic possibility itself.

Such revision is difficult, though. Scientists who model e.g. epistemic how-possibly explanations will plausibly have something more demanding in mind, i.e. scientific practice works with stronger notion(s) of epistemic possibility, on which *p* is epistemically possible just in case the truth of *p* is in some sense *supported by* a given body of epistemically prioritized (e.g. known, believed, justified) propositions (compare Przyjemski 2017). On the other hand, modal modeling—including the modeling of epistemic possibilities—is often a crucial part of *exploratory* science, where existing knowledge is scarce or put into question. Such practices arguably work with a less demanding, weak notion of epistemic possibility—but even the epistemic possibility space(s) of exploratory practices are arguably constrained in some ways that go beyond the standard weak formulation. It is plausible that science requires both weaker and stronger notions of epistemic possibility. But the notions of epistemic possibility currently available (as well as their formal axiomatization in epistemic modal logic) are arguably misaligned with these practices and the conceptual needs that arise from them.

Another methodological problem arises when a properly justified modal claim is irrelevant to the purpose at hand because constraints other than those relevant to the purpose were applied in the modeling process. This is of particular relevance for objective possibility claims, which are often distinguished according to such differing constraints—with logical, mathematical, physical, biological, or economic possibilities being examples. An illustrative example of this is the criticism that general equilibrium models are overly "formalistic" (Blaug 2003) or that it is just an empty piece of mathematics (Rosenberg 1992). One way to understand these worries is that the models provide true objective possibility claims, but not of a relevant modality. In particular, critics claim that these models show general equilibrium to be mathematically possible, but that such a mathematical possibility is irrelevant to the epistemic goals of economics. Other authors have argued against this, stating that these models are not an exercise in pure mathematics because many assumptions had an "economic interpretation," i.e. they were consistent with stylized economic facts and background theory (Hands 2016). Both sides agree that the models establish true claims of objective possibility. However, for some, it is a 'mere' mathematical possibility, whereas for others it is a stronger modal claim, for example an economic possibility. This suggests two potential points of contention concerning the epistemic value of general equilibrium modeling. One is about which sort of modal claims economics should seek to establish. Can claims of mathematical possibility ever be relevant for economics? Another is about the sort of claims general equilibrium models support: do the models establish mathematical or economic possibility? (Grüne-Yanoff and Verreault-Julien 2021).

To conclude, the analysis of modal modeling practices raises some methodological problems. On the one hand, given standard notions of epistemic possibility, assertions that models justify modal claims are often too facile. On the other hand, without proper regard for the purpose at hand, even legitimately justified modal conclusions might be irrelevant if the modeling process is not properly aligned with those purposes. Such methodological problems show the need to provide tools for reliable normative assessment of modal modeling.

5. The epistemic question for modal modeling

The fact that models are used to support modal claims raises what Sjölin Wirling and Grüne-Yanoff (2021) call the *epistemic question for modal modeling* (see also Tan 2022). This is really a two-part question, asking (i) *under what conditions* models are reliable tools for making justified modal claims, and (ii) *in virtue of what* are models, under the conditions specified in (i), reliable tools for modal justification. The first part of the question can be sufficiently answered by giving a descriptive account, whereas the second part asks for a deeper explanation of *why* models are reliable tools for modal justification under those conditions, or with those characteristics, specified in the descriptive account. It should be noted—especially in light of Section 3 above—that one should not expect just *one* answer to the epistemic question. As modeling may involve different kinds of modalities (either epistemic or some kind of objective possibility), and different modalities place different epistemic constraints on the modeling practices in question, the epistemic question will likely require different answers depending on which type of modality is at issue. The conditions under which an epistemic modal claim is plausibly supported by a model will presumably differ from the conditions under which an objective modal claim is plausibly supported, and one can also expect variation between different flavors of objective modality.

In approaching the epistemic question, philosophers of science can gain insight from considering work in the epistemology of modality: the branch of philosophy that focuses on knowledge and justification of modal claims. In particular, that field has seen a strong recent trend toward *modal empiricism*—the view that our modal knowledge derives from experience and/or (non-modal) empirical knowledge, rather than being *a priori*—which makes its findings more amenable to the philosophy of science. Some examples of empiricist modal epistemologies are Bueno and Shalkowski (2014), Dohrn (2021), Fischer (2017), Roca-Royes (2017), Ruyant (2020), Strohminger (2015), Vetter (forthcoming), Williamson (2007). In fact, among the few but notable existing attempts to address the epistemic question for modal modeling (especially its first part), there are already several interesting parallels to modal epistemology (Sjölin Wirling and Grüne-Yanoff 2021).

To give just one example, according to Michela Massimi (2019), some *exploratory* models—such as models of hypothetical particles in physics and Maxwell's honeycomb model—can give scientists knowledge of what is objectively possible because they involve what she calls "physical conceivability." To physically conceive of *p* is to manage to imagine *p* while holding fixed what one knows about the laws of nature. The answer to part (i) of the epistemic question given here is: model *m* supports the claim that *p* is possible (presumably in the sense of *physically* possible) if *m* prompts scientists to successfully physically conceive of *p*. Appealing to conceivings or imaginings as a way of justifying modal claims is one of the most venerable strategies in modal epistemology (Kung 2010; Yablo 1993). Moreover, the need to somehow *constrain* imagination in order for it to provide justification since it is clear that one can easily imagine impossible things is widely recognized in that literature (Kind and Kung 2016; Mallozzi, Vaidya, and Wallner 2021). One way to constrain it is to do as Massimi does: specify that the imagining must be compatible with (knowledge of) the general principles that constrain the relevant possibility space. In such cases, the justification is provided not by the imagination but by the background knowledge that constrains it. More generally, and in the same vein, modal epistemologists have suggested that justification for particular possibility claims is downstream from justification for *theories* (Bueno and Shalkowski 2014; Fisher 2017).

6. Roles for models in modal justification

That the justificatory strategies underlying modal modeling are continuous with the strategies people rely on in modal thinking in general is, to some extent, just what one should expect. However, since these strategies do not refer to scientific models, as described in the modal epistemological literature, this can be taken as an indication that models are a transient and dispensable part of modal thinking in science. But this does not correspond to scientific practice, where models often appear to be indispensable tools for obtaining certain (but not all) modal knowledge in science. A question that philosophers interested in modal modeling should therefore ask themselves is: what role do models play in modal modeling, and why are they important for drawing modal inferences?

Although there is not much work explicitly addressing this issue, here are two hypotheses that seem to be good starting points for thinking about it. First, given the great diversity among modal modeling practices that has been indicated throughout this text—in terms of disciplines, the nature of the models, the types of modal claims, and the justificatory strategies apparently underlying them—it seems reasonable to expect pluralism with respect to the roles that models can play in modal justification. Second, it would seem wise to draw on what the existing philosophical literature says about the role of models in (non-modal) reasoning more generally.

For instance, it has been suggested that models can be used to probe scientific theories, e.g. by conceptualizing phenomena in ways that make theoretical principles applicable to them (Cartwright 1997) or mediate between theories and the states of the world that the theory applies to (Morgan and Morrison 1999). Thus, in modeling contexts where empirically well-grounded theory or knowledge of the relevant laws are available—as must be the case if Massimi's physical conceivability strategy is to be employed—one could expect the role of the model to be a way to tease out more particular modal implications of the theory or the laws (compare Fischer 2017). Indeed, in Massimi's physical conceivability account, the model appears to provide a 'testing ground', a concrete situation in which some prospective possibility p is true and in which the relevant laws are implemented together, in order to check whether p is compatible with the laws or the theoretical principles.

However, in many contexts where modal modeling occurs, no established background theory can provide the relevant justification. This also suggests that there must be other roles for models to play in modal justification. Again, a more general philosophy of modeling can point the way here. Many philosophers of science have suggested that models perform their epistemic function insofar as they are relevantly *similar* to the target system(s) they afford knowledge of (e.g. Weisberg 2013; Giere 2010). In this view, scientists are directly comparing phenomena in the world with models for similarity, and insofar as the model is similar to its target phenomenon, it can be relied on for conclusions about the target. The role of the model here is to function as a surrogate system that enables study, manipulation, and comparison with the specific target individual. Presumably, something analogous could be going on in modal modeling. This is especially interesting in light of how the idea that one can draw modal conclusions about one individual on the basis of what one knows to be the case with another relevantly similar individual has been the subject of much attention in modal epistemology of late (Roca-Royes 2017; Dohrn 2019; Hawke 2010; Schoonen 2022). While the focus of similarity-based accounts of modeling is typically not on *modal modeling*, there is no principled obstacle to extending it to account for some such cases. Sjölin Wirling (2022) suggests as much. The idea is that if one wants to know if a target system T can possibly be F, one can find out by constructing a model system M, trying to realize F in M, and then compare M with T, to see if they are substantially similar in relevant ways. If this is the case, there seems to be reason to believe that T can possibly be F. This would be especially useful to *extend* one's epistemic reach in cases where there are no *actual* systems that are relevantly similar to T and that are known to be F, or in which F can be realized. In this kind of similarity-based reasoning, the role of the model is to allow for surrogative study and informative comparison with targets. In particular, they stand in for real, relevant individuals that one could have used for comparison had they been available.

Finally, as was already noted, it is common among modal epistemologists to ascribe *some* role to the imagination. Some take it to be an independent source of justification for modal claims (Yablo 1993; Kung 2010), whereas others allow it merely as a useful tool in assessing what really matters for justification, such as compatibility with background knowledge (e.g. Fischer 2017). Interesting views of the former variety take some modes of imagination, especially those that have an "involuntary" character, to be reliable means to true beliefs because they by design "develop in a reality-oriented way" (Williamson 2016, 118; see also Balcerak Jackson 2018; Byrne 2005). This is, like much else in this literature, controversial, but if it is correct, then one might expect modal modelers in science also to sometimes rely on the imagination in this sense. Is there a role for models to play in such imaginative reasoning to modal conclusions? Here is just one way in which this thought might be explored. Consider how Nersessian (1992) connects thought experiments in science with what she calls "mental models". In thought experimenting, which is a form of imagining, these mental models serve as means by which thought experimenters represent their scenarios. The imagining agent uses the mental model to represent the scenario that she is experimenting on, so to speak. She draws together the elements of the scenario from various forms of chiefly non-propositional (e.g. sensory) knowledge and seeks to create a coherent whole—that is the mental model. Perhaps scientific models more generally—not just "mental" models—can play something akin to this role of recording an imagined scenario, and thereby make it available to the imaginer for further probing and imagining, in order to draw modal conclusions. This should be especially useful in complex, temporally extended uses of the imagination, which might proceed in several stages.

7. Lacunae in the literature

Once modal modeling is on the table as an interesting type of scientific practice, several philosophical questions arise. In the three sections immediately preceding this one, we have presented some issues that face philosophers of science who take an interest in modal modeling, and then either reviewed existing work in response to them or sketched the shape that attempts to handle them might take. In this last section, we bring up some further interesting challenges that model modeling presents philosophers with, but which have not yet been addressed or acknowledged in the literature. Hopefully, this can stimulate further work on these questions.

First, existing responses to the epistemic question for modal modeling tend to focus only on its first part. That is, while there are some attempts to outline the various conditions under which models can be good guides to modal truth, there has so far been little attention paid to the follow-up question of *why* or *in virtue of what* such-and-such models should be expected to be good guides to modal truth. Some philosophers of science may consider this a hopeless or unnecessary question. But it relates very closely to a number of *normative* questions concerning justificatory strategies and roles for models in modal reasoning, e.g. about whether scientists are indeed right to trust a particular kind of modeling practice to justify a particular type of modal claims, what type of justificatory strategy (and what type of role for the model) is appropriate for probing this or that type of scientifically interesting modal question, and so on.

Second and very much relatedly, relatively little has been said in response to the question of how to understand what grounds or constrains the modal spaces *relevant to scientific inquiry*. There are some relatively detailed accounts of the nature of physical/nomological possibility (e.g. Wilson 2020), and some attempts to develop a single empiricist-friendly[4] notion of possibility (Ismael 2017; Norton, 2022), but there is very little discussion of what constrains or determines other objective modal notions relevant to, e.g. biology or economics. Yet these questions of what it *is* to be possible in such-and-such a sense are crucial to any attempt to evaluate whether a particular modeling practice or justificatory strategy is plausibly reliable (Sjölin Wirling 2021). Moreover, and also noted in Section 3, there is a need for more thorough work on the notion(s) of *epistemic* possibility relevant to various scientific practices.

Third and finally, one may also suspect that the philosophy of modal modeling faces something of a delineation problem. Given how modal modeling is currently defined, its scope is rather broad, and one might worry that a lot of perfectly "ordinary" modeling about the actual world will now suddenly count as modal modeling. For instance, if toy models worked the way Nguyen (2020) proposes that they do, namely by ascribing dispositional properties to real-world targets, they would not seem to be examples of 'modal modeling' in a particularly interesting sense of that term. The issue also arises in light of a reasonable ambition among philosophers of science to locate the targets of modal models in the actual world, rather than to see them as merely possible entities or locate them in possible worlds (Verreault-Julien, 2022). The worry is that if modal modeling is too liberally defined, it will not be a very novel or interesting way of classifying models, and will not allow for anything like a unified treatment. While pluralism as such in the treatment of modal modeling might be unproblematic and to be expected, there is also reason to think that this worry is worth taking seriously. There may well be interesting distinctions to be drawn here between modal modeling *proper* and modal modeling in a wider sense, but it has not to date been addressed in the literature.

8. Concluding remarks

Scientific models are often used to infer and justify modal claims. This deserves more attention from philosophers of science, who until recently largely ignored modal modeling practices. For such an analysis, work done in modal epistemology and the philosophy of modality more broadly might be helpful, both for the delineation of different notions of possibility, as well as for accounting for various justificatory strategies. However, important issues specific to modal modeling remain, in particular regarding questions on how to avoid apologetic modal modeling, concerning the normative bases of modal claim justification, and identifying the specific roles models might play in such justifications.

Acknowledgement

Ylwa Sjölin Wirling received funding from the Swedish Research Council grant no. 2019-00635. Till Grüne-Yanoff acknowledges funding from the Swedish Research Council, grant no. 2018-01353.

Notes

1 Note that not all instances of minimal/toy/exploratory modeling need be instances of, or understood as, modal modeling.
2 Some modeling practices that may also be said to be "modal" involve models of target systems that are *impossible* in some sense or other. This chapter will not say more on the issue, but we note that this relates to the role of "counterpossibles" in scientific reasoning more generally, a subject of some discussion in recent years (see e.g. Jenny 2018; Tan 2019; McLoone 2020).
3 In epistemic logic, all worlds that are logically compatible with what is known by some agent counts as epistemically possible for that agent.
4 "Empiricist" in the regular philosophy of science sense, not in the more liberal sense used in "modal empiricism" in the epistemology of modality to just demarcate a difference from rationalists who take modal knowledge to be *a priori*.

References

Akerlof, George A. 1970. "The Market for 'Lemons'.: Quality Uncertainty and the Market Mechanism" *The Quarterly Journal of Economics* 84(3): 488–500.

Balcerak Jackson, Magdalena. 2018. "Justification by Imagination." In *Perceptual Imagination and Perceptual memory*, edited by Fiona McPherson and Fabian Dorsch, 209–226. Oxford: Oxford University Press.

Blaug, Mark. 2003. "The Formalist Revolution of the 1950s." *Journal of the History of Economic Thought* 25(2): 145–156.

Bokulich, Alisa. 2014. "How the Tiger Bush Got Its Stripes: 'How Possibly' vs. 'How Actually' Model Explanations." *The Monist* 97(3): 321–338.

Bueno, Otavio and Scott Shalkowski. 2014. "Modalism and Theoretical Virtues: Toward an Epistemology of Modality." *Philosophical Studies* 172(3): 671–689.

Byrne, Ruth. 2005. *The Rational Imagination.* Cambridge: MIT Press.

Cartwright, Nancy. 1997. "Models: The Blueprints for Laws." *Philosophy of Science* 64: S292–S303.

Chalmers, David. 2011. "The Nature of Epistemic Space." In *Epistemic Possibility*, edited by Andy Egan and Brian Weatherson, 60–107. New York: Oxford University Press.

Dohrn, Daniel. 2019. "Modal Epistemology Made Concrete." *Philosophical Studies* 176(9): 2455–2475.

———. 2021. "A Humean Modal Epistemology." *Synthese* 199(1): 1701–1725.

Edgington, Dorothy. 2004. "Two Kinds of Possibility." *Supplement to the Proceedings of the Aristotelian Society* 78(1): 1–22.

Fischer, Bob. 2017. *Modal Justification via Theories.* Berlin, Heidelberg, New York: Springer International.

Gelfert, Axel. 2016. *How to Do Science with Models: A Philosophical Primer*. Berlin, Heidelberg, New York: Spring International.

Giere, Roland N. 2010. "An Agent-based Conception of Models and Scientific Representation." *Synthese* 172: 269–281.

Grüne-Yanoff, Till. 2009. "Learning from Minimal Economic Models." *Erkenntnis* 70(1): 81–99.

———. 2013. "Appraising Models Nonrepresentationally." *Philosophy of Science* 80(5): 850–861.

Grüne-Yanoff, Till and Philippe Verreault-Julien. 2021. "How-Possibly Explanations in Economics: Anything Goes?" *Journal of Economic Methodology* 28(1): 114–123.

Hands, Wade D. 2016. "Derivational Robustness, Credible Substitute Systems and Mathematical Economic Models: The Case of Stability Analysis in Walrasian General Equilibrium Theory." *European Journal for Philosophy of Science* 6(1): 31–53.

Hawke, Peter. 2010. "Van Inwagen's Modal Skepticism." *Philosophical Studies* 153(3): 351–364.

Ismael, Jenann. 2017. "An Empiricists Guide to Objective Modality." In *Metaphysics and the Philosophy of Science: New Essays*, edited by Matthew Slater and Zanja Yudell, 109–125. New York: Oxford University Press.

Jenny, Matthias. 2018. "Counterpossibles in Science: The Case of Relative Computability." *Noûs* 52(3): 530–560.

Khalifa, Kareem. 2017. *Understanding, Explanation, and Scientific Knowledge*. Cambridge: Cambridge University Press.
Kind, Amy and Peter Kung, eds. 2016. *Knowledge through Imagination*. Oxford: Oxford University Press.
Kment, Boris. 2021. "Varieties of Modality." *The Stanford Encyclopedia of Philosophy* (Spring 2021 Edition) edited by Edward N. Zalta, https://plato.stanford.edu/archives/spr2021/entries/modality-varieties/
Knuuttila, Tarja and Rami Koskinen. 2021. "Synthetic Fictions: Turning Imagined Biological Systems into Concrete Ones." *Synthese* 198: 8233–8250.
Knuuttila, Tarja and Andrea Loettgers. 2022. "(Un)Easily Possible Synthetic Biology." *Philosophy of Science*, *89*(5): 908–917.
Koskinen, Rami. 2017. "Synthetic Biology and the Search for Alternative Genetic Systems: Taking How-Possibly Models Seriously." *European Journal for the Philosophy of Science* 7(3): 493–506.
Kung, Peter. 2010. "Imagining as a Guide to Possibility." *Philosophy and Phenomenological Research* 81(3): 620–663.
Mallozzi, Antonella, Anand Vaidya and Michael Wallner. 2021. "The Epistemology of Modality." *The Stanford Encyclopedia of Philosophy* (Fall 2021 Edition), edited by Edward N. Zalta, https://plato.stanford.edu/archives/fall2021/entries/modality-epistemology/
Massimi, Michela. 2019. "Two kinds of exploratory models." *Philosophy of Science* 86(5): 869–881.
Maynard Smith, John and George R. Price. 1973. "The Logic of Animal Conflict." *Nature* 246(5427): 15–18.
McLoone, Brian. 2020. "Calculus and Counterpossibles in Science." *Synthese* 198: 12153–12174.
Morgan, Mary S. and Margaret Morrison. 1999. *Models as Mediators*. Cambridge: Cambridge University Press.
Nersessian, Nancy. 1992. "In the Theoretician's Laboratory: Thought Experimenting as Mental Modeling." *PSA: Proceedings of the Biennial Meeting of the Philosophy of Science Association 1992*, 2: 291–301.
Nguyen, James. 2020. "It's Not a Game: Accurate Representation with Toy Models." *British Journal for the Philosophy of Science* 71(3): 1013–1041.
Norton, John D. 2022. "How to Make Possibility Safe for Empiricists." In *Rethinking the Concept of Laws of Nature: Natural order in the Light of Contemporary Science*, edited by Yemima Ben-Menahem, 129–159. Berlin, Heidelberg, New York: Springer International.
Przyjemski, Katrina. 2017. "Strong Epistemic Possibility and Evidentiality." *Topoi* 36(1): 183–195.
Reutlinger, Alexander, Dominik Hangleiter, and Stephan Hartmann. 2018. "Understanding (with) Toy Models" *British Journal for the Philosophy of Science* 69(4): 1069–1099.
Roca-Royes, Sónia. 2017. "Similarity and Possibility: An Epistemology of de re possibility for Concrete Entities." In *Modal Epistemology After Rationalism*, edited by Bob Fischer and Felipe Leon, 221–245. Berlin, Heidelberg, New York: Springer International.
Rosenberg, Alexander. 1992. *Economics--Mathematical Politics or Science of Diminishing Returns?* Chicago: University of Chicago Press.
Ruyant, Quentin. 2020. "The Inductive Route towards Necessity." *Acta Analytica* 35(2): 147–163.
Schoonen, Tom. 2022. "Possibility, Relevant Similarity, and Structural Knowledge." *Synthese* 200: Article 39.
Sjölin Wirling, Ylwa. 2021. "Is Credibility a Guide to Possibility? A Challenge for Toy Models in Science." *Analysis* 81(3): 470–478.
———. 2022. "Extending Similarity-based Modal Epistemology with Models." *Ergo* 8(45): 570–594.
Sjölin Wirling, Ylwa and Till Grüne-Yanoff. Forthcoming. "Epistemic and Objective Possibility in Science." *British Journal for the Philosophy of Science*. doi: 10.1086/716925
Sjölin Wirling, Ylwa and Till Grüne-Yanoff. 2021. "The Epistemology of Modal Modeling." *Philosophy Compass* 16(10): e12775.
Strohminger, Margot. 2015. "Perceptual Knowledge of Nonactual Possibilities." *Philosophical Perspectives* 29(1): 363–375.
Sugden, Robert. 2011. "Explanations in Search of Observations." *Biology and Philosophy* 26: 717–736.

Tan, Peter. 2019. "Counterpossible Non-Vacuity in Scientific Practice." *Journal of Philosophy* 116(1): 32–60.

———. 2022. "Two Challenges Regarding Hypothetical Modeling." *Synthese* 200: Article 448.

Verreault-Julien, Philippe. 2019. "How Could Models Possibly Provide How-Possibly Explanations?" *Studies in History and Philosophy of Science Part A* 73: 22–33.

———. 2022. "Representing Non-Actual Targets?" *Philosophy of Science* 89 (5): 918–927.

Vetter, Barbara. 2015. *Potentiality*. Oxford: Oxford University Press.

———. Forthcoming. "An Agency-based Epistemology of Modality." In *The Epistemology of Modality and Philosophical Methodology*, edited by Duško Prelevic and Anand Vaidya. New York: Routledge.

Weatherson, Brian and Andy Egan. 2011. "Epistemic Modals and Epistemic Modality." In *Epistemic Modality*, edited by Andy Egan and Brian Weatherson, 1–18. New York: Oxford University Press.

Weisberg, Michael. 2013. *Simulation and Similiarity: Using Models to Understand the World*. Oxford: Oxford University Press.

Williamson, Timothy. 2007. *The Philosophy of Philosophy*. Malden: Blackwell Publishing.

———. 2016a. "Modal Science." *Canadian Journal of Philosophy* 46(4–5): 453–492.

———. 2016b. "Knowing by Imagining." In *Knowledge through Imagination*, edited by Amy Kind and Peter Kung, 113–123. Oxford: Oxford University Press.

Wilson, Alastair. 2020. *The Nature of Contingency*. Oxford: Oxford University Press.

Yablo, Stephen. 1993. Is conceivability a guide to possibility?. *Philosophy and Phenomenological Research*, 53(1), 1–42.

Ylikoski, Petri and Emrah N. Aydinonat. 2014. "Understanding with Theoretical Models." *Journal of Economic Methodology* 21(1): 19–36.

24
SCIENTIFIC MODELS AND THOUGHT EXPERIMENTS

Rawad El Skaf and Michael T. Stuart

1. Introduction

Thought experiments (TEs) and models are devices at the heart of modern science with a history of usage long predating their modern names. They are created, interpreted, reinterpreted, published in research, and used in pedagogy. It is possible to tell the whole story of science via either of them.

Recently, philosophers have been drawing attention to their similarities. El Skaf and Imbert (2013) argue that in some cases TEs and (computational) models could be treated as functionally, but not epistemically, substitutable. Arcangeli (2018) distinguishes between the different processes of mental simulation that play a role in TEs and computer simulations, which she understands as implemented models. Salis and Frigg (2020) argue that the same fictionalist epistemological framework can be applied to TEs and models insofar as they employ the same kind of imagination. Stuart (2022) also categorizes TEs and models together by putting them under the same consequentialist epistemological framework.

The above contributions highlight similarities and differences between TEs and models, but there is still much more to be said about this connection. Following Frigg and Hartmann (2020), the discussion in this entry is divided into three categories: ontology, semantics, and epistemology. In each category, the relevant work on TEs and models is summarized, pointing out cases where insights about one kind of device can be extended to the other. It will also turn out that a sharp separation between ontology, semantics, and epistemology can only be achieved with lots of gymnastics, seeing that each is informed by and builds on the others.

2. Ontology

2.1 Scientific models

What, exactly, are models? A popular option for discussing them is pluralism: i.e., models are not one single kind of thing (Callender and Cohen 2006; Suárez 2004; Swoyer 1991). So, what are the different kinds of models?

 DOI: 10.4324/9781003205647-29

Some models are material; they can be found in the world, not (just) in the mind. Some material models are scale models, e.g., a model of a ship in water. Some material models are used expressly because of the material similarities between model and target. However, others share almost no relevant material properties with their targets, like Watson and Crick's model of DNA, which was made of metal sheets, rods, and clamps, not nucleotides. In some cases, destroying some particular material construction would also destroy the model. For example, destroying a scale model of an airplane destroys that model. In other cases, destroying a specific material token would not destroy the model because the model is a *type*: destroying a particular fruit fly would not destroy drosophila as a model organism, though destroying *all* drosophila might.

Models can also be non-material. One important starting point when analyzing non-material models is to differentiate between model descriptions and model systems. It is the *model description* that we find in textbooks and papers, typically in the form of equations, text, or code. These descriptions define, specify, or constrain the *model system*. For example, there are simple population growth models in ecology that are given by the logistic equation. In such cases, the model system is the population whose growth is described by that equation. But what is such a population? More generally, what are model systems?

There are many possibilities. They might be Meinongian (or neo-Meinongian) objects, possible objects, abstract entities (Giere 1988), Platonic forms, set-theoretic structures (da Costa and French 2003), abstract cultural artifacts (Thomasson 2020), imagined concrete objects (Godfrey-Smith 2006), or entities that only exist inside a fiction. This last option, fictionalism, is now quite popular. It is really a family of different views, many of which are based on Walton's (1990) pretense theory of fiction. Briefly stated, the idea is that models involve model descriptions, which prescribe that certain model systems are to be imagined as described. This has typically been an anti-realist position, in that model systems only live in scientists' imaginations (Fine 1998; Frigg 2010). In any case, there are two variants of fictionalism that are importantly different with respect to ontology. The first commits itself only to model descriptions and denies the existence of model systems (Levy 2012; Toon 2012). The second commits itself ontologically to both model descriptions and model systems. Salis' "new fiction view" draws on both by reconceptualizing models as "complex objects constituted by model-descriptions and model-contents" (Salis 2021).

Finally, there is an "artifactualist" approach, according to which models are human-made tools that fulfill certain purposes (Knuuttila 2011; 2017; 2021; Sanches de Oliveira 2021; 2022; Parker 2020). On this kind of view, a model could be either abstract or concrete. What makes it the thing it is, is its purpose or function. The most radical version of this approach (Sanches de Oliveira 2022) denies that non-material models exist. This kind of artifactualist provides a unified deflationist answer to the ontology of models: all models are (just) material tools. Less radical artifactualists are open to non-material "representational modes," but remain committed to the materiality of "representational media" (Knuuttila 2011). This kind of artifactualism continues to portray models primarily as epistemic tools, but allows that those tools can be partially non-material. In either case, identifying models with tools only pushes back the ontological question, until we know what tools are.

2.2 *Scientific thought experiments*

One interesting difference between TEs and models is that ontological issues have not historically played much of a role in discussions of TEs. Given this, metaphysical views about

TEs tend to remain implicit. Another difference is that ontological views about TEs tend to be less pluralistic than views about models.

What are TEs? According to the argument view, pioneered by John D. Norton, TEs are just picturesque arguments (Norton 1991). While equating TEs with arguments is clearly an ontological claim, this move is not typically characterized ontologically. But it could be. We tend to think of arguments as being "made of" inferences and propositions. What are propositions? There is a long history of debate about this, with positions ranging from Fregean thoughts, senses of (declarative) sentences, predicated subjects, "pictures" of the world, sets of possible worlds, properties, and abstract mind-independent entities (King 2017). What are inferences? Norton does not want to say that these are mental actions (Norton 2021, 20). Instead, inferences seem to be something like a *rule* (when the argument is deductive) that describes a logical connection, or a transformation of propositions, or a *fact* (when the argument is inductive) that licenses an expansion of the domain of reference. Thus, for Norton, deductive TEs appear to be hylomorphic duos of form and content (Stuart 2020, El Skaf 2021), while inductive TEs can perhaps be reduced entirely to facts (Stuart 2020).

Another monolithic ontology is extractable from the "mental models account" of TEs (Miščević 1992; Nersessian 1992, 2007). The main thing to note here is that TEs are portrayed not as facts, rules, or propositions, but rather as a combination of mental states and processes, including mental actions. "Mental model" is a term of art taken from psychology and cognitive science, and it refers to a structure in the mind. Different accounts adopt different definitions of what mental models are, but they all share several common ideas: TEs have a narrative form that enables us to construct, and reason upon, mental models. Instead of focusing on the TE itself, the focus shifts to *reasoning through a TE*, which is a non-propositional *activity* aimed at building and manipulating mental models and "seeing" what happens in those models. The ontology of this account is the ontology of (mental) action, beliefs, knowledge, memory, imagination, and imagery.

A third option is to portray TEs as actual experiments (Sorensen 1992; Buzzoni 2008; Stuart 2016b). But what is an experiment? It seems there are at least two options: an experiment is a set of actions that people perform, or it is a set of instructions for actions that people *could* perform. TEs can be interpreted as experiments in either way. On these views, then, the ontology of TEs plausibly reduces to the ontology of actions, or of instructions. Focusing on actions, many thorny problems arise, concerning, e.g., how to differentiate between actions and events, how actions relate to intentions, whether an action is the same under different descriptions (e.g., the moving of a trigger finger vs. the firing of a gun vs. the killing of a person), and whether actions exist in space-time and if so how to say where and when an action begins/ends. Focusing on instructions, different options exist, e.g., depending on how we characterize the "could" in "instructions for actions that people *could* perform." Specifically, should we require (or expect) that the scenario of a good scientific TE will *not* include instructions for actions that are theoretically/nomologically impossible or indeterminate? If that is a necessary criterion for TEs to be portrayed as (a limiting case of) actual experiments, then it seems that some interesting case studies can not be counted as "successful" TEs (El Skaf 2017).

A fourth option is akin to the artifactualist and deflationist approaches to models, in which we define TEs by their function or purpose. For more details on these functions, see Section 4.2.

A fifth option is to adopt a fictionalist view of TEs (Meynell 2014; Salis and Frigg 2020; Sartori 2023). As with fictionalism about models, these accounts adopt Kendall Walton's pretense view such that TEs are real-world props (e.g., some text on a page), which, in combination with implicit and explicit rules, prescribe imaginings in a game of make believe.

2.3 Similarities, differences, and new possibilities

Accounts of the ontology of models and TEs overlap considerably. For example, Norton's argument view of TEs mirrors views about models that portray them as sets of inferences (Beisbart 2018; Suárez 2004). There are also fictionalist views of both models and TEs, and while there are no explicitly artifactualist accounts of TEs, many authors do treat TEs as epistemic tools in a way that accords with artifactualism about models.

However, there are also some important differences. In the literature on models, ontological pluralism was accepted relatively quickly: there are different kinds of models, which are "made of" different kinds of stuff. The literature on TEs seems to be more essentialist: whatever TEs are, they are all "made of" the same kind of thing (e.g., arguments, mental actions, fictions) or they are a single thing with a blurry definition (McComb 2013). Perhaps this is because there is more inherent variety among models or less among TEs. But arguments would have to be given to substantiate such a claim.

We also note that while there are material models, there are no material TEs. TEs can become "real" experiments when actualized, but models remain models whether they are material or non-material. Also, models can be quite general (e.g., a mere analogy or calculation device), while TEs seem always to focus on specific particular situations.

Comparing the views on models and TEs can be helpful for inspiring a number of potentially interesting new positions. One is applying the (admittedly no-longer very popular) view of models as abstract entities to TEs. Another would be to apply the mental models view of TEs (i.e., that TEs are mental actions on mental structures) to models. On such a view, we downplay the thing-like nature of models in favor of an emphasis on the kinds of mental actions they afford (for a start, see Boesch 2019; Brewer 2001; Nersessian 1999; 2008; 2022). Finally, we could apply the experimentalist view of TEs to models. There are views that are similar to this already, e.g., Morrison (2009) has argued that models can function as measuring instruments and simulating a model in a computer can count as an experimental measurement. However, this only portrays models as important parts of experiments, without yet claiming that models *are* experiments (see Knuuttila and Loettgers 2021 for a discussion of models as experiments).

3. Semantics

There are several ways to think about the semantics of things like models and TEs. We begin by separating two questions: what kinds of things are proper objects of semantic analysis, and what are semantic properties themselves? With respect to the first, we want to keep our options open in order to maximize potentially interesting applications of insights from the philosophy of language. Thus, we will consider words, concepts, sentences, propositions, texts, and actions, as well as models and TEs themselves, as potential carriers of semantic content.

The second issue is about what makes the above entities "semantic," or, in other words, what it means to say that something has meaning or makes reference. This is highly contested, to say the least. We might think of an entity's "meaning" as merely the experiences that gave rise to it (as the early British empiricists allegedly did). One wrinkle here is that many words refer to things that are not experienced, and others to things that could not be experienced. Following Frege, philosophers have tackled this issue by separating an entity's intention/sense/connotation from its extension/reference/denotation. This distinguishes between the more subjective, cognitive significance of an entity, and what it "points to" in the real world, which allows for meaning even in the absence of reference to real-world entities.

What is important for present purposes is that there are many semantic questions we can ask about models and TEs other than how they refer. The reference question is of special interest in light of the epistemic question of how we learn from models. But asking about the semantic content of (parts of) models and TEs can be a fruitful way of analyzing these two scientific tools, beyond the question of how they represent, which is the question we will mainly tackle in the following subsections.

3.1 Scientific models

Different kinds of models and targets exist, be they actual or merely possible, general or particular. So, how do models represent their targets? This has become the main semantic concern in the literature on models, especially since it is taken to solve, among other things, a central (epistemic) problem; that of surrogative reasoning. Surrogative reasoning enables one to draw inferences about the target system based on investigating the model (Swoyer, 1991; El Skaf et al., 2022). Models are thus tools that generate (explanatory, explorative) hypotheses, as well as predictions about target systems, to name a few functions of models. But what surrogate inferences are licensed is an epistemological issue. The semantics underlying epistemic uses of models is usually understood as follows: we (arguably, according to representationalists see Section 4.1) are justified in our surrogative inferences as long as the model represents its target. In addition, models being representational devices could also be understood as an ontological claim (Sanches de Oliveira 2022). In this section, the focus will be only on semantics.

There are different ways to cash out how a model represents its target (for an extensive discussion, see Frigg and Nguyen 2020). One account claims that models represent by stipulative fiat (Callender and Cohen 2006). That is, a model represents whatever a scientist says it does. Another possibility is that a model represents its target in virtue of being relevantly similar, and similar enough, to it (e.g., Giere 1988). There are also structuralist accounts of representation (e.g., da Costa and French 2003; Bueno, French, and Ladyman 2002). On these accounts, models are set-theoretic structures that represent their targets by having (some of) their elements mapped onto elements of the target. These mappings might be monomorphic, isomorphic, homomorphic, or partially isomorphic. There is also an "inferential" view, according to which a model represents its target if its users can draw inferences about that target from the model (see, e.g., Suárez 2004), and an "interpretational" view, according to which a model represents its target if the model is interpreted in terms of that target (Contessa 2007; 2011).

Proponents of the fictionalist accounts of models have also developed theories of scientific representation. Roughly, the postulation of the usefulness of fictional model systems has divided the fictional view into direct representationalism (Toon 2012; Levy 2012; 2015) and indirect representationalism (Frigg and Nguyen, 2016; 2020). The former denies the existence and even the utility of postulating a fictional model system that lives in scientists' imaginations, and argues that the model description prescribes imaginings that are directly about some real-world target, while the latter calls upon a fictional model system to stand between the model description and the target.

The best-developed indirect view is the DEKI account (Frigg and Nguyen 2016; 2020). On this view, what a model is "about" is determined by an act of denotation. This is the

"D" in DEKI (the rest: Exemplification, Keying-up, and Imputation). For material models, the model system is a material entity, but for non-material models, the model system is a fiction. In both cases, the model system represents its target if: it denotes the target, exemplifies certain features, there is a key that associates those features to a new set of features, and at least one of those features in the new set is imputed to the target.

3.2 Scientific thought experiments

It is rare to find an analysis of TEs in terms of representation between a source and a target. There are, however, some exceptions. In the mental models literature, Nersessian (1992) argues that it is the representation relation (usually a structural similarity) between the mental model and the real-world phenomena that does the justificatory work in TEs. And Sartori (2023) applies both Frigg's fictionalist approach to the ontology of models and Frigg and Nguyen's DEKI model of scientific representation to epistemically analyze TEs.

Despite there not being a lot of direct discussion about the semantic properties of TEs, we can extrapolate somewhat. If TEs are arguments, we should equate the semantic content of TEs with the semantic content of their underlying arguments. Norton defines TEs in a way that makes it necessary that they contain imaginative "particulars" which are not relevant to the generality of the conclusion and are thus eliminable from the reconstructed argument (see Norton's elimination thesis in Norton 1991). It is hard to say what "particulars" are in Norton's analysis, but think about experimental details that appear in TEs, such as the material make-up of Galileo's falling bodies, or the details of the weighing procedure in Einstein's photon box. Indeed, in Norton's reconstructions of TEs into arguments, many of these particulars do not appear, and when they do, they are absent from the (more general) conclusion. This makes sense: Norton's claim is an epistemological one, according to which TEs can be reconstructed into arguments without epistemic loss. It is not a semantic claim about what the content of a TE is. Adherents of the argument view can perhaps allow for extra semantic content in the TE that is not in the argument.

If TEs are a kind of real experiment, then their semantic content consists either of actions that could be performed, or they are actions. This makes it hard to say what their semantic content might be. Of course, actions can be interpreted as having semantic content. For example, while driving on a country road at night, someone might flash their car headlights to communicate police presence up ahead. For experiments to have semantic content, they must likewise be performed with a communicative or representational intention. This is possible in some pedagogical contexts, for example, where an experiment demonstrates something to a classroom of students. We think that in the majority of cases, TEs are not performed (merely or mainly) to communicate some definite semantic content, but rather to aid in exploring something from a first-person perspective.

However, Buzzoni adapts the Kantian dictum about concepts and experience to thought and real experiments, such that "TEs without real experiments are empty, and real experiments without TEs are blind" (Buzzoni 2018, 327). TEs are required to give meaning to experience, but they also get their content from previous experience. This point has been extended in an explicitly semantic direction, such that many famous TEs have been reinterpreted as something like Kantian schemata that help scientists and students of science to "fill in" the semantic content of new theoretical structures when they are first introduced (Stuart 2016a; 2017; 2018).

Finally, for a Platonist, we might expect platonic TEs not to have semantic content, given that they are *routes* to knowledge, and routes do not, on their own, have semantic content.

However, for Brown (the main defender of Platonism about TEs), TEs are not *merely* routes. Brown provides two different interpretations of TEs, one wide and one narrow (2007, 158). On the narrow interpretation, the TE is the mental experience we undergo. Here, the TE has the semantic content of the relevant mental states. On the wide interpretation, the TE encompasses the theory and background assumptions, plus the mental experience, and then also the theoretical interpretation of the experience. In this case, the TE has the semantic content of the mental experience but also whatever content the theory, background, and interpretation have.

3.3 Similarities, differences, and new possibilities

It should be possible to extend the work done on the semantic content of TEs to models. For example, if TEs sometimes have semantic aims, perhaps models do as well. In other words, if at least some models are created and used to increase our understanding of the "meaning" of some bit of theory or reality, then we can perhaps explain successful models in the same way that we explain successful TEs that have the same aim—in terms of imaginatively supplying and exploring possible experiences that we would have in a given scenario, drawing on tacit knowledge, background knowledge, and previous experience. Then, rather than judging a model on how well it increases knowledge about a target system, it could be judged in terms of how much useful semantic content it tends to make accessible.

What about going in the other direction? One frequently gets the impression from the literature on models that if we only knew the nature of scientific representation, we would be able to answer all the other questions about models, including about their ontology and epistemology. One does *not* get this impression from the literature on TEs. Why is this? Perhaps one reason is that models were commonly framed as relations between symbols/structures and the world, whereas TEs were seen from the start as arguments or mental activities. Indeed, insofar as the literature on TEs has touched on representation, it is mostly about *mental* representation, instead of scientific representation. Perhaps TEs involve first-personal, subjective mental representations, while models involve or require some kind of intersubjective representation. But whatever the differences, representation *is* part of thought experimentation, and so, perhaps the literature on TEs could benefit from the well-developed discussion on representation in models. For example, the literature on models shows that similarity is not necessarily the best way to understand representation. A mapping account might be preferred, or something like the DEKI account, which was designed to solve issues about representation. For example, Elgin argues that TEs teach us about the world by instantiating features of interest. But a TE cannot really instantiate features like mass or movement, so the instantiation must be "metaphorical." One motivation for the DEKI account was to avoid postulating metaphorical instantiation, and so those working on TEs who are attracted to Elgin's account but want to avoid metaphorical instantiation can perhaps do so by adopting the view of representation we find attached to the DEKI account.

One obstacle to directly applying insights from the representational literature on models to TEs is that representation might not have the same function in TEs and models: that of extrapolating from the model system to some target system. Let us suppose, following El Skaf and Imbert (2013), that both TEs and models are functionally similar, i.e., they both unfold scenarios and arrive at an output. Now, the outputs of TEs seem different from the outputs of models. The former are often propositions like the following: two objects fall both faster and slower (Galileo's falling bodies TE), Maxwell's demon separates fast from slow molecules without expenditure of work, Schrödinger's cat is dead and alive at the same time, Langvin's twin is both younger and older than his brother, the heat of matter lowered from a lab near a black hole could be converted to work with 100% efficiency (Geroch's engine TE), and the total entropy of the universe may have decreased when we throw two cups of tea into a black hole (Wheeler's demon TE). These outputs present apparent inconsistencies (El Skaf 2021; El Skaf and Palacios 2022). Certainly, TEs' narratives represent this and that, but the output does not seem to represent anything real about the world, since the world is (arguably) not inconsistent. This is not what we find with models. In modeling, the output often tends to be a specific claim about the model system, which is then extrapolated – via some theory of scientific representation – as a claim about actual or possible real-world target systems. To see things more clearly, consider Galileo's TE about falling bodies (El Skaf 2018), and "Malileo's" model of the same scenario using classical or relativistic mechanics (Salis and Frigg 2020). Despite unfolding the same sort of scenario, one difference between them is that Malileo's seems designed to produce predictions with precise values of the rate of fall of these and other falling bodies, while Galileo's is not. Another is that Galileo uses the TE to criticize the dominant framework by revealing an inconsistency, while Malileo *applies* the dominant framework.

4. Epistemology

There are three main epistemic issues about models and TEs: (1) Do they produce epistemic good(s)? (2) If so, which? And (3), if so, how? There are interesting similarities and differences between the answers given to these questions in the literatures on models and TEs.

4.1 Scientific models

It is generally agreed that models do provide some epistemic good(s). Even a cursory glance at the history and practice of science shows that models are important, if not central, to scientific progress, and respect for this fact motivates philosophers to accept that models provide some epistemic good(s).

But which epistemic good(s) do they provide? There are a number of ways to answer. For example, Alexandrova (2008) argues that idealized deductive models are best understood as contributing causal hypotheses. Others claim that models produce knowledge about their target systems, but only if the model accurately represents the target and there are no "defeaters" present that would invalidate inferences from the model to the target. For example, a model in economics might cause us to infer that if the price of a commodity increases, demand will decrease. That might be correct, *as long as* the price of competing commodities does not also increase.

Still others focus on the less-obviously epistemic properties of models, for example, their status as a means by which theory can be applied to particular cases (Morgan and Morrison 1999), as a means of theory-building (Hartmann 1995), or as a vehicle of explanations (De Regt 2017). Given the important role of models in explanations, there is also a case to be made that (good) models increase scientific understanding, whether this is in addition to, as part of, or as opposed to, increasing scientific knowledge (Elgin 2017; Dellsén 2020; Potochnik 2017; Sullivan 2022; Stuart and Nersessian 2019). Echoing Section 2, pluralism is generally a popular option, such that some models produce one kind of epistemic good, and others produce others.

In answer to the third question, the philosophical literature on models seems to be roughly divided into two camps: one representational and the other non-representational, with subdivisions in each. In the representational camp, an epistemological account can sometimes be drawn from the details of a given account of representation, and sometimes this is made explicit. For example, proponents of a similarity account can claim that things learned about the model will also hold in the target if the model and target are relevantly similar. Structuralists can claim that things learned about the structure of a model can be extrapolated to the target by means of an appropriate mapping relation. Inferentialists define models as things that license inferences about targets, and the question then becomes one about defining "correct" inferences. Fictionalists have different ways of answering the question, but those who follow Walton's pretense view of fiction will claim that the model is a prop in a game of make believe that we explore while constrained by implicit and explicit rules, to see what is true in the fiction. On the direct view, what is true in the fiction can be true about the target because the fiction was always "about" the target. On the indirect view, what is true in the fiction requires keying up and imputation to the target and can be true about the target depending on how the fiction's features are chosen, interpreted, and keyed up.

The non-representational camp, as we understand it, combines different approaches under the umbrella of artifactualism. They share the idea that an analysis of "scientific models in general will, at best, be limited" (Sanches de Oliveira 2022, 6). The epistemic contribution of models should be assessed on a case-by-case basis. In a series of papers, Knuuttila (2011; 2017; 2021) argues that what and how we can learn from models depends on the way the model is constructed to explore a particular scientific question. This question can be general in nature or address only what is possible or impossible. In line with fictionalist and other indirect representation accounts, Knuuttila distinguishes between internal representation and external representation: what is represented within a model does not yet make the model a representation of some determinable social or natural target system. However, the artifactual account also pays attention to the epistemic affordances of the specific representational modes and media used in model construction.

Parker (2020) also emphasizes the importance of problem-solving. She develops a view of models that evaluates them for their adequacy for a purpose, not in terms of representational accuracy. While this is usually assessed on a case-by-case basis, Parker claims in general "what is required is that the model stands in a suitable relationship with a target, (type of) user, (type of) methodology, (type of) circumstances, and purpose jointly. Put differently, the model must constitute a 'solution' in a kind of problem space" (Parker 2020, 475).

In addition to epistemic concerns about targets, philosophers also raise epistemic questions about how we learn about models themselves. Learning about material models raises questions akin to those of laboratory experiments: we manipulate models, subject them to tests, interpret the results, and so on. Learning about non-material model systems is a different story. It has been suggested (by Frigg and Hartmann 2020) that we learn about some abstract models by doing TEs. On this view, abstract models and TEs are complementary tools: scientists write down the description of a model and use a TE to mentally manipulate the fictional system described. However, other models are more easily unfolded by implementation in a computer simulation. In the case where a model is unfolded by a TE, the epistemology of TEs is part of the epistemology of models. Where a model is unfolded by a computer simulation, the epistemology of simulations is part of the epistemology of models. There are still further ways of thinking about the epistemology of models, e.g., as (or as including) metaphors (Camp 2020; Levy 2020; Stuart and Wilkenfeld 2022), analogies

(Hesse 1966; Nersessian 2015), diagrams (Sheredos and Bechtel 2020), and idealizations (Cassini and Redmond 2021).

4.2 Scientific thought experiments

Like models, TEs are mostly accepted as being epistemically profitable. What sorts of epistemic good(s) do TEs provide? Here, there is just as much pluralism as with models. TEs might generate new knowledge (Brown 2011; Norton 2004; Nersessian 2018; Miščević 2022) or understanding (Brown 2014; Lipton 2009; Murphy 2020a; Stuart 2016a; 2018), new theoretical possibilities (Stuart 2021, El Skaf 2021), reveal and resolve inconsistencies (El Skaf 2021; El Skaf and Palacios 2022; Sorensen 1992; Häggqvist 2009; 2019), give examples, illustrate a claim (Brown 1991; Schabas 2018; Peacock 2018), demonstrate pursuitworthiness (Miller 2002; Šešelja and Straßer 2014; El Skaf 2021), control variables (Sorensen 1992), exemplify features (Elgin 2014), give "hypothetical explanations" (Schlaepfer and Weber 2018), and test a theory's non-empirical virtues (Bokulich 2001).

How do TEs produce the epistemic goods they do? Norton (e.g., 1991; 1996; 2004) argues that TEs can always be reconstructed as deductive or inductive arguments. This means that the new insight that TEs provide depends on the type of argument that underlies the TE. If the argument constructed from a TE is deductive, the TE would just serve to rearrange our existing knowledge without adding any new knowledge. If the argument is inductive, the TE could extend our knowledge to new cases, in the same way as inductive arguments do.

Brown (1991) has defended a different approach. In contrast to Norton, he does not identify TEs with arguments, and provides a detailed taxonomy of the different types of TEs, which are associated with different epistemic functions of TEs, such as constructive, conjectural and "platonic." The most controversial are the platonic TEs, which, according to Brown, can provide us with a priori access to the laws of nature, without the need for any new empirical data. They do this by producing mental phenomena which serve as evidence for claims about connections between universals. If the Dretske-Tooley-Armstrong account of laws of nature is correct that laws of nature are relations between universals, and Brown is correct that TEs give us insights about universals, then platonic TEs are capable of providing us with knowledge of laws of nature.

Defenders of the mental model account of TEs (e.g., Miščević 2022; Nersessian 2018) have rejected the view that the justificatory power of TEs can be reduced to the logical structure of their propositional content and that the experimental details are irrelevant and eliminable. Nersessian, for instance, argues that we acquire new knowledge about the real-world target system by mentally modeling a structural analog of that system and not (only) by mentally reasoning through a set of logically related propositions.

Those who portray TEs as genuine members of the experiment family understand the epistemology of TEs in the same way as the epistemology of experiments. Thus, a TE will be epistemologically good insofar as it meets the conditions of a good experiment, such as Franklin's five criteria (1986): the experimental system must be well-isolated, experimental bias must be eliminated, sources of error must be identified and accounted for, instruments must be calibrated as well as possible, and there should be a theory of our instruments (see Stuart 2016b).

El Skaf (2021) and El Skaf and Palacios (2022) argue that many TEs, both from the history of physics and from ongoing physics such as black hole thermodynamics, aim at revealing and resolving inconsistencies. These two functions have different epistemic forces and are justified differently: while the revelation of the inconsistency could be analyzed as conclusive knowledge, its resolution is only conjectural.

The above epistemological accounts of TEs mostly deal with the question of how TEs produce new knowledge. Different accounts might be necessary to explain how TEs can produce other epistemic goods. For example, Stuart (2018) has argued that TEs are capable of producing all three of the major types of understanding: explanatory, objectual, and practical, and the way they do this might be different in each case.

4.3 *Similarities, differences, and new possibilities*

One interesting thing to note in comparing the epistemologies of models and TEs is that in both cases, most of the work is offloaded onto accounts of more traditional "ways of knowing," including logical inference, pure reason, metaphor, analogy, representation, experiment, and storytelling. Another point of consilience is that in both literatures, the epistemological issue is usually phrased as concerning how models and TEs produce new knowledge, even though in practice what philosophers discuss is much more varied, and perhaps not all the epistemic goods produced can or should be reduced to knowledge. A third point of agreement is that constraints play a major role in explaining how epistemic goods are generated. These might be reduced to two kinds of constraints: logical constraints on valid reasoning, and representational constraints on accurate reasoning.

Another similarity concerns the use of imagination, which appears to be at the root of both models and TEs (Salis and Frigg 2020; Stuart 2022). This explains the fact that there are fictionalist views about both models and TEs. But it raises the following question: Can imagination produce new knowledge or understanding, or does it only mediate that production? This has been called the question of the "epistemic generativity of imagination" (Miyazono and Tooming 2022), and it will be crucial moving forward to see whether a positive answer can address skepticism about the epistemic power of imagination (for discussion, see, e.g., Kinberg and Levy 2022; Myers 2021; Stuart 2019; 2022).

Nevertheless, there are also interesting differences. Unlike with models, the question of how we can learn *about* TEs is not asked. Perhaps it should be. Also, as we noted in Section 3.3, the literature on TEs has not focused as much on representation as the literature on models. Perhaps some of the insights about representation in models could be used in the case of the epistemology of TEs. Although, if non-representationalists about the epistemology of models are correct, perhaps not.

Interestingly, and contrary to TEs, the epistemological question of how we learn from models was only a derivative concern in the philosophical literature on models, given the large consensus (before artifactualism) that an account of representation is all we need. Put differently, semantics took center stage in the epistemological literature on models, unlike in TEs.

Additionally, scientists often learn about and from models by intervening on them numerically. That is not exactly the case with how scientists engage with TEs: thought experimental scenarios are manipulated by playing around with the theoretical statements or with some qualitative and technical experimental detail, not with numerical values of parameters and variables. Probably this difference explains some of the differences in their respective epistemologies.

Finally, there are tantalizing opportunities for epistemic "cross-pollination" between the literatures. The artifactualist view about models could surely be applied to TEs in more detail. The Kantian, phenomenological (Hopp 2014; Wiltsche 2018), unfolding-based, understanding-based, and experimentalist perspectives on TEs could also be applied to models.

5. Conclusion

This entry has summarized work done on the ontology, semantics, and epistemology of both models and TEs, pointing out similarities and differences, and hinting at new philosophical possibilities. Other comparative lenses could have been taken up as well, such as the aesthetics of models and TEs. Do scientists employ different standards of aesthetic value for these? Are there different ways that aesthetic features relate to non-aesthetic (e.g., semantic or epistemic) features? Another potentially interesting lens is social epistemology: usually, models are team-built, and TEs can also be understood as social uses of imagination (Molinari 2022), even though they are usually conceived by a single scientist. A third lens is ethics. Models play a key role in justifying scientific claims, which then go on to justify ethically relevant actions, e.g., concerning climate change and pandemic lockdowns. The ethical features of models are gaining attention (Winsberg and Harvard 2022), however, the ethics of TEs is not yet a topic of much discussion (except in jest, see Lerner 2010, Norton 2010). A fourth lens is functional. What functions are common to both, and which are not? Can we find a more general function that both TEs and models all perform? One has been suggested by El Skaf and Imbert, who argue that all tools that unfold scenarios are "composed of functionally similar parts" (2013, 3455). They call the set of these parts a "CUI pattern of inquiry" where this stands for the Construction of a scenario in the context of an inquiry, Unfolding of the scenario, and Interpretation of the result. Thus, instead of focusing on ontological, semantic, or epistemological differences, TEs and models could be pragmatically analyzed as functionally similar in that they share the CUI pattern, and they are both tools that unfold scenarios, though also different in the sense that they often have different kinds of outputs.

Another way to take the discussion further would be to expand what the lenses focus on. We talked about models and TEs, but many scholars have drawn connections between both of these and simulations, and laboratory experiments, analogy, metaphor, and much else. Many who discuss different tools of scientific reasoning discuss two or three of these, but there have been few attempts to bring all their literatures together to find points of resonance and dissonance.

Finally, it could be worthwhile to analyze the underlying cognitive nature of models and TEs to see how they compare. One question concerns what kind of imagination fuels both. Salis and Frigg (2020) argue that we need only talk about *propositional* imagination. This cannot be reconciled with the work of philosophers of mind, who argue that imagination is fundamentally sensory, or imagistic (Kind 2001; Nanay forthcoming). For this and other reasons, Murphy (2020b) argues we should be pluralists about what kind of imagination is relevant for TEs. If correct, this will affect discussions of the epistemology of imagination in both models and TEs.

References

Alexandrova, Anna. 2008. "Making Models Count." *Philosophy of Science* 75: 383–404.

Arcangeli, Margherita. 2018. "The Hidden Links between Real, Thought and Numerical Experiments." *Croatian Journal of Philosophy* 18(1): 3–22.

Beisbart, Claus. 2018. "Are Computer Simulations Experiments? And If Not, How Are They Related to Each Other?" *European Journal for Philosophy of Science* 8(2): 171–204.

Boesch, Brandon. 2019. "The Means-End Account of Scientific, Representational Actions." *Synthese* 196(6): 2305–2322.

Bokulich, Alisa. 2001. "Rethinking Thought Experiments." *Perspectives on Science* 9(3): 285–307.

Brewer, William F. 2001. "Models in Science and Mental Models in Scientists and Nonscientists." *Mind & Society* 2(2): 33–48.

Brown, James Robert. 1991. *The Laboratory of the Mind: Thought Experiments in the Natural Sciences*. New York: Routledge.

———. 2007. "Counter Thought Experiments." *Royal Institute of Philosophy Supplement* 61 (October): 155–177.

———. 2014. "Explaining, Seeing, and Understanding in Thought Experiments." *Perspectives on Science* 22(3): 357–376.

Bueno, Otávio, Steven French, and James Ladyman. 2002. "On Representing the Relationship between the Mathematical and the Empirical." *Philosophy of Science* 69(3): 497–518.

Buzzoni, Marco. 2008. *Thought Experiment in the Natural Sciences: An Operational and Reflexive-Transcendental Conception*. Würzburg: Königshausen & Neumann.

———. 2018. "Kantian Accounts of Thought Experiments." In *The Routledge Companion to Thought Experiments*, edited by Michael T Stuart, Yiftach Fehige, James Robert Brown, 327–341. London: Routledge.

Callender, Craig, and Jonathan Cohen. 2006. "There Is No Special Problem about Scientific Representation." *Theoria. Revista de Teoría, Historia y Fundamentos de La Ciencia* 21(1): 67–85.

Camp, Elisabeth. 2020. "Imaginative Frames for Scientific Inquiry: Metaphors, Telling Facts, and Just-So Stories." In *The Scientific Imagination*, edited by Arnon Levy and Peter Godfrey-Smith, 304–336. Oxford University Press.

Cassini, Alejandro, and Juan Redmond, eds. 2021. *Models and Idealizations in Science: Artifactual and Fictional Approaches*. Cham: Springer International Publishing.

Contessa, Gabriele. 2007. "Scientific Representation, Interpretation, and Surrogative Reasoning." *Philosophy of Science* 74(1): 48–68.

———. 2011. "Scientific Models and Representation." In *The Continuum Companion to the Philosophy of Science*, edited by Steven French and Juha Saatsi, 120–137. London: Continuum Press.

da Costa, Newton, and Steven French. 2003. *Science and Partial Truth: A Unitary Approach to Models and Scientific Reasoning*. Oxford: Oxford University Press.

de Regt, Henk. 2017. *Understanding Scientific Understanding*. New York: Oxford University Press.

Dellsén, Finnur. 2020. "Beyond Explanation: Understanding as Dependency Modelling." *The British Journal for the Philosophy of Science* 4: 1261–1286.

El Skaf, Rawad. 2017. "What Notion of Possibility Should We Use in Assessing Scientific Thought Experiments?" *Lato Sensu: Revue de La Société de Philosophie Des Sciences* 4(1): 19–30.

———. 2018. "The Function and Limit of Galileo's Falling Bodies Thought Experiment." *Croatian Journal of Philosophy* 18(1): 37–58.

———. 2021. "Probing Theoretical Statements with Thought Experiments." *Synthese* 199(3–4): 6119–6147.

El Skaf, Rawad, Laura Feline, Patricia Palacios, and Giovanni Valente, eds. 2022. "Surrogative Reasoning in the Sciences [Topical Collection]." *Synthese* 203(105) https://doi.org/10.1007/s11229-024-04518-x.

El Skaf, Rawad, and Cyrille Imbert. 2013. "Unfolding in the Empirical Sciences: Experiments, Thought Experiments and Computer Simulations." *Synthese* 190(16): 3451–3474.

El Skaf, Rawad, and Patricia Palacios. 2022. "What Can We Learn (and Not Learn) from Thought Experiments in Black Hole Thermodynamics?" *Synthese* 200(6): 434.

Elgin, Catherine. 2014. "Fiction as Thought Experiment." *Perspectives on Science* 22(2): 221–241.

———. 2017. *True Enough*. Cambridge: MIT Press.

Fine, Arthur. 1998. "Fictionalism." In *Routledge Encyclopedia of Philosophy*, edited by Edward Craig, 3:667–668. London: Routledge.

Franklin, Allan. 1986. *The Neglect of Experiment*. Cambridge: Cambridge University Press.

Frigg, Roman. 2010. "Models and Fiction." *Synthese* 172(2): 251–268.

Frigg, Roman, and Stephan Hartmann. 2020. "Models in Science." In *The Stanford Encyclopedia of Philosophy*, edited by Edward N. Zalta, Spring 2020. https://plato.stanford.edu/cgi-bin/encyclopedia/archinfo.cgi?entry=models-science

Frigg, Roman, and James Nguyen. 2016. "The Fiction View of Models Reloaded." *The Monist* 99(3): 225–242.
———. 2020. *Modelling Nature: An Opinionated Introduction to Scientific Representation*. Cham: Springer Synthese Library.
Giere, Ronald N. 1988. *Explaining Science: A Cognitive Approach*. Chicago: University of Chicago Press.
Godfrey-Smith, Peter. 2006. "The Strategy of Model-Based Science." *Biology and Philosophy* 21(5): 725–740.
Häggqvist, Sören. 2009. "A Model for Thought Experiments." *Canadian Journal of Philosophy* 39(1): 55–76.
———. 2019. "Thought Experiments, Formalization, and Disagreement." *Topoi* 38 (4): 801–10. https://doi.org/10.1007/s11245-017-9491-7.
Hartmann, Stephan. 1995. "Models as a Tool for Theory Construction: Some Strategies of Preliminary Physics." In *Theories and Models in Scientific Processes*, edited by William Herfel, Władysław Krajewski, Ilkka Niiniluoto, and Ryszard Wójcicki, 49–67. Boston: Brill.
Hesse, Mary. 1966. *Models and Analogies in Science*. Notre Dame: University of Notre Dame Press.
Hopp, Walter. 2014. "Experiments in Thought." *Perspectives on Science* 22(2): 242–263.
Kinberg, Ori, and Arnon Levy. 2022. "The Epistemic Imagination Revisited." *Philosophy and Phenomenological Research* 107(2): 319–336.
Kind, Amy. 2001. "Putting the Image Back in Imagination." *Philosophy and Phenomenological Research* 62(1): 85–109.
King, Jeffrey C. 2017. "The Metaphysics of Propositions." In *The Oxford Handbook of Topics in Philosophy*. Oxford: Oxford University Press. https://doi.org/10.1093/oxfordhb/9780199935314.013.26
Knuuttila, Tarja. 2011. "Modelling and Representing: An Artefactual Approach to Model-based Representation." *Studies in History and Philosophy of Science Part A*, Model-based Representation in Scientific Practice, 42(2): 262–271.
———. 2017. "Imagination Extended and Embedded: Artifactual versus Fictional Accounts of Models." *Synthese* 198: 5077–5097.
———. 2021. "Epistemic Artifacts and the Modal Dimension of Modeling." *European Journal for Philosophy of Science* 11(3): 65.
Knuuttila, Tarja, and Loettgers Andrea, 2021, "Biological Control Variously Materialized: Modeling, Experimentation and Exploration in Multiple Media". *Perspectives on Science* 29(4): 468–492.
Lerner, Berel Dov. 2010. "My Evening with Mr. Wang." *Think* 10(27): 83–93. https://doi.org/10.1017/S1477175610000382
Levy, Arnon. 2012. "Models, Fictions, and Realism: Two Packages." *Philosophy of Science* 79(5): 738–748.
———. 2015. "Modeling without Models." *Philosophical Studies* 172(3): 781–798.
———. 2020. "Metaphor and Scientific Explanation." In *The Scientific Imagination*, edited by Arnon Levy and Peter Godfrey-Smith, 280–303. Oxford: Oxford University Press.
Lipton, Peter. 2009. "Understanding without Explanation." In *Scientific Understanding: Philosophical Perspectives*, edited by H. W. de Regt, S. Leonelli, and K. Eigner, 43–63. Pittsburgh: University of Pittsburgh Press.
McComb, Geordie. 2013. *Thought Experiment, Definition, and Literary Fiction*, edited by Mélanie Frappier. London: Routledge.
Meynell, Letitia. 2014. "Imagination and Insight: A New Account of the Content of Thought Experiments." *Synthese* 191(17): 4149–4168.
Miller, Arthur I. 2002. "Inconsistent Reasoning toward Consistent Theories." In *Inconsistency in Science*, edited by Joke Meheus, 35–41. Origins. Dordrecht: Springer Netherlands.
Miščević, Nenad. 1992. "Mental Models and Thought Experiments." *International Studies in the Philosophy of Science* 6(3): 215–226.
———. 2022. *Thought Experiments*. Cham: Springer International Publishing.
Miyazono, Kengo, and Uku Tooming. 2022. "On the Putative Epistemic Generativity of Memory and Imagination." In *Philosophical Perspectives on Memory and Imagination*, edited by Anja Berninger and Íngrid Vendrell Ferran, 127–145. London: Routledge.
Molinari, Daniele. 2022. "Thought Experiments as Social Practice and the Clash of Imaginers." *Croatian Journal of Philosophy* 22(2): 229–247.

Morgan, Mary, and Margaret Morrison. 1999. *Models as Mediators*. Cambridge: Cambridge University Press.
Morrison, Margaret. 2009. "Models, Measurement and Computer Simulation: The Changing Face of Experimentation." *Philosophical Studies* 143(1): 33–57.
Murphy, Alice. 2020a. "The Aesthetic and Literary Qualities of Scientific Thought Experiments." In *The Aesthetics of Science: Beauty, Imagination and Understanding*, edited by Milena Ivanova and Steven French, 146–166. London: Routledge.
———. 2020b. "Toward a Pluralist Account of the Imagination in Science." *Philosophy of Science* 87(5): 957–967.
Myers, Joshua. 2021. "The Epistemic Status of the Imagination." *Philosophical Studies* 178: 3251–3270.
Nanay, Bence. forthcoming. "Against Imagination." In *Contemporary Debates in the Philosophy of Mind*. 2nd Edition. Edited by Jonathan Cohen and Brian McLaughlin. Oxford: Blackwell.
Nersessian, Nancy J. 1992. "In the Theoretician's Laboratory: Thought Experimenting as Mental Modeling." *PSA: Proceedings of the Biennial Meeting of the Philosophy of Science Association* 1992: 291–301.
———. 1999. "Model-based Reasoning in Conceptual Change." In *Model-Based Reasoning in Scientific Discovery*, edited by Lorenzo Magnani, Nancy J. Nersessian, and Paul Thagard, 5–22. Boston, MA: Springer.
———. 2007. "Thought Experiments as Mental Modelling: Empiricism without Logic." *Croatian Journal of Philosophy* VII: 125–161.
———. 2008. *Creating Scientific Concepts*. Cambridge: MIT Press.
———. 2015. "The Cognitive Work of Metaphor and Analogy in Scientific Practice." *Philosophical Inquiries* 3(1): 133–156.
———. 2018. "Cognitive Science, Mental Modeling, and Thought Experiments." In *The Routledge Companion to Thought Experiments*, edited by Michael T Stuart, Yiftach Fehige, James Robert Brown, 209–226. London: Routledge.
———. 2022. *Interdisciplinarity in the Making: Models and Methods in Frontier Science*. Cambridge: MIT Press.
Norton, John D. 1991. "Thought Experiments in Einstein's Work." In *Thought Experiments in Science and Philosophy*, edited by Tamara Horowitz and Gerald J. Massey, 129–147. Savage, MD: Rowman & Littlefield.
———. 1996. "Are Thought Experiments Just What You Thought?" *Canadian Journal of Philosophy* 26(3): 333–366.
———. 2004 "On Thought Experiments: Is There More to the Argument?" *Philosophy of Science* 71 (5): 1139–1151.
———. 2010. "Ethics of Imaginary Research." *Goodies*. 2010. https://sites.pitt.edu/~jdnorton/Goodies/ethics_te/ethics_te.html.
———. 2021. *The Material Theory of Induction*. Calgary: BSPS Open.
Parker, Wendy S. 2020. "Model Evaluation: An Adequacy-for-Purpose View." *Philosophy of Science* 87(3): 457–477.
Peacock, Kent A. 2018. "Happiest Thoughts: Great Thought Experiments of Modern Physics." In *The Routledge Companion to Thought Experiments*, edited by Michael T Stuart, Yiftach Fehige, James Robert Brown, 211–242. London: Routledge.
Potochnik, Angela. 2017. *Idealization and the Aims of Science*. Chicago: University of Chicago Press.
Salis, Fiora. 2021. "The New Fiction View of Models." *The British Journal for the Philosophy of Science* 72(3): 717–742.
Salis, Fiora, and Roman Frigg. 2020. "Capturing the Scientific Imagination." In *The Scientific Imagination*, edited by Arnon Levy and Peter Godfrey-Smith, 17–50. Oxford: Oxford University Press.
Sanches de Oliveira, Guilherme. 2021. "Representationalism Is a Dead End." *Synthese* 198(1): 209–235. https://doi.org/10.1007/s11229-018-01995-9.
———. 2022. "Radical Artifactualism." *European Journal for Philosophy of Science* 12(2): 1–33.
Sartori, Lorenzo. 2023. "Putting the 'Experiment' Back into the 'Thought Experiment.'" *Synthese* 201: 34.

Schabas, Margaret. 2018. "Thought Experiments in Economics." In *The Routledge Companion to Thought Experiments*, edited by Michael T Stuart, Yiftach Fehige, James Robert Brown, 171–182. London: Routledge.
Schlaepfer, Guillaume, and Marcel Weber. 2018. "Thought Experiments in Biology." In *The Routledge Companion to Thought Experiments*, edited by Michael T Stuart, Yiftach Fehige, James Robert Brown, 243–256. London: Routledge.
Šešelja, Dunja, and Christian Straßer. 2014. "Epistemic Justification in the Context of Pursuit: A Coherentist Approach." *Synthese* 191(13): 3111–3141.
Sheredos, Benjamin, and William Bechtel. 2020. "Imagining Mechanisms with Diagrams." In *The Scientific Imagination: Philosophical and Psychological Perspectives*, edited by Arnon Levy and Peter Godfrey-Smith. Oxford University Press.
Sorensen, Roy. 1992. *Thought Experiments*. Oxford: Oxford University Press.
Stuart, Michael T. 2016a. "Taming Theory with Thought Experiments: Understanding and Scientific Progress." *Studies in History and Philosophy of Science Part A* 58 (August): 24–33.
———. 2016b. "Norton and the Logic of Thought Experiments." *Axiomathes* 26(4): 451–466.
———. 2017. "Imagination: A Sine Qua Non of Science." *Croatian Journal of Philosophy* XVII (49): 9–32.
———. 2018. "How Thought Experiments Increase Understanding." In *The Routledge Companion to Thought Experiments*, edited by Michael T Stuart, Yiftach Fehige, James Robert Brown, 526–544. London: Routledge.
———. 2019. "Towards a Dual Process Epistemology of Imagination." *Synthese* 198: 1329–1350.
———. 2020. "The Material Theory of Induction and the Epistemology of Thought Experiments." *Studies in History and Philosophy of Science Part A* 83: 17–27, https://doi.org/10.1016/j.shpsa.2020.03.005.
———. 2021. "Telling Stories in Science: Feyerabend and Thought Experiments." *HOPOS: The Journal of the International Society for the History of Philosophy of Science* 11(1): 262–281.
———. 2022. "Sharpening the Tools of Imagination." *Synthese* 200(6): 451.
Stuart, Michael T., and Nancy J. Nersessian. 2019. "Peeking Inside the Black Box: A New Kind of Scientific Visualization." *Minds and Machines* 29(1): 87–107.
Stuart, Michael T., and Daniel Wilkenfeld. 2022. "Understanding Metaphorical Understanding (Literally)." *European Journal for Philosophy of Science* 12(3): 49.
Suárez, Mauricio. 2004. "An Inferential Conception of Scientific Representation." *Philosophy of Science* 71(5): 767–779.
Sullivan, Emily. 2022. "Understanding from Machine Learning Models." *The British Journal for the Philosophy of Science* 73(1): 109–133.
Swoyer, Chris. 1991. "Structural Representation and Surrogative Reasoning." *Synthese* 87(3): 449–508.
Thomasson, Amie L. 2020. "If Models Were Fictions, Then What Would They Be?" In *The Scientific Imagination*, edited by Arnon Levy and Peter Godfrey-Smith, 51–74. Oxford: Oxford University Press.
Toon, Adam. 2012. *Models as Make-Believe: Imagination, Fiction and Scientific Representation*. New Directions in the Philosophy of Science. London: Palgrave Macmillan.
Walton, Kendall L. 1990. *Mimesis as Make-Believe: On the Foundations of the Representational Arts*. Cambridge, MA and London: Harvard University Press.
Wiltsche, Harald A. 2018. "Phenomenology and Thought Experiments: Thought Experiments as Anticipation Pumps." In *The Routledge Companion to Thought Experiments*, edited by Michael T Stuart, Yiftach Fehige, James Robert Brown, 342–365. London: Routledge.
Winsberg, Eric, and Stephanie Harvard. 2022. "Purposes and Duties in Scientific Modelling." *J Epidemiol Community Health* 76(5): 512–517.

25
MODELS AND MAPS

Rasmus Grønfeldt Winther

1. Introduction

Generatively ambiguous, the concept of a *map* finds its natural home in cartography. The geographer John Andrews archived 321 definitions of the term published between 1649 and 1996. The single characterization dominating all others is "a representation… in a plane… of all or part of the earth's surface" (Andrews 1996, 1). This is a map as a representational cartographic object. In the first and wonderfully philosophical chapter of their book *The Nature of Maps*, the cartographers Arthur Robinson and Barbara Petchenik provide the following definition: "a map is a graphic representation of the milieu" (1976, 16). The *Oxford English Dictionary* catalogs further instances of this tradition of dramatic cartographic representationalism.

In contrast, and in line with dialectic tensions and perennial discussions in the philosophy of science, some cartographers and geographers beg for a more practice-based conceptualization. Geographers Rob Kitchin and Martin Dodge argue, "that cartography is profitably conceived as a processual, rather than representational, science" (2007, 331). J.B. Harley worried about the relation between "cartographic rules" and "the cultural production of the map": "In the map itself, social structures are often disguised beneath an abstract, instrumental space, or incarcerated in the coordinates of computer mapping" (1989, 4–5). Finally, Denis Wood portrays maps as "weapons" wielded by those with power – the state, the military, or the corporate elite (1992; 2012).

The contrast between representation and theory on the one hand and process and practice on the other is familiar to cartographers as well as to philosophers of science, showing one way that maps and mapping raise questions about models and modeling in general. In this chapter, I archive map discourse in the founding generation of philosophers of science (Section 2) and the subsequent generation (Section 3). In focusing on these two original framing generations of philosophy of science, I intend to remove us from the heat of contemporary discussions to see, in a more distant and neutral light, the many productive ways in which maps can stand in analytically for scientific theories and models. I also expand on what I take to be the map analogy – i.e., *a scientific theory is a map of the world* (Section 4) – illustrating its fruitfulness for understanding abstraction, representation, and practice in science.

 DOI: 10.4324/9781003205647-30

2. Archive I: the founding generation of philosophy of science and map discourse

Maps and mapping provide ubiquitous inspiration and intuition pumps, as it were, for the philosophy of science literature on representation and models. To name just a few examples, I will consider how maps are deployed as analogies for scientific representations by four figures from the founding generation of professionalized philosophy of science: Rudolf Carnap (b. 1891, PhD. 1921), Nelson Goodman (b. 1906, PhD. 1941), Stephen Toulmin (b. 1922, PhD. 1948), and Thomas Kuhn (b. 1922, PhD. 1949). Of particular interest here is the extensive use of the map analogy made both by the structuralist Carnap and the pragmatist Goodman.[1]

Turn first to Rudolf Carnap's 1928 *Aufbau* (1967/2003). According to Michael Friedman, the "fundamental aim" of the *Aufbau* was "the articulation and defense of a radically new conception of objectivity" (1987, 526). For Carnap, objectivity was intimately linked to "logical form or structure" (526). This form amounted to a system of "structural definite descriptions," a rich and enormous network of scientific concepts, within which each unique scientific concept finds its place. This central aim is developed in §§12–15 of the *Aufbau*, including the single-longest "concrete example" in the book, a map of "the Eurasian railroad network." This example explores how we can identify and distinguish each node of the total global structure – i.e., each station or each scientific concept – by examining the number of edges of each node, and of the nodes connected to it. As in identifying each station node by topology and connectivity within a railroad structure, an important step toward scientific objectivity is finding the location of different concepts within the unified, deductive logical structure of a "constructional system" (Konstitutionssysteme).[2]

In his 1963 commentary on Carnap's *Aufbau*, pragmatically oriented philosopher Nelson Goodman deployed the map analogy to show how the philosopher is a map-making meta-scientist. Experience is the "territory" of the constructionalist philosopher's map-making enterprise: "the function of a constructional system is not to recreate experience but rather to map it" (552). Philosophers can even construct "alternative schemes" using cues from *Aufbau* (553).[3]

With the map analogy in hand, Goodman defends Carnap against two critics: the "anti-intellectualist" (e.g., Henri Bergson, whom Goodman mentions by name) and the "verbal analyst" or "ordinary language" philosopher (552–554). Contrary to the anti-intellectualist who decries a constructional system or map because it does "not recreate experience," Goodman argues that "the relevant question about a system or a map" is not a choice "between misrepresentation and a meticulous reproduction," but "whether [a map] is serviceable and accurate in the way intended" (553). Goodman implores: "let no one accuse the cartographer of merciless reductionism if his map fails to turn green in the spring" (553). The map is not the territory. Anti-intellectualists, Goodman believes, are disingenuously indicting Carnap for conflating map and world, something Carnap was not doing.

Concerning the verbal analyst, Goodman admits that "verbal analysis is a necessary preliminary and accompaniment of systematic construction" but finds it counterproductive for the verbal analyst to be hostile to the constructionalist mapper (554). Although they are presented in an "artificial language" (like constructional systems), maps have "advantages." They are "consistent, comprehensive, and connected," "reveal unsuspected routes," "rectify misconceptions," and give "an organized overall view that no set of verbal directions and no experience in travelling can provide unaided" (553). The verbal analyst,

Goodman argues, need not perceive Carnap and other constructionalist mappers (including Goodman) as competitors or foes. A constructional definition is not privative. Rather than implying that there "is nothing more than" the map and its elements, the map has a critical self-awareness built in so that it should be read as making the careful claim only of "is here to be mapped as" (554).[4] In a loose sense, reality has a one-to-many relationship with all of the legitimate maps that may be made of it. In short, Goodman interprets the constructionalist as wishing neither to conflate nor to confuse map and territory, nor as claiming to have an absolute, total representation.

Carnap approved. In the 1961 preface to the second edition of the *Aufbau*, Carnap admiringly noted that Goodman's constructional system had "essentially the same goal as my own" (1967, x). In his 154-page response to his critics in *The Library of Living Philosophers* volume dedicated to him, he also commends Goodman's "comparison ... of construction with the drawing of a map" since it "clears up misunderstandings which are the basis of many criticisms of constructionism." According to Carnap, Goodman "emphasizes correctly" that "a total language is not intended to copy or picture reality either as a whole, or in part, or on a diminished scale, but to represent the relations among the objects in question by an abstract schema" (1963, 940). Carnap's structuralism and constructionalism, which he believed reflected a new "style of thinking and doing... which demands clarity everywhere" distinguished (linguistic) abstractions from reality (1967, xviii; this preface to the first edition is from 1928.).

A third example from this generation is Stephen Toulmin's analysis of "the analogy between physical theories and maps" as found in chapter 4 of his *Philosophy of Science: An Introduction* (1953/1960, 105–139). Toulmin's pioneering discussion is worth considering in detail.

First, Toulmin considered the scientist – especially the physicist – a "surveyor of phenomena" (110). Cartography involved empirically grounded inferential uniformity: "from a limited number of highly precise and well-chosen measurements and observations, one can produce a map from which can be read off an unlimited number of geographical facts of almost as great a precision." Such uniformity also obtained, Toulmin thought, in science: "a limited number of highly accurate observations on [physical] systems" allowed one to formulate a theory, which then underwrote "an unlimited number of inferences of comparable accuracy" (111).

Second, according to Toulmin, an important scientific project was to derive more context-bound, "refined" theories from "fundamental" theories. He drew explicitly on maps: "the relation between geometrical optics [i.e., the refined theory] and the wave-theory [i.e., the fundamental theory] is not unlike that between a road map and a detailed physical map" (115). The latter sort of map Toulmin characterized as "the fundamental map on which the Ordnance Survey might record all the things which it is their ambition to record." That is, geometrical optics and road maps are derived, respectively, from wave theory and a fundamental map through "selection and simplification" (116). Abstraction permits the production of evermore contextual and purpose-specific scientific theories – or, I would add, models – and maps.

Finally, maps negotiate and synthesize the truth and correctness of representations with their use and implementation. In these efforts, both precision and conventions are essential:

> Cartographers and surveyors have to choose a base-line, orientation, scale, method of projection and system of signs, before they can even begin to map an area. They make these choices in a variety of ways, and so produce maps of different types. But the fact

> that they make a choice of some kind does not imply in any way that they falsify their results. For the alternative to a map of which the method of projection, scale and so on were chosen in this way, is not a truer map—a map undistorted by abstraction: the only alternative is no map at all.
>
> *(127)*[5]

Maps (and theory) must distort (and be distorted), and choices about how and what to represent must be made. It is only through such choices that something – i.e., a representation – exists to which "facts" "can be true to *or* falsify" (127).

In short, and without claiming to exhaust Toulmin's many uses of the map analogy, Toulmin draws on basic cartography to usefully describe theory construction as involving the surveying of phenomena; to capture the relation between fundamental and refined theory; and to negotiate truth and use of representation.

This section has attempted to archive at least some of the key uses of the map analogy by first-generation professional philosophers of science. Already, perennial themes of the philosophy of science can be seen to emerge: the use of cartographic objects to illustrate logical structure and conceptual topology; the importance of distinguishing representation (map; theory, model) from world (territory; object, target); and the necessity of negotiating the content and abstraction with the development and application of scientific representations. We see this last point, especially in the single place in *The Structure of Scientific Revolutions* where Thomas Kuhn used the map analogy, observing: "paradigms provide scientists not only with a map but also with some of the directions essential for map-making. In learning a paradigm, the scientist acquires theory, methods, and standards together, usually in an inextricable mixture" (1970, 109).[6]

3. Archive II: the second philosophy of science generation sharpens the map analogy

The second generation of philosophers of science consistently relies on the map analogy to accentuate the purpose- and scale-relative nature of scientific representation, as well as highlight the importance of the partiality and creativity of scientific models.[7] Their efforts increasingly turned to the actual work theories and models do in the world (the so-called "practice turn"), as opposed to the first generation's concerns with rationally reconstructing the structure of physical and biological theories.

For this archivist project, the focus will be on three figures, first sketching Philip Kitcher's uses of the map analogy to elaborate a kind of pragmatic realism and then using my concept of *contextual objectivity* as a conceptual umbrella to explore Helen Longino's and Bas van Fraassen's analyses of the map analogy.

In his 2001 book, Philip Kitcher devotes an entire chapter, titled "Mapping Reality," to philosophical cartography. Very much in line with the third feature of the Toulmin discussion above, Kitcher cares about interpenetrating accuracy and application of representations, whether cartographic or scientific. There is no trade-off between accuracy and convention, nor are they mutually exclusive. There is a single complex world, Kitcher insists, noting that realism is "perfectly compatible with recognizing the fact that human interests change and, in consequence, maps are drawn with very different reading conventions" (2001, 58). In map-making, we divide "the world into things and kinds of things," "depending on our capacities and interests" (59). Analogously, in scientific knowledge production, we classify the parts and properties of the world in various ways and identify and favor different sorts of regularities, causes, and laws (of the world) according to questions of concern (72). The world can be cut in various convention-dependent manners.

Kitcher draws a surprising lesson from the map analogy. He goes so far as to argue that, given the conventions, "the map of the [London] Underground is not *approximately* accurate. It is exact" (59). This is because once we have specified what he calls the *intended content* – i.e., "the region and the types of entities and properties that the map intends to portray" – as well as the *reading conventions* – i.e., the conventions that "link items in the visual display to those [physical] entities and also specify which features of the display do not correspond to any aspect of nature [e.g., the Underground tunnels are not literally colored as in the map]" – then the map is exactly accurate (57). Does the same hold for scientific representation? We are not exactly told, but it would seem so. The important lesson is that accuracy and convention require one another and are both necessary for appropriate and useful cartographic and scientific representation.[8]

In *When Maps Become the World*, I drew on two members of the second generation of philosophy of science to develop my concept of *contextual objectivity*, or "the quality resulting from good and proper application of a representation" (2020a, 95). Accurate bike maps of Copenhagen, Amsterdam, or San Francisco are contextually objective for biking purposes. However, such maps are neither precise, informative, nor useful – i.e., not in any way objective – for a geologist who wishes to know about the kinds of soils, minerals, and fossils that might be found in these cities. Ditto in science, where accurate theories and models are also highly contextually objective when used for the particular ends for which they were designed. As biologists Richard Levins and Richard Lewontin (1985) write, "the problem for science is to understand the proper domain of explanation of each abstraction rather than become its prisoner" (149–150).

Could a bike map like those mentioned above be considered true, approximately true, or even true only for certain local purposes, without being true in general? This might appear like an odd question. But different aspects and elements of the map fit, are accurate, and capture the world in distinct ways. *Truth* seems too generic a success term to capture such varieties of fit. One option for addressing myriad concerns about fit and accuracy, whether in map-, model-, or theory-making (e.g., confirmation) or map-, model-, and theory-use (e.g., explanation and understanding), is a more pluralist strategy about modes of epistemic success. That is: permit a plethora of success terms, depending on the epistemic and pragmatic aims and values of the scientist, scientific community, or public at large.

In *The Fate of Knowledge*, Helen Longino draws on map discourse to develop a contextualist proposal of reference pertinent to scientific theories and models: the *conformation* account. Conformation is a capacious concept delineating a family of epistemic success terms (2002, 117). There is no single, monist principle of justification or truth:

> Maps fit or conform to their objects to a certain degree and in certain respects. I am proposing to treat conformation as a general term for a family of epistemological success concepts including truth, but also isomorphism, homomorphism, similarity, fit, alignment, and such notions. Classical truth is a limiting concept in a category of evaluation that in general admits of degree and requires the specification of respects.
> *(117)*

Different cartographic criteria of fit can be deployed. For instance, is the exact location of relevant features or objects necessary, or is their relative topology sufficient? Finished maps generated by the second criterion (e.g., the famous – not to say clichéd – London Tube map) will be justified differently (and look different) than those generated by the first (e.g., a London street map).[9] If empirically verified by its own criteria, each map can be relevantly precise and accurate – i.e., conformational – for different users and uses, as it is with scientific models. Extracting further from the analogy between map conformation and model (and idealization)[10] conformation, Longino continues: "like maps, models must be sorted out into grades of adequacy in multiple categories, rather than into a single binary category" (118). Thus, Longino deploys the map analogy to argue that a variety of representation relations and criteria of representational accuracy are at play in cartography and science in general. The pragmatic context is critically important in choosing among these and concretizing any one of them in particular.

This contextualism holds not only for the representational relations of mapping and modeling but also for evidence as such. "The pluralist philosopher," Longino says, holds that "it makes no sense to detach measurements and data descriptions from the contexts in which they are generated, or that, as soon as one does, one creates a new context relative to which they are to be assessed and understood" (201). No neutral observation language is necessary or possible for confirmation. After all, different approaches may use commensurable data to produce distinct representations and knowledge of the same system, "each of which conforms to that system differently as both Mercator and Peters projections produce two-dimensional maps that conform, but differently, to the topography of the spherical planet Earth" (201).

Conformation is a broad concept of "epistemological success" marking the appropriate use of an abstraction or representation. I interpret conformation as a component of contextual objectivity: Longino's concept helps us understand the context-dependency and epistemic specificity of partially objective representations and their components.

The map analogy demands an explicit acknowledgment of the simultaneous role of the objective and the subjective. Diverse subjects with locally situated purposes and politics produce public cartographic abstractions representing the (objective) world. This is also the case with scientific theory, as seen in the book *Scientific Representation: Paradoxes of Perspective*. On the one hand, Bas van Fraassen holds that scientific theories, with their model structures, can "be written in coordinate free, context-independent form" (2008, 82). That is, scientific theories are detached, public, and express a "view from nowhere" – they are objective.[11] On the other hand, in order to test or apply information contained in scientific theories, the scientific community must situate the user in the context of the theory (82).[12] Our theories are *also* personal, biased, and express a "view from the inside," as it were.

Van Fraassen insists on the simultaneous importance of subjectivity and objectivity in cartographic endeavors, and, by extension, in science. The map analogy strongly motivates his attempts to show that scientific theories and models are context-independent as well as user-specific – objectivity and subjectivity reach a synthesis. He draws from Immanuel Kant, who discusses the necessity of having both "a map of the heavens" and knowing how "my hands" are positioned relative to it if one wishes to infer where on the horizon the rising sun will appear (1768/1992). In a section called "Mapping and Perspectival Self-Location," Van Fraassen develops the "inevitable indexicality of application" (2008, 80) in light of Kant's map example. Through the concept of the *essential indexical*, van Fraassen argues that maps and scientific theories are context-independent in their universality, detachment, and public availability (i.e., their objectivity) as well as user-specific and therefore biased in their application (i.e., their subjectivity). But what is this essential indexical (2008, 3, 83, 88)?[13]

In order to use a map, we must know where we are *on* it. In this moment of application, we take the map's context-independent information and make a context-bound location judgment, and perhaps even an itinerary that allows us to get from Point A to Point B. And since "models" and "maps" are equivalent "metaphors," according to van Fraassen, it is also the case that "we must locate ourselves with respect to that model" (83). That is, the act of application requires subjective indexicality in scientific modeling as much as in mapping. Precisely because science is "use[d]," we have to let in "consciousness and agency." And to those who would seek to banish subjectivity from science, van Fraassen says, "We will just have to admit a non-pejorative sense of 'subjective', if the essential indexical has to be labeled as something subjective" (83).

In his 1992 presidential address to the Philosophy of Science Association, van Fraassen counters critiques of his anti-foundationalist theory and epistemology of science.[14] The map analogy drives the argument in the first three sections. Van Fraassen concludes that those who dream of a non-theoretical observation language are wrong to relinquish a contextual role for experience in models, maps, and language: "[in] maps and language equally [,] we need, and aim to have, accuracy only in *relevant* respects - inaccuracy elsewhere does not pre-empt the criteria of correctness of self-location with respect to them" (1992, 14). We also learn that "the topic of self-ascription belongs to pragmatics and not to semantics" (7). Pragmatics is necessary to understand the application of science in designing and building technology.

In short, subjectivity and objectivity, accuracy and inaccuracy, pragmatics and semantics, are all required for a full understanding of experience in science. It would be a grave mistake, van Fraassen (1992; 2008) argues – and I concur – to throw the fallible and contextual observation baby of actual science out with the theory-neutral and unified experience bathwater of the positivists. Van Fraassen's (and Perry's) essential indexical is an analytical component of contextual objectivity. The essential indexical highlights the centrality of the user of representations, and also the creator of new representations based on old ones.[15]

There is a strong pragmatic streak in Kitcher's realism, Longino's concept of conformation, and van Fraassen's concept of the essential indexical. They draw on map discourse – and on the map analogy – to illuminate the non-binary nature of scientific theorizing and modeling: accuracy and convention, objectivity and subjectivity, and context-independence and context-dependence are dialectical poles of different spectra that are simultaneously important.[16] Moreover, a plurality of representational and epistemic success relations is necessary for a full comprehension of how scientific representation works.

4. Expanding the map analogy

A pattern of reasoning emerges: as in cartography, so in the philosophy of science. One might say that cartography is the source domain, while philosophy of science is the target domain. When thinking or reasoning analogically, one item or feature from one type of domain, field, or case is compared to – and, hopefully, found in – another domain, field, or case. When the same object or characteristic is found across domains, we say we have or have found a *positive* analogy; when the analogy fails and we do not have the item or feature of the source domain in the target domain, the analogy is *negative* – some might say we have a *disanalogy*; and when we do not know, the analogy is *neutral*.[17] Isaac Newton found positive analogies between fast projectiles and planets in orbit, and Alfred Wegener analogized icebergs on water to continents floating on Earth's hot, inner geological fluid (Newton 1728; Wegener 1966). Here is the central, basic map analogy (Winther 2020a, 29; compare Sismondo 1998; 2004):[18]

A scientific theory is a map of the world.

Both theories and maps are simplifications and idealizations imposing counterfactual assumptions. Both portray only a small subset of the properties and processes of their respective targets – world and territory – in purpose-dependent manners. And both, I have argued, can all too easily be confused and conflated with their target – a phenomenon I call *pernicious reification* (Winther 2014; 2020a; 2020b; compare Dupré and Leonelli 2022.). Indeed, "If pernicious reification is an epistemic and practical failure, contextual objectivity is a knowledge-enhancing and concrete success" (2020a, 90). As is always the case with analogical reasoning, the map analogy breaks down in places. But it is fruitful and beautifully pervasive, as we have also seen above.

In many respects, there is continuity and similarity between the concepts of theory and model. The first two generations of philosophers of science primarily spoke in terms of theories, viewing models either as specific physical instantiations or "analogies" to theories (e.g., Mary Hesse) or as formal offshoots or pieces, as it were, of theory. Nancy Cartwright forced a "modeling turn" in the philosophy of science in the 1980s when the philosophy of science increasingly focused on models. We are still coming to terms with this shift (Cartwright, 1983).

In 2010–2011, I sent a survey to 20 eminent scientists and received 16 responses. This survey included the question "What do you think is the difference between theory and model, if there is any?" The respondents distinguished these two in varied ways. Common distinctions included that theories were quite general and broad and covered many potential and actual phenomena, while models were more local and built-to-purpose. Regardless, it was obvious that both were deemed important, and they were taken to interact. (Of course, today, a decade later, answers might differ.)

In consideration of the above, the potential utility of a distinct analogy for models became evident. A few are on offer already. Cartwright, Shomar, and Suárez proposed the *toolbox* view of science, in which theory was but one input to making a model:

> real things… are represented by models, models constructed with the aid of all the knowledge and technique and tricks and devices we have. Theory plays its own small important role here. But it is a tool like any other; and you can not build a house with a hammer alone.
>
> *(1995, 140)*

In contrast, Marcel Boumans tells us that "model building is like baking a cake without a recipe. The ingredients are theoretical ideas, policy views, mathematisations of the cycle, metaphors and empirical facts" (1999, 67). Both the toolbox and the baking analogies have strengths. So does, I think, a model map analogy, which replaces "theories" with "models" in the analogy above. Each has strengths and illuminates different features of models.

Therefore, I would like to add the following model analogy to the mix:

A scientific model is a vehicle for understanding.

Let us take seriously the play on the term "vehicle," precisely because it does seem to capture important analogies between the physical and the phenomenological, as well as between the objective and subjective. After all, as already Lakoff and Johnson (1980) taught, language captures important correspondences between bodily features and cognitive or moral properties. In its simplest meaning, a vehicle is a train, bike, boat, and, of course, a car. It helps you get from Point A to Point B. Because of work, family, pleasure, or curiosity, we often need – or just wish – to get to a new physical place and space. A vehicle, then, is necessary for satisfying our needs and desires to move our bodies (and minds) to new places.

This sense of movement, I believe, also helps capture what a scientific model can do. It can help us "move" from a state of ignorance or incomprehension to a state of understanding. Since models are somewhat concrete, local, and idealized scientific constructs, we can play with them and draw out lessons about climate, alleles in gene pools of populations, and gravitational waves. In their specificity, models transport us from confusion to understanding. Theories can also do this, but the modeling turn has taught that models are much more concrete playthings helping us along in understanding and intervening in the world.

Consider for a moment an electric vehicle, both literally and metaphorically. Literally, an electric car has new technologies questioning our assumptions about fossil fuel consumption (but of course worries regarding the extractivist mining of rare earth metals abound, and perhaps *reducing* consumption – and *degrowing* our economies – in general would be better). Metaphorically, an electric car qua vehicle is a collection of collective and norm-driven processes interacting with physical technology, that permits us to travel efficiently and (arguably) sustainably from Point A to Point B. We travel or are moved from incomprehension or ignorance to understanding. This version of a vehicle thus analogically captures the locality, complexity, and epistemic value of models.

But a vehicle qua transportation is not the only vehicle possible. I suggest that just as we can broaden the cartographic object from a standard topographical map to a political or military map to, for instance, a geological or extreme-scale or state-space map (Winther 2020a; 2024), so can we expand the notion of vehicle metaphorically, to be the *apparatus* needed to satisfy our aims of interacting with the world. Thus, the vehicle for a scuba diver includes diving gear and air tanks. The vehicle for a hiker includes all the hiking gear. The vehicle for a scientist, all the instruments, lab spaces, computers, etc. This is a version of the *epistemic artifacts* view of models by Tarja Knuuttila and Natalia Carrillo (Knuuttila 2011; Carrillo and Knuuttila 2021). *Models move us towards scientific understanding and they are the scaffolding we require for understanding.*[19]

And, importantly, models do so in constant feedback with general theories. At the risk of being repetitive: if a scientific theory is a map of the world, then a scientific model is a vehicle for understanding. And the analogies are dialectical – we need abstract/general maps and concrete/artifactual vehicles, in interaction, both cognitively and socially, to achieve understanding. The map "points," and the vehicle "moves." The theory-map analogy and the model-vehicle analogy illuminate the interrelation and back-and-forth of models and theory. Models are not "models of theory," but they require theories and theoretical components as one aspect of their structure, development, and use.

5. Conclusion

To be fair, not all philosophers of science have embraced map discourse and the map analogy. Karl Popper was skeptical: "the familiar analogy between maps and scientific theories [is] a particularly unfortunate one." For him, maps were only descriptive and "non-argumentative"; in contrast, theories were "argumentative systems of statements" that could explain and describe deductively (1982, 86). Admittedly, Popper's deductive, normative falsificationism does not on the surface articulate well with a pragmatic reading of the map analogy, but I suspect it might upon further exploration (e.g., maps have a normative ontology).[20]

The analogy – or set of analogies – between maps, mapping, and cartography on the one hand and scientific theories and models, theorizing and modeling, and science on the other has been extensively explored by philosophers of science. In this chapter, I have reviewed some uses of the map analogy in the founding generation of philosophers of science as well as the second generation. Especially the latter interpreted the map analogy in pragmatic ways, while the former was perhaps more exploratory. Thinking cartographically allows us to think in non-dualistic and dialectical manners about structure and practice, representation and world, and truth and convention in the philosophy of science.

Notes

1 Other philosophers of science in the founding generation include Paul Feyerabend (b. 1924, PhD. 1951), C.G. Hempel (b. 1905, PhD. 1934), Mary Hesse (b. 1924, PhD. 1948), Ernst Nagel (b. 1901, PhD. 1931), Karl Popper (b. 1902, PhD. 1928), Patrick Suppes (b. 1922, PhD. 1950), and J.M. Ziman (b. 1925, PhD. 1952). My discussion here significantly expands my earlier too-brief discussion on Carnap and Goodman (Winther 2020a, 46–47).

2 For a discussion of Carnap's project and its rich cultural context, see Daston and Galison (2007, chapter 5 "Structural Objectivity," esp. pages 289–296 and Fig. 5.7 "Structural Map," page 292). See also Leitgeb and Carus (2022), whose "Main Point and Motivation of the *Aufbau*" section summarizes a telling 1929 popular lecture Carnap gave. This lecture contrasted "critical intellect" and "imagination," claiming that human culture started with the latter, but developed the former through "the discovery of *one* [single] *comprehensive space*." Furthermore, critical intellect eventually abstracted this physical space into "an all-comprehending conceptual space" (Carnap's own terms, as presented by Leitgeb and Carus 2022).

3 On Goodman's own constructionalism, see Goodman (1951).

4 In an analogous manner, William James critiques "vicious abstractionism," which interprets concepts as involving "nothing but" definitions. See James (1909); Winther (2014; 2020a; 2020b).

5 In my 2020 book, I present a compressed version of this quote (footnote 24, 96), and a too-brief discussion of Toulmin's deployment of the map analogy (ibid and footnote 1, 60).

6 As explored in Winther (2020a, 195–196), Kuhn also used the map analogy in an essay, "Possible Worlds in History of Science" (2000b), addressing matters of translating and interpreting lexica or vocabularies (alternatively: ontologies or taxonomies) of historical paradigms into later scientific languages. This essay resonated with an earlier essay's themes about incommensurability, translation manuals, and "taxonomic categories of the world" (Kuhn 2000a, 52).

7 This generation includes Nancy Cartwright (b. 1944, PhD. 1971), John Dupré (b. 1952, PhD. 1981), Michael Friedman (b. 1947, PhD. 1973), Ronald Giere (b. 1938, PhD. 1968), Helen Longino (b. 1944, PhD. 1973), Thomas Ryckman (b. 1950, PhD. 1986), and Bas van Fraassen (b. 1941, PhD. 1966), most of whom received their PhDs in the 1970s. Aptly, Ian Hacking (b. 1936, PhD. 1962) falls between this generation and the first.

8 These particular Kitcherian lessons of the map analogy are not discussed in Winther (2020a). Others are.

9 Ziman distinguishes four maps of London: a highway map, "a street directory," a bus route map, and the underground map. He observes: "these four maps all cover the same region on much the same scale, and in spite of various simplifications are all essentially 'truthful'" (2000, 129).
10 Longino notes that idealizations also have various criteria of appropriateness: "Like maps, they are useful just because they do not represent any particular situation, but rather make salient a feature common to a family of similar situations, and in particular because they make salient a feature in which the law's users are interested" (2002, 117).
11 Elisabeth Lloyd critically reviews "four distinct meanings of 'objective' and 'objectivity' that are currently in broad use in contemporary philosophy" (1995, 353), as well as different forms of contrast between objectivity and subjectivity. Nagel (1986) stands as one defense of objectivity as a "view from nowhere."
12 Toulmin drew on the map analogy to motivate non-exclusive distinctions between science and technology, and representing and intervening (compare Hacking 1983).
13 Perry (1979) influentially developed this concept in a philosophy of language context.
14 Interestingly, this 1992 address contains language identical to van Fraassen (2008) on the "self-ascription of location" in maps and in models and is a piece worth examining on its own terms (see, e.g., van Fraassen 1992, 7).
15 Winther (2020b) urges caution with exaggerating a centralized, "world navel" point of view.
16 See Winther (2021a) for the analysis of dialectics used in this chapter.
17 Hesse (1966; 1967) developed this language; compare Bartha (2010).
18 See Winther (2021b) for a general analysis of scientific theory, and of shifting understandings of theory in the philosophy of science.
19 On the philosophy of science of understanding, see de Regt, Leonelli, and Eigner (2009) and Grimm, Baumberger, and Ammon (2017). In instructive conversations, James Griesemer reminds me of the importance of compasses and "navigationism" as supplements to maps and "representationalism."
20 For a "multiple representations account" of the "ontological layer" of maps and models, see Chapter 5 of Winther (2020a). For a philosophical analysis reversing the map analogy – i.e., maps-as-models rather than models-as-maps – see Frigg and Nguyen (2020) and Nguyen and Frigg (2023).

References

Andrews, John H. 1996. "What Was a Map? The Lexicographers Reply." *Cartographica* 33(4): 1–11.

Bartha, Paul. 2010. *By Parallel Reasoning: The Construction and Evaluation of Analogical Arguments*. New York: Oxford University Press.

Boumans, Marcel. 1999. "Built-In Justification." In *Models as Mediators: Perspectives on Natural and Social Science*, edited by Mary S. Morgan and Margaret Morrison, 66–96. Cambridge: Cambridge University Press.

Carnap, Rudolf. 1963. "Replies and Systematic Expositions." In *The Philosophy of Rudolf Carnap*, edited by Paul Arthur Schilpp, 859–1013. La Salle, IL: Open Court.

———. (1967) 2003. *The Logical Structure of the World and Pseudoproblems in Philosophy*. Translated by Rolf A. George. Peru, IL: Carus Publishing. Originally published as *Der logische Aufbau der Welt*, 1928.

Carrillo, Natalia, and Tarja Knuuttila. 2021. "An Artifactual Perspective on Idealization: Constant Capacitance and the Hodgkin and Huxley Model." In *Models and Idealizations in Science: Fictional and Artefactual Approaches*, edited by Alejandro Cassini and Juan Redmond. Logic, Epistemology, and the Unity of Science, vol 50. Cham: Springer. https://doi.org/10.1007/978-3-030-65802-1_2

Cartwright, Nancy. 1983. *How the Laws of Physics Lie*. New York: Oxford University Press.

Cartwright, Nancy, Towfic Shomar, and Mauricio Suárez. 1995. "The Tool Box of Science: Tools for the Building of Models with a Superconductivity Example." In *Theories and Models in Scientific Processes* (Poznan Studies in the Philosophy of the Sciences and the Humanities, Volume 44), edited by William E. Herfel, Władysław Krajewsk, Ikka Niiniluoto, and Ryszard Wojcicki: 137–149. Amsterdam: Rodopi.

Daston, Lorraine J., and Peter Galison. 2007. *Objectivity*. Brooklyn: Zone Books.

de Regt, Henk W., Sabina Leonelli, and Kai Eigner. 2009. *Scientific Understanding: Philosophical Perspectives*. Pittsburgh, PA: University of Pittsburg Press.

Dupré, John, and Sabina Leonelli. 2022. "Process Epistemology in the COVID-19 Era: Rethinking the Research Process to Avoid Dangerous Forms of Reification." *European Journal for Philosophy of Science* 12(20). https://doi.org/10.1007/s13194-022-00450-4

Friedman, Michael. 1987. "Carnap's *Aufbau* Reconsidered." *Noûs* 21(4): 521–545.

Frigg, Roman, and James Nguyen. 2020. *Modelling Nature: An Opinionated Introduction to Scientific Representation*. Cham: Springer.

Goodman, Nelson. 1951. *The Structure of Appearance*. Cambridge, MA: Harvard University Press.

———. 1963. "The Significance of *Der logische Aufbau der Welt*." In *The Philosophy of Rudolf Carnap*, edited by Paul Arthur Schilpp, 545–558. La Salle, IL: Open Court.

Grimm, Stephen R., Christoph Baumberger, and Sabine Ammon. 2017. *Explaining Understanding: New Perspectives from Epistemology and Philosophy of Science*. New York: Routledge.

Hacking, Ian. 1983. *Representing and Intervening: Introductory Topics in the Philosophy of Natural Science*. Cambridge: Cambridge University Press.

Harley, John B. 1989. "Deconstructing the Map." *Cartographica* 26(2): 1–20.

Hesse, Mary 1966. *Models and Analogies in Science*. Notre Dame: University of Notre Dame Press.

———. 1967. "Models and Analogy in Science." In *The Encyclopedia of Philosophy* (Volume 5), edited by Paul Edwards, 354–359. New York: Macmillan.

Kant, Immanuel. 1992. "Concerning the Ultimate Ground of the Differentiation of Directions in Space." In *Theoretical Philosophy, 1755–1770: The Cambridge Edition of the Works of Immanuel Kant*, edited and translated by David Walford, in collaboration with Ralf Meerbote, 361–372. Cambridge: Cambridge University Press. Originally published as "Von dem ersten Grunde des Unterschiedes der Gegenden im Raume," 1768.

Kitcher, Philip. 2001. *Science, Truth, and Democracy*. New York: Oxford University Press.

Kitchin, Rob, and Martin Dodge. 2007. "Rethinking Maps." *Progress in Human Geography* 31(3): 331–344.

Knuuttila, Tarja. 2011. "Modelling and Representing: An Artefactual Approach to Model-Based Representation." *Studies in History and Philosophy of Science, Part A* 42(2): 262–271.

Kuhn, Thomas S. 1970. *The Structure of Scientific Revolutions*. 2nd ed. Chicago: University of Chicago Press.

———. 2000a. "Commensurability, Comparability, Communicability." In *The Road since Structure: Philosophical Essays, 1970–1993, with an Autobiographical Interview*, edited by James Conant and John Haugeland, 33–57. Chicago: University of Chicago Press.

———. 2000b. "Possible Worlds in History of Science." In *The Road since Structure: Philosophical Essays, 1970–1993, with an Autobiographical Interview*, edited by James Conant and John Haugeland, 58–89. Chicago: University of Chicago Press.

Lakoff, George and Mark Johnson. 1980. *Metaphors We Live By*. Chicago, IL: University of Chicago Press.

Leitgeb, Hannes and André Carus. 2022. "Rudolf Carnap," *The Stanford Encyclopedia of Philosophy* (Fall 2022 Edition), edited by Edward N. Zalta, forthcoming https://plato.stanford.edu/archives/fall2022/entries/carnap/

Levins, Richard, and Richard C. Lewontin. 1985. *The Dialectical Biologist*. Cambridge, MA: Harvard University Press.

Lloyd, Elisabeth A. 1995. "Objectivity and the Double Standard for Feminist Epistemologies." *Synthese* 104(3): 351–381.

Longino, Helen E. 2002. *The Fate of Knowledge*. Princeton, NJ: Princeton University Press.

Nagel, Thomas. 1986. *The View from Nowhere*. New York: Oxford University Press.

Newton, Isaac. 1728. *A Treatise of the System of the World*. London: F. Fayram. Originally published as *De mundi systemate*, 1728.

Nguyen, James, and Roman Frigg. 2023. "Maps, Models, and Representation." In *Scientific Understanding and Representation: Modeling in the Physical Sciences*, edited by Kareem Khalifa, Insa Lawer, and Elay Shech. New York: Routledge, 19 pp.

Perry, John 1979. "The Problem of the Essential Indexical." *Noûs* 13(1): 3–21.

Popper, Karl 1982. *Unended Quest. An Intellectual Autobiography*. Revised edition. La Salle, IL: Open Court Publishing Company.

Robinson, Arthur H., and Barbara B. Petchenik. 1976. *The Nature of Maps: Essays toward Understanding Maps and Mapping*. Chicago: University of Chicago Press.

Sismondo, Sergio 1998. "The Mapping Metaphor in Philosophy of Science." *Cogito* 12(1): 41–50.

———. 2004. "Maps and Mapping Practices: A Deflationary Approach." In *From Molecular Genetics to Genomics: The Mapping Cultures of Twentieth-Century Genetics*, edited by Jean-Paul Gaudillière and Hans-Jörg Rheinberger, 203–209. London: Routledge.

Toulmin, Stephen E. 1953/1960. *The Philosophy of Science: An Introduction*. Reprint. New York: Harper & Row.

van Fraassen, Bas C. 1992. "From Vicious Circle to Infinite Regress, and Back Again." In *PSA: Proceedings of the Biennial Meeting of the Philosophy of Science Association*, edited by David L. Hull, Micky Forbes, and Kathleen Okruhlik, vol. 2, 6–29.

———. 2008. *Scientific Representation: Paradoxes of Perspective*. New York: Oxford University Press.

Wegener, Alfred. 1966. *The Origin of Continents and Oceans*. Translated by John Biram. New York: Dover. Originally published as *Die Entstehung der Kontinente und Ozeane*, 1929.

Winther, Rasmus Grønfeldt. 2014. "James and Dewey on Abstraction." *The Pluralist* 9(2): 1–28.

———. 2020a. *When Maps Become the World*. Chicago, IL: University of Chicago Press.

———. 2020b. "Cutting the Cord: A Corrective for World Navels in Cartography and Science." *The Cartographic Journal* (British Cartographic Society) 57(2): 147–159.

———. 2021a. "Lewontin as Master Dialectician: Rest in Power, Dick." In *Science for the People*, edited by The Editorial Collective. https://magazine.scienceforthepeople.org/lewontin-special-issue/lewontin-as-master-dialectician/

———. 2021b. "The Structure of Scientific Theories." In *The Stanford Encyclopedia of Philosophy* (Spring 2021 Edition), edited by Edward N. Zalta. https://plato.stanford.edu/archives/spr2021/entries/structure-scientific-theories

———. 2024. "Map Thinking across the Life Sciences." In *The Routledge Handbook of Geospatial Technologies and Society*, edited by Alexander J. Kent and Doug Specht, 600–612. New York: Routledge.

Wood, Denis. 1992. *The Power of Maps*. With John Fels. New York: Guilford Press.

———. 2012. "The Anthropology of Cartography." In *Mapping Cultures: Place, Practice, Performance*, edited by Les Roberts, 280–303. Basingstoke: Palgrave Macmillan.

Ziman, John M. 2000. *Real Science: What It Is, and What It Means*. Cambridge: Cambridge University Press.

26
METAPHORS, ANALOGIES, AND MODELS

Sergio F. Martínez

1. Introduction

In the 21st century, the close relationship between metaphors, analogies, and models is an important topic of discussion in some scientific disciplines and in the philosophy of science. But there are different ways of thinking about this relationship. In this chapter, some of this diversity is presented and discussed. The different approaches have implications for traditional ways of framing questions about models, and in particular, about how they earn their epistemic (or pragmatic) value. The chapter starts by reviewing some basic concepts and delimiting their scope. The topic is too vast and complex to deal with in an entire book, let alone in a single chapter. However, some insights can be gained by focusing attention on a few key questions. The second section presents a brief outline of the historical context in which metaphors and analogies in science as tools of scientific inquiry were discussed during the 20th century. The third section reviews the concept of the mathematical model developed from the mid-19th century to the mid-20th century, culminating in the highly influential approach of scientific structuralism.[1] Scientific structuralism, focusing on the semantic interpretation of models, has had a major role in setting the agenda for philosophical discussions about models. Variants and spin-offs of the program have shaped the study of modeling since then. But such variants and spin-offs tend to inherit presuppositions, which lead to focusing on the question of realism, or on formalizing the inferential and representational practices of scientists. Often, questions are inherited, which, even if important, conceal the neglect of other relevant issues, for example, the significant contribution of metaphors in shaping conceptual change. In the fourth section, relevant work in the cognitive sciences on the changing meaning of analogy, metaphor, and model (and their relation) will be presented. Also in this section, the work of Max Black and Mary Hesse will be presented. Hesse argued in dozens of publications for almost five decades about the importance of rethinking the relation between metaphors, analogies, and models, and above all, about the need to unveil epistemic presuppositions in traditional discussions about models.

Hesse developed different views about what metaphor is in science and how figurative language in general relates to the question of the relation between models, analogies, and metaphors. However, only some key points will be focused on here.[2] Hesse claims that the

DOI: 10.4324/9781003205647-31

distinction between literal and metaphoric language is only a pragmatic, not a semantic distinction, and thus, all language in the relevant sense is metaphoric. Serious consideration of this view has important implications even today for how to account for the cognitive role of metaphors and models in science. In the fifth section, examples are presented of the diversity of ways in which the relation between metaphors and models is being developed in contemporary philosophy of science; it concludes by pointing to the relevance of current work in embodied cognition and, in particular, to work on ecological theories of metaphor for advancing the study of the relation between metaphor and models in the context of philosophy of science.

2. The lost connection between metaphoric language and philosophy

The close connection between metaphoric language and philosophy was a main ingredient of the earliest philosophical traditions. Pre-Socratic philosophy arose as a refinement of folk theories and metaphors; this also seems to be the case in other philosophical traditions. In classical Greek philosophy and Western philosophy, the tendency has been to separate sharply between logic, good reasoning, and truth on the one hand, and the importance of metaphor in Art and Rhetoric on the other. Reviel Netz has shown how the development of the "Euclidean tool-box" leads to "the shaping of deduction" (Netz 1999, 222), which extends this rift to a more general opposition between science and philosophy on one side, and the humanities on the other.

Aristotle regarded metaphor highly, as a valuable tool in all linguistic communication. He considered the mastery of metaphor a mark of artistic genius, yet he warned about the use of metaphors in philosophy and science. This view follows easily from the Greeks' view of science as structured in deductive arguments. For Aristotle, the key role of metaphor is to allow for analogical comparison. A metaphor is converted into a simile and is then interpreted by comparing the respects in which two things are similar. Thus, the perception of similarity grounds the understanding of the metaphor and its use. Since this relation of similarity cannot be reduced to deductive reasoning, it follows that metaphors should be kept away from scientific reasoning. Such a view of the role of metaphor in reasoning continues to be influential throughout the history of philosophy, even today. Classical philosophy abounds in warnings about the use of metaphors in the pursuit of epistemic ends. Bacon and mainstream empiricist philosophy later fought against the use of metaphors in science as part of their crusade against ambiguous language. Locke's idea that metaphors lead us astray by moving the passions and misleading judgment (Locke 1690) is advice closely followed by empiricist philosophers even to this day. Often, this advice is formulated by saying that metaphors do not have a cognitive role or function.

In Anglo-Saxon philosophy of mind and language, the existence of a literal language having a special epistemic significance has been taken for granted. Scientists should strive to formulate claims to knowledge as close to this literal language as possible, and that means to formulate knowledge in a logical and mathematical form as much as possible. Logical empiricism is a well-known example of this kind of approach. The warning against the use of metaphors is no less strong in the history of science and even in today's scientific writings. When Darwin presented his theory of evolution by natural selection, most critics questioned its scientific status by pointing to the metaphorical nature of his proposal.[3] Pierre Duhem famously questions the metaphorical nature of physical theory based on the use of mechanical models. He accepts that models drawn from familiar mechanical devices might be a psychological aid in picturing theory, but he claims that such models have no epistemic value.[4] After all, dreams, astrological beliefs, or palm reading might play a role in discovery, but such things have no lasting significance and should not be considered part of science. Duhem thinks that the world is ultimately logically ordered and that science is the search for this order. Accepting models (metaphors) as explanations just distracts us from the only important task in science, the search for logical order. Mary Hesse famously presented an imaginary dialog between Pierre Duhem and N.R. Campbell. Campbell defends the view that analogies are not aids to the establishment of theories; instead, he thinks that they are an essential part of theories (Campbell 1920). Hesse makes clear that even though the discussion between Duhem and Campbell revolves around the importance of mechanical models, there are important points of the discussion that were relevant when she wrote the book and are still relevant today (see Hesse 1966 and Section 4 below).

Warning about metaphors has been and still is common today among scientists and philosophers (Pauwels 2013; Boudry and Pigliucci 2013). Nevertheless, metaphors are increasingly recognized as a key ingredient in scientific inquiry, not only for having heuristic value but also for being important in the development of scientific understanding (Olson, Arroyo-Santos, and Vergara-Silva 2019; Reynolds 2022; Keller 2002; McCloskey 1990).

3. On the history of the concept of model in relation to the changing views on metaphor

In the philosophy of science, the history of scientific modeling is often said to have started in the 19th century, in the scientific writings of Kelvin, Thompson, and Maxwell. There are indeed good reasons to see the beginning of contemporary discussions on modeling in these authors' work, but we should keep in mind the close connection between analogy, metaphor, and model presupposed by Kelvin and contemporaries, and that the idea of model in question was not the one that became the canonical notion in the twentieth century. Maxwell and Kelvin do not use the term model in the semantic sense but in the rather narrow sense of prototype, scale model, or mechanical model (which is a kind of prototype model). In addition, it should be kept in mind that there are also significant differences in how Maxwell and Kelvin understood the term model. For Kelvin, one may claim to understand a phenomenon if one can make a mechanical model of it. Maxwell also thought highly of mechanical models and of their important place in science, but only as parts of a useful epistemic tool. For him, models in this narrow, mechanical sense are part of a methodology that can lead to understanding. This methodology is what Maxwell calls "the method of physical analogy" or the method of "truly scientific illustration." A truly scientific illustration is grounded in the metaphor of the "Book of Nature":

> The Book of Nature, in fact, contains elementary chapters, and, to those who know where to look for them, the mastery of one chapter is a preparation for the study of the next.
>
> *(Maxwell 1890a, 338)*

The metaphor points to the importance of what we already know to what we want to know. Thus, a truly scientific illustration, i.e., a good physical metaphor, exploits to the fullest a comparison between a mathematical description and a physical hypothesis, and this requires identifying a similarity between two sciences. Maxwell formulates the idea as saying that "partial resemblance between the laws of a science and the laws of another science" allows us to use one science to "illustrate" the other (Maxwell 1890b).[5] The comparison leads to a new meaning metaphorically expressed. For Maxwell, a physical analogy is "science forming" in the sense that the physical analogy orients us in the direction in which experimental and conceptual work can make the analogy more precise, also by showing the limits and possible relations with other metaphors and concepts.

The sense in which the method of physical analogy is "science forming" of course requires elaboration, and, as it turns out, there is not only one way in which analogy is "science forming". More about this point later on. For now, it is only important to be clear that the method of physical analogy is not meant to be a version of the well-known argument by analogy, nor can it be identified with the usual claim that a relation of analogy can be reduced to an isomorphism between two structures. In an argument by analogy, the fact that two sets of phenomena A and B have several properties (say x, y, z) that are similar, invites us to conclude (non-deductively) that if A has another property w, then B should also have it. Clearly (contrary to Duhem) this is not what Maxwell is doing. But then how do we understand a process of analogical reasoning that would be genuinely "science forming" as Maxwell claims?

One important answer that arises from the development of the formal concept of the mathematical model is a byproduct of structuralism in mathematics (and mathematical logic in particular) that played a key role in logical empiricism. Structuralism led to the view that this question could be answered in a rather straightforward way. The formalization of the semantics of a theory in terms of "models," in the specialized sense developed in mathematical logic, could be extended from the formal to the "non-formal" sciences. This extension of the formal notion of model to all sciences would provide a clear and definite normative account of the methodological unity of all sciences and, in the process, dispense with the need to account for metaphors or analogies as having a cognitive role. Also, the related distinction between the context of discovery (of interest for psychology) and the context of justification (of interest for philosophy) implied that the construction of models involved psychological processes that were not relevant to the philosophical task of understanding the structure and advancement of science. Consequently, metaphors and analogies were largely overlooked in the subsequent discussions in the philosophy of science.

Freudenthal (1961), for example, reunites scientists and philosophers of the mid-twentieth century, aiming to show how the formal study of models could be extended into the "non-formal sciences." As Apostel puts it in the first chapter of the book, "*the concept of model will be useless if we cannot deduce from its function a determinate structure, thereby providing a 'rational reconstruction' of the informal use of models by scientists*" (Apostel 1961, emphasis added). The consequence of such a view is clear. Metaphors and analogies in science do not play a role in such a "rational reconstruction" and thus, even if they can be part of the context of discovery, they do not play an epistemic role. The cognitive role of models has to be expressed in formal semantics. The plausible and common-sense (but difficult to spell out) connection of models with analogies and metaphors is cut off. The use of analogies and metaphors as part of methodologies like Maxwell's "method of physical analogy" should be understood via a "rational reconstruction" (in a formal model of inference), and any further discussion could be relegated to historians or psychologists.

One consequence of this project is that analogical reasoning should be formalizable. Carnap spent decades attempting to carry out this task to no avail, but the idea that analogical inference can be reduced to Bayesian inference (a kind of formal reasoning) is quite common even to this day (Bartha 2010). From the perspective of scientific (semantic) structuralism, the philosophy of science has little to gain from a search to better understand the contribution of metaphors to the meaning or explanatory value of scientific statements (as this is done in the humanities and the empirical sciences).[6] Metaphors and models are two different kinds of things. Whereas metaphors are purely linguistic devices, models are cognitive tools playing an epistemic role (in explanation, prediction, or understanding). Whatever meaning metaphors have arises in the context of everyday language, whereas the meaning of models is given by its relation to truth in the formal Tarskian sense.[7] It was soon realized that Tarskian models were not sufficient to account for the way a formal theory (and by extension, a non-formal theory seen from the semantic structuralist perspective) related to the world. Such relations required more than a Tarskian notion of truth, and this led to the introduction of the notion of representation to fill out the gap (Giere 1988; Van Fraassen 1980). Another way of accounting for the relation between a model (of a theory) and the world was the one proposed by Mary Hesse. She suggested that the way to answer this challenge was to develop a "family resemblance theory of meaning" which led to the view that all language is metaphorical and brought back the discussion about the cognitive role of metaphors and analogies in the philosophy of science. From Hesse's perspective, the discussion of the role of analogies and metaphors in science was only a special case of a more general theory of analogy. Such theory would require the development of a non-Tarskian notion of truth, which should be congruent with work in the cognitive sciences about the kinds of cognitive processes involved in analogical reasoning.[8]

Hesse's theory initially started as an important modification of the *interaction theory* of Max Black (1962), according to which metaphorical statements can generate new knowledge by changing the way in which a system designated as a primary subject relates to a second system designated as a secondary or subsidiary subject. For Black, such a change of relationships between the primary and the secondary subject constitutes a cognitive function of metaphors. As Hesse and other philosophers pointed out, it is far from clear why such interactions have cognitive value. How are we supposed to relate such interactions, for example, to Gentner's structural mappings (see Section 4), or to some other credible empirical theory of metaphors' cognitive content?

Hesse frames the problem of characterizing the cognitive role of metaphors in science as a more general philosophical problem. How do metaphors and analogies contribute to our understanding of art, music, religion, and science? Hesse believed we should search for a way of characterizing metaphoric meaning in general, and then we could, as a special case, explain the cognitive role of metaphors in science.

Hesse's adoption of Wittgenstein's family resemblance analysis of concepts leads her to her central thesis that all language is metaphorical but distinguishes acceptable from unacceptable schemes of categorization following Rosch's work on prototypes structured by cue-properties within a category (2002). A concept then is represented by an interconnected set of constraints, and the modification of constraints or the generation of new constraints leads to conceptual change. For Hesse, science can be regarded as a special case of such a general theory of categories.

Maybe the most important consequence that Hesse derives from the claim that all language is metaphoric is that metaphors can be used as descriptions and can also play a normative-evaluative role that influences our actions as constrained by certain values and expectations. As she puts it, "metaphor is concerned with action as well as description" (Hesse 1988).

Hesse has been influential in suggesting ways in which work on metaphors and in general work on figurative language could be useful in philosophical discussions about models and, in particular, for accounting for the explanatory value of (some) models. Nonetheless, her work has serious limitations in that she thinks of models paradigmatically as models of theories and her focus is on the language of science. Nowadays, the focus is on the practices of science and models as part of such practices. Of course, models have important relations to theories, but such relations can be multifaceted, contingent, and transitory. After all, the concept and theory of models also have changing meanings in different disciplines and throughout the history of science. The following section provides examples of these more contemporary approaches to the relation between models and metaphors and provides examples of the variety of ways in which metaphors are considered to play a cognitive role.

4. The changing meaning of metaphors, analogies, and models

The traditional view of metaphor in literary studies and philosophy has been that metaphor is "a poetically or rhetorically ambitious use of words, a figurative as opposed to literal use" (Hills 2022). Since Aristotle, there have been many attempts to characterize the different kinds of metaphors one finds in linguistic practice. Aristotle distinguishes four kinds of metaphor, but he considered metaphor supported by analogy as the most important kind. This way of subordinating metaphors to analogy (or likeness) has been quite influential in literary studies and philosophy. It is often accompanied by the assumption that there is a distinction between literal and metaphorical language, and thus the question of the cognitive content of a metaphor is reduced to the question of whether likeness can be a source of knowledge. Most often in the philosophy of science metaphors are introduced as subordinated to models since models are understood in the structuralist tradition as models of theories. A well-known article by Richard Boyd, "Metaphor and theory change: what is 'metaphor' a metaphor for?" (1993), suggests metaphors work as assistants to models, as useful ways of illustrating how models work. But this way of looking at metaphors assumes a fundamental distinction between what the role of metaphors in science is, and what the role of metaphors in non-specialized human cognition is.

Contemporary work on the relation between metaphors and models abandons the idea that models are models of theory and, thus, are required to account for the source of epistemic support of a model; empirical sciences and the cognitive sciences in particular enter to fill this gap.

In cognitive psychology, there has been a lot of work on the relation between metaphors and analogies. Dedre Gentner talks of metaphors as a special kind of analogy, in that the source and the target domains are semantically distant (1982), which explains the well-known observation that metaphors are more sensitive to the semantic context; metaphors are often combined or take the role of other figures of speech. For example, the image of a flea in Hooke's famous book (1665) stands metonymically for the whole microscopic world, but it can also function as a metaphor in the sense that the image focuses on the possibility of a new kind of knowledge anchored in scientific instrumentation and methods (for an elaboration of this point, see Martínez 2023).

Maybe the most common account of the relation between metaphors and analogies in cognitive psychology is the idea that metaphor can be characterized as *structure mapping*, which at the same time can serve as a theory of analogy. Gentner has written extensively on this topic. In an early paper written together with Michael Jeziorski, they question the view that the faculty for analogical reasoning is an innate part of human cognition. Using examples from the history of science throughout different epochs, they show that there are important changes in what is considered a good use of analogy (Gentner and Jeziorski 1993). Examples from the history of science lead them to claim that the importance of metaphors has given space to the preeminence of analogy based on the (often implicit) acceptance of several principles that constrain the way analogical reasoning is carried out. The most important of those principles is that the human processing of analogy is carried out through structure mapping. This basic idea has been elaborated by Gentner and many other psychologists, and it is maybe the most influential account of the relation between metaphor and analogy. An analogy elaborates the similarity present in a metaphor by providing a more systematic and focused formulation in terms of structural mappings (which can be characterized at least in principle as isomorphisms). In a more recent paper, Gentner and Bowdle argue that figurative statements begin as novel comparison statements and evolve gradually into category-inclusion statements as the vehicle terms develop a metaphorical abstraction (2008). From this perspective, metaphor and analogy only differ along the axis of conventionality. The mappings in metaphor can be activated automatically or with little effort, which amounts to their conventionality.

The theory of structure mapping has been very influential in the cognitive sciences as a way of unifying discussions about metaphors and analogies. There is no doubt that structure mapping is important in metaphor processing and analogical reasoning at the individual level, but as we shall see, there is important recent work suggesting that metaphor processing (and the construction of analogies based on such processing) is also done at an embodied and collective level, and this can be particularly important for understanding the way conceptual change takes place and metaphors and scientific models relate to each other (see the next section). Besides, thinking of metaphors and analogies in terms of structure mapping does not help us in accounting for the tension (mentioned in Section 2) that is quite relevant for making sense of the cognitive roles of metaphors. On the one hand, there are a lot of warnings about the use of metaphors in science, on the other it is a fact that metaphors are widespread in all human communication and in science and that many scientists consider metaphorical thought, if not synonymous with, at least closely related to, creative thought. Finally, as we have seen in the case of Maxwell, there is more to the role of metaphors and analogies in "science forming" than mere structure mapping. Metaphors can play their cognitive (normative) role in conceptual change, in the evaluation of methodologies, or in their contribution to the narrative form of explanations. Should we expect that such a cognitive role is the same in different contexts? If there are different roles in different contexts, should we not take the aim of different models into consideration, or their ecologies, in clarifying the normative role of metaphors in modeling?

5. On the diversity of ways in which models and metaphors can connect

In a text in which he goes back to defend and modify his interaction view of metaphors (published almost 15 years earlier), Max Black says the following:

> I am now impressed, as I was insufficiently so when composing Metaphor, by the tight connections between the notions of models and metaphors. Every "implication complex" supported by a metaphor's secondary subject, I now think, is a model of

> the ascriptions imputed to the primary subject: every metaphor is the tip of a submerged model.
>
> *(Black 1977, 444)*

Black seems to have been convinced that ultimately, structuralism was on the right track and that the cognitive role of metaphors should be explained in terms of models. That was 1977.

Nowadays, the iceberg seems to have flipped over. Morgan says that "turning a metaphor, which begins as a figure of speech and idle likeness, into an analogical model involves both cognitive and imaginative work" (2012, 173). On the next page, she mentions Bowman's way of describing how a metaphor becomes a model. For Bowman, a metaphor is one dimensional like "money is liquid": the development of its various possibilities or dimensions transforms the metaphor into a model. In the discipline of economics, there is a long tradition of thinking of models as metaphors, starting at least with Henderson (1982). To say that (economic) models are metaphors, points toward viewing models as artifacts, in the sense that the artifactual perspective on models highlights the tool aspect of modeling, as well as the role of imagination (Knuuttila 2021). It also points toward the cognitive-behavioral dimension of metaphors and models. Metaphors, as Hesse said, are concerned with action as well as description.

Another way of characterizing the relation between metaphors and models in science is what Nancy Nersessian calls a "cognitive-historical method" (1987; 2008). Her approach of focusing on the practices that supported the creation of stable explanatory models elaborates on Hesse's idea that models, are a sort of analogy constructed in what is often a complex process. However, such processes do not often fit the traditional idea of interaction from a secondary to a primary subject. Nersessian points out *that models are often built explicitly to serve as analogical sources*, which leads to the conclusion that, in order to understand the cognitive role of models we require an account of analogical reasoning. It also often requires the construction of intermediate models that embody the features and constraints of both the source and the target of the analogy (or metaphor). Such reconstruction of a process of intermediate models, is an important and often neglected kind of analogical reasoning (Nersessian 2015). See also Morgan and Morrison (1999); Morgan and Knuuttila (2012); Knuutila and Loettgers (2014).

There are many other senses in which models are metaphors. Jordi Cat argues that metaphors in Maxwell often function as illustrations of the relation between abstractions and the concrete conceptions that characterize a privileged representation of our interactions with the world (Cat, 2001), a representation which in turn guides our actions therein. Here the metaphor is not "one dimensional," rather, the metaphor is the model, but in a very special sense. As Cat argues, for Maxwell, the concreteness in question is not reducible to geometrical imagery, but involves the embodiment of such abstractions in what (following Otto Sebum and Michael Polanyi) he calls "embodied understanding."

Cat's discussion makes clear that the cognitive role of metaphors is not something that can be identified independently of a particular time and tradition of inquiry and that it is not readymade. This is a result that resonates with recent theories of metaphor that in different ways promote the importance of understanding metaphors as embodied. Lakoff and Johnson's theory (Lakoff and Johnson 1980) distinguishes between metaphor as a communicative device, metaphor as a linguistic phenomenon, and metaphor as an embodied cognitive tool allowing the construction of abstract concepts from other more basic or concrete concepts. People's use of metaphorical language points to the presence of underlying conceptual metaphors that support the sense of understanding of metaphorical language. Instead of Black's mechanism of interaction between primary and secondary subjects, Lakoff and Johnson proposed the concept of metaphorical transfer, which involves whole domains and not isolated concepts (1987, 288). Another important distinction they introduce is between orientational, ontological, and structural metaphors. Ontological metaphors are the kind of metaphors that, as in Maxwell's example, allow us to construct abstractions. Structural metaphors represent a more complex kind of transfer in which metaphors serve to organize or decompose a concept in terms of another concept. "Money is liquid" leads to "money slips through your fingers". Orientational mappings are metaphors that organize a whole system of concepts with respect to one another. For example, happy is up/sad is down leads to "I am low" or "I am feeling up". Lakoff and Johnson's theory has been very influential and also strongly criticized.

More recent theories understand metaphors as embodied processes, not merely as figures of language, but in different senses of embodiment. In what are called ecological (or dynamical) theories of metaphor (Gibbs 2019; Jensen and Greve 2019), cognition is no longer assumed to take place only in the head, but in a cognitive niche. Gibbs argues that *metaphorical performances* are always part of dynamical ecological cognition, processes that do not just take place in the head but involve collective actions and organized practices. For Gibbs, as with several promoters of the view of metaphor as embodied cognition, metaphor "is a dynamical constraint on action that is distributed across brains, bodies, and real-world ecologies" (Gibbs 2019; Thibodeau and Boroditsky 2011).

This way of formulating the embodied nature of metaphors, focusing more on metaphors as cognitive processes than as mere figures of language, is particularly well suited to characterize the way historians and philosophers of science often talk about metaphors (and their relation to modeling). Evelyn Fox-Keller and Mary Morgan, for example, talk of "central metaphors" in the structuring of mastering narratives, which have an important role in explaining the development of science (see Keller 2002; 2015; Morgan 2012; Morgan, Hajek, and Berry 2022). The metaphors grow and stabilize their meaning through the process of structuring a master narrative. In a recent publication, Carrillo and Martínez (2023) show the importance of tracing the historical lineages of key metaphors because they articulate criteria of explanatory relevance. In this way, they claim, investigating how metaphors evolve historically and their relation to the culture at the time they emerge is key to making explicit the grounds for abstractions that articulate research programs.

Nancy Nersessian has emphasized for several decades that the sort of experimental studies in cognitive psychology that have been used to study analogical reasoning is not adequate to understand the role of metaphors and analogies in science, at least not in cases in which important conceptual changes are involved (Nersessian 2002; 2015). In everyday language as well as in theories of metaphorical reasoning common in cognitive psychology, it is assumed that there is a ready-at-hand solution to the source problem which is transferred to the target problem.

What Nersessian shows with several important case studies is that the source analogy itself needs to be constructed. In the same vein, Knuutila and Loettgers (2014) also emphasize the importance of negative analogies for analogical reasoning and compare analogical reasoning to template-based model transfers (Knuuttila and Loettgers 2020). Such reconstruction processes of analogies involve ethnographic, historical, as well as cognitive methodological resources. This methodology is what Nersessian calls the cognitive-historical method for the study of analogical inference in science. More generally, there are different ways in which models metaphors and narratives mutually constrain each other and shape conceptual change and scientific inference. Metaphors as models are cognitive tools that are undergoing constant retooling processes.

6. Conclusion

The philosophy of scientific modeling has been centered until recently on the question of representation and inference following the initial framing of the question by scientific structuralism. Such an approach does not make enough space for the significant cognitive role of metaphors and analogies in science beyond the role that analogical reasoning can play in scientific inference. Scientific inference is, of course, quite an important topic, but as we have seen, the cognitive role of metaphors and analogies in science should not be reduced to its role in a theory of scientific inference. The question of the cognitive role of metaphors has been an important topic in the history of science for decades, in a thriving subfield of research in cognitive psychology, and more recently in other areas of the cognitive and social sciences.

Recent work by philosophers and historians of science on modeling practices in different scientific disciplines shows the importance of understanding the role of metaphors and analogical models in the dynamics and organization of science (including its epistemic organization), which requires going beyond the linguistic role of metaphors. Metaphors and analogies form part of a web of cognitive and epistemic artifacts scaffolding the development of abstract concepts, models, and narratives, which are "science forming," as Maxwell said.

Acknowledgement

This work has been supported by Proyecto IN400027 UNAM.

Notes

1 Scientific structuralism refers to several approaches in philosophy of science which, in different ways, elaborate an analogy with mathematical structuralism, a highly successful account of mathematics as a collection of structures. For mathematical structuralism, mathematical objects are ultimately "positions" of the structure (see, e.g., Benacerraf 1965, 70). This has led to the idea that a mathematical theory can be characterized by its models (see Landry and Marquis 2005). Analogously, scientific theories can be characterized as a collection of models sharing a common structure. For discussion on the different kinds of scientific structuralism and its problems, see Van Fraassen (2007); Brading and Landry (2006); Lorenzano (2013).

2 More in-depth discussions about Hesse's proposals can be found in Helman (1988). French (2017) introduced a virtual issue of the *British Journal for the Philosophy of Science* on the work of Mary Hesse. Another important reference for the discussion of the work of Mary Hesse is the special issue: *Philosophical Inquiry* (3)1 (2015), which discusses the work of Hesse on the question of how to understand the cognitive role of metaphors. Rentetzi (2005) elaborates on Hesse's view on explanation and shows the relevance of this proposal for contemporary discussions on models of explanation.

3 Reynolds shows how the evidence gathered by Darwin and others for decades in favor of the theory of evolution by natural selection was not readily accepted because for most of his contemporaries, the battle was of "fundamental metaphors" (Reynolds 2022). Keller talks of "global narratives" which are articulated by different metaphors (Keller 2002), see Section 2. See also (White, Hodge and Radick 2021).

4 A famous example of the use of metaphors and analogies in the construction of scientific knowledge is what Maxwell calls "the method of physical metaphor" or "physical analogy" which he claimed led him to his famous results on electromagnetism. Duhem claimed that Maxwell in fact found these results by other means and the analogy in question was formulated after the fact (1954, 98). For a detailed argument showing the credibility of Maxwell's claims see Nersessian (2002) and Cat (2001).
5 But to read such "partial resemblance between laws" as isomorphism would be anachronic. See Nersessian (2002); Cat (2001).
6 The importance of metaphors in the construction of public opinion on key issues, as well as the different interpretations of the same situation or fact depending on the use of different metaphors is an important topic in science communication and in theories of reasoning (Semino and Demjén 2017; Thibodeau and Boroditsky 2011).
7 A model is said to satisfy a sentence if the interpretation of that sentence within the model makes the sentence true.
8 In this chapter, the question of how to characterize analogical reasoning will be discussed no further. For a discussion of this issue see Bartha (2010; 2022); Gentner, Holyoak, and Kokinov (2001); Holyoak and Thagard (1996).

References

Apostel, Leo. 1961. "Towards the Formal Study of Models in the Non-Formal Sciences." In *The Concept and the Role of the Model in Mathematics and Natural and Social Sciences*, edited by Hans Freudenthal, 1–37. Dordrecht: D. Reidel Publishing Company.

Bartha, Paul. 2010. *By Parallel Reasoning*. New York: Oxford University Press.

———. 2022. "Analogy and Analogical Reasoning." *The Stanford Encyclopedia of Philosophy* (Summer Edition), edited by Edward N. Zalta. https://plato.stanford.edu/archives/sum2022/entries/reasoning-analogy/

Benacerraf, Paul. 1965. "What Numbers Could Not Be." *Philosophical Review* 74(1): 47–73.

Black, Max. 1962. *Models and Metaphors: Studies in Language and Philosophy*. New York: Cornell University Press.

———. 1977. "More about Metaphor." *Dialectica* 31(3–4): 431–457.

Boudry, Maarten, and Massimo Pigliucci. 2013. "The Mismeasure of Machine: Synthetic biology and the Trouble with Engineering Metaphors." *Studies in History and Philosophy of Science Part C* 44(4): 660–668.

Boyd, Richard. 1993. "Metaphor and Theory Change: What Is 'Metaphor' a Metaphor for?" In *Metaphor and Thought*, edited by Andrew Ortony, 481–532. Cambridge: Cambridge University Press.

Brading, Katherine, and Elaine Landry. 2006. "Scientific Structuralism: Presentation and Representation." *Philosophy of Science* 73(5): 571–581.

Campbell, Norman R. 1920. *Physics, the Elements*. Cambridge: Cambridge University Press.

Carrillo, Natalia and Sergio Martinez. 2023. "Scientific Inquiry: From Metaphors to Abstraction." *Perspectives on Science* 31(2): 233–261.

Cat, Jordi. 2001. "On Understanding: Maxwell on the Methods of Illustration and Scientific Metaphor." *Studies in History and Philosophy of Science Part B*: 32(3): 395–441.

Duhem, Pierre. 1954. *The Aim and Structure of Physical Theory*. Princeton, NJ: Princeton University Press.

French, Steven. 2017. "Models and Meaning Change: An Introduction to the Work of Mary Hesse." *British Journal for the Philosophy of Science, Special Virtual Issue on the Work of Mary Hesse.*

Freudenthal, Hans, ed. 1961. *The Concept and the Role of the Model in Mathematics and Natural and Social Sciences*. Dordrecht: D. Reidel Publishing Company.

Gentner, Dedre. 1982. "Why Nouns Are Learned before Verbs: Linguistic Relativity versus Natural Partitioning. In *Language Development Vol. 2: Language, Thought and Culture*, edited by Stan Kuczaj, II. 301–334. Hillsdale: Lawrence Erlbaum.

Gentner, Dedre, and Brian Bowdle. 2008. "Metaphor as Structure-Mapping." In *The Cambridge Handbook of Metaphor and Thought*, edited by Raymond W. Gibbs, Jr., 109–128. New York: Cambridge University Press.

Gentner, Dedre, and Michael Jeziorski. 1993. "The Shift from Metaphor to Analogy in Western Science." In *Metaphor and Thought*, edited by Andrew Ortony, 447–480. Cambridge: Cambridge University Press.

Gentner, Dedre, Keith J. Holyoak, and Boicho N. Kokinov, eds. 2001. *The Analogical Mind: Perspectives from Cognitive Science*. Cambridge: The MIT Press.

Gibbs Jr., Raymond W. 2019. "Metaphor as Dynamical–Ecological Performance." *Metaphor and Symbol* 34(1): 33–44.

Giere, Ronald N. 1988. *Explaining Science. A Cognitive Approach*. Chicago: The University of Chicago Press.

Helman, David H. ed. 1988. *Analogical Reasoning: Perspectives of Artificial Intelligence, Cognitive Science, and Philosophy*. London: Springer.

Henderson, Willie. 1982. "Metaphor in Economics." *Economies* 18: 147–153.

Hesse, Mary B. 1966. *Models and Analogies in science*. Indiana: Notre Dame Press.

———. 1988. "The Cognitive Claims of Metaphor." *Journal of Speculative Philosophy* 2(1): 1–16.

Hills, David. 2022. "Metaphor." *The Stanford Encyclopedia of Philosophy (Fall Edition)*, edited by Edward N. Zalta & Uri Nodelman. https://plato.stanford.edu/archives/fall2022/entries/metaphor/.

Holyoak, Keith J., and Paul Thagard. 1996. *Mental Leaps: Analogy in Creative Thought*. Cambridge: Bradford Books.

Hooke, Robert. 1665. *Micrographia, Some Physiological Descriptions of Minute Bodies Made by Magnifying Glasses with Observations and Inquiries Thereupon*. London: Royal Society.

Jensen, Thomas W., and Linda Greve. 2019. "Ecological Cognition and Metaphor." *Metaphor and Symbol* 34(1): 1–16.

Keller, Evelyn F. 2002. *Making Sense of Life: Explaining Biological Development with Models, Metaphors, and Machines*. Cambridge, MA: Harvard University Press.

———. 2015. "Cognitive Functions of Metaphor in the Natural Sciences." *Philosophical Inquiries* 3(1): 113–132.

Knuuttila, Tarja and Andrea Loettgers. 2014. "Varieties of Noise: Analogical Reasoning in Synthetic Biology." *Studies in History and Philosophy of Science* 48: 76–88.

———. 2020. "Magnetized Memories. Analogies and Templates in Model Transfer." In *Philosophical Perspectives on the Engineering Approach in Biology*, edited by. Sune Holm and Maria Serban, London: Routledge. 123–140.

Knuuttila, Tarja and Natalia Carrillo. 2021. "An Artifactual Perspective on Idealization: Constant Capacitance and the Hodkin and Huxley Model." In Alejandro Cassini & Juan Redmont (eds.), [*Models and Idealizations in Science: Artifactual and Fictional Approaches*], Springer Verlag. pp. 51–70.

Lakoff, George, and Mark Johnson. 1980. *Metaphors We Live By*. Chicago: The University of Chicago Press.

———. 1987. *Women, Fire, and Dangerous Things: What Categories Reveal about the Mind*. Chicago: The University of Chicago Press.

Landry, Elaine and Jean Pierre Marquis. 2005. "Categories in Context: Historical Foundational, and Philosophical." *Philosophia Mathematica* 13(1): 1–43.

Locke, John. 1690. *An Essay Concerning Human Understanding*. London.

Lorenzano, Pablo. 2013. "The Semantic Conception and the Structuralist View of Theories: A Critique of Suppe's Criticisms." *Studies in History and Philosophy of Science Part A* 44(4): 600–607.

Martínez, Sergio F. 2023. "What Makes a Good Metaphor in Science." *Signo* 48(91): 23–30.

Maxwell, James C. 1890a. "An Essay on the Mathematical Principles of Physics. By the Rev. James Challis, M.A., &c. (Review)." In *The Scientific Papers of James Clerk Maxwell Vol. 2*, edited by William D. Niven, 338–342. Cambridge: Cambridge University Press.

———. 1890b. "On Faraday's Lines of Force" In *The Scientific Papers of James Clerk Maxwell Vol. 1*, edited by William D. Niven, 155–229. Cambridge: Cambridge University Press.

McCloskey, Deirdre N. 1990. *If You're So Smart: The Narrative of Economic Expertise*. Chicago: The University of Chicago Press.

Morgan, Mary S. 2012. *The World in the Model*. Cambridge: Cambridge University Press.

Morgan, Mary S., and Tarja Knuuttila. 2012. "Models and Modelling in Economics." *Handbook of the Philosophy of Science* 13: 49–87.

Morgan, Mary S., and Margaret Morrison, eds. 1999. *Models as Mediators: Perspectives on Natural and Social Science*. Cambridge: Cambridge University Press.
Morgan, Mary S., Kim Hajek, and Dominic Berry, eds. 2022. *Narrative Science: Reasoning, Representing and Knowing since 1800*. Cambridge: Cambridge University Press.
Nersessian, Nancy J. 2002. "Maxwell and "the Method of Physical Analogy": Model-based Reasoning, Generic Abstraction, and Conceptual Change." In *Reading Natural Philosophy: Essays in the History and Philosophy of Science and Mathematics*, edited by David B. Malament, 129–166. Illinois: Open Court.
———. 2008. *Creating Scientific Concepts*. Cambridge: The MIT Press.
———. 2015. "The Cognitive Work of Metaphor and Analogy in Scientific Practice." *Philosophical Inquiries* 3(1): 113–132.
———, ed. 1987. *The Process of Science: Contemporary Philosophical Approaches to Understanding Scientific Practice*. Amsterdam: Kluwer Academic Publishers.
Netz, Raviel. 1999. *The Shaping of Deduction in Greek Mathematics: A Study in Cognitive History*. Cambridge: Cambridge University Press.
Olson, Mark E., Alfonso Arroyo-Santos, and Francisco Vergara-Silva. 2019. "A User's Guide to Metaphors in Ecology and Evolution." *Trends in Ecology & Evolution* 34(7): 605–615.
Pauwels, Eleonore. 2013. "Mind the Metaphor." *Nature* 500: 523–524.
Rentetzi, Maria. 2005. "The Metaphorical Conception of Scientific Explanation: Rereading Mary Hesse." *Journal for General Philosophy of Science* 36: 377–391.
Reynolds, Andrew S. 2022. *Understanding Metaphors in the Life Sciences*. Cambridge: Cambridge University Press.
Rosch, Eleanor. 2002. "Principles of Categorization." In *Foundations of Cognitive Psychology: Core Readings*, edited by Daniel. J. Levitin, 251–270. Cambridge: The MIT Press.
Semino, Elena, and Zsófia Demjén, eds. 2017. *The Routledge Handbook of Metaphor and Language*. London: Routledge.
Thibodeau, Paul H., and Lera Boroditsky. 2011. "Metaphors We Think With: The Role of Metaphor in Reasoning." *PLoS ONE* 6(2): e16782.
Van Fraassen, Bas C. 1980. *The Scientific Image*. New York: Oxford University Press.
———. 2007. "Structuralism(s) about Science: Some Common Problems." *Proceedings of the Aristotelian Society, Supplementary* 81: 45–61.
White, Roger M., Michael J. Hodge, and Gregory Radick. 2021. *Darwin's Argument by Analogy: From Artificial to Natural Selection*. Cambridge: Cambridge University Press.

27
NARRATIVE AND MODELS

Mary S. Morgan

1. Narrative and models: good companionships[1]

It may be taken for granted by some commentators on science that models are scientific objects and narratives are humanist constructions. But in many respects, models and narratives function as good companions and, in some aspects and some cases, narratives are constitutive in the core of models. Before getting into this agenda of companionship, some preliminary remarks are needed.

First, Models are understood here as objects designed by scientists to help them investigate some part, or some characteristics and behaviours, of their science's phenomena that they do not fully understand, and cannot access directly or consistently.[2] These model-objects are *representations of* how they think their phenomena behave and may be in mathematical, statistical, graphical, or diagrammatic formats (or perhaps even in verbal accounts, though they are mostly non-linguistic entities). They may even be real living objects, specially chosen to be *representatives of* biological life *for* certain purposes such as the use of the model organism: fruit flies for genetics, the lab mice used in medical science, or the lab rats of psychology.[3] But the important point here is their function: models—in construction and usage—provide scientists with *a means of enquiry into* the theoretical and conceptual accounts they have of the world, and *with* the model into the real world.[4]

Second, there are multiple definitions of what a narrative is in narrative theory, probably many more than the different notions of a model within the sciences and offered by philosophers of science. The most basic, and very helpful, way to think about narratives in science is that they provide an account of how things are related together (see Morgan 2017). Narrativising could involve relating events over time, across space, between social groups, or within individual behaviour, and so forth. It is important that a narrative is always *more* than a chronicle (a simple ordering of events), but how much more and what is involved is rather open, so the definition is focussed, but still relatively unrestricted.[5] An important point about using this definition is that narrative-making does not just place things into an order (e.g. according to their time sequence or spatial arrangements) but configures them: it brings together more or less disparate elements into an account that indicates or makes claims about, their relationships, and in this way 'makes sense' of the scientific phenomena

 DOI: 10.4324/9781003205647-32

of interest. That is, narrativising (or narrative-making) provides *a sense-making technology* for scientists (see Morgan 2022).[6]

Finally, the discussion here will *not* be concerned with the rhetoric of the sciences: namely, the important ways in which the use of expressive forms—e.g. clever metaphors in literary analysis, elegance in mathematical proofs, good design in diagrams—matter to the way ideas are presented and understood. This is a perfectly good agenda, for rhetoric is never 'mere' and always matters; but is not the issue here. Rather, the agenda is concerned with how narratives and models work together in various ways in scientific work. Two main and very different sources of examples: the natural historical sciences, and the more technically oriented social sciences (particularly economics), are used here to explore and explain the relationships between models and narratives, that is, to see how *models as a means of enquiry*, and *narrative as a sense-making technology* collaborate in certain sites of science and in certain of their practices. The companionships of models and narratives are discussed first for situations where narrative is constitutive in a science's theoretical explanatory framework and appears so in its models. The discussion then turns to other reasons and other ways in which narrative is involved in the construction and usage of scientific models.

2. Where narratives are constitutive in science and its models

2.1 Narratives in the core of models

In certain sites in the sciences, narrative is constitutive in the core of the scientific account, not just in sense-making descriptions but often in reasoning and explanation (see Olmos 2022). This is pretty obviously so in the natural historical sciences, where accounts of how the natural world changes over time seem 'almost naturally' to take a narrative form—as we can see at several levels in evolutionary biology. At the most all-encompassing theoretical level, the general theory of biological evolution has a narrative structure telling of the adaptation of species to their environment, or the role of random mutations, or of both. When applied at a broad level, that narrative structure is used in giving an account of, or explaining, how major groups of living things in the world developed over time: e.g. plants, fish, mammals, insects. Then, that narrative structure remains similar in discussing more specific individual evolutionary changes—for explaining, for example, how some kinds of fish became flat fish, and even down to the most particular level, such as the turbot. Thus, narrative is constitutive in such accounts of evolution that run from the most general to the most particular. At all levels, these scientists make sense of and explain what happened by telling narrative accounts, and this close relation between narrative and explanation can be found throughout the natural historical sciences of evolutionary biology, geology, palaeontology, cosmology, and so forth.

Where do models fit into these natural-historical scientific narratives? In general, there is no one standard way that models fit into a science; different sciences use models in different places and for different purposes, wherever scientists have found them useful in doing science. In the natural historical sciences, these differences are illustrated by three specific examples of the ways narrative and models come together and fit into the levels of evolutionary biology arguments as suggested above.

First, at the general level of evolutionary change, two major theorists of the 1920s—Sewall Wright and Ronald A. Fisher—produced competing accounts of the main drivers of evolutionary change. According to Rosales' analysis (2017, 7), they agreed on the mathematical elements but held different narrative accounts of the evolutionary processes. Fisher's narrative privileged 'natural selection as the driving force'; Wright's narrative involved 'drift, migration and selection'. The point was that they used these different qualitative narratives to integrate their mathematical elements together. Otto and Rosales (2020) argue that both narratives and mathematics have played key roles in the development of evolutionary theorising, that they are interactive, and, going further, that maths may be embedded in the narratives rather than be the main carriers of theory.

At the second level in this evolutionary biology domain, modelling can be found in the ubiquitous use of 'phylogenetic tree' diagrams representing the evolution of species, making sense of their evolutionary changes over time and space, and placing them into relationships with each other just as genealogical kin diagrams do. Starting with Darwin's famous 'tree of life' diagram, these 'trees' are found not just in museums and children's books but in serious scientific work tracking the evolutionary development of the phenomena of our natural world. For example, such a tree diagram tracing the evolutionary spread of marsupials (the kangaroo family) from South America to Australasia over time offers a similar structure of narrative history as found in our own human-family kinship diagrams (see Kranke 2022). They chart the narrative of both generational descent and spread, and they can be found in a variety of vertical and horizontal forms. These tree diagrams are not habitually thought of as 'models', but, as with many other diagrammatic devices in other fields, they function as just such representations, i.e. as shorthand, artefactual accounts, that express knowledge claims about relationships in a non-linguistic way, and notably here with a narrative structure. According to Priest, such tree diagrams form the 'scaffolding' for explanation in the field in which narrative remains constitutive (see Priest, 2018; and his entry XI in *Narrative Science Anthology II*).

The third and most particular level of evolutionary narrative might also be depicted with diagrammatic modelling to unravel the possible order of adaptation of some fish to become flat fish (see Beatty 2022). Although it is known that flounders (the generic term for flat fish) began their evolutionary lives upright, their evolutionary narrative from living their lives vertically to living lives horizontally is not known. The possibilities of such paths of adaptation can be modelled as a branching tree sequence to suggest alternative 'back stories', e.g. that first these fish had become bottom layers, then their fins had become side flaps, and finally their eyes had moved over the top. But in the absence of the relevant fossil record, any other order seems just as plausible. Each possible pathway or ordering in the branching tree betokens a narrative account created from following different adaptation routes along the branches of the tree.

This narrative 'following' process (which could be done backwards or forwards), is also used with the tree diagrams in the social sciences of psychology and economics to model sequences of decision-making in human life—for example, in the use of 'game theory' in economics and in political science. These diagrammatic models also have similarities with the chemical reaction diagrams depicting possible routes to a successful synthesis in chemistry, showing not adaptive evolutionary moves nor human decision processes, but narrating the possible processes of chemical reactions in the formation of complex molecules. For example, Paskins (2022) shows two chemical syntheses 'equations' for a particular molecule: tropinone. One from the early 20th century offers a narrative 'recipe', telling scientists how to make that molecule, while a more recent one from the early 21st century is understood to represent the relevant chemical reaction processes that occur in such a synthesis.[7] This example also nicely illustrates the useful distinction delineated by Meunier (2022) between the 'research narrative' of the scientist's research work, and the 'narrative of nature': namely what is thought to happen in the natural process.

It is perhaps worthwhile to draw out the related implications of this broad argument. One aspect is the relationship between historical and philosophical notions of explanation. In the natural historical sciences: laws, concepts, and theories address fundamental changes over time using a narrative structure, and in this sense, narratives are constitutive in the core of scientific accounts and so in their explanations of their phenomena. To get closer to explaining and understanding the details of historical changes in the world, scientists in those fields need to adopt and adapt such narratives to more particular levels. In contrast, then, to the normal divide drawn in philosophy—that scientific explanations rely on laws or on causal mechanisms that hold generally while historical explanations can only be about particulars—the claim here is rather different. Rather, these narratives of evolution, from their most general down to their most particular level (of the turbot), remain as much scientific narratives as historical narratives for they depend on, or they embed, or they are driven by, the general scientific laws or the mechanisms envisaged in their discipline and so remain narratives of adaptive evolution or random mutation, or both.

This raises the question: How do these natural scientific laws and causal mechanisms appear in such narratives and models of the natural historical sciences? Hopkins (2022) argues of the equivalent base-level geological laws that the forces of deposition and erosion tightly constrain the narratives of geological change, though the policing by these laws may perhaps remain hidden; the laws lurk in the narratives rather than being found explicit. This lurking is indeed how they appear in the models and accompanying narrative texts in Hopkins' examples of the narratives and diagrams/models that appear in geology.[8]

This suggests the following reflection on a certain useful similarity in the relation of the base theories/laws/mechanisms/concepts of a scientific field and both its models and its narratives. A model is not a scientific law, or general theory, or a concept, but institutes some aspect of these into its representational qualities. The same can be said of the narratives of a science. How might this similarity of character of models and narratives be understood? The main message of the 'models as mediators' account (Morrison and Morgan 1999) was to point out that models do not sit neatly as a sandwich filling between laws/theories and empirical evidence, but rather that they are independent representational objects, artefacts designed (or chosen, as in model organisms) to embed elements of those theories and a field's realities in such a way that the model can be used to explore both realms. That is, they are not simply derivative copies of either laws or world descriptions. In this respect, scientific narratives are like scientific models, they take some relevant elements of the scientific laws/theories of their field into their constructed accounts. This is how narratives fit onto, or into, scientists' explanatory accounts, making use of their sciences' concepts, ideas, framings, and so forth in more generic, or more particular, accounts of how things happen.

In this framing then, models and narratives, can *both* be understood as representations used by scientists: they have much the same qualities, and have similarities in status, with respect to the sciences' explanatory frameworks and phenomena. Regardless of how models are fashioned and framed, they always provide thinner, smaller, and less comprehensive accounts of the phenomena of the world than the world itself, by definition and purpose; and they are usually accounts in a different medium from the phenomena they model. The narrative accounts of science have very similar characteristics.

2.2 *Narrative motivations vs narrative at the core*

While narratives come in 'almost naturally' in the natural historical sciences because of their general commitment to understanding how the natural world changes, there are other sciences where narrative has a less obvious or less secure relationship to a main thesis of a field and their depictions in models. Consider the account that economists make of peoples' decision-making behaviour when they make choices. In that account, consumers are assumed to make decisions that 'maximise their utility', and they do so by preferring more to less, and by making their preference choices consistent amongst several goods. These economists' axioms about utility (that is, the values humans associate with the outcomes of the decisions they make) translate neatly into a mathematical description that can be applied to many (every?) decision(s). But that base-level account of model man's choices is rather thin, empty even. It has descriptive content and may offer predictive outcomes, but it lacks, within the model itself, the agency of decision-making. Narratives may be told by scientists about such human actors' decision-making to give reasons for the scientists' choices for the depictions in their model, yet those presumed narratives may not be recognised within the model. The general question here is whether the narratives are constitutive of the model or merely give an account of such motivations, whose rationality hinges on something else than those narratives. We can examine this in the history of how economists modelled this 'choosing' problem.

When a group of economists in the late 19th century began their utility theorising, they motivated their accounts by telling lots of small stories, imagining how people (including themselves) thought through their choices, and how they made valuations and decisions based on their preferences. These various forward-looking motivational accounts about how people would behave were used in justifying the particular details of three different versions of these theories, expressed in three different model forms: mathematical, graphical, and tabular (laid out in Morgan 2020, 248), but it is notable that the human actors' stories were not really built into the models, rather they were verbal accompaniments to motivate and explain, *ex ante*, the behavioural habits and rules that lay behind the economists' choice of model representation. The human actors' stories were not represented in the models themselves.

By the early 20th century, economists had mostly given up these initial attempts to link decision-making back to psychology, and so, no longer relied on these back stories about how people think about their preferences to 'explain' their choices. One of these three models, developed from Jevons' original geometrical and algebraic representations of 1871, grabbed the mainstream, and his theory of choices was developed into a more general mathematical model account of rational economic man's behaviour by the mid-20th century with little narrative accompaniment (see Morgan, 2006 for a fuller account).

A few decades later, that mathematical model was found to lack traction when applied in laboratory and field experimental work (where people behaved unexpectedly with respect to the theory) and in less than straightforward set-ups (where outcomes were uncertain). These findings produced a set of patches onto the basic theory, and thence into a proliferation of versions of that basic model. Significantly, each of these late 20th century versions of the basic model was again motivated by 'small stories' told by the economists about people's behaviour, but this time *post hoc* to make sense of those experimental findings about how people acted in those situations or reasoned about the valuations and decisions they made. People were understood to have made their decisions by thinking forwards about 'prospects', or by thinking backwards about 'regret' (and so forth) in their decision-making. These narrativised accounts by economists after the events (rather than in the earlier 19th century stories of motivations beforehand) of how and why people had made the decisions they made created extensions or versions of the original theory model. It is again difficult to see exactly how far these more focussed narratives became constitutive into the model, but they were a critical input into the fashioning of the new generation of models.

A well-used model where a sense-making narrative seems to be more central within the model was offered by Hirschman (1970) who was interested in characterising the situation that prompted the three-way choice decision that people made in organisations between exiting a situation they found uncomfortable, or expressing their disquiet ('whistle-blowing'), or staying and keeping quiet. This 'exit, voice or loyalty' decision model grew out of an anecdote, a small story he told of a puzzling experience he had in Nigeria about their railway system. Working away to make sense of his puzzle, he came up with this multipurpose model that could be applied in lots of circumstances: to a firm in an industry, a person in an organisation, a country in a trade alliance, that is, to any individual unit in a larger set facing this three-way choice. Because the model situation was more complex, and the possibilities of various options needed more content, the narrative connections became more central in the model. And because this model situation is a generic kind of situation (i.e. neither completely particular, nor completely general), the model-narrative works as a generic tool for exploring many different situations: that is, narrative sense-making in the model could be applied regardless of the relative details of the person, situation, and choice descriptions.

And, more recently, two sets of commentators have argued that such narratives of decision-making are more than devices for economists in explaining how people make choices before or after the event, but must actually be constitutive within these models because narrative reasoning is constitutive in human decision-making processes (as indeed, seems consistent with the experimental and field evidence mentioned above). That is, narrative is not part of the scientific rationale offered by the scientist in supporting the use of such a model, but rather the model must embed narrative processes because narrative is constitutive of human reasoning. Thus, Tuckett and Nikolic (2017) draw on *cognitive* psychology to show how people make decisions in situations of radical uncertainty to develop an account that relies on narrative reasoning on the part of those people. Bianchi and Patalano (2017) draw on *developmental* psychology to explain how people reason from their current situation to the outcome they hope to reach by creating narratives linking those situations. Both accounts depend on narrative being a core element of human reasoning in decision-making and so, they argue, should be constitutive (even if not fully identifiable) in economists' models of people's decision-making.

3. Narrative in constructing and using models

3.1 Narrative configuring in making models

Although model-making varies across time and subject fields, and histories and philosophies of science may throw up other different kinds of relationships of models with narratives, there are useful comparisons to be drawn that give insight into their relationships. As suggested above, both narratives and models can be understood as forms of scientific representation: representations of how scientists think the world is and how it works. But it is also worth noticing that the *nouns* of narratives and models are the outcome of different practices of scientific reasoning that can be described in the *verbs* of narrativising (or narrative sense-making) and model-making (or modelling). So, the relationship between narrativising as a sense-making activity and the narratives that result parallel those of model-making and the models that result. Such modelling and narrativising practices may also be connected or conjoint. This relationship is most evident when models are the outcome of sense-making processes that involve 'narrativising' an account of their phenomena of interest (as occurred in the exit-voice-loyalty economic choice example above).

Narrative sense-making (the verb) can be found in constructing both theoretical models and empirical models: narrative sense-making may inform, drive, or be more or less strongly instantiated into the relational representations that form either kind of model. In the 1920s and 1930s, for example, economists were deeply concerned with understanding the relatively new phenomena of 'the business cycle'. Some were dealing with the evidential trails of business cycle data in fashioning empirical models while others, at the same time, were creating nascent theoretical models of how an economy as a whole (the 'macro-economy') might generate such cycles. Jan Tinbergen was one of the special group of economists who worked on both kinds of modelling and used narrative sense-making in both. At the former site, he used narrative-making in configuring sets of different statistical data trails of the economic-cycle phenomena to fit together into an empirical model. At the latter site, he constructed theoretical, aggregate-level, models out of a variety of elements relying on narrative-making to configure the causal relations of the cycle to provide both for their cyclical dynamics and their variabilities. Marcel Boumans (1999) used this historical example to motivate his account of model-making as a practice that picks out and integrates a set of ingredients to produce a model (a kind of lego project that then relied on mathematical moulding to configure the parts to fit together). He pointed out how economists' meta-narrative about how the aggregate economic system worked involved them not only pulling together a narrative of causal chain parts but also drawing in a small narrative that functioned as a key ingredient in this theoretical modelling of the cycle.[9] This little narrative—of a child hitting a rocking horse—instantiated the dynamic role of randomness into the mathematical model. Thus, narrative sense-making was important in creating both Tinbergen's mathematical model structure and his statistical-econometric model.

The natural historical sciences offer especially instructive examples of how narrative-making may be integral in model formation. Those concerned with understanding the extinction of the dinosaurs have suggested two alternative models, with different associated narrative forms, applied to the same data (see Huss 2022). For one group of scientists, that particular extinction is understood as just one of many similar such events in a recurring pattern of such mass extinctions, to be pinned down by revealing a cyclical pattern (of 26 million years' periodicity) in the long data series of the timing of mass extinction events. Their narrative is rather thin for it is descriptive rather than explanatory, though speculation suggests a regular 'cause' narrated in the events of cosmology (which might then contribute to explanatory reasoning with the model). For another group of scientists, that particular extinction—of the dinosaurs—is understood as one amongst the set of different such cases, each of which has its own set of causes. The challenge for the latter is to make sense out of a messy evidential domain and configure the set of causes into a model with data that would support a narrative explanation for that one event.

3.2 Narrative in the mediating role: sense-checking the model

The above examples relied on narrativising (sense-making) in configuring sets of elements to help create viable structures and pinpoint relevant relationships in a model. But narratives are at least as important in providing a sense-checking device for models once they are in usage as a means of enquiry. In this context of a collaborative account of models and narratives, how models are used might be at least as important and interesting as how they are made, and particularly how scientists reason with narrative in using their models and what they learn from that usage. In the 'models as mediators' account (Morgan and Morrison 1999), models mediate between theories and the world, having reference to them both and being able to mediate between them by being partially of them both. Mari and Giordani (2014) re-describe these mediating possibilities when they suggest that a 'model is used both as a theoretical tool for interpreting our concepts and as an operational tool for studying the corresponding portion of the world' (83).[10]

One important place where these mediating possibilities are used together is in exploring models through conducting simulations, where narratives associated with the simulations provide a means of potentially exploring both the theoretical qualities and possible empirical validity of models, and so offer a form of double quality control for scientists in working with models. How does this work? In this 'what happens if' reasoning, the model can be simulated and the resulting narratives of this explorative usage provide one of the criteria scientists use to validate their models, and so inform what scientists take to be a 'good' model for their purposes. This narrative usage enables the scientist to enquire into the theoretical world of the model, to suggest domains where it might be usefully applied (a kind of 'applied theory') or into the applicability of the model to the world, either in rather general form or in closer fit to the kinds of phenomena experienced in the world. These narrative 'tests' of validity are qualitative: concerned with *kinds* of outcomes (rather than quantitative in the sense of dealing with domains of uncertainty or error as associated with statistical kinds of testing regimes) and so provide a kind of quality control testing. This is how the theoretical tool and operational tool of mediating using narratives can be seen working together, as in the next two cases.

As a tool of theoretical investigation, this exploratory reasoning mode of model-generated narratives is used to see what kinds of outcomes might be possible, plausible, or implausible according to the model. For example, a mathematical model seen to embody a particular theory might be run either informally in a kind of thought experiment, or via a computer simulation with different starting values, or with different parameter values, according to different assumptions about the world depicted in the model. These exploratory/reasoning modes effectively use the model to 'tell' narratives about the possible paths and outcomes of events, or the predicted outcomes of these models under various settings. One early example is given by the first, hand-cranked simulation of a very small mathematical model of the aggregate economy in 1939 by economist Paul Samuelson. He 'ran' his model for a few nominal 'years' forwards to see how the patterns changed by choosing different parameter values in the model. Each 'run' produced a sequence of model outcomes that provided paths over nominal units of time. Sometimes these narrative paths were rather plausible, with outcomes that were not too big, not too small, and with regular cycles. In other words, they seemed to make sense in terms of being consistent with observed variability in the economy. Other parameter sets produced implausible outcomes: economies that expanded towards infinity within a small number of iterations, other runs that showed no effects, and others that created expanding cycles. Samuelson concluded that almost anything can happen in the world depicted in the model: the 'model-world' of his theorising (see Morgan 2012 for a full analysis). His simulation narratives offered insight into the models' theoretical pretensions but had less to say about the qualities of the real world that the model might be compatible with, and indeed, they were not intended to have this kind of representational verisimilitude. Nevertheless, negative information is often as useful as positive in showing where and when a model is useful to 'explore' a particular phenomenon. Narrative explorations are a way of showing the limitations of the theory that the model is designed to capture, and/or in application to things in the world. What these narrative explorations do is to suggest to the scientist what it cannot be applied to, which might be as important as telling what it might apply to, both in developing the conceptual or theoretical domain of the model and in suggesting constraints on that model's usefulness in the empirical domain.

Another example, from the natural science domain, is found in the model-based, computer simulation of snowflakes (see Wise 2017). Snowflake formation can be modelled with a relatively simple mathematical model based on water droplets falling through the atmosphere with changing temperatures at different elevations. Each of these simulations charts out an individual snowflake's life history, a narrative told through the successive changes in that snowflake's shape and size. As this simple theoretical model simulation process goes on, the computer generates simulated snowflakes into a surprisingly different set of visual shapes, not at all the kind of standard six-sided shapes that were long presumed by scientists (and still cherished by children drawing them at Xmas). This might suggest a rejection of the model on implausibility grounds, but, surprisingly, these outcomes from the simple mathematical model are consistent with the observed evidence of snowflakes: which come in a huge variety of shapes. That is, in simulations, this simple model created highly varied, but empirically valid, outcomes in snowflake shapes equivalent to those seen arriving on the ground—each one separately narrated by the mathematical model simulation.

Samuelson's macro-economic model was also very simple in terms of the structure of the mathematical model, and it too generated a variety of different narrative outcomes, but unlike the snowflakes case, only some of these were empirically plausible. In contrast to these simple model scenarios, Beck (2014) rightly argues that one could not tell narratives with simulations from climate science models because the latter are just too complex to be able to understand the processes in any kind of narrative format. That is, he might argue, as an operational tool for models in climate science, narrative exploration just does not work. That may well be, and the large-scale macro-economic models developed in the 1960s would probably be equally problematic in simulation checks in the theoretical domain. And since the macro-economy is a large open system subject to shocks (as in climate science), empirical sense-checking using narratives would also probably be equally unrewarding. So, it is important to distinguish between a small-scale, simple core model (as in Samuelson's model), and a large-scale more detailed and complex overall model. These will be quite different objects, used differently, and narrative exploration might not be feasible or helpful with a large-scale or complex model in any science (not just climate science). There are probably no obvious or easy general statements about the relations between the simplicity/complexity of a scientific model and the associated variability of the narratives associated with model simulations and with their empirical referents.

Yet, as implied already, for some circumstances this exploratory aspect of model narratives can provide a kind of quality control tool: Does this model provide sensible narratives when you ask a sensible question (either in the context of theorising or in empirical domains)? If it does not, ummmm! In the philosophy of science framing, this use of narrative as a quality indicator for the model is thinly characterised and might seem extremely non-rigorous. But in the context of a scientific community, it may look much more reasonable. For any given community of expert knowledge, a verdict that something makes sense, or makes nonsense, is likely quite a good criterion, for it is about the extent to which this model-narrative knowledge matches up with all the other elements of knowledge these scientists hold (theoretical and empirical) about the phenomena in their domain, a 'coherence' quality not to be underrated (see Currie and Sterelny 2017). And this plausibility facet overlaps with the explanatory services offered by such narratives with or without models. The narratives, by reasoning through the linkages depicted in the model, proffer answers to scientists' questions about how their model-world works, and so offer explanatory possibilities. This quality of the model-narrative nexus reappears again next.

3.3 Narrative closure, opening, and transfer

A third focus for the models-narrative collaboration comes in adopting the notion of 'closure' from narrative theory into scientific uses of narrative. The classic example of closure is found in detective novels, where narrative sense-making requires that by the end of the story, there should be no bits of knowledge left out or leftover or that do not fit the narrative, otherwise the narrative is not closed satisfactorily. An equivalent reasonable quality test for models might suggest that a model is complete for the task at issue if all bits thought to be important are fitted together, there are no essential holes in the account, and anything not considered relevant is omitted.[11] These ideas of closure are not so well formed in philosophy of science discussions of models, but they make a strong appearance in narrative theory and might be applied to the use of narrative in collaboration with models as another aspect of sense-making with models and associated modes of assessing model quality.[12]

One case investigated by Biddle (2023) was the problem of modelling disequilibria in agricultural markets, such as what happens with the introduction of hybrid corn, or the use of fertiliser. Economists' models of markets typically focus on the equilibrium conditions and outcomes in a model, but how this equilibrium comes about over time, and what moves were involved in agricultural markets, were not easily or well modelled. These gaps in the model became very evident when it was applied to the data and evidence of particular markets for particular times. In some of the early days of such empirical modelling, economists sought out farmers and market participants to hear their narrative accounts of what happened in these markets, and in particular how they adjusted their behaviour, and how fast they adapted when things in the market changed. The economists used these narrative inputs to fill the information gap, and so plug the holes in the model with relevant adjustment factors so that it could make sense of the phenomena under study. This is reminiscent of Rosales' account of the evolutionary biology mathematical modelling gap (discussed above), which had to be closed with narrative.

Tinbergen's macro-models (also discussed above) also used the idea of closure. He wanted to develop macro-models that could be used in policymaking. The original models (theoretical and econometric or statistical) involved not just closure of the equation system (where everything that needed explanations had an explanatory equation), but also closure in the sense of enabling all the equations to hold simultaneously in the system. This was essential, for it was difficult to use the model to figure out policy actions unless the whole model was 'closed': all the inter-relations within the model need to be tied up; otherwise, the narratives told in using the model would have loose ends and the policy analysis would fail.

These narrative notions of closure parallel the literature on legal case narratives, which require that all evidence is taken into account and that the evidence fits together in a consistent way, and so forth (see MacCormick 2005). Going further, Campbell (1975), argued that in case study work, when one piece of evidence fails to fit an account that has been built up from lots of different elements, the account—effectively a case narrative—is 'infirmed' (in contrast to 'confirmed'). These closure notions of narrative seem to reflect Tinbergen's ideas about working with mathematical models, and perhaps the failures of such 'closure' are best, or most easily, revealed in failures in the narrative accounts that might be told with the model.

Sometimes narratives play a role not in model closure, but in opening models up. Sharon Crasnow (2017; 2022) details how political scientists open up their statistical models that come from a set of data on political phenomena by going into narrative 'process-tracing' for a few particular individual cases chosen from the data set. Here narrative comes in, not to 'test' the model statistically, but to test in another sense: namely, to explore alternative hypotheses and bring evidence and explanation together in a very different way than with the use of statistics. Such process tracing of political events to tell viable narrative accounts is designed to provide causal stories and perhaps reveal causal mechanisms. For example, the 'democratic peace hypothesis', when tested with statistics, suggests that democracies don't go to war with each other. But: 'How does the democratic peace hypothesis work in practice? What is the process by which war is avoided?' These questions cannot easily be answered in statistical models but can be in case work, taking individual cases and filling in the account; which is in turn dependent on narrative-making as the tool for guiding the analysis, joining up disparate pieces of information at different times to answer the questions and so make sense of that case. For example, Crasnow (2017) examines political scientists' investigation of the Fashoda Incident, when French and Anglo-Egyptian forces came to a standoff over the boundaries of their colonial power in Africa in 1898. Process tracing involved narrative sense-making at the evidential level, but working through conceptual categories and ideas of the political science field to understand why the two forces did not go to war.

This use of narrative in opening up models, rather than closure, may actually be rather common in science. Regardless of how scientists get to their models (and of the accounts philosophers give to these processes), there is an important moment when scientists seek to go on from their theoretical models to try to make them fit the materials from the world. Narratives emerge in the process of using the model to speak directly about particular situations, cases and contexts in the world (as in the example above). For some philosophers of science, this is called model application, for others it might be termed de-idealisation (Knuuttila and Morgan 2019), but the process of making the model fit a particular world is much the same, and narratives are a significantly useful tool in doing so.

Looking at the histories of particular models, it has often been found easier to incorporate additional elements to the base model rather than go back to start anew, and narrative can play a key role here. Much modelling work involves the application of an existing model, with some revisions, to another problem in the field, often motivated by a narrative rationale for such application. This might be especially true of model transfers between fields. For example, Quack and Herfeld (2023) trace how the problem of understanding political coalitions involved the transfer of game-theoretic models from economics into political science, where that transfer depended on narratives (both thinner and thicker) of empirical cases to justify the relevance and fit of the model transfer. To what extent this transfer relies on the fact that stories may be constituted in the core of game theory (in the narrative 'rules of the game', see Morgan 2007b) is one interesting question here. Another might be to look for the role of narrative in arguably the most famous historical transfer of game theory—from mathematics and social sciences into evolutionary biology models. As in the process tracing of Crasnow's example (above), narrative emerges as an indispensable companion to model work, not just in historical moments, but as part of everyday practice found in a variety of sites in the sciences.

4. Conclusion

Both model-making and narrative-making are part of the creative practices of science. Model-making and -using offer means of *enquiry into both theories and the phenomena* of the world that the theory is about. Narrative-making and -using offer ways of *making sense of those phenomena* by configuring disparate elements together and exploring their implications. Narratives provide inputs to model creation: they are sometimes constitutive in the scientific laws/theories/mechanisms embedded in the model; they sometimes feature as connectors or closers in model-making. They are often used and developed to explore the models' possibilities; to help develop models in a field; and as suggestive quality controllers with associated criteria (that is, complementary to formal testing devices). Narratives are not models, and models are not narratives, but in usage, they have similarities as representational devices, and in explanation and reasoning about scientific phenomena. They have synergies of practice which create many areas of collaboration for scientists using them together: they function as good companions.

Acknowledgements

I thank the editors of this volume for their support in putting this chapter together, participants at various seminars over past years with whom I have discussed the many intersections of modelling and narratives going back to my first paper on that intersection (Morgan, 2001), and now particularly the participants at seminars in May 2023 (at the LSE (Information Systems Seminar and at the Institute for Cultural Inquiry Berlin). This account of narrative draws on research undertaken for the Narrative Science Project at LSE that received funding from the European Research Council under the European Union's Horizon 2020 research and innovation programme (grant agreement No. 694732): see www.narrative-science.org/. I thank the many research team members involved in this wide-ranging project.

Notes

1 This account of narrative draws on research undertaken for the Narrative Science Project at LSE: www.narrative-science.org/ and references many of the book chapters of Morgan et al (2022) and *Anthology* resources of that project; and before that for Morgan and Wise (2017).

2 This account is sometimes called the 'artefactual' account of models (see Knuuttila 2011), but much of what is argued here might be just as relevant for other understandings of the nature and role of models in science.

3 For the distinction on how models represent, see Morgan (2007a); for an extensive account of model organisms, see Ankeny and Leonelli (2020).

4 This focus on the role/function of models as a means of enquiry comes from the 'models as mediators' account (Morgan and Morrison, 1999).
5 For example, definitions within literary theory and narratology may require many more conditions: e.g., a 'beginning, middle, and end'; a 'change of state between beginning and end'; the 'role of human agency'; etc.
6 This account of narratives as 'a general purpose technology' for sense-making in science draws on the many workshops, projects, and collaborators involved in the ERC project referenced above in note 1.
7 Further narrative examples from the chemistry of making things can be found in extracts and commentaries by Mat Paskins in *Narrative Science Anthology II* (entries XIX and XXVI) on recipe narratives; and by Sabine Baier in *Anthology I* (entry VIII) on narratives as a navigation tool.
8 See Hopkins 2022, and *Narrative Science Anthology II*, XXVII and XXVIII.
9 A parallel small story usage in physics is given in Hartmann's (1999) account of the development of the 'MIT Bag model'.
10 An alternative framing that makes use of the 'stories' element is suggested by Cartwright (2010) who suggests that models are 'fables' in their relation to scientific laws but 'parables' in relation to the empirical world.
11 These qualities can be framed in philosophy of science as equivalent to fulfilling the full set of *ceteris paribus* conditions on a model (see Boumans and Morgan 2001) but are rarely portrayed as a critical test of model completeness.
12 See Hajek (2022) on narrative closure in science; and Carroll (2007) which engages with both philosophy and narrative on the issue of closure; see also Anand (2023) and Morgan and Stapleford (2023).

References

Anand, Ibanca. 2023. "Resisting Narrative Closure: The Comparative and Historical Imagination of Evsey Domar." In Morgan and Stapleford, 497–522.

Ankeny, Rachel and Sabina Leonelli. 2020. *Model Organisms*. Cambridge: Cambridge University Press.

Beatty, John. 2022. "Narrative Solutions to a Common Evolutionary Problem." In Morgan, Hajek and Berry, 405–423.

Beck, M. Bruce. 2014. "Handling Uncertainty in Environmental Models at the Science-Policy-Society Interfaces." In *Error and Uncertainty in Scientific Practice*, edited by Marcel Boumans Giora Hon and Arthur C. Petersen, 97–136. London: Pickering & Chatto.

Bianchi, Marina, and Roberta Patalano. 2017. *Storytelling and Choice*. Rounded Globe: https://roundedglobe.com/books/1dd3bc40-0775-4182-87a1-d20e0529076f/Storytelling%20and%20Choice/

Biddle, Jeff. 2023. "Narratives and Empirical Strategies in Zvi Griliches Early Research." In Morgan and Stapleford. 447–470.

Boumans, Marcel. 1999. "Built in Justification." In *Models as Mediators*, edited by Mary S. Morgan and Margaret Morrison, 66–96. Cambridge: Cambridge University Press.

Boumans, Marcel and Mary S. Morgan. 2001 "*Ceteris Paribus* Conditions: Materiality and the Application of Economic Theories." *Journal of Economic Methodology* 8(1): 11–26.

Campbell, Donald T. 1975. "'Degrees of Freedom' and the Case Study." *Comparative Political Studies* 8(2): 178–193.

Carroll, Noël. 2007. "Narrative Closure." *Philosophical Studies: An International Journal for Philosophy in the Analytic Tradition* 135(1): 1–15.

Cartwright, Nancy. 2010. "Models: Parables v Fables." In *Beyond Mimesis and Convention: Representation in Art and Science*, edited by Roman Frigg and Matthew Hunter, 19–31. New York: Springer.

Crasnow, Sharon L. 2017. "Process Tracing in Political Science: What's the Story?" *Studies in History and Philosophy of Science* 62(April): 6–13.

———. 2022. "Process Tracing and Narrative Science." In *Narrative Science: Reasoning, Representing and Knowing since 1800*, edited by Mary S. Morgan, Kim M. Hajek, and Dominic J. Berry, 229–245. Cambridge: Cambridge University Press.

Currie, Adrian and Kim Sterelny. 2017. "In Defence of Story-Telling." *Studies in History and Philosophy of Science*. 62(April): 14–21.

Hajek, Kim M. 2022. "What Is Narrative in Narrative Science? The Narrative Science Approach." In *Narrative Science: Reasoning, Representing and Knowing since 1800*, edited by Mary S. Morgan, Kim M. Hajek, and Dominic J. Berry, 31–57. Cambridge: Cambridge University Press.

Hartmann, Stephan. 1999. "Models and Stories in Hadron Physics." In *Models as Mediators*, edited by Mary S. Morgan and Margaret Morrison, 326–346. Cambridge: Cambridge University Press.

Hirschman, Albert O. 1970. *Exit, Voice, and Loyalty*. Cambridge, MA: Harvard University Press.

Hopkins, Andrew. 2022. "The Narrative Nature of Geology and the Rewriting of the Stac Fada Story." In *Narrative Science: Reasoning, Representing and Knowing since 1800*, edited by Mary S. Morgan, Kim M. Hajek, and Dominic J. Berry, 82–103. Cambridge: Cambridge University Press.

Huss, John E. 2022. "Mass Extinctions and Narratives of Recurrence." In *Narrative Science: Reasoning, Representing and Knowing since 1800*, edited by Mary S. Morgan, Kim M. Hajek, and Dominic J. Berry, 61–81. Cambridge: Cambridge University Press.

Knuuttila, Tarja. 2011. "Modelling and Representing: An Artefactual Approach to Model-based Representation." *Studies in History and Philosophy of Science* 42(2): 262–271.

Knuuttila, Tarja and Mary S. Morgan. 2019. "Deidealization: No Easy Reversals." *Philosophy of Science* 86(4): 641–661.

Kranke, Nina. 2022. "The Trees' Tale: Filigreed Phylogenetic Trees and Integrated Narratives." In *Narrative Science: Reasoning, Representing and Knowing since 1800*, edited by Mary S. Morgan, Kim M. Hajek, and Dominic J. Berry, 206–226. Cambridge: Cambridge University Press.

MacCormick, Neil. 2005. *Rhetoric and the Rule of Law: A Theory of Legal Reasoning*. Oxford: Oxford University Press.

Mari, Luca and Alessandro Giordani. 2014. "Modelling Measurement: Error and Uncertainty." In *Error and Uncertainty in Scientific Practice*, Marcel Boumans, Giora Hon and Arthur C. Petersen, 79–96. London: Pickering & Chatto.

Meunier, Robert. 2022. "Research Narratives and Narratives of Nature in Scientific Articles: How Scientists Familiarize Their Communities with New Approaches and Epistemic Objects." In *Narrative Science: Reasoning, Representing and Knowing since 1800*, edited by Mary S. Morgan, Kim M. Hajek, and Dominic J. Berry, 247–266. Cambridge: Cambridge University Press.

Morgan, Mary S. 2001. "Models, Stories and the Economic World." *Journal of Economic Methodology* 8(3): 361–384; Reprinted in *Fact and Fiction in Economics* (2002) edited by Uskali Mäki (Cambridge University Press, 178–201) [Research Memorandum in History and Methodology of Economics, University of Amsterdam, 1999].

———. 2006. "Economic Man as Model Man: Ideal Types, Idealization and Caricatures." *Journal of the History of Economic Thought*, 28:1–27.

———. 2007a. "Reflections on Exemplary Narratives, Cases, and Model Organisms." In *Science Without Laws: Model Systems, Cases, Exemplary Narratives*, edited by Angela Creager, M. Norton Wise, and Elizabeth Lunbeck, 264–274. Duke University Press.

———. 2007b. "The Curious Case of the Prisoner's Dilemma: Model Situation? Exemplary Narrative?" In *Science Without Laws: Model Systems, Cases, Exemplary Narratives*, edited by Angela Creager, M. Norton Wise, and Elizabeth Lunbeck, 157–185. Duke University Press.

———. 2012. *The World in the Model*. Cambridge: Cambridge University Press.

———. 2017. "Narrative Ordering and Explanation." *Studies in History and Philosophy of Science*, 62: 86–97.

———. 2020. "Inducing Visibility and Visual Deduction." *East Asian Science, Technology and Society*, Special Issue 14: 22552.

———. 2022. "Narrative: A General Purpose Technology for Science." In *Narrative Science: Reasoning, Representing and Knowing since 1800*, edited by Mary S. Morgan, Kim M. Hajek, and Dominic J. Berry, 3–30. Cambridge: Cambridge University Press.

Morgan, Mary S. and Margaret Morrison, eds. 1999. *Models as Mediators*. Cambridge: Cambridge University Press, *Ideas in Context* Series.

Morgan, Mary S. and Thomas A. Stapleford. 2023. "Narrative in Economics: A New Turn on the Past." In *Narrative in Economics: Historical Experiences. History of Political Economy*, edited by Mary S. Morgan and Thomas A Stapleford, 395–422. Duke University Press.

Morgan, Mary S. and M. Norton Wise. 2017. "Narrative Science and Narrative Knowing: Introduction to Special Issue on Narrative Science." In *Narratives in Science*. Special Issue of *Studies in History and Philosophy of Science* Vol 62, edited by Mary S. Morgan and M. Norton Wise, 1–5. Elsevier Ltd.

Morrison, Margaret and Mary S. Morgan. 1999. "Models as Mediating Instruments." In *Models as Mediators*, edited by Mary S. Morgan and Margaret Morrison, 10–37. Cambridge: Cambridge University Press.

Narrative Science Anthologies I and *II* at https://www.narrative-science.org/anthology-of-narrative-science.html

Olmos, Paula. 2022. "Just-So What?" In *Narrative Science: Reasoning, Representing and Knowing since 1800*, edited by Mary S. Morgan, Kim M. Hajek, and Dominic J. Berry, 424–444. Cambridge: Cambridge University Press.

Otto, Sarah P. and Alirio Rosales. 2020. "Theory in the Service of Narratives in Evolution and Ecology." *The American Naturalist* 195(2): 290–299.

Paskins, Mat. 2022. "Thick and Thin Chemical Narratives." In *Narrative Science: Reasoning, Representing and Knowing since 1800*, edited by Mary S. Morgan, Kim M. Hajek, and Dominic J. Berry, 267–286. Cambridge: Cambridge University Press.

Priest, Greg. 2018. Diagramming Evolution: The case of Darwin's Trees." *Endeavour* 42(2–3):157–171.

Quack, Alexandra and Catherine Herfeld. 2023. "The Role of Narratives in Transferring Rational Choice Models into Political Science." In *Narrative in Economics: Historical Experiences. History of Political Economy*, edited by Mary S. Morgan and Thomas A Stapleford, 549–576. Duke University Press.

Rosales, Alirio. 2017. "Theories that Narrate the World: Ronald A Fisher's Mass Selection and Sewall Wright's Shifting Balance." In *Narratives in Science*. Special Issue of *Studies in History and Philosophy of Science* Vol 62, edited by Mary S. Morgan and M. Norton Wise, 1–5. Elsevier Ltd.

Tuckett, David and M. Nikolic. 2017. "The Role of Conviction in Decision-making under Radical Uncertainty." *Theory & Psychology* 27(4): 501–503.

Wise, Norton M. 2017. "On the Narrative Form of Simulations." In *Narratives in Science*. Special Issue of *Studies in History and Philosophy of Science* Vol 62, edited by Mary S. Morgan and M. Norton Wise, 74–85. Elsevier Ltd.

28
MODELS AND VALUES

Kristina Rolin

1. Introduction

It is widely recognized that non-epistemic values can legitimately enter practices whereby scientific hypotheses and theories are evaluated, and ultimately, either rejected or accepted, communicated to others, and used in practical decision-making. That non-epistemic values often influence such practices does not mean that they are allowed to play just any role. Philosophers of science aim to specify what the proper roles of non-epistemic values are in the assessment of hypotheses and theories. While traditional debates on science and values have sidestepped models and modeling practices, the past decade has witnessed a growing interest in models and values (for an introduction to an early special issue on the topic, see Peterson and Zwart 2014). In the literature on models and values, one can find modified versions of three main arguments against the value-free ideal of science, an argument from inductive risk, an argument from value-laden background assumptions, and an argument from a plurality of theoretical virtues. These three arguments aim to show that in most cases, the value-free ideal is not feasible without seriously distorting the image of scientific knowledge – and even when the ideal is feasible, it is not a standard for good science because it lacks epistemic or moral/social justification. It lacks epistemic justification, as it is not necessary for achieving the epistemic aims of scientific inquiry, and moral/social justification because it obscures scientists' moral and social responsibility. These arguments have led many philosophers to reject the value-free ideal, the view that non-epistemic values are not allowed to play any role in the core practices of scientific inquiry where scientific knowledge is accepted, either by individual or collective epistemic agents.

This chapter is organized in the following way. Each section discusses one of the three arguments against the value-free ideal of science and explains how it has been applied to models. Each argument against the value-free ideal is also coupled with a normative approach aiming to answer the question of which principles should replace the value-free ideal of science.

At this point, it is worth clarifying how the distinction between epistemic and non-epistemic values is understood. By "epistemic values," it is often meant "values that promote the attainment of truth," either intrinsically or extrinsically (Steel 2010, 18). As Steel

DOI: 10.4324/9781003205647-33

defines them, epistemic values are intrinsic when they constitute an attainment of truth or they are necessary for truth, and they are extrinsic when they promote the attainment of truth without themselves being an indicator or a requirement of truth (2010, 18). While this understanding of epistemic values is widely shared in the philosophy of science, the emerging debate on models and values poses a challenge to it. Models are not candidates for truth, and therefore, the task is to make sense of the epistemic/non-epistemic distinction without necessarily appealing to the notion of truth. There have been different attempts to make sense of the epistemic role of models, acknowledging that the notion of truth is not applicable to models or that models can include false assumptions (see, e.g., Wimsatt 2007; Giere 2004; Teller 2008; Mäki 2011; Knuuttila 2011; Weisberg 2013). For example, Giere (2004) argues that models are tools that scientists use to represent aspects of the world for specific purposes. In his view, it is similarities between models and aspects of the world that make it possible to use models in this way. Mäki (2011) argues that models sometimes include assumptions that can be true. However, for models to function as tools for representation, not all assumptions need to be true. Insofar as models contain truths, such truths are merely partial (2011, 62).

To acknowledge different accounts of models and truths, this chapter adopts an extended definition of epistemic values. An extended definition recognizes that models can have a variety of epistemic functions in addition to representing aspects of the world. According to an extended definition, epistemic values are those values that promote epistemic goals, which may involve truth or the empirical adequacy of models but do not have to. The empirical adequacy of models refers to their ability to account for empirical data in a way that is seen as adequate or good enough given the purposes that models are used for. An extended definition of epistemic values is a starting point that can be modified and revised as philosophers' understanding of models and their various uses evolves.

Given an extended definition of epistemic values, it is not surprising that there is a "borderland area" between epistemic and non-epistemic values (Rooney 2017). A borderland area arises for two reasons. First, it stems from the contingent nature of epistemic values. Even if a candidate for an epistemic value leads scientists toward their preferred epistemic goal under some circumstances, it does not follow that it does so under other types of circumstances. Second, the borderland area arises because the line between epistemic and other goals is often blurred. For example, if scientists use models to represent aspects of the world, it is hard, if not impossible, for them to separate this epistemic goal from moral, pragmatic, and social/political goals. This is because the latter goals define in part for what purpose models are used to represent aspects of the world. As this chapter shows, these difficulties have led some philosophers to doubt whether epistemic values can be distinguished from non-epistemic ones at all (e.g., Winsberg 2012).

For the purposes of this chapter, it is assumed that even if the distinction between epistemic and non-epistemic values is not sharp, it is still a useful conceptual tool in debates about models and values. Its usefulness lies in its capacity to clarify how the proper roles of values in the construction and evaluation of models depend on the type of values in question. While epistemic values are concerned with the epistemic purposes for which models are constructed (e.g., representation), moral and social/political values are concerned with specific purposes such as respect for human rights, the well-being of human beings, and desirable social arrangements (e.g., democracy, civil liberties, and social justice), and pragmatic values are concerned, for instance, with the costs of model building and the execution time for computer simulations.

2. Inductive risk argument and models

An argument from inductive risk is perhaps the best-known argument against the value-free ideal of science. The argument claims that the value-free ideal is not desirable because non-epistemic values, especially moral and social ones, are needed to decide how much uncertainty is tolerable when scientific knowledge is accepted, communicated to others, and used as a reason for action (Douglas 2009; see also Elliott and Richards 2017). The inductive risk argument is based on the observation that accepting a hypothesis typically involves some degree of uncertainty. When uncertainty is unavoidable, scientists must decide whether the evidence at hand is sufficiently strong to warrant acceptance (Rudner 1953). Such decisions require a moral or social value judgment because scientists must identify and evaluate the harm that erroneous beliefs could cause to human beings or society. When such harm could be serious, it is wise to raise the threshold for the acceptance of a hypothesis, thereby reducing the risk of error. Also, when the evaluation of the risks involved in error precedes political decision-making, it should be informed by social and political considerations, and involve the consultation of various stakeholders (Intemann 2015).

The inductive risk argument recognizes that scientists can err in two ways. Not only can they make the mistake of accepting a false hypothesis, but they can also make the mistake of rejecting a true one. The latter mistake is likely to be harmful when the hypothesis in question is part of a much-needed remedy to a problem (e.g., a vaccine to protect human beings from a dangerous disease). Thus, not only can acting on the basis of false beliefs have severe consequences, but refraining from acting can also be damaging when there is an urgent need to find a solution. The core idea of the inductive risk argument is that moral, social, and political values are necessary to assess the consequences of potential errors in all those sciences that are expected to produce knowledge for use. To insist on the value-free ideal of science would be morally and socially irresponsible.

As the inductive risk argument has led philosophers to reject the value-free ideal of science, it has given rise to the question of which normative principles should replace it. Douglas (2009) proposes one such principle. Her view appeals to the distinction between *indirect* and *direct* roles for non-epistemic values in scientific reasoning. Non-epistemic values play an indirect role when they act as reasons to accept a certain level of uncertainty about a hypothesis, and they would play a direct one if they acted in the same way as evidence. In Douglas' view, non-epistemic values are allowed to play an indirect role in the assessment of hypotheses, but they are not allowed to play a direct one. An indirect role is acceptable because scientists, like other human beings, are morally responsible for their actions and the foreseeable consequences of their actions. A direct role is forbidden because it would undermine the epistemic integrity of scientific research (Douglas 2009, 156). For example, political ideologies would play a direct role if they led a person to overlook empirical evidence of the benefits of preschool daycare for children and accept the view that caring for small children at home is always the best for children. Political values are not allowed to play a direct role because in this role they lead a person to ignore existing evidence or believe that evidence is not relevant to the issue at hand.

The argument from inductive risk has given rise to a lively debate about the role of non-epistemic values in the construction and evaluation of models. Biddle and Winsberg (2009) apply the argument to climate models and argue that non-epistemic values play an inevitable role in the estimation of acceptable uncertainty in these models. They distinguish among three sources of uncertainty. First, *structural model uncertainty* is uncertainty about the basic structure that climate models ought to have (e.g., which equations are part of the model). This type of uncertainty is difficult to avoid due to the complex nature of the target system that climate models are meant to represent. Second, *parameter uncertainty* is uncertainty about what the best value for many parameters is. While parameter values are expected to match with empirical data, they nevertheless involve uncertainty in various measurements. Third, *data uncertainty* is uncertainty about evidence concerning the past climate. This type of uncertainty stems from limitations in past practices of data gathering and varies from one case to another.

Based on this tripartite analysis of uncertainty, Biddle and Winsberg (2009) argue that non-epistemic values can legitimately enter decisions concerning an acceptable level of uncertainty in numerous modeling decisions. Moral and social values are relevant to such decisions because it would be morally, socially, or politically worse to err in one way rather than another. To use Douglas' (2009) terms, non-epistemic values can play an indirect role, but they are not allowed to play a direct one, for example, by leading scientists to ignore empirical data when they define parameter values. As Biddle and Winsberg emphasize, their analysis of uncertainty should not be taken as a reason for skepticism about climate change. Eventually, Biddle and Winsberg argue that the three types of uncertainty amount to uncertainty about predictions that are based on climate models. Estimation of acceptable predictive uncertainty is of great moral, social, and political importance because predictions of future climate change are potentially relevant to numerous policy decisions.

In addition to applying the inductive risk argument to model-based science, Biddle and Winsberg (2009) argue that non-epistemic values can legitimately enter the assessment of climate models in another way. Climate models are evaluated based on their ability to predict and retrodict certain tasks well, and non-epistemic values can influence decisions to prioritize certain predictive or retrodictive tasks over others. That non-epistemic values can play such a role in climate models should not be taken to mean that there is no consensus about a causal connection between fossil fuel emissions and global climate change. The point is rather that what human beings care about can legitimately guide scientists' assessments of which predictive and retrodictive tasks are seen as important (e.g., extreme weather events, sea level rise).

Winsberg (2010) aims to refute one possible objection to the application of the inductive risk argument to climate models. According to the objection, scientists could refrain from making non-epistemic value judgments by merely assigning probabilities to each hypothesis in a value-neutral fashion. If scientists succeeded in suspending moral, social, and political value judgments, then it would be up to those who use knowledge to decide whether uncertainties are not too high. Against this objection, Winsberg argues that "Scientists cannot assign probabilities to hypotheses about climate change - or, more specifically, estimate the uncertainties of climate predictions - in a manner that is free from non-epistemic considerations, because non-epistemic considerations invariably influence the choices of prediction tasks, and the choices of prediction tasks invariably influence the estimation of both structural model uncertainty and parameter uncertainty" (2010, 119).

In "Values and Uncertainties in the Predictions of Global Climate Models" (2012), Winsberg takes a closer look at the role of non-epistemic values in methods of uncertainty quantification (UQ). The purpose of UQ is to provide quantitative estimates of the degree of uncertainty associated with the predictions of global and regional climate models. As Winsberg explains, UQ is meant to be a tool for communicating knowledge from experts to policymakers in a way that is seemingly free from the influence of moral, social, and political values (2012, 111). Against the value-free interpretation of UQ, he argues that UQ methods cannot be used to assess the probability of certain events in a value-free fashion because climate models are too complex for any attempt to separate a purely epistemic value from a non-epistemic one. Moreover, climate models are based on an extensive division of epistemic labor among numerous scientists. This means that past choices about acceptable uncertainty are embedded in current models, and consequently, it is difficult to extract non-epistemic value influences from models. Winsberg concludes that "The bits of value-ladenness lie in all the nooks and crannies; they might very well have been opaque to the actors who put them there, and they are certainly opaque to those who stand at the end of the long, distributed, and path-dependent process of model construction" (2012, 132).

While Parker (2014) does not deny that non-epistemic values can legitimately influence the construction and evaluation of climate models, she raises two objections against Winsberg's view. One objection is that not all pragmatic and subjective considerations in modeling involve non-epistemic values (see also Morrison 2014). Parker admits that Winsberg is right to claim that modeling practices involve unforced methodological choices, but she does not accept the claim that "no unforced methodological choice can be defended in a value vacuum" (Winsberg 2012, 130). In contrast to Winsberg, Parker argues that methodological choices are often influenced by pragmatic factors which are not the same as moral, social, and political values (2014, 27). An example of a pragmatic factor is the cost of developing a model as well as the time reserved for such work. Relatively expensive technologies and software licenses are available to some modelers, whereas others will have to work with less expensive ones.

Another objection advanced by Parker (2014) is that even when non-epistemic values influence decisions concerning acceptable uncertainty, Winsberg (2012) tends to exaggerate their impact. Parker argues that uncertainties can be represented in many ways, not only as precise probabilities assigned to parameters or predictions. In her view, Winsberg's emphasis on precise probabilities is misplaced. This is because climate model uncertainties are typically represented with coarse estimates. An example of a coarse estimate would be a claim that a hypothesis about future climate change has a probability between 90% and 100%. An estimate is coarse precisely because it defines a range for a probability and not a precise value. Parker argues that the influence of non-epistemic values on the assessment of acceptable uncertainty will be smaller when uncertainty is represented with a coarse estimate than when it is represented with a precise probability (e.g., 95%). She does not deny that non-epistemic values play a role in assessing an acceptable level of uncertainty in climate models and, especially, in accepting a certain level of uncertainty in predictions that are drawn from climate models. But in her view, the argument from inductive risk has been used to overstate the influence of non-epistemic values.

While Parker (2014) is critical of the way the inductive risk argument has been applied to climate models, she advances her own modified version of the inductive risk argument. She argues that non-epistemic values have a legitimate role to play in decisions concerning second-order uncertainty (2014, 28). Second-order uncertainty means uncertainty about estimates of uncertainty. As Parker explains, "any decision to offer a particular estimate of uncertainty implies a judgment that this second-order uncertainty is insignificant/unimportant; but such a judgment is a value judgment, as it is concerned with (among other things) how bad the consequences of error (inaccuracy) would be; hence even decisions to offer coarser uncertainty estimates at least implicitly reflect value judgments (see, e.g., Douglas, 2009, p. 85)" (2014, 29).

In sum, the inductive risk argument has been applied to models to argue that model construction and evaluation cannot be free from non-epistemic values. Philosophers identify numerous sources of uncertainty in modeling decisions, including structural model uncertainty, parameter uncertainty, data uncertainty, predictive uncertainty, and second-order uncertainty. While some decisions concerning an acceptable range of uncertainty are based on pragmatic considerations, some others require moral, social, and political value judgments.

3. An argument from value-laden background assumptions and models

An argument from value-laden background assumptions claims that the value-free ideal is not feasible because non-epistemic values can legitimately influence the choice of background assumptions that are necessary for evidential reasoning (Longino 1990; see also de Melo-Martín and Intemann 2016). The argument is part of a broader view that Longino (1990) labels contextual empiricism. As a form of empiricism, contextual empiricism emphasizes the importance of empirical data, and as contextual it holds that empirical data can have a bearing on scientific knowledge only in the context of background assumptions. A context of background assumptions is necessary to establish the relevance of a particular data set to a hypothesis or a theory (1990, 43–44). Without a link between a data set and a hypothesis or a theory, data do not have the status of evidence. While background assumptions do not always "encode" non-epistemic values, they often do so (1990, 216). An established body of scientific knowledge typically provides scientists with several plausible background assumptions or theories. Given the plurality of background assumptions, scientists must decide which assumptions or theories they rely on in their evidential reasoning. Non-epistemic values can legitimately guide such choices if there are no epistemic reasons to reject a background assumption or a theory.

Longino (1990) uses research on human evolution to illustrate the argument from value-laden background assumptions. An androcentric "man-the-hunter" framework functions as a set of background assumptions, which connect data to a theory about human evolution. In this framework, male hunting is seen as the main activity that has favored the development of distinctly human forms of intelligence and sociability (1990, 107). An alternative and equally value-laden set of background assumptions comprises a "woman-the-gatherer" framework, which assigns a major role to the changing behavior of females (1990, 107). In the gynocentric framework, the development of human intelligence and sociability is understood as a function of female food-gathering activities. Changing female gathering activities go hand in hand with longer infant dependency and increasing brain size (1990, 108). In both cases, a set of background assumptions is necessary to guide the way in which various types of data (e.g., fossils, bits of bone or teeth, tools, footprints) are interpreted as evidence in support of a theory. The value-free idea is not feasible because it is hard to see how relevant judgments could be made without value-laden background assumptions.

Like the argument from inductive risk, the argument from value-laden background assumptions gives rise to the question of which normative principles should replace the value-free ideal of science. Longino (2002) proposes a social value management view, which recommends that the role of non-epistemic values in scientific inquiry be analyzed, criticized, and judged as either acceptable or unacceptable by a scientific community that satisfies certain conditions. Scientific communities should act in accordance with four norms: the norm of "publicly recognized venues," "uptake of criticism," "shared standards," and "tempered equality of intellectual authority" (Longino 2002, 129–131). When scientific communities follow these norms, it is more likely that the often-tacit influence of moral, social, and political values will be revealed and brought into a critical discussion than otherwise. Along similar lines, Anderson (2004) argues that the value-laden nature of background assumptions is not a problem in and of itself. It becomes a problem when it gives rise to dogmatism. According to Anderson, scientific communities need to ensure that "value judgments do not operate to drive inquiry to a predetermined conclusion" (2004, 11). When scientists cannot avoid moral, social, and political value judgments, they should make such judgments and their reasons explicit (see also Elliott 2017). Scientific communities play an important role in articulating non-epistemic value influences as individuals are not always aware of the value-ladenness of their assumptions.

Peschard and van Fraassen (2014) apply the argument from value-laden background assumptions to models. They argue that modeling often starts with abstract concepts aiming to capture the phenomenon under investigation and proceeds to a concrete model that can be tested for its empirical accuracy. A concrete model specifies a data-generating procedure that needs to be realized to produce the testing data (2014, 4). To specify a data-generating procedure, modelers must make judgments about what kind of data is relevant. As Peschard and van Fraassen point out, "differences in relevance judgments will lead to differences in the concrete models of the phenomenon" (5). And they add that "these judgments are not empirical; they are normative judgments and through them norms and values are incorporated to the modeling process" (5).

Parker and Winsberg (2018) develop a novel version of the argument from value-laden background assumptions. Like Longino, they recognize that evidential reasoning (e.g., Bayesian reasoning) typically takes place against background knowledge that helps scientists estimate a prior probability to a hypothesis. In many cases, the challenge is to deal with background knowledge that is too complex to be fully incorporated into reasoning. To manage the complexity of background knowledge, scientists can use models as proxies for background knowledge. Evidential reasoning is easier to handle in the context of models because models are necessarily limited in nature due to idealizing assumptions. A model embodies some background assumptions, while it leaves out some others. By sacrificing some information, models enable scientists to create a context for evidential reasoning. As Parker and Winsberg explain, models can be used as "surrogates" for background knowledge that is impossible to manage otherwise (2018, 141). When models function as surrogates for background knowledge, evidential reasoning can be non-epistemically value-laden in several ways. First, non-epistemic values can influence the selection of which models suit this purpose well. Second, models themselves can be laden with non-epistemic values as they are shaped by past decisions concerning the aims and purposes they are expected to serve. Third, the selection of background assumptions to be included in models can reflect non-epistemic values.

In sum, the argument from value-laden background assumptions has been applied to models to argue that moral, social, political, and pragmatic value judgments can be implicit in decisions concerning what data are relevant for testing models and which background assumptions are included in models.

4. An argument from plurality of theoretical virtues and models

An argument from a plurality of theoretical virtues claims that the value-free ideal is not feasible because non-epistemic values can legitimately enter the selection of theoretical virtues (Elliott and McKaughan 2014; Kuhn 1977; Longino 1995; Solomon 2001). They can do so because the set of theoretical virtues includes a variety of criteria and desiderata, and such virtues typically cannot be realized simultaneously. For example, scientists may have to strike a balance between accuracy and broad scope, between emphasizing the depth of empirical evidence or its breadth. Given the plurality of theoretical virtues, non-epistemic values help scientists decide which virtues are given priority over others. The main idea of the argument from a plurality of theoretical virtues is that while non-epistemic values should not replace epistemic ones, they are allowed to guide the use of epistemic values. While the role of epistemic values is to protect the epistemic integrity of scientific inquiry, the role of non-epistemic ones is to ensure that scientific inquiry serves valued moral and social goals.

Like the argument from inductive risk and the argument from value-laden background assumptions, the argument from a plurality of theoretical virtues gives rise to the question of which normative principles should replace the value-free ideal of science. Intemann argues that value judgments are legitimate when they promote democratically endorsed epistemological and social aims of research (2015, 218). In her view, scientists should not have disproportionate power in deciding which non-epistemic values ought to be endorsed. According to Intemann, the aims of model construction should be informed by the epistemic and non-epistemic values of stakeholders (2015, 227). This principle gives rise to the question of which stakeholders should be heard and how the hearings are to be organized.

It is easy to see how the argument from a plurality of theoretical virtues can be applied to models. Due to the cognitive limitations of human beings and the complexity of the world, the construction of models involves trade-offs between different desiderata. Models cannot at once maximize generality, realism, and precision (Levins 1966, 422). A decision to sacrifice one desideratum in favor of another can be value-laden morally and socially. Alternatively, modelers can adopt a multiple-model strategy without expecting that any single model will offer a complete story (Weisberg 2007, 646). Such a strategy has been applied, for instance, in ecology, where scientists are dealing with highly complex phenomena. Diekmann and Peterson argue that non-epistemic values are not only secondary values that become important just in case epistemic values leave some issues open (2013, 208). In their view, non-epistemic values are as important as epistemic ones because their role is to help scientists and engineers envision the best models of a process or problem (2013, 208). On a somewhat similar note, Sterrett (2014) argues that non-epistemic values, especially moral ones, are at the core of model-making in engineering sciences because models are not merely models of representation; they are also models of intervention, and as such, they guide the actions of scientists, engineers, and knowledge users.

Like Giere (2004), Parker and Winsberg (2018) argue that models are constructed and evaluated with a set of purposes in mind, some of which are more important than others. While such purposes are related to knowledge, they stem in part from various human interests and non-epistemic values. Non-epistemic values can play a legitimate role in decisions concerning which features of the target system are represented and how accurate these representations are. Like many other philosophers, Parker and Winsberg emphasize that model building involves idealizations and simplifications. Purposes help scientists see when distortions caused by idealizations and simplifications are justifiable. Purposes tell scientists which modeling results should be compared with empirical data and what counts as a "good enough fit with observations" (2018, 128).

If purposes are necessary for constructing and evaluating models, one might wonder whose purposes they are. The answer to this question is more complex than one might expect. According to Parker and Winsberg (2018), it is not only the purposes of current modelers that are relevant but also the purposes of past modelers. This is because models are often built on previous models, and purposes from their earlier times continue to exert influence (2018, 129). Jebeile and Crucifix argue that the dependency of climate models on scientists' purposes can give rise to "epistemic inequality" (2021, 120). By epistemic inequality, they mean "the risk that models might more accurately represent the future climates of the geographical regions and sources of concern prioritised by the values of the modellers, thus making some people better informed than others" (120). They propose Longino's (2002) social value management approach as one way to prevent or correct epistemic inequalities.

In "Model Evaluation: An Adequacy-for-Purpose View" (2020), Parker sets out to develop a more systematic account of the role of purposes in model-making. Like Giere (2004), Parker argues that models should be assessed with respect to their adequacy or fitness for a particular purpose. As she explains, "Such a view can be contrasted with one on which model quality is (just) a matter of how accurately and completely a model represents a target, where the ideal limit is a perfect and complete representation" (2020, 458). An adequacy-for-purpose view is meant to do justice to the widespread view that models are not just representations of a target system; they are more appropriately understood as tools or "epistemic artifacts" (Knuuttila 2011), which are selected and used for epistemic and practical purposes (Parker 2020, 459). As practical purposes often stem from non-epistemic values, such values are at the very core of model evaluation.

While Giere's (2004) pragmatic account of models has already considered the user and the purpose of the model, Parker introduces additional constraints on modeling. She argues that for a model to be adequate for purpose, it must stand in a suitable relationship not just with a target T but with a type of user U, methodology W, circumstances B, and goal P jointly (2020, 464). While Parker prefers to use the term "user," some other philosophers propose that the adequacy of a model depends on an "audience" (e.g., Mäki 2011, 55). In either case, a model's usefulness or persuasiveness may vary from one user or audience to another. Parker encourages philosophers to think of T, U, W, B, and P as dimensions of a problem space that is constituted by a goal (P) and a set of constraints (T, U, W, B) on how the desired goal can be achieved (2020, 464). In Parker's view, models can be seen as "solutions" in a kind of problem space (2020, 475). There is often more than one way to construct a model that is adequate for a particular purpose (472). Parker stresses that evaluating a model's adequacy-for-purpose involves a kind of holism that is absent when someone evaluates merely a model's representational accuracy (471).

Lusk and Elliott (2022) analyze further the implications of the adequacy-for-purpose account of model evaluation for debates on science and values. In their view, the emphasis in these debates should be shifted from the question of how non-epistemic values can figure in scientific assessment to the question of how scientific assessment can accommodate non-epistemic values (2022, 9). They argue that if philosophers reject the value-free ideal of science, they also need to modify scientific assessment so that it incorporates other goals in addition to the pursuit of truth. Lusk and Elliott suggest that the adequacy-for-purpose account provides a general framework for describing the assessment of scientific knowledge in a way that goes beyond purely epistemic considerations. By scientific assessment, they mean the appraisal of the outcomes of scientific inquiry to determine whether they are acceptable.

Lusk and Elliott (2022) argue that the adequacy-for-purpose framework improves debates on values and science in three ways. First, it provides a better way of describing the purpose-relativity of many scientific hypotheses. The illusion that hypotheses can be assessed in a value-free manner is created by a myopic focus on a "plain hypothesis" (2022, 12). The illusion disappears as soon as one sets out to assess "adequacy-for-purpose-style hypotheses" that refer to particular purposes, including practical ones (2022, 13). Second, the adequacy-for-purpose framework provides a better way to examine how the assessment of a hypothesis or a model depends on the potential consequences of accepting or rejecting one, an insight that stems from the inductive risk argument. As many philosophers acknowledge, it is difficult to foresee all the consequences of accepting or rejecting a hypothesis or a model, and hence, scientists can merely be expected to predict the consequences as well as they can. The adequacy-for-purpose framework helps clarify what it takes for scientists to consider consequences to the best of their knowledge. The framework tells scientists to focus especially on the consequences that the use of a hypothesis or a model might have in a particular context. Third, the adequacy-for-purpose framework provides a better way to describe the interplay between epistemic and non-epistemic values in scientific assessment. Given the framework, it turns out to be misleading to ask whether epistemic values should trump non-epistemic ones (or vice versa). The main lesson to be drawn from the framework is that scientific assessment calls for the joint satisfaction of epistemic and non-epistemic values (2022, 17). A hypothesis or a model can be rejected if any of the various criteria for adequacy are not met (2022, 17).

Potochnik (2015) argues that model-based science should change the way philosophers think about the aims of science (see also Potochnik 2012). While some philosophers believe that the aims of science are best captured by the notion of "significant truth" (Kitcher 1993), Potochnik argues that "understanding" is a more appropriate way to describe the aims of science. As she explains: "I suggest that continuing, widespread idealization calls into question the idea that science aims for truth" (2015, 72). She adds that "If instead science aims to produce understanding, this would enable idealizations to directly contribute to science's epistemic success" (2015, 72). Potochnik holds the view that science has a wide variety of aims, both epistemic and non-epistemic, and this explains why the aims of science can be served by different kinds of scientific products (2015, 72).

In "Idealization and Many Aims" (2020), Potochnik sets out to give a general account of understanding. In her view, understanding has a dual nature. As she explains, "Understanding is at once a cognitive state and an epistemic achievement" (2015, 72). In virtue of being a cognitive state, understanding depends on the psychological characteristics of those who seek to understand. Representations that incorporate idealizations can provide understanding to various epistemic agents because the phenomena that are of interest to science are complex, whereas the powers of human cognition and action are limited. Given the dual nature of understanding, non-epistemic values can shape what kind of understanding is sought and offered. However, the influence of non-epistemic values is kept in check by the requirement that scientific models be "true enough" (2015, 78). Idealizations are acceptable when they do not diverge from truth in significant ways, taking into account their role in the representation and the epistemic purpose to which that representation is put (2020, 937).

To summarize, the argument from a plurality of theoretical virtues has been applied to models to argue that non-epistemic values can legitimately help scientists navigate trade-offs between different desiderata that models cannot maximize simultaneously (e.g., generality, realism, and precision). The construction and evaluation of models need to appeal to the purposes models are expected to serve. The description of modeling purposes often mixes epistemic, pragmatic, moral, and social values in a way that makes it difficult to perceive how model construction and evaluation could proceed without non-epistemic values.

5. Conclusion

This chapter has explained how three arguments against the value-free ideal of science, the argument from inductive risk, the argument from value-laden background assumptions, and the argument from a plurality of theoretical virtues, have been applied to models. The inductive risk argument states that the construction and evaluation of models cannot be free from non-epistemic values because moral, social, political, and pragmatic values are necessary to decide how much uncertainty is acceptable in modeling decisions. The argument from value-laden background assumptions states that model-making cannot be free from non-epistemic values because value-laden background assumptions are necessary for relevance judgments. Relevance judgments tell modelers what kind of data are relevant for testing a model's empirical adequacy. The argument from a plurality of theoretical virtues states that model construction cannot be free from non-epistemic values because it involves trade-offs between different desiderata. As models cannot at once maximize generality, realism, and precision, non-epistemic values help modelers decide which desiderata should guide model construction. Non-epistemic values can play a legitimate role in decisions concerning which features of the target system are represented in models and how accurate these representations are. Non-epistemic values can provide justification for idealizations and simplifications in models. Models are constructed and evaluated with a set of purposes in mind, and such purposes are often a mixture of epistemic, moral, and social values. The literature on models and values poses a challenge to the traditional definition of epistemic values as values that promote the attainment of truth. As models are not candidates for truth, the definition of epistemic values needs to be extended to include those values that serve other epistemic purposes that models are used for.

References

Anderson, Elizabeth. 2004. "Uses of Value Judgments in Science: A General Argument, with Lessons from a Case Study of Feminist Research on Divorce." *Hypatia* 19(1): 1–24.

Biddle, Justin, and Eric Winsberg. 2009. "Value Judgments and the Estimation of Uncertainty in Climate Modeling." In *New Waves in the Philosophy of Science*, edited by P. D. Magnus and Jacob Busch, 172–197. New York: Palgrave MacMillan.

de Melo-Martín, Inmaculada, and Kristen Intemann. 2016. "The Risk of Using Inductive Risk to Challenge the Value-Free Ideal." *Philosophy of Science* 83: 500–520.

Diekmann, Sven, and Martin Peterson. 2013. "The Role of Non-Epistemic Values in Engineering Models." *Science and Engineering Ethics* 19: 207–218.

Douglas, Heather. 2009. *Science, Policy, and the Value-free Ideal*. Pittsburgh: University of Pittsburgh Press.

Elliott, Kevin C. 2017. *A Tapestry of Values: An Introduction to Values in Science*. New York. Oxford University Press.

Elliott, Kevin C., and Daniel J. McKaughan. 2014. "Nonepistemic Values and the Multiple Goals of Science." *Philosophy of Science* 81(1): 1–21.

Elliott, Kevin C., and Ted Richards, eds. 2017. *Exploring Inductive Risk: Case Studies of Values in Science*. New York. Oxford University Press.
Giere, Ronald. 2004. "How Models Are Used to Represent Reality." *Philosophy of Science* 71(5): 742–752.
Intemann, Kristen. 2015. "Distinguishing between Legitimate and Illegitimate Values in Climate Modeling." *European Journal for Philosophy of Science* 5: 217–232.
Jebeile, Julie, and Michel Crucifix. 2021. "Value Management and Model Pluralism in Climate Science." *Studies in History and Philosophy of Science* 88: 120–127.
Kitcher, Philip. 1993. *The Advancement of Science: Science without Legend, Objectivity without Illusions*. New York and Oxford: Oxford University Press.
Knuuttila, Tarja. 2011. "Modeling and Representing: An Artefactual Approach." *Studies in History and Philosophy of Science Part A* 42(2): 262–271.
Kuhn, Thomas. 1977. "Objectivity, Value Judgment, and Theory Choice." In *The Essential Tension: Selected Studies in Scientific Tradition and Change*, 320–339. Chicago: University of Chicago Press.
Levins, Richard. 1966. "The Strategy of Model Building in Population Biology." *American Scientist* 54 (4): 421–431.
Longino, Helen. 1990. *Science as Social Knowledge*. Princeton, NJ: Princeton University Press.
———. 1995. "Gender, Politics, and the Theoretical Virtues." *Synthese* 104(3): 383–397.
———. 2002. *The Fate of Knowledge*. Princeton, NJ: Princeton University Press.
Lusk, Greg, and Kevin C. Elliott. 2022. "Non-epistemic Values and Scientific Assessment: An Adequacy-for-Purpose View." *European Journal for Philosophy of Science* 12: 35.
Mäki, Uskali. 2011. "Models and the Locus of their Truth." *Synthese* 180(1): 47–63.
Morrison, Margaret. 2014. "Values and Uncertainty in Simulation Models." *Erkenntnis* 79: 939–959.
Parker, Wendy. 2014. "Values and Uncertainties in Climate Prediction, Revisited." *Studies in History and Philosophy of Science Part A* 46: 24–30.
———. 2020. "Model Evaluation: An Adequacy-for-Purpose View." *Philosophy of Science* 87: 457–477.
Parker, Wendy, and Eric Winsberg. 2018. "Values and Evidence: How Models Make a Difference." *European Journal for Philosophy of Science* 8: 125–142.
Peschard, Isabelle F., and Bas C. van Fraassen. 2014. "Making the Abstract Concrete: The Role of Norms and Values in Experimental Modeling." *Studies in History and Philosophy of Science Part A* 46: 3–10.
Peterson, Martin, and Sjoerd D. Zwart. 2014. "Introduction: Values and Norms in Modeling." *Studies in History and Philosophy of Science Part A* 46: 1–2.
Potochnik, Angela. 2012. "Feminist Implications of Model-Based Science". *Studies in History and Philosophy of Science Part A* 43: 383–389.
———. 2015. "The Diverse Aims of Science". *Studies in History and Philosophy of Science Part A* 53: 71–80.
———. 2020. "Idealization and Many Aims." *Philosophy of Science* 87(5): 933–943.
Rooney, Phyllis. 2017. "The Borderlands between Epistemic and Non-epistemic Values." In *Current Controversies in Values and Science*, edited by Kevin C. Elliott and Daniel Steel, 31–45. New York and London: Routledge.
Rudner, Richard. 1953. "The Scientist qua Scientist Makes Value Judgments." *Philosophy of Science* 20(1): 1–6.
Solomon, Miriam. 2001. *Social Empiricism*. Cambridge: MIT Press.
Steel, Daniel. 2010. "Epistemic Values and the Argument from Inductive Risk." *Philosophy of Science* 77(1): 14–34.
Sterrett, Susan. 2014. "The Morals of Model-Making." *Studies in History and Philosophy of Science Part A* 46: 31–45.
Teller, Paul. 2008. "Of Course Idealizations Are Incommensurable!" In *Rethinking Scientific Change and Theory Comparison*, edited by Léna Soler, Howard Sankey, and Paul Hoyningen-Huene, 247–264. Dordrecht: Springer Netherlands.
Weisberg, Michael. 2007. "Three Kinds of Idealization." *Journal of Philosophy* 104(12): 639–659.
———. 2013. *Simulation and Similarity. Using Models to Understand the World*. New York. Oxford University Press.

Wimsatt, William C. 2007. "False Models as Means to Truer Theories." In *Re-Engineering Philosophy for Limited Beings: Piecewise Approximations to Reality*, 96–132. Cambridge, MA: Harvard University Press. Originally published in 1983 in M. Nitecki and A. Hoffman, eds., London: Oxford University Press, 23–55.
Winsberg, Eric B. 2010. *Science in the Age of Computer Simulation*. Chicago and London: The University of Chicago Press.
———. 2012. "Values and Uncertainties in the Predictions of Global Climate Models." *Kennedy Institute of Ethics Journal* 22(2): 111–137.

29

INTERDISCIPLINARITY THROUGH MODELLING

Mieke Boon

1. Introduction

Over the last few decades, research organizations such as the National Academy of Sciences (2005) have emphasized the importance of interdisciplinary research and education (see also Tuana 2013). Research policymakers often acknowledge that interdisciplinary research is challenging for numerous reasons, such as the organization and funding of research, political obstacles, the complexity of interdisciplinary research, and the difficulty of communication within a multidisciplinary team (see Jacobs and Frickel 2009 for a critical evaluation). However, hardly any attention has been paid to the epistemological, methodological, and conceptual barriers and cognitive constraints of working across disciplinary domains (MacLeod 2018). In the philosophy of science, Nancy Nersessian, Miles MacLeod, Uskali Maki, and Michiru Nagatsu have done pioneering work in studying the strategies (esp. modeling strategies) of researchers in interdisciplinary scientific practices.

Thus, while the philosophy of science initially focused on questions of the nature, ontology, and representational properties of models, analyses of research into complex problems include the cognitive, epistemological, methodological, and pragmatic aspects related to modelers and model-users. Analyzing the cognitive complexity of modeling complex problems thereby offers new insights into the nature of models and modeling practices. When focusing on the nature of the intellectual work researchers accomplish through building and using models, cognitive processes becomes an inherent part of these studies, introducing new notions, such as *model-based understanding*, *model-based reasoning*, *model-based explanation*, *modeling strategies*, *mental models*,[1] and *models as cognitive artifacts* (Nersessian 2009; 2022, see also Magnani and Bertolotti 2017; Mattila 2005; O'Malley and Soyer 2012; MacLeod 2018). By including cognitive processes in philosophical analyses of models and modeling practices, other notions that emerge are: *inferential reasoning*, *model-users* and *competent cognitive agents* (Suárez 2004; Giere 2010);[2] *epistemological responsibility* (Van Baalen and Boon 2015); *epistemology of models and modeling* (Boon and Van Baalen 2019); and *researchers having disciplinary perspectives* (Boon 2020b). Additionally, this turn of focus provides crucial insights for *education* in interdisciplinary research (e.g., Boon 2020a; Boon et al. 2022; Nersessian 2022; Van den Beemt et al. 2020) and modeling

 DOI: 10.4324/9781003205647-34

strategies in *science-based policy* (e.g., MacLeod 2018; MacLeod and Nagatsu 2018; Nagatsu and Ruzzene 2019; Nagatsu et al. 2020; Frisch 2013; Inkpen and DesRoches 2020). Furthermore, when the *epistemic usefulness* of models in practical applications such as science-based policy is taken into account, where models are considered *epistemic tools* (Boon and Knuuttila 2009; Knuutila and Boon 2011) for problem-analysis, forecasting, and scenario studies, still other features of modeling become prominent, which have implications for philosophical views on models, in particular regarding their representational characteristics. For example, Elliot and McKaughan (2014) argue that scientific representations should also be evaluated on their suitability for the practical and epistemic purposes of model users, which requires including non-epistemic values. Similarly, in the context of climate modeling, Parker (2020) proposes an adequacy-for-purpose view on models. Studying interdisciplinary research practices thus leads to new themes and research questions for the philosophy of science (see Mäki 2016).

The topic of this chapter – interdisciplinarity through modeling in research, science-based policy, and education – connects two subjects that are often treated separately within the philosophy of science: interdisciplinarity and models. Section 2 addresses the why, what, and how of interdisciplinary research, and the role of models and modeling therein. To this end, scholarly, policy-related, and philosophical literature on interdisciplinary research has been surveyed. Section 3 discusses accounts of models and modeling strategies and provides an outline of epistemological and methodological issues of interdisciplinary research practices. Use is made of both scientific literature on methodologies in interdisciplinary research and philosophy of science literature on the role of models in this. Section 4 concludes with a brief overview of issues to be addressed in a *philosophy for interdisciplinary modeling practices*.

2. Interdisciplinarity

2.1 *Definition of interdisciplinary research*

Interdisciplinarity is studied in scholarly domains ranging from science policy studies, governance studies, STS (science, technology, and society), science education, cognitive sciences, philosophy of science, and social epistemology. One of the scholarly aims is a correct *definition* (e.g., Klein 1990; Aboelela et al. 2007; Repko 2008; Newell and Gagnon 2013). Three characteristics are usually found in definitions of interdisciplinary research: (I) the *rationale* for interdisciplinary research is solving a problem, or addressing a topic that is *too broad or complex* to be dealt with adequately by a single discipline or profession (cf. Newell and Gagnon 2013); (II) the *epistemic purpose* of interdisciplinary research is (a) to advance *fundamental understanding* of a phenomenon, or (b) to develop knowledge and understanding for *solving (complex) problems*; and (III) the crucial role of *integration* of (a) *knowledge* (or, more broadly, epistemic resources such as data, concepts, laws, and theories), (b) *instruments* (including methods and technologies), or even (c) *disciplinary perspectives*.[3] An example is the oft-cited definition by *The National Academy of Science* (2005): "Interdisciplinary research (IDR) is a mode of research by teams or individuals that integrates information, data, techniques, tools, perspectives, concepts, and/or theories from two or more disciplines or bodies of specialized knowledge to advance fundamental understanding or to solve problems whose solutions are beyond the scope of a single discipline or area of research practice" (National Academy of Science et al. 2005, 2).

2.2 *Interdisciplinarity in scientific research, higher education, and science-based policy*

Research policy documents from leading organizations, institutes, and research councils emphasize the critical importance of interdisciplinary research (e.g., NSF, NRC, NAS, ESF, ERC,[4] GRC,[5] NWO, Van Noorden 2015). Three arguments are often made in favor of interdisciplinary research (Rylance 2015). First, the grand challenges facing society – energy, water, climate, food, health – are not amenable to single-discipline investigation; they often require many types of expertise across the biological, physical, and social disciplines (see also Frodeman 2016; De Grandis and Efstathiou 2016; Nagatsu et al. 2020). Second, discoveries are said to be more likely on the boundaries between fields, where the latest techniques, perspectives, and insights can reorient or increase knowledge. Third encounters with others benefit single disciplines, extending their horizons. Moreover, the proliferation of disciplines in the twentieth century increasingly calls for bridging them and transcending the scope of single disciplines on complex problems, i.e., for interdisciplinary research (e.g., Allwood et al. 2020).

Similarly, higher education policy documents assume that interdisciplinarity is increasingly becoming the hallmark of contemporary knowledge production and professional life (Mansilla 2005).[6] Graduate students and their training programs are recognized as essential to increasing interdisciplinary research capacity (Borrego and Newswander 2010; Spelt et al. 2009; Tripp and Shortlidge 2019; Nersessian 2022). An example of this move towards interdisciplinary research and education is an AAAS vision report (2009)[7] on developments in biology research and education that are becoming increasingly interdisciplinary. However, scientific research into teaching and learning in interdisciplinary higher education, for example regarding necessary research and thinking skills, is still limited and exploratory (Spelt et al. 2009; Van den Beemt 2020; Boon et al. 2022).

Additionally, there is a strong interest in promoting and funding collaboration between scientific disciplines to support science-based policy. For example, between ecologists, economists, sociologists, civil engineers, and atmospheric scientists working on an integrated understanding of environmental problems in which social, economic, ecological, and climate systems are causally intertwined (MacLeod and Nagatsu 2018; see also Inkpen et al. 2020), or on assessment models that assist in climate policies (e.g., Frisch 2013; Goodwin 2015; Parker 2018). Similar examples are the interdisciplinary modeling of an ecosystem management approach to marine social-ecological systems (Starfield and Jarre 2011; see also Levontin et al. 2011; Niinimäki et al. 2012; Kelly et al. 2013; Strasser et al. 2014; Ni et al. 2020). Other examples of the importance of interdisciplinary research to policy and management are chronic disease management (e.g., Bardhan et al. 2020) and the policy and management of risk (e.g., Zinn and Taylor-Gooby 2006).[8]

2.3 *Cognitive and epistemological challenges of interdisciplinary research*

Interdisciplinarity scholars also propose models of the interdisciplinary research process (e.g., Klein 1990; Repko 2008; Menken and Keestra 2016; Repko and Szostak 2017) drawing on literature in cognitive science and social psychology. These authors assume *integration* (of the research question, theoretical frameworks, method, results, and conclusions) as a crucial aspect of interdisciplinary research. They recommend step-by-step research processes that closely resemble common models of research processes, with the addition that *finding or creating common ground* is recommended as a way to achieve *integration between disciplines*. This approach thus relies heavily on communication between the disciplines but disregards the fundamental cognitive and epistemological challenges of communication and integration between disciplines (cf. MacLeod 2018). Integration (or connecting, or fitting together) of epistemic resources and methodologies from different disciplines is challenging because they are embedded in a tightly-knit network of scientific concepts, theories, fundamental principles, epistemic and pragmatic values, as well as techniques, procedures, routines, and modeling strategies that form the discipline, to the effect that disciplines or their content cannot be put together in a straightforward manner (Boon 2020b; Nersessian 2022). Moreover, the mentioned scholarly studies do not assign an explicit role to models and modeling in achieving integration between disciplines, while modeling is standard practice in existing interdisciplinary research. So, despite scholarly studies to create strategies and plans for doing interdisciplinary research, there is still a lack of proper articulation and testing of interdisciplinary research approaches (cf. Nagatsu et al. 2020, 1810; see also Grüne-Yanoff 2016; Mäki 2016).

2.4 *Interdisciplinary research in practice*

Scientific disciplines are not closed silos but develop, among other things, through the transfer and implementation of aspects from other disciplines. Grüne-Yanoff and Mäki (2014) provide a systematic overview of types of exchanges between disciplines. Elaborate examples of such exchanges are described in the ethnographic studies conducted by Nersessian (2009; 2022), MacLeod (2016), MacLeod and Nersessian (2013; 2015; 2016; 2018), and MacLeod and Nagatsu (2016). Exchange includes elements such as: knowledge about specific phenomena; experimental methods to create and investigate phenomena; measurement equipment and techniques; scientific concepts (e.g., 'conservation principles,' 'operations,' 'mechanisms,' 'energy,' 'equilibrium,' 'dynamics,' 'threshold,' 'saturation,' 'buffer,' 'reversibility,' 'hysteresis,' 'evolution,' 'ecology,' 'ecosystem'); mathematical and statistical methods to find structure in data and establish meaningful, quantifiable phenomena or patterns in data; mathematical templates (Humphreys 2019); model templates (e.g., Knuuttila and Loettgers 2016; Houkes and Zwart 2019); computer simulation methods to estimate unknown parameters or to link different types of models and study the dynamics of a system; the combination of different types of (quantitative and qualitative) research methods into mixed methods that expand research designs; and modeling strategies (e.g., from engineering sciences to molecular or systems biology).[9] Section 3 explains that these types of (heterogeneous) elements (exchanged between disciplines) are built into scientific models (Boumans 1999; Boon and Knuuttila 2009; Knuutila and Boon 2011). Interdisciplinarity is thus achieved through modeling, whereby integration of the mentioned elements takes place in modeling (i.e., models as integrators) and the resulting models become epistemic tools. As a result of these dynamics between research practices, some of these aspects are no longer discipline-specific but are shared cross-disciplinarily and embedded in multiple disciplines.

New disciplines emerge when researchers collaborate on problems or systems that are considered to consist of *causally interacting sub-systems* investigated in distinct disciplines. The sub-systems and their interactions are often investigated in experimental models and represented and interconnected by means of conceptual models, mathematical models, computer simulations (Nersessian 2022), and diagrammatic models (Boon 2008). Traditional examples are specialized disciplines in the engineering, agricultural, and biomedical sciences (e.g., Nerssessian and Patton 2009; Nersessian 2009; 2022). More recent examples are nuclear physics, systems biology (Coveney and Fowler 2005; O'Malley and Soyer 2012; Green 2013; MacLeod and Nerssessian 2013; 2015; 2016; 2018), neurosciences (e.g., Fagan 2017), computer sciences, geo- and climate sciences (e.g., Parker 2018; MacLeod and Nagatsu 2018). Interdisciplinary research, therefore, does not always take place through *integration* in the sense of the aforementioned definition of interdisciplinary research (cf. Grüne-Yanoff 2016) but is often a matter of cross-fertilization through transfer and exchange between disciplines.

A major motivation for promoting interdisciplinary research is to contribute to problems or opportunities outside science, such as those addressed in so-called applied sciences (the engineering, agricultural and biomedical sciences), and more generally, "real-world" problems related to new industrial opportunities, complex policy issues in society, and the UNESCO's sustainability goals. In these application contexts, interdisciplinary research projects usually focus on developing technologies,[10] computer simulations, scenario designs, and other types of tools for epistemic purposes, such as measurement, diagnosis, exploration, forecasting, and scenario investigation.

The distinction between interdisciplinary research within academic disciplines focused on *true knowledge about (fundamental) aspects of the world* versus interdisciplinary research focused on *actionable epistemic tools that make it possible to address real-world problems* (e.g., in science-based policy contexts) implies different epistemic and pragmatic criteria for research quality (cf. Elliot and McKaughan 2014; Brister 2016; De Grandis and Efstathiou 2016; Parker 2020),[11,12] as well as epistemologies, methodologies, and modeling strategies to meet these various criteria.

3. Models and modeling in interdisciplinary research practices

3.1 Models as integrators

In research practices, models and modeling are standard practices to achieve integration. Boumans' (1999) study on business cycles in the seminal collection *Models as Mediators* (Morrison and Morgan 1999) shows that models are *constructed by integrating many heterogeneous "ingredients,"* such as analogies, metaphors, theoretical notions, mathematical concepts, mathematical techniques, stylized facts, empirical data and finally relevant policy views, whereby the correctness of the resulting scientific model is partly justified by the scientifically sound choices researchers make in the modeling process. This approach to modeling in scientific practices is also studied by ethnographic studies. For example, Nersessian and Patton (2009), have studied biomedical engineering laboratories and argue that mental, physical, and computer models function as hubs that enable the integration ("interlocking") of biological and engineering concepts, methods, and materials. These models, in turn, are mental and external representations that enable model-based inferences that support research and learning about the system (see also Nersessian 2022).

In this view, modeling thus plays a role in integration processes, with *models as integrators* of not only the "ingredients" mentioned by Boumans, but also, as will be illustrated below with examples from practice, of sub-models that represent sub-systems within interdisciplinary research.

3.2 *How the construction of scientific models facilitates interdisciplinary research*

This process towards philosophical accounts of models and modeling that includes the cognitive, epistemological, methodological, and pragmatic aspects related to modelers and model-users in research practices, is further elaborated by Boon and Knuuttila (2009; see also Knuuttila and Boon 2011), who propose considering *models as epistemic tools*. They thereby build on Knuuttila's (2005) notion of models as *epistemic artefacts*, which explicitly deviates from the idea that our understanding of modeling should be reduced to models representing some external target systems – for models are not only representative artefacts, but also productive artefacts in, for example, model-based reasoning about the target system. Boon (2020a) elaborates on how models are constructed, namely by determining the heterogeneous "ingredients" that are usually built into the model (cf. Boumans 1999). Boon (2020b) provides further epistemological substantiation for this account, which also emphasizes the choices that researchers have to make in the construction of a model. Researchers can be held accountable for these choices, which is captured by the concept of *epistemological responsibility* (cf. Van Baalen and Boon 2015). Additionally, scientific models are *justified* and tested in at least three ways that complement each other, namely: (i) by justifying the relevance, physical plausibility, and adequacy of aspects built into the model; (ii) by assessing whether the model meets relevant epistemic and pragmatic criteria; and (iii) by empirical or experimental testing against reality, e.g., by comparing *model-outcomes* and experimental results (cf. Boon 2020b).

But the construction of models is also determined by "the specificities of a discipline," each with its own concepts and specific modeling strategies, which makes interdisciplinary collaboration (including integration and transfer between disciplines) difficult (cf. MacLeod 2018). Boon and Van Baalen (2019) and Boon (2020b) analyse this problem of interdisciplinary research in terms of *disciplinary perspectives* and argue that these are not necessarily opaque. Instead, disciplinary perspectives should be made explicit and explained in interdisciplinary research projects. Based on Kuhn's notion of disciplinary matrices and the aforementioned epistemology of model construction, they develop a framework for analyzing disciplinary perspectives that can be used by individual researchers (recognizing that researchers may have slightly different perspectives even within a discipline), which facilitates interdisciplinary understanding and communication.

On a more fine-grained practical level, model construction in interdisciplinary research involves a broad spectrum of *modeling strategies*, which raise additional epistemological, methodological, and ethical issues, for example:

- How to connect models from different disciplines, for which researchers use the notion of *coupling* (e.g., Coveney and Fowler 2005; Kremling and Saez-Rodriguez 2007; MacLeod and Nersessian 2013; MacLeod and Nagatsu 2016).
- How to deal with connecting models of dynamic physically related systems at *different time – and length-scales* as in: systems biology (e.g., Coveney and Fowler 2005; Kremling and Saez-Rodriguez 2007; MacLeod and Nersessian 2015; 2016); integrated assessment of agricultural production systems (Antle and Stoorvogel 2006); or integrated environmental assessment and management (Kelly et al. 2013).
- How to connect models of different kinds in the natural and engineering sciences, such as mechanistic and mathematical models, for which *diagrammatic models* are proposed (cf. Boon 2008).

- How to connect models from the natural sciences (broadly interpreted as sciences that concern natural and physical processes) and social sciences, e.g., in climate modeling to support policy decisions, for which *integrated assessment models* are proposed (e.g., Frisch 2013; also see Strasser et al. 2014; Parker 2006; 2011).
- How to assess the *reliability* of (*complex multiscale*) models that result from interdisciplinary research as in climate models (e.g., Goodwin 2015; Parker 2006).
- How to deal with the *uncertainty* of (e.g., complex multiscale) models and their predictions that result from interdisciplinary research as in climate models (e.g., Parker 2011).
- How to achieve an integrated treatment of complex societal issues, e.g., by integrating stakeholders, models of dynamic processes, different scales, and societal considerations into *integrated environmental assessment models* for management decisions under uncertainty (cf. Kelly et al. 2013; see also Strasser et al. 2014; Inkpen et al. 2020).

3.3 Modeling strategies in interdisciplinary research practices

Practicing researchers have developed several modeling and integration strategies to address the issues mentioned. This is illustrated with a number of examples, ranging from modeling in systems biology to models that support the management of complex systems.

Kremling and Saez-Rodriguez (2007) propose an engineering approach to *systems biology*, for which they adopt a *modeling framework based on network theory*. Network theory considers all processes a connection of *components* and *coupling elements*. Components represent physical quantities like energy, mass, (bio)chemical substances, or momentum. That is, the time- and location-dependent amounts of these components in the physical system are (conceptually and mathematically) represented as time- and location-dependent variables in the model while coupling elements describe the physical fluxes of components. In other words, the physical amount of components flowing into or out of a location is (conceptually and mathematically) represented as changes in the time- and location-dependent variable values in the model. Additionally, components and coupling elements can be defined on different *hierarchical modeling levels*, which enable the aggregation of systems of components and coupling elements into a single component on a higher level.

Similarly, Coveney and Fowler (2005) explain, "from the perspective of a physicist," the role of *multiscale models* in connecting models of systems at different time- and length-scales. Their case study also resides in systems biology. Their ultimate epistemic goal is to construct a *whole-organ heart model* (for example, to study the dynamics of the heart or circadian rhythms), by coupling models that represent processes at the molecular and cellular scale. Hence, (conceptual and mathematical) models of processes at the molecular biological level must be connected (i.e., integrated) with models of processes at the cellular level, in order to represent (conceptually and mathematically) interactions between dynamical systems that are physically related. One of the challenges they aim to solve by the coupled multiscale approach is to account for the role of feedback, i.e., to build into the model changes on the larger length-scale that affect behavior at the smaller length-scale.

Antle and Stoorvogel (2006) study vulnerable *agricultural* (or, agro-eco) *production systems*. They view these as complex and dynamic systems that result from interacting physical, biological, and human decision-making processes and many internal feedbacks. Their goal is a *computer simulation model* of the system describing the interacting bio-physical and economic decision-making subsystems on compatible spatial and temporal scales. Their modeling strategy is a *modular model-coupling* approach, in which models of subsystems are coupled by using a subset of (spatially and temporally varying) state variables from one subsystem as inputs into another subsystem. According to these authors, advantages to the modular approach are that the disciplines involved develop (modular) models of subsystems, which, when coupled, are kept in their original (perhaps simplified) form. This warrants the transparency of models and makes it easier for researchers to build and test the models. In a case study of a vulnerable agricultural system, they illustrate the importance of a *modular model-coupling* approach that includes the dynamics and spatial heterogeneity in the analysis of the agro-eco behavior of the production system. For example, the economic problem facing farmers is deciding which crop to grow. This is where the computer simulation of the agricultural system in their area can assist by showing the long-term impacts, such as soil depth falling below a critical threshold due to erosion, which can be prevented if farmers opt for crop rotation.

Ni et al. (2020) developed a hybrid model aimed at an *accurate and reliable forecasting model* for water resource planning and management. Their *hybrid model* is based on the *principle of modular modelling*, in which a complex problem is divided into more simple sub-models. The epistemic and pragmatic purpose of these types of models is accurate and reliable streamflow (low and high) forecasting to provide information for water resource management and timely warning of natural disasters, such as droughts and floods.

Levontin et al. (2011) use *Bayesian belief networks* (BBN) to integrate the findings of separate biological, economic, and sociological studies, to be used as a decision-support tool for the interdisciplinary evaluation of potential Baltic salmon management plans. Their epistemic and pragmatic aim is to evaluate the robustness of management decisions to different priorities and various sources of uncertainty. The BBN can thus be considered a model constructed as an epistemic tool to represent interactions and responses to policy decisions.

Kelly et al. (2013) present a comprehensive review of five common modeling approaches in environmental sciences that have the capacity to integrate knowledge – that is, modeling approaches that can accommodate multiple issues, values, scales (e.g., time- and length-scales) and uncertainty considerations, as well as facilitate stakeholder engagement. These modeling approaches are *systems dynamics*, *Bayesian networks*, *coupled component models*, *agent-based models*, and *knowledge-based models* (as in expert systems). Additionally, Kelly et al. use their analysis to develop a framework to help modelers and model-users select an appropriate modeling approach for their integrated environmental assessment and management applications and enable more effective learning in interdisciplinary settings.

Starfield and Jarre (2011) propose a set of recommendations for conducting interdisciplinary research – which in their case focuses on *interdisciplinary modeling for an ecosystem approach to management in marine social-ecological systems* – emphasizing that "Interdisciplinary work needs to be constrained by clear system objectives. The emphasis is on the word 'system' because it is a mistake to define objectives from the viewpoint of the disciplines themselves. …. It is essential to use a modeling paradigm that focuses on objectives and leads to a balanced contribution from each discipline" (Starfield and Jarre 2011, 217–218). They consider *frame-based modeling* suitable as a modeling paradigm for addressing long-term changes in social-ecological systems. Notably, the emphatic premise of letting the overarching epistemic and pragmatic goal take precedence (rather than the epistemic goals of the disciplines) may conflict with "the advantages of the *modular model-coupling* approach" recommended by Antle and Stoorvogel (2006).

Strasser et al. (2014) develop a *coupled component model* to facilitate an integrative assessment of the impact of climate change on snow conditions and skiing tourism in a typical Austrian ski resort. They use this as a case study for the design of *interface tools* to enable the integration between disciplinary sub-models. Importantly, their focus on interfaces to enable integration of quantitative and qualitative knowledge– that is, values, from relevant natural and social science disciplines-such as *variables* from climate and weather sciences, and *indicators* and *threshold values* from economy and ecology. These interface tools were jointly developed by scientists (in climate, snow hydrology, economy, and tourism) and the decision-makers responsible for the skiing industry and regional tourism development. The authors emphasize that "the joint model development and interface design are core elements of integration, and can be regarded as a mutual learning and negotiation process where understanding continuously develop" (Strasser et al. 2014, 186; see also Antle and Stoorvogel 2006, Kelly et al. 2013). Similarly, De Sandes-Guimarães et al. (2022) argue that for this type of problem, policymakers should take part in the interdisciplinary research project, thus making it a process of *knowledge coproduction* aimed at supporting policy decisions for complex problems (see also De Grandis and Efstathiou 2016).

3.4 Philosophical accounts of modeling practices in interdisciplinary research

These kinds of examples from interdisciplinary research practices are analyzed by philosophers of science to uncover epistemological, methodological, and ethical aspects of interdisciplinary scientific research (cf. Mäki 2016). The practice examples show that the same concepts are used to characterize the nature of a target-system across a wide range of scientific disciplines, such as: "complex systems," "dynamical systems," "sub-systems," "physically (or otherwise causally) related processes," "feedbacks," "processes at different time- and length-scales," and "variables." The same applies to the concepts used by researchers in different research areas to describe modeling strategies, such as "integration," "modularity," "model coupling," "coupled-component models," "multi-scale modeling," "hierarchical modeling," "hybrid modeling," "networks," "systems dynamics," and "interfaces between models." In the scientific literature, these concepts are used to explain interdisciplinary research strategies and methodologies.

Philosophical analyses of existing scientific research practices show that scientific researchers in a wide range of scientific disciplines generally follow the same strategy when developing *conceptual models* (cf. Boon 2020a; also see MacLeod and Nersessian 2013; Nersessian 2022). The similarity of research strategies enables integration between disciplines (Boon 2020b). An example is the way researchers develop an integrated model of a more complex system, by representing the system as (causal) interactions between relevant (often dynamic) processes or subsystems (typically represented in space-time diagrams, cf. Boon 2008). Usually, each of those subsystems is the subject of a separate scientific discipline. In this strategy, the relevant (discipline-specific) measurable and calculable *variables and parameters* are determined for each subsystem. Based on this, a *mathematical sub-model* can be constructed for each subsystem. Integration then takes place by constructing a mathematical model that connects the mathematical sub-models via the time- and space-dependent variables (also called state variables), namely as input and output variables between the sub-models. Finally, these mathematical models form the basis for the construction of *computer simulation models.*

These examples also show that models across a wide range of complex systems are usually aimed at a specific epistemic purpose, e.g., the closer study of the system in terms of its dynamic behavior, the effects of interventions, and the determination of unknown parameter values (e.g., through computer simulations), or as an aid in policy decisions using the model in scenario studies or forecasting (e.g. Kelly et al. 2013; Ni et al. 2020). Altogether, this implies that models created in the specific research contexts can be interpreted as *epistemic artifacts* and *tools* built for use by researchers and other stakeholders in understanding, handling, or intervening with complex systems (cf. Knuuttila 2005; see also Parker 2020; Nersessian 2022).

It is worth mentioning separately that some modeling strategies also aim at incorporating social, economic, and sustainability values (cf. Elliott and McKaughan 2014; Parker 2020) and mapping the vulnerability of the dynamic system in relation to them, which is built into the model, for example, via threshold values (e.g., Strasser et al. 2014). These practice examples, therefore, illustrate how models can simultaneously play a role in exploring the ethical implications of (postponing) interventions in or (lack of) decisions about a complex system.

In ethnographic studies, philosophers stay close to first aiming at a rich and detailed description of these practices and making explicit salient features. Ethnographic methods have thus been used (cf. Nersessian and MacLeod 2022; Nersessian and Patton 2009; MacLeod 2016; Nersessian 2009; 2022; MacLeod and Nersessian 2013; 2015; 2016; 2018; MacLeod and Nagatsu 2016) to make *modeling strategies* in concrete interdisciplinary research practices explicit and to analyze critically their epistemological approach, inventions, and quality (e.g., Mattila 2005; Parker 2006; 2011; Nersessian and Patton 2009; Grüne-Yanoff 2016; MacLeod 2018; MacLeod and Nagatsu 2016; 2018; Nagatsu et al. 2020; Inkpen and DesRoches 2020; Nersessian 2022). Some examples are:

Green's (2013) analysis of modeling practices by a case study on network modeling in systems biology, shows that engineering approaches are applied to the study of biological systems. Based on this case study, she argues that *the use of engineering principles* affords a conceptualization of biological functions in language from control- and graph theory, which can open a *new epistemic space for understanding biological function*.

MacLeod and Nagatsu's (2016) ethnographic study of the collaboration of economists and ecologists in the resource economy aims to analyze the role of *model-building frameworks and strategies* that can play a role in overcoming the inherent difficulties of interdisciplinary research. They distill various features of how models are put together and show how a *coupled-model framework* is used to coordinate and combine background models from ecology and economics.

Nersessian's (2022) book-long study analyses research on the epistemic practices of interdisciplinary research in laboratories of biomedical engineering (BME) and integrative systems biology (ISB). She argues that interdisciplinary modeling in BME uses *engineering design methods and principles* to *understand basic biological phenomena* in order to control disease processes or create interventions for specific medical disorders. ISB aims at an *integrative analysis* of the behavior of complex *(nonlinear) biological systems at all levels*, from intracellular interactions to ecosystem processes, to investigate *how higher-level functionality emerges* from myriad interactions at lower levels. To this end, ISB modeling practices *integrate* computation, applied mathematics, engineering concepts and methods, and biological experimentation (see also MacLeod and Nersessian 2016).

In addition to ethnographic studies that provide rich and detailed descriptions of interdisciplinary modeling practices, philosophers also aim at targeting epistemological and ethical aspects. Some examples are: Elliott and McKaughan (2014) on the role of *non-epistemic values*, Andersen and Wagenknecht (2013) on the role of *epistemic dependence and trust* in interdisciplinary research, Andersen (2016) on the tension between interdisciplinarity and *quality control*, and MacLeod and Nagatsu (2018) who propose *categorizing* four different integrative modeling strategies. Green (2013) argues that the use of *multiple representational means* is an essential part of the dynamic of knowledge generation because the diversity of constraints of different interlocking epistemic means creates a *potential for knowledge production*. Parker (2006) shows how *incompatible climate models* are used together in *multi-model ensembles* and explains why this practice is *reasonable*, given scientists' inability to identify a "best" model for predicting the future climate. Finally, Frisch (2013) argues that integrated assessment models used in climate policies involve highly conjectural (non-evidenced), simplified (unjustified), and intrinsically normative *assumptions*.

4. Philosophy for interdisciplinary modeling practices

The knowledge of epistemological and methodological challenges of interdisciplinary research and the role of modeling therein is far from complete. The presented overview highlights a number of aspects. First, representational accounts of models are problematized because the construction of models is enabled by the specificities of the scientific disciplines (i.e., the disciplinary perspective) so that discipline-specific theoretical, conceptual, instrumental, and strategic features determine the model content. This explains why crucial characteristics of interdisciplinary research, namely *transfer* and *integration* (e.g., of epistemic resources and methodologies), encounter epistemological, methodological, and conceptual barriers. It also means that models function as integrators (hubs) of heterogeneous aspects and, in interdisciplinary research, of sub-models. Another aspect arises from the advocacy of interdisciplinary research focused on epistemic utility, which implies that models are seen as epistemic tools that must meet epistemic and pragmatic criteria relevant to the intended epistemic purpose, and in the case of science-based policy also ethical criteria, e.g., model-based reasoning or computer simulations for the analysis, prediction, or scenario-study of complex target-systems. Researchers do cope with the mentioned epistemological, methodological, and cognitive issues and barriers, as illustrated by the aforementioned real-world examples of interdisciplinary modeling practices.

The *philosophy of scientific modeling* that targets interdisciplinary research practices, science-based policy, and higher education, should therefore study epistemologies and methodologies of modeling strategies aimed at understanding complex systems, including the critical roles of human cognition and responsibility therein (cf. Boon et al. 2022; Nersessian 2022, 283).

Acknowledgments

This work is financed by an Vici-Aspasia grant (409.40216) of the Dutch National Science Foundation (NWO) for the project *Philosophy of Science for the Engineering Sciences* and by the work package *Interdisciplinary Engineering Education* at the 4TU-CEE (Centre Engineering Education) in The Netherlands. I would like to thank Meghan Bohardt and Rami Koskinen for their helpful suggestions for the clarity of this chapter.

Notes

1 The cognitive scientist, Barbara Tversky (2017) offers a concise explanation of mental models, in which models as representations are interpreted from cognitive sciences perspective: "representations are internalized perceptions. However, representations cannot be copies, they are highly processed. They are interpretations of the content that is the focus of thought. They may select some information from the world and ignore other information, they may rework the information selected, and they may add information, drawing on information already stored in the brain. In this sense, representations are models" (Tversky 2017, VI–VII).

2 Suárez (2004) proposes an inferential conception of representation, which entails the idea that "[the internal structure of the representation, e.g., a model] A allows competent and informed agents to [correctly] draw specific inferences regarding [the target] B" (Suárez 2004, 773).

3 For a more comprehensive review of aspects addressed in definitions of interdisciplinary science, see Tripp and Shortlidge (2019).

4 E.g., Speech by ERC President Prof. Jean-Pierre Bourguignon (2019).

5 Gleed and Marchant (2016) *Interdisciplinarity Survey Report for the Global Research Council 2016 Annual Meeting*. Also see: Global Research Council (n.d.) *Statement of Principles on Interdisciplinarity*.

6 For example: National Academy of Sciences et al. (2005). National Academy of Engineering (2005). National Science Foundation (2008). National Academies of Sciences et al. (2018, Chapter 3). Witchel (2022) and Psychological Society (2021). Craciun et al. (2023). Moser et al. (2022).

7 American Association for the Advancement of Science AAAS. (2009). *Vision and change in Undergraduate Biology Education: A Call to Action, Final Report*. Washington, DC. Retrieved January 3, 2023. This report is no longer available online; see Woodin et al. (2010).

8 Chronic disease management requires an integrated care approach to managing illness that includes screenings, check-ups, monitoring, and coordinating treatment, and patient education (cf. Bardhan et al. 2020). Policy and management of risk (e.g., by governments, insurance companies, and industries) requires interdisciplinary research that combines technical risk analysis (focusing on the controllability, safety, and reliability of technical systems and processes, and analysis of how failure can occur) or epidemiological and toxicological risk analysis (focusing on probability and seriousness of illness due to toxic compounds or medicines) with studies into public perception of risk (e.g., conceptualizing and studying social processes influencing risk perception) and risk communication (Zinn and Taylor-Gooby 2006).

9 These kinds of (heterogeneous) elements —that are exchanged between disciplines— are built-into models, as in models-as-integrators and models-as-epistemic-tools. More elaborate accounts of knowledge transfer between disciplines can be found in a special issue on this topic edited by Herfeld and Lisciandra (2019).

10 Van Baalen (2019) provides an example of interdisciplinary biomedical research to develop a diagnostic technology. She conducted an ethnographic study to analyse reasoning and decision-making processes within a multidisciplinary research team —consisting of a clinician, a radiologist (specialized in thorax imaging), a radiographer and an MRI engineer— who collaboratively developed a new clinical MRI imaging technique for the non-invasive diagnosis of respiratory diseases.

11 Recognizing different epistemic goals is also crucial to interdisciplinary research within academia (c.f. Green 2013). See also Parker (2020). Love and Brigand (2017) push for a shift in focus *from metaphysics to epistemology*. Philosophers should approach conceptual problems in science (such as the problem of biological individuality) by paying attention to the variety of *epistemic goals* underlying successful scientific practice.

12 Notable, pragmatic and epistemic criteria relevant to the research project at hand, should also guide the assessment of the quality of interdisciplinary work in educational settings (cf. Mansilla 2005).

References

Aboelela, Sally W., Elaine Larson, Suzanne Bakken, Olveen Carrasquillo, Allan Formicola, Sherry A Glied, Janet Haas, and Kristine M Gebbie. 2007. "Defining interdisciplinary research: Conclusions from a critical review of the literature." *Health Services Research* 42(1): 329–346. https://doi.org/10.1111/j.1475-6773.2006.00621.x

Allwood, Jens, Olga Pombo, Clara Renna, and Giovanni Scarafile, eds. 2020. *Controversies and Interdisciplinarity: Beyond Disciplinary Fragmentation for a New Knowledge Model.* Vol. 16. John Benjamins Publishing Company. https://doi.org/10.1075/cvs.16

Andersen, Hanne. 2016. "Collaboration, interdisciplinarity, and the epistemology of contemporary science." *Studies in History and Philosophy of Science Part A* 56: 1–10. https://doi.org/10.1016/j.shpsa.2015.10.006

Andersen, Hanne, and Susann Wagenknecht. 2013. "Epistemic dependence in interdisciplinary groups" *Synthese* 190: 1881–1898. https://doi.org/10.1007/s11229-012-0172-1

Antle, John M., and Jetse J. Stoorvogel. 2006. "Incorporating systems dynamics and spatial heterogeneity in integrated assessment of agricultural production systems." *Environment and Development Economics* 11(1): 39–58. https://doi.org/10.1017/S1355770X05002639

Bardhan, Indranil, Hsinchun Chen, and Elena Karahanna. 2020. "Connecting systems, data, and people: A multidisciplinary research roadmap for chronic disease management." *MIS Quarterly* 44(1): 185–200.

Boon, Mieke. 2008. "Diagrammatic models in the engineering sciences." *Foundations of Science* 13(2): 127–142. https://doi.org/10.1007/s10699-008-9122-2

———. 2020a. "Scientific methodology in the engineering sciences." *The Routledge Handbook of the Philosophy of Engineering,* edited by Diane P. Michelfelder and Neelke Doorn. New York, Routledge: 80–94. https://doi.org/10.4324/9781315276502-8

———. 2020b. "The role of disciplinary perspectives in an epistemology of scientific models." *European Journal for Philosophy of Science* 10(3): 1–34. https://doi.org/10.1007/s13194-020-00295-9

Boon, Mieke, and Sophie Van Baalen. 2019. "Epistemology for interdisciplinary research-shifting philosophical paradigms of science." *European Journal for Philosophy of Science* 9(1): 1–28. https://doi.org/10.1007/s13194-018-0242-4

Boon, Mieke, and Tarja T. Knuuttila. 2009. "Models as epistemic tools in engineering sciences: a pragmatic approach." In *Philosophy of Technology and Engineering Sciences*, edited by Anthonie Meijers, 687–720. Handbook of the philosophy of science (Vol. 9). Elsevier/North-Holland. https://www.sciencedirect.com/book/9780444516671/philosophy-of-technology-and-engineering-sciences

Boon, Mieke, Mariana Orozco, and Kishore Sivakumar. 2022. "Epistemological and educational issues in teaching practice-oriented scientific research: Roles for philosophers of science." *European Journal for Philosophy of Science* 12(1): 16. https://doi.org/10.1007/s13194-022-00447-z

Borrego, Maura, and Lynita K. Newswander. 2010. "Definitions of interdisciplinary research: Toward graduate-level interdisciplinary learning outcomes." *The Review of Higher Education* 34(1): 61–84. https://doi.org/10.1353/rhe.2010.0006

Boumans, Marcel. 1999. "Built-in justification." In *Models as Mediators - Perspectives on Natural and Social Science*, edited by Mary S. Morgan and Margaret Morrison, 66–96. Cambridge: Cambridge University Press.

Bourguignon, Jean-Pierre. 2019. "Supporting interdisciplinarity, a challenging obligation." Accessed January 3, 2023. https://erc.europa.eu/news/supporting-interdisciplinarity-challenging-obligation.

Brister, Evelyn. 2016. "Disciplinary capture and epistemological obstacles to interdisciplinary research: Lessons from central African conservation disputes." *Studies in History and Philosophy of Science Part C: Studies in History and Philosophy of Biological and Biomedical Sciences*, 56: 82–91. https://doi.org/10.1016/j.shpsc.2015.11.001

Craciun, Daniela, Frans Kaiser, Andrea Kottmann, and Barend van der Meulen. 2023. *Research for CULT Committee - The European Universities Initiative: First Lessons, Main Challenges and Perspectives.* https://www.europarl.europa.eu/RegData/etudes/STUD/2023/733105/IPOL_STU(2023)733105_EN.pdf

Coveney, Peter V., and Philip W. Fowler. 2005. "Modelling biological complexity: A physical scientist's perspective." *Journal of the Royal Society Interface* 2(4): 267–280. https://doi.org/10.1098/rsif.2005.0045

de Grandis, Giovanni, and Sophia Efstathiou. 2016. "Grand challenges and small steps. Introduction to the special issue 'Interdisciplinary integration: The real grand challenge for the life sciences?'" *Studies in History and Philosophy of Science Part C: Studies in History and Philosophy of Biological and Biomedical Sciences* 56: 39–47.

de Sandes-Guimarães, Luisa Veras, Raquel Velho, and Guilherme Ary Plonski. 2022. "Interdisciplinary research and policy impacts: Assessing the significance of knowledge coproduction." *Research Evaluation* 31(3): 344–354.

Elliott, Kevin, and Daniel McKaughan. 2014. "Nonepistemic values and the multiple goals of science." *Philosophy of Science* 81(1): 1–21. https://doi.org/10.1086/674345

Fagan, Melissa B. 2017. "Explanation, multiple perspectives, and understanding." *Balkan Journal of Philosophy*, 9(1): 19–34. https://www.pdcnet.org/bjp/content/bjp_2017_0009_0001_0019_0034

Frisch, Mathias. 2013. "Modeling climate policies: A critical look at integrated assessment models." *Philosophy and Technology*, 26: 117–137. https://link.springer.com/article/10.1007/s13347-013-0099-6

Frodeman, Robert. 2016. "Interdisciplinarity, grand challenges, and the future of knowledge." *Studies in History and Philosophy of Science Part C: Studies in History and Philosophy of Biological and Biomedical Sciences* 56: 108–110. https://doi.org/10.1016/j.shpsc.2015.11.011

Giere, Ronald N. 2010. "An agent-based conception of models and scientific representation." *Synthese* 172(2): 269–281. https://doi:10.1007/s11229-009-9506-z.

Gleed, Alasdair, and David Marchant. 2016. *Interdisciplinarity. Survey Report for the global Research Council 2016 Annual Meeting.* https://www.jsps.go.jp/english/e-grc/data/5th/Survey_Report_on_Interdisciplinarity_for_GRC_DJS_Research.pdf

Global Research Council. n.d. "Statement of principles on interdisciplinarity." https://globalresearchcouncil.org/fileadmin/documents/GRC_Publications/Statement_of_Principles_on_Interdisciplinarity.pdf

Goodwin. W. M. 2015. "Global climate modeling as applied science." *Journal for General Philosophy of Science* 46: 339–330. https://link.springer.com/article/10.1007/s10838-015-9301-0

Green, Sara. 2013. "When one model is not enough: Combining epistemic tools in systems biology." *Studies in History and Philosophy of Science Part C: Studies in History and Philosophy of Biological and Biomedical Sciences* 44(2): 170–180. https://doi.org/10.1016/j.shpsc.2013.03.012

Grüne-Yanoff, Till. 2016. "Interdisciplinary success without integration." *European Journal for Philosophy of Science* 6(3): 343–360. https://doi.org/10.1007/s13194-016-0139-z

Grüne-Yanoff, Till and Uskali Mäki. 2014. "Introduction: Interdisciplinary model exchanges." *Studies in History and Philosophy of Science Part A*, 48: 52–59. https://doi.org/10.1016/j.shpsa.2014.08.001

Herfeld, Catherine, and Chiara Lisciandra, eds. (2019). "Knowledge transfer and its contexts." *Studies in History and Philosophy of Science Part A* 77: 1–10.

Houkes, Wybo, and Sjoerd D. Zwart. 2019. "Transfer and templates in scientific modelling." *Studies in History and Philosophy of Science Part A* 77: 93–100. https://doi.org/10.1016/j.shpsa.2017.11.003

Humphreys, Paul. 2019. "Knowledge transfer across scientific disciplines." *Studies in History and Philosophy of Science Part A* 77: 112–119. https://doi.org/10.1016/j.shpsa.2017.11.001

Inkpen, S. Andrew, and C. Tyler DesRoches. 2020. "When ecology needs economics and economics needs ecology: Interdisciplinary exchange during the anthropocene." *Ethics, Policy and Environment* 23(2): 203–221. https://doi.org/10.1080/21550085.2020.1848182

Jacobs, Jerry A., and Scott Frickel. 2009. "Interdisciplinarity: A critical assessment." *Annual Review of Sociology* 35: 43–65. https://doi.org/10.1146/annurev-soc-070308-115954

Kelly, Rebecca. A., Anthony J. Jakeman, Olivier Barreteau, Mark E. Borsuk, Sondoss ElSawah, Serena H. Hamilton, and Hans Jørgen Henriksen, et al. 2013. "Selecting among five common modelling approaches for integrated environmental assessment and management." *Environmental Modelling and Software* 47: 159–181. https://doi.org/10.1016/j.envsoft.2013.05.005

Klein, Julie T. 1990. *Interdisciplinarity: History, Theory, and Practice.* Detroit, MI: Wayne State University Press.

Kremling, A., and J. Saez-Rodriguez. 2007. "Systems biology-an engineering perspective." *Journal of Biotechnology* 129(2): 329–351. https://doi.org/10.1016/j.jbiotec.2007.02.009

Knuuttila, Tarja. 2005. *Models as Epistemic Artefacts: Toward a Non-representationalist Account of Scientific Representation.* PhD thesis. University of Helsinki.

Knuuttila, Tarja, and Mieke Boon, 2011. "How do models give us knowledge? The case of Carnot's ideal heat engine." *European Journal for Philosophy of Science* 1(3): 309–334. https://doi:10.1007/s13194-011-0029-3

Knuuttila, Tarja, and Andrea Loettgers. 2016. "Model templates within and between disciplines: From magnets to gases-and socio-economic systems." *European Journal for Philosophy of Science* 6: 377–400. https://doi.org/10.1007/s13194-016-0145-1

Levontin, Polina, Soile Kulmala, Päivi Haapasaari, and Sakari Kuikka. 2011. "Integration of biological, economic, and sociological knowledge by Bayesian belief networks: The interdisciplinary evaluation of potential management plans for Baltic salmon." *ICES Journal of Marine Science* 68(3): 632–638. https://doi.org/10.1093/icesjms/fsr004

Love, Alan C., and Ingo Brigandt. 2017. "Philosophical dimensions of individuality." In *Biological Individuality: Integrating Scientific, Philosophical, and Historical Perspectives*, edited by Scott Lidgard, Lynn Nyhart, 318–348. Chicago: University of Chicago Press.

MacLeod, Miles, 2018. "What makes interdisciplinarity difficult? Some consequences of domain specificity in interdisciplinary practice." *Synthese* 195(2): 697–720. https://doi.org/10.1007/s11229-016-1236-4

MacLeod, Miles, and Michiru Nagatsu. 2016. "Model coupling in resource economics: Conditions for effective interdisciplinary collaboration." *Philosophy of science*, 83(3): 412–433. https://doi.org/10.1086/685745

———. 2018. "What does interdisciplinarity look like in practice: Mapping interdisciplinarity and its limits in the environmental sciences." *Studies in History and Philosophy of Science Part A* 67: 74–84. https://doi.org/10.1016/j.shpsa.2018.01.001

MacLeod, Miles, and Nancy J. Nersessian. 2013. "Coupling simulation and experiment: The bimodal strategy in integrative systems biology." *Studies in History and Philosophy of Science Part C: Studies in History and Philosophy of Biological and Biomedical Sciences* 44(4): 572–584. https://doi.org/10.1016/j.shpsc.2013.07.001

———. 2015. "Modeling systems-level dynamics: Understanding without mechanistic explanation in integrative systems biology." *Studies in History and Philosophy of Science Part C: Studies in History and Philosophy of Biological and Biomedical Sciences* 49: 1–11.

———. 2016. "Interdisciplinary problem-solving: Emerging modes in integrative systems biology." *European journal for philosophy of science* 6(3): 401–418. https://doi.org/10.1007/s13194-016-0157-x

———. 2018. "Modeling complexity: Cognitive constraints and computational model-building in integrative systems biology." *History and Philosophy of the Life Sciences* 40: 1–28. https://doi.org/10.1007/s40656-017-0183-9

Magnani, Lorenzo, and Tommaso Bertolotti, eds. 2017. "Preface." *Springer Handbook of Model-based Science*, XI–XIII. Springer Dordrecht Heidelberg London New York. https://link.springer.com/book/10.1007/978-3-319-30526-4

Mäki, Uskali. 2016. "Philosophy of interdisciplinarity. What? Why? How?" *European Journal for Philosophy of Science* 6(3): 327–342. https://doi.org/10.1007/s13194-016-0162-0

Mansilla, Veronica Boix. 2005. "Assessing student work at disciplinary crossroads." *Change: The Magazine of Higher Learning* 37(1): 14–21. https://doi.org/10.3200/CHNG.37.1.14-21

Mattila, E. 2005. "Interdisciplinarity "in the making": Modeling infectious diseases." *Perspectives on Science*, 13(4): 531–553. https://doi.org/10.1162/106361405775466081

Menken, Steph, and Machiel Keestra, eds. 2016. *An Introduction to Interdisciplinary Research: Theory and Practice.* Amsterdam: Amsterdam University Press. https://www.aup.nl/en/book/9789463724692/an-introduction-to-interdisciplinary-research

Morgan, Mary S., and Margaret Morrison, eds. 1999. *Models as Mediators.* Cambridge: Cambridge University Press.

Moser, Peter, Susanne Feiel, and Volkmar Kircher. 2022. "The (R) evolution of European education policy: European higher education alliances." *BHM Berg-und Hüttenmännische Monatshefte* 167(10): 457–461.

Nagatsu, Michiru, and Attilia Ruzzene, eds. 2019. *Contemporary Philosophy and Social Science: An Interdisciplinary Dialogue.* Bloomsbury Academic. https://www.bloomsbury.com/uk/contemporary-philosophy-and-social-science-9781474248754/

Nagatsu, Michiru, Taylor Davis, C. Tyler DesRoches, Inkeri Koskinen, Miles MacLeod, Milutin Stojanovic, and Henrik Thorén. 2020. "Philosophy of science for sustainability science." *Sustainability Science* 15(6): 1807–1817. https://doi.org/10.1007/s11625-020-00832-8

National Academy of Engineering. 2005. *Educating the Engineer of 2020: Adapting Engineering Education to the New Century*. Washington, DC: National Academies Press.

National Academy of Sciences, National Academy of Engineering, and Institute of Medicine. 2005. *Facilitating Interdisciplinary Research.* Washington, DC: The National Academies Press. https://doi.org/10.17226/11153.

National Academies of Sciences, Engineering, and Medicine. 2018. *Graduate STEM Education for the 21st Century*. National Academies Press. https://nap.nationalacademies.org/read/25038

National Science Foundation. 2008. *Impact of Transformative Interdisciplinary Research and Graduate Education on Academic Institutions*. Workshop Report. Carol Van Hartesveldt, and Judith Giordan.

Nersessian, Nancy J. 2009. "How do engineering scientists think? Model-based simulation in biomedical engineering research laboratories." *Topics in Cognitive Science* 1(4): 730–757. https://doi.org/10.1111/j.1756-8765.2009.01032.x

———. 2022. *Interdisciplinarity in the Making: Models and Methods in Frontier Science*. MIT Press. https://mitpress.mit.edu/9780262544665/interdisciplinarity-in-the-making/

Nersessian, Nancy J., and Miles MacLeod. 2022. "Rethinking ethnography for philosophy of science." *Philosophy of Science* 89(4): 721–741. https://doi.org/10.1017/psa.2022.8

Nersessian, Nancy J., and Christopher Patton. 2009. "Model-based reasoning in interdisciplinary engineering." In *Handbook of the Philosophy and Engineering Sciences*, edited by Anthonie Meijers, 687–718. Amsterdam: Elsevier.

Newell, William H., and Pauline Gagnon. 2013. "The state of the field: Interdisciplinary theory." *Issues in Interdisciplinary Studies* 31: 22–43. https://our.oakland.edu/handle/10323/4478

Ni, Lingling, Dong Wang, Jianfeng Wu, Yuankun Wang, Yuwei Tao, Jianyun Zhang, and Jiufu Liu. 2020. "Streamflow forecasting using extreme gradient boosting model coupled with Gaussian mixture model." *Journal of Hydrology* 586: 124901. https://doi.org/10.1016/j.jhydrol.2020.124901

Niinimäki, Sami, Olli Tahvonen, and Annikki Mäkelä. 2012. "Applying a process-based model in Norway spruce management." *Forest Ecology and Management* 265: 102–115. https://doi.org/10.1016/j.foreco.2011.10.023

O'Malley, Maureen A., and Orkun S. Soyer. 2012. "The roles of integration in molecular systems biology." *Studies in History and Philosophy of Science Part C: Studies in History and Philosophy of Biological and Biomedical Sciences* 43(1): 58–68. https://doi.org/10.1016/j.shpsc.2011.10.006

Parker, Wendy S. 2006. "Understanding pluralism in climate modeling." *Foundations of Science* 11(4): 349–368. https://doi.org/10.1007/s10699-005-3196-x

———. 2011. "When climate models agree: The significance of robust model predictions." *Philosophy of Science* 78(4): 579–600. https://doi.org/10.1086/661566

———. 2018. "Climate science." In *The Stanford Encyclopedia of Philosophy*, edited by E.N. Zalta. https://plato.stanford.edu/archives/sum2018/entries/climate-science/

———. 2020. "Model evaluation: An adequacy-for-purpose view." *Philosophy of Science* 87(3): 457–477. https://doi.org/10.1086/708691

Psychological Society, the. 2021. "The Future of Interdisciplinary Research beyond REF 2021." https://www.physoc.org/policy/research-landscape-and-funding/interdisciplinary-research/.

Repko, Alan F. 2008. *Interdisciplinary Research*. Thousand Oaks, CA: Sage.

Repko, Alan F., and Rick Szostak. 2017. *Interdisciplinary Research: Process and Theory*. 3rd edition. Los Angeles: Sage.

Rylance, Rick. 2015. "Grant giving: Global funders to focus on interdisciplinarity." *Nature* 525: 313–315. https://doi.org/10.1038/525313a

Spelt, Elisabeth J.H., et al. 2009. "Teaching and learning in interdisciplinary higher education: A systematic review." *Educational Psychology Review* 21: 365–378. https://doi.org/10.1007/s10648-009-9113-z

Starfield, Anthony M., and Astrid Jarre. 2011. "Interdisciplinary modeling for an ecosystem approach to management in marine social-ecological systems." *World Fisheries: A Social-Ecological Analysis*, edited by Rosemary E. Ommer, R. Ian Perry, Kevern Cochrane, Philippe Cury 105–119. https://doi.org/10.1002/9781444392241.ch6

Strasser, Ulrich, et al. 2014. "Coupled component modelling for inter-and transdisciplinary climate change impact research: Dimensions of integration and examples of interface design." *Environmental Modelling and Software* 60: 180–187. https://doi.org/10.1016/j.envsoft.2014.06.014

Suárez, Mauricio. 2004. "An inferential conception of scientific representation." *Philosophy of Science* 71: 767–779.

Tripp, Brie, and Erin E. Shortlidge. 2019. "A framework to guide undergraduate education in interdisciplinary science." *CBE-Life Sciences Education*, 18(2): es3. https://doi.org/10.1187/cbe.18-11-0226

Tuana, Nancy. 2013. "Embedding philosophers in the practices of science: Bringing humanities to the sciences." *Synthese* 190: 1955–1973. https://doi.org/10.1007/s11229-012-0171-2

Tversky, Barbara. 2017. "Foreword." In *Springer Handbook of Model-Based Science*, edited by Lorenzo Magnani, and Tommaso Bertolotti. Springer International Publishing. https://link.springer.com/book/10.1007/978-3-319-30526-4

Van Baalen, Sophie. 2019. "Developing an imaging tool for clinical practice." In *Knowing in Medical Practice: Expertise, Imaging Technologies and Interdisciplinarity*. PhD Thesis University of Twente: 103–144. https://doi.org/10.3990/1.9789036546935

Van Baalen, Sophie, and Mieke Boon. 2015. "An epistemological shift: From evidence-based medicine to epistemological responsibility." *Journal of Evaluation in Clinical Practice* 21(3): 433–439.

Van den Beemt, Antoine, et al. 2020. "Interdisciplinary engineering education: A review of vision, teaching, and support." *Journal of Engineering Education* 109(3): 508–555. https://doi.org/10.1002/jee.20347

Van Noorden, Richard. 2015. "Interdisciplinary research by the numbers." *Nature* 525(7569): 306–307.

Witchel, Harry J. 2022. "Interdisciplinary research in physiology: Insights into the physiological society's (UK) report on "The future of interdisciplinary research beyond REF 2021."" *The FASEB Journal* 36. https://doi.org/10.1096/fasebj.2022.36.S1.R2687

Woodin, Terry, V. Celeste Carter, and Linnea Fletcher. 2010. "Vision and change in biology undergraduate education, a call for action-initial responses." *CBE-Life Sciences Education* 9(2): 71–73.

Zinn, Jens O., and Peter Taylor-Gooby. 2006. "Risk as an interdisciplinary research area." *Risk in Social Science*. Edited by Peter Taylor-Gooby and Jens O. Zinn, 20–53. Oxford: Oxford University Press.

30
THE LEARNING OF MODELING

K. K. Mashood and Sanjay Chandrasekharan

1. Introduction

The learning of modeling is now a key thread in science education research (Etkina, Warren, and Gentile 2006; Matthews 2007; Gilbert and Boulter 2012; Passmore, Gouvea, and Giere 2014; Jung and Newton 2018; Cheng, Wu, and Lin 2021; Rost and Knuuttila 2022). This trend is driven by multiple factors, including new insights from philosophy of science and cognitive science analyses. For instance, the 'practice-turn' in philosophy of science highlighted the ineffectiveness of focusing on the final products of scientific inquiry (Duschl 1990; Lehrer and Schauble 2012; Passmore, Gouvea, and Giere 2014). This insight resulted in a call to provide students with opportunities to engage in authentic scientific practices, similar to the ones that allow professional practitioners of science to solve open-ended problems. In parallel, cognitive science case studies of scientific discovery, based on analyses of the thinking of scientists (such as Maxwell and Carnot) outlined how the building of models led to scientific discoveries (Nersessian 1992; Knuuttila and Boon 2011; Bokulich 2015). Such cognitive studies provided an operational understanding of modeling. They also highlighted implications for science education, at times explicitly (Nersessian 1992). These discussions provided a clear way to operationalize the call for authentic practices in science education – through an increased focus on models and modeling.

In a different approach from the above studies – addressing scientific practice and its cognitive dimension – Giere (1988) analyzed textbooks, to characterize how scientists understood theory and models. This approach linked the learning of modeling to the practice of modeling and was premised on the assumption that most scientists form their first impressions of theory and models from textbooks and associated lectures during their education. Pedagogy designers were motivated by this discussion, as it closely related to issues they had faced in science classrooms. For example, student difficulties in managing and organizing the large pile of content they encounter is a pervasive problem (Van Heuvelen 1991; Malone 2008). This pile accumulates over time, and students then resort to unproductive learning strategies like memorization and recall, as they are unaware of deeper conceptual structures that could help them to coherently organize the knowledge (Hestenes 1992; Jackson, Dukerich, and Hestenes 2008). Giere's study suggested a way out of this problem,

DOI: 10.4324/9781003205647-35

through the restructuring of textbook content (and thus also instruction) around a few core models. This focus on models naturally demanded pedagogies centered around modeling. Students' poor understanding of the nature of scientific thinking is another pervasive problem in science education (Redish, Saul, and Steinberg 1998; Abd-El-Khalick 2006). This is particularly stark in classrooms that follow the traditional lecture format, where discussion is restricted to formalisms, as presented in textbooks. Pedagogies centered around modeling provide a more sophisticated understanding of the scientific thinking process.

Modeling-based approaches can also help pedagogy designers integrate and interweave insights from the history and philosophy of science and cognitive science, to develop a new perspective on the learning of formalisms. This integration capability of modeling approaches has gained more relevance recently, with educational policies across the globe now advocating a radical transition to curricula promoting interdisciplinary and computational thinking (NRC 2013; MHRD 2020). This is a major challenge, as developing a smooth transition to these novel pedagogical approaches from the existing practices and media which currently structure science education, is quite difficult. As modeling underlies both classical/existing science practices and emerging ones, it provides a unifying framework to address this challenging transition (Mashood et al., 2022). The transition is in its infancy, and it builds on a series of science education approaches that draw from philosophical discussions on models and modeling. Some of these approaches are described in the next section.

2. Modeling-centered approaches in science education – some illustrative examples

There are many pedagogical implementations based on models and modeling. This entry focuses specifically on approaches that draw on theoretical discussions in the philosophy of science and cognitive science.

2.1 Prelude from Giere

Giere's (1988) analysis of textbooks is one of the earliest philosophical discussions that provided clear directions for pedagogies based on models and modeling. A primary goal of his analysis was a naturalistic account of how scientists develop their understanding of theory and models. Assuming that textbooks and associated lectures play a crucial role in this process, he analyzed the way mechanics textbooks presented and structured content. He found that a standard mechanics textbook consists of a range of models, such as the simple harmonic motion (formed by the conjoining of Newton's second law $F = ma$ with particular force functions, $F = -kx$ in this case). These models were posited as intermediaries that mediate the relationship between theoretical statements, such as Newton's laws of motion, and the real world (Giere 1999). Based on this analysis, Giere concluded that textbook content is organized around a limited set of clusters of models, and not axiomatically. This view paved the way for the claim that scientists understood theory in terms of clusters of models. Further, he corroborated his model-centered view by invoking studies of expertise in chess and problem-solving in physics. In these cases, expert performance is considered to involve the retrieval and deployment of patterns or models. Giere's analysis ends with the speculation that physics textbooks have evolved to reach their present form by adapting to support human cognitive operations.

Giere argued that a similar analysis is possible in domains other than classical mechanics. For example, quantum mechanics could be considered to be structured around exemplar models, such as *particles in a potential well*. The structuring of quantum mechanics around a cluster of models was taken up in detail by Develaki (2007). Follow-up discussions sought to extend this cluster model approach further, to build an epistemological foundation for science education (Grandy 2003; Aduriz-Bravo and Izquierdo-Aymerich 2005). The next section discusses Modeling Instruction (MI), a widely known model-based science education approach that is centered around ideas similar to the ones proposed by Giere.

2.2 Modeling instruction

Modeling Instruction (MI) is an approach that seeks to provide students with opportunities to build, test, deploy, and revise models (Hestenes 1992; Halloun 2004; Brewe 2008). The approach was pioneered by David Hestenes, a theoretical physicist with interests in philosophy of science and cognition, with two graduate-level physics education researchers (Ibrahim Halloun and Malcom Wells; the latter was also a committed high school teacher). MI advocates a curriculum and instruction that are centered around models. It was implemented at the high school and university levels.

In the MI approach, scientific knowledge is considered to consist of factual knowledge and procedural knowledge (Hestenes 1987). The former includes models, theories, and interpreted empirical data. The latter involves heuristics, strategies, and other procedures used by practitioners, to develop and validate factual knowledge. MI highlights the poor and inadequate treatment of procedural knowledge in typical textbooks. To address this problem, MI advocates a reformulation of textbook content (the factual knowledge) around 'models' and 'theory'. This explicit treatment directly provides the modeling perspective to students. In the earlier structure, students were expected to decipher this perspective on their own, similar to practicing scientists, who managed to figure it out through many examples and practice problems in physics. Traditional courses and pedagogy focus on teaching the formalisms of the subject. They fail to provide students with a proper appreciation of modeling, theory, and their connection to reality.

MI practice involves organizing the content of physics courses around a small set of core models (Hestenes 1987; Wells, Hestenes, and Swackhamer 1995; Halloun 2004). For example, mechanics content is organized around models such as the harmonic oscillator, motion involving constant velocity, and constant acceleration. Students then analyze physical phenomena and situations, going through 'modeling cycles', often in small groups, starting with the problem of explaining or making predictions about a phenomenon (Brewe 2008). This is often done through laboratory activities, in which the students first explore physical systems phenomenologically, and then by generating representations. For example, if the constant velocity model is the target of instruction, students will be moving around and experimenting with motion detectors. They will be encouraged to generate different representations, like motion maps and graphs, and coordinate them. Once students are sufficiently familiar with the different quantitative representations, they are given problems that can be solved using the constant velocity model. Standard textbook problems are altered to make them less structured; such that the problems can be solved only by invoking and engaging with the model under consideration. A set of such semi-open problems are designed, so that students get sufficient experience with the procedural aspects of modeling. This experience is followed by a discussion of the general characteristics of the problems, to help students develop a sense of the situations where the model under consideration would be valid and could be productively deployed. The focus then shifts to incrementally developing the model further, and connecting it to other models.

In his later work, Hestenes has augmented this approach with recent insights from cognitive linguistics, to provide MI a stronger theoretical footing (Hestenes 2006; 2010). The proposed approach seeks to build a modeling theory that can account for cognition in everyday life, science, and mathematics. Some of the claims of the theory are also validated using empirical studies on students' misconceptions. The modeling theory also provides a framework for concept inventories (CIs), which are a key thread in physics education research. CIs are sets of carefully constructed multiple-choice questions, designed to elicit and diagnose student conceptions related to various physics concepts (Mashood 2014). The Force Concept Inventory (FCI), developed by Hestenes and colleagues – arguably the most popular and widely used CIs – has a more robust theoretical footing in modeling, compared to others (Hestenes, Wells, and Swackhamer 1992). Hestenes argues that CIs that lack this theoretical depth, particularly related to assumptions concerning cognition, tend to devolve into question banks, which may not reveal much about student cognition and learning related to modeling.

2.3 *Critiques of MI*

Approaches centered on core models aim simultaneously to teach students extant canonical content in the subject, as well as impart procedural knowledge and authentic practices. This dual objective generates some tension, as there is a huge difference between the context of learning and the context of discovery (Guy-Gaytán et al. 2019). In the former, the learning of existing models is often the priority, whereas in the latter, the emphasis is on modeling for constructing new knowledge. The intertwining of content (to be learned) with practices (to be engaged in) – or learning the finished products of scientific inquiry, and doing science – generates an artificial process. This tension has been highlighted by several authors (Manz, 2015; Miller et al., 2018; Elby, 2019).

This tension is not easily reconcilable and often results in a subtle domination of content-related aspects. In approaches like MI, this domination manifests as a shift in focus – toward the structural aspects of the model, and their representational role. Other epistemic functions and purposes of models, and the modeling process, are sidelined. The canonical model gets decoupled from the web of practices with which it is entangled, for the purpose of making sense of phenomena. The models are then taught and learned as standalone abstract entities. Gouvea and Passmore (2017) elaborate on this point using biology examples (e.g., a model of DNA). They provide a heuristic – models *of* and models *for* – to clearly distinguish between these aspects.

The above critique, which points at a drift – where the effort to teach the modeling process in general devolves into the learning of specific models – may also involve deep-rooted structures related to the epistemology of physics, which likely play a constraining role in the drifting process. In the edifice of physics, there are a limited number of core models, which are highly generalizable. Also, experiments are highly controlled, and they are usually considered to validate – and thereby follow – theory. These two structural aspects of physics lead to the core models enjoying a special status, and their learning consequently dominates pedagogical approaches.

In contrast, biology education permits relatively more open-ended inquiry, with field experiments, which are situated in the real world, changing theory significantly. Computational approaches such as agent-based modeling also provide a similar open-ended and exploratory structure. This structure allows such models to connect well with the practices of emerging interdisciplinary fields, such as engineering sciences. An example of a modeling-based implementation that takes these factors into account, including criticisms leveled against MI, is discussed in the next section. This approach explicitly tries to foreground the practices that are shadowed over by MI's focus on content and formalisms.

2.4 *Practice-centered pedagogies*

There are many practice-centered approaches in science education, particularly within the tradition of inquiry-based learning. The discussion below focuses on pedagogical approaches that are informed or inspired by philosophical discussions.

2.4.1 *Philosophical and theoretical underpinnings*

Practice-centered modeling approaches for learning draw insights from the analysis of historical case studies of iconic discovery episodes (such as Maxwell's model of electromagnetism) and in situ ethnographic studies of scientists at work (Chandrasekharan and Nersessian, 2015; 2021). These philosophical reconstructions of historical discovery episodes make explicit the dynamic and extended nature of the modeling process and provide insights into science cognition. Such accounts foreground situated and distributed cognitive processes, and the role of model-based reasoning in discoveries made by scientists. These include the way scientists articulate their epistemic aim, how phenomena are conceptualized in ways that are amenable to existing cognitive and mathematical operations, how new models are subsumed under existing scientific concepts, and the role of analogical and imagistic reasoning (Nersessian 1992; Knuuttila and Boon 2011). To make this case, the reasoning modes, representations, and practices used by scientists are analyzed closely, based on scientists' original writings, notes, and historical records. Such rich analyses provide critical pushback against minimalist accounts (Thagard 2012), which focus only on the final acts of the discovery process, or on just the representational role of models (see Chandrasekharan 2013 for a discussion).

An influential approach in this tradition is cognitive-historical analysis, developed by Nersessian, who used cognitive science theories to interpret historical episodes of radical knowledge construction, particularly by scientists such as Faraday, Maxwell, and Einstein (Nersessian 1985; 2012). In a similar vein, Knuuttila and Boon (2011) discuss how Carnot constructed his ideal engine. They interweave this historical account with a philosophical discussion of the development of new concepts, representations, and theoretical principles. Both of these studies are premised on modeling as the quintessential practice of science, leading to the generation of new knowledge. Related to this work, but based on sociological frameworks, in situ ethnographic studies of scientists at work focus on understanding science as a way of making sense of the world, by a community of practitioners situated in a particular social, cultural, political, and economic context (Latour 1999; Lynch and Woolgar 1990; Pickering 2010). The nature of modeling activities engaged in by scientists, and the artifacts involved in the process, are center-stage in such accounts as well.

As the human mind and human practices are key players in such rich narratives, these accounts are psychologically and socially realistic and offer specific directions for education and learning. One educational direction that follows from such studies is the focus on practices as part of a community. The canonical content – the core models, which are the focus in implementations such as MI – plays a subsidiary role. This view leads to a reorientation of the role of the learner – from a recipient of core models to an active participant in a social enterprise, with specific epistemic goals. It is in stark contrast to the dominant model of science learning, which focuses on learning content topics and associated skills, in isolation, or in unrealistic and contrived settings. The pedagogical approach discussed below provides an illustrative example of the focus on practice and learning communities.

2.4.2 A situated modeling design

Lehrer, Schauble, and colleagues have developed a pedagogical approach where the process of modeling does not begin with canonical models as in MI (Lehrer and Schauble 2004). Rather, the physical world and inquiry related to problems therein are the starting points of modeling. This situated inquiry is followed by attempts to find possible solutions through quantification. In this design, students are required to make decisions about what data is to be collected, how to collect it, how to represent it, and how to make inferences (Lehrer and Schauble, 2004; 2007; 2012).

An illustrative example of such a situated modeling investigation is the study of the ecology of retention ponds by grade 6 students, facilitated by their teacher (Lehrer, Schauble, and Lucas 2008). Students are asked to study the factors affecting the health of two ponds near their school. The modeling process begins with students visiting the pond multiple times, documenting its structure and changes. They also pose and answer simple questions related to the living things inhabiting the pond. As students' knowledge about the pond grows, the questions get revised, toward comparing the diversity of animal life in pond 1 and pond 2. The process of questioning, and identifying the characteristics of good quality research questions, are part of the discussion between the students and the teacher. Students then design and develop microcosm models of the ponds, using gallon jars. This, in turn, spurs related investigations, such as studying the effect of pH value on the growth of plants, and how oxygen in water affects the life of fish. Sustaining the microcosm model is a struggle, and this pushes students toward investigating in detail the interactions among different components in the microcosm. For example, when some students noted that their fish are not doing well, a recovery proposal is made, based on their knowledge about dissolved oxygen. The proposal is to transfer the fish to a jar with higher levels of oxygen. Such activities organically interleave content and process, blurring the dichotomy between the two. This structure leads to a more involved engagement in the process of modeling.

2.5 Agent-based modeling: a computational extension of the practice-centered approach

A recent thread extends the practice-centered modeling approach to computational modeling (Sengupta and Wilensky 2009; Dickes and Sengupta 2013; Farris, Dickes, and Sengupta 2019). One of the key characteristics of this approach is the use of microworlds, which are computational environments that embed the phenomenon to be learned. In these microworlds, the learner interacts with agents whose behavior is defined by simple rules. By varying the different parameters, using the interface or through code, the behavior of the systems to be learned can be explored, and predictions can be made and tested. Multiple representations, such as graphs, are also linked to the system, which allows students to track the macro effects of their manipulations in real time, and thus make general inferences about the phenomenon under study. Similar to the practice-centered modeling approach discussed earlier, the agent-based modeling (ABM) approach seeks to replicate modeling practices used by contemporary science practitioners. As the learning approach is based on exploration, the learner is encouraged to discover the canonical models – through active manipulation and data models. Students are also encouraged to extend these to address new problems, rather than just apply pre-given models to standardized problems. ABM also provides opportunities to engage in authentic practices, such as measurements, recognition of patterns, and building interpretations. This approach thus deviates significantly from the dominant science education approaches, where most of the focus is on memorizing canonical models.

The ABM approach allows introducing computational modeling – and modeling in general – very early in the curriculum, including at the lower primary school level (Farris, Dickes, and Sengupta 2019). Recent educational policies across the globe advocate the introduction of computational thinking at all educational levels (NRC 2013, MHRD 2020). While the ABM approach allows the introduction of important contemporary modeling practices into science education, it has some limitations in terms of practical implementation, particularly in developing country contexts. In countries like India, most students do not have access to computational hardware, and education is predominantly driven by textbooks. To address these concerns, an approach was developed, to facilitate a smoother transition to computational modeling, where teachers build on existing content, media, and classroom practices (Mashood et al. 2022). The next section discusses in detail this augmentation approach.

3. Building performative models (BPM): modeling as building of symbol-based systems that can enact/emulate state changes in the world

In response to the fast-changing practices in contemporary science, technology, engineering, and mathematics (STEM), recent educational policy initiatives have focused on developing computational thinking and interdisciplinary model-building skills at the high school and undergraduate levels. However, education systems around the world are struggling to implement this much-needed systemic change, as there are no clear operational models that illustrate effective ways to transition from existing practices to interdisciplinary ones. To address this issue, a new pedagogical approach was designed, where the central operational focus is *augmenting* existing curricula, and *compensating* for its limitations, rather than replacing existing pedagogical practices with completely new ones (Mashood et al., 2022).

The development of the Building Performative Models (BPM) pedagogy started with a process analysis of the modeling practices present (though rarely enacted) in existing curricula in India, and most other developing countries. Derivations in physics were identified as a practice that could be augmented, to support computational modeling. Some of the popular derivations at the higher secondary and undergraduate levels were then deconstructed, to highlight the core modeling decisions, moves, and practices that went into their construction. Drawing on a recent educational approach based on conceptual blending (Redish and Kuo 2015), the process of derivation was recast as a process of 'loading' reality into mathematics, to build equations that 'enact' state changes in the world. The following four key intertwined operational steps involved in the loading process were identified: Physical phenomena → Structural diagram or schematic → Geometrical model → Algebraic model (see Mashood et al. 2022 for further details). Based on this conceptual structure, derivations are presented as starting from the real world. An idealized schematic or external representation is then used to identify the key state changes. The idealization also allows a certain distancing and decoupling from the sensorimotor representation of the physical world. Judicious omissions, the possibility of mathematization, and theoretical considerations, are invoked in the construction of the idealized schematic. This schematic is then moved to a coordinate grid, which allows changes in the world to be mapped to changes in quantities. The relevant variables and parameters that allow this mapping are then conjoined using operators, to develop an algebraic expression. This equation is then conjoined with a general equation (like Newton's Second law or Maxwell's equation). The resulting differential equation, on solving, will 'act out' the behavior of the real-world system, using quantity changes as a proxy for physical state changes. Figure 30.1 shows the above four steps in (1) the derivation of the motion of objects on an inclined plane and (2) the derivation of the wave equation. The derivation of the wave equation was further developed into an interactive learning system (see Interactive Derivations, HBCSE-LSR 2022).

A key distinguishing feature of the BPM approach is the identification of the four-step, topic-independent, *process structure* that is common to many derivations in physics. This process structure presents a generic modeling practice, which can help students organize physics knowledge better. In contrast with MI, which is a *structural* and content-centered analysis of derivations or canonical models, the reality-loading approach allows distinct derivations to be conceived as similar, based on the *process* of loading, with core commonalities.

The process view embedded in the BPM approach allows for paving a path from derivations to computational modeling. For this, a set of bridge simulations were developed

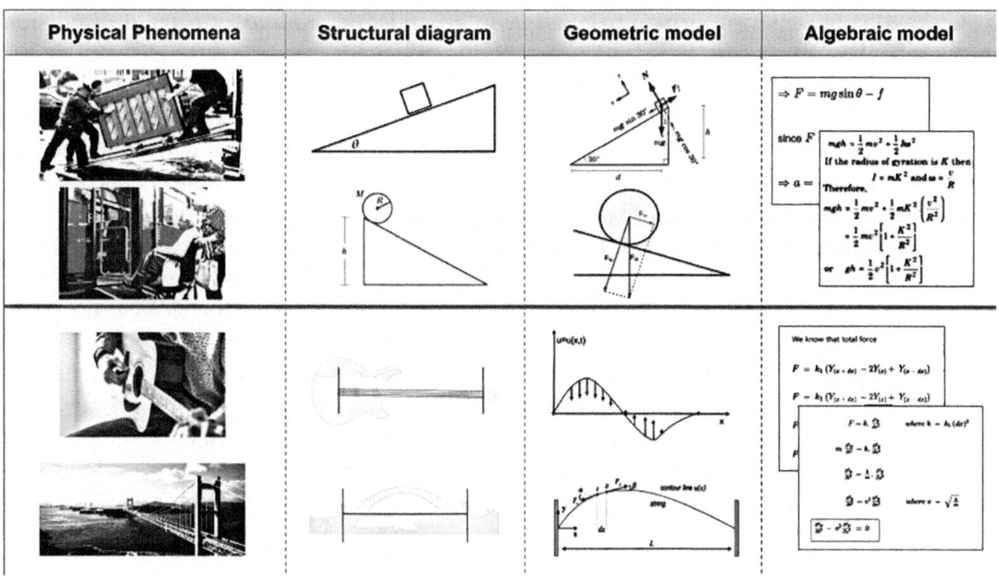

Figure 30.1 The four key steps in the derivations of (1) the motion of objects along an inclined plane and (2) that of a string fixed at both ends (wave equation). These illustrative examples show a topic-independent conceptual structure underlying many derivations in physics, which is a feature of the Building Performative Models (BPM) approach.

(see Pendulum and Piecewise Oscillator in Manipulable Simulations, HBCSE-LSR 2022), which allowed textbook-based derivation models to be turned into fully manipulable interactive models. These can be accessed directly from textbook discussions, through QR codes. The bridge simulations, which are publicly available, interconnect physical phenomena (such as oscillations), their equations, and their graphs. Augmented textbooks based on these simulations allow students to actively manipulate (and thus enact) formal systems. This process enables learners to appreciate the dynamic nature of formal models. The simulations also allow learners to integrate the multiple representations used in science learning and discovery, in a coherent way.

A second key design element of the BPM approach is a smoother transition from physics derivations to interdisciplinary modeling, using numerical solutions as the key boundary-crossing space. The BPM approach emphasizes numerical ways of solving equations, in contrast to the predominantly analytical approaches practiced in current physics classrooms. Numerical approaches involve thinking in terms of numbers and difference equations, which provides a smooth segue into computational thinking. This structure also allows connecting multiple equations using logical operators, which makes it possible to interleave theory and data from different disciplines, thus effectively facilitating interdisciplinary modeling. In contrast, the analytical approach to solving equations entails algebraic thinking and differential equations, and a rather exclusive reliance on the equality operator and linear models.

The teacher training program, which was developed based on the BPM structure, included an interactive learning system that shows the structure of numerical methods (see Piecewise Oscillator in Manipulable Simulations, HBCSE-LSR 2022). The training program also introduced teachers to easily available technological tools, such as the free version of WolframAlpha, which allows equations to be solved based on numerical approaches. A NetLogo simulation of virus transmission in a pandemic, and an associated discussion of the modeling of the problem, were used to illustrate the commonalities, as well as the differences, underlying derivation models and interdisciplinary models.

The integrative pedagogical framework focuses on modeling as a process of 'building performative models' (BPM), as both derivations and computational models are presented as 'acting out' the real-world phenomena. This framework smoothly connects derivations, bridge simulations, computational thinking, and interdisciplinary modeling. The narrative of the workshops, based on existing standard themes (oscillation, heat) captures the systematic evolution of mathematical model building, moving from simple derivation systems (solved analytically) to complex systems (solved numerically). The BPM framework thus allows teachers to subsume, and significantly advance, a large section of the content they currently teach, and shift to an integrative and process-based teaching and learning of mathematical and computational modeling.

Based on these design elements, teacher training modules were developed, and a series of two-day workshops were conducted. The workshops were initially focused on undergraduate teachers, but they were then extended to the higher secondary level (grades 11–12) as well. See Mashood et al. (2022) for further details of this design, along with the iterative process involved in the development of the modules.

In terms of underlying theory, the above design draws on and integrates a range of philosophy of science and cognitive science discussions related to modeling (see Figure 30.2), particularly recent accounts of the way building computational models leads to new knowledge (Chandrasekharan and Nersessian 2015; 2021; Chandrasekharan 2009; 2014). The BPM

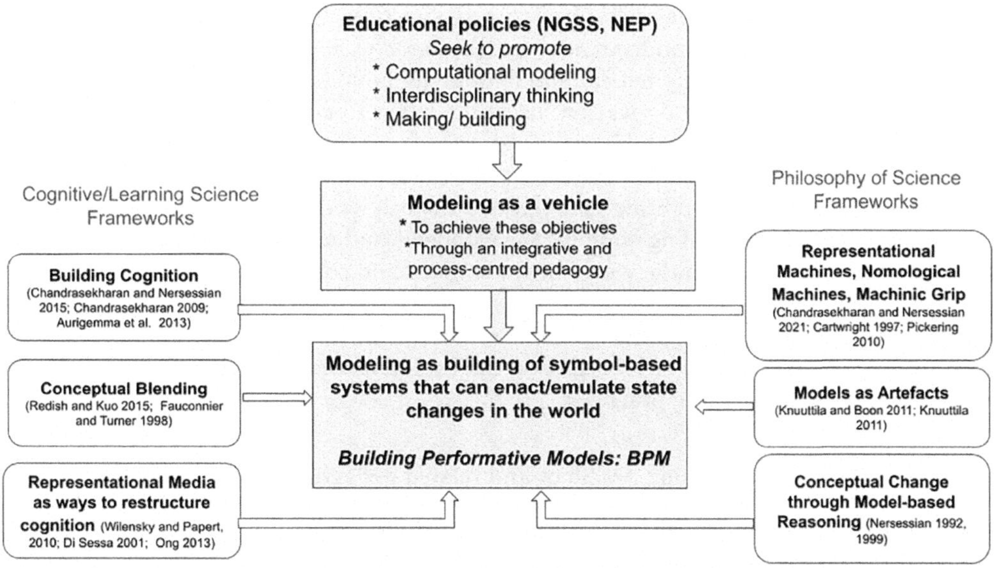

Figure 30.2 A schematic showing how the Building Performative Models (BPM) approach operationalizes the recommendations of new educational policies like NGSS and NEP (NRC 2013, MHRD 2020). The different theoretical frameworks BPM draws on, from cognitive/learning science and philosophy of science, are also shown.

design also relies on earlier discussions on model building and discovery (Nersessian 1992; Knuuttila and Boon 2011) and the way final models tend to erase (or 'vanish') the instrumental, analogical, and imagination/reasoning structures that lead to new empirical and theoretical advances (Pickering 2010). The design also draws on embodied cognition approaches that characterize equations (Majumdar et al. 2014) and mathematics in general (De Freitas and Sinclair 2014; Abrahamson and Sánchez-García 2016; Rahaman et al. 2018) as performative structures. Finally, the design builds on discussions in learning sciences on media form factors – particularly how cognition, learning, and discovery change with transitions from old to new representational and computational media (DiSessa 2001; Wilensky and Papert 2010; Ong 2013). The following are the key features of the design (for details see Mashood et al., 2022):

1 The explicit characterization of model building as a process of 'loading' reality into symbols, and the illustration of the different stages of this process using interactive systems, provide students with a topic-independent perspective on the model-building process.
2 The treatment of equations as 'acting out' state changes in the world – using magnitude changes in a coordinate space as a proxy – provides students with a new way to understand equations, particularly through enaction (with the help of new enactive computational media systems, such as virtual and augmented reality). This pedagogical approach is supported by recent theoretical discussions in embodied and enactive cognition.
3 The proposal that the curriculum overemphasizes analytical (closed form) solutions – a common feature of physics pedagogy across the world – and the tracing of the roots of this emphasis to difficulties in doing extensive numerical computational operations

during the development of early physics, provide students and teachers with a better perspective on the transition from analytical approaches to numerical solutions.

4 The characterization of numerical simulations as boundary-crossing spaces provides a smooth transition from derivation-based modeling to computational modeling, and also the building of interdisciplinary models, which are central to contemporary science practice.

5 The use of textbook linked bridge simulations – which work as user-friendly entry points into computational modeling – allows even students and teachers from resource-limited countries to participate in the world of simulations and computational modeling.

Current work expands this approach to data modeling, which is an emerging trend in contemporary science practice. The next section outlines some of the challenges related to this transition and other emerging practices.

4. Emerging modeling practices and related pedagogy design challenges

The recent success of a machine learning (ML) system in solving the protein folding problem has established ML as a significant approach to developing models based on integrating the massive amounts of data that are now available in many domains. However, this approach raises many complex questions and challenges, including the 'black box' nature of ML models; the high levels of computational infrastructure needed to develop such models; the related issue of the increasing role played by private players (the protein folding system was developed by DeepMind, which is owned by Google) in basic science research areas (such as quantum information processing, computational biology, and drug design); the limited creative roles played by scientists in the development and use of such models; and complex ethics questions related to these practices. As of now, there are no systematic characterizations of this modeling practice in the philosophy of science literature.

A second emerging trend is the rapidly vanishing distinction between engineering and science, particularly in the emerging engineering sciences (such as robotics, nanotechnology, bioengineering, and systems biology) where 'building' practices are used to address basic research questions. This practice, termed 'building to discover' (Chandrasekharan 2009; Chandrasekharan and Nersessian 2015; 2021), is not fully characterized and understood, as it has only recently attracted the attention of philosophers of science and technology.

Finally, there is an ongoing transition toward studying – and managing – very large natural systems that overlap significantly with society, such as the climate, environmental sustainability, and pandemics. These studies also closely align with efforts to develop sustainable technologies and practices. This area, broadly termed 'sustainability science', requires highly interdisciplinary models that capture very complex interactions between natural systems, society, and technology. The modeling approaches used here are quite eclectic, ranging from measurements, surveys, ethnography, and design. There are very few systematic studies of such modeling practices in the philosophy of science and technology.

These three trends, and the lack of clear philosophical and cognitive theories on their nature, pose very complex challenges for science education, as these practice-turns – particularly in concert – are highly disruptive, with the capacity to make much of contemporary science education practice obsolete. Related to this disruption, there is the influential view that computation needs to be understood as a massive cognitive transition, similar to humanity's shift to the use of writing (literacy) from just speech (DiSessa 2001; Wilensky and Papert 2010; Ong 2013; Chandrasekharan 2014). This theoretical view emerged in response to the 'deductive' digital computing revolution. However, ongoing ML modeling practices represent a new 'inductive' computing revolution, with quantum computing and reservoir computing waiting in the wings. If these representational, computational, and practice transitions also require, and bring about, revolutionary changes in cognition (similar to literacy), it would be very difficult for science education to catch up – particularly because these disruptions are occurring in 'internet time', compared to the meandering 10,000-year transition to literacy.

These trends and related theoretical views suggest that the design strategy we have outlined – starting with philosophical characterizations, and drawing on them to develop new pedagogy designs – might no longer work. Practices are changing too quickly for designers to wait for philosophical accounts. Also, the fast-moving changes require complete rethinking of the curricula, rather than changing parts of it in a piecemeal fashion based on analytical frames that are external to science education. Pedagogy designers now need to actively collaborate with science practitioners, philosophers of science, cognitive scientists, learning scientists, and new media designers, to develop systematic and integrative analyses of frontier practices, to develop pedagogy designs that smoothly support the emerging scientific practices, starting from existing pedagogies. For this, such interdisciplinary teams need to immerse themselves in novel practices such as ML and the building of physical models, as well as develop new pedagogies in tandem with changing science practices. Given the rapid pace of change, waiting for frontier practices to slowly sediment into pedagogies is no longer an option for science education.

Acknowledgements

We thank Harshit Agrawal, Ambar Narwal, and Aamir Sahil for their help in developing the learning systems and related discussions. We acknowledge the support of the Govt. of India, Department of Atomic Energy under Project Identification No. RTI4001.

References

Abd-El-Khalick, Fouad. 2006. "Over and Over and Over Again: College Students' Views of Nature of Science." In *Scientific Inquiry and Nature of Science*, 389–425. Dordrecht: Springer.

Abrahamson, Dor, and Raúl Sánchez-García. 2016. "Learning Is Moving in New Ways: The Ecological Dynamics of Mathematics Education." *Journal of the Learning Sciences* 25(2): 203–239.

Aduriz-Bravo, Agustin, and Merce Izquierdo-Aymerich. 2005. "Utilising the '3P-Model'to Characterise the Discipline of Didactics of Science." *Science & Education* 14(1): 29–41.

Bokulich, Alisa. 2015. "Maxwell, Helmholtz, and the Unreasonable Effectiveness of the Method of Physical Analogy." *Studies in History and Philosophy of Science Part A* 50: 28–37.

Brewe, Eric. 2008. "Modeling Theory Applied: Modeling Instruction in Introductory Physics." *American Journal of Physics* 76(12): 1155–1160.

Chandrasekharan, Sanjay. 2009. "Building to Discover: A Common Coding Model." *Cognitive Science* 33(6): 1059–1086.

———. 2013. "The Cognitive Science of Feynmen." *Metascience* 22: 647–652.

———. 2014. "Becoming Knowledge: Cognitive and Neural Mechanisms that Support Scientific Intuition." In *Rational Intuition: Philosophical Roots, Scientific Investigations*, edited by Lisa M. Osbeck and Barbara S. Held, 307–337. Cambridge University Press.

Chandrasekharan, Sanjay, and Nancy J. Nersessian. 2015. "Building Cognition: The Construction of Computational Representations for Scientific Discovery." *Cognitive Science* 39(8): 1727–1763.

———. 2021. "Rethinking Correspondence: How the Process of Constructing Models Leads to Discoveries and Transfer in the Bioengineering Sciences." *Synthese* 198(21): 1–30.

Cheng, Meng-Fei, Tsung-Yu Wu, and Shu-Fen Lin. 2021. "Investigating the Relationship between Views of Scientific Models and Modeling Practice." *Research in Science Education* 51(1): 307–323.

De Freitas, Elizabeth, and Nathalie Sinclair. 2014. *Mathematics and the Body: Material Entanglements in the Classroom*. Cambridge: Cambridge University Press.

Develaki, Maria. 2007. "The Model-based View of Scientific Theories and the Structuring of School Science Programmes." *Science & Education* 16(7): 725–749.

Dickes, Amanda Catherine, and Pratim Sengupta. 2013. "Learning Natural Selection in 4th Grade with Multi-agent based Computational Models." *Research in Science Education* 43(3): 921–953.

DiSessa, Andrea A. 2001. *Changing Minds: Computers, Learning, and Literacy*. Cambridge: The MIT Press.

Duschl, Richard Alan. 1990. *Restructuring Science Education: The Importance of Theories and their Development*. New York: Teachers College Press.

Elby, Andrew. 2019. "Did the Framework for K-12 Science Education Trample Itself? A Reply to 'Addressing the Epistemic Elephant in the Room: Epistemic Agency and the Next Generation Science Standards'." *Journal of Research in Science Teaching* 56(4): 518–520.

Etkina, Eugenia, Aaron Warren, and Michael Gentile. 2006. "The Role of Models in Physics Instruction." *The Physics Teacher* 44(1): 34–39.

Farris, Amy Voss, Amanda C. Dickes, and Pratim Sengupta. 2019. "Learning to Interpret Measurement and Motion in Fourth Grade Computational Modeling." *Science & Education* 28(8): 927–956.

Giere, Ronald N. 1988. *Explaining Science: A Cognitive Approach*. Chicago: University of Chicago Press.

———. 1999. *Science without Laws*. Chicago: University of Chicago Press.

Gilbert, John K., and Carolyn Boulter, eds. 2012. *Developing Models in Science Education*. Berlin: Springer Science & Business Media.

Gouvea, Julia, and Cynthia Passmore. 2017. "'Models of' Versus 'Models for'." *Science & Education* 26(1): 49–63.

Grandy, Richard E. 2003. "What Are Models and Why Do We Need Them?" *Science & Education* 12(8): 773–777.

Guy-Gaytán, Candice, Julia S. Gouvea, Chris Griesemer, and Cynthia Passmore. 2019. "Tensions between Learning Models and Engaging in Modeling." *Science & Education* 28(8): 843–864.

Halloun, Ibrahim. 2004. *Modeling Theory in Science Education*. Dordrecht: Kluwer.

HBCSE-LSR. 2022. Learnware, Accessed November 14, 2022. https://lsr.hbcse.tifr.res.in/interactive-derivations/, https://lsr.hbcse.tifr.res.in/teaching-material/

Hestenes, David. 1987. "Toward a Modeling Theory of Physics Instruction." *American Journal of Physics* 55(5): 440–454.

———. 1992. "Modeling Games in the Newtonian World." *American Journal of Physics* 60(8): 732–748.

———. 2006. "Notes for a Modeling Theory." In *Proceedings of the 2006 GIREP Conference: Modeling in Physics and Physics Education*, edited by Ed van den Berg, Ton Ellermeijer, and Onne Slooten, 31, 27–55. Amsterdam: University of Amsterdam.

———. 2010. "Modeling Theory for Math and Science Education." In *Modeling Students' Mathematical Modeling Competencies*, edited by Richard Lesh, Peter L. Galbraith, Christopher R. Haines, and Andrew Hurford, 13–41. Boston, MA: Springer.

Hestenes, David, Malcolm Wells, and Gregg Swackhamer. 1992. "Force Concept Inventory." *The Physics Teacher* 30(3): 141–158.

Jackson, Jane, Larry Dukerich, and David Hestenes. 2008. "Modeling Instruction: An Effective Model for Science Education." *Science Educator* 17(1): 10–17.

Jung, Hyunyi, and Jill A. Newton. 2018. "Preservice Mathematics Teachers' Conceptions and Enactments of Modeling Standards." *School Science and Mathematics* 118(5): 169–178.

Knuuttila, Tarja, and Mieke Boon. 2011. "How Do Models Give Us Knowledge? The Case of Carnot's Ideal Heat Engine." *European Journal for Philosophy of Science* 1(3): 309–334.

Latour, Bruno. 1999. *Pandora's Hope: Essays on the Reality of Science Studies*. Cambridge: Harvard University Press.

Lehrer, Richard, and Leona Schauble. 2004. "Modeling Natural Variation through Distribution." *American Educational Research Journal* 41(3): 635–679.

———. 2007. "A Developmental Approach for Supporting the Epistemology of Modeling." In *Modelling and Applications in Mathematics Education*, edited by Blum, Werner, Peter Galbraith, Hans-Wolfgang Henn and Mogens Niss, Boston: Springer, 153–160.

———. 2012. "Seeding Evolutionary Thinking by Engaging Children in Modeling its Foundations." *Science Education* 96(4): 701–724.

Lehrer, Richard, Leona Schauble, and Deborah Lucas. 2008. "Supporting Development of the Epistemology of Inquiry." *Cognitive Development*, 24: 512–529.

Lynch, Michael, and Steve Woolgar. 1990. *Representation in Scientific Practice*. Cambridge: The MIT Press.

Majumdar, Rwitajit, Aditi Kothiyal, Ajit Ranka, Prajakt Pande, Sahana Murthy, Harshit Agarwal, and Sanjay Chandrasekharan. 2014. "The Enactive Equation: Exploring How Multiple External Representations are Integrated, Using a Fully Controllable Interface and Eye-Tracking." In *2014 IEEE Sixth International Conference on Technology for Education*, edited by Kinshuk and Sahana Murthy, Amritapuri, India, 233–240.

Malone, Kathy L. 2008. "Correlations among Knowledge Structures, Force Concept Inventory, and Problem-Solving Behaviors." *Physical Review Special Topics-Physics Education Research* 4(2): 020107.

Manz, Eve. 2015. "Resistance and the Development of Scientific Practice: Designing the Mangle into Science Instruction." *Cognition and Instruction* 33(2): 89–124.

Mashood, K. K. 2014. *Development and Evaluation of a Concept Inventory in Rotational Kinematics*. Mumbai: Tata Institute of Fundamental Research.

Mashood, K. K., Kamakshi Khosla, Arjun Prasad, Sasidevan V., Muhammed Ashefas CH, Charles Jose, and Sanjay Chandrasekharan. 2022. "Participatory Approach to Introduce Computational Modeling at the Undergraduate Level, Extending Existing Curricula and Practices: Augmenting Derivations" *Physical Review Physics Education Research* 18(2): 020136.

Matthews, Michael R. 2007. "Models in Science and in Science Education: An Introduction." *Science & Education* 16(7): 647–652.

Miller, Emily, Eve Manz, Rosemary Russ, David Stroupe, and Leema Berland. 2018. "Addressing the Epistemic Elephant in the Room: Epistemic Agency and the Next Generation Science Standards." *Journal of Research in Science Teaching* 55(7): 1053–1075.

Ministry of Human Resource Development (MHRD). 2020. *National Education Policy*. New Delhi: MHRD.

National Research Council (NRC). 2013. *NGSS Lead States, Next Generation Science Standards: For States, By States*. Washington, DC: National Academies Press.

Nersessian, Nancy J. 1985. "Faraday's Field Concept." *In Faraday Rediscovered: Essays on the Live and Work of Michael Faraday*, Editors: David Gooding and Frank James Publisher: Palgrave: London 175–187.

———. 1992. "How do Scientists Think? Capturing the Dynamics of Conceptual Change in Science." *Cognitive Models of Science* 15: 3–44.

———. 2012. *Faraday to Einstein: Constructing Meaning in Scientific Theories*. Dordrecht: Kluwer Academic Publishers.

Ong, Walter J. 2013. *Orality and Literacy*. London: Routledge.

Passmore, Cynthia, Julia Svoboda Gouvea, and Ronald Giere. 2014. "Models in Science and in Learning Science: Focusing Scientific Practice on Sense-Making." In *International Handbook of Research in History, Philosophy, and Science Teaching*, edited by Michael R. Matthews, 1171–1202. Dordrecht: Springer.

Pickering, Andrew. 2010. *The Mangle of Practice: Time, Agency, and Science*. Chicago: University of Chicago Press.

Rahaman, Jeenath, Harshit Agrawal, Nisheeth Srivastava, and Sanjay Chandrasekharan. 2018. "Recombinant Enaction: Manipulatives Generate New Procedures in the Imagination, by Extending and Recombining Action Spaces." *Cognitive Science* 42(2): 370–415.

Redish, Edward F., and Eric Kuo. 2015. "Language of Physics, Language of Math: Disciplinary Culture and Dynamic Epistemology." *Science & Education* 24(5): 561–590.
Redish, Edward F., Jeffery M. Saul, and Richard N. Steinberg. 1998. "Student Expectations in Introductory Physics." *American Journal of Physics* 66(3): 212–224.
Rost, Marvin, and Tarja Knuuttila. 2022. "Models as Epistemic Artifacts for Scientific Reasoning in Science Education Research." *Education Sciences* 12(4): 276.
Sengupta, Pratim, and Uri Wilensky. 2009. "Learning Electricity with NIELS: Thinking with Electrons and Thinking in Levels." *International Journal of Computers for Mathematical Learning* 14(1): 21–50.
Thagard, Paul. 2012. *The Cognitive Science of Science: Explanation, Discovery, and Conceptual Change*. Cambridge: The MIT Press.
Van Heuvelen, Alan. 1991. "Learning to Think Like a Physicist: A Review of Research-based Instructional Strategies." *American Journal of Physics* 59(10): 891–897.
Wells, Malcolm, David Hestenes, and Gregg Swackhamer. 1995. "A Modeling Method for High School Physics Instruction." *American Journal of Physics* 63(7): 606–619.
Wilensky, Uri, and Seymour Papert. 2010. "Restructurations: Reformulations of Knowledge Disciplines through New Representational Forms." *Constructionism* 17: 1–15.

PART 5

Modeling in the wild

31

STATISTICAL MECHANICAL MODELS OF FINANCE

Patricia Palacios and Jennifer S. Jhun

1. Introduction

The last 30 years have seen an increase in the number of approaches to economic modeling inspired by analogies from statistical mechanics and other areas of physics. Work in this tradition has come to be known as *econophysics*. Despite the apparent empirical success of some models in econophysics, the field has been widely criticized. The arguments offered by the critics of these approaches are mainly based on the physical or material disanalogies between physical and economic systems. For instance, it has been said that the absence of conserved quantities in economic systems makes the whole project of using physics in the context of economics untenable (Gallegati et al. 2006, 5). It has also been pointed out that human behavior is not as stable and predictable as physical phenomena (Lo and Mueller 2010) and that economic systems are deprived of "universal empirical regularities" of a sort amenable to predictive mathematical modeling (Gallegati et al. 2006, 2). Some authors have also criticized specific econophysics models by stressing particular disanalogies. For example, Arioli and Valente (2021) have criticized the applicability of quantum mechanics to economics by focusing on a formal disanalogy between quantum mechanics and finance.

On the other hand, advocates of econophysics models usually justify the validity of these models by arguing that they best explain the empirical data and the so-called "stylized facts," which correspond to regularities characterizing economic phenomena (Johansen et al. 1999, 2000; Rickles 2007, 2011; Jhun et al., 2018).

In this chapter, we analyze the extent to which the empirical adequacy of econophysics models suffices to justify the analogies involved in econophysics modeling. To that end, we focus on two models: the Johansen-Ledoit-Sornette (Johansen et al. 2000) model, which posits that financial crashes occur at "critical points," and Jakimowicz and Juzwiszyn's (2015) model, which postulates that financial data exhibit "turbulence."[1] In both cases, we contend that the empirical adequacy of these models serves to justify a formal rather than a material analogy. However, we will point out that while the formal analogy may be sufficient to allow for descriptive and some predictive power, it does not endow the model with explanatory power (in a causal sense). In other words, we will argue that the formal analogies cannot yield information about what causal relations make up the system of

 DOI: 10.4324/9781003205647-37

interest, nor which interventions might be effective. We will point out that in order for this model to fulfill that epistemic role, the scientist must construct a material analogy alongside the formal one, which may require invoking additional and important idealizations about the target system. We will stress that, whereas Johansen et al. (2000) explicitly offer such a material analogy, Jakimowicz and Juzwiszyn (2015) do not. Finally, taking a cue from Bradley and Thébault's (2019) analysis, we suggest that the case studies we consider should be interpreted as cases of *model migration* rather than cases of *model imperialism*. We will suggest that in the cases in which a material analogy can be built alongside a formal one, robustness analysis can work as a potential avenue for the justification of the transfer of causal mechanisms from physics to economics.[2]

2. Analogical reasoning

Econophysics models typically focus on modeling financial time-series data and may have one or more of the following goals: (a) to reproduce descriptively stylized facts, such as the occurrence of bubbles and crashes; (b) to predict phenomena; and (c) to explain phenomena. For instance, they may seek to explain why such events occur (and perhaps how to disrupt them). In order to achieve these goals, most econophysics models rely on analogies between physical and economic systems, which motivate the transfer of mathematical formalisms as well as causal mechanisms from physics to economics.

Such a transfer raises the following questions: What types of analogies are involved in econophysics models? To what extent are these analogies justified? In order to answer these questions, it is useful to recall some important distinctions drawn by Hesse (1966).[3] The most important one is between *formal* and *material analogies*. According to Hesse's terminology, formal analogies are different "interpretations of the same formal theory" (1966, 68). An example is that between the flow of electric current in a wire, which is described by Ohm's law, and fluid in a pipe, which is described by Poiseuille's law. In this case, the analogy is formal because Ohm's law has the same mathematical form as Poiseuille's law, but it is interpreted differently. More precisely, Ohm's law is described by the following expression: $\Delta v = iR$, where Δv is the voltage difference along a wire, i is the current and R is a constant resistance. This has the same mathematical form as Poiseuille's law: $\Delta p = Vk$, where Δp is the pressure difference along a pipe, V is the volumetric flow rate and k is a constant. On the other hand, material analogies are regarded by Hesse as similarities between "observables" in some pre-theoretical sense. An example of the latter are the similarities between fins on a fish and wings on a bird. In the case of material analogies, we would like the causal structures that are observed in one system to be observed in the other. For instance, in the previous example, fins on a fish and wings on a bird serve, in some sense, the same causal role for their possessors, namely, they are manipulated by the organisms in order to propel themselves through some medium.

Hesse's distinctions have been extended elsewhere. In particular, Bartha (2010) provides additional terminology that enables us to more precisely characterize the econophysics project at hand. First, he broadens Hesse's conception of the material analogy by introducing the notion of physical *analogy*, based simply on physical similarities. As Fraser (2020) points out, material analogies can be interpreted as a sub-category of physical analogies, in that the former require the causal relations in the target and source systems to be of the same kind, whereas the latter do not.

Bartha (2010) also loosens Hesse's conception of the formal analogy: "[t]wo features are formally similar if they occupy corresponding positions in formally analogous theories" (195). This contrasts with Hesse's stipulation that a formal analogy is "one-to-one correspondence between different interpretations of the same formal [i.e., uninterpreted] theory" (1966, 68). In order to distinguish between Hesse's strict notion of formal analogy and Bartha's liberal version, Fraser (2020) uses the term *strict formal analogy* to denote the former and *liberal formal analogy* to denote the latter. In this chapter, we adopt Fraser's terminology and distinguish between "liberal formal analogies" and "strict formal analogies" as well as between "material" and "physical analogies."

For many econophysics models, there is often some loose material or physical analogy being used as inspiration, which is meant to later account for the explanatory power of the model. However, it is largely the (liberal) formal analogy that initially anchors econophysics models. Moreover, our analysis of two different econophysics models will suggest that data-fitting can at best justify the use of formal but not of material or physical analogies. Therefore, we will argue that in order to justify the material or physical analogies, one needs something more than empirical adequacy. In particular, one needs to justify the idealizations involved in the migration of physics methods and properties to economics.

3. The JLS model

3.1 Details of the model

An empirically adequate model which has also captured the attention of philosophers of science (e.g., Jhun et al. 2018; Yee forthcoming) is the Johansen-Ledoit-Sornette model (henceforth "JLS model"), developed by Johansen et al. (2000). This model is motivated by an analogy between financial crashes, which are characterized by many traders executing "sell" orders at the same time, and critical phase transitions, such as the transition of a magnet from its paramagnetic to ferromagnetic state. During this transition, the material spontaneously goes from a phase where all spins are pointing in random different directions to a phase in which all those spins align in one direction.

In this section, we look at the JLS model in some detail. The starting point of the model is to derive the expression for the price dynamics for the period prior to the crash:

$$\log\left[\frac{p(t)}{p(t_0)}\right] = \kappa \int_{t_0}^{t} h(t)dt \tag{31.1}$$

where κ is the percentage by which the price drops, $p(t)$ is the price of the asset at a particular time, and $h(t)$ is the *hazard rate*, defined as the instantaneous rate of change of the probability of an event (i.e., the crash) happening at time t, given that it has not occurred yet. A higher hazard rate implies that the asset price will increase more quickly.

In order to make sense of the hazard rate $h(t)$ in Equation (31.1), Johansen et al. appeal to an analogy with critical phase transitions in physical systems. More precisely, given that in magnetic materials the magnetic susceptibility can be interpreted as the tendency of the system to change suddenly its global state (i.e., magnetization) under a very small perturbation, they posit that the hazard rate behaves in the same way. In other words, they postulate that the hazard rate can be described by an equation that is *formally analogous* to the equation that characterizes the magnetic susceptibility. In magnetic materials, the magnetic susceptibility follows power law behavior and diverges at the critical point:

$$\chi \approx A\left(t_c - t\right)^{-\gamma} \tag{31.2}$$

where γ is the critical exponent, A is a positive constant, and t_c is the critical temperature. They postulate that the hazard rate has the same mathematical form, namely:

$$h(t) \approx B\left(t_c - t\right)^{-\alpha} \tag{31.3}$$

where, t_c is the most probable time for a crash, B is a constant, and α is the critical exponent, which is between 0 and 1. Furthermore, Johansen et al. (2000) argue that the critical exponent must be complex, based on the observation that prices exhibit accelerating oscillations as they approach a crash. They derive then the general solution for the power law governing $h(t)$:

$$h(t) \approx B\left(t_c - t\right)^{-\alpha} + B_1\left(t_c - t\right)^{-\alpha}\cos[\omega\log(t_c - t) + \psi] \tag{31.4}$$

where B, B_1, ω and ψ are real constants. Finally, plugging this back into Equation (31.1), they derive the equation for the evolution of the price before the crash:

$$\log\left[p(t)\right] \approx \log\left[p_c\right] - \frac{\kappa}{\beta}\left\{B_0\left(t_c - t\right)^{\beta} + B_1\left(t_c - t\right)^{\beta}\cos[\omega\log(t_c - t) + \psi]\right\} \tag{31.5}$$

where $\beta = 1 - \alpha \in (0, 1)$, $p_c = p(t_c)$ is the price at the critical time leading up to the crash, and ϕ is a phase constant. The second term at the right-hand side of the equation describes log-periodic oscillations that accompany the power law behavior specified by the hazard rate.

Note that although this derivation is inspired by a physical analogy (the crash occurs, analogously as a phase transition, when all agents decide to sell at the same time), this setup makes no reference to the underlying material constitution of either system and relies instead on the formal analogy between Equations (31.2) and (31.3), which are phenomenological equations for macroscale variables.[4]

What justifies this particular formal analogy is the fact that the model gives a descriptive account of the stylized facts observed in the data. In fact, the authors show that the log-periodic oscillations predicted by the model were present before the crashes of 1929, 1962, and 1987 on Wall Street, the 1997 crash on the Hong Kong Stock Exchange, and the Russian market collapse of 1997–1998 (Johansen et al. 1999).

However, it is important to point out that the formal analogy does not allow us to tell the whole story about stock market crashes. It may allow for some qualitative predictions in the sense that accelerating log-periodic oscillations provide a signature of approaching criticality and an attendant crash, but it does not help explain why markets crash or what we can do to stop them, nor does it allow for quantitative predictions. To do this, we must answer the following question: by what mechanism do individuals in a network suddenly manage to organize a coordinated sell-off? Because the formal analogy is mute on this, we need to draw on yet another analogy.

To this end, Johansen et al. (2000) rely on a *material* analogy: Analogous to the behavior of a magnet, they posit, "a crash may be caused by *self-reinforcing imitation* between noise traders" (Johansen et al. 2000, 219). More specifically, they assume that a local imitation procedure that propagates hierarchically and with discrete scale invariance is responsible for the crash.[5] This material analogy between the mechanisms responsible for a phase transition in magnetic materials and the mechanisms responsible for a financial crash is what causally explains how macro-level coordination can arise from micro-level imitation (Jhun et al. 2018).

It is important to note that this material analogy, which is responsible for the causal power of the model, requires that the noisy traders are representable as a lattice network and behave like magnetic spins, responding to the state of their surrounding neighbors, such as in the Ising model (Johansen et al. 2000). In other words, in order for the material analogy to hold, the interactions must be in some way "local." The question that arises is how we ought to interpret "locality" in the context of stock market crashes.

Interpreted literally, the locality assumption implies that agents look to their spatial geographical neighbors for cues as to what to do. However, data fitting alone does not help justify this assumption, and worse, even observations of herding behavior do not serve to justify the assumption that imitation is literally "local" (Bikhchandani and Sharma 2000). To the contrary, the interactions between traders today appear to be anything but "local." In fact, we know that agents are globally interconnected through the internet, television, and other communication and social media. Furthermore, the model aims to describe different crashes throughout the history starting from the crash of 1929, and we know that communication media has changed dramatically since the 1920s. It seems, therefore, that the "locality assumption" needs to be interpreted as an idealization. How we can justify this relevant idealization is the question to which we now turn.

3.2 *Justifying the analogy*

A natural way of justifying the idealization of "locality" in the JLS model is to assume that the term is open to different interpretations and that it is consistent with the fact that the population underlying a crash may look any number of ways. This means that we need to demonstrate that the JLS model is *robust* against different interpretations of locality and that it remains a suitable enough representation over a range of different conditions.

What makes us believe that the model is in fact robust in respect to different interpretations of "locality" is that JLS themselves (Johansen et al. 2000, 6) postulate that the evolution of the hazard rate is:

$$\frac{dh}{dt} = Ch^{\delta}, \tag{31.6}$$

where the exponent $\delta > 1$ quantifies the effective number equal to $\delta - 1$ of interactions felt by a typical trader and C is a positive constant. Like spins on a lattice, agents interact with their neighbors. Appealing to this analogy and to the use of mean field theory, $h(t)$ captures the "collective result of the interactions between [traders] (sic)" (2008, 6). Since they assume that a typical trader must be connected to more than one other trader, they allow for δ to be within the interval: $2 < \delta < +\infty$. Furthermore, they assume that this does not determine at the microscale the number of neighbors for particular individuals. So, depending on the structure of the network they live on, different agents may have different numbers of neighbors. It seems, therefore, that the model is in fact robust upon different interpretations of locality, which could serve to justify the assumption of locality in the model.

There is, however, a potential difficulty. As stated by Jhun et al. (2018), the JLS model not only seeks to explain the occurrence of stock market crashes, but also aims to help visualize possible avenues for intervention. The problem is that according to the JLS model, some possible interventions that would help prevent market crashes, such as trading curbs, seem to be targeted at disrupting the cascade or herding behavior by forcing market participants to reflect on their subsequent actions in specific ways (whether this is successful or not is another story, see Jhun et al. 2018 for a critical discussion of possible interventions). That is, interventions that are meant to induce a specific type of behavior at the level of individuals.

However, if this is the case, there is a tension between robustness and intervention goals. The more robust the model is, the easier it would be to justify certain material analogies, but the harder it would be to find specific policies that would make a difference for the behavior that is heading towards a crash. A possible solution to this worry is to formulate *robust policies*, which are policies that perform well over a range of circumstances.[6] This is especially important when scientists face model uncertainty, which describes cases in which they are not quite sure if their model of a target system has all the relevant details right.[7] Typically, this means trying to minimize the worst-case scenarios for all salient scenarios, relative to other potential policies. In fact, if what we want is to keep a crash from occurring, we would like our intervention to be sufficiently robust such that it does not depend on the specific micro-configuration of agents. Whether any of the policies deployed by, for instance, the US Securities and Exchange Commission would plausibly qualify as robust remains to be seen.

Taking stock of what has been said in this section: we believe that the assumption of locality, which we consider to be a material analogy, can be transferred from physics to economics only if we interpret locality differently than in physics, where it represents spatial contiguity. In fact, we argued that the assumption of locality in the context of finance could be justified only if one proves the model to be robust upon different interpretations of this assumption. Although this stresses an important difference between the causal mechanisms underlying a physical phase transition and the ones underlying a crash, this difference, as we will explain next, should be taken simply as a feature rather than a bug of econophysics modeling.

3.3 The JLS as a case of model migration

Bradley and Thébault (2019) draw an important distinction between "model migration" and "model imperialism," which can also be useful for understanding the export of analogous causal mechanisms from physics to economics. In model migration, a model moves from one discipline to another by a radical reinterpretation of its representational function. Since there is reinterpretation, the idealizations deployed in the new model need a different justification from the ones used to ground the idealizations in the originating model. On the other hand, model imperialism occurs when the domain of validity of models in one discipline extends to include other target systems previously described by different disciplines.

Very recently, Yee (forthcoming) discusses different models in econophysics, including the JLS model, and criticizes the "imperialist" attitude of econophysicists by pointing out manifest differences in the causal mechanisms in physics and economics. He concludes from this that econophysics, so far, has only legitimately exported the mathematical methods from physics to economics and has not succeeded in exporting causal mechanisms. This can also be associated with a recent discussion on the use of cross-disciplinary templates, which occurs when computational templates are transferred from one discipline to another and adjusted to fit the field of application (e.g., Humphreys 2004; Knuuttila and Loettgers 2016, 2021). Knuuttila and Loettgers (2016) worry about the prospects of cross-disciplinary template transfer, as the attempt may involve the template "[losing] its associated theoretical and methodological toolbox that provided its justification and empirical content in the original field of application" (380). In these cases, they argue, the model in the new discipline, say the socio-economic context, will not have the same predictive value, and empirical and theoretical grounding as in the original discipline, say physics. The analogy between, for instance, the socio-economic and physical cases may be best interpreted as a "thin analogue model" (380) instead of as a case of robust template transfer.

We think that our account clarifies what it is that the economist must provide in order for the analogical inference to hinge on more than a "thin" analogy. As explained above, the economist needs to justify independently the material/physical analogy, and such justification is usually distinct from the justification for the formal analogies. We believe, furthermore, that interpreting econophysics models as a result of *model migration* instead of imperialism allows for a more charitable interpretation of the transfer of causal mechanisms from one discipline to the other. More specifically, it allows us to extrapolate causal mechanisms without further requiring that the rationale for the new model's idealizing assumptions must be the same. While justifying the idealizations *is* required for the target system—in this case, the econophysics model—the justification of the idealizations can look different from the ones deployed for the source system—in this case, the physics model. For instance, in the JLS model, we need to justify (we suggest by robustness analysis) the locality assumption, but the justification of this idealization has little to do with how locality was justified for models of magnetic materials.

This allows for a more optimistic attitude towards econophysics and its ability to export not only mathematical methods but also analogous causal mechanisms used in physics for the study of economic phenomena. In fact, for the econophysics model to be successful, all we need to do is justify whether it represents its own target well with these particular idealizations. This may not be a simple task, but in the case of the JLS model, robustness analysis appears to be a promising strategy to cope with this.

4. Jakimowicz and Juzwiszyn turbulence model

4.1 Details of the model

Another class of econophysics models are based on analogies with turbulence in fluid mechanics (e.g., Mantegna and Stanley 1996; 2000; Ghosh and Kozarević 2018). An example is the model developed by Jakimowicz and Juzwiszyn (2015), which proposes an analogy between the Reynolds number in fluid dynamics with a financial Reynolds number for stock market volatility. This analogy is built in a number of steps.

The starting point of this model is the expression for the Reynolds number in fluid mechanics, which helps predict flow patterns in different fluid flow situations:

$$R_e = \rho u L/\mu \qquad (31.7)$$

where R_e is the Reynolds number, ρ is the density of the fluid, u is the speed of the flow, L the characteristic length, and μ the fluid's (dynamic) viscosity.

If we are concerned with water flowing through a pipe, a Reynolds number of less than about 1,000 indicates laminar (stable) flow, between 1,000 and 2,000 indicates an unstable transitional infraregime with both turbulent and laminar flows, and above 2,000 indicates a flow with turbulent rapid mixing that is unpredictable.

They then try to derive a formally analogous expression for the context of the stock market, with the goal of distinguishing between different phases in the market. Deriving this expression was not a trivial task, since they noted a *negative* analogy between fluid mechanics and financial behavior. Namely, something that looks like fluid viscosity is not initially found in financial mathematics. To make the equations sensible from an economic standpoint, they use Frenkel's fluid equation and the Smoluchowski/Einstein equation to derive (with some additional simplifying assumptions) a *liberal formal analogy* of the form:

$$R_e = \rho\sigma^2\upsilon^2 \tag{31.8}$$

where ρ is the density (constant number of companies making up the stock exchange index), v is stock market flow velocity, and σ^2 is a volume variance.

With this formal analogy, the model aims mainly: (a) to determine the threshold values of the Reynolds rate for various types of markets, i.e., laminar (stable) and unpredictable turbulent markets, and (b) to identify transition points, which could be informative for investment decisions. The hope is that such a financial Reynolds number could be helpful as a warning rate that indicates when the market will exhibit turbulent dynamics. In particular, high values of the Reynolds number would indicate a turbulent regime.

However, insofar as these aims are concerned, Jakimowicz and Juzwiszyn need more than this formal analogy. More specifically, they propose modeling the dynamics of the WIG (Poland) stock exchange index in a space of three dimensions:

$$R^3_+ = P \times Q \times T, \tag{31.9}$$

where P is the index value, Q is volume, and T is time.

They note that the type of movement made by the market in R^3_+ around a *hypothetical line of balance* is analogous to the rotational movement of fluids (Figure 31.1). In fact, they find that the chaotic vibrations of the "stock exchange particles," which are the data vectors in the three-dimensional space, move in a similar way as liquid particles in a pipe. They call this a *logical homology*, referring to the structural similarity between the spiraling movement of the stock market data and that of material particles in vortices. When the stock market helices take such a spiral shape, they claim that this can usefully make forecasts (in the short run) of how the economic market vectors will develop. In particular, they associate the transition from rotational to spiral movement with the transition from a laminar phase into a turbulent phase. Furthermore, when the financial Reynolds number reaches a high number, the previously laminar flow becomes turbulent. In this way, knowing the Reynolds rate may help predict market behavior.

Although they suggest that there may be a physical analogy between the movement made by the market in R^3_+ and the movement exhibited by hydrodynamic systems, it is somewhat unclear whether this analogy can be interpreted as a physical one, since we are not comparing the physical properties of two analogous systems. Instead, we believe that this should be interpreted as another *liberal formal analogy* between the movement of the fluid and the movement of the data.

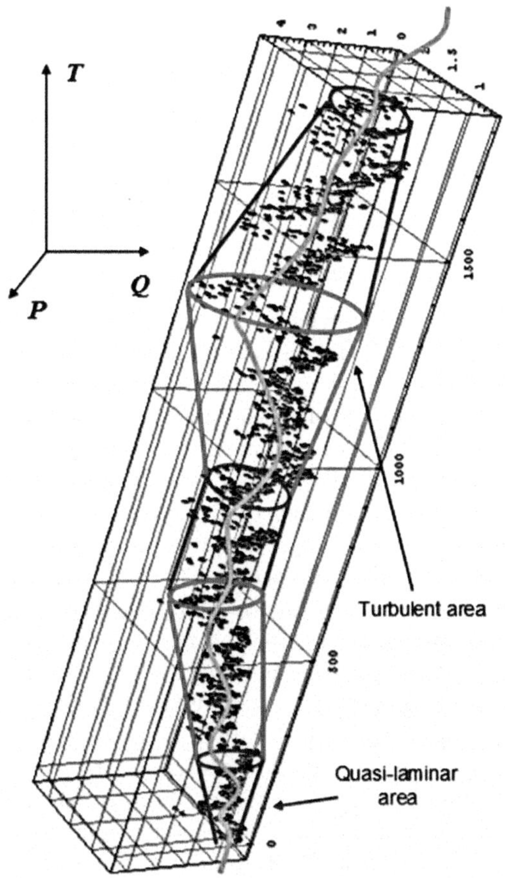

Figure 31.1 Three-dimensional rotational trajectory of WIG stock exchange index listed on the Warsaw Stock Exchange. (Jakimowicz and Juzwiszyn 2015, A-80. Reproduced with permission.)

By analyzing the time series for two periods, one in 1994 and one in 1995, of the Warsaw Stock Exchange, the authors confirm that in both cases, a change in trend was associated with the Reynolds number exceeding its maximum value R_e, which suggests, for them, that the financial Reynolds number could serve as an adequate warning value. It appears, therefore, that this model has at least *some* predictive power and that it is to some extent empirically adequate, at least for the cases under investigation. Since the model consists mainly of describing the flow of data vectors (i.e., the data are the ones that make up the economic analogue to fluid flow), the reliability with which we can predict turbulence using the financial Reynolds number helps to justify, at least *partially*, the liberal formal analogies mentioned above. Interestingly, this is similar to the case of the JLS model, in which we argued that empirical adequacy (in the sense of data-fitting) could only justify the formal analogy and not the material analogy. However, in contrast to the JLS model, Jakimowicz and Juzwiszyn's Turbulence Model does not seem to rely on a material analogy that may account for the causal power of the model. On the contrary, the model is only predictive and does not explain why the market moves from one regime to another. Indeed, as explained above, they intend to use this model mainly for technical analysis of financial data, which focuses on forecasting using patterns of financial data rather than uncovering substantial information about companies themselves behind those stocks.

We will argue in the next section that a complete justification for the most important analogies involved in the turbulence model requires more than empirical adequacy. In particular, it requires justifying core assumptions (or idealizations) in the econophysics model.

4.2 *Justifying the analogy*

In order to justify the analogies involved in the turbulence model, it is important to emphasize that the most important analogy is not material in the intuitive sense. In fact, the similarity that they identify between the two systems is between the kind of movement the entities in question display, where in one case the relevant entity is actually the shape of *the data*. Recall that material analogies require the same causal relations in both target and source systems. However, this particular analogy is built upon the structural features of the data rather than the underlying mechanisms that generate the data. Thus, we believe that it is not appropriate to interpret this analogy as a material one. One particularly interesting feature of this analogy is that it does not seem to fit well with the categories introduced. Data, for instance, has no causal power (nor is it a physical entity). The two cases are thus not two physical interpretations of the same formalism—only one is. Furthermore, there is, strictly speaking, no material analogy between the data set and fluid movement (the former lacks causal structure altogether). No material analogy exists for us to extrapolate causal structure from the source right to the target. Nonetheless, we hesitate to say there is no physical analogy at all, in Bartha's and Fraser's sense. After all, the structure of the data vectors really does look similar to that of swirling fluid. It may very well be the case that the analogy is both a loose formal and physical analogy, suggesting that perhaps it is a sort of hybrid analogy, or that these particular distinctions do not cut cleanly.

We argued above that the empirical adequacy of the model for certain cases, such as the Polish stock market, could *partially* justify the most important formal analogies between fluid dynamics and data dynamics. However, a fuller justification would also require validating the main idealizations involved in the model. More precisely, in order to represent the dynamics in three-dimensional stock market kinematics and extract a financial Reynolds number, Jakimowicz and Juzwiszyn postulate a "hypothetical line of dynamic balance" (2015, 81) around which the market vectors rotate. They continue that this dynamic balance line is "a virtual construction, as it is never achieved, and it only provides a frame of reference for actual movements of economic vectors" (81). Importantly, it is this hypothetical line, which can be interpreted as an idealization, which helps the authors distinguish between laminar and turbulent areas. Given the crucial role that this idealization plays for the empirical adequacy of the model, it needs to be adequately justified.

Part of the justification for the hypothetical line of balance is theoretical. Jakimowicz and Juzwiszyn postulate that the line of balance is meant to be the equilibrium of supply and demand. To give a more satisfactory justification for this assumption, we believe that these arguments need to be supplemented with robustness arguments. In fact, robustness tests can potentially demonstrate that the hypothetical line of balance is involved in different successful reconstructions of the turbulent data flow that generate Reynolds numbers, which reliably indicate changes in regime. If so, then we have evidence that the hypothetical posit is either a harmless idealization, or even justifiable as part of a model that itself is, and must be, a holistic distortion that pervasively misrepresents its target (Rice 2019).

Robustness tests can also play an important role in establishing the generality of the model. Indeed, if this model successfully helps establish a warning Reynolds number for more data sets, one could infer the validity of the model outside the Polish stock market. This evidence, however, is limited. As Ghosh et al. (2020) point out: "Though the WSE is rated as one of the larger Central European stock markets, globally, they are not part of the 16 stock markets spread across three continents that make up the US 1$ trillion club, accounting for 87% of all market capitalization" (Ghosh et al 2020; 209). This would mean that further robustness tests are required to see if similar data patterns appear in other regimes, although there are some extensions of this work to the CNX Nifty Regular and CNX Nifty High Frequency Trading domains (Ghosh et al, 2018).

We would like to conclude by returning to Bradley and Thébault's (2019) distinction between model imperialism and migration. Like the JLS model, the turbulence model suggested by Jakimowicz and Juzwiszyn appears to be a case of model migration rather than imperialism. Recall that model migration means that the justificatory moves one may make for idealizations in a certain system will not carry over to the other because the model's representational function changes in its new context. That the reinterpretation is quite radical is obvious. In fact, data vectors (or agents, or even market prices themselves) are very different from the constituents of fluids. Therefore, this should not be considered simply as an extension of physics to the economic domain. On the contrary, the model relies on a liberal formal analogy between the physical properties of a fluid and features of the data.[8] The physical interpretation of the formalism (the swirling data vectors, which accord with a particular "flow" equation) requires its own idealizing assumption (the logical homology) that in turn further requires its own justification, independently of the fact that the two systems are analogous.

If one were further interested in policy interventions, discovering the underlying causal mechanism (in terms of interacting agents) generating the data would be of tantamount importance. Indeed, Jakimowicz and Juzwiszyn suggest, though do not develop the idea, that "In three-dimensional stock market kinematics, vortices (helices) emerging in a micro-scale (time scale) cause vortices in higher, meso- and macro scales" (82). One might then propose a material interpretation, for instance, as Ghasghaie et al. (1996) do, that these price dynamics are driven by information cascades (so that information is thought to be analogous to energy in hydrodynamic models). In the Jakimowicz and Juzwiszyn case, even though there was a formal analogy between the dynamics of the data and that of the fluid, modeling the data this way requires further justification. One needs, for example, to justify the particular idealizations invoked independently of how the model of the source system was justified. Our remarks regarding model migration would also apply to a material interpretation in case causal structure was in question; such an interpretation would itself need independent investigation and confirmation. One possible route is indeed by a kind of robustness investigation: if other time-series data could be modelled as if it were fluid-like in the same manner, it would give us some reason to think that the approach is not misguided.

This perhaps gives the econophysicist room to defend their project against skeptical naysayers even if the justification for the idealizations made has nothing to do with the ontological (read: material or physical) unification between the systems under consideration. In fact, the justification can be entirely new in the new context and tied to the material system at hand, rather than to the analog model (or system).

5. Conclusion

We have diagnosed the precise nature of the seemingly unsatisfactory nature of econophysics models by examining two empirically successful models. After all, we find such model transfers to be successful precisely because the model can reproduce some stylized fact of interest.[9] We have concluded that while they may achieve predictive or descriptive accuracy via data-fitting, and their equations may seem analogous to those we find in physics, in order to extrapolate the causal mechanisms from one discipline to another, one needs something more. In particular, one needs to justify the idealizations in the economic context. We have suggested, for instance, robustness analysis as one potential avenue to cope with the problem of justifying the idealizations in econophysics models. We have also relied on Bradley and Thébault's (2019) distinction between *model imperialism* and *migration* to claim that the econophysics models under investigation should be interpreted as cases of migration. So interpreted, we can find room for econophysics models that differ vastly from their physical analogues.

Notes

1 While we acknowledge that these models are not representative of the entire field of econophysics, we do believe that they may capture salient features of many econophysics models.

2 This work fits into the larger literature concerning the use of *model templates* more broadly, usage of which includes, specifically, their application via analogical reasoning as we describe here in our case studies. For details on this kind of use, see Knuuttila and Loettgers (2016; 2021). Knuuttila and Loettgers (2021) distinguish that the literature on analogical reasoning and that on template-based reasoning in model construction and transfer have largely progressed without interfacing with one another, and argue that they should be seen as complementary projects. The language of templates originates from Humphreys' (2004; 2019) use of the concept of computational template.

3 The others, which we do not consider in detail, are as follows. The second distinction is between horizontal and vertical relations. While horizontal relations are identities/similarities or differences between the source and the target systems, vertical relations are causal ones within a system. The last distinction drawn by Hesse is between positive, negative, and neutral analogies. Positive analogies consist of those properties that we can attribute to both target and source systems; negative analogies are those properties that they do not share; and neutral analogies are ones that we have not yet established but that can allow for novel predictions.

4 Note that since Equations (31.2) and (31.3) are merely phenomenological, they are not necessarily attached to microscopic theories and therefore they are not attached to the material analogy that we were discussing.

5 We can interpret this as offering a causal explanation (Jhun et al. 2018).

6 See, for instance, Brainard's (1967) paper discussing the case when one has a single model but the economy lies in the vicinity of it. See Levin et al (1999) on robust monetary policy rules and Hansen and Sargent (2001) on robust control.

7 Using robustness as a strategy to combat model uncertainty has been addressed in the philosophical literature by, for instance, Lloyd (2015) and Parker (2011) in the context of climate science and Kuorikoski et al. (2010) in economics.

8 Notice that here the source and the target systems should not be interpreted as different interpretations of the *same* formalism because even the formal analogies are sometimes not strict.

9 Thanks to Tarja Knuuttila for clarification on this point.

References

Arioli, Gianni, and Giovanni Valente. 2021. "What is really quantum in quantum econophysics?" *Philosophy of Science* 88: 665–685.

Bartha, Paul. 2010. *By Parallel Reasoning*. Oxford: Oxford University Press.

Bikhchandani, Sushil, and Sunil Sharma. 2000. "Herd behavior in financial markets: a review." *International Monetary Fund*, WP/00/48.

Bradley, Seamus, and Karim P. Thébault. 2019. "Models on the move: Migration and imperialism." *Studies in History and Philosophy of Science Part A* 77: 81–92.

Brainard, William. 1967. "Uncertainty and the effectiveness of policy." *American Economic Review (Papers and Proceedings)* 57(2): 411–425.

Fraser, Doreen. 2020. "The development of renormalization group methods for particle physics: Formal analogies between classical statistical mechanics and quantum field theory." *Synthese* 197: 3027–3063.

Gallegati, Mauro, Steve Keen, Thomas Lux, and Paul Ormerod. 2006. "Worrying trends in econophysics." *Physica A: Statistical Mechanics and its Applications* 370(1): 1–6.

Ghasghaie, Shoaleh, Wolfgang Breymann, Joachim Peinke, Peter Talkner, and Yadolah Dodge. 1996. "Turbulent cascades in foreign exchange markets." *Nature* 381: 767–770.

Ghosh, Bikramaditya, Corlise Le Rouk, and Anjali Verma. 2020. "Investigation of the fractal footprint in selected EURIBOR panel banks." *Banks and Bank Systems* 15(1): 185–198.

Ghosh, Bikramaditya and Emira Kozarević. 2018. "Identifying explosive behavioral trace in the CNX Nifty Index: A quantum finance approach." *Investment Management and Financial Innovations* 1: 208–223.

Hansen, Lars Peter, and Thomas J. Sargent. 2001. "Robust control and model uncertainty." *American Economic Review* 91(2): 60–66.

Hesse, Mary. 1966. *Models and Analogies in Science*. Notre Dame, IN: University of Notre Dame Press.

Humphreys, Paul. 2004. *Extending Ourselves. Computation Science, Empirical and Scientific Method*. Oxford: Oxford University Press.

Humphreys, Paul. 2019. Knowledge transfer across scientific disciplines. *Studies in History and Philosophy of Science* 77: 112–119.

Jakimowicz, Aleksander, and Jan Juzwiszyn. 2015. "Balance in the turbulent world of economy." *Acta Physica Polonica A* 127(3a): A-78–A-85.

Jhun, Jennifer, Patricia Palacios, and James O. Weatherall. 2018. "Market crashes as critical phenomena? Explanation, idealization, and universality in econophysics." *Synthese* 195(10): 4477–4505.

Johansen, Anders, Olivier Ledoit, and Didier Sornette. 2008. "Crashes as critical points." *International Journal of Theoretical and Applied Finance* 3(2): 219–255.

Johansen, Anders, Didier Sornette, and Olivier Ledoit. 1999. "Predicting financial crashes using discrete scale invariance." *Journal of Risk* 1(4): 5–32.

Knuuttila, Tarja, and Andrea Loettgers. 2016. "Model templates within and between disciplines: From magnets to gases-and socio-economic systems." *European Journal for Philosophy of Science* 6: 377–400.

———. 2021. "Magnetized memories: Analogies and templates in model transfer." In *Philosophical Perspectives on the Engineering Approach in Biology*, edited by Sune Holm and Maria Serban, 123–140. Abingdon, Oxon: Routledge.

Kuorikoski, Jaakko, Aki Lehtinen, and Caterina Marchionni. 2010. "Economic modelling as robustness analysis." *The British Journal for the Philosophy of Science* 61(3): 541–567.

Levin, Andrew T., Volker Wieland, and John Williams. 1999. "Robustness of simple monetary policy rules under model uncertainty." In *Monetary Policy Rules*, edited by John B. Taylor, 263–318. Chicago, IL: University of Chicago Press.

Lloyd, Elizabeth A. 2015. "Model robustness as a confirmatory virtue: The case of climate science." *Studies in History and Philosophy of Science Part A* 49: 58–68.

Lo, Andrew W., and Mark T. Mueller. 2010. "Warning: Physics envy may be hazardous to your wealth!" *Journal of Investment Management* 8(2): 13–63.

Mantegna, Rosario N., and H. Eugene Stanley. 1996. "Turbulence and Financial markets." *Nature* 383: 587–588.

———. 2000. *Introduction to Econophysics: Correlations and Complexity in Finance*. Cambridge: Cambridge University Press.

Parker, Wendy S. 2011. "When climate models agree: The significance of robust model predictions." *Philosophy of Science* 78(4): 579–600.

Rice, Collin. 2019. "Models don't decompose that way: A holistic view of idealized models." *The British Journal for the Philosophy of Science* 70: 179–208.

Rickles, Dean. 2007. "Econophysics for philosophers." *Studies in History and Philosophy of Science Part B* 38(4): 948–978.

———. 2011. "Econophysics and the complexity of financial markets." In *Philosophy of Complex Systems, Handbook of the Philosophy of Science Vol 10*, edited by Cliff Hooker, 531–565. Kidlington, Oxford: North Holland.

Yee, Andrew K. Forthcoming. "Econophysics (making sense of a Chimera)." *European Journal for the Philosophy of Science* 11. https://doi.org/10.1007/s13194-021-00413-1

32
CLIMATE MODELS

Ilkka Pättiniemi and Rami Koskinen

1. Introduction

Given the observed changes in global climate, especially in global mean surface temperature in the period after the 1850s, there has been a growing interest for scientists to understand the cause of said changes and to quantify key climate processes in all their complexity.[1] The main tools developed for gaining an understanding of the Earth's climate system are *climate models*. Climate models are used not only to give insight into the complex dynamics of the climate system and the drivers of climate change (such as the increase in greenhouse gas emissions), but more specifically, they are also used to derive predictions of future weather events, including rainfall, hurricanes, draughts, and so on. That is, climate models are used to study the past and future global and local effects of changes in climate and the drivers of said changes.

This chapter introduces the basic concepts of climate science and climate modeling, gives the bare bones of what a climate model consists of, and discusses the philosophical (mainly epistemological) issues concerning climate models and their use. Also provided is a discussion of the application of climate models to understand and predict extreme weather-related events.

It is important to note that not all climate science consists of practices best characterized as modeling. Understanding climate systems could hardly advance without carefully conducted data collection and assimilation, the designing and building of various instruments and measurement devices, and the theoretical work on poorly understood natural processes (although this is often difficult to distinguish from non-climate-oriented work in physics, chemistry, and other fields). Still, climate modeling forms arguably the central part of contemporary climate science, and is a constant topic of interest both in public debate and in philosophical discussion on climate and climate change. Indeed, in the eyes of the public as well as professional experts, climate science is one of the paradigmatic examples of contemporary model-based science.

Before climate models can be characterized, however, we need to have at least a preliminary understanding of what these models are used to study, that is, of climate and climate system(s). Section 2 provides a short characterization of the notion of a climate system

 DOI: 10.4324/9781003205647-38

by introducing common definitions, and explaining the basic vocabulary. Section 3 then turns to various climate models and climate modeling strategies that have been especially prominent in shaping climate science for the past few decades. The models are introduced in roughly increasing order of complexity, starting with relatively simple versions of energy balance models (EBMs) and proceeding to models of various intermediate and higher levels of complexity, including different strains of global circulation models (GCMs) and regional climate models (RCMs). The basic formal properties of important models are also quickly explained, but the overall treatment is qualitative and focused on more conceptual features. Important concepts like *parametrization* and *tuning* of climate models are also touched upon, as is the idea of *projections*. Section 4 draws from recent work in the philosophy of science to discuss some epistemological issues surrounding climate models. The *reliability* and *robustness* of various climate models are assessed. Section 5 zooms in on an important and recently debated special case of *extreme event attribution*. Finally, Section 6 concludes the entry.

2. What is climate?

The definitions of "climate," and of "climate system," are somewhat controversial, in that there is no general consensus on how to best define the terms. It is useful to start by citing the definition of the terms provided by the Intergovernmental Panel on Climate Change (IPCC) before tackling any controversies. The IPCC special report *Managing the Risks of Extreme Events and Disasters to Advance Climate Change Adaptation* (IPCC 2012) defines "climate" thus:

> Climate in a narrow sense is usually defined as the average weather, or more rigorously, as the statistical description in terms of the mean and variability of relevant quantities over a period of time ranging from months to thousands or millions of years. The classical period for averaging these variables is 30 years[.] [...] The relevant quantities are most often surface variables such as temperature, precipitation, and wind. Climate in a wider sense is the state, including a statistical description, of the climate system.
>
> *(IPCC 2012, 557)*

A "climate system" on the other hand is:

> the highly complex system consisting of five major components: the atmosphere, the oceans, the cryosphere, the land surface, the biosphere, and the interactions between them. The climate system evolves in time under the influence of its own internal dynamics and because of external forcings such as volcanic eruptions, solar variations, and anthropogenic forcings such as the changing composition of the atmosphere and land use change.
>
> *(IPCC 2012, 557)*

Some notes are, however, in order. Since arguably humanity is a part of the biosphere, the first part of this definition will render anthropogenic forcing[2] as a part of the normal variability of the climate system. So, there is *some* tension here, but it will suffice to note that the distinction between internal variability and external forcing is a pragmatic one.[3] We are, after all, interested in the effects of human behavior on the climate, as it is a forcing we are (in principle) able to change by reducing (or increasing) greenhouse gas emissions, altering our land use, and so on. At the same time, it is important to note that there can be considerable changes in climate because of purely internal factors alone, due to so-called *internal variability* in the climate system. Understanding climate even without the effect of major external forcings is a highly non-trivial feat.

The concepts of climate science are a subject of debate among scientists and philosophers of science, and quite rightly so. Here we will sidestep most of these definitional debates since the primary interest of this chapter is climate models. For those interested in these questions, a good place to start is Katzav and Parker (2018).

3. What are climate models?

With some basic terminology in hand, we are now ready to step into the world of climate models. At its heart, a climate model is a set of (often differential) equations derived from basic and well-understood mechanics and thermodynamics. There are several different kinds of climate models with differing scopes and complexities. The simplest of these are the so-called *energy balance models* (EBMs).[4] As their name suggests, these models describe situations where incoming and outgoing radiation (i.e., energy) is balanced; $E_{in} = E_{out}$. Now, combining this with some rudimentary thermodynamics, geometry, and observational data, a prediction for the Earth's temperature can be derived in the following way: Let the incident solar radiation that the Earth receives per square meter be S_0. Then the total energy that the Earth receives will be $\pi R^2 S_0$ (πR^2 is the area of the Earth presented to the sun). The Earth reflects a portion α of the incident energy out, called the Earth's *albedo*.[5] So the total incident energy absorbed will be $E_{in} = (1-\alpha)\pi R^2 S_0$. From the Stefan-Boltzmann law, we get for outgoing radiation $E_{out} = 4\pi R^2 \sigma T^4$, where T is the temperature in Kelvin and σ is the Stefan-Boltzmann constant. Here S_0 and α are empirically determined, and T can be easily calculated from the equation $E_{in} = E_{out}$. Thus, we get:

$$T = \left(\frac{(1-\alpha)S_0}{4\sigma} \right)^{\frac{1}{4}}$$

Plugging in the relevant numbers, one gets $T \approx 255$ K (≈ -18.5 °C). For the Earth's surface temperature (≈ 15 °C), this is way off, but it is a good approximation for vertically averaged *atmospheric* temperature (Jeevanjee 2018, 3).

The above zero-dimensional energy balance model is likely the simplest case of a climate model. One can build better energy balance models by considering atmospheric layers and the fact that some of the radiation from the surface will be reflected back. But even those models will end up rather lacking if one wishes for a better understanding of the Earth's climate. For a better model, one will wish to include the movement of air, the oceans, the effect of various greenhouse gases (GHG), the formation of clouds, air moisture, precipitation, and so on. These more advanced models are of several differing kinds, all with different scopes and details. The following three types are the most common. First, there are *Earth system models of intermediate complexity* (EMICs). These models represent the components of the climate system and geography in a simplified and coarse-grained manner. They have a relatively low computational cost but suffer from inaccuracies due to the simplifications made. The second type of model are *global climate models* or *general circulation models* (GCMs). These models have a more detailed representation of the components of the climate system, higher resolution (more on resolution and computation later) than EMICs, and explicit representation of many atmospheric and oceanic processes. *Earth system models* are GCMs that also take into account the biosphere via the representation of biochemical processes. Finally, there are *regional climate models* (RCMs) that have higher resolutions than GCMs but cover only a portion of the globe rather than the whole Earth (Katzav and Parker 2015). For a listing of different types of climate models by increasing complexity, see Table 5.2 of Neelin (2011, 175).

Let us now take a brief look at the physics and mathematics that will be involved in a climate model. Building the core equations for a climate model is rather straightforward, namely, one accounts for the dynamics and thermodynamics of parcels of air and water.

We will start with Newton's second law of motion, also known as the law of acceleration, and consider the forces acting on a parcel. For horizontal movement, we get (considering forces per unit mass):

$$\frac{dv_x}{dt} = fv_y - \frac{1}{\rho}\frac{\partial p}{\partial x} + F_{dr}^x \tag{32.1}$$

$$\frac{dv_y}{dt} = -fv_x - \frac{1}{\rho}\frac{\partial p}{\partial y} + F_{dr}^y \tag{32.2}$$

Here, v_i are the components of velocity, f is the Coriolis parameter, ρ the density of the parcel, p the air or ocean pressure, and F_{dr}^i the turbulent drag on the flow. The terms on the right-hand side correspond to the Coriolis force, the pressure gradient force, and the drag caused by turbulence. For the vertical direction, things are even easier, as one only has to consider pressure, height, and density:[6] $\frac{\partial p}{\partial z} = -\rho g$, where g is gravitational acceleration.

To bring temperature (and some other factors) into play, climate models need to account for the thermodynamics of the situation. That is, the equations of state for the atmosphere and the oceans are needed. For the atmosphere, one can simply use the ideal gas law $\rho = \frac{p}{RT}$, where R is the ideal gas constant. The oceans are a lot trickier, as water density is temperature-dependent in a non-trivial way (water gets denser as it gets colder until ~4 °C after which it gets less dense), and water density is also affected by salinity S. The equation of state will then be of the form $\rho = \Upsilon(T, S, p)$.[7]

Now, while the equations given above seem rather innocuous, the system—or model—they form does have some problematic properties. The dynamics of the system are *non-linear*, which can already be seen from the two coupled differential Equations (32.1) and (32.2). From non-linearity, it follows that in general, the system will not be analytically solvable, thus necessitating the use of *numerical methods* usually implemented on computers. The system is also *chaotic*, which means that even a slight change in initial conditions can result in a drastic difference when the system evolves forward in time—the so-called butterfly effect. More on this second problem later.

The numerical solution of a (group of) differential equation(s) involves *discretizing* them. For climate models, this means dividing the atmosphere, the oceans, and so on into a grid of discrete chunks reminiscent of a giant layered chessboard pattern. Given the sheer size of the climate system, the *cell size* of the discretized model will, for reasons of computer power and time/cost-effectiveness, be rather large, typically in the order of 100 km per side for global climate models. This large cell size makes it necessary to *parametrize* some phenomena that occur at smaller scales rather than having them directly represented in the model. These phenomena include clouds and cloud formation in the atmosphere and eddies in the oceans.

A computerized model will include dynamics of some parts of the climate system and parameterizations of phenomena that happen at small scales or that are otherwise too difficult to include in the dynamics. Indeed, all sub-grid phenomena of note need to be parameterized. These include, but are not limited to, cloud formation, ocean eddies, small-scale turbulent exchanges of fluid (i.e., air or water) parcels near the ocean's surface, convection, the changes in sea ice and surface snow, evapotranspiration (the processes that move water from land surface to the atmosphere), and the area covered by leaves. Roughly put, a parameterization consists of representing a process via a set of equations that includes some parameters whose values are usually empirically determined, as opposed to being inherited from the laws of physics from which the model is derived (Neelin 2011; Hourdin et al. 2017).

Unfortunately, the parameters used in different parameterizations are often not easy to derive from observation. This leads to the *tuning* of the parameters—a process by which modelers seek to reduce the discrepancy between model predictions and observations. This tuning can be done by hand, relying mainly on the modeler's expertise, automated multi-parameter tuning utilizing fast optimization, or Bayesian methods called *uncertainty quantification* (Frisch 2015; Hourdin et al. 2017).

Simulation models are used to make predictions of the behavior of (parts of) the climate system. These models involve systems of equations that are introduced into a computer program. The simulation offers the possibility of virtual experimentation, in the sense that one can ask the computer program what the future states of the system would be once some initial conditions are specified.[8] Conditional predictions—say those predicated on a certain greenhouse gas emission scenario—are called *projections* (see, e.g., Parker 2013; Werndl 2019).

When a simulation model starts its run, there will be a relatively long period of simulation time, called *model spin-up*, for the model to reach equilibrium. The amount of spin-up time required depends on the kind of model in question. Atmospheric models make do with shorter spin-ups (about a year) than ocean models do (decades for the upper ocean, up to millennia for the deep ocean) (Neelin 2011, 177–178). This is because heat transfer in the oceans is very slow, which means that it takes more simulation time for the model to reach equilibrium (Neelin 2011, 178; Jeevanjee 2018). Also, ice sheet models can have spin-ups of over a thousand years (Lofverstrom et al. 2020). Regional models will have a shorter spin-up time than global models (from days up to a year) (Lavin-Gullon, Milovac, García-Díez, and Fernández 2022). Global circulation models take roughly the same spin-up time as ocean models do. Spin-up times can make certain model kinds prohibitively (computationally) costly.

Having looked at computerized, or simulation models, it is good to note that, in addition to simple energy balance models, some rudimentary climate models can be run on pen and paper and that doing so can be a good way of getting some insight into some features of the climate system (Jeevanjee 2018). Also, it is worth mentioning that the models examined in this chapter belong primarily to the so-called *theory-based* (or *physics-based*) family of models. This means that they start from generally accepted principles and try to achieve an understanding through a process of iterative accommodation of more and more detailed, and typically computationally costly, processes.[9] Most climate models in use today are of this type. However, besides theory-based models, climate scientists have also developed so-called *empirical* (or *data-driven*) climate models which rely on machine learning and statistical analysis of huge datasets to predict climate development (see Knüsel and Baumberger 2020; Song et al. 2023).

Before turning to the epistemological assessment of climate models and climate modeling strategies, it is good to say a few words about *weather models*. What is the relationship between climate models and numerical weather prediction? Both climate models and the models used in numerical weather prediction are based on the same physical principles, and both exhibit non-linear, chaotic dynamics. Since regional models usually suffice for weather prediction, they can have a much smaller cell size compared to general circulation models—typically in the range of 2–100 km in width and several hundred meters to a few kilometers in height. In most areas, there is abundant weather data available, from radiosondes, satellites, weather stations, and so on, and thus there can be high confidence in initial values. Model ensembles can attain model skill (i.e., reliable prediction) for up to ten days in the future. However, the reliability of the models drops dramatically after this period (Lynch 2008; Finnish Meteorological Institute 2022). The contrast here with climate models is sharp: climate models are bad at predicting localized and short-term phenomena but have high skill when it comes to larger-scale phenomena. In short, climate models and weather models are based on the same physical principles but have different strengths and limitations due to the different phenomena they seek to study.

4. Epistemological assessment of climate modeling

There are several sources of uncertainty when it comes to climate models, which give rise to epistemological worries of differing severity. A minor worry is that future external forcing on the climate system is hard to predict. From a modeling perspective, this is easily tackled: one produces conditional predictions, or projections, for differing forcing scenarios. For example, one could run a model for scenarios in which human-induced CO_2 emissions rise, are kept constant, or decrease during the immediate future and estimate the differential effect of these scenarios for climate in the coming decades. The adequacy of these projections will then depend on the overall adequacy of the climate model used.

A second worry is the effect of uncertainty about initial and boundary conditions. As mentioned above, the dynamics of the climate system are chaotic, meaning highly sensitive to perturbations in initial conditions. This is a smaller worry than one might think since long-term projections show little overall sensitivity to initial conditions (Knutti et al. 2010; see Werndl 2019 for an opposing view). Things do get tricky, however, if one wishes to make projections at relatively small scales; for example, will a certain area suffer from floods or droughts due to climate change? Indeed, chaos theory has its very roots in weather research, as is famously codified in the provocative question from Edward Lorenz from 1972: could a butterfly flapping its wings in Brazil produce a tornado in Texas? (Lorenz 1993, 14). This issue will be considered in more detail in the next section.

Next, there is a group of worries that have to do with the models themselves. As stated above, due to the cell size of climate models, certain phenomena must be parameterized. It is, however, not clear what exact values these parameters should take. This gives rise to *parametric uncertainty* (Knutti et al. 2010; Hourdin et al. 2017). As noted before, parameters need to be tuned in order to get the models to match up to observations. There are several ways of doing this, all of them leaning, at least in part, on expert opinion, making parameter tuning at least in some respects subjective (Hourdin et al. 2017). This is not the main epistemological worry here, however. The main risk in tuning a single parameter by hand is that one might end up with a suboptimal parameter value, since the optimal value might require varying other parameter values. This can be alleviated by using computerized methods for finding an optimal value for a set of several parameters. The problem here is that they may lead to *overtuning*, that is, they may lead to unphysical behavior of some of the untuned processes (Hourdin et al. 2017).[10]

There is another worry concerning parameterization, namely that the parameter tunings are *ad hoc* and without empirical support; that is, they are only made to make the model results fit observations (Leuschner 2015). Yet another worry is that because some processes are completely left out or only approximately included in climate models, the models will not agree with observations; the model *might* simply be inadequate. This worry about the adequacy of the model is called *structural uncertainty*. Wendy Parker sums up this uncertainty succinctly as "uncertainty about the form the modeling equations should take and how they should be solved computationally" (Parker 2013, 215).

Testing for model skill is also problematic. The problem here is that there is only one Earth system. How much does a model's skill in replicating the climate system for the past century warrant trust in its further success? For instance, warmer air has a higher saturation point for water vapor, meaning that climate warming leads to a higher concentration of water, a known greenhouse gas, in the atmosphere. This, in turn, leads to more clouds, and thus more cloud formation. This might lead to current parametrizations being wrong, leading to models that give incorrect results. Worries of this kind can be called *response uncertainty*, which is uncertainty about what response the climate system will have to a given forcing scenario (Parker 2013).

The usual way to deal with the above epistemological worries is to use *model ensembles*. Model ensembles are of two main types: perturbed physics ensembles (PPEs) and multimodel ensembles (MMEs). Perturbed physics ensembles use the same basic model but with perturbed parameters. Multimodel ensembles consist of a collection of *different* models of the same general type (e.g., they are all global circulation models) (Knutti et al. 2010; Parker 2013). Both types of ensembles can also include variations in initial conditions and thus help to address worries related to the chaotic nature of climate dynamics.

Perturbed physics ensembles are used to deal with parameter uncertainty. The same model is run parallelly with differing plausible values for the parameters of the model. Here, plausibility is determined by expert opinion. If all or most models in the ensemble give the same prediction, say a certain rise in mean temperature in some region by the year 2040, the projection will be *robust* with regards to parameter value (for philosophical discussion on robustness, see Lloyd 2015; Wimsatt 2007). This alleviates the worry of parameter uncertainty but does nothing to lessen worries about structural uncertainty. Indeed, this convergence might even be a symptom of some structural failing of the model.

Multimodel ensembles try to deal with structural uncertainty. Here the idea is that if different models (with different dynamics, parameters, and so on) starting with the same initial conditions and forcing scenarios give the same predictions for some event, then the prediction is not an artifact of the models' imperfections (Knutti et al. 2010; Parker 2013). Indeed, there is *some* robustness to models with well-aligning predictions, but does this imply *reliability*? Elisabeth Lloyd argues that what she calls *model robustness* can be confirmatory in nature: that is, such robustness can indeed imply reliability. Model robustness is achieved when members of the same model type yield the same prediction (or retrodiction) for some phenomenon. According to Lloyd, climate scientists do take this attitude when it comes to robust predictions from climate models (Lloyd 2015). There are those who are critical of this line of argument, however. It is a fact that the success of, say, global circulation models is based on both empirically true assumptions and assumptions known to be *false* (Katzav 2013). Therefore, the achieved robustness might be an artifact of the false assumptions, leading us to suspect the reliability of the results.

We already saw that in the case of perturbed physics ensembles, the ensemble does not help with structural uncertainty, but what is the case for multimodel ensembles? The multimodel case seems promising in that there *is* robustness in some results. Since we are using distinct models in the ensembles, such robustness could only be an artifact if *all* the models were erroneous in the same way. Here, there is actually a cause for concern since the simulation models in use are not as independent of each other as one might wish. As Wendy Parker reports in one case of a multimodel ensemble, CMIP3, "the two dozen or so … models behave like a set of only 5–10 statistically independent models" (Parker 2013, 219). This is because the models have not been developed fully independently. Indeed, most models will even share parts of computer code, sometimes dating back several years (Baumberger, Knutti, and Hirsch Hadorn 2017). In general, there seems to be considerable *generative entrenchment* when it comes to productive climate models and both their formal and artefactual makeup (Lenhard and Winsberg 2010; see also Wimsatt 2007). A part of a model may become fixed because so many other things depend on it that it would be practically risky to try to change it anymore.

There is a neighboring question of whether convergence of predictions is a reason to think that climate models themselves will converge towards one correct model of the climate system. Lenhard and Winsberg argue that it is highly unlikely that even if we had a good theoretical understanding of small-scale phenomena like cloud formation and the behavior of aerosols, we would be able to include them in a global climate model (Lenhard and Winsberg 2010). If this is indeed the case, then we will still need a plurality of climate models to deal with parameterizations. This *convergence skepticism* is no reason to doubt the usefulness of climate models, however. It merely states that while model robustness can give good reasons to accept model predictions, it is not a good reason to think that we might one day have a singular "best" climate model for all purposes. Indeed, there is evidence that despite the advancements in climate modeling, there has not been a reduction in the range of model predictions—or in the so-called *model spread*—in climate projections. This might not be fatal, however, as reducing model spread might not be as important a goal as ensuring the independence of models used in multimodel ensembles (Jebeile and Barberousse 2021).

There is a further worry about multimodel ensembles. How does one know whether the models at hand are a representative sample of all possible climate models? Is such a set even well-defined? (Parker 2013; Baumberger, Knutti, and Hirsch Hadorn 2017). It might well be that current ensembles sample a highly unrepresentative part of the model space, leading to systematic error.

Given all this, are there still good reasons to trust climate models? Yes, there are. Climate models are based on well-known physical principles, such as the conservation of energy and momentum, the laws of thermodynamics and radiation, and so on. Climate models are also able to "retrodict" current and past climates. Of special import is the models' capacity to reproduce paleoclimate data, since in the distant past the Earth's climate was drastically different from the recent past (Knutti et al. 2010). Interestingly, a recent general circulation model developed for simulating Saturn's climate has been successful in reproducing at least some of the planet's climate phenomena (Cabanes, Spiga, and Young 2020). This goes some way toward alleviating worries related to the evaluation of model skill.

Finally, it is good to note that while scientifically and philosophically the increased understanding of the climate system is of high importance, one of the main motivations for using model ensembles is to produce quantifiable results for policymakers. The next section will zoom in on a particular climate modeling strategy of extreme event attribution that makes use of policy-oriented language and more localized methodological choices.

5. Extreme event attribution

One particularly important use of climate models is for *extreme event attribution*. After all, knowing facts about general things like the *mean* surface temperature of the Earth in the future is not necessarily going to be very helpful in guiding what to do in practice in local situations around the globe. Recently, there have been an unprecedented number of floods, droughts, hurricanes, and other extreme weather-related events. At least some of these are likely due to changes in the Earth's climate. So, we would like to know which events have indeed been caused by climate change on the one hand, and on the other, we would like to be able to predict future extreme events caused by climate change in order to better prepare for them (Parker 2010; Shepherd 2014; Shepherd et al. 2018; Lloyd and Oreskes 2018).

How does one go about figuring out whether a particular event, say a flood, occurred *because* of climate change? One way to do so would be to run, say, a regional climate model for the area where the extreme event occurred, with and without human-caused forcing. Then one could see whether such events occur with the forcing scenario but do not occur without it. If this is the case, the event in question can be attributed to climate change. However, if one uses only a single model and not an ensemble, any results might be caused by structural problems in the model. So, a multimodel approach is necessary. But there is a worry here. Different models project different extreme events, and it is difficult to give weight to different models. If the models are averaged over without weighing them, then extreme events are likely to average out (Shepherd 2014). This means that both attributing and predicting extreme events will be difficult with the usual modeling approach.

A recent proposal to aid in extreme event attribution is the so-called *storyline approach*. In the storyline approach one constructs a *physically plausible* model scenario—or storyline—and checks whether, under this scenario, a certain past event occurs. Let us say that one wishes to ascertain whether a certain flood occurred because of the regional rise in temperature caused by climate change. Then one constructs a scenario *S* which includes a rise in temperature, and checks whether a flood occurs under *S*. If a flood occurs under *S*, but not under a scenario where the regional temperature has *not* risen, one can conclude that the flood was indeed caused by climate change (Shepherd et al. 2018). The proponents of the storyline approach claim that the standard modeling approach not only underestimates the past and future impact of climate change but also is unable to take into account the unique nature of extreme events (Shepherd et al. 2018).

The storyline approach does suffer from its own epistemological worries. The main constraint for the scenarios is that they must be physically plausible, but it is not clear what physical plausibility entails. At a minimum, physical plausibility requires physical possibility, a fact that is in line with the recent call for a *possibilistic* interpretation of climate modeling practices in the philosophy of science (Katzav 2023). Indeed, at least prima facie, the storyline approach to climate modeling seems to be a candidate for a form of *modal modeling* (Sjölin Wirling and Grüne-Yanoff 2021; Koskinen 2023). In the philosophy of science literature, it is customary to distinguish so-called *how-possibly models* from how-actually (or, especially in earlier discourse, from why-necessarily) models. While it is natural to interpret the storyline approach in terms of how-possibly modeling, several factors make this epistemically and methodologically problematic.

One problem is that, at least without further specification, physical possibility, and possibility in general, is rather cheap, as all that is required for something to be (physically) possible is that it be consistent with certain rules (e.g., the laws of physics, some initial/boundary conditions, etc.) (Maudlin 2020; Hirvonen, Koskinen, and Pättiniemi 2021). Indeed, Theodore Shepherd and his co-authors discuss a storyline where plausibility is taken to be consistency with available climate model ensembles and external (to climate, in the sense of Section 2) factors known to be significant for the event at hand (Shepherd et al. 2018). Also, in the context of climate modeling, there is arguably a kind of mixing of various notions of possibility. Sometimes the usage of modal language may simply carry an "apologetic" function (Grüne-Yanoff and Sjölin Wirling 2021) that can be used to soften the success conditions of model-based claims as well as to guard against disappointment on the part of policymakers and the general public in the event the predicted possibility fails to occur.

Another problem with the storyline approach is that it is possible to come up with several differing storylines for a *future* extreme event. For instance, an increase in surface temperature will increase water evaporation, which in some regions R might lead to a drought. At the same time, warmer air has a higher saturation point for water vapor, leading to more precipitation. Now, this increased rainfall might happen at R, causing flooding instead of a drought. So, we could easily have two storylines for differing events in R, both based on physical possibility. Storylines are meant to be an aid to policymakers, but in a situation where we cannot point out a singular outcome and local resources are limited, it will be hard to say whether policymakers should opt for water storage systems or measures against flooding. This seems like a real problem since, according to Shepherd et al. "No a priori probability of the storyline is assessed; emphasis is placed instead on understanding the driving factors involved, and the plausibility of those factors" (2018, 555). This is not such a huge problem in using storylines for *past* events, since here the event itself is fixed, and thus, a constraint on the possible storylines. However, usually the most important questions concern future events.

6. Conclusions

Climate science is one of the paradigmatic model-based sciences of our current age. By utilizing an impressive array of tools and techniques that rely both on basic science and data-intensive computation, scientists are able to understand the behavior of complex climate systems and even predict their future development. However, climate modeling is not without its epistemic and methodological challenges, a fact that has received an increasing amount of interest from philosophers of science. For example, even though a general consensus exists about anthropogenic climate change and its overall qualitative direction, there is a lot of uncertainty and dissent when it comes to the exact nature of future climate conditions, especially in the case of precise localized predictions. Different kinds of robustness-probing, multimodel ensemble approaches go some way towards alleviating the situation, but they also have methodological problems of their own. Nevertheless, climate models are our foremost window into the future climate, and at their best, they provide invaluable tools to help us better characterize, explain, and predict climate systems of various levels of scale and detail. Indeed, as Elisabeth Lloyd aptly puts it: "Climate models should not be judged primarily on the basis of what they are weak at; if we approached other scientific theories or models this way, we would never accept any of them" (Lloyd 2010, 982).

Acknowledgments

This work has received funding from the European Research Council (ERC) under the European Union's Horizon 2020 research and innovation program (Grant Agreement No. 818772).

Notes

1 For an important early achievement in climate science, see Arrhenius (1896). Building on the work of Tyndail, Fourier, and others, Arrhenius gave the first quantitative prediction of the influence of CO_2 on the greenhouse effect and speculated on its contribution to long-term variation in climate.
2 In climate science terminology, a *forcing* is any (natural or human-induced) factor that drives the climate system to change.
3 On the philosophical issues surrounding the definition of a climate, and its complex relation to climate modeling, see Werndl (2016).
4 Here, mainly the full names of model-types are used rather than their acronyms. However, common acronyms are given to familiarize the reader with the terminology, which features heavily in practically any article dealing with climate science and climate models.
5 More generally, *albedo* is the measure, ranging from 0 to 1, that indicates how much sunlight a given body reflects. The average *albedo* of the Earth is estimated to be about 0.3, while pure snow would have significantly higher and oceans clearly lower values.
6 This is true only for large enough scales. If one wishes to model smaller-scale phenomena (e.g., thunderstorms) an acceleration term in the vertical direction should be retained.
7 There will of course be more phenomena to consider. The present chapter introduces the central equations and principles involved in modeling the climate. For a more thorough look at the physics of climate models, see Chapter 3 of Neelin (2011) and Chapter 3 of Winsberg (2018).
8 On the philosophical novelty and methodologically central role of computer simulations, see Humphreys (2009). He argues that "Computational science introduces new issues into the philosophy of science because it uses methods that push humans away from the centre of the epistemological enterprise" (Humphreys 2009, 616). Climate model simulations are no exception here.
9 Jebeile and Roussos (2023) argue that climate modeling has been characterized by a physics-first approach, which puts emphasis on the primacy of physics at the expense of the social and life sciences in understanding climate change and its effects. However, they think that in order for climate models to provide usable information for a wider range of stakeholders (for example, the public health sector), the physics approach should be more strongly coupled with environmental, ecosystemic, and socioeconomic dimensions.
10 The *tuned* parameters/processes will be fine, since the allowed parameter space will be constrained to be physically possible (*modulo* expert opinion).

References

Arrhenius, Svante. 1896. "On the Influence of Carbonic Acid in the Air upon the Temperature of the Ground." *Philosophical Magazine and Journal of Science* 41: 237–276.

Baumberger, Cristoph, Reto Knutti and Gertrude Hirsch Hadorn. 2017. "Building Confidence in Climate Model Projections: An Analysis of Inferences from Fit." *WIREs Climate Change* 8: e454.

Cabanes, Simon, Aymeric Spiga and Roland M. B. Young. 2020. "Global Climate Modeling of Saturn's Atmosphere. Part III: Global Statistical Picture of Zonostrophic Turbulence in High-Resolution 3D-Turbulent Simulations." *Icarus* 45: 113705.

Finnish Meteorological Institute. 2022. "Numerical Weather Prediction." https://en.ilmatieteenlaitos.fi/numerical-weather-prediction, accessed 25.5.2023.

Frisch, Mathias. 2015. "Predictivism and Old Evidence: A Critical Look at Climate Model Tuning." *European Journal for Philosophy of Science* 5: 171–190.

Grüne-Yanoff, Till and Ylwa Sjölin Wirling. 2021. "The Possibilistic Interpretation of Climate Model Ensembles". Unpublished conference paper, presented at PSA20/21, Baltimore, MD, 11.11.2021.

Hirvonen, Ilmari, Rami Koskinen and Ilkka Pättiniemi. 2021. "Modal Inferences in Science: A Tale of Two Epistemologies." *Synthese* 199: 13823–13843.

Hourdin, Frédéric, Thorsten Mauritsen, Andrew Gettelman, Jean-Christophe Golaz, Venkatramani Balaji, Qingyun Duan, Doris Folini, Duoying Ji, Daniel Klocke, Yun Qian, Florian Rauser, Catherine Rio, Lorenzo Tomassini, Masahiro Watanabe and Daniel Williamson. 2017 "The Art and Science of Climate Model Tuning." *Bulletin of the American Meteorological Society* 98(3): 289–602.

Humphreys, Paul. 2009. "The Philosophical Novelty of Computer Simulation Methods." *Synthese* 169: 615–626.

IPCC. 2012. "Glossary of Terms." In: *Managing the Risks of Extreme Events and Disasters to Advance Climate Change Adaptation. A Special Report of Working Groups I and II of the Intergovernmental Panel on Climate Change (IPCC)*, edited by C.B. Field, V. Barros, T.F. Stocker, D. Qin, D.J. Dokken, K.L. Ebi, M.D. Mastrandrea, K.J. Mach, G.-K. Plattner, S.K. Allen, M. Tignor, and P.M. Midgley, 555–564. Cambridge: Cambridge University Press.

Jebeile, Julie and Anouk Barberousse. 2021. "Model Spread and Progress in Climate Modelling." *European Journal for Philosophy of Science* 11: 66.

Jebeile, Julie and Joe Roussos. 2023. "Usability of Climate Information: Toward a New Scientific Framework." *WIREs Climate Change* 14(5): e833. https://doi.org/10.1002/wcc.833

Jeevanjee, Nadir. 2018. "The Physics of Climate Change: Simple Models in Climate Science." https://doi.org/10.48550/arXiv.1802.02695

Katzav, Joel. 2013. "Hybrid Models, Climate Models and Inference to the Best Explanation." *The British Journal for the Philosophy of Science* 64(1): 107–129.

———. 2023. "Epistemic Possibilities in Climate Science: Lessons from Some Recent Research in the Context of Discovery." *European Journal for Philosophy of Science* 13:57.

Katzav, Joel and Wendy S. Parker. 2015. "The Future of Climate Modeling." *Climate Change* 132: 475–487.

———. 2018. "Issues in the Theoretical Foundations of Climate Science." *Studies in History and Philosophy of Modern Physics* 63: 141–149.

Knüsel, Benedikt and Christoph Baumberger. 2020. "Understanding Climate Phenomena with Data-Driven Models". *Studies in History and Philosophy of Science* 84: 46–56.

Knutti, Reto, Reinhard Furrer, Claudia Tebaldi, Jan Cermak and Gerard A. Meehl. 2010. "Challenges in Combining Projections from Multiple Climate Models." *Journal of Climate* 23(10): 2739–2758.

Koskinen, Rami. 2023. "Kinds of Modalities and Modeling Practices." *Synthese* 201: 196.

Lavin-Gullon, Alvaro, Josipa Milovac, Markel García-Díez and Jesus Fernández. 2023. "Spin-Up Time and Internal Variability Analysis for Overlapping Time Slices in a Regional Climate Model." *Climate Dynamics* 61: 47–64.

Lenhard, Johannes and Eric Winsberg. 2010. "Holism, Entrenchment, and the Future of Climate Model Pluralism." *Studies in History and Philosophy of Modern Physics* 41(3): 253–262.

Leuschner, Anna. 2015. "Uncertainties, Plurality, and Robustness in Climate Research and Modeling: On the Reliability of Climate Prognoses." *Journal for General Philosophy of Science* 46: 367–381.

Lloyd, Elisabeth A. 2010. "Confirmation and Robustness of Climate Models." *Philosophy of Science* 77(5): 971–984.

———. 2015. "Model Robustness as a Confirmatory Virtue: The Case of Climate Science." *Studies in History and Philosophy of Science* 49: 58–68.

Lloyd, Elisabeth A. and Naomi Oreskes. 2018. "Climate Change Attribution: When Is It Appropriate to Accept New Methods?" *Earth's Future* 6: 311–325.

Lofverstrom, Marcus, Jeremy G. Fyke, Katherine Thayer-Calder, Laura Muntjewerf, Miren Vizcaino, William J. Sacks, William H. Lipscomb, Bette L. Otto-Bliesner and Sarah L. Bradley. 2020. "An

Efficient Ice Sheet/Earth System Model Spin-up Procedure for CESM2-CISM2: Description, Evaluation, and Broader Applicability." *Journal of Advances in Modeling Earth Systems* 12(8): 1–23. https://doi.org/10.1029/2019MS001984
Lorenz, Edward N. 1993. *The Essence of Chaos*. Seattle, WA: The University of Washington Press.
Lynch, Peter. 2008. "The Origins of Computer Weather Prediction and Climate Modeling." *Journal of Computational Physics* 227: 3431–3444.
Maudlin, Tim. 2020. "A Modal Free Lunch." *Foundations of Physics* 50(6): 522–529.
Neelin, J. David. 2011. *Climate Change and Climate Modeling*. Cambridge: Cambridge University Press.
Parker Wendy S. 2010. "Comparative Process Tracing and Climate Change Fingerprints." *Philosophy of Science* 77: 1083–1095.
———. 2013. "Ensemble Modeling, Uncertainty and Robust Predictions." *WIREs Climate Change* 4: 213–223.
Sjölin Wirling, Ylwa and Till Grüne-Yanoff. 2021. "The Epistemology of Modal Modeling." *Philosophy Compass*. https://doi.org/10.1111/phc3.12775
Shepherd, Theodore G. 2014. "Atmospheric Circulation as a Source of Uncertainty in Climate Change Projections." *Nature Geoscience* 7: 703–708.
Shepherd, Theodore G., Emily Boyd, Raphael A. Calel, Sandra C. Chapman, Suraje Dessai, Ioana M. Dima-West, Hayley J. Fowler, Rachel James, Douglas Maraun, Olivia Martius, Catherine A. Senior, Adam H. Sobel, David A. Stainforth, Simon F. B. Tett, Kevin E. Trenberth, Bart J. J. M. van den Hurk, Nicholas W. Watkins, Robert L. Wilby and Dimitri A. Zenghelis. 2018. "Storylines: An Alternative Approach to Representing Uncertainty in Physical Aspects of Climate Change." *Climatic Change* 151: 555–571.
Song, Jiechen, Tong, Guanchao, Chao, Jiayou, et al. 2023. "Data Driven Pathway Analysis and Forecast of Global Warming and Sea Level Rise." *Nature Scientific Reports* 13: 5536.
Werndl, Charlotte. 2016. "On Defining Climate and Climate Change." *The British Journal for the Philosophy of Science* 67: 337–364.
———. 2019. "Initial-Condition Dependence and Initial-Condition Uncertainty in Climate Science." *The British Journal for the Philosophy of Science* 70: 953–976.
Wimsatt, William C. 2007. *Re-Engineering Philosophy for Limited Beings: Piecewise Approximations to Reality*. Cambridge, MA: Harvard University Press.
Winsberg, Eric. 2018. *Philosophy and Climate Change*. Cambridge: Cambridge University Press.

33

EPISTEMIC IMPLICATIONS OF MACHINE LEARNING MODELS IN SCIENCE

Stefan Buijsman and Juan M. Durán

1. Introduction

Machine learning models are quickly gaining ground in scientific practice. A particular story of success is the use of the deep learning model AlphaFold 2 to predict protein folding (Jumper et al. 2021), but examples abound. There is, for example, the usage of deep learning in climate models (Rasp et al. 2018), astronomy (Agarwal et al. 2012), and materials science (Schmidt et al. 2019). Furthermore, a wide range of deep learning models are used in computational neuroscience (e.g., Zhuang et al. 2021; Güçlü and van Gerven 2017). This increased usage of machine learning techniques in scientific research raises important philosophical questions regarding the epistemic implications of such tools. Most prominently, the issue is that many machine learning models fail to represent a target system with a set of equations, as is the case in other types of (process-based) models. To see this, consider the workings of deep learning models such as random forest models. These models, a type of neural network, consist of a large number of artificial neurons that have a (standardly non-linear) activation function determining the output value of the neuron based on the input values. These artificial neurons are then ordered into (a large number of) layers, with connections from neurons in one layer to neurons in the next layer. It is those connections that matter, as the weights on them—how much the output of a neuron counts toward the input of the next neuron—are adjusted based on training data. Typically, a machine learning model has millions of weights, and the largest neural network models have trillions of such weights that are adjusted in training.

A number of differences from traditional theoretical models and modeling have already become apparent from this very brief description of machine learning models. First and foremost, there are no (explicit) representations of physical quantities in such models. This differentiates machine learning models from other statistical models, where regression based on data may be used, but representations of physical quantities are still present in the model. Furthermore, the adjustment of the weights in machine learning models happens automatically, based on a training set. There are too many such weights to monitor this process directly, nor can the final model be easily inspected to understand its exact functioning. It follows that it is incredibly difficult to tell which patterns the model uses to arrive

DOI: 10.4324/9781003205647-39

at predictions. As a result, machine learning models have a high degree of epistemic opacity, defined as (see also Durán and Formanek 2018; Beisbart 2021):

> [A] process is epistemically opaque relative to a cognitive agent X at time t just in case X does not know at t all of the epistemically relevant elements of the process
> *(Humphreys 2009, 618)*

These two differences, and the increased epistemic opacity that results, raise the philosophical question: what is the scientific value of using machine learning models? The answer depends somewhat on the scientific field. In the case of neuroscience, for instance, artificial neurons might be seen as (idealized) representations of physical neurons. Thus, neural networks can be seen to yield scientific understanding in these contexts and can give way to new specifications of functionalism in the philosophy of mind (Section 2). In the other sciences, their status is much more contentious. The statistical nature of machine learning is a more serious concern here, as is the opacity of models and the absence of representations. Can machine learning models yield scientific explanations and, possibly via those explanations, understanding? Are we justified in believing their predictions? As we will see (Section 3), views range from pessimistic, seeing machine learning models as substantially different from other kinds of models, to more optimistic, where epistemic opacity is not necessarily an issue and machine learning models are on the same scale of explainability as other models used in science.

Does this entail new ways of doing science and, as such, novel issues for the philosophy of science? Some answers to these questions are found in the literature on computer simulations. Although one can find some early skepticism about the *scientific* novelty of computer simulations (e.g., Teichroew and Lubin 1966, 724), the general feeling is that computer-based methodologies extend the class of tractable mathematics and representation, thus broadening the range of modeling phenomena (Frigg and Reiss 2009). Fewer agreements are found, however, on the *philosophical* novelty of computer-based research. Famously, Frigg and Reiss (2009) club together four skeptical arguments against "a new metaphysics, epistemology, semantics, and methodology" (595) for the philosophy of science. Humphreys, however, alerts us that an anthropocentric epistemology is no longer viable and that we are required to understand and evaluate the world through "computationally based scientific methods that transcend our own abilities" (Humphreys 2009, 617), as opposed to representations tailored to human cognitive capacities. Within a non-anthropocentric epistemology emerge diverse philosophical issues that, according to Humphreys, have not been addressed by a more familiar philosophy of science. Perhaps the most famous of all is the problem of *epistemic opacity* mentioned earlier. Having said that, the epistemic and methodological implications of using machine learning models are still heavily debated. However, their successful use in the sciences shows that they certainly have a role to play.

2. Neural networks and neuroscience

The application of machine learning in neuroscience is a special case. As opposed to other sciences, neural networks (but no other machine learning techniques) can be argued to contain explicit representations in neuroscience. Artificial neurons represent actual neurons, the weights in neural networks represent the strength of connections between neurons, and so on. There are, of course, a number of differences between neural networks and actual neurons, as López-Rubio (2018) enumerates: backpropagation is unrealistic for the brain, artificial neurons are far simpler than biological neurons, the brain isn't structured as neatly as neural networks are, activation functions differ, and so on. Despite such differences, López-Rubio considers it plausible that neural networks are representative of the brain, according to a similarity view of model representation:

> [f]rom the current state of research, it is likely that the similarities among biological and artificial features extend from the highest level of description, i.e., the overall inputs and outputs, to a certain intermediate level of description, while the lowest levels such as the electrical signals in the biological synapses do not match well with their artificial counterparts.
>
> *(681)*

In virtue of this similarity between neural networks and the brain, López-Rubio holds that we can formulate an updated version of computationalism he terms neural computational functionalism:

Neural computational functionalism (NCF): the mind is the set of synaptic weights of the brain.

> This is to be interpreted in the sense that: (a) the brain stores synaptic weights in its neural structures, (b) some of those neural structures are organized in a hierarchy of layers, (c) those synaptic weights determine the computation of significant features of progressively higher level as we traverse the neural hierarchy, (d) those features ultimately determine behavior.
>
> *(682–683)*

Neural networks can then clearly function as models of the brain in much the same way that other types of models work. In line with that idea, neural network models would be able to offer explanations of the functioning of the brain. Piccinini makes a concrete suggestion of what such explanations would look like:

> An *explanation by synaptic weights* of a capacity C possessed by a (biological or artificial) neural system S is a set of weights W for C such that S possesses C because S operates according to its stored weights W.
>
> *(Piccinini 2010, 277)*

Neural networks can then clearly offer explanations of the functioning of the brain in this proposal. Such ideas are more widespread, as Miłkowski (2013) and Stinson (2018) similarly argue that neural network models can offer (mechanistic) explanations of the brain. Buckner (2018) even argues that the functioning of (convolutional) neural networks gives an important insight into the way the brain handles concepts. They illustrate a process he calls *transformational abstraction*, where complexity is reduced by iterative transformations into simplified (abstract) representations. This type of abstraction, which occurs in neural networks in order to detect later layers, e.g., the presence of a chair or shovel, is in Buckner's (2018) eyes, also a fitting solution to the question of how humans manage to acquire such concepts from experience. Can these neural networks then function as mechanistic explanations for the visual cortex of our brains?

There are a number of problems laid out by Buckner (2018), such as the fact that neural networks are prone to adversarial examples: Small changes to the input image can cause the model to yield a wildly different output classification. An image of a panda in which a few select pixels are changed might be classified as showing a gibbon, for example. Such adversarial examples are hard to eliminate in neural networks, and yet our brains are clearly not susceptible to them. Buckner (2018, 5367) does not see this as too problematic and argues instead that neural networks are best seen as mechanism sketches (Piccinini and Craver 2011) or as generic mechanisms of the kind Stinson (2018) suggests. Taking the limitations of neural networks into consideration, they still have an explanatory role to play: "DCNNs show that the generic kind of neural mechanism found in mammalian perceptual cortex can learn and deploy abstract category representations using only domain-general mechanisms—vindicating a key theme of empiricism" (Buckner 2018, 5369). Yet, at the same time, neural networks are far worse than we are at generalizing (see Section 3.1) and make very different mistakes in image classification and other tasks than humans. Such substantial differences call for caution when using neural networks as models of, e.g., human concept formation. Further attention to these functional differences is needed before we can see neural networks as explaining our actual higher-level cognitive functions.

Machine learning models, in conclusion, might provide an idealized representation of biological neurons and synapses, and neural networks can act as (mechanistic) explanations of their functioning on an appropriate level of abstraction. There is more work to do on the exact nature of these representations and idealizations and the effect this has on the conclusions that can be drawn from the models. Can neural networks explain higher-level cognitive functions, or do they only provide how-possibly explanations? Does their limited generalizability imply a limit to their role in how-actually explanations? Or will neural networks become one of the dominant modeling tools for neuroscience? This requires further reflection, but there is little doubt that neural networks have an explanatory role to play. That is quite different when applications of machine learning models are considered in other sciences. We, therefore, turn to those other applications now.

3. Machine learning in the other sciences

As mentioned in the introduction, machine learning models present us with difficulties in the other sciences, as they do not contain explicit representations of the physical quantities involved and are epistemically opaque. That is, we typically do not understand why a machine learning model yields a particular prediction as opposed to a different one. Consequently, it is tempting to hold that such models do not yield (scientific) explanations for the phenomena they are trained to predict. Srećković et al. (2022), for example, argue that machine learning complicates the obtaining of two types of explanations: Process explanations and phenomenon explanations. This is uncontroversial for process explanations, which would be explanations of the process that led to a specific model prediction. Machine learning models are typically too complex to survey, and it is a serious challenge to obtain explanations for their outputs. This is widely studied under the name explainable AI (XAI; see Das and Rad 2020 for a review) and is considered an ethical issue for the application of AI.

The more crucial question for the use of machine learning methods in science is whether this also means that explanations of the scientific phenomena that are predicted are not forthcoming. Srećković et al. (2022, 6) consider such explanations to be unforthcoming due to the lack of causal relations underlying the model predictions: the problem is "the associativity of the method, which involves searching solely for correlations between the features in the data without a theoretical back-up to provide causal relationships, traditionally considered crucial for explanations." The lack of process explanations exacerbates this issue, as it obscures the correlations used by the model to make predictions. These underlying correlations, as a result, cannot be extracted from the model, and so no experiments can be designed to find causal relations. In short, machine learning models, as Srećković et al. (2022) argue, cannot be used to arrive at causal relations linking inputs to outputs, and so do not yield causal explanations. They can be used for (highly accurate) predictions, but not for understanding. However, it is not even clear that predictions of machine learning models will have a similar, or better, epistemic status as those of process-based models. Thus, before diving deeper into the question of explanations, we turn first to the predictions of machine learning models.

3.1 Epistemic status of machine learning predictions

Machine learning models are usually associated with high accuracy. In the case study that Kawamleh (2021) looks at, the model predictions for parametrizations in climate models are reported to be of high accuracy (Rasp et al. 2018). Despite this success on the test set with which the model was evaluated, Kawamleh (2021) argues that the machine learning model fails to generalize to new situations. This limited generalizability of machine learning models is a known problem, as machine learning models often perform badly when presented with input that is (in our view slightly) different than that present in the test set. For example, object recognition systems become highly inaccurate when objects are presented in unusual locations (Rosenfeld et al. 2018), or when they are rotated into an unusual pose (Alcorn et al. 2019). A similar situation occurs in the machine learning model that predicts parameters for climate models. These are trained on input-output pairs generated by a physical model (that is much more computationally intensive to use for long-term climate change modeling). If the situation deviates too much from these training pairs, which one might expect when modeling climate change, then the machine learning model loses its accuracy. As Rasp et al. (2018, 9687) state, "the neural network cannot handle temperatures that exceed the ones seen during training," in this case, an increase of sea-surface temperatures of more than 4 Kelvin. They blame this on overfitting, but as Kawamleh (2021) shows, no machine learning models have managed to generalize on this task to date (and as pointed out above, it is, in fact, a common feature of such models).

Where does this lack of generalizability come from? Kawamleh (2021) blames the lack of representations of physical processes:

> Traditional and cloud resolving parameterizations represented processes directly or indirectly and this process representation has added an irreducible value for the reliability of model predictions because it provides (a) physical/empirical constraints and (b) facilitates forms of model development and evaluation which guard against overfitting.
>
> *(1019)*

Machine learning models do not have this protection against overfitting and instead rely purely on correlations present in the data set generated from running the process-based model on a chosen set of training cases. The upshot is that:

> the trained NNP [machine learning model] fails to learn convection and generalize beyond its training data because it fails to represent the causal convective processes which relate the climate variables of interest. [...] The very *representation* of processes adds significant and irreplaceable value for the reliability of climate model *predictions*.
>
> *(Kawamleh 2021, 1019, emphasis in original)*

This matches with explanations of the incredible performance of machine learning models in protein folding, where AlphaFold 2 is largely considered to have 'solved protein folding' because it gives accurate predictions of the folding for almost all protein specifications. Note, however, that "[t]he key to why AF2 works is the fact the library of single domain protein structures is essentially complete" (Skolnick et al. 2021, 4827). It is this lack of outliers compared to the training set that has led to a uniformly strong performance. If it were not for that completeness, there would likely be the same issues with generalizability (and indeed, issues do occur when more than one fold is possible). For:

> AlphaFold has not learned from ligands and is actually not aware of the actual energy minima that are essential for folding in real life. In reality, AlphaFold has not solved the folding problem as it would occur in solution or in a cell, but it has provided a practical solution: It has learned the results of folding at the amino acid residue contact level and can, therefore accurately predict a single-chain hemoglobin fold that would never exist on its own or in the absence of the heme cofactor in nature.
>
> *(Perrakis and Sixma 2021, 2–3)*

So, does this issue with the generalizability of machine learning models affect the epistemic status of their predictions? It need not, depending on one's views of justification from machine learning models. We only give a brief overview of the options here. These range from more liberal views, such as that of Beisbart (2017), who holds that one is justified to believe the predictions of a computer simulation (here generalized a bit to machine learning) if one is justified to believe that the computer program works as intended. Verification of this will be very difficult for machine learning models, however, due to their epistemic opacity, so when are we justified to believe that the program works as intended? One can also wonder which intentions are relevant, as intending that the model predicts the phenomenon accurately for a test set is easy to verify, but too limited to be justified in believing its results generally speaking.

Durán and Formanek (2018) are more detailed matters about justification, though from a more externalist standpoint. They hold that one is justified to believe the output of a computer simulation if the model is sufficiently reliable, in their account of *Computational Reliabilism* (which can be generalized to machine learning models):

> (CR) if S's believing p at t results from m, then S's belief in p at t is justified.
>
> where S is a cognitive agent, p is any truth-valued proposition related to the results of a computer simulation, t is any given time, and m is a reliable computer simulation.
>
> *(Durán and Formanek 2018, 654)*

Reliability here is to be understood as more than simply that the model produces correct predictions sufficiently often. Instead, it is a more complex notion where the reliability of a model can be supported by reliability indicators such as verification and validation methods, robustness analyses, a history of (un)successful implementations, and expert knowledge. The account does not tell how these factors fit together, and thus, how to determine when a model is reliable and for what range of cases (e.g., only temperature variations under 4 Kelvin). Such details would need to be delivered by applying computational reliabilism to specific cases.

Finally, Symons and Alvarado (2019) take this idea somewhat further, holding that justifications for the results of computer simulations (i.e., machine learning model predictions) in scientific contexts come with high demands. They are, consequently, fairly pessimistic about machine learning models, as they argue that "trust in simulations should be grounded in empirical evidence, good engineering practice, and established theoretical principles. Without these constraints, computer simulation risks becoming little more than unmoored speculation" (Symons and Alvarado 2019, 57–58). Such grounding is difficult, though what it exactly entails is left unclear. Still, Kawamleh (2021) can be read as an argument that scientific grounding is lacking for those machine learning models, so justified beliefs might be hard to come by. The epistemic status of machine learning predictions is thus a matter of active debate, and there is a need for more specific accounts that can adjudicate specific cases. The lack of representations in these models presents a problem for their generalizability and grounding in established theoretical principles. That, in turn, affects the epistemic status of their predictions. Does it also rule out any hope for scientific explanations and understanding?

3.2 *Explanations from machine learning models?*

The statistical nature of machine learning models, combined with the opacity of the precise correlations they rely on, are for a number of philosophers good reason to be skeptical of their explanatory prospects. We have already discussed the arguments of Srećković et al. (2022), but López-Rubio and Ratti (2021) make a similar point in the context of molecular biology. They focus on the prospect of mechanistic explanations, the standard account for molecular biology, resulting from machine learning models. They, too, are skeptical that such explanations can be obtained: " If you do molecular biology with machine learning techniques, and if you want to have the best machine learning performances, then you cannot even in principle elaborate fully-fledged mechanistic explanations" (López-Rubio and Ratti 2021, 3152). Not because of technological limitations, but because "the more the size of the model increases, the less the human mind is able to organize the model's components into a causal narrative, which forms the backbone of any mechanistic description with explanatory force" (López-Rubio and Ratti 2021, 3152). As machine learning models rely on a vast number of parameters to achieve high accuracy, the argument goes, that they hinder the formulation of a causal narrative and, thus, of a mechanistic explanation. Here it is the associativity, i.e., the lack of a clear causal link between inputs and model predictions, in addition to the complexity that hinders understanding.

Yet other philosophers do not consider it a given that there are no scientific explanations to extract from machine learning models (primarily seen as involving causal relations, though, importantly, not all philosophical accounts of explanation give a central role to causation). They hold that, at least in some cases, it is possible to acquire these kinds of explanations.

Sullivan (2022) started this line of thought, defending that machine learning models can yield understanding despite their epistemic opacity. She holds that the implementation of a model is often irrelevant to the explanations that can be generated from that model and gives the example of Schelling's model, used to study the causes of segregation. This model holds that a person will move if more than 70% of her neighbors belong to a different group than she does. In a situation with two groups present, this simple rule ultimately leads to a segregated situation, irrespective of the starting situation. How that model is implemented, however, whether on a checkers board (as originally the case) or on a computer, is irrelevant for extracting scientific explanations. What matters for us to obtain explanations of segregation in the real world is whether the model shows a process that actually occurs. In other words, what matters is whether people in real life tend to move when they belong to the minority in a specific neighborhood. If the model links to such a real-world process, then it can provide explanations. If it does not, then it fails to yield scientific explanations. The real problem, according to Sullivan, then, is what she calls *link uncertainty*, where "link uncertainty constitutes a lack of scientific and empirical evidence supporting the link connecting the model to the target phenomenon" (Sullivan 2022, 21). Note, however, that the explanation resulting from the model here also crucially relies on us knowing the process implemented by the model: that people move when 70% of their neighbors are in a different group is built-in in the model (Räz and Beisbart 2022). However, as discussed in the context of epistemic opacity, the knowledge of the implemented process is difficult to obtain from machine learning models. Sullivan (2022), however, is optimistic that, in some simple cases, one can still know enough about the implemented process and reduce the link uncertainty sufficiently to obtain explanations from machine learning models.

Sullivan argues that this is the case for a skin lesion classifier, where a machine learning model classifies moles based on their visual appearance. As there is a strong scientific basis for a link between visual appearance and the type of mole it is (e.g., whether it is a kind of cancer or requires a biopsy), the reasoning goes that the link uncertainty, therefore, is low. The model also receives the input information that is scientifically known to be relevant to the decision, and thus, correlations found based on that information are of interest. Perhaps they do not correspond to causal relations, but Sullivan maintains that such (new) correlations "can further understanding, especially once these newly discovered patterns undergo further investigation" (2019, 24). While she does not discuss how the correlations the machine learning model uses would be identified, deal with the worry that they may be too complex, or how link uncertainty is reduced, the idea that available scientific background information can make machine learning models explanatory has been picked up and developed in further detail by others.

Knüsel and Baumberger (2020) do so in the context of climate change modeling - not for parametrizations, but for models that try to determine if the rise in average temperature is due to human actions. In such a case, they consider it possible for machine learning models to provide understanding. The condition here is that:

> for data-driven models to be useful for understanding phenomena, researchers should be in a position to argue from the coherence of the model with background knowledge to its representational accuracy. This can for example be achieved if important bivariate relationships are known. This sort of reasoning provides exactly the kind of evidence that reduces the link uncertainty discussed by Sullivan
>
> *(Knüsel and Baumberger 2020, 47)*

How does this background knowledge help in modeling historical changes in temperature? First of all, it is a setting where we can approximate the situation quite well using the energy-balance model, consisting of a single differential equation. It coheres with background knowledge, has decent empirical accuracy, is robust, and is easily graspable (as it is only a single differential equation). As such, it can be used to show that human actions are the cause of the temperature rise, as that rise only comes out of the model if the effects of human actions are taken into account. Filter them out, and the average temperature predicted by the model remains stable. We can then explain why the average temperature has risen (and why human actions are the culprit).

Knüsel and Baumberger (2020) then compare this process-based model to a machine learning model making the same predictions. This machine learning model shows the same difference whether human actions are included or not and has similar empirical accuracy. They argue that it is robust because outputs are similar to the process-based model, in that it is coherent with background knowledge and because the outputs are consistent with the known physical laws (though recall Kawamleh (2021) that robustness and coherence are more complicated), and that manipulating the model and studying the feature importance makes it somewhat graspable. Therefore, they hold that this machine learning model can also be used to explain why the average temperature has risen in the last hundred years. Machine learning models may do worse on all these scales of explanation except for empirical accuracy, but they can still do well enough in some cases to provide explanations. The argument, however, focuses on whether certain input values (human factors) are relevant to the outcome. More interesting, and problematic, given the associativity and opacity of machine learning models, is *why* human actions cause a rise in temperature. Knüsel and Baumberger (2020) do not discuss that question. In addition, it is unclear if the link uncertainty can be reduced sufficiently without a transparent process-based model being available. Only if that is possible would machine learning models add new explanations.

A similar shortcoming can be seen in the work of Jebeile et al. (2021), who look at yet another type of machine learning in climate modeling to argue that said models are on a continuous scale along with other types of models. They argue that their empirical accuracy is often better, but they do worse on intelligibility, representational accuracy, coherence with background knowledge, and assessment of the domain of validity. In some cases, however, we might know enough about the domain that we can give a sufficiently confident assessment of machine learning models' coherence with background knowledge. In those cases, they can explain the phenomena they are trained to predict. Yet, what kind of explanations can be obtained if the processes the models implement remain unclear?

Meskhidze (2023) tries to provide more substantive answers here. She argues that machine learning models in cosmology (predicting cosmological parameters in large simulations of the formation of galaxies) answer some why-questions, but do not help us understand "why phenomena of this general type occur across a variety of circumstances" (Meskhidze 2023, 1901). The reason is their lack of physical representations; they do not adhere to physical laws and so are not suited to explain such questions about the unfolding of physical processes. This is not a problem, though, for such machine learning models to help us understand "why, for example, our universe has the particular distribution of matter it does. By filling out the parameter space of interest, such methods can point cosmologists to the relevant values of the cosmological parameters that led to a particular distribution of matter" (Meskhidze 2021, 1906). The argument seems to be that if the outputs of the machine learning models correspond to the actual values, then this can be explanatory of the actual distribution of matter. However, scientific explanations are typically thought to require a covering rule or mechanism sketch. The machine learning model does not seem to provide that overarching process, which is instead given by physics-based N-body models. As such, the machine learning model does not seem to answer the question of why our universe has the particular distribution of matter that it does on its own. It thus remains unclear what the explanatory value of the machine learning models is exactly.

Despite widespread optimism, no clear answers have emerged on how machine learning models lead to (novel) explanations, even if the link uncertainty is reduced. At the same time, the pessimists might be too hasty to dismiss the extraction of causal relations from machine learning models, as there is a burgeoning literature connecting causal inference to machine learning (Pearl 2019). Buijsman (2023) connects this literature to machine learning techniques for causal inference to argue that in a few specific cases we can get (causal) scientific explanations from machine learning models. However, he also argues that this is unlikely to work for predictive machine learning models due to inherent biases in these models. Furthermore, causal accounts of explanation are not the only option. Other epistemic accounts of explanation are likewise viable; for example, Durán (2017, 2021) approaches scientific explanation and machine learning from a unificationist perspective. Such alternative accounts deserve more attention in the debate on scientific explanations from machine learning. The central challenge of formulating how explanations arise from machine learning models (if at all), remains an open question and calls for both a broader look at explanations and more in-depth case studies.

Let us finally note that the current debate on understanding machine learning largely happens in light of explanation. This is either because explanations are seen by many as a one-solution-fits-all (e.g., it reduces opacity, increases transparency, provides trustworthy machine learning, and adds to our understanding of the system) or because it is the standard philosophical pathway to understanding. Take, for instance, the objections raised by Räz and Beisbart (2022) to Sullivan's uncertainty link. To these authors, Sullivan's view depends on which notion of understanding is at play, and a strong notion would require explanatory understanding. Although they do not adopt a specific definition of explanatory understanding, they accept de Regt's (2017) and Khalifa's (2017) as suitable interpretations for their purposes. In this context, the overall strategy of Räz and Beisbart consists of showing that understanding ML comes in close connection with explanations. But not just any form of explanation. In particular, Sullivan's how-explanations strike them as unconvincing: "She writes that the deep patient model can answer the question of 'how it is possible to predict disease development for a range of diseases'" (Sullivan 2022, 123). As pointed out by Räz and Beisbart, "This is not a request for a how-possibility explanation of phenomena in the target system, it is a question about the possibility of predictive modeling itself" (2022).

Some authors have taken a somewhat different path in the connection between explanation and understanding. Páez (2019), for instance, claims that the search for explainable AI must be formulated in terms of the broader project of offering a pragmatic and naturalistic account of understanding. The result is the same: the analysis of explanations is in light of understanding. But is there a way to address understanding without resorting to explanation (and vice versa)? Räz and Beisbart think so. They suggest that machine learning can produce some degree of objectual understanding, here taken to be:

> the understanding of a domain of things; it is often taken to imply some knowledge of this domain and the grasp of connections between items in the domain. These connections may be explanatory, but need not be; they may be merely logical or probabilistic.
>
> *(Räz and Beisbart 2022)*

Examples of objectual understanding have been discussed in the philosophical literature. Gijsbers (2013) shows that some classifications, such as those used in biology, can effectively enhance our understanding of, say, species without providing explanations. Based on these, Räz and Beisbart suggest that "ML models can lead to some objectual understanding, e.g., by establishing correlations, or by simply adding to knowledge of a domain of things" (Räz and Beisbart 2022).

4. Conclusion

What are the epistemic implications of machine learning models in the sciences? In the case of neuroscience, these epistemic implications are fairly clear. Neural networks, a type of machine learning model, can be seen as representing (parts of) the brain, and elements of neural networks can be linked to elements of biological neurons and synapses. Questions remain on the limitations of neural networks as models of the brain, e.g., due to their limited ability to generalize, but it is clear that they play a role in understanding the functioning of the brain.

When Machine Learning functions as a tool, in other sciences, its contribution to understanding is far less clear. Since machine learning models do not contain physical representations, they are harder to link to the actual situation they model. Furthermore, their epistemic opacity makes it difficult to extract causal relations and even to determine the reliability and robustness of their predictions. As a result, it is unclear when scientists are justified to believe the predictions made by such models, and more work is needed on specifying exactly what conditions hold for justification in these contexts. Furthermore, it is unclear whether and what explanations (and understanding) can be gained from machine learning models. The discussion so far has focused on causal accounts of explanation but has not yet yielded examples of causal explanations that are clearly obtained from the machine learning model. Both a broader look at accounts of explanations and more detailed case studies are needed to determine the explanatory role of machine learning models in science. Such models are here to stay due to their benefits of higher empirical accuracy and lower computational costs, as the range of case studies has shown.

Acknowledgments

Juan M. Durán has received support from the EU program under the scheme "INFRAIA 2020-2024-SoBigData++: European Integrated Infrastructure for Social Mining and Big Data Analytics," grant agreement 871042. He has also received support from the EU program under the scheme "ICT48 Humane AI Net," grant agreement 952026. Their support is gratefully acknowledged.

References

Agarwal Shankar, Filipe B. Abdalla, Hume A. Feldman, Ofer Lahav, and Shaun A. Thomas. 2012. "PkANN - I. Non-Linear Matter Power Spectrum Interpolation through Artificial Neural Networks." *MNRAS* 424: 1409–1418.

Alcorn, Michael A., Qi Li, Zhitao Gong, Chengfei Wang, Long Mai, Wei-Shinn Ku, and Anh Nguyen. 2019. "Strike (with) a Pose: Neural Networks are Easily Fooled by Strange Poses of Familiar Objects." In *Proceedings of the IEEE/CVF Conference on Computer Vision and Pattern Recognition*, Long Beach, CA, USA, 4845–4854.

Beisbart, Claus. 2017. "Advancing Knowledge through Computer Simulations? A Socratic Exercise." In *The Science and Art of Simulation I*, edited by Michel Resch, Andreas Kaminski, and Petra Gehring, 153–174. Berlin: Springer.

———. 2021. "Opacity Thought Through: on the Intransparency of Computer Simulations." *Synthese* 199: 11643—11666.

Buckner, Claus. 2018. "Empiricism without Magic: Transformational Abstraction in Deep Convolutional Neural Networks." *Synthese* 195(12): 5339–5372.

Buijsman, Stefan. (2023). Causal scientific explanations from machine learning. *Synthese*, 202(6), 202.

Das, Arun, and Paul Rad. 2020. "Opportunities and Challenges in Explainable Artificial Intelligence (XAI): A Survey." *arXiv preprint* arXiv:2006.11371.

de Regt, Henk W. 2017. *Understanding Scientific Understanding*. New York: Oxford University Press.

Durán, Juan M. 2017. "Varying the Explanatory Span: Scientific Explanation for Computer Simulations." *International Studies in the Philosophy of Science* 31(1): 27–45.

———. 2021. "Dissecting Scientific Explanation in AI (sXAI): A Case for Medicine and Healthcare." *Artificial Intelligence* 297: 103498.

Durán, Juan M., and Nico Formanek. 2018. "Grounds for Trust: Essential Epistemic Opacity and Computational Reliabilism." *Minds and Machines* 28(4): 645–666.

Frigg, Roman, and Julian Reiss. 2009. "The Philosophy of Simulation: Hot New Issues or Same Old Stew?" *Synthese* 169(3): 593–613.

Gijsbers, Victor. 2013. "Understanding, Explanation, and Unification." *Studies in History and Philosophy of Science Part A* 44.3: 516–522.

Güçlü, Umut, and Marcel J. van Gerven, M. A. 2017. "Modeling the Dynamics of Human Brain Activity with Recurrent Neural Networks." *Frontiers in Computational Neuroscience* 11: 7.

Humphreys, Paul W. 2009. "The Philosophical Novelty of Computer Simulation Methods." *Synthese* 169(3): 615–626.

Jebeile, Julie, Vincent Lam, and Tim Räz. 2021. "Understanding Climate Change with Statistical Downscaling and Machine Learning." *Synthese* 199(1): 1877–1897.

Jumper, John, Richard Evans, Alexander Pritzel, Tim Green, Michael Figurnov, Olaf Ronneberger, Kathryn Tunyasuvunakool, Russ Bates, Augustin Žídek, Anna Potapenko, et al. 2021. "Highly Accurate Protein Structure Prediction with AlphaFold." *Nature*, 596: 583–589.

Kawamleh, Suzanne. 2021. "Can Machines Learn How Clouds Work? The Epistemic Implications of Machine Learning Methods in Climate Science." *Philosophy of Science* 88(5): 1008–1020.

Khalifa, Kareem. 2017. *Understanding, Explanation, and Scientific Knowledge*. Cambridge: Cambridge University Press.

Knüsel, Benedikt, and Christoph Baumberger. 2020. "Understanding Climate Phenomena with Data-Driven Models." *Studies in History and Philosophy of Science Part A* 84: 46–56.

López-Rubio, Ezequiel. 2018. "Computational Functionalism for the Deep Learning Era." *Minds and Machines* 28(4): 667–688.

López-Rubio, Ezequiel, and Emanuelle Ratti. 2021. "Data Science and Molecular Biology: Prediction and Mechanistic Explanation." *Synthese* 198(4): 3131–3156.

Meskhidze, H. (2023). Can machine learning provide understanding? How cosmologists use machine learning to understand observations of the universe. *Erkenntnis*, 88(5): 1895–1909.

Miłkowski, Marcin. 2013. *Explaining the Computational Mind*. Cambridge: MIT Press

Páez, Andrés. 2019. "The Pragmatic Turn in Explainable Artificial Intelligence (XAI)." *Minds and Machines* 29: 441—59.

Pearl, Judea. 2019. "The Seven Tools of Causal Inference with Reflections on Machine Learning." *Communications of the ACM* 62(3): 54–60.

Perrakis, Anastassis, and Titia K. Sixma. 2021. "AI Revolutions in Biology: The joys and Perils of AlphaFold." *EMBO Reports* 22(11): e54046.

Piccinini, Gualtiero. 2010. "The Mind as Neural Software? Understanding Functionalism, Computationalism, and Computational Functionalism." *Philosophy and Phenomenological Research* 81(2): 269–311. https://doi.org/10.1111/j.1933-1592.2010.00356.x

Piccinini, Gualtiero, and Craver, Carl. 2011. Integrating psychology and neuroscience: Functional analyses as mechanism sketches. *Synthese*, *183*, 283-311.

Rasp, Stephan, Michael S. Pritchard, and Pierre Gentine. 2018. "Deep Learning to Represent Subgrid Processes in Climate Models." *PNAS* 115(39): 9684–9689.

Räz, Tim, and Claus Beisbart. 2022. "The Importance of Understanding Deep Learning." *Erkenntnis*. https://doi.org/10.1007/s10670-022-00605-y

Rosenfeld, Amir, Richard Zemel, and John K. Tsotsos. 2018. "The Elephant in the Room." *arXiv preprint*. arXiv:1808.03305.

Schmidt, Jonathan, Mario R. Marques, Silvana Botti, and Miguel A. L. Marques. 2019. "Recent Advances and Applications of Machine Learning in Solid-State Materials Science." *npj Computational Materials* 5(1): 1–36.

Skolnick, Jeffrey, Mu Gao, Hongyi Zhou, and Suresh Singh. 2021. "AlphaFold 2: Why It Works and Its Implications for Understanding the Relationships of Protein Sequence, Structure, and Function." *Journal of Chemical Information and Modeling* 61(10): 4827–4831.

Srećković, Sanja, Andrea Berber, and Nenad Filipović. 2022. "The Automated Laplacean Demon: How ML Challenges Our Views on Prediction and Explanation." *Minds and Machines* 32: 159–183.

Stinson, Catherine. 2018. "Explanation and Connectionist Models." In *The Routledge Handbook of the Computational Mind*, edited by Matteo Colombo and Mark Sprevak, 120–133. New York: Routledge.

Sullivan, Emily. 2022. "Understanding from Machine Learning Models." *British Journal for the Philosophy of Science* 73(1): 109–133.

Symons, John, and Ramón Alvarado. 2019. "Epistemic Entitlements and the Practice of Computer Simulation." *Minds and Machines* 29(1): 37–60.

Teichroew, Daniel, and John F. Lubin. 1966: "Computer Simulation Discussion of the Technique and Comparison of Languages." *Communications of the ACM* 9(10): 723–741.

Zhuang, Chengxu, Siming Yan, Aran Nayebi, Martin Schrimpf, Michel C. Frank, James J. DiCarlo, and Daniel Yamins. 2021. "Unsupervised Neural Network Models of the Ventral Visual Stream." *PNAS* 118(3): e2014196118.

34

IN VITRO ANALOGIES

Simulation modeling in biomedical engineering sciences

Nancy J. Nersessian

1. Introduction

A major epistemic practice in biomedical engineering sciences (hereafter, BMES) is to use engineering concepts, theories, methods, and materials to create living in vitro models, composed of cells or tissues and engineered materials, that serve as epistemic tools (Knuuttila 2011) for probing and learning about the behaviors of selected system components under controlled experimental conditions. Such models are epistemically and ontologically hybrid. These in vitro simulation models (often called "devices") provide BMES researchers with a means to investigate the dynamics of normal and disease processes in biological systems. They aspire to understand the phenomena sufficiently to enable medical, clinical, and pharmaceutical researchers to develop treatments to mitigate or prevent disease processes. Frontier biomedical engineers formulate problems with respect to phenomena that customarily have not been investigated by bioscientists, such as the effects of forces of blood flow on cardiovascular cells and tissues or network learning in neurons. These are systems for which there are no general biological theories of the phenomena under investigation that can provide a resource from which to begin research, so bioengineers frame the problem from an engineering perspective, and models are built from the ground up with the aid of engineering concepts, theories, materials, and methods. Modeling the dynamics of such systems comprises iterative and incremental processes of design, construction, evaluation, experimentation, and redesign, that is, cycles of *building* models to discover (Chandrasekharan and Nersessian 2015; Nersessian 2022).

A central epistemic aim of the practice of in vitro simulation modeling is to build models that provide the basis for inference about the target system, that is, to build an analogical source. BME researchers aim to build models that allow them to transfer inferences that derive from experiments they conduct with in vitro models to target in vivo phenomena as candidate understandings and hypotheses. As a researcher explained about her model:

> We typically use models to predict what is going to happen in a system [in vivo]. Like people use mathematical models to predict… what's going to happen in a mechanical system? Well, this is an experimental model that predicts what would happen – or you hope that it would predict – what would happen in real life.

 DOI: 10.4324/9781003205647-40

Such prediction is a form of analogical transfer. Building is a bootstrapping process in which models are developed toward becoming an analogical source. Once developed and justified, in vitro models provide structural, behavioral, or functional analog systems through which researchers can reason not only about the model but also about the real-world system by transferring inferences as hypotheses. Developing the warrant for such transfer is an important dimension of model building. This chapter will focus on the analogical dimension of modeling, which has largely been overlooked in the philosophical literature on models (Bailer-Jones 2009; Harré 1970; Black 1962; Hesse 1963 are notable exceptions). That literature has tended to focus on the representational nature of models, but when we look at models, generally, from the perspective of iterative and incremental investigative tools for building understanding, the dimension of analogical inference comes to the fore, which, of course, has representational implications as well.

My research group's 12-year cognitive-ethnographic investigation of model-building practices in two pioneering BMES university research laboratories – one in tissue engineering and one in neuroengineering – has provided a wealth of data on numerous dimensions of the nature of these epistemic practices. Importantly, in collecting field observations, interviews, and archival data (grant proposals, paper drafts, PowerPoint presentations, and so forth) over a sustained period, we were able to track the formation of problems and goals; to log the various methods, steps, and iterations of building; to ascertain specific concepts, theories, methods, and materials in use; to probe the decisions and judgments behind the development and alteration of a specific model; to examine how and what kind of inferences an experimental simulation with such models enables; and to note interactions among researchers relevant to the problem-solving process (Nersessian 2022). Here, I draw on that material to examine the analogical nature of models. In particular, I focus on an important aspect of analogical reasoning that has been overlooked in both the philosophy of science and cognitive science but is widespread in frontier science: building the analogical source (Nersessian 2008). In these laboratories, most of the reasoning we have observed is focused on the model, especially its capabilities and limitations, as well as on how to make it a better analogical source, which requires researchers to think not only about the biological target but also about the resources available for building, including the constraints of the materials and methods. In both laboratories, cellular systems are seen as providing design possibilities that feed into various design options. Research in both laboratories revolves around engineering living cell cultures into simulation models – wet "devices" with experimental potential that is constrained both by the cellular systems and by the engineered artifacts with which they interlock. Since it builds the easiest bioengineered model for the non-specialist to comprehend, I focus on the tissue engineering laboratory.

2. Building in vitro models in a tissue engineering laboratory

The tissue engineering laboratory (lab A) had been in existence for nearly 20 years when we entered. The director and the graduate student researchers had backgrounds in mechanical engineering, and the students were working toward degrees in bioengineering or in the biomedical engineering educational program under development. We conducted interviews, field observations, and collected archival data intensely for two years, and then continued to follow the graduate student researcher projects for another five years. The director was, by then, a widely recognized pioneer in tissue engineering and BMES. His research program started with a problem he had encountered as a mechanical engineer conducting aeronautics research. NASA tapped him to help them understand the effects of the forces of launch and re-entry ("pogo stick vibration") on the cardiovascular systems of the astronauts. He reported not knowing "*anything about biology and medicine*," but that he felt an obligation to try to help them, and the problem was interesting. He discovered that no one had examined the effects of even the natural physical forces of blood flow through the cardiovascular system. He came to suspect that the mechanical forces, in the first instance, shear, would most likely impact the endothelium – the innermost layer of cells in a blood vessel. In our initial interview with him, he formulated the insight he had then that would transform his research into a biomedical engineering program as follows:

> characteristics of blood flow [mechanical stress/strain forces] actually were influencing the biology of the wall of a blood vessel. And even more than that... the way the blood vessel is designed is... it has an inner lining, the endothelium. It's a monolayer – it's the cell layer in direct contact with flowing blood. So, it made sense to me that, if there was this influence of flow on the underlying biology of the vessel wall, that somehow that cell type had to be involved.

Lab A director's research, thus, started with an engineering framing of a biological problem and a goal to understand complex biological processes of the cardiovascular system in terms of mechanical engineering concepts and methods. The hypothesis that mechanical forces were "*influencing the biology*" was radical at a time when the nascent field of vascular biology was focused on biochemical processes, and biologists initially rejected it. His statement also reveals the design perspective of an engineer on biology, which pervaded his investigative program. This engineering framing provided a means to manage the complex biological problem of the nature and effects of the dynamical processes within blood vessels by reducing it to understanding the effects of the flow (mechanical forces) of blood on a specific cell type. The director proposed a novel hybrid "placeholder" concept (Carey 2009), "*arterial shear*," that is, the frictional force of blood on the endothelium as it flows in the parallel plane through the lumen (the inner space of the arterial tube), and the aim of articulating various dimensions of this concept drove the research for over 40 years. His research began with using cows as animal models to investigate the effects of stenosis that researchers induced surgically in their arteries. However, that mode of research did not allow sufficient controls and also required the animals to be sacrificed. He decided to see if it would be possible to "*take the research in vitro*," by which he meant launching a program to study the impact of shear stress flow on cultures of endothelial cells in bioengineered models. Such models would isolate and control the relevant features of the target cells, and blood flow, while (hopefully) producing a relevant and useful understanding of the processes and effects of their interactions. Building in vitro simulation models would open the possibility of controlled experimental studies, amenable to qualitative and quantitative analysis, of the impact of both normal and pathological flow processes on cardiovascular cells and tissues. In this section, I provide an overview of the development of the main in vitro model-systems lab A researchers developed to instantiate features they deemed to be relevant – and feasible – of such processes, including some of the reasons and justifications researchers advanced for specific design decisions.

2.1 The flow-loop—cells-on-slides model-system

At the outset, researchers need to determine what abstractions might be feasible from a biological perspective for designing an in vitro model while yielding relevant and significant information about the dynamical processes of interest. As one laboratory member expressed, the design process:

> as engineers we try to emulate that environment [in vivo], but we also try to eliminate as many extraneous variables as possible, so we can focus on the effect of one or perhaps two, so that our conclusions can be drawn from the change of only one variable.

In one major abstraction, the director decided, in line with his initial insight, to isolate and study only the endothelial cells and not include other components of the blood vessel. The researchers reasoned that this abstraction is warranted because these cells line the inner blood vessels, and so are in direct contact with the blood flow forces, and bear the brunt of the frictional force. Further, as one researcher justified the choice, "*cell culture is not a physiological model; however, it is a model where biologic responses can be observed under carefully designed and well-defined laboratory conditions.*" This fact enabled them to derive reliable quantitative measures. Another important abstraction was to begin with studying laminar flow, which is steady and uniform in contrast to in vivo blood flow, which is turbulent and pulsatile along much of its pathway. The in vitro model system is, thus, greatly simplified, but to investigate just the response of endothelial cells to laminar flow would at the very least provide baseline information on the biological responses of cells to fluid forces.

An in vitro model of the target system requires, at a minimum, that it can replicate the shear forces of blood on the cells. The channel flow device ("*flow loop*") is a functional model of that process, which enables controlled experimentation directly on endothelial cell cultures, thus creating what the researchers call a model system. A specific model system can be the locus of an experiment or just one step in a multi-model experimental process. The flow loop in use at the time of our investigation was the result of several iterations of the design. The important modeling parts of the flow loop comprise a peristaltic pump, a liquid, and a channel in which the liquid flows over cells. The speed at which the pump operates reflects a range of potential blood flow in vivo, and the pulse dampener allows control over the constancy of the flow; for instance, it can turn pulsating flow into laminar flow. Both normal and abnormal flows can, in principle, be studied. The channel through which an incompressible fluid flows over the endothelial cell cultures on slides is engineered to exact geometrical specifications in a physiologically meaningful range. The liquid medium has the viscosity of blood, a cell-friendly Ph, and other in vivo features. The flow loop in use was the product of iterations of design and redesign dating back 20 years.

The initial flow loop was designed in 1981 with the capacity only to produce laminar (steady, uniform) flow. However, the mechanical features of blood flow in vivo vary with the distance from the heart, as well as with respect to the topological features of the arteries, especially constrictions. The flow loop was redesigned in 1989 to allow "*studies in which fluid mechanic conditions can be systematically varied,*" which include pulsatile and oscillatory flows, in order "*to determine the extent of any such flow effects*" that can occur in vivo. It was a large bench-top device, and contamination was a constant problem because the viability of cell cultures requires that they are maintained at appropriate CO_2 levels and in a specific temperature range, which was impossible with this model. Over 50% of their experiments failed because of contamination.

In 1995, new technology made possible a significant redesign of the flow loop to address the contamination issue. In an interview, a recent graduate of the laboratory chronicled the process (see, Kurz-Milcke, Nersessian, and Newstetter 2004) of "*model-revising this design to go into the incubator*," which made long-term experiments possible. This was important because it takes 24 hours for the effects of flow on the cells to be seen, and contamination increases with time. "*Model-revising*" entailed a redesign of the model to replace the heating function of the coils with the incubator and to use a pump rather than a pressure difference to derive flow. The revision also made the components sufficiently decomposable to allow for independent redesign if needed as the research program advanced. In fact, minor modifications continued to take place throughout our investigation. The redesigned flow loop could be assembled under a sterile hood, operated in an incubator, and had an integrated peristaltic pump. The geometry of the flow channel, where cells-in-culture interface with mechanical parts, was left unchanged. This redesign of the flow loop device was central to its function in the model system because its viability as an in vitro model is totally dependent on the ability of the endothelial cell cultures to resist contamination. To determine the response of the cells to the applied forces, the researchers remove them from the chamber and examine them with various instruments, including the Coulter counter and confocal microscope, which provide information about proliferation, alignment, alive/dead status, morphology, migration, and so forth in the form of numerical and visual (graphical, diagrammatic, color-coded) representations. This information can be directly related to the controlled shear stresses and quantified.

The flow loop, then, is a dynamical model that, when in operation, has the possibility to simulate normal and pathological forces of blood flow, laminar and pulsatile, through the lumen of an artery. In most experiments, however, the process instantiates the shear forces of a steady (constant speed), laminar (straight streamlines) flow over a flat surface (cells on slides). The flow is two-dimensional and unidirectional. The researchers listed all of these features as contributing to their assessment that the model system "*emulates*" in vivo shear to a "*first-order approximation... as blood flows over [sic] the lumen.*"[1] They argued that instantiating this process with only the characteristics of first-order flow is justified because it provides a "*way to impose a very well-defined shear stress across a very large population of cells such that their aggregate response will be due to*" it and enables them to "*base... conclusions on the general response of the entire population.*" Experiments flowing cell cultures on slides continued to be conducted throughout the period of our investigation, but investigations with a more complex vascular wall model, the construct device, were, by then, the focal point of laboratory research.

2.2 *The flow-loop—construct model-system*

Improving the devices and model systems so as to instantiate additional relevant features is an ongoing part of the research. Although an extended discussion is not possible here, this process can be understood as one of "de-idealization," especially involving processes of reformulation and concretization as discussed by Tarja Knuuttila and Mary Morgan (2019). As they argue for immaterial models, processes of de-idealizing a material model are not a straightforward reversal of any prior idealization process. Lab A's research began with building a simplified model that focused on one causal interaction that "made sense" to the director as the most important for his purposes. Reformulating a model to instantiate (make concrete) additional features, in the case of in vitro models, depends on an assessment both of what other features might be causally relevant and of what it is feasible to do with the biological materials, the engineering materials, and the technologies at hand, which change over time.

Simulations with the endothelial cells in isolation from other components of arterial tissue enable a basic, provisional understanding of cell response to shear, but the researchers were fully aware that "*cell culture is not a physiological model*" of the blood vessel wall. It leaves out many features of the blood vessel and, thus, produces a limited understanding of their target problem of the effects of mechanical forces on the blood vessel wall, which has other components. In a first attempt to add relevant features, they created a "*co-culture*" of endothelial and smooth muscle cells, but the limitations remained much the same since it does not capture their structural relations in the tissue of a blood vessel. Specifically, as the director noted,

> putting cells in plastic and exposing them to flow is not a very good simulation of what is actually happening in the body. Endothelial cells... have a natural neighbor, smooth muscle cells. If you look within the vessel wall you have smooth muscle cells and then inside the lining is [sic] the endothelial cells, but these cell types communicate with one another. So, we had an idea: let's try to tissue-engineer a better model-system for using cell cultures.

Their aim became "*to use this concept of tissue engineering to develop better models to study cells in culture*;" that is, to work toward building "*a more physiological model*" – one that would instantiate more features and have the functionality, eventually, of an in vivo vessel along mechanical, physical, and biochemical dimensions. With this more complex model, they could study the effects of shear on more components of the blood vessel wall, as well as the interactions of different cell types. But the " *big gamble*" the laboratory took to try to build a model that instantiated more of the relevant features of a blood vessel wall was only possible because new tissue-engineering techniques and materials had been developed. If successful, building the construct model could also open up a novel application potential: to turn the model into a vascular graft to repair diseased arteries in vivo. Within the laboratory, this tissue-engineered model was referred to, variously, as "*the construct*" device, the "*tissue-engineered blood vessel wall model*," and, underscoring its application potential, the "*tissue-engineered vascular graft*."

An in vivo blood vessel is tubular in shape and comprises several layers: the lumen where the blood flows; a first, mono-layer of endothelial cells that sit on collagen; an internal elastic lamina; a second layer of smooth muscle cells, collagen, and elastin; an external elastic lamina; and an additional layer of loosely connected fibroblasts. The construct is grown on a specially designed structure that comprises tiny silicon tubes, which allow cells to attach and grow on them and then be slipped off (Figures 34.1a and b). To function as a model of the target arterial system, the materials used to grow them must coalesce in ways that mimic the properties of native tissues, and the cells that are embedded in the scaffolding material must replicate the capabilities and behaviors of native cells so that their higher-level tissue functions can be achieved. Depending on the goals of an experiment, the in vitro model can be constructed to instantiate some or all of the in vivo features. It is possible, for example, to use only collagen and not add elastin, or to seed it with either endothelial cells or smooth

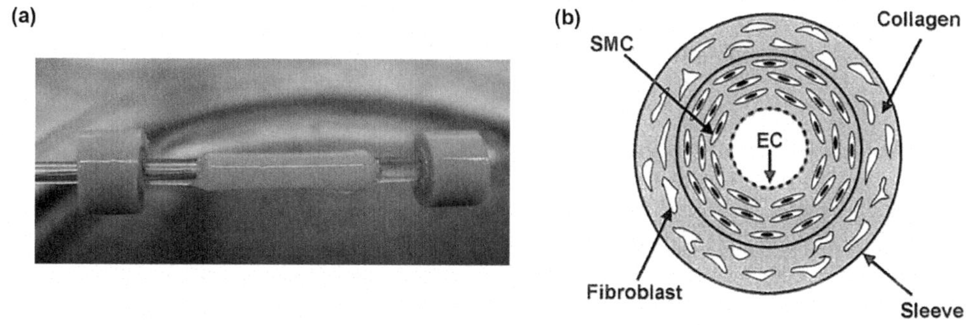

Figure 34.1 (a) A construct seeded onto a mandrel. (b) Cross-section of a construct. (A Teflon sleeve is used to strengthen it for a specific experiment).

muscle cells. Thus, the construct forms a *family of models* that can be designed for different experimental purposes.

The flow loop – construct model-system provides insight into how multiple in vitro models can interlock in experimental simulations. Because of the geometry of the flow chamber, the researchers would have needed to undertake a major redesign of it to accommodate the tubular shape of the construct. Instead, they decided to cut open the construct so it would lie flat in the existing chamber. They justified using the flat constructs by arguing that since the cells are so small with respect to the construct, the shear forces they experience would be the same as if they were in a curved vessel. As one researcher explained their reasoning, from the "*cell's perspective,*" a cut-open construct is not an approximation because

> the cell sees basically a flat surface. You know, the curvature is maybe one over a centimeter, whereas the cell is like a micrometer.... It's like ten-thousandth the size, so to the cell – it has no idea that there's actually a curve to it.

That is, flowing the fluid over a flat construct instantiates the force the cell experiences in vivo. Because the cell is so small with respect to the arterial wall, the cell's in vivo experience is as though it lives in a flat world.

As with all in vitro models, the iterative and incremental construct design is based on what is understood at the time of the biological environment of endothelial cells and vascular biology, on the kinds of materials available, and on the tissue engineering techniques developed in the laboratory and the field thus far. The laboratory's ongoing research sought to advance all these aspects through numerous iterations. Thus, with the move to tissue engineering, the laboratory's major research question became:

> The big, big question is how do our constructs act like a modeling tool, how do they respond to – or biological markers respond to – mechanical stimulation. So is there a certain correlation to the stress and strain and the distribution being applied to these constructs to certain biological markers... Does it respond in the same manner? That's the big, big question.

To "*respond in the same manner*" means, among other things, that it expresses the in vivo proteins and genetic markers, and possesses the in vivo mechanical properties.

For any device to perform as a "*modeling tool*," researchers must understand both how it represents in vivo phenomena (device qua model) and how it is an object in its own right (device qua device), an environment for biological experimentation with constraints and affordances due to the nature of the design, the materials, and the engineering challenges. Also, with respect to the former, although these modeling tools are highly specific in the details of their construction, they are understood to represent generic biological systems – systems of that kind (e.g., cardiovascular systems) – rather than specific systems. All these factors need to be taken into account when researchers plan experiments, evaluate outcomes, and make inferences about what to transfer as hypotheses to the in vivo target system.

2.3 *In vitro models as built analogical sources*

In vitro models are the primary means through which BME researchers gain epistemic access to complex biological phenomena. As we have seen in the brief overview of two model systems, researchers build epistemic warrants for a model through the principled decisions and rationalizations they make in the processes of building it. Researchers design and perform in vitro simulation experiments with devices in processes they claim to "*parallel*" or "*mimic*" salient aspects of in vivo situations. The warrant for using these kinds of models as epistemic tools is connected to how the models function as dynamic representations, that is, how they are built to instantiate and simulate in vivo features. What I consider now is that, to fathom how the practice can achieve its epistemic aims through model-based reasoning, we need to understand the epistemic affordances of the models as built analogies.

2.3.1 *Building the analogical source*

The BME epistemic practice of building devices and model systems is, fundamentally, an analogical practice. The researchers aim to design models to provide analogical sources that can enable them to gain an understanding and control of complex biological systems. This analogical practice is quite unlike any considered in customary philosophical and cognitive science literature. Usually, analogy is cast as a process of making sense of what we do not understand (target) in terms of what we do (source). Here, little is understood about either the source (model) or target (real-world phenomena) at the outset. Customarily, in analogical problem solving, the reasoner retrieves a previously solved problem that provides a source analogy, determines a mapping between source and target, transfers features from source to target, and evaluates inferences with respect to the target domain. Mary Hesse (1963), whose account has been most influential in both philosophy and cognitive science, called the features that form the mapping "positive," if they match the target, "negative," if they do not match, and "neutral," if their status is unknown.[2] On her account, the neutral features provide a resource for further development. Recently, Tarja Knuuttila and Andrea Loettgers (2014) have shown that for retrieved analogical sources used in synthetic biology, negative features can also lead to further development. With built analogies in BMES, as illustrated above, researchers are well aware at the outset of negative features not instantiated in a model, and these negative analogies, indeed, provide opportunities for development.

Although models have pride of place in the contemporary philosophy of science, scant attention has been directed toward the analogical dimension of models. I venture that this stems from the fact that the literature focuses on models derived, at least partially, from theories, which draws attention to the traditional "realism" issues associated with thinking about theories. However, starting from the other direction, that of building models "from the ground up" in the absence of a theory of the phenomena under investigation, underscores how models and analogies are tightly bound (see, e.g., Nersessian 1992; 2008; 2022). This analogical relationship has importance, thus far not addressed, for models derived from theories since analogical inference provides a means to transfer prediction, explanation, and understanding from model to world. Additionally, as Hesse (1963) pointed out, the source model is always a "false" representation in that it cannot accurately or adequately represent all the features of the target. With her, I contend that "true" and "false" are not the appropriate categories for thinking about models. However, in the case of in vitro models, neither is Hesse's nor the customary notion of similarity. These models must instantiate features with the same biological properties and functions as those features selected from the real-world system in order to perform properly as a living system, and so the notion of similarity is also not an appropriate category for the representational relation between in vitro simulation models and the in vivo target. In general, BMES researchers strive to design in vitro models that both refer to and instantiate features of the in vivo biological system germane to their epistemic goals. As will be discussed in the next section, the notion of exemplification, advanced by Nelson Goodman (1968) and extended to scientific practices by Catherine Elgin (2018), can best capture this representational relation between source and target.

Although what we customarily understand as analogy occurs in science, for frontier research problems, there is often no pre-existing analogical source. Rather, the source itself needs to be created in interaction with the goals and constraints of the target problem – a bootstrapping process that furthers the articulation of the problem as well as its solution. There are several sources of data on scientific problem-solving, including historical, think-aloud protocol, and ethnographic (see, e.g., Nersessian 1984, 2008, 2022), that provide evidence of this important *representation-building* aspect of analogy. My analyses of data from all these sources provide a list of features that are relevant to understanding in vitro models as built analogies:

- building processes are goal-directed
- building processes are iterative and incremental
- interaction between source and target is ongoing in the building process
- elements used in building analogies can derive from more than one domain ("hybrid analogies")
- various abstractive processes are used in selecting features and merging target, source, and model constraints
- mappings are established during the building processes, so in most cases, mappings develop/evolve over time
- models are built toward instantiating features germane to the epistemic goals
- models are evaluated based on whether they exemplify features germane to the problem
- features not exemplified can provide a resource for further development
- analogical transfer requires that a model instantiate relevant features, and leave out nothing essential to that inference

It is important to note that although the word "abstraction" is commonly used for a separate process alongside "idealization" and other abstractive notions, this is confusing. It is better to reserve "abstraction" for a comprehensive notion comprising various processes, including idealization, approximation, simplification, omission, limiting case, and generic modeling. All of these abstractive processes can play a role in model building.

As we saw, BME researchers aim to build physical simulation models to the degree of specificity they believe is sufficient to examine an aspect of in vivo phenomena in a cognitively tractable manner. This goal is informed by an assessment both of the current state of understanding of the phenomena and of how the available materials and technologies constrain and enable design possibilities. Given the frontier nature of the research, all of these factors change over time; thus, the building process is incremental, as a satisfactory representation is developed. Additionally, models are hybrid bioengineered constructions, and there is tension between the constraints on the design and functionality of a device that derive from biology and those that derive from engineering. Some selections are made in order to merge these constraints and need to be considered in assessing the warrant for transferring any inferences.

During our investigation in both laboratories, the researchers' concerns about a model's relation to the real-world system informed decisions about design and redesign, as well as their evaluations of experimental simulation outcomes. Importantly, in vitro models are dynamic systems, and a model needs to instantiate those features that enable the cells and tissues to behave in an experimental simulation as they would in the in vivo phenomena under those conditions. A major epistemic task, therefore, is to determine what those features might be and whether or not any abstractions made can impact behavior, and how they might do so. For instance, a flow-loop simulation instantiates first-order (laminar) blood flow. This is a counterfactual situation because there are always higher-order effects in vivo, but for their initial epistemic goal of understanding in what ways forces can affect the morphology and proliferation of endothelial cells, the researchers argued that there is no need to capture the full complexity of the in vivo blood flow at the outset. The reasons researchers gave for this choice included such considerations as there are places of laminar flow in the circulatory system as the flow gets further away from the heart, laminar flow enables them to impose a well-defined shear on a population of cells, and if indeed the cells functioned differently in significant ways in vivo (e.g., gene expression), the device design affords (or can be redesigned to) the possibility to simulate higher-order effects. These reasons (in order) are of the following sort: the model instantiates a germane feature of a part of the in vivo system of interest, the model achieves an important engineering goal that reduces the complexity of the analysis, and the model can be made to instantiate other features of the system if in vitro biological function is importantly different. They did use the flow loop's capacity to simulate higher-order effects in later research when it became technologically possible to examine gene expression, which made it worthwhile to investigate these effects.

Importantly, redesign is ongoing with in vitro model building. Some redesigns have to do with improving the engineering, and others are made for practical purposes, such as enhancing the viability of cells. The most important redesigns, however, are to improve the nature of the parallelism to the biological phenomena of interest, if only in minor ways, as they are made to provide better or different exemplifications. Redesign can be driven by a change in the understanding of the phenomena or of the problem or by a change in technological and material capabilities as the research progresses. At any point in time, in vitro models are in different stages of development. Thus, exemplification, here, is a historical process. During the period of our investigation, the flow loop was quite stable, but the construct was still undergoing design changes aimed mostly at improving its mechanical strength. Once the kinks have been worked out of a design and the researchers assess that it has met their epistemic goals, change is largely incremental. In vitro systems are meant to be sites of long-term investment so as to enable systematic experimentation.

Negative analogies are a major driver of model development and redesign. For instance, in the design of the flow loop, researchers were aware of a negative analogy from the outset: flow loop simulations are "*something very abstract because there are many in vivo environments and many in vivo conditions within that environment.*" Things change constantly in human bodies over the day and over their lifetimes, including physiological flow rates. These changes had been a significant problem in the director's earlier animal studies and motivated his move to in vitro. The first flow loop could produce only laminar flow, but when redesigned, it had the capability to produce a range of flow rates. Flow-loop simulations could instantiate higher-order effects if there were reasons to do so, such as "*if there's a whole different pattern of genes that are upregulated in pulsatile shear.*" In this instance, however, for many years, there was no way to investigate possible salient differences in gene regulation. That potential came quite late in the research program when gene array technology was developed. The prior basis for a partial comparison of their results was provided by studies of morphology and proliferation in vascular biology and whatever biological markers were available from biochemical studies. The possibilities for comparing a model with biological research are always fluid and incomplete.

Two other negative analogies were important to furthering the laboratory's research program. First, the flow loop model exemplifies only one of the in vivo mechanical forces: shear stress. This is the force with the greatest impact on the endothelial cells. Blood vessels are also subject to strain forces from the blood pressing on the vessel wall, but to instantiate this force requires a model system that instantiates the topology of the vessel. A second negative analogy concerns the use of slides with endothelial cells in culture in flow loop simulations. The researchers recognized that this model system does not provide "*a physiological model.*" This simulation does not instantiate some of what they knew to be relevant mechanical and biochemical features of blood flow through the lumen of an artery and thus limits the understanding obtained. For one thing, endothelial cells have a "*natural neighbor*," smooth muscle cells. It was not until the technologies to engineer complex tissues started to develop in the 1990s that it became feasible for the laboratory to attempt to build a blood vessel wall model that could also instantiate smooth muscle cells and other in vivo components, i.e., build the construct family of models. These models enabled simulations and assessments of the shear forces of blood flow that more closely mimic the in vivo system. They also led to researchers building other model systems to investigate the forces of pressure and strain, the other negative analogy (see Nersessian 2022, chap. 4).

2.3.2 *Analogy and exemplification*

In the words of our researchers, devices are designed to "*parallel*" or "*mimic*" selected features of the in vivo phenomena. Their expressions can be interpreted to mean that in vitro physical simulation models are built to provide structural, behavioral, or functional analog representations of selected dimensions of complex in vivo biological systems. They provide a way to get a grip on the behavior of a biological system by creating a parallel or virtual world through which to conceptualize, control, and experimentally probe aspects of that complex dynamic system. Such models can only function as epistemic tools if they have been designed with an appropriate representation of what is understood about the biological facts.

Importantly, unlike computational virtual worlds, in vitro models are composed in part of biological materials, so the cells and tissues have biological functionality that needs to be maintained as they interface with engineered materials and perform under greatly simplified conditions, all of which figure into how they function epistemically. And, to add a level of complexity, most model systems are *nested* analogies, that is, analogies within an analogy (Nersessian and Chandrasekharan 2009). For example, the flow loop provides an analogy to hemodynamics, the construct provides an analogy to the blood vessel wall, and the model system they constitute provides an analogy to blood flow in an artery. So, the considerations in play need to be not only about each model but also about how the model system fits together.

What enables the researchers to have some assurance they are on a productive path with a device or model-system design? Despite their complexity, in vitro models are missing much of the in vivo target system. What we found in our data is that researchers were continually asking the question that can be phrased generically as: "Is the model of *the same kind as* the in vitro system along the relevant dimensions?" That is, are the features instantiated such that the researcher is warranted to infer that the behaviors of the model belong, along specified dimensions, to the same class of phenomena as those of the in vivo biological system? Answering that question requires an assessment of the relevance of both the features that are instantiated in the model to its behavior and those that have been left out. The best way to interpret that question is by asking whether the built analogy *exemplifies* the features relevant to the research.

In the sense advanced by Goodman (1968) and Elgin (2018), "X exemplifies Y" means "X instantiates relevant features of Y and refers to Y by means of that instantiation." The notion of exemplification captures the representational relation the researchers aim for as they build models to "parallel" or "mimic" in vivo phenomena. The flow loop, in performing, not only refers to shear stress forces in a process of blood flow through the endothelial cells in a blood vessel, but it also produces those shear stress forces. The liquid has what the researchers judge to be relevant fluid-dynamic features of blood as it flows over the endothelial cells, cultures, or the construct device that has been designed to have relevant features of the blood vessel wall. The in vitro models, then, are successful exemplifications if, indeed, they possess the features of the in vivo phenomena germane to the problem at hand, and much of the research is directed toward determining if this is the case. Such determination requires the researcher to consider both the relevance of what is and what is not instantiated to the behavior of the system. What is not instantiated provides a potential resource for further development (negative analogy). Building in vivo models toward exemplifying features is an incremental, and thus, historical process in which models are *built towards* exemplifying the features determined to be relevant to the functioning of a target biological system – features that can change as research progresses.

Models that are satisfactory exemplifications provide the researchers with a warrant for the analogical transfer of hypotheses based on experimental findings, but with the proviso that what inferences are justified depends on the historical state of the model. So, *analogy and exemplification work together in model-based reasoning.*

3. Conclusion

In vitro simulation modeling is a significant epistemic practice in BMES. It has become even more widespread with the advent of the "next generation" tissue-engineered "organ on a chip emulation model." These are in vitro simulation models the size of a memory stick, which instantiate the requisite structure and functionality of in vivo organs to be used in experiments aimed at understanding disease mechanisms or evaluating the therapeutic effects of drugs (Ingber 2022). Although the ethnographic investigations we conducted into the practice of in vitro simulation modeling ended several years ago, the data are still relevant to the fundamental epistemic issue: what justifies researchers transferring inferences from in vitro simulation models to in vivo systems? As I have argued, these kinds of models are built to instantiate epistemically relevant features of the target system in order to serve as source analogies. Exemplification, then, provides the criteria for assessing the affordances and limitations of an in vitro model – at a particular stage in its development – as an analogical source through which to investigate the target in vivo phenomena. These assessments enable researchers to determine for which inferences about the behaviors of the in vitro model there is epistemic warrant to transfer as hypotheses to the in vivo system.

Notes

1 That the researchers all use "*over*" instead of "*through*" the lumen (which is tubular in in vivo) is an interesting slip. I suspect they made the mistake because they were thinking about the phenomenon in terms of the in vitro simulation, in which, as we will see, the tubular constructs are cut open and laid flat in the flow chamber.

2 Hesse called these features "properties," but "features" is a better expression to use since it captures the notion that properties, relations, and relational structures can be mapped.

References

Bailer-Jones, Daniella M. 2009. *Scientific Models in Philosophy of Science*. Pittsburgh: University of Pittsburgh.

Black, Max. 1962. *Models and Metaphors*. Ithaca, NY: Cornell University Press.

Carey, Susan. 2009. *The Origin of Concepts*. Oxford: Oxford University Press.

Chandrasekharan, Sanjay, and Nancy J. Nersessian. 2015. "Building cognition: The construction of external representations for discovery." *Cognitive Science* 39: 1727–1763.

Elgin, Catherine Z. 2018. *True Enough*. Cambridge: MIT Press.

Goodman, Nelson. 1968. *Languages of Art*. Indianapolis: Hackett.

Harré, Rom. 1970. *The Principles of Scientific Thinking*. London: Macmillan.

Hesse, Mary. 1963. *Models and Analogies in Science*. London: Sheed and Ward.

Ingber, Donald E. 2022. "Human organs-on-chips for disease modelling, drug development and personalized medicine." *Nature Reviews Genetics* 23: 467–491.

Knuuttila, Tarja. 2011. "Modelling and representing: An artefactual approach to model-based representation." *Studies in History and Philosophy of Science Part A* 42(2): 262–271.

Knuuttila, Tarja, and Andrea Loettgers. 2014. "Varieties of noise: Analogical reasoning in synthetic biology." *Studies in History and Philosophy of Science Part A* 48: 76–88.

Knuuttila, Tarja, and Mary S. Morgan. 2019. "Deidealization: No easy reversals." *Philosophy of Science* 86(4): 641–661.

Kurz-Milcke, Elke, Nancy J. Nersessian, and Wendy Newstetter. 2004. "What has history to do with cognition? Interactive methods for studying research laboratories." *Journal of Cognition and Culture* 4: 663–700.

Nersessian, Nancy J. 1984. *Faraday to Einstein: Constructing Meaning in Scientific Theories*. Dordrecht: Martinus Nijhoff/Kluwer.

———. 2008. *Creating Scientific Concepts*. Cambridge: MIT Press.

———. 2022. *Interdisciplinarity in the Making: Models and Methods in Frontier Science*. Cambridge: MIT Press.

Nersessian, N. J. (1992). How do scientists think? Capturing the dynamics of conceptual change in science. In R. Giere (Ed.), *Minnesota Studies in the Philosophy of Science* (pp. 3–45). Minneapolis: University of Minnesota Press.

Nersessian, Nancy J., and Sanjay Chandrasekharan. 2009. "Hybrid analogies in conceptual innovation in science." *Cognitive Systems Research* 10: 178–188.

35
SYNTHETIC MODELS IN BIOLOGY

Tarja Knuuttila and Andrea Loettgers

1. Introduction

The first two synthetic models, the genetic toggle switch (Gardner, Cantor, and Collins 2000) and the repressilator (Elowitz and Leibler 2000), were published in the same issue of *Nature* independently from one another. Many practitioners within synthetic biology consider these models constitutive of the field of synthetic biology, although various kinds of synthetic constructs have a long lineage within the biological sciences. The research on the genetic toggle switch and the repressilator can be traced back to the work of Jacob and Monod on the *lac* operon in *E. coli* that was instrumental to the idea of assembling new regulatory systems from molecular components (Cameron, Bashor, and Collins 2014; Jacob and Monod 1961). The implementation of these ideas had to wait for a host of technological developments and discoveries such as molecular cloning and polymerase chain reaction (PCR), automated DNA sequencing, and green fluorescent proteins. Armed with molecular toolkits, engineers, physicists, and computer scientists migrated at the end of the 1990s to molecular biology, combining the construction of synthetic systems in the wet lab with mathematical and computational modeling.

Synthetic models in biology are artifacts constructed from biological components, such as genes and proteins, to form biological parts or wholes (Bensaude Vincent 2013; Bursten 2019). They have the same materiality as natural biological systems but are not the result of natural evolutionary processes. The early synthetic models were genetic circuits, whose design was inspired by circuit engineering, with the genetic toggle switch obviously referring to toggle switches, while the repressilator mimics a ring oscillator, which is an electronic circuit composed of an odd number of NOT gates in a ring, whose output oscillates between two voltage levels. While most of the hype and hope around synthetic biology at the beginning of the 2000s focused on possible new applications—novel materials, synthesized medicines and therapies, biofuels, and environmental solutions (Singh 2022)—the genetic toggle switch and the repressilator represented a basic-science approach to synthetic biology.

In this entry, we focus on synthetic genetic circuits using the repressilator as an example. Apart from being experimental systems, they can also be considered models in contrast

DOI: 10.4324/9781003205647-41

to many other synthetic biology constructs. We discuss how the construction of synthetic models has enabled researchers to probe gene regulation and, more generally, biological organization. The construction of synthetic genetic circuits assumes that biological organization is governed by some general design principles (Section 2), based on various kinds of feedback systems also utilized in engineering. Another foundation for synthetic biology and the construction of synthetic genetic circuits is the enterprise of achieving standardized biological parts, also called BioBricks,[1] from which the synthetic circuits can be assembled (Section 3). In Section 4, we discuss the repressilator model more in-depth, as well as the research program on the functional meaning of noise it gave rise to, while Section 5 focuses on the question of what distinguishes synthetic models from other kinds of synthetic constructs.

2. Probing biological design principles

Organization in biology has traditionally been addressed by mathematical modeling—employing concepts from engineering and physics—and by performing experiments on model organisms. Using engineering principles, such as feedback loops to model biological systems, goes back to cybernetics, which claims that the same mechanisms can describe engineered and biological systems (Green and Wolkenhauer 2013; Wiener 1961; Bertalanffy 1969). The idea of such general organizational principles is shared by most synthetic and systems biologists. As systems biologist Uri Alon states:

> [...] studies led to the discovery that one can, in fact, formulate general laws that apply to biological networks. Because it has evolved to perform functions, biological circuitry is far from random or haphazard. It has a defined style, the style of systems that must function. Although evolution works by random tinkering, it converges again and again onto defined set of circuit elements that obey general design principles.
>
> *(Alon 2007, XV)*

When talking about general laws, Alon is not drawing an analogy to laws in physics. Rather, laws in biology are more like general design principles by which network structures and dynamics become related to functions. Such design principles define "generic features of a class of systems that operates under a similar set of constraints" (Green 2015, 631). Consequently, they are assumed to be independent from the specific biological context, which would make them multiply realizable (Koskinen 2019). However, instead of taking such universal constraints for granted, general design principles are a subject of investigation in synthetic biology. Thus, one can ask why scientists introduced engineering principles to biology in the first place.

Interestingly, many pioneers of synthetic biology have a background in physics, although they were applying engineering approaches to biology. Concepts from physics, which were successfully used, for instance, in modeling neural networks (Hopfield 1982), did not seem suitable for modeling biological functions. Such a realization was one of the key arguments of Hartwell et al. in a programmatic article (1999) that was published shortly before the introduction of the first synthetic models (Gardner, Cantor, and Collins 2000; Elowitz and Leibler 2000). Hartwell et al. (1999) claim that biology is different from physics due to the phenomena of survival and reproduction, related to the importance of the notion of function. Instead of physics, they draw analogies to synthetic sciences such as engineering and computational sciences: "Just as electrical engineers design circuits to perform specific functions, modules have been evolved to perform biological functions (C49)." Hartwell et al. assume that selection and evolutionary constraints shape biological design principles and that simplifying, higher-level models are needed for understanding the functioning of different biological modules. Toward the end of their article, they mention synthetic biology noticing that "[s]eeing how well the behaviour of such modules matches our expectations is a critical test of how well we understand biological design principles" (C52).

The statements of Gardner et al. and Elowitz and Leibler in their pioneering works are in line with Hartwell et al. (1999). Gardner et al. (2000) write that the construction of their synthetic model provides proof of concept of the proposal "[...] that gene-regulatory circuits with virtually any desired property can be constructed from networks of simple regulatory elements" (339). Such properties include metastability and oscillatory behavior observed in organisms such as bacteriophage lambda and Cyanobacteria, with the genetic toggle switch instantiating the flipping between two bistable states. Consequently, such a network design could also be a general design principle in biology. Design principles such as the genetic toggle switch could then become part of the repertoire in constructing genetic networks to be used in "biotechnology, biocomputing, and gene therapy" (339).

Elowitz and Leibler (2000) explicitly declare their aim for probing biological design principles: "Networks of interacting biomolecules carry out many essential functions in living cells, but the 'design principles' underlying the functioning of such intracellular networks remain poorly understood, despite intensive efforts including quantitative analysis of relatively simple systems" (335). They draw inspiration from the theoretical biologists, René Thomas and Richard d'Ari, who develop in their book, *Biological Feedback*, a formal methodology for analyzing dynamic systems (see also Thomas 1998). The book studies different types of general regulation mechanisms, describing their architecture, interaction, and dynamics (Thomas and D'Ari 1990).

Even though the probing of "design principles" is the main motivation for the construction of synthetic models for Gardner et al. (2000) and Elowitz and Leibler (2000), the engineering aim of designing novel behaviors is part and parcel of their agenda as well. Elowitz and Leibler write: "Such 'rational network design' may lead both to the engineering of new cellular behaviours and an improved understanding of naturally occurring networks" (2000, 335). Indeed, synthetic biology provides a good example of science in which basic science and engineering aims are difficult to tell apart, given that synthetic biology studies theoretical ideas by constructing synthetic systems that are supposed to realize them. Cookson et al. characterize this kind of theoretical work as "basic science by engineering" (Cookson, Tsimring, and Hasty 2009). Knuuttila and Loettgers (2013a) discuss how such a basic-science-oriented engineering program has paradoxically made it clearer that synthetic biology should become more biology-inspired. While the program is premised on drawing analogies between biological and engineered systems, not all analogies are positive, but turn out negative (Knuuttila and Loettgers 2014). The research on the functional meaning of noise is a case in point; in biology, noise acquires a functional meaning that it does not have in engineering (see Section 4.2).

3. Standardized biological parts and the BioBrick initiative

Apart from general design principles, the construction of synthetic genetic circuits relies on standardized biological parts. In the 1990s, the Human Genome project started to accumulate vast part lists of different organisms, cataloged in several repositories (e.g., iGEM Registry of Biological Parts,[2] [Madsen et al. 2016]). Such repositories enable scientists to select parts from a catalog and construct from them circuits or pathways in a chassis microbe such as *E. coli* (Kendig and Bartley 2019). Arkin and Endy (1999) recognized that without standardization it would be difficult to design new biochemical circuitry: the new circuits would likely be restricted to the casual successes of researchers to "choose" suitable biochemical parts that fulfill some criteria. Consequently, standardization of biological parts would be a crucial step in making biology easier to engineer. BioBrick parts are DNA sequences that can be used to design and construct synthetic circuits from individual parts and their combinations with defined biological functions. They were introduced by Tom Knight (2003). The analogy to Lego Bricks invokes the idea, and the ideal, of biological components becoming standardized and combinable. This engineering-based conceptualization of biological parts led to the foundation of the BioBrick initiative, a registry founded in 2006.[3] The registry itself is an open resource to which synthetic biologists are supposed to contribute standardized DNA sequences but also can make use of those available in the registry.

Drew Endy, one of the founders of the BioBricks organization[4] and one of the most public representatives of synthetic biology, argued that biology should implement foundational technologies of engineering (Endy 2005). Instead of considering biology difficult to engineer because of its complexity, bioengineers should consider the possibility that "biology remains complex because we have never made it simple" (449). The construction of buildings provides Endy with an example of how synthetic biologists should proceed. The construction work relies on a limited set of predefined and ready-to-order materials, rules for their combination, and skilled workers, who have the knowledge and means to apply them (450).

Endy does recognize, however, the challenges faced by engineering biology, limiting any synthetic biology endeavors. Such challenges, according to Endy, are as follows: "(1) an inability to avoid or manage biological complexity, (2) the tedious and unreliable construction and characterization of synthetic biological systems, (3) the apparent spontaneous physical variation of biological system behavior, and (4) evolution" (450). Endy's answer for managing or ameliorating the first three problems consists of employing the principles of standardization, decoupling, and abstraction, which are central to engineering.

Standards, Endy notes, "underlie most aspects of the modern world" (Endy 2005, 450). Decoupling entails the idea of disassembling complicated problems into simpler ones (451). Abstraction, in turn, should allow for identifying a hierarchy in the complex architecture of biological systems that starts from the DNA level, proceeding to parts, devices, and system levels in an ascending abstraction. Such an abstraction hierarchy would enable scientists to work independently at each level.

Behind Endy's engineering approach, there is a greater vision of how synthetic biology could contribute to a better life on our planet: "The potential is for civilization-scale flourishing, a world of abundance, not scarcity, supporting a growing global population without destroying it" (Lohr 2021). It is important to note, however, that Endy's engineering-oriented synthetic biology program goes far beyond the construction of synthetic genetic circuits. Instead of constructing networks of a specific topology and implementing and studying them within the biological environment of a cell, the organisms themselves are envisioned as microbial production units, and the engineering takes place directly in the organisms. Kendig calls synthetic biology a "platform technology" because "it generates highly transferrable theoretical models, engineering principles, and know-how that can be applied to create potential products in a wide variety of industries" (Kendig 2014, 1695). The construction of synthetic models is not at the center of such an engineering-oriented approach, where the "know-how" of engineering and manipulating organisms is more central.

The engineering-oriented visions of synthetic biology have largely shaped the outside perception of the field as one that focuses on designing and redesigning organisms for various purposes. Examples include vaccines (Ro et al. 2006)(Ro et al. 2006), biofuels (Bond-Watts, Bellerose, and Chang 2011) and cancer-cell-killing bacteria (Anderson et al. 2006). An early success story was the synthetic anti-malaria drug artemisinin for the production of which *E. coli* and *S. cerevisiae* (baker's yeast) were engineered using synthetic biology tools (Paddon and Keasling 2014). The artemisinin is not a result of the rational parts-based engineering approach, however. Many scientists and philosophers have indeed remained skeptical of the "parts-based engineering approach" because "the key question synthetic biologists have to address is what properties these parts should have so that they give a predictable output even when they are used in different contexts" (Güttinger 2013, 202). Most of the constructs of the application-oriented branch of synthetic biology are best described by the notion of kludging, the word *kludge* referring to a workaround solution that is *k*lumsy, *l*ame, *u*gly, *d*umb, but *g*ood *e*nough (O'Malley 2011). As the notion of kludge was used in engineering long before the emergence of synthetic biology proper, it does not contest the engineering agenda of synthetic biology but rather questions the possibilities of the rational design of biological parts and systems. Indeed, such a conclusion was imminent also within the construction of synthetic genetic circuits that perhaps come closest to the idea of rational engineering in synthetic biology.

4. Synthetic models

Since the introduction of the genetic toggle switch and the repressilator, circuit engineering has provided synthetic biologists with various kinds of templates for forward-engineering genetic circuits from the continuously expanding inventories of molecular parts. The typical construction process of a genetic circuit consists of the following steps. First, some network design is chosen, typically inspired by the repertoire of various kinds of feedback systems studied in engineering and determined by the question to be investigated or the function to be realized. Next, a mathematical model is constructed to study the dynamics of the network, also informing the choice of the component genes of the network. Special software exists for choosing the genes, and the actual synthetic circuits are constructed by companies specialized in DNA synthesis. In the last step, the resulting genetic circuit in the form of a plasmid is implemented within a living organism, frequently *E. coli* bacteria, to study its dynamic. The next section concentrates on the repressilator model, which is perhaps the best-known synthetic genetic circuit, and has also generated philosophical discussion (e.g., Knuuttila and Loettgers 2013a; 2013b; Weber 2014; Green 2022). Moreover, it has led to a rich research program addressing noise in biological systems, making use of other synthetic constructs and even an electronic version of the original synthetic repressilator.

4.1 *The repressilator model*

The original aim of the construction of synthetic genetic circuits was to gain understanding of the organization and related functions in biological systems. A classic example is the circadian clock, a biochemical oscillator that regulates organisms' sleep/wake cycles, body temperature, and metabolic processes. This particular oscillator has been intensively studied by mathematical modeling (Asgari-Targhi and Klerman 2019; Winfree 2001) and genetic screening and experiments in molecular biology on model organisms, such as *Drosophila melanogaster* (Konopka and Benzer 1971; Sehgal 2015). Mathematical biologist Brian Goodwin introduced one of the first models of a biochemical oscillator (Goodwin 1963). The structure of the model is a feedback loop that agrees in its structure and function with the control mechanism in engineering but, interestingly, not in how control is facilitated. In engineering, the oscillations emerging in the feedback loop are unwanted disturbances, whereas in biology, they have been hypothesized to be the primary way of control (Bechtel and Abrahamsen 2011).

The first synthetic model of an oscillatory system is the repressilator constructed by Michael Elowitz together with Stanislas Leibler (Elowitz and Leibler 2000). The repressilator is a network of three genes, arranged as a ring, repressing each other's expression. As explained in the introduction of this entry, it is a molecular analog of the ring oscillator in electronics. The transcription factors (proteins) of each gene bind to the transcription site of its neighboring gene and repress the production of its transcription factor. This arrangement leads to oscillations in protein production.

Characteristic for the construction of synthetic models, such as the repressilator, is making use of a combinational strategy, whereby scientists triangulate experimentation on model organisms, mathematical modeling, and synthetic modeling (Knuuttila and Loettgers 2011). This combinational strategy is depicted in the upper part of Figure 35.1. The lower part on the left-hand side shows in a schematized form the present understanding of the "natural gene-regulatory circuit" of the circadian clock in *Drosophila melanogaster* (the fruit fly), consisting of interacting genes and proteins. The right-hand side, in turn, depicts a synthetic model of the circadian clock, the repressilator. The representation on the right-hand side of the diagram indicates that the natural system exhibits a much higher degree of complexity than the repressilator. From the perspective of modeling, this is only to be expected. Models typically are highly simplified in comparison to the natural or social systems they are studying. Interestingly, however, the synthetic circuit has been designed by using different genes and proteins than the natural *Drosophila* circuit, or any other known circadian clock circuit. It does not aim to even partially replicate the circadian clock of the *Drosophila melanogaster*, only to mimic some of its behavior, i.e., oscillation.

Synthetic models have different advantages vis-à-vis model organisms and mathematical models. They are less complex than model organisms and therefore easier to control and investigate. In contrast to mathematical models, synthetic models are of the same materiality as biological systems and are subject to the same constraints as biological systems even though they are constructed from different genes than those occuring in natural systems. Choosing the "right" genes is crucial for optimizing the properties of interest in the model, such as the strength of the oscillations in protein level and the goal of making the genetic circuit as independent as possible from the rest of the cell.

When it comes to the actual construction process, the first step usually consists of choosing the circuit design and building a mathematical model, the latter often referred to by synthetic biologists as the "blueprint." The notion of a blueprint may be misleading, however, giving the impression of an easy transition from a mathematical to a synthetic model. This is not the case due to the manifold construction decisions, assumptions, and technologies involved as well as the particularities of biochemical material and biological organization.

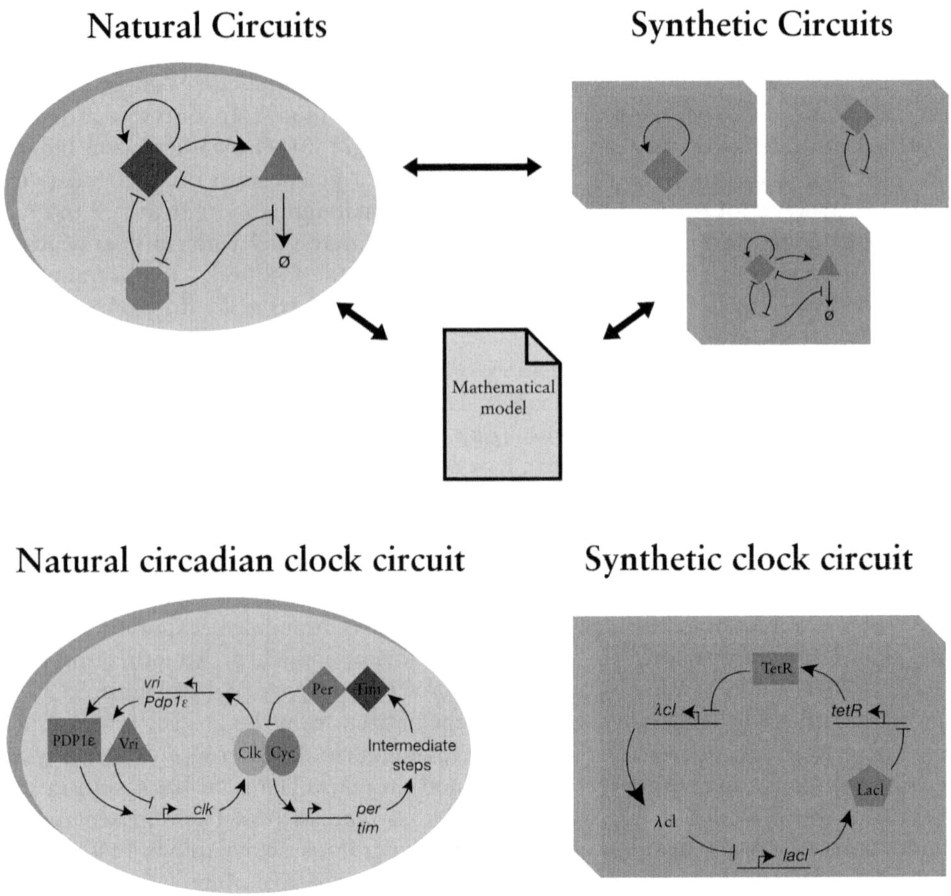

Figure 35.1 Combinational modeling according to Sprinzak and Elowitz (2005, 444). The upper part of the diagram depicts the combinational modeling strategy. The lower part compares the natural gene regulatory network (the fruit fly) and a synthetic one (the repressilator).

The mathematical model of the repressilator is based on kinetic equations and it is of the following form:

$$\frac{dm_i}{dt} = -m_i + \frac{\alpha}{\left(1 + p_j^n\right)} + \alpha_0$$

$$\frac{dp_i}{dt} = -\beta\left(p_i - m_i\right)$$

$$\text{With:} \begin{pmatrix} i = lacI, tetR, cl \\ j = cl, lacI, tetR \end{pmatrix}$$

In this set of equations p_i stands for the concentrations of the proteins suppressing the function of the neighbor genes (where i stands for *lacI*, *tetR*, or *cl)* and m_i are the corresponding concentrations of mRNA. There are six molecule species (three proteins functioning as repressors and three genes), each of them taking part in transcription, translation, and degradation reactions. In general, there are no analytical solutions to such non-linear coupled differential equations; therefore, Elowitz and Leibler performed computer simulations based on this mathematical model. The main purpose of these computer simulations was the identification of relevant experimental parameters as well as the different possible states that could be exhibited by the system. There are two such states: a steady state, and a state in which the system performs limit-cycle-oscillations. As they were interested in biological control, Elowitz and Leibler focused on limit-cycle oscillations and the particular experimental parameters that were critical for attaining stable oscillations. The simulations showed that such oscillations require, for example, strong promoters and tight transcriptional repression, influencing which genes and proteins were chosen.

In the next step, the synthetic repressilator was constructed in the lab and implemented within a living bacterial cell. The lab-constructed repressilator plasmid was transferred into *E. coli* bacteria by making use of the ability of *E. coli* bacteria to take up extra-chromosomal DNA from the environment. Furthermore, to make the oscillations observable, a green fluorescent protein (GFP) was fused to one of the genes, functioning as a "reporter." The oscillations in the protein level of the gene thus became visible through fluorescence microscopy.

Being constructed from biological components and integrated into the bacteria, the repressilator system was clearly closer to biology than models constructed in other media, such as the original mathematical model. Although the biochemical interactions in the cell are largely unknown, this embedment, as Waters (2012) has pointed out, "avoids having to understand the details of the complexity, not by assuming that complexity is irrelevant but by incorporating the complexity in the models."

Positively surprising its authors, the repressilator was able to produce oscillations, but they turned out noisy (in contrast to what the underlying mathematical model predicted). Figure 35.2 shows the oscillations of the repressilator, both in the growing bacteria colony and in single-sibling bacteria.

The Elowitz lab took films of the blinking bacteria, which revealed that the oscillations made visible by the reporter were not synchronized (Figure 35.2). This non-synchronization is manifest in the lower diagrams (a–c), showing the fluorescence of two sibling cells. Here one (red line in the online version of this book) line is a reference line representing the oscillations of the whole bacteria colony and the two other ones (blue and green lines in the online version of this book) show the oscillations of sibling cells. The diagrams show that the amplitudes of the oscillations of the sibling cells change over time, indicating a difference in the amount of proteins produced over time by the reporter gene. Secondly, the phases of the oscillations in the two bacteria shift over time. In other words, the sibling cells show some individual behavior (phase shift), but there is also some variability in this individual behavior (changes in amplitude). (The graph (d) presents oscillations obtained in different experiments, and (e–f) are the result of negative control experiments.)

4.2 The research on the functional role of noise

The observed individual behavior of cells, as shown by the phase shifts and fluctuations provided a first clue that the fluctuations could be of a stochastic nature. Elowitz and the members of his lab assumed that most probably they were caused by the limited number of molecules in cells. To explore the noisy behavior exhibited by the repressilator, the

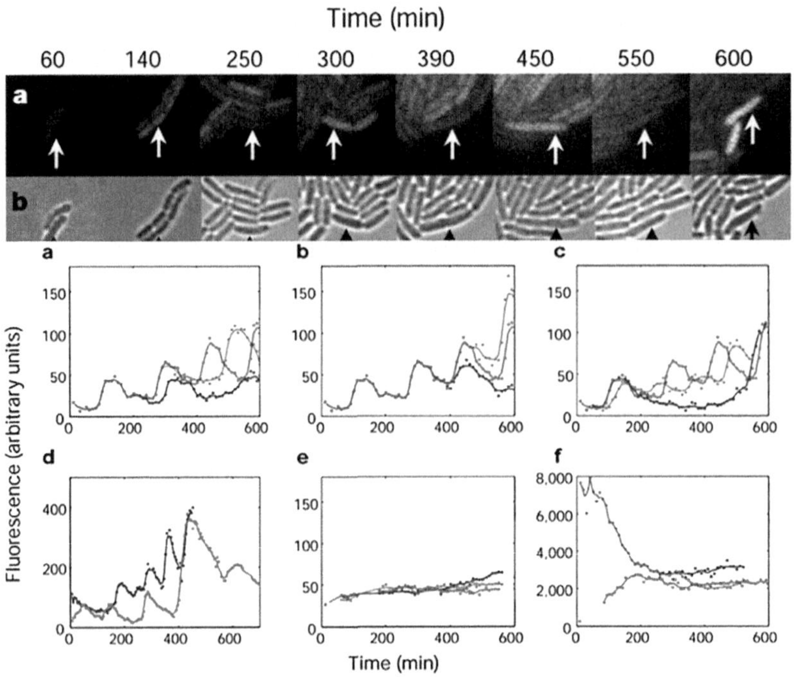

Figure 35.2 The upper snapshots from the growing colony of bacteria show the oscillations of the repressilator in the living bacteria. In the diagrams below, (a–c) show the repressilator dynamics of sibling cells, which exhibit great variability in period and amplitude; (d) presents oscillations obtained in other experiments, and (e–f) are the result of negative control experiments (Elowitz and Leibler 2000, 336).

researchers performed computer simulations of a stochastic version of the initial mathematical model that appeared to confirm the stochastic nature of the observed fluctuations. Two related questions appeared. First, how are regular oscillations possible at all in the stochastic environment of a cell, and second, how are stochastic fluctuations related to other sources of noise that occur independently from the observed stochastic fluctuations? Both questions were explored by further models and synthetic constructs.

To explore stochastic fluctuations within individual *E. coli* bacteria, the Elowitz group developed a synthetic intracellular measuring device.[5] To distinguish between extrinsic and intrinsic sources of noise, they integrated into the chromosomes of the bacteria cyan *cfp* and yellow *yfp* alleles of green fluorescent proteins. With extrinsic sources of noise, they referred to noise that is independent of the gene, such as the stage of the cell cycle or cell environment fluctuations. The two proteins were put under the control of identical promoters (Elowitz et al. 2002; Swain, Elowitz, and Siggia 2002). In the absence of intrinsic noise, the two reporter genes that are located in the same cell are only exposed to extrinsic noise that is the same for each of the genes. In this case, the cells have the same amount of each protein and have the same color (yellow in the online version of this book) as shown in the upper part of Figure 35.3a. However, in the presence of intrinsic noise, the proteins produced by the genes fluctuate in an uncorrelated fashion since the two genes are uncorrelated. This gives rise to a population of cells, in which some cells express more of one fluorescent protein than

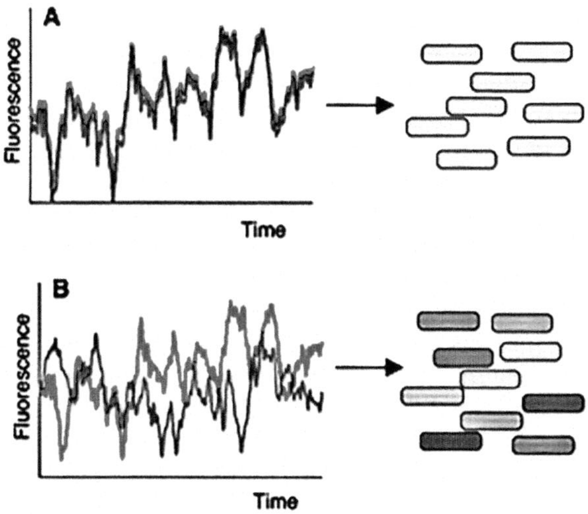

Figure 35.3 Fluctuations due to the extrinsic (a) and intrinsic (b) noise and the corresponding variations in fluorescence (Elowitz et al. 2002, 1185).

the other. As a consequence, the cells in such populations appear in different colors (such as yellow, orange, red and green in the online version of this book) (Figure 35.3, lower part b).

In their experiments with different E. coli strains, the researchers were able to attain differently colored bacteria colonies. These strains differed from each other in how strongly the genes of the regulator sequences to which the reporter genes were fused were transcribed.

As already mentioned, at the beginning of the construction of the measuring device, Elowitz and his co-workers considered stochastic fluctuations as perturbations, and the question was how to change the design of the repressilator in such a way that it would lead to more robust performance. What then started to intrigue the researchers was whether the stochastic fluctuations observed could also have a functional role (Loettgers 2009). Stochastic noise caused by variations in the gene expression of different cells need not be a perturbation but could rather be an essential feature of cellular organization. For example, stochastic fluctuations allow cells to transition between different states (Süel et al. 2006).

Interestingly, in addition to the synthetic model, an electronic version of the repressilator was also constructed (Buldú et al. 2007). Such an electronic network provides a good analog of the repressilator for the study of robust oscillations, since, as the researchers put it, "this system is subject to electronic noise and time delays associated with its operation, and since its parameters depend on the actual values of capacitances and resistors [...]" (Mason et al. 2004, 709). The researchers found out that in the electronic repressilator robust oscillations were possible in the presence of noise.

5. The model-like character of synthetic genetic circuits

In a lecture at the Kavli Institute of Theoretical Physics, Michael Elowitz discussed the novel epistemic possibilities that synthetic genetic circuits provide (Kavli Institute for Theoretical Physics 2017). The traditional approach of biology would rely on deconstruction or perturbation. One conventional method is to grow a population in a test tube, slice them, and then analyze biochemically what was happening on average in those cells before they were killed. The synthetic biology approach makes it possible to study genetic control within single living cells. Moreover, in contrast to experimentally perturbing very complicated natural systems, synthetic biologists construct simple circuits that they put within a living cell. These simple circuits do not aim to represent any natural circuit. Instead, they are optimized for the purpose of attaining some chosen function.

Within the philosophical discussion, models are typically taken as representations of some real-world target systems. Moreover, it is also often assumed that they are implemented in a different medium than their supposed target systems (Morgan 2012; Rheinberger 2015). Rheinberger (2015) distinguishes what he calls "preparations" from models based on the medium; while preparations "participate in the materiality of the object of knowledge in question" models do not (325). However, there is no need to treat models as representations of some determinable real-world target systems (Knuuttila 2011; Weisberg 2013). Nor to suppose that synthetic systems could not be regarded as models since they are implemented in the same materiality as the biological systems they are used to explore.

First, synthetic circuits are like many other models in being tightly constrained and self-contained constructions that study some dependencies in order to answer pending theoretical and empirical questions (Knuuttila 2011; 2021b). It is instructive to compare the repressilator to the synthetic measuring device that the Elowitz lab constructed. This noise sensor was not constructed to fulfill some specific biological function or to function as a partially independent synthetic module. In contrast to the repressilator, the noise sensor did not have a dynamic of its own. As a mere measuring device, it was supposed to be responsive to various conditions produced within the cell. Consequently, although the repressilator and the noise sensor were both synthetic genetic constructs, considering both preparations would miss their different characters and roles in the research program of the Elowitz lab (Knuuttila and Loettgers 2021).

Second, the repressilator does not aim to *re*present some naturally evolved genetic circuit, however partially. The Elowitz lab's research, and that of synthetic biology more generally, focuses on hypothetical designs that would apply both to actual and possible, non-actual, biological systems. Elowitz and Lim envision synthetic biology as "the expansion of biology from a discipline that focuses on natural organisms to one that includes potential organisms" (Elowitz and Lim 2010, 889). While arguably many models study possibilities, the ability of synthetic biology to materially realize these designs is crucial for its modal character (Koskinen 2019; Ijäs and Koskinen 2021; Knuuttila 2021a).

6. Synthetic biology becoming more biology-inspired

The construction of the repressilator was motivated by the question of whether feedback systems, familiar from physics and engineering and already theorized since the early 1960s (e.g., Jacob and Monod 1961), could be realizable in biological organisms. Synthetic biology finally provided a means for scientists to study such possible general principles of biological organization within living cells. Biological systems have turned out to be challenging to engineer, however, and the actual biological designs are unintuitive from the engineering perspective. Biological designs are easier to conceive abstractly, while synthetic biologists must balance multiple constraints. Some constraints are technological and independent of biological capabilities. Other constraints include the standardizability and combinability of biochemical parts and the synthetic systems' dependence on the cellular environment and metabolic system. Another problem is that of relating the molecular features and interactions to mathematical models, which are used to design synthetic systems and study their behavior.

The main challenges of current synthetic biology are the stochasticity of biological processes, and the increasing complexity of synthetic systems once the research has advanced from bacteria to diverse cell types and multicellular creatures. Experimental and mathematical modeling have shown that stochastic fluctuations may help bacteria decide their cellular fate (Eldar and Elowitz 2010). On the other hand, transporting synthetic circuits into cells is difficult in the case of multicellular organisms (Gao et al. 2020). While synthetic biologists can leverage bacteria's ability to take up DNA from their environment, other cell types require alternative methods. Synthetic biologists have engineered viruses as "cargo" systems (Nayerossadat, Maedeh, and Ali 2012). Preventing circuit integration into the host genome is crucial here since viruses make use of the chromosomes of the host to reproduce.

Apart from the quest to tackle more complex "unintuitive" designs, the basic assumption of modular architecture has been questioned. Early synthetic biology research focused on isolating modules, but it is now becoming more obvious that interactions within the cell environment may make them more robust (Cookson, Tsimring, and Hasty 2009). However, in complicated synthetic systems with multiple circuits, their components may interact and disturb their functions. As one solution, synthetic biologists employ synthetic "cells" to isolate circuits and achieve modular organization (Adamala et al. 2017). Even though these synthetic cells are not live cells, they can read DNA and make proteins. Finally, the turn from bacteria to multicellular organisms has prompted researchers to study how cells "communicate" in a timely and accurate manner. The Elowitz laboratory, along with other research groups, has addressed this problem in terms of information processing. Their methodology involves investigating the encoding and decoding of information using experimental techniques, as well as mathematical and synthetic modeling methods (Li and Elowitz 2021).

Perhaps paradoxically, then, the rational engineering part of synthetic biology, which is based on modularity and electrical engineering design principles, has gathered evidence questioning these very assumptions, becoming more biology-inspired. Thus the engineering approach to biology has led to an increased appreciation of the differences between engineering and biology. Synthetic modeling has been central to this process.

Acknowledgments

This chapter and Handbook were produced thanks to funding from the European Research Council (ERC) under the European Union's Horizon 2020 research and innovation program (Grant Agreement No. 818772). They are also funded by the John Templeton Foundation, project "Pushing the Boundaries" (Grant ID: 62581).

Notes

1 http://dspace.mit.edu/handle/1721.1/21168.
2 https://technology.igem.org/registry.
3 https://biobricks.org.
4 https://biobricks.org.
5 For more extensive discussion of the Elowitz lab's research on noise, see Knuuttila and Loettgers (2021, 2014).

References

Adamala, Katarzyna P., Daniel A. Martin-Alarcon, Katriona R. Guthrie-Honea, and Edward S. Boyden. 2017. "Engineering Genetic Circuit Interactions within and between Synthetic Minimal Cells." *Nature Chemistry* 9(5): 431–439. https://doi.org/10.1038/nchem.2644.

Alon, Uri. 2007. *An Introduction to Systems Biology: Design Principles of Biological Circuits*. Boca Raton, FL: Chapman & Hall/CRC Press.

Anderson, J. Christopher, Elizabeth J. Clarke, Adam P. Arkin, and Christopher A. Voigt. 2006. "Environmentally Controlled Invasion of Cancer Cells by Engineered Bacteria." *Journal of Molecular Biology* 355 (4): 619–27. https://doi.org/10.1016/j.jmb.2005.10.076.

Arkin, Adam P., and Drew Endy. 1999. "A Standard Parts List for Biological Circuitry." Working Paper. DARPA White Paper. https://dspace.mit.edu/handle/1721.1/29794.

Asgari-Targhi, Ameneh, and Elizabeth B. Klerman. 2019. "Mathematical Modeling of Circadian Rhythms." *WIREs Systems Biology and Medicine* 11(2): e1439. https://doi.org/10.1002/wsbm.1439.

Bechtel, William, and Adele Abrahamsen. 2011. "Complex Biological Mechanisms: Cyclic, Oscillatory, and Autonomous." In *Philosophy of Complex Systems. Handbook of the Philosophy of Science*, edited by C. A. Hooker, 10: 257–285. Oxford: Elsevier. https://www.sciencedirect.com/science/article/pii/B9780444520760500092.

Bensaude Vincent, Bernadette. 2013. "Between the Possible and the Actual: Philosophical Perspectives on the Design of Synthetic Organisms." *Futures* 48(April): 23–31. https://doi.org/10.1016/j.futures.2013.02.006.

Bertalanffy, Ludwig von. 1969. *General System Theory: Foundations, Development, Applications*. New York: G. Braziller. https://doi.org/10.1109/TSMC.1974.4309376.

Bond-Watts, Brooks B., Robert J. Bellerose, and Michelle C. Y. Chang. 2011. "Enzyme Mechanism as a Kinetic Control Element for Designing Synthetic Biofuel Pathways." *Nature Chemical Biology* 7(4): 222–227. https://doi.org/10.1038/nchembio.537.

Buldú, Javier M., Jordi García-Ojalvo, Alexandre Wagemakers, and Miguel a. F. Sanjuán. 2007. "Electronic Design of Synthetic Genetic Networks." *International Journal of Bifurcation and Chaos* 17(10): 3507–3511. https://doi.org/10.1142/S0218127407019275.

Bursten, Julia R. S. 2019. *Perspectives on Classification in Synthetic Sciences: Unnatural Kinds*. London: Routledge.

Cameron, D. Ewen, Caleb J. Bashor, and James J. Collins. 2014. "A Brief History of Synthetic Biology." *Nature Reviews Microbiology* 12(5): 381–390.

Cookson, Natalie A., Lev S. Tsimring, and Jeff Hasty. 2009. "The Pedestrian Watchmaker: Genetic Clocks from Engineered Oscillators." *FEBS Letters* 583(24): 3931–3937. https://doi.org/10.1016/j.febslet.2009.10.089.

Eldar, Avigdor, and Michael B. Elowitz. 2010. "Functional Roles for Noise in Genetic Circuits." *Nature* 467(7312): 167–173. https://doi.org/10.1038/nature09326.

Elowitz, Michael, and Stanislas Leibler. 2000. "A Synthetic Oscillatory Network of Transcriptional Regulators." *Nature* 403(6767): 335–338. https://doi.org/10.1038/35002125.

Elowitz, Michael, Arnold J. Levine, Eric D. Siggia, and Peter S. Swain. 2002. "Stochastic Gene Expression in a Single Cell." *Science* 297(5584): 1183–1186. https://doi.org/10.1126/science.1070919.

Elowitz, Michael, and Wendell A. Lim. 2010. "Build Life to Understand It." *Nature* 468 (7326): 889–890. https://doi.org/10.1038/468889a.

Endy, Drew. 2005. "Foundations for Engineering Biology." *Nature* 438(7067): 449–453. https://doi.org/10.1038/nature04342.

Gao, Xiaojing J., Lucy S. Chong, Michaela H. Ince, Matthew S. Kim, and Michael B. Elowitz. 2020. "Engineering Multiple Levels of Specificity in an RNA Viral Vector." bioRxiv. https://doi.org/10.1101/2020.05.27.119909.

Gardner, Timothy S., Charles R. Cantor, and James J. Collins. 2000. "Construction of a Genetic Toggle Switch in Escherichia Coli." *Nature* 403(6767): 339–342. https://doi.org/10.1038/35002131.

Goodwin, Brian C. 1963. *Temporal Organization in Cells; a Dynamic Theory of Cellular Control Processes*. London: Academic Press. https://doi.org/10.5962/bhl.title.6268.

Green, Sara. 2015. "Revisiting Generality in Biology: Systems Biology and the Quest for Design Principles." *Biology & Philosophy* 30(5): 629–652. https://doi.org/10.1007/s10539-015-9496-9.

———. 2022. "Philosophy of Systems and Synthetic Biology." In *The Stanford Encyclopedia of Philosophy*, edited by Edward N. Zalta, Summer 2022. Metaphysics Research Lab, Stanford University. https://plato.stanford.edu/archives/sum2022/entries/systems-synthetic-biology/.

Green, Sara, and Olaf Wolkenhauer. 2013. "Tracing Organizing Principles: Learning from the History of Systems Biology." *History and Philosophy of the Life Sciences* 35(4): 553–576.

Güttinger, Stephan. 2013. "Creating Parts That Allow for Rational Design: Synthetic Biology and the Problem of Context-Sensitivity." *Studies in History and Philosophy of Science Part C: Studies in History and Philosophy of Biological and Biomedical Sciences, Philosophical Perspectives on Synthetic Biology* 44(2): 199–207. https://doi.org/10.1016/j.shpsc.2013.03.015.

Hartwell, Leland H., John J. Hopfield, Stanislas Leibler, and Andrew W. Murray. 1999. "From Molecular to Modular Cell Biology." *Nature* 402(6761): C47–C52. https://doi.org/10.1038/35011540.

Hopfield, J. J. 1982. "Neural Networks and Physical Systems with Emergent Collective Computational Abilities." *Proceedings of the National Academy of Sciences* 79(8): 2554–2558. https://doi.org/10.1073/pnas.79.8.2554.

Ijäs, Tero, and Rami Koskinen. 2021. "Exploring Biological Possibility through Synthetic Biology." *European Journal for Philosophy of Science* 11(2): 39. https://doi.org/10.1007/s13194-021-00364-7.

Jacob, François, and Jacques Monod. 1961. "Genetic Regulatory Mechanisms in the Synthesis of Proteins." *Journal of Molecular Biology* 3(3): 318–356. https://doi.org/10.1016/S0022-2836(61)80072-7.

Kavli Institute for Theoretical Physics, dir. 2017. *Michael Elowitz: Life at the Single Cell Level*. https://www.youtube.com/watch?v=NxPcIQsscoE.

Kendig, Catherine, and Bryan A. Bartley. 2019. "Synthetic Kinds: Kind-Making in Synthetic Biology." In *Perspectives on Classification in Synthetic Sciences: Unnatural Kinds*, edited by Julia R. S. Bursten, 78–91. London: Taylor & Francis.

Kendig, Catherine, Paul B. Thompson, and David M. Kaplan. 2014. "Synthetic Biology and Biofuels." In *Encyclopedia of Food and Agricultural Ethics*, edited by Kaplan, David M. and Paul B. Thompson, 1695–1703. Dordrecht: Springer. https://doi.org/10.1007/978-94-007-0929-4_124.

Knight, Thomas. 2003. "Idempotent Vector Design for Standard Assembly of Biobricks." MIT Synthetic Biology Working Group. MIT Artificial Intelligence Laboratory. https://dspace.mit.edu/handle/1721.1/21168.

Knuuttila, Tarja. 2011. "Modelling and Representing: An Artefactual Approach to Model-Based Representation." *Studies in History and Philosophy of Science Part A*, Model-Based Representation in Scientific Practice 42(2): 262–271. https://doi.org/10.1016/j.shpsa.2010.11.034.

———. 2021a. "Epistemic Artifacts and the Modal Dimension of Modeling." *European Journal for Philosophy of Science* 11(3): 65. https://doi.org/10.1007/s13194-021-00374-5.

———. 2021b. "Imagination Extended and Embedded: Artifactual versus Fictional Accounts of Models." *Synthese* 198(21): 5077–5097. https://doi.org/10.1007/s11229-017-1545-2.

Knuuttila, Tarja, and Andrea Loettgers. 2011. "Causal Isolation Robustness Analysis: The Combinatorial Strategy of Circadian Clock Research." *Biology & Philosophy* 26(5): 773–791. https://doi.org/10.1007/s10539-011-9279-x.

———. 2013a. "Basic Science through Engineering? Synthetic Modeling and the Idea of Biology-Inspired Engineering." *Studies in History and Philosophy of Science Part C: Studies in History and Philosophy of Biological and Biomedical Sciences*, Philosophical Perspectives on Synthetic Biology 44(2): 158–169. https://doi.org/10.1016/j.shpsc.2013.03.011.

———. 2013b. "Synthetic Modeling and Mechanistic Account: Material Recombination and Beyond." *Philosophy of Science* 80(5): 874–885. https://doi.org/10.1086/673965.

———. 2014. "Varieties of Noise: Analogical Reasoning in Synthetic Biology." *Studies in History and Philosophy of Science Part A* 48(December): 76–88. https://doi.org/10.1016/j.shpsa.2014.05.006.

———. 2021. "Biological Control Variously Materialized: Modeling, Experimentation and Exploration in Multiple Media." *Perspectives on Science* 29(4): 468–492. https://doi.org/10.1162/posc_a_00379.

Konopka, Ronald J., and Seymour Benzer. 1971. "Clock Mutants of Drosophila Melanogaster." *Proceedings of the National Academy of Sciences of the United States of America* 68(9): 2112–2116.

Koskinen, Rami. 2019. "Multiple Realizability and Biological Modality." *Philosophy of Science* 86(5): 1123–1133. https://doi.org/10.1086/705478.

Li, Pulin, and Michael B. Elowitz. 2021. "Communication Codes in Developmental Signaling Pathways | Development | The Company of Biologists." *Development* 146(12): 1–12.

Loettgers, Andrea. 2009. "Synthetic Biology and the Emergence of a Dual Meaning of Noise." *Biological Theory* 4(4): 340–356. https://doi.org/10.1162/BIOT_a_00009.

Lohr, Steve. 2021. "Can Synthetic Biology Save Us? This Scientist Thinks So." *The New York Times*, November 23, 2021, sec. Business. https://www.nytimes.com/2021/11/23/business/dealbook/synthetic-biology-drew-endy.html.

Madsen, Curtis, James Alastair McLaughlin, Göksel Mısırlı, Matthew Pocock, Keith Flanagan, Jennifer Hallinan, and Anil Wipat. 2016. "The SBOL Stack: A Platform for Storing, Publishing, and Sharing Synthetic Biology Designs." *ACS Synthetic Biology* 5(6): 487–497. https://doi.org/10.1021/acssynbio.5b00210.

Mason, Jonathan, Paul S. Linsay, J. J. Collins, and Leon Glass. 2004. "Evolving Complex Dynamics in Electronic Models of Genetic Networks." *Chaos: An Interdisciplinary Journal of Nonlinear Science* 14(3): 707–715. https://doi.org/10.1063/1.1786683.

Morgan, Mary S. 2012. *The World in the Model: How Economists Work and Think*. Cambridge: Cambridge University Press. https://doi.org/10.1017/CBO9781139026185.

Nayerossadat, Nouri, Talebi Maedeh, and Palizban Abas Ali. 2012. "Viral and Nonviral Delivery Systems for Gene Delivery." *Advanced Biomedical Research* 1(July): 27. https://doi.org/10.4103/2277-9175.98152.

O'Malley, M. A. 2011. "Exploration, Iterativity and Kludging in Synthetic Biology." *Comptes Rendus Chimie* 14: 406–412.

Paddon, Chris J., and Jay D. Keasling. 2014. "Semi-Synthetic Artemisinin: A Model for the Use of Synthetic Biology in Pharmaceutical Development." *Nature Reviews. Microbiology* 12(5): 355–67. https://doi.org/10.1038/nrmicro3240.

Rheinberger, Hans-Jörg. 2015. "Preparations, Models, and Simulations." *History and Philosophy of the Life Sciences* 36(3): 321–334. https://doi.org/10.1007/s40656-014-0049-3.

Ro, Dae-Kyun, Eric M. Paradise, Mario Ouellet, Karl J. Fisher, Karyn L. Newman, John M. Ndungu, Kimberly A. Ho, et al. 2006. "Production of the Antimalarial Drug Precursor Artemisinic Acid in Engineered Yeast." *Nature* 440 (7086): 940–943. https://doi.org/10.1038/nature04640.

Sehgal, Amita. 2015. *Circadian Rhythms and Biological Clocks Part A*. New York: Academic Press.

Singh, Vijai. 2022. *New Frontiers and Applications of Synthetic Biology*. New York: Academic Press.

Sprinzak, David, and Michael B. Elowitz. 2005. "Reconstruction of Genetic Circuits." *Nature* 438(7312): 443–448. https://doi.org/10.1038/nature04335.

Süel, Gürol M., Jordi Garcia-Ojalvo, Louisa M. Liberman, and Michael B. Elowitz. 2006. "An Excitable Gene Regulatory Circuit Induces Transient Cellular Differentiation." *Nature* 440(7083): 545–550. https://doi.org/10.1038/nature04588.

Swain, Peter S., Michael B. Elowitz, and Eric D. Siggia. 2002. "Intrinsic and Extrinsic Contributions to Stochasticity in Gene Expression." *Proceedings of the National Academy of Sciences* 99(20): 12795–12800. https://doi.org/10.1073/pnas.162041399.

Thomas, Rene. 1998. "Laws for the Dynamics of Regulatory Circuits." *International Journal Developmental Biology* 42: 479–485.

Thomas, Rene, and Richard D'Ari. 1990. *Biological Feedback*. 1st edition. Boca Raton, FL: CRC Press.

Waters, Kenneth C. 2012. "Experimental Modeling as a Form of Theoretical Modeling." Paper presented at the Philosophy of Science Association 23rd Meeting, San Diego.

Weber, Marcel. 2014. "Experimental Modeling in Biology: In Vivo Representation and Stand-Ins as Modeling Strategies." *Philosophy of Science* 81(5): 756–769. https://doi.org/10.1086/678257.

Weisberg, Michael. 2013. *Simulation and Similarity: Using Models to Understand the World*. Reprint Edition. Oxford: Oxford University Press.

Wiener, Norbert. 1961. *Cybernetics: Or Control and Communication in the Animal and the Machine*. 2nd edition. Cambridge: MIT Press Ltd.

Winfree, Arthur T. 2001. *The Geometry of Biological Time*. New York: Springer Science & Business Media.

36
MODELLING THE DEEP PAST

Adrian Currie

1. Introduction

Palaeoscientists adopt 'methodologically omnivorous' strategies to mitigate the epistemic scarcity generated by millions of years of information decay. This chapter focuses on one set of those strategies: the use of models to understand the deep past. Standard philosophical accounts of historical science fail to capture the importance and role of modelling, examining why and how such scientists use models is revelatory of both their methods and the nature of knowledge in the palaeosciences.

Knowledge of the past is built upon knowledge of possibility, flirting with the idea that the historical sciences are less about temporality than about modality. *Contra* accounts of modelling that emphasize the independence of models from empirical data; many models of the deep past are intimately connected to historical evidence. This 'phenomena-driven' modelling strategy ensures the possibilities the models explore are relevant possibilities. Models in the palaeosciences are tools for understanding what is (or was) possible.

Section 2 discusses some preliminaries: characterizing 'trace-based' accounts of historical reasoning, sketching an account of modelling practice, and touring philosophical work on modelling in the palaeosciences. In Section 3, two examples are introduced: the application of ecological models to Ediacaran ecosystems, and robotics to dinosaur aquatic propulsion. In Section 4, the role of model-explored possibility for understanding the past is examined, before the conclusion in Section 5.

2. Models and traces

This section zeroes in on a (non-technical) dilemma: many accounts of palaeoscientific reasoning de-emphasize modelling, yet models are ubiquitous in palaeoscience. This suggests the nature of that reasoning should be rethought, as well as the place of modelling within it.

2.1 *Trace-based reasoning*

How do we know anything about the deep past? Past events, entities, and processes have reach into the present, that is, they sometimes leave traces. The bodies of long-dead

DOI: 10.4324/9781003205647-42

organisms, as well as their tracks and burrows, fossilize, and they live on via descendants and environmental effects. If we understand processes of preservation, have a good grip on, say, fossilization processes, then we can infer from traces to the past. *Trace-based reasoning*, then, involves (1) a set of traces, (2) a set of theories about trace formation together enabling (3) inferences to past states of affairs (Currie 2017, Currie 2018a, chaps. 3–5; Currie 2019a). This inferential strategy forms the foundation of many philosophical accounts of historical scientific methods.

Carol Cleland's account is grounded in trace-based reasoning (Cleland 2002; 2011; 2013). For her, proto-typical historical science (as opposed to experimental science) distinctively proceeds via *smoking gun reasoning*. This involves, first, identifying a set of surprising correlations between present observations, second, generating a set of hypotheses about the past which, if true, would explain those correlations, and third, searching for further traces which, if found, would discriminate between those hypotheses. In smoking gun reasoning, historical scientists navigate between diverse traces in order to develop and test common-cause explanations (see also Tucker's treatments: 2004; 2011; Kleinhans et al. 2005; 2010). Forber and Griffith (2011) develop a similar view, emphasizing independent evidential convergence. By these accounts, historical reasoning begins and ends with the discovery, analysis, and inferential use of, traces.

Derek Turner's defence of anti-realism is also grounded in trace-based reasoning (Turner 2004; 2005; 2007; 2009a). Turner argues that, compared to experimental science, knowledge of the deep past is systematically underdetermined. That is, traces will be insufficient to discriminate between many past hypotheses. First, historical background knowledge(s) provides councils of despair, that is, information destruction is characteristic of history: traces decay. Second, unlike in experimental cases, our incapacity to intervene in the past leaves us unable to mitigate information loss. Therefore, Turner's arguments concern trace-based reasoning's lack of power.

Although much recent philosophical analysis of palaeoscience has expanded our conception beyond trace-based reasoning[1] (see below), suffice to say, it is still (perhaps rightly) central to our conceptions. However, that picture does not seem to leave room for the surrogative reasoning characteristic of modelling.

2.2 *Modelling as strategy*

As this volume attests, philosophers of science have been much concerned with models, characterizing them and their role in science, asking what models are and how they might be revelatory of nature. I will follow one strand of this literature which tackles models not via their content (being idealized, say), but in terms of a particular scientific strategy. This account serves as a useful contrast for how historical scientists often use models.

What counts as a model? Both Michael Weisberg and Peter Godfrey-Smith have answered by asking us not to look at the features of the model itself, but rather at how models are developed and the use scientists put them to (Weisberg 2007; Godfrey-Smith 2006). At base, they understand modelling as a kind of indirect strategy: rather than looking at our target system or systems, we examine some model system.

Weisberg captures the modeller's strategy by distinguishing it from what he calls *abstract direct representation*, or ADR. In ADR, we build a representation via the use of empirical data generated from our target system. By contrast, the modeller's strategy takes inspiration from a target system or phenomena, constructing a comparatively transparent, easy-to-manipulate-and-study proxy, investigating *that*, and then comparing it to the phenomenon downstream. A classic set of studies from the dawn of the 'palaeontological revolution' illustrates the modeller's strategy applied to the deep past.

From the mid-1960s, palaeontology became palaeobiology, that is, palaeontologists began answering a richer set of questions concerning extinct organisms (Sepkoski and Ruse 2009). One crucial aspect of this involved interrogating large-scale macroevolutionary patterns. By careful analysis of typically invertebrate fossils, palaeontologists were able to infer palaeontological phenomena from large data sets. Perhaps the most well-known of these are patterns of mass extinction (Raup and Sepkoski 1982; Dresow 2021; Bocchi et al. forthcoming). Raup and Sepkoski use careful collation and analysis, characterise a pattern of 'spikes' in extinction across deep time: periods when against a background extinction rate, improbably large numbers of taxa disappear. An abstract representation is come to *directly,* in Weisberg's terms, if they infer from the data to the relevant phenomenon.

The availability of invertebrate data sets enabled palaeontologists to characterise various large-scale patterns of life: radiation, speciation, extinction, and so forth. These patterns raised further questions concerning their underlying processes: do evolutionary events occur via gradual increments; does speciation require geographic separation; is natural selection required to explain mass extinctions, radiations, and other macroevolutionary events? One way of tackling these questions involves an indirect strategy, that is, model building.

In the early 1970s, a group of scientists meeting at the Woods Hole Marine Biology Lab developed a computer programme designed to simulate large-scale evolutionary patterns. This model eventually became known as the "MBL" model (Raup et al. 1973; Huss 2009; Turner 2011, 60–64). At its most simple, the MBL model is extraordinarily so. Beginning with a single lineage, at each time step it can: split into two lineages (speciate), remain as it is, or go extinct. By running the simulation over multiple time steps, and re-running these, the MBL group hoped to generate qualitatively similar patterns seen in the record. The model was stochastic: the probability of each event occurring had equal probability (with an equilibrium measure added to stop the phylogeny number from exploding beyond computational capacity). These probabilities can be manipulated, and deterministic elements added as required. Crucially, the stochastic model does not include fitness: there is no feature of a lineage itself, nor external influence, that makes a difference to its chances of speciating, surviving, or becoming extinct. As such, Raup et al. took the model as a way of asking: *what macroevolutionary patterns or events require fitness to generate them*? The strategy was to use the simple model as a kind of baseline with which to identify those macro-evolutionary events which required more complexity to model:

> If it appears, for example, that a given evolutionary event (such as a mass extinction) cannot be simulated by the simple random model, we will abandon certain of the stochastic elements in favor of additional deterministic constraints. The power of the model will then be its ability to specify the minimum departure from randomness necessary to produce a satisfactory replica of the real world situation.
>
> *(Raup et al. 1973, 527)*

For our purposes, the initial MBL model illustrates the modeller's strategy according to Weisberg's account. As he summarizes:

> In the first stage, a theorist constructs a model. In the second, she analyzes, refines, and further articulates the properties and dynamics of the model. Finally, in the third stage, she assesses the relationship between the model and the world if such an assessment is appropriate. If the model is sufficiently similar to the world, then the analysis of the model is also, indirectly, an analysis of the properties of the real-world phenomenon. Hence, modeling involves indirect representation and analysis of real-world phenomena.
>
> *(Weisberg 2007, 209–210)*

Raup et al. (1973) began by building a single stochastic branching model. They then examined the model's parameters, investigating which patterns emerged, and which deterministic elements were required to generate other patterns. Only after this process did they explicitly consider the relationship between model and macroevolutionary pattern.

Thus, we can understand modelling as a kind of scientific strategy involving the indirect representation of a target system. Instead of sticking to the phenomena, modellers take a step back, constructing and investigating a relevantly similar proxy. This is only one way of understanding models: there are scientific tools and representations that philosophers and scientists call 'models' which are much more intimately concerned with data. For instance, models are very often used in inferences from data sets to patterns and phenomena, and indeed, representations of those phenomena might rightly be considered 'data-models' (Suppes 1966; Leonelli 2019; Antoniou 2021). The point of introducing this particular account is as a contrast: as will be discussed in Sections 3 and 4, modelling in palaeontology is often phenomena-led in a way that differs from the strategy identified here.

2.3 *The philosophy of palaeoscientific modelling*

Many philosophers have emphasized the role of models in the sciences of the deep past (see Bokulich and Oreskes 2017 for a systematic tour focused on geoscience, Sepkoski 2012 for an insightful discussion of palaeobiological models). Some of that work is sketched below.

Philosophers have responded to Cleland's trace-centric account by appealing to the use of models in the historical sciences. Derek Turner, for instance, highlights the use of computer simulations in estimating gaits in long-extinct therapods (Turner 2009b). Functional morphologists build models representing an animal's maximal speed and gait based on its anatomical properties. These simulations are tested on living analogues and, based on fossil reconstructions, generate gait and speed by running iterative competitions between varying gaits, where variations are generated from the best performers to form the basis of the next bout. I have pointed to the use of simple geometric models to explore and test models of long-extinct echinoderm development (Currie 2018a, chap. 9). With very simple assumptions, palaeontologists are able to generate geometric forms very similar to those seen in echinoderm fossils, and these results were put to work in exploring differences between early and late echinoderm development.

Functional morphology provides many examples of modelling strategies, which philosophers have discussed. Like Turner, I appealed to gait simulations, providing an analysis of how models and simulations might provide surprising results (Currie 2018b). In this case, an exploration of sauropod gait, simulations suggested that sauropods walked with a gait unseen—novel and surprising—in extant animals. Marco Tamborini explores the use of robotic models—in a sense, concrete versions of the *in silico* examples Turner and I examine—to emphasize the entanglement of technology and nature in historical reasoning (Tamborini 2020; 2021).

Alison Wylie has argued that archaeology should be understood as a fundamentally model-based discipline, suggesting understanding "...archaeological practice as a genre of empirically grounded, investigative reasoning with and through models" (2017a, 3). Adopting a broad conception of models, Wylie includes abstract characterizations of assemblage patterns (e.g., the use of 'Clovis' arrow-heads to identify cultural spread across the Americas); models representing patterns in spatial distribution (e.g., the positioning and orientation of grave goods); chronological models representing the appearance and disappearance of assemblages, as well as various models from related fields (carbon dating, ethnographies, etc.). Often, models in archaeology guide interpretation and aid in understanding the various forces, which form and transform the archaeological record—thus crucial for trace-based reasoning—see also Nyrup (2020). Wylie also covers models closer to the modelling strategy identified above.

Wylie discusses computational models of subsistence activities, such as Flannery and Reynold's (1986) simulation of agricultural spread based on data from Gila Naquitz cave. Simulation runs begin by evolving a diverse set of foraging activities, followed by the introduction of agricultural strategies. The model aimed to distinguish between hypotheses concerning what drove the emergence of agriculture and includes parameters aimed at capturing features revealed in actual data. Wylie argues that such models work as 'scaffolding' for various archaeological interpretations: they provision interpretations and their results drive further tests (see also Routledge 2021). This involves, as she notes, navigating between understanding *'how actually'* and *'how possibly'*—a feature emphasized downstream.

Modelling work also plays a crucial role in re-integrating and re-using old (or 'legacy') data. Both Wylie (2017b) and I (Currie 2021a) have emphasized how modelling strategies can aid in the resurrection of apparently lost data, and in bringing old data into renewed contact with up-to-date evidence and questions. As such, modelling is a way of bringing data into contact with hypotheses about the past beyond the discovery of new traces.

Like its neontological cousin, palaeoclimatology relies on a combination of empirical measurements and often-complex coupled computational systems that represent factors such as global temperature, albedo, ice-pack coverage, continental positions, and so forth. I have appealed to simulation studies of Snowball Earth to defend the notion of a 'surrogate experiment' and to suggest that idealisation strategies in the historical sciences are geared towards representing phenomena at the right grain for trace data to be relevant (Currie 2018b, chap. 10). Wilson and Boudinot (2022) highlight the use of vicarious controls, particularly highly localised models of measurement processes, to make sense of the diversity of proxy measurements palaeoclimatology relies upon.

Alisa Bokulich discusses the use of models to make sense of historical data, describing how models correct and represent fossil data in estimating biodiversity. This grounds further exploration of the nature of data models more generally (Bokulich 2021a). Expectations about biodiversity, represented by models, are brought into dialogue with data from the fossil record both to correct for the latter's (infamous) unreliability—often through vicarious post-hoc modelling work—but also to further refine the former. Bokulich also appeals to conceptual and table-top models representing the dynamics of river channels to explore how the 'tyranny of scales' is navigated in the geosciences (Bokulich 2021b). In brief, variations between conceptual and concrete models—sometimes inconsistent ones!—are used to identify threshold effects across scales and to tailor various models towards specific purposes.

Modelling is ubiquitous in historical science and is plausibly crucial for historical scientists to mitigate epistemic scarcity. This suggests that accounts of historical reasoning that emphasize trace-based reasoning misapprehend both historical method and its epistemic prowess.

3. Two models of the deep past

This section introduces two case studies. Each is picked to emphasize particular features of historical reasoning, which enrich our picture of the strategies historical scientists pursue.

3.1 Dinosaur tails

The first model consists of a set of .93-millimetre-thick pieces of plastic, laser-cut into similar shapes and (scaled) proportions as the tails of various reptiles, including small therapods, extant newts, crocodiles, and the study's central topic, *Spinosaurus*, the largest therapod dinosaur known. These 'tails' are attached to a robotic controller that can generate left-to-right or up-and-down movement and submerged in a water flume. As the plastic waggles in the flume, Ibrahim et al. (2020) are able to calculate the tails' thrust and efficiency:

> Our experimental results show that the Spinosaurus tail shape was capable of generating more than 8 times the thrust of the tail shapes of other theropods, and achieved 2.6 times the efficiency.
>
> *(Ibrahim et al. 2020, 69)*

Ibrahim et al. draw on the results to argue that the *Spinosaurus* was an aquatic pursuit predator. That is, the dinosaur used to hunt aquatic prey by, well, chasing them:

> [...] the vertically expanded tail shape of *Spinosaurus* imparts a substantial positive benefit to aquatic propulsion relative to the long and narrow tails of terrestrial theropods, supporting the inference that *Spinosaurus* used tail-propelled swimming. (69)

In 2005, the same palaeontological team discovered a remarkably complete specimen of *Spinosaurus aegyptiacus*. Their 2014 report (Ibrahim et al. 2014) recounted careful extraction and a functional morphological analysis in favour of *S. aegyptiacus* being semi- or fully aquatic. The long snout, for instance, is reminiscent of piscivores, while the reduced hindlimbs (compared to bipedal therapods) suggested unusual motion. These conclusions led to a flurry of objections (e.g. Hone and Holtz 2021; Brusatte 2021). In their 2020 paper, Ibrahim et al. respond, focusing on the tail.

In addition to the model, Ibrahim et al. defend their morphological reconstruction. They argue that the disarticulated fossil remains came from the same individual and provide a more detailed reconstruction of the tail. For instance, where standard theropods have stiff tails, providing balance for bipedal stances, *S. aegyptiacus'* was much more flexible, suggesting to Ibrahim et al. that the tail's morphology "... allowed it to function as a propulsive structure for aquatic locomotion" (69). Ibrahim et al.'s model was constructed as a way of testing a hypothesis: could, in fact, the tail produce the relevant thrust required by the specified function? They think yes.

3.2 Avalonian ecological communities

Our second model consists of a Spatial Point Process Analysis (SPPA), a set of equations that compare the spatial distribution of a set of objects to a stochastic null. These statistical analyses start from the notion of a *point process* or *field*, a collection of randomly assigned locations across a Euclidean space. By generating a set of randomly assigned locations (where the probability of a location being occupied is equal), we can generate a kind of reference—a spatially random distribution—against which various non-random distributions can be compared. In ecological contexts, Euclidean space is taken to represent—you guessed it—physical space, and the points are interpreted as taxa or ecotype locations.

By comparing the actual distribution of taxa to the random distribution generated by the model, ecologists determine whether that distribution can be understood stochastically (where taxa placement is independent) or whether there are further factors determining the spatial pattern. The stochastic distribution acts as a pseudo-null, and to the extent that the actual spatial distribution diverges from it, the more that spatial distribution is non-stochastic.[2] Mitchell et al. (2019) make use of SPPA to examine whether ecosystems are structured according to niche (deterministic/predictable) or neutral (stochastic) processes.

Most ecological systems are structured by *niche*: competition for resources leads to different species living in different environments or niches, which explains the spatial distribution of differing taxa. This niche model can be compared to *neutral* structures, where taxa are distributed stochastically. If we know the spatial distribution of a set of immobile taxa in an ecosystem, we can use a model to generate a null, basically, the kind of distribution to be expected if taxa did not compete. By comparing the actual distribution to the null, it can be determined whether taxa are distributed neutrally or by niche differentiation. Mitchell et al. use the model to analyse the oldest metazoan ecosystems.

Metazoan life arose in the Ediacaran approximately 575 million years ago (Narbonne 2005, Liu, Kenchington, and Mitchell 2015). Its earliest are the Avalonian assemblages found in Newfoundland and the UK, a series of sessile, benthic communities. These are dominated by rangeomorphs, frond-like creatures attached to the seafloor (of unknown metabolic, developmental, phylogenetic, and ecological status), and a thick, leathery microbial mat coating the seafloor. These assemblages were fossilised via volcaniclastic events and, due to the properties of the mat, and the lack of large, complex, mobile life, *in situ* position was likely preserved. Thus, we have the closest we can hope to of a *census community*: a snapshot of the various taxa and their spatial locations. If Mitchell and company can work out taxa membership, then, SPPA can be undertaken.

Mitchell and her team used laser-line probe technology to scan fossil beds (Mitchell and Butterfield 2018). This data was used to produce spatial and taxonomic data models that formed the basis of the distribution analysis. Mitchell et al. claim that the analysis provides "... strong evidence that neutral process dominated Avalonian assemblage communities" (Mitchell and Butterfield 2018). As SPPA is also used to study extant ecosystems, the model enables the comparison of Avalonian palaeocommunities to communities from other times, places, and ecological makeup:

> These neutral-process-dominated community dynamics contrast with those observed in the modern marine realm, where neutral processes are typically rare [...] This stark difference raises the question of whether Ediacaran early animal paleocommunities had fundamentally different community dynamics [...] to those of the present day.
>
> *(Mitchell et al. 2019, 2034)*

This point of difference potentially holds between Avalonian and Phanerozoic communities, Jackson and Blois (2015), for instance, found niche-structured communities in the Quaternary. What could explain this stark difference? Mitchell et al. point to a set of further differences, suggesting that the conditions of the Avalonian limited the development of the dynamics that underwrite niche differentiation:

> The studied Ediacaran paleocommunities have comparatively small populations, experienced frequent disturbance events, and include many taxa with short dispersal ranges, so within this framework we should expect neutral processes to dominate.
> *(2035)*

Additionally, Mitchell et al. emphasize that resource limitations were significantly fewer in the Avalonian. Thus, they argue that, given the relevant conditions, Avalonian communities do not challenge our ideas of how ecosystem structure should work. As opposed to fundamentally differing community dynamics, the same basic set of dynamics structure these communities the differences being due to the (ecologically) impoverished conditions of the Avalonian.

4. Modality and modelling in historical science

We are now in a position to provide a positive characterization of the role of models in historical reasoning. First, models are crucial for historical science because understanding possibility is crucial for historical science. Second, the 'phenomena-driven' strategy often adopted by historical scientists differs somewhat from the modelling strategy identified by Weisberg and Godfrey-Smith. While both involve surrogative approaches, the phenomena-driven modelling strategy is in deep conversation with historical data and phenomena.

4.1 Possible pasts

Historical science is as much about what is possible—what may have happened—as what actually happened. Even when historical scientists are primarily interested in understanding the actual past, this requires a rich understanding of possibility. We get to the actual by situating our data within possibility. First, trace-based reasoning requires an understanding of the regularities by which traces form (Jeffares 2008). Second, tests of hypotheses about the past do not only involve hunting for further traces, but also confirming *capacity-hypotheses*. Can certain kinds of structures in fact behave as posited—as seen in Ibrahim et al.'s robotic dinosaur tail? Third, the application of models across temporal contexts allows us to confirm and test more general regularities—as we see in the application of ecological models to the Avalonian.

The initial argument that *Spinosaurus* is an underwater pursuit predator leant on functional morphology and analogy. Some of *Spinosaurus*' morphological traits plausibly indicate a partially aquatic lifestyle, and others are common amongst fish-eaters. Such evidence, when coming from multiple sources, can make a strong case. However, it does not speak directly to particular *capacities*. For instance, the hypothesis is presaged on the idea that *Spinosaurus* could swim fast. This does not directly concern whether in fact it did: this is not a question of actuality, but a question of capacity or possibility.

Ibrahim et al. are reasonably well understood as performing a *Kon-Tiki Experiment*, named for Thor Heyerdahl's voyages from South America to the Pacific (Novick et al. 2020).

Heyerdahl sought to prove possible his preferred hypothesis for Pacific migrations: from the West via drift voyaging. Although his hypothesis has been rightly rejected, as the Pacific was purposefully settled from the East via extremely sophisticated navigational knowledge and technology (Holton 2004), it is plausible that his journey proved that drift voyaging from the Americas to the Pacific is possible. Ibrahim et al.'s experiment similarly seeks to show that a tail like *Spinosaurus'* could produce a significant amount of propulsion, at least compared to other large therapods. This establishes possibility, but not actuality.

How does establishing possibility aid Ibrahim et al.? It does not in itself establish that *Spinosaurus* was capable of fast underwater pursuit: these animals were not disembodied tails, but enormous carnivorous dinosaurs with great big sails on their backs. Rather, the various tail experiments explore a particular possibility space, one concerning various tail morphologies (or at least various plastic proxies of tail morphology) and their ability to produce underwater propulsion. The model departs from the real world:

> Some limitations of this study are the simplicity of the robotic structure design (i.e. a flat plastic tail profile mounted on an undulating rack) and the fact that motion, although set to an amplitude and speed informed by living undulatory swimmers (salamanders of the genus *Amblystoma* and the American alligator) did not account for specific anatomical constraints (i.e., vertebral motion ranges, flexibility, muscle configuration).
> *(Gutarra and Rahman 2022, 20)*

To make sense of the capacity hypothesis concerning the tail, it must be contextualised together with other features of *Spinosaurus*. The modal space must then be *integrated* into a larger picture of the animal (see Currie 2019b).

Ibrahim et al.'s study, then, can be understood as a test of the aquatic-pursuit hypothesis: had it rendered a negative result—if the tail could not support sufficient propulsion—then the aquatic-pursuit picture would be seriously undermined. However, the model goes further: it also provides crucial modal information about the relationship between tail morphology and aquatic propulsion.

Mitchell et al.'s work is not plausibly considered a Kon-Tiki experiment—they are not testing for some capacity required by some hypothesis—however, their application of SPPA models goes beyond inferring the neutrality of Avalonian metazoan communities. They also situate those communities in comparison to others and use it as a test case for the generality of the ecological regularities underlying SPPA models. Thus, past instances are used to establish regularities (Page 2021). As we saw above, the surprising neutrality-dominated ecosystems of the Avalonian raised the possibility that they obeyed 'fundamentally different dynamics', as Mitchell et al. put it. The model's predictions were a kind of anomaly: either ecosystems were different in the Ediacaran such that the model does not apply, or the model is getting it right and we need to explain its results further.

Mitchell et al. opt for the latter: once we take Avalonian conditions into account, our usual understanding of ecological communities will expect neutrality. Pointing to factors such as a lack of ecological maturity, resource competition, dispersal, etc., and comparing the Avalonian to analogues under similar conditions (which do tend towards neutrality), they show how the anomalous case fits within our understanding of ecosystems after all:

> While the dominance of neutral processes within these paleocommunities differs substantively from the majority of the modern marine realm, the underlying dynamics

> are entirely consistent with models of assembly which include both niche and neutral processes, and are similar to those of modern communities subject to the same conditions.
>
> *(Mitchell et al. 2019, 2015)*

While some philosophers—Cleland in particular—have argued that historical scientists are primarily interested in understanding the token, actual past (Cleland 2011), others have denied this on several fronts—and we see these in both of our modelling examples. First, trace-based reasoning relies crucially on regularities about how traces form (Jeffares 2008). Second, reconstructions of the past require an understanding of the various capacities posited in hypotheses about the past. If *Spinosaurus* was an underwater pursuit predator, it must have been capable of aquatic pursuit. Third, as Page (2021) has compellingly argued, sometimes the past is the key to the future: models of regularities can be tested, refined, and confirmed against the historical record. Indeed, historical investigation is often interested in 'fragile regularities' (Currie 2018a, chap. 7), regularities which only hold, often imperfectly, under certain conditions. With care, Mitchell et al. are able to further understand the ecological regularities SPPA models explore—their fragilities, and indeed how they might potentially work into the future—via their application to the past.

Therefore, historical sciences are not simply sciences of the past, but sciences of modality. Models are well suited for representing, exploring, and testing the contours of modal spaces. This concern with the possible explains the ubiquity of modelling practices in the historical sciences.

4.2 *Phenomena-driven modelling*

Many scientific pursuits, particularly theoretical ones, can be understood as exercises in possibility exploration. Indeed, philosophers have long recognized the critical importance of modal knowledge for prediction, explanation, and understanding. However, palaeoscience often explores possibility in a particular way: in intimate dialogue with historical data. Ibrahim et al.'s proxy exploration of thrust generated by pieces of plastic are constructed in light of, and justified through, comparisons with reconstructions of extinct organisms. Mitchell et al.'s use of SPPA requires complex data analysis and careful application to the Avalonian context. These modelling pursuits can be characterised as *phenomena-driven* (as opposed to *theory-driven*) investigations, and two things follow. First, it implies a different modelling strategy to that identified by Weisberg and Godfrey-Smith. Second, it helps us understand how fossil and other data ensure the relevance of the possibilities that historical scientists examine.

Many model-based sciences are *theory-driven* (Currie 2019b). In theory-driven investigation, the relevance of evidence is determined by the relationship between data and the target theory. Although many investigations of the deep past concern theories—Mitchell et al. for instance are interested in, explore, and employ ecological theories about community spatial composition and its relationship to the dynamics of those communities—they proceed primarily through the identification, articulation and explanation of particular empirical phenomenon (Dresow 2021). In *phenomena-driven* investigations, evidential relevance is determined by the phenomenon at hand. Mitchell et al. do not explore ecological community dynamics via further model-based explorations. The models they employ, how they employ them, and what they conclude are intimately tied to phenomena they use the model to identify: the neutral structure of Avalonian communities. An extraordinary amount of effort goes into characterizing these communities, including data collection, analysis, and application.

A major difference between modelling in theory-driven versus phenomenon-driven contexts concerns *relevance*. All modelling practices explore possibility, but what are these modal spaces good for? In particular, for what are they relevant? In theory-driven contexts (some contenders being game-theoretic and signalling models seeking to understand the emergence of co-operation, language, multi-cellularity, and so on) the distance between modelling results—as sophisticated and fascinating as they might be—and the real past, is potentially minimal (Currie and Sterelny 2017). There is little constraint provided by actual instances of the historical phenomena these modelling activities are supposed to speak to. Moreover, when actual history is brought into discussion, theoretical models are often shown to be barking up the wrong modal trees. Overall, models best understand the actual past when in intimate dialogue with evidence pertaining to that past and that dialogue reassures us that the possibility spaces examined are of relevance to the phenomenon of interest.

Phenomena-driven modelling differs from the strategy described by Weisberg. There, modellers take inspiration from natural phenomena, build, and examine a model capturing the basic contours of the phenomenon; eventually comparing modelling results to the original phenomena. In building the MBL model—at least at first—Raup et al. were interested in how patterns, qualitatively similar to those in the fossil record, could be generated in a highly simple, abstract, 'bare-bones' system. Phenomena-driven modelling is significantly more iterative, as features of the model are continually checked against the relevant properties of the phenomena in question. Ibrahim et al. are not simply interested in differences in broad tail-shape as captured by pieces of plastic, but go to some lengths to demonstrate that the relative proportions match the fossil reconstructions in relevant ways. The model does not afford an indirect strategy for the understanding of theory in the abstract, but is tailor-made in accordance with specific features of empirical data towards exploring highly specific questions.

In phenomena-driven modelling, idealisation is a tool that enables both tractability and relevance. For Mitchell et al., the spatial and taxic information that was retrievable from Avalonian deposits afforded the application of SPPA models. While for Ibrahim et al., a model needed to be constructed matching the specifics of the epistemic situation, that is, one that captured sufficiently the relevant features of therapod tails while also being amenable to experimentally establishing the relevant capacities. That models may be idealised and de-idealised to fit various levels of description is crucial for phenomena-driven modelling: with much care, modellers can ensure their representational tools are at the right grain for the available data to be relevant to them (see Currie 2018a, chap. 10).

5. Conclusion

Scientists interested in what was are just as interested in what could have been. Reconstructing token past events requires situating them in possibility space, past patterns and events are powerful sources of modal knowledge. Models are, if nothing else, tools for understanding modality: they explore possibility (Bokulich 2014; Massimi 2019; Knuuttila 2021; Wirling and Grüne-Yanoff 2021). Thus, the prevalence of modelling practices in the sciences of the deep past is no surprise. Nevertheless, historical scientists are not interested in any-old possibility: past phenomena guide their exploration, ensuring relevance. In such work, then, modelling practices are intimately concerned with, and are in iterative contact with, historical data. Knowing the deep past requires modelling the deep past.

Notes

1 For instance: both myself (2017) and Alisa Bokulich (2020) have emphasized the role of coherency-testing in historical science, while Ben Jeffares (2008), Meghan Page (2021) and Daniel Swaim (2019) emphasize the role, generation and testing of regularities.
2 The use of 'pseudo-null' hypotheses is not without controversy: Bausman (2018), Bausman and Halina (2018).

References

Antoniou, Antonis. 2021. "What is a data model?" *European Journal for Philosophy of Science*, 11(4): 1–33.
Bausman, William. 2018. "Modeling: neutral, null, and baseline." *Philosophy of Science*, 85(4): 594–616.
Bausman, William and Marta Halina. 2018. "Not null enough: pseudo-null hypotheses in community ecology and comparative psychology." *Biology & Philosophy*, 33(3): 1–20.
Bocchi, Frederica, Alisa Bokulich, Leticia Brache, Gloria Grand-Pierre and Aja Watkins. Forthcoming. "Are we in a sixth mass extinction? The challenges of answering and value of asking." *British Journal for the Philosophy of Science.*
Bokulich, Alisa. 2021a. "Using models to correct data: paleodiversity and the fossil record." *Synthese*, 198(24): 5919–5940.
———. 2021b. "Taming the tyranny of scales: models and scale in the geosciences." *Synthese*, 199(5): 14167–14199.
———. 2020. "Calibration, coherence, and consilience in radiometric measures of geologic time." *Philosophy of Science*, 87(3): 425–456.
———. 2014. "How the tiger bush got its stripes: 'How Possibly' vs. 'How Actually' model explanation." *The Monist*, 97(3): 321–338.
Bokulich, Alisa and Naomi Oreskes. 2017. "Models in geosciences." In *Springer Handbook of Model-Based Science*, edited by Lorenzo Magnani and Tommaso Bertolotti, 891–911. Dordrecht: Springer.
Brusatte, Stephan. 2021. "Spinosaurus." *Current Biology*, 31(20): R1369–R1371.
Cleland, Carol. 2013. "Common cause explanation and the search for a smoking gun." *Geological Society of America Special Papers*, 502: 1–9.
———. 2011. "Prediction and explanation in historical natural science." *The British Journal for the Philosophy of Science*, 62(3): 551–582.
———. 2002. "Methodological and epistemic differences between historical science and experimental science." *Philosophy of Science*, 69(3): 447–451.
Currie, Adrian and Kim Sterelny. 2017. "In defence of story-telling." *Studies in History and Philosophy of Science Part A*, 62: 14–21.
Currie, Adrian. 2021a. "Stepping forwards by looking back: underdetermination, epistemic scarcity and legacy data." *Perspectives on Science*, 29(1): 104–132.
———. 2019a. *Scientific Knowledge and the Deep Past: History Matters.* Cambridge: Cambridge University Press.
———. 2019b. "Mass extinctions as major transitions." *Biology & Philosophy*, 34(2): 1–24.
———. 2018a. *Rock, Bone, and Ruin: An Optimist's Guide to the Historical Sciences.* New York: MIT Press.
———. 2018b. "The argument from surprise." *Canadian Journal of Philosophy*, 48(5): 639–661.
———. 2017. "Hot-Blooded Gluttons: dependency, coherence, and method in the historical sciences." *British Journal for the Philosophy of Science*, 68: 929–952.
Dresow, Max. 2021. "Explaining the apocalypse: the end-Permian mass extinction and the dynamics of explanation in geohistory." *Synthese*, 199(3): 10441–10474.
Flannery, Kent and Robert Reynolds. 1986. "Simulating Foraging and Early Agriculture in Oaxaca." *Gila Naquitz: Archaic Foraging and Early Agriculture in Oaxaca, Mexico*, edited by Kent V. Flannery, 433–508. New York: Academic Press.
Forber, Patrick and Eric Griffith. 2011. "Historical reconstruction: Gaining epistemic access to the deep past." *Philosophy and Theory in Biology*, 3(201306): 1–19.

Godfrey-Smith, Peter. 2006. "The strategy of model-based science." *Biology & Philosophy*, 21(5): 725–740.
Gutarra, Susanna and Imran Rahman. 2022. "The locomotion of extinct secondarily aquatic tetrapods." *Biological Reviews*, 97(1): 67–98.
Holton, Graham. 2004. "Heyerdahl's Kon Tiki theory and the denial of the indigenous past." *Anthropological Forum* 14(2): 163–181.
Hone, David and Thomas Holtz Jr. 2021. "Evaluating the ecology of Spinosaurus: Shoreline generalist or aquatic pursuit specialist?" *Palaeontologia Electronica* 24(1): a03. https://doi.org/10.26879/1110
Huss, John. 2009. "The shape of evolution: the MBL model and clade shape." In *The Paleobiological Revolution: Essays on the Growth of Modern Paleontology*, edited by David Sepkoski and Michael Ruse, 326–345. Chicago: Chicago University Press.
Ibrahim, Nizar, Simone Maganuco, Cristiano Dal Sasso, Matteo Fabbri, Marco Auditore, Gabrielle Bindellini, David Martill, et al. 2020. "Tail-propelled aquatic locomotion in a theropod dinosaur." *Nature*, 581(7806): 67–70.
Ibrahim, Nizar, Paul Sereno, Cristiano Dal Sasso, Simon Maganuco, Matteo Fabbri, David Martill, Samir Zouhri, Nathan Myhrvold and Dawid A. Iurino. 2014. "Semiaquatic adaptations in a giant predatory dinosaur." *Science*, 345(6204): 1613–1616.
Jackson, Stephen and Jessica Blois. 2015. "Community ecology in a changing environment: perspectives from the quaternary." *Proceedings of the National Academy of Sciences*, 112(16): 4915–4921.
Jeffares, Ben. 2008. "Testing times: regularities in the historical sciences." *Studies in History and Philosophy of Science Part C*, 39(4): 469–475.
Kleinhans, Maarten, Chris Buskes and Henk de Regt. 2010. "Philosophy of earth science." In *Philosophies of the Sciences*, edited by Fritz Allhoff, 213–236. Oxford: Wiley-Blackwell.
———. 2005. "Terra incognita: explanation and reduction in earth science." *International Studies in the Philosophy of Science*, 19: 289–317.
Knuuttila, Tarja. 2021. "Epistemic artifacts and the modal dimension of modeling." *European Journal for Philosophy of Science*, 11(3): 1–18.
Leonelli, Sabina. 2019. "What distinguishes data from models?" *European Journal for Philosophy of Science*, 9(2): 1–27.
Liu, Alexander, Charlotte Kenchington, and Emily Mitchell. 2015. "Remarkable insights into the paleoecology of the Avalonian Ediacaran macrobiota." *Gondwana Research*, 27(4): 1355–1380.
Massimi, Michaela. 2019. "Two kinds of exploratory models." *Philosophy of Science*, 86(5): 869–881.
Mitchell, Emily, Simon Harris, Charlotte Kenchington, Philip Vixseboxse, Lucy Roberts, Catherine Clark, Alexandra Dennis, Alexander G. Liu and Philip Wilby. 2019. "The importance of neutral over niche processes in structuring Ediacaran early animal communities." *Ecology Letters*, 22(12): 2028–2038.
Mitchell, Emily and Nicholas Butterfield. 2018. "Spatial analyses of Ediacaran communities at Mistaken Point." *Paleobiology*, 44(1): 40–57.
Narbonne, Guy. 2005. "The Ediacara biota: neoproterozoic origin of animals and their ecosystems." *Annual Review of Earth and Planetary Science*, 33, 421–442.
Novick, Rose, Adrian Currie, Eden McQueen and Nathan Brouwer. 2020. "Kon-tiki experiments." *Philosophy of Science*, 87(2): 213–236.
Nyrup, Rune. 2020. "Three uses of analogy: a philosophical view of the archaeologist's toolbox." In *Interarchaeologia 6: Archaeology and Analogy*, edited by Marko Marila, Marja Ahola, Kristiina Mannermaa and Mika Lavento, 12–31. Helsinki: Department of Cultures, University of Helsinki.
Page, Meghan. 2021. "The role of historical science in methodological actualism." *Philosophy of Science*, 88(3): 461–482.
Raup, David, Stephen Gould, William Schopf and Daniel Simberloff. 1973. "Stochastic models of phylogeny and the evolution of diversity." *The Journal of Geology*, 81(5): 525–542.
Raup, David and Jack Sepkoski Jr. 1982. "Mass extinctions in the marine fossil record." *Science*, 215(4539): 1501–1503.
Routledge, Bruce. 2021. "Scaffolding and Concept-Metaphors: Building Archaeological Knowledge in Practice." In *Explorations in Archaeology and Philosophy*, edited by Anton Killin and Sean Allen-Hermanson, 47–63. Cham: Springer.
Sepkoski, David. 2012. *Rereading the Fossil Record: The Growth of Paleobiology as an Evolutionary Discipline*. Chicago: University of Chicago Press.

Sepkoski, David and Michael Ruse, eds. 2009. *The Paleobiological Revolution: Essays on the Growth of Modern Paleontology*. Chicago: University of Chicago Press.
Suppes, Patrick. 1966. "Models of data." *Studies in Logic and the Foundations of Mathematics*, 44, 252–261.
———. 2019. "The roles of possibility and mechanism in narrative explanation." *Philosophy of Science*, 86(5): 858–868.
Tamborini, Marco. 2021. "The material turn in the study of form: from bio-inspired robots to robotics-inspired morphology." *Perspectives on Science*, 29(5): 643–665.
———. 2020. "Technoscientific approaches to deep time." *Studies in History and Philosophy of Science Part A*, 79: 57–67.
Tucker, Aviezer. 2011. "Historical science, over- and underdetermined: A study of Darwin's inference of origins." *British Journal for the Philosophy of Science*, 62(4): 805–829.
———. 2004. *Our Knowledge of the Past: A Philosophy of Historiography*. Cambridge: Cambridge University Press.
Turner, Derek. 2011. *Paleontology: A Philosophical Introduction*. Cambridge: Cambridge University Press.
———. 2009a. "How much can we know about the causes of evolutionary trends?" *Biology & Philosophy* 24(3): 341–357.
———. 2009b. "Beyond detective work: Empirical testing in paleontology." In *The Paleobiological Revolution: Essays on the Growth of Modern Paleontology*, edited by Sepkoski, David and Michael Ruse, 201–214. University of Chicago Press.
———. 2007. *Making Prehistory: Historical Science and the Scientific Realism Debate*. Cambridge: Cambridge University Press.
———. 2005. "Local underdetermination in historical science." *Philosophy of Science* 72(1): 209–230.
———. 2004. "The past vs. the tiny: historical science and the abductive arguments for realism." *Studies in History and Philosophy of Science Part A*, 35(1): 1–17.
Weisberg, Michael. 2007. "Who is a modeler?" *British Journal for the Philosophy of Science*, 58 (2): 207–233.
Wilson, Joseph and Garrett Boudinot. 2022. "Proxy measurement in paleoclimatology." *European Journal for Philosophy of Science*, 12(1): 1–20.
Wirling, Yilwa and Till Grüne-Yanoff. 2021. "The epistemology of modal modeling." *Philosophy Compass* 16(10): e12775.
Wylie, Alison. 2017a. "Representational and experimental modeling in archaeology." In *Springer Handbook of Model-Based Science*, edited by L Magnani, 989–1002. Cham: Springer.
———. 2017b. "How archaeological evidence bites back: strategies for putting old data to work in new ways." *Science, Technology, & Human Values* 42(2): 203–225.

37

MODELS AND MEASUREMENT OF INEQUALITY

Alessandra Basso and Chiara Lisciandra

1. Introduction

Inequality is a broad and multifaceted concept. Generally speaking, we talk of inequality whenever the distribution of a certain resource in a population departs from a desirable state of equity. Inequality is typically thought of in relation to economic goods, such as income, wealth, or consumption. Handbooks and reports for policymaking often define it in relation to poverty: both concepts are concerned with the distribution of economic resources among a population; however, while poverty focuses only on the lower end of the distribution, inequality refers to disparities within the distribution (Rohwerder 2016; UN 2013; McKay 2002; Haughton and Khandker 2009).

The scientific investigation of inequality relies on both modeling and measurement. Economists build theoretical and statistical models of inequality that represent the distribution of income within a population (McGregor et al. 2019). Moreover, inequality is a topic of empirical research that aims to measure inequality in such a way that it can be compared across regions and tracked over time. This chapter focuses on the relationship between models and measurements of income inequality and highlights the importance of their interdependence.

The relationship between models and measurement has recently been emphasized by *model-based accounts of measurement* (Tal 2017; 2020; this volume; Boumans 2006, 2015; this volume; Morgan 2001). According to model-based accounts, measurement involves two interrelated levels: a concrete procedure to gather empirical indications and a model of the measurement process, constructed from simplifying assumptions about the object being measured, the instrument, and the surrounding environment. In this view, models and measurement are highly intertwined. On the one hand, models need measurement to assign values to the models' parameters. On the other hand, measurement needs models for the interpretation of empirical data.

In order to show the interplay of models and measurement of income inequality, Section 2 starts by introducing one of the most common models to represent income inequality in society, the Lorenz curve. The Lorenz curve is a theoretical model of income distribution that provides a graphical illustration of how income is distributed across a population.

 DOI: 10.4324/9781003205647-43

In this section, we emphasize that this model is closely related to the Gini index, which is a statistical model that summarizes the dispersion of inequality in a population. Today, the Gini index is one of the most commonly employed statistics to measure inequality. Within a representational view of measurement, Gini index values represent inequality with numbers (Vessonen 2021). For instance, when measuring national inequality, relations between Gini index values represent relations between countries in terms of inequality.

To gain empirical significance, the Lorenz curve and the Gini index need data. By looking at recent debates about global inequality, Section 3 provides examples of how empirical data from household surveys, tax records, and national accounts are employed to produce inequality outcomes based on models like the Lorenz curve and the Gini index. The role of models in measurement, however, is not limited to providing tools to elaborate empirical data (Tal 2025, this volume). Model-driven considerations are also required to interpret and understand inequality outcomes. At the same time, the measurement process can serve as a test for the models' underlying assumptions and can reveal important insights into how to refine the modeling instruments themselves.

2. Models of income inequality

This section discusses the Lorenz curve and the Gini index. The first of these models can be considered a theoretical model, while the latter is a statistical one. The formulation of these models exemplifies a dominant practice in the economists' study of inequality. First, inequality models are built on a set of theoretical principles or criteria. Then, as we will see in the next section, such models are used as "instruments" to measure properties of interest and interpret their outcomes (Boumans 2001).

2.1 Lorenz curve

The Lorenz curve is a model of income distribution that provides a graphical illustration of how income is distributed across a population. In the process of building a model of inequality, social scientists start with a set of desiderata that the model should satisfy.[1] Typically, the following theoretical principles apply to a model of income inequality:

1 *Anonymity principle*
2 *Population principle*
3 *Relative income principle*
4 *Transfer principle*

The *anonymity principle* states that income inequality only depends on the actual income of the individuals, and not on any other identifying features. For instance, when comparing the income of Alice and Bob, no features of Alice and Bob matter—whether Alice is more virtuous or older than Bob—other than their income.

The *population principle* states that income inequality is relative to the size of the population whose inequality is measured. This way, populations with different sizes in absolute numbers can be compared with each other with respect to their inequality.

The *relative income principle* states that inequality is relative to total income and to the currency in which the income is calculated. This way, populations whose income differs in absolute terms, or whose income is calculated in different currencies can still be compared with each other independently of such factors.

The *transfer principle* states that when a transfer of income occurs from a richer individual to a poorer individual, inequality decreases—unless the transfer is so large that inequality flips across individuals. In this way, a distribution A is more equal than a distribution B, when part of what is owned by the richer in distribution B is owned by the poorer in distribution A, all other things being equal.

These principles fulfill different functions: the first principle ensures that the analysis focuses on income inequality only, excluding other kinds of inequality, such as inequality of endowments, and how inequality comes about, such as whether it depends on merit or luck. The second and third principles require that inequality normalize over population and income. Finally, the fourth principle serves as a sort of testbed: if a model of income inequality violates the principle, it does not qualify as an adequate model, because it goes against the principle that a transfer from the poorer to the richer increases inequality.

Based on these criteria, we can define a function that determines the inequality of the distribution of income in a population. The *Lorenz curve* is a graphical representation of income distribution in a population, that satisfies the four criteria stated above (Lorenz 1905). Figures 37.1 and 37.2 below show a series of Lorenz curves that exhibit different levels of inequality.

To see how to build a Lorenz curve, the following steps apply: first, a population is divided into sub-groups that are arranged in ascending order of income. In the specific examples represented below, a population is divided into quintiles. Second, a graph reports on the horizontal axis *cumulative* population groups, where the first segment is the poorest

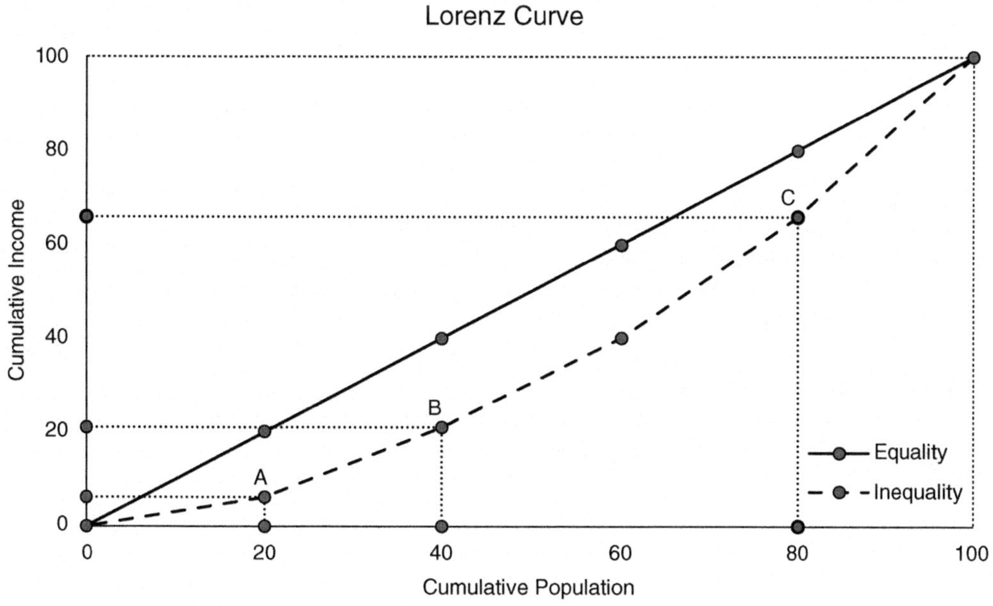

Figure 37.1 Lorenz curve.

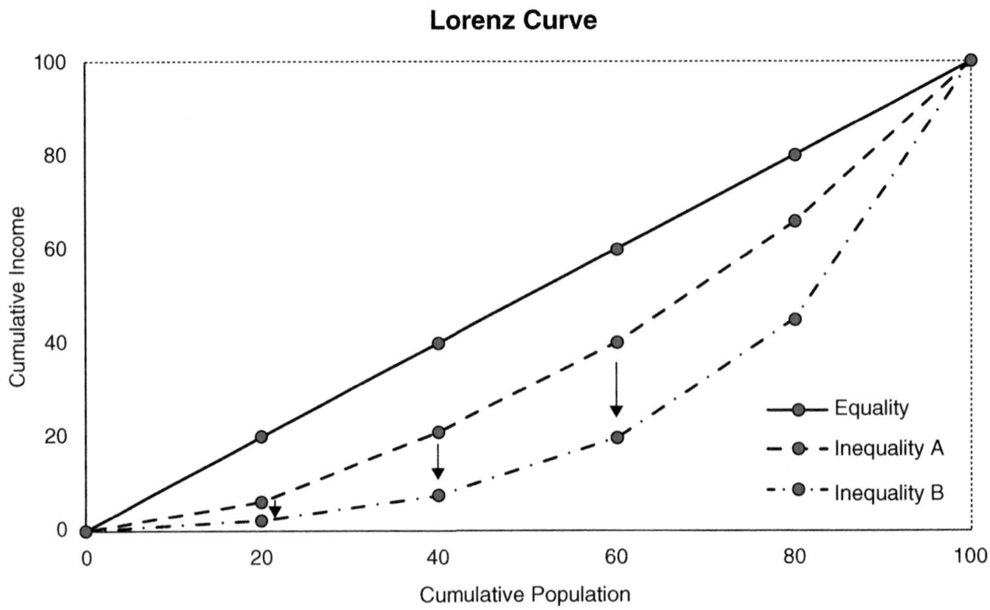

Figure 37.2 Lorenz curve and transfer principle.

20% of the population; the first and second segments together are the poorest 40% of the population, and so on, up to 100% of the population. Similarly, the vertical axis reports *cumulative* income groups.

In a population where income is distributed equally, the Lorenz curve corresponds to the 45-degree line. This is because the lowest income group owns 20% of the overall income; the second income group owns 20% of the overall income; and so on up to the entire population. Each group has the same share of overall income. The opposite situation, i.e., where inequality is highest, displays a curve that coincides with the horizontal axis almost entirely and then with the vertical axis at the very end.

In Figure 37.1, the curve to the right of the equality line represents an unequal distribution of income. To see this, note that the poorest quintile has less than 20% of the overall income (point A on the curve). The first and second poorest quintiles have less than 40% of the overall income (point B on the curve). This means that their income is below what they would have if the distribution were equal. Moreover, it is possible to see that the richest segment of the population owns all that comes roughly after 60% of cumulative income (point C on the curve). In other words, the richest 20% of the population owns roughly 40% of the overall income, which is the difference between the income of the entire population and all that is owned by the rest of the population.

Two properties of the Lorenz curve are particularly relevant. First, Lorenz curves always fall below the 45-degree line, and they are concave, facing up and increasing. Lorenz curves cannot bend in the other direction because population segments that come earlier on the horizontal axis have lower incomes than those who come later. Secondly, as the curves go more and more to the right, i.e., as the distance from the 45-degree line increases, the level of inequality increases.

To see this, consider the situation in Figure 37.2. Take the curve in the graph that is labeled "Inequality B," which is closer to the x-axis. It is possible to see that for "Inequality B," the first segments of the population have a lower income share, as compared to the curve labeled "Inequality A." At the same time, the richest segment of the population starts earlier, having around 40% of total income. This means that the richest 20% now own roughly 60% of overall income. To put it more simply, to derive the second curve B from the first curve A, a transfer of income has occurred from a poorer segment of the population to a higher segment; therefore, following the *transfer principle*, inequality has increased in the population.

One final consideration about the Lorenz curve concerns cases where curves cross paths (Figure 37.3). This happens when transfers occur both from richer segments of the population to poorer segments (also called *progressive transfers*) and vice versa, from poorer segments to richer segments (also called *regressive transfers*). As Figure 37.3 shows, going from the curve labeled "Inequality A" to the curve labeled "Inequality B," both kinds of transfers occur. In situations like these, it is not possible to use the Lorenz curves to conclude that one distribution is more equal than the other; such cases remain ambiguous. In other words, while the *transfer principle* states that regressive transfers determine an increase in inequality, it leaves it open to compare cases where income distribution changes for both *regressive and progressive transfers.*

To conclude, the Lorenz curve is a model of income distribution that starts from a set of theoretical principles, or criteria, that the model should satisfy. The same four principles underlie another model of income inequality, the Gini index, whose main properties we will discuss in the next section.

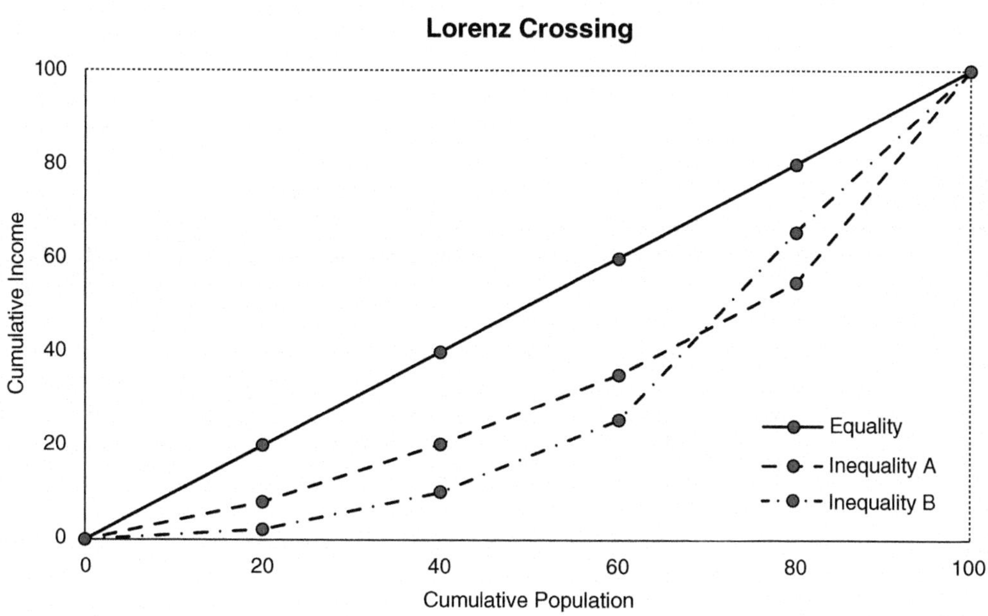

Figure 37.3 Lorenz crossing.

2.2 The Gini index

Lorenz curves are helpful tools to compare inequality graphically. However, it would often be convenient to express levels of inequality in numbers rather than via diagrams. In the literature, several statistical models have thus been built to quantify inequality levels. Such models aim to find appropriate ways to calculate inequality as a property of the data that reflects the dispersion and concentration of resources in a group. Among these, the Gini index, which is named after the statistician who formulated it, is the most extensively used statistic to measure inequality (Gini 1914).

The Gini index considers the difference between each income group and the others by means of pairwise comparisons. The main intuition behind this index is that inequality in a population increases as the sum of the distances between income groups increases.[2]

More formally, the *Gini index* is calculated as follows

$$G = \frac{1}{2n^2\mu}\sum_{j=1}^{m}\sum_{k=1}^{m} n_j n_k \left|y_j - y_k\right|$$

where n stays for the individuals in a population; y for income and j for income groups, with $j = 1, \ldots, m$; μ is the average income in the population.

The index is the sum of the difference in income between each group $\left|y_j - y_k\right|$ for all the members $n_j n_k$ of such groups divided by twice the population square and average income.

It is easy to see that the Gini index is consistent with the four principles stated above. It is consistent with the *anonymity principle* as only the income of the individual matters to their ordering. It satisfies the *population principle* and *the relative income principle* as the measure normalizes over population size n and average income μ.

Moreover, the Gini index is consistent with the *transfer principle*. Intuitively, one way to grasp this is to consider that the sum of the difference in income between income groups can only increase if the distance between two of them gets larger, i.e., if a transfer occurs from the poorer to the richer. More specifically, take an income group—call it I_p—that becomes poorer because of a transfer. After the transfer, I_p gets closer to the other poor income groups; thus, their distance d decreases by a certain amount—call it λ; at the same time, I_p gets further away from the richer income groups; thus, their distance increases, and it increases exactly by the same amount λ. While these changes cancel each other out, the only amount that effectively increases is that between the two income groups that have transferred income from one to the other. Since the poorer get poorer and the richer get richer, the distance from each other increases and, accordingly, the Gini index goes up.

With respect to the relation between the Gini index and the Lorenz curve, it is possible to show that the Gini index is the numerical version of the graphical representation of the Lorenz curve. To see this, consider Figure 37.4. Gini himself (1914) showed that we can use the Lorenz curve to calculate a *ratio of concentration*, which is measured by the area between the Lorenz curve and the 45-degree line—the area called *a* in Figure 37.4—divided by the area of the triangle *BCD*. This ratio is the numerical version of the Lorenz curve and is equivalent to the Gini index. Given that curves of higher inequality cover larger and larger areas as they move to the right of the 45-degree line, the ratio will increase, and the Gini index will increase correspondingly as the overall difference in income grows too. This shows that the Gini index and the Lorenz curve offer two alternative models to represent income inequality in a population (Schneider 2021).

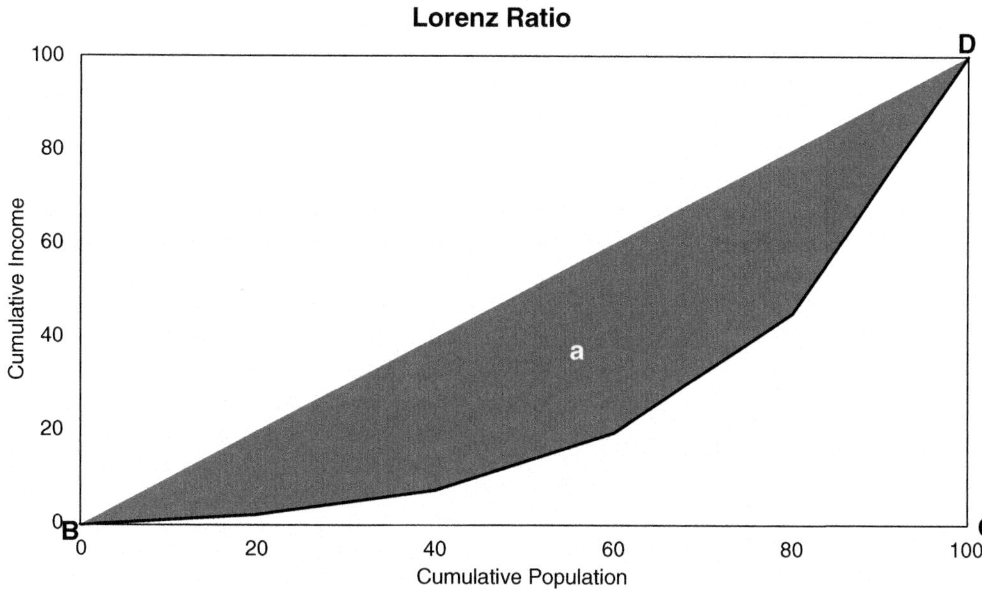

Figure 37.4 Lorenz ratio.

To gain empirical content, the Lorenz curve and the Gini index need measurement. Social scientists rely on a variety of data sources to measure income inequality, which is elaborated on the basis of these theoretical and statistical models. Nevertheless, the role of models in measurement is not limited to providing the underlying tools for constructing an income distribution and summarizing its inequality into a single statistic. The interpretation of the measurement outcomes, too, requires considerations that appeal to the underlying theoretical and statistical models that made measurement possible. The next section illustrates this with an example.

3. Measurements of income inequality

National statistical institutes employ the Gini index to measure income inequality within countries based on household surveys, tax records, and national accounts. These measurements allow researchers to compare income inequality across countries, but they do not provide information regarding the distribution of income across the world. How unequal is the world? Is global inequality on the rise, or is it declining? As long as we use the Gini index—or the Lorenz curve—to measure inequality within single countries, we cannot answer these questions. For example, measuring inequality within countries allows us to say that country *x* has higher inequality than country *y*, but it does not give us information about overall inequality, i.e., about whether a country is richer or poorer overall. To figure out which countries have the largest share of global income and how that share has evolved over time, we need to look at the distribution of global income. When measuring global inequality, the world is treated as a single entity, and the Gini index summarizes the level of inequality in the world as a whole.

Global inequality is more difficult to measure than national inequality because it requires aggregating national incomes into a single distribution (cf. Basso forthcoming).

This is different from the case of national inequality. Currencies do not matter when comparing national inequality across countries. The reason is that, thanks to the *relative income principle* (see Section 2.1), statistical tools like the Gini index depend on relative income differences and are therefore invariant to the absolute level of such incomes or the currency in which they are expressed. The investigation of global inequality, instead, treats the world as a single entity and therefore requires constructing an aggregate income distribution. As a consequence, national currencies matter again because, when aggregating income from multiple countries, the currency in which income is reported varies.

To investigate global income inequality, researchers use a variety of approaches. Some investigate inequality of income among all people in the world, independently of where they live (e.g., Ferreira and Ravallion 2009; Lakner and Milanović 2013; 2016). Others focus on the inequality of aggregated national income across countries (e.g., Piketty 2014). In addition, there are a variety of methods to combine incomes that are expressed in different currencies and to account for differing living standards across nations. Alternative methods, however, produce different and sometimes conflicting results. One method suggests that the world has entered a phase in which rich and poor countries are converging in income (Piketty 2014). Another method indicates instead that global income inequality has remained rather constant in the past decades (Lakner and Milanović 2013; 2016). The question arises as to which interpretation in correct. With no external reference to test the outcomes, how can researchers tell between reliable and unreliable findings?

There is no simple answer to this question. Researchers have developed strategies to detect errors and improve their measurements, but global inequality is a multifaceted phenomenon that cannot be summarized by a single index producing unambiguous results (Piketty 2014, 66). Instead, it is possible that when considering certain dimensions, global inequality appears to be declining, but when considering others, it seems to have remained constant through time.

When confronted with contrasting findings, however, researchers must decide whether they are looking at erroneous outcomes or at distinct but complementary dimensions of global inequality. Because of this ambiguity in the interpretation of alternative measurements, the literature on global inequality provides us with an opportunity to highlight a tension between measurement errors and the definition of the parameter being measured via the model. Making progress in measurement is not simply a matter of intervening in a procedure that makes mistakes; it also involves testing the underlying theoretical assumptions about the parameter under measurement, and refining them when needed.

To evaluate and improve inequality measurement, researchers compare fallible measurements of inequality to each other. Alternative measurements can capture distinct conceptualizations of inequality and rely on different data sources. Consequently, alternative methods are prone to different sources of error. Comparing a measurement to another that is not subjected to the same sources of error can help estimate these errors and devise ways to remedy them. When measuring global inequality, researchers are aware of specific methodological weaknesses that could compromise the reliability of their findings. For instance, the exchange rates that are used to combine incomes in different currencies have significant margins of error. In this case, researchers can compare measurements that are based on different exchange rates to evaluate the extent to which their findings depend on the exchange rate used. Household surveys, on the other hand, tend to underestimate top incomes—they often fail to correctly report the incomes of the very rich. By comparing household surveys to other measurements that are not subject to this source of error, researchers are able to estimate the amount of unreported income.

These kinds of comparisons, however, only work under the assumption that alternative measurements are about the same parameter of interest, so that

- Agreement among them can be taken as a sign that the outcomes are not influenced by the methodological differences,
- Disagreement can be taken as a sign of error.

However, the interpretation of discrepancies is ambiguous because they could also indicate that the alternative measurements capture distinct concepts of global inequality (Tal 2017; 2019). Conceptual discrepancies are not mistakes in themselves but are rather related to differences in what precisely is being measured (Blanchet, Chancel, and Gethin 2019). Depending on the underlying theoretical and statistical models, existing measurements of global inequality may focus on different types of income and distinct population units, but researchers sometimes assume that they are roughly about the same broad inequality concept. When comparing these measurements to each other, therefore, disagreement can be a sign of error, but could also be due to conceptual discrepancies. Consequently, the assessment of global inequality is not solely a matter of comparing measurement results but rather involves an evaluation and refinement of the underlying representation of inequality (Basso 2017). In the assessment of measurement, modeling and procedural considerations are interdependent.

The following subsections provide examples of assessment practice in global inequality measurement. While both examples are successful in improving the accuracy of measurements by comparing alternative measurements to each other, their findings are in contrast. Piketty (2014) suggests that global income inequality is on the decline, while Lakner and Milanović (2013; 2016) suggest instead that it has remained constant in the last decades. This raises the question of whether their disagreement is a sign of inaccuracy or conceptual discrepancies. Only a combination of measurement and model-driven considerations can provide a solution to this dilemma and reconcile seemingly contradictory results.

3.1 Converging national incomes

Piketty's (2014) work on global income inequality highlights a converging trend among countries' national incomes. After a period in which the global production of goods and services was concentrated in Europe and America, the world appears to have entered a phase of convergence. The share of global income of wealthy countries has been declining since the 1970s, while the national income of poor countries has been steadily on the rise.

When discussing this result, Piketty warns us that the margin of error is considerable. In particular, he is concerned about the uncertainty associated with the exchange rates used to combine national incomes expressed in multiple currencies. The most commonly employed method for combining different currencies relies on Purchasing Power Parities (PPP) exchange rates, which adjust for the differing purchasing powers of national currencies based on a set of assumptions. Exchange rates, however, are rather uncertain. There is more than one way to calculate PPP, and there are several vexing issues with their employment.[3] Moreover, global inequality measurements vary significantly when using different exchange rates. For example, Piketty notes that global inequality would be markedly higher if he used current exchange rates rather than PPP rates. Because of the uncertainties surrounding exchange rates, Piketty tests the sensitivity of his main results to the choice of exchange rates. To do that, he compares the global inequality outcomes produced using current exchange rates and PPP rates. He observes that the choice of exchange rates has a significant influence on global inequality measurements, but the orders of magnitude remain the same. The historical trend in global income inequality is robust across alternative exchange rates, and this strengthens confidence in this particular robust result

> "Still, the orders of magnitude remain the same, as does the fact that the share of income going to the wealthy countries has been declining steadily since the 1970s. Regardless of what measure is used, the world clearly seems to have entered a phase in which rich and poor countries are converging in income."
>
> *(Piketty 2014, 67)*

The confidence in the robust result is increased by showing that it does not depend on the choice of exchange rate. Alternative measurement methods, however, highlight different historical trends.

3.2 *Stable inequality of individual incomes*

Using an alternative method of measuring global inequality, which is subject to different sources of error, Lakner and Milanović (2013; 2016) suggest instead that global inequality has remained rather stable in the past decades.

Lakner and Milanović study income inequality between people around the world, regardless of where they live. Their work is based on a different model and conceptualization of inequality than Piketty's (2014). The main difference with Piketty's approach is that Lakner and Milanović study inequality among people rather than among nations. Therefore, they rely on different data sources. Piketty's work relies on national account data, which provides information about national aggregates such as the gross domestic product and its main components. These data have the advantage of being available for extended periods of time and covering a broad range of countries. Because their focus is on personal income, Lakner and Milanović cannot rely on the aggregate data of national accounts but employ household surveys instead. Household surveys provide information about individual income, but they are subject to additional sources of error, like the underreporting of top incomes.

The underreporting of top incomes is a well-known source of error in the measurement of income inequality on the basis of household surveys (Atkinson and Piketty 2010; Atkinson, Piketty, and Saez 2011). The sample size of a typical household survey is too small to capture the incomes of the very rich. In addition, extreme incomes in the survey data are top-coded or eliminated as outliers. The very rich are also less likely to participate in surveys and more likely to understate their own income (Lakner and Milanović 2013; 2016). As a result, the estimates of individual income inequality are likely to be downward biased.

The underreporting of top incomes can be estimated and corrected only by comparing the outcomes based on household surveys to a measurement that is not affected by this source of error. Recent works on the size of top income shares relative to the rest of the distribution have been using tax records and national accounts to correct for this source of error (Atkinson and Piketty 2007; 2010). Although tax data and national accounts still suffer from misreporting, especially at the bottom of the distribution, they are considered more reliable than surveys for top incomes.

As part of their work on the evolution of global inequality, Lakner and Milanović (2013; 2016) use the gap between national accounts and household surveys as a rough estimate of top income under-reporting. Because national accounts provide aggregate national income rather than individual incomes, they can be used to estimate the overall amount of unreported income, but this amount must be allocated across the population. After estimating the amount of unreported income, Lakner and Milanović allocate it across the income distribution, imputing it more to the top decile than to the rest of the distribution.[4] In order to check the robustness of the outcomes across alternative allocation methods, their analysis considers alternative scenarios, which impute different weights to the top tail. Without correction, the measurement outcomes show a decrease in global inequality from 1988 to 2008. However, with a top-heavy allocation of the gap, this result almost entirely dissipates.[5] In other words, once corrected, the outcomes no longer exhibit a decreasing trend.[6] According to the authors, this result "further supports a more cautious view about the decline in global inequality: if indeed surveys tend to underreport incomes at the very top, it could well be that global inequality, measured by the Gini index, has not gone down during the twenty-year period considered here" (Lakner and Milanović 2013, 39).[7]

The question arises of wheather this conclusion is at odds with Piketty's finding of a converging trend in national (rather than individual) income, or whether this discrepancy is attributable to conceptual differences between national income inequality and individual income inequality. Global income inequality appears to fall or be steady at the same time, depending on the way we look at it. Based on a combination of measurement and model-based considerations, the World Inequality Report 2022 provides a model of global inequality that reconciles these contrasting findings (Chancel et al. 2022, 11). The representation of global inequality reproduced in Figure 37.5 combines two components of global inequality – between and within-country inequality – and highlights their respective contributions to global inequality. This model is able to reconcile findings based on different ways of conceptualizing and measuring global inequality. Between-country inequality reflects the trend of national income inequality as measured by Piketty (2014), while within-country inequality conveys the discrepancies in individual incomes. Since 1980, inequality between countries has declined because of the economic catch-up and the strong growth in some emerging countries. At the individual level, however, this effect is mitigated by a sharp increase in inequality within most countries, making global inequality – as measured by Lakner and Milanović – steadily high. While national incomes are becoming more equal, they are becoming more and more unequally distributed among individuals within each country. Therefore, the decline in inequality between countries is contrasted by an opposite trend within countries and, as a result, the distribution of individual income across the world's population remains highly unequal, as found by Lakner and Milanović (2013). The lesson to be learned is that the growth of national income brings about a reduction of inequality between individuals only if the revenue is equally distributed among the country's population. If, instead, the rising national income is concentrated in the hands of a few, inequity between individuals increases rather than falls.

To reconcile seemingly contrasting findings, the model provided in the World Inequality Report brings out the empirical and conceptual relationship between two aspects of global inequality that jointly contribute to determining its trends. In this sense, measurement can

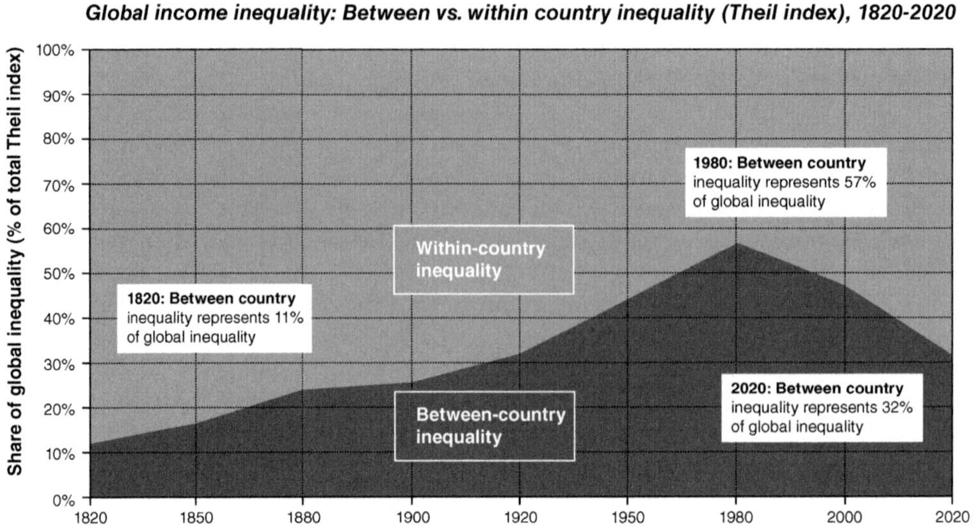

Figure 37.5 Global income inequality: between vs. within country inequality. From World Inequality Report 2022 (Chancel et al. 2022 13).

serve as a test for the models' underlying assumptions: discordant measurement outcomes reveal conceptual discrepancies and encourage a refinement of how inequality is represented. On the other hand, the model refines the conceptualization of global inequality by identifying its distinct components.

Overall, understanding trends in global inequality requires both models and measurements. Neither alone is sufficient to answer our questions. Theoretical and statistical models guide the elaboration of empirical data. In the assessment of measurement, moreover, contrasting measurement findings are not necessarily mistaken unless model-driven considerations indicate that they are about the same broad concept of global inequality. Model-driven considerations can also clarify conceptual discrepancies prompted by the outcomes of measurement procedures. Either way, the modelling of global inequality contributes to advancing the understanding of the measured parameter and its components.

4. Conclusion

In the investigation of income inequality, models and measurement are interdependent. The Lorenz curve and the Gini index respond to the theoretical principles that are used to model the distribution of income across a population. On the basis of models like the Lorenz curve and the Gini index, social scientists elaborate income data from household surveys, national accounts, or tax records to provide outcomes that summarize the inequality of the income distribution across a population. The interpretations of these outcomes, however, also require model-driven considerations. Contrasting outcomes about global income inequality can be reconciled once they are interpreted on the basis of a more complex model able to clarify the relations between complementary components of global inequality.

Notes

1 For a more detailed review, see Ray (1998) and Todaro and Smith (2012).
2 Current data show that countries with low inequality levels have a Gini index between 0.20 and 0.35, while countries on the higher side of the inequality spectrum have Gini index between 0.50 and 0.70 (Todaro and Smith 2012).
3 For instance, it is important to emphasize that PPP is based on price indices that measure different aspects of social life. The price of energy measures purchasing power of energy, while the price of health services measures purchasing power in that area. It is difficult to combine and weigh price indices in different areas to reflect people's purchasing needs/habits across the world. For a discussion of issues related to estimation of price indices see Reiss (2008, chap. 2).
4 This involves making assumptions about the corrected shape of the income distribution. Lakner and Milanovic assume a Pareto upper-tailed distribution of top incomes.
5 Lakner and Milanović found that the gap between national accounts and surveys has risen over the considered period. In other words, the under-reporting of top income has increased. If top income underreporting increases, we get the (misguided) impression that inequality is decreasing.
6 Using a more sophisticated method to improve the comparability across national survey data, Blanchet, Chancel and Gethin (2019) do not observe a reduction in income inequality in Europe since the early 1980s. This result is consistent with Lakner and Milanovic (2016)'s analysis.
7 Other results, instead, are robust: "Using the income reported in surveys we concluded above that the global income distribution had moved from a twin-peak to a single-peak. This also holds for the global distribution of income which adjusts for missing top incomes" (Lakner and Milanovic 2013, 40).

References

Atkinson, Anthony B., and Thomas Piketty, eds. 2007. *Top Incomes over the Twentieth Century: A Contrast between Continental European and English-Speaking Countries*. Oxford; New York: Oxford University Press.

———. 2010. *Top Incomes: A Global Perspective*. Oxford; New York: Oxford University Press.

Atkinson, Anthony B., Thomas Piketty, and Emmanuel Saez. 2011. "Top Incomes in the Long Run of History." *Journal of Economic Literature* 49(1): 3–71. https://doi.org/10.1257/jel.49.1.3.

Basso, Alessandra. 2017. "The Appeal to Robustness in Measurement Practice." *Studies in History and Philosophy of Science Part A*, The Making of Measurement, 65–66 (October): 57–66. https://doi.org/10.1016/j.shpsa.2017.02.001.

———. forthcoming. "The Comparison of Inequality Measurements across Countries and Time." *The British Journal for the Philosophy of Science*, February. https://doi.org/10.1086/719568.

Blanchet, Thomas, Lucas Chancel, and Amory Gethin. 2019. "How Unequal Is Europe? Evidence from Distributional National Accounts, 1980–2017." *World Inequality Working Papers*, https://wid.world/document/bcg2019-full-paper/.

Boumans, Marcel. 2001. "Measure for Measure: How Economists Model the World into Numbers." *Social Research* 68(2): 427–553.

———. 2006. "The Difference between Answering a 'Why' Question and Answering a 'How Much' Question." In *Simulation: Pragmatic Construction of Reality*, edited by Johannes Lenhard, Günter Küppers, and Terry Shinn, 107–124. Dordrecht: Springer.

———. 2015. *Science Outside the Laboratory: Measurement in Field Science and Economics*. Oxford; New York: Oxford University Press.

Chancel, Lucas, Thomas Piketty, Emmanuel Saez, Gabriel Zucman, Felix Bajard, Francois Burq, Rowaida Moshrif, Theresa Neef, and Anne-Sophie Robilliard. 2022. 'World Inequality Report 2022.' World Inequality Lab. https://wir2022.wid.world/

Ferreira, Francisco H. G., and Martin Ravallion. 2009. "Poverty and Inequality: The Global Context." In *The Oxford Handbook of Economic Inequality*, edited by Brian Nolan Wiemer Salverda, and Tim Smeeding. Oxford: Oxford University Press. https://doi.org/10.1093/oxfordhb/9780199606061.013.0024.

Gini, Corrado. 1914. "Sulla misura della concentrazione e della variabilità dei caratteri," *Atti del R.Istituto Veneto di Scienze, Lettere ed Arti* 73(2): 1203–1248.

Haughton, Jonathan Henry, and Shahidur R. Khandker. 2009. *Handbook on Poverty and Inequality*. Washington, DC: World Bank.
Lakner, Christoph, and Branko Milanović. 2013. "Global Income Distribution: From the Fall of the Berlin Wall to the Great Recession." *World Bank, Policy Research Working Paper No. 6719*.
———. 2016. "Global Income Distribution: From the Fall of the Berlin Wall to the Great Recession." *World Bank Economic Review* 30(2): 203–232.
Lorenz, Max. 1905. "Methods of Measuring the Concentration of Wealth." *Publications of the American Statistical Association* 9(70): 209–219.
McGregor, Thomas, Brock Smith, and Samuel Wills. 2019. "Measuring Inequality." *Oxford Review of Economic Policy* 35(3): 368–395.
McKay, Andrew. 2002. "Defining and Measuring Inequality." Briefing Paper No 1. UK Department for International Development by the Economists' Resource Centre.
Morgan, Mary S. 2001. "Making Measuring Instruments." In *The Age of Economic Measurement. History of Political Economy. Annual Supplement*, edited by Judy L. Klein, and Mary S. Morgan, 235–251. London: Duke University Press.
Piketty, Thomas. 2014. *Capital in the Twenty-First Century*. Harvard, MA: Harvard University Press.
Reiss, Julian. 2008. *Error in Economics. Towards a More Evidence–Based Methodology*. London: Routledge.
Ray, Debraj. 1998. *Development Economics*. Princeton, NJ: Princeton University Press.
Rohwerder, Brigitte. 2016. *Poverty and Inequality: Topic Guide*. Birmingham: GSDRC, University of Birmingham.
Schneider, Michael. 2021 "The Discovery of the Gini Coefficient: Was the Lorenz Curve the Catalyst?" *History of Political Economy* 53(1): 115–141.
Tal, Eran. 2017. "A Model-Based Epistemology of Measurement." In *Reasoning in Measurement*, edited by Nicola Mößner and Alfred Nordmann, 233–253. London: Routledge.
———. 2019. "Individuating Quantities." *Philosophical Studies* 176(4): 853–878. https://doi.org/10.1007/s11098-018-1216-2.
———. 2020. "Measurement in Science." In *The Stanford Encyclopedia of Philosophy*, edited by Edward N. Zalta. Metaphysics Research Lab, Stanford University, https://plato.stanford.edu/. https://plato.stanford.edu/archives/fall2020/entries/measurement-science/.
———. 2025. "Models and Measurement." In *The Routledge Handbook of Philosophy of Scientific Modeling*, edited by Tarja Knuuttila, Natalia Carrillo, and Rami Koskinen. Routledge.
Todaro, Michael, and Stephen Smith. 2012. *Economic Development*. London: Pearson.
UN. 2013. "Inequality Matters. Report on World Social Situation 2013." Report. United Nations' Department of Economic and Social Affairs.
Vessonen, Elina. 2021. "Representation in Measurement." *European Journal for Philosophy of Science* 11(3): 76. https://doi.org/10.1007/s13194-021-00365-6.

38

FORMAL LANGUAGE THEORY AND ITS INTERDISCIPLINARY APPLICATIONS

Chia-Hua Lin

1. Introduction

Formal language theory (FLT) studies mathematically defined languages. In FLT, a language is a set of strings composed of a given alphabet based on a set of rules. As a branch of mathematics and a key component of theoretical computer science, FLT deals with the description, manipulation, and classification of formal languages. In this chapter, the basic elements of formal language theory will be presented, as well as its application to linguistics, computer science, and animal cognition.

Grammars and automata are formal systems used in FLT for the analysis and classification of languages, as well as serving as means to produce sentences within a language and even as a way to check whether certain strings are syntactically valid according to a given language. These formal systems are explained in Section 2. Section 3 then examines how these models are used to determine a hierarchy of formal languages according to how the language relates to these models. These systems, as well as a scheme that classifies them called the Chomsky hierarchy, have then served as models to investigate diverse phenomena including natural languages, computer code written in programming languages, and the cognitive infrastructure of humans and non-human animals (Section 3). In these interdisciplinary applications of FLT, the Chomsky hierarchy serves as a model template used to understand the relation between different languages in computing science and to describe the difference between different cognitive infrastructures in animal cognition. A model template refers to any model that was successful in a field and is then applied to a different system, or even in a different discipline. These model templates have been characterized as formal structures (Humphreys 2019) or as something closer to a formal-conceptual complex (Knuuttila and Loettgers 2016, 2022). Examples of models that have been used as templates in other disciplines include the Ising model and the harmonic oscillator. This chapter discusses the role of the Chomsky hierarchy as a model template, and some philosophical implications (Section 4).

 DOI: 10.4324/9781003205647-44

2. The basic components of formal language theory

In FLT, a formal language is a subset of the set of all strings that can be arranged using a number of symbols (including zero) of a pre-declared alphabet according to a set of rules. All such well-formed strings are called sentences of a language, whereas strings that do not follow the rules are syntactically invalid with respect to the language. For instance, consider an alphabet of only two symbols: *a* and *b*. The set that consists of strings such as {*ab, abab, ababab*} can be a formal language, and so is the set {*ab*, *aabb*, *aaabbb*, *aaaabbbb*}. These two examples show that despite sharing an alphabet, two languages can be distinguished by the distinctive patterns among their sentences.

Using set notations to describe the pattern is one of three ways to define a formal language. Take the two examples above, for instance. One writes $\{(ab)^n: 1 \leq n \leq 3\}$ to refer to the language {*ab*, *abab*, *ababab*}. The notation indicates that, of the language in question, a sentence is composed of a number of *a*'s and *b*'s in pair concatenation, up to three such pairs. Similarly, the notation $\{a^nb^n: 1 \leq n \leq 4\}$ refers to the language {*ab*, *aabb*, *aaabbb*, *aaaabbbb*}, whose sentences are strings of a number of *a*'s followed by the same number of *b*'s. Any string violating either pattern disqualifies as a sentence of the respective language. Formal languages can be infinite in size while still holding a pattern. For instance, $\{(ab)^n\}$ refers to {ε, *ab*, *abab*, *ababab*, ...}, a set that contains an empty string, i.e., when $n = 0$, denoted by ε, as well as other strings of the pattern $(ab)^n$ denoted by "..."

In FLT, lowercase letters are reserved for *terminal* symbols. Much like the 26 letters in the English alphabet, terminal symbols are the symbols that can appear in a sentence. In contrast, upper case letters are typically used to denote *non-terminal* symbols. Much like names of sentence structure types in English, such as *Subject*, *Verb*, and *Noun Phrase*, non-terminals indicate different parts of a sentence. They serve as placeholders for terminals during the process of sentence formation, and, as such, a sentence is always free of non-terminals.

With non-terminal symbols, one can define languages with more sophistication. For instance, $\{(AB)^n: n \geq 1\}$ refers to an infinite language whose sentences are pairs of an *A*-class symbol and a *B*-class symbol. Let the *A*-class be the set of all odd numbers between 1 and 9, and let the *B*-class be the set of all even numbers in the same range. Then the strings *12*, *3892*, *723498* qualify as sentences of said language, whereas strings *2*, *43*, *02374* do not. The patterns $(AB)^n$ and A^nB^n have been applied in the study of animal cognition for preparing experimental stimuli, an episode to be discussed in Section 3.3.

In addition to set notations, one may describe a language by designing an abstract machine that does the following: it accepts as input all sentences of the language and rejects the input when the string is syntactically invalid with regard to the language. Such an abstract machine, also called a decision program, operates on the basis of checking whether an input string conforms to a set of predefined patterns. To illustrate, consider an automaton that has been programmed to recognize $\{(AB)^n: n \geq 1\}$. Upon receiving a string as input, the automaton will read the first symbol of the string. In the case in which the symbol is *not* an odd number between 1 and 9, the automaton enters a "rejecting" state, writes "no" as output, and then halts the program. Otherwise, i.e., in the case in which the symbol *is* an odd number between 1 and 9, the automaton moves to read the second symbol of the string. In the case in which the second symbol is *not* an even number between 1 and 9, the automaton again moves to the "rejecting" state, writes "no" as output, and then halts the program. Otherwise, the automaton moves on to read the third symbol of the string if it has one. In the case in which the string does not have a third symbol, the automaton moves to the "accepting" state and writes "yes" as output before halting the program. Otherwise, the automaton continues. It checks as it did with the first two symbols, whether the third and the fourth symbols are a pair of an odd number and an even number between 1 and 9, and halts the program when the string is either rejected or accepted.

The third, and final way to define a language is to articulate a finite set of rules that dictate how symbols of an alphabet should be arranged to generate a sentence. A set of such rules, called grammar, also displays patterns. By placing increasingly more restrictive requirements on the patterns that grammar can take, Chomsky (1959) created a hierarchy of four grammar types. From the most restricted to the least, they are Type 3 regular grammars, Type 2 context-free grammars, Type 1 context-sensitive grammars, and finally Type 0 unrestricted grammars.

Chomsky's creation of the grammar hierarchy became of tremendous practical significance when it comes to programming physical computing machines. The more restricted a grammar is, the less complex a language it generates. However, the lower the degree of complexity a language has, the less powerful automaton it requires to recognize the language. This observation results in a hierarchy of four automaton types, each of which is distinguishable by the nature of its memory store, and is matched with a grammar type for the complexity of the formal languages that they express or recognize. The hierarchy of four formal language types is, from the most complex to the least, as follows: Type 0 recursively enumerable languages, Type 1 context-sensitive languages, Type 2 context-free languages, and finally Type 3 regular languages. See Figure 38.1 for a summary of the Chomsky hierarchy.

At the outermost level of the hierarchy, Type 0 recursively enumerable languages contain all languages that can be recognized by a particular kind of automaton called a Turing machine. A formal language is recursively enumerable if it is logically possible to design a decision program that, when presented with any of its sentences as input, will halt and accept the input, reject the input, or otherwise loop it forever (the possibility of looping forever, which is important in computer science, does not concern our present purpose). A Turing machine is an abstract model that can simulate any algorithm, including any Type 0 unrestricted grammars. When a Turing machine is programmed to recognize a Type 0 recursively enumerable language, it is functionally equivalent to the Type 0 unrestricted grammars that can generate the same language. This principle applies to other automaton types throughout the hierarchy. Only Types 2 and 3 of the Chomsky hierarchy are directly applied in the studies relevant to this entry, but because other automata are conceptual extensions of the Turing machine, it is instrumental to take a closer look at it.

A Turing machine (Turing 1936) consists of a "head," a "tape," and a program that dictates its behavior. The tape of a Turing machine is a one-dimensional, infinite memory store divided into discrete cells that can hold a symbol from a given alphabet. The behaviors of the head include: reading the symbol in a cell on the tape, erasing or writing a symbol, while the head either remains in its current position or moves along the tape in either direction.

Type 1 context-sensitive languages are the subset of Type 0 languages that can be recognized by linear-bounded automata, a less powerful automaton type than the Turing machine. Unlike the infinite tape in a standard Turing machine, a linear-bounded automaton operates on a tape whose length is bounded by a function of the input size. Type 1 languages are called "context-sensitive" because the grammars that generate them consider the surrounding symbols when determining whether a string is permissible.

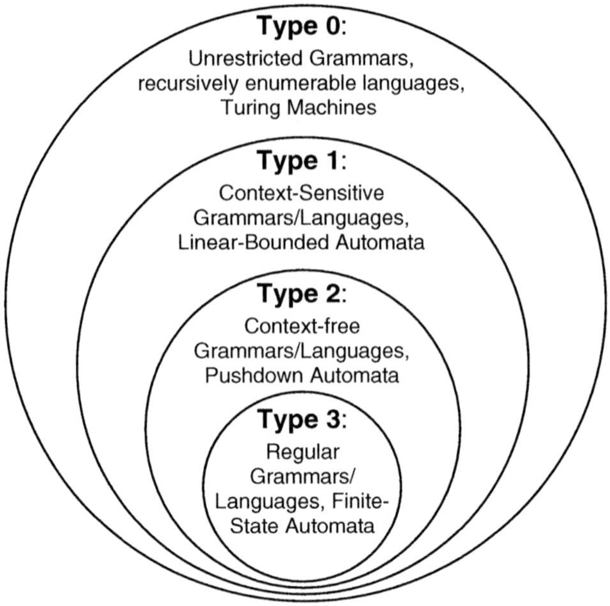

Figure 38.1 Formal language theory.

Type 2 context-free languages are the subset of Type 1 languages that can be generated by pushdown automata, a less powerful automaton type than the linear-bound automata. Unlike the tape in a linear-bound automaton, a pushdown automaton's memory store is a *last-in-first-out* stack. It pushes items down the stack and pops the latest one out before an earlier item may be retrieved. A pushdown automaton accepts the input string if, after processing the entire input, it reaches an accepting state and the stack becomes empty. If the stack is not empty or the automaton reaches a non-accepting state, the input is rejected. Moreover, Type 2 context-free languages are called "context-free" because the grammars that generate them do not need to meet the context requirement: during sentence formation, the next symbol that can be generated by context-free grammar does not depend on the previous symbols that have been generated. Nonetheless, Type 2 context-free grammars are more restricted than Type 1 context-sensitive grammars in other ways, making them a less expressive type of sentence generator than Type 1 context-sensitive grammars.

Finally, Type 3 regular languages are the subset of Type 2 languages that can be generated by finite-state automata, the simplest automaton type in the hierarchy. Such abstract machines operate without an explicit memory store. Like other types of automata, a finite-state automaton includes an initial state and one or more accepting states and rejecting states. Type 3 regular languages are generated by Type 3 regular grammars, the most restricted and thus, the least expressive of the four grammar types. Because finite-state automata can be programmed as sentence acceptors for Type 3 regular languages, with proper programming, a finite-state automaton can be functionally equivalent to the Type 3 regular grammars that generate Type 3 languages.

Crucial to applying the Chomsky hierarchy is the idea of restricted expressive power: each type of grammar or automaton can only be programmed to generate or recognize up to a language of its corresponding type. For instance, a Type 3 regular grammar or a finite-state automaton is too weak to generate or recognize a "supra-regular" language, i.e., any language beyond Type 3. Similarly, a Type 2 context-free grammar or a pushdown automaton cannot be made to generate or recognize a language beyond context-free, and so forth. This principle is implemented in the study of animal cognition.

3. Applying the Chomsky hierarchy as a classification scheme

Chomsky, the linguist responsible for developing the hierarchy of grammars, is the first to apply it in science as a scheme for classifying systems into different types. In his work (1956; 1959) on the syntax of natural language (Section 3.1), Chomsky shows that English is beyond Type 3 because Type 3 finite-state automata cannot model the syntax of English. Chomsky's use of the hierarchy influenced two other applications of FLT: First, shaping compiler design in computer science in which computer code is classified into different language types (Section 3.2), and subsequently, informing the study of animal cognition in which cognitive infrastructure across species is classified into different automaton types (Section 3.3).

3.1 Modeling the syntax of natural languages

Chomsky (1956) argues that finite-state automata lack adequate power to model the syntax of English. His argument hinges on two linguistic phenomena of English: that there is no upper bound for the length of an English sentence, and that there are pairs of words that are in what he calls "long-distance dependence."

For instance, consider the following English ***sentence: "If*** it rains, ***then*** the ground will be wet." There is no upper bound on the length of sentences of this sort because following the word "***if***," one may add infinitely many clauses—such as "Mary knows that," "Tom believes that," "the man whose jacket is black thinks that," and so on—without resulting in an ungrammatical sentence. That is, despite changing meanings, and while the resulting sentence will quickly become incomprehensible to typical English users as more clauses are added on, at no point would modifications of this sort yield an ungrammatical string of words. This observation indicates that English is open-ended. In contrast, one can immediately undermine the grammaticality of the initial sentence by replacing, for instance, "***if***" with "***either***" without also replacing "***then***" with "***or***." This suggests that the word-pair "***if***" and "***then***," and for that matter "***either***" and "***or***," are in long-distance dependence. One consequence of these observations is that to model the syntax of English using an automaton, the abstract machine would need to handle indefinite lengths of symbols between pairs of words in long-distance dependence.

In order to handle both features, an automaton needs either a flexible number of states or an explicit memory store. However, by definition, a finite-state automaton has a fixed number of states and lacks an explicit memory store. Thus, the features of long-distance dependency and open-endedness in English will result in sentences that are admissible by native English speakers but escape finite-state automata. Hence, no finite-state automata can adequately model English.

Chomsky's application of FLT gave rise to the hierarchy that bears his name and the idea of supra-regularity: The hierarchy became a theoretical pillar of computer science (Section 3.2), whereas the idea of supra-regularity later inspired the experimental classification of animal cognition (Sections 3.3). Because both regular grammars and finite-state automata are Type 3, Chomsky's work is interpreted as arguing that English is *supra-regular*, and consequently, to model the syntax of English, one would need a *supra-regular* grammar, a formal system with more expressive power than a Type 3 regular grammar.

3.2 Designing compilers for computer code

The significance of FLT to computer science was widely acknowledged after it was found that the syntax of one of the early programming languages (ALGOL) is Type 2. This theoretical discovery would shape the programming practice of compiler design for years to come (Hyman 2010). Unlike English, programming languages, such as C++, Python, or machine language, are artificial, i.e., they are the products of deliberate design. It follows that applying FLT in computer science is not about "revealing" the syntax of a language, as it was in linguistics, but about anticipating the type of language of computer input: when a computer needs to handle code that belongs to a Type 2 language, e.g., to parse said code, it needs to be functionally equivalent to, or more powerful than, a Type 2 automaton.

In software engineering, programmers use what are called high-level languages, such as Python, Java, or C++ to write code. Before executing the code, a computer first checks whether the code is syntactically valid for the programming language. If it is, the code is then translated into a machine language. A compiler is a program that performs these two tasks. To ensure performance, a compiler programmer must make sure that the program rejects all syntactically invalid code and accepts all syntactically valid code. In other words, this part of the compiler is an automaton that carries out the decision program.

To code a decision program, the compiler designer first classifies the code that the compiler will be dealing with and then chooses a corresponding automaton type for the task. The underlying rationale is that when the syntax of a programming language belongs to a particular Type n (where $0 \leq n \leq 3$) on the Chomsky hierarchy, then all code written in that language will be correctly recognized by a properly programmed automaton of Type n.

Moreover, because programming languages are products of design, one can specify the syntax of a programming language to be, say, Type 2. This step is much like in the study of mathematically defined languages when one defines a language by specifying a set of rules that dictate how a set of symbols may be arranged to generate well-formed strings. When the syntax of a programming language is deliberately designed to be Type 2, all code properly written in that programming language, which amounts to well-formed strings of a Type 2 grammar, will be Type 2. Consequently, for a computer to recognize any code written in that language, its compiler would need to be programmed with a Type 2 pushdown automaton or beyond.

In other words, applying the Chomsky hierarchy in compiler design amounts to utilizing the principle that because a Type n grammar generates a Type n language, programming a compiler for a Type n language requires the implementation of at least a Type n automaton. Applying this principle, along with the idea of supra-regularity (Section 3.1), scientists designed artificial languages to investigate animal cognition, which we turn to next.

3.3 Modeling the linguistic gap between humans and other animals

With a brand-new interpretation of the formal systems in FLT, cognitive biologist Tecumseh Fitch and comparative psychologist Marc Hauser (2004) implemented the Chomsky hierarchy to articulate the evolutionary gap in linguistic behaviors: an organism with a Type 3 cognitive infrastructure cannot recognize a supra-regular language. Fitch and Hauser (2004) used an alphabet consisting of two classes of syllables, *A*-class and *B*-class, to design artificial languages in the patterns of $(AB)^n$ and A^nB^n, respectively. The organisms participating in the experiment, including humans, are exposed to either language, and the tests of their learning outcomes are compared across groups. The experimenters found that while humans can recognize both patterns $(AB)^n$ and A^nB^n, tamarin monkeys can only recognize the pattern $(AB)^n$. Based on these results, Fitch and Hauser suggest that tamarin monkeys lack something that humans have, either in their internal program or in their brain, or both, which they conclude "would minimally be a push-down stack" (378).

One key premise of Fitch and Hauser's experiment of artificial grammar learning (AGL) is that the participants that cannot recognize the pattern A^nB^n but can recognize the pattern $(AB)^n$ are limited to learning only Type 3 grammar, whereas the participants that recognize both patterns are able to learn up to Type 2 grammars. The experimenters support this premise by elaborating on how they designed the experiment to test for supra-regularity in their participants.

First, Fitch and Hauser chose the patterns $(AB)^n$ and A^nB^n to generate the experimental stimuli because the string set $\{(AB)^n\text{: } n \geq 1\}$ is a Type 3 regular language, while the string set $\{A^nB^n\text{: } n \geq 1\}$ Type 2 is a context-free language with the feature of long-distance dependence (Section 3.1). For instance, the shared alphabet consists of, as terminals, two sets of voice-recorded syllables: *A*-terminals = {**ba, di, yo, tu, la, mi, no, wu**} and *B*-terminals = {pa, li, mo, nu, ka, bi, do, gu}. The audio recordings of these syllables, voiced by a female and a male speaker, respectively, are then used to form sentences of two different languages. For example, the strings "**no** li **ba** pa" and "**la** pa **wu** mo **no** li" belong to $\{(AB)^n\text{: } n \geq 1\}$ because the former is an instance of $(AB)^2$, i.e., two pairs of *A*-terminal followed by a *B*-terminal in concatenation. Similarly, the string "**la** pa **wu** mo **no** li" is an instance of $(AB)^3$. In contrast, strings such as "**yo la** pa do" and "**ba la tu** li pa ka" belong to $\{A^nB^n\text{: } n \geq 1\}$ because they are instances of A^2B^2 and A^3B^3, respectively.

Moreover, the language $\{A^nB^n\text{: } n \geq 1\}$ has the feature of long-distance dependence. To correctly recognize strings that belong to $\{A^nB^n\text{: } n \geq 1\}$, one needs to at least be able to compare the number of *A*-terminals and the number of *B*-terminals in a given string. This requires participants to commit the *A*-terminal syllables to their memory store at the same time as an input string is revealed to them. The items stored previously would then be retrieved upon deciding the string's membership. For this reason, Fitch and Hauser consider the language $\{A^nB^n\text{: } n \geq 1\}$ supra-regular and use it in their AGL experiment to test supra-regularity across species.

A typical AGL experiment includes a training phase and a brief re-familiarization phase followed by a test phase at the end. In Fitch and Hauser's experiment, 20 cotton-top tamarin monkeys (*Saguinus Oedipus*) of varying age and sex were evenly divided into two groups of ten. In the training phase, all ten participants were exposed to 20 minutes of repeated playback of 60 grammatical strings in random order. The group that was assigned to learn the pattern $(AB)^n$ was exposed to training strings in patterns such as AB, ABAB, or ABABAB. Similarly, the group assigned to learn the pattern A^nB^n was exposed to training strings in the patterns of AB, AABB, or AAABBB. In both cases, the *A*s and *B*s are implemented by voice recordings with randomly chosen *A*-terminals and *B*-terminals from their shared alphabet.

After a brief re-familiarization, the test phase began while the participants' behavior was monitored and videotaped. When the participating animal was both looking down and away from the loudspeaker in the space, the experimenter played a sequence of testing stimuli.

Between the two groups of participants, the test stimuli consisted of eight strings in total: four strings consistent with $(AB)^n$, while the other four were consistent with A^nB^n. All eight strings were novel to both groups because the strings were excluded from their training stimuli.

The two groups of monkeys responded systematically differently to the test stimuli. In the group trained with the pattern $(AB)^n$, nine out of the ten looked more often toward the loudspeaker when it played recordings that violated the pattern. That is, the participants seemed to pay more attention to the A^nB^n strings than to the strings consistent with their training set. To Fitch and Hauser, this suggests two things. First, the monkeys could distinguish between the *A* and *B* classes of terminals in their training languages. Second, and more importantly, the monkeys are sensitive to the grammar that generates their training set, meaning that they are capable of learning Type 3 grammar. This second point is a standard inference from the so-called "familiarization-novelty" experiment protocol. It is common to conclude whether a pre-linguistic subject recognizes a pattern in the training stimuli based on the subject's reaction toward novel violations. An important precursor to Fitch and Hauser's work can be found in Saffran, Aslin, and Newport (1996). The assumption is that subjects show more interest in novelty when they are sufficiently familiarized with a pattern in the training material. Thus, looking more frequently toward novel violations is considered similar to verbal feedback as it indicates that the subjects have learned the pattern in the training string set.

In contrast, the group trained with the pattern A^nB^n showed no statistically significant difference in their looking behavior throughout the test phase. Fitch and Hauser (2004) interpret this indifference to novelty to indicate that this group of monkeys failed to recognize the pattern in their training language. In particular, they point out that all extraneous factors in the experiment were consistent between the two training languages. For example, the stimuli used across groups are of the same length and loudness. Subjects can perceive that there are two classes of terminals, *A* and *B*, as shown in their recognition of the pattern of $(AB)^n$. The duration of exposure, testing, and evaluation procedures were the same. Moreover, as the authors argue, earlier work with this species using the same experimental procedure has demonstrated that these animals can store and recall up to three separate stimuli and compare them with subsequent strings. Fitch and Hauser (2004) thus conclude that the tamarins' inability to learn supra-regular grammar must be due to a constraint on their cognitive infrastructure.

Yet in stark contrast, all 20 adult human participants "showed rapid learning of either" language ... and were easily able to discriminate grammatical from non-grammatical stimuli for both grammars (Fitch and Hauser 2004, 379). For Fitch and Hauser, these responses suggest that the human participants are able to recognize both Type 3 and Type 2 grammars, whereas the tamarin monkeys are only able to recognize Type 3 grammars.

Over the course of ten years, the subject of AGL research in comparative psychology has grown from understanding the evolution of human language faculty to looking for the neural substrate of the pushdown stack (Fitch and Friederici 2012). Based on the findings of the AGL experiments, Fitch (2014) suggests that this neural substrate, once found, will explain how the human brain processes linguistic structures like the automata process non-terminal elements in supra-regular grammars. A likely location for the pushdown stack could be at the "inferior frontal gyrus (IFG, comprising of Broca's area and its neighbors), with sensory and association regions in the temporal and parietal lobes" (2014, 355). In Fitch's words:

> [R]everberations in the fronto-sensory feedback loop would play the role of the stack in the pushdown automaton implementing a context-free grammar.... [T]he IFG

> would thus have an additional storage mechanism into which intermediate results (and in particular unfinished structural computations) could be placed for later retrieval.
>
> *(355)*

In other words, the IFG to Fitch is the structure in the brain that "serves as a kind of 'abstract scratchpad'" much like a memory store in a compiler involved in processing non-terminals before all the symbols turn terminal.

4. Interdisciplinary model application

The progression of the three applications of FLT as discussed in this chapter shows that, including the Chomsky hierarchy, the basic components of FLT are given diverse interpretations across linguistics, computer science, and cognitive biology, appropriate to the phenomenon investigated in each respective discipline.

Modeling practices involving a mathematical construct striding an interdisciplinary trajectory, much like that of the Chomsky hierarchy, have been studied by philosophers in terms of the transfer of a model template (Knuuttila and Loettgers 2016; 2022). A model template is a conceptual framework embedded within a mathematical form. The notion of a model template is developed to explain why certain mathematical models or computational methods that have similar forms can successfully be applied across disciplines to study phenomena in different domains, even when the connections between these domains are not immediately apparent. Using the transfer of the Chomsky hierarchy as an example, Lin (2019) argues that by introducing FLT to the AGL experiment, Fitch and Hauser (2004) turned the experimental procedure into a supra-regularity detector. In other words, supra-regularity, an idea that came from Chomsky's work on the syntax of natural language, served as the organizing idea, or a model template, behind Fitch and Hauser's use of the Chomsky hierarchy, and is arguably key to the success of the transfer.

The Chomsky hierarchy's importance is not only as an example of model transfer—the study of each of the transfers has allowed philosophers to understand the features of model templates more generally. For instance, examining the interdisciplinary use of the Chomsky hierarchy leads Lin (2022) to develop the notion of *spillovers* to illustrate the truth-functional or justificatory dependence between two modeling efforts that share a mathematical construct. The notion of spillover captures *how* and *when* the transfer of a template across disciplines becomes epistemically consequential, allowing us to better understand the nature of the model transfer practice. A spillover is a knowledge claim from a prior use of a mathematical construct that is essential for justifying a knowledge claim for a subsequent use of the same mathematical construct. As Lin (2022) argues, by juxtaposing the notions of a model template and a spillover, one may distinguish between two kinds of justificatory dependency between modeling efforts that stem from the same template. A prior modeling effort—such as Chomsky's work on the syntax of natural language which introduces the idea of supra-regularity—serves as a model template for a subsequent modeling effort when the latter is conceptually dependent on the former. In contrast, a prior modeling effort provides a spillover for a subsequent modeling effort when the latter is truth-functionally dependent on the former. An example of a spillover can be found in Fitch and Hauser's (2004) use of the Chomsky hierarchy. In it, the principle from compiler design—parsing input of a Type n language requires an implementation of at least a Type ***n*** automaton (discussed in Sections 3.2)—is indispensable for justifying the claim that supra-regularity is absent in tamarin monkeys.[1] This dialog between philosophy and Formal Language Theory also illustrates the way that studying specific models closely allows us to understand modeling practices more generally.

Note

1 To better understand the notion of a spillover through this example, see Lin (2022, Sections 4.2.1 and 5).

References

Chomsky, Noam. 1956. "Three Models for the Description of Language." *IRE Transactions on Information Theory* 2(3): 113–124. https://doi.org/10.1109/TIT.1956.1056813

———. 1959. "On Certain Formal Properties of Grammars." *Information and Control* 2: 137–167. https://doi.org/10.1075/bjl.1.08mil

Fitch, W. Tecumseh. 2014. "Toward a Computational Framework for Cognitive Biology: Unifying Approaches from Cognitive Neuroscience and Comparative Cognition." *Physics of Life Reviews*, 11(3): 329–364. https://doi.org/10.1016/j.plrev.2014.04.005

Fitch, W. Tecumseh, and Angela D. Friederici. 2012. "Artificial Grammar Learning Meets Formal Language Theory: An Overview." *Philosophical Transactions of the Royal Society B: Biological Sciences*, 367: 1933–1955. https://doi.org/10.1098/rstb.2012.0103

Fitch, W. Tecumseh, and Marc D. Hauser. 2004. "Computational Constraints on Syntactic Processing in a Nonhuman Primate." *Science* 303(5656): 377–380. https://doi.org/10.1126/science.1089401

Humphreys, Paul. 2019. "Knowledge Transfer across Scientific Disciplines." *Studies in History and Philosophy of Science Part A* 77: 112–119. https://doi.org/10.1016/J.SHPSA.2017.11.001

Hyman, D. Malcolm. 2010. "Chomsky between Revolutions." In *Chomskyan (R)evolutions*, edited by Douglas A. Kibbee, 265–298. John Benjamins Publishing Company, Amsterdam/Philadelphia. https://doi.org/10.1075/z.154.09hym

Knuuttila, Tarja, and Andrea Loettgers. 2016. "Model Templates within and between Disciplines: From Magnets to Gases – and Socio-Economic Systems." *European Journal for Philosophy of Science*, 6(3): 377–400. https://doi.org/10.1007/s13194-016-0145-1

———. 2022. "Model Templates: Transdisciplinary Application and Entanglement." *Synthese* 201: 200. https://doi.org/10.1007/s11229-023-04178-3

Lin, Chia-Hua. 2019. "Tool Migration: A Framework to Study the Cross-disciplinary Use of Mathematical Constructs in Science." PhD diss. University of South Carolina.

———. 2022. "Knowledge Transfer, Templates, and the Spillovers." *European Journal for Philosophy of Science* 12(1): 1–30.

Saffran, Jenny R., Richard N. Aslin, and Elissa L. Newport. 1996. "Statistical Learning by 8-Month-Old Infants." *Science* 274(5294): 1926–1928.

Turing, Alan. M. 1936. "On Computable Numbers, with an Application to the Entscheidungsproblem." *Proceedings of the London Mathematical Society*, 2(42): 230–265. https://doi.org/10.2307/2268810

39

HOW NETWORK MODELS CONTRIBUTE TO SCIENCE[1]

Charles Rathkopf

1. Introduction

Network models are surprisingly easy to construct. There are at least two reasons for this. First, the construction process typically requires rather little theoretical guidance. Network models represent empirical objects as graph nodes and relations between objects as graph edges. The objects in question are usually discrete and, at least when viewed within the relevant scientific context, easily individuated. Examples include people, corporations, power plants, academic publications, tree species, and protein types. Moreover, the rules that govern the mapping between the graph and the data are straightforward. On the mathematical side, the nodes and edges are simple mathematical objects, mostly devoid of internal structure. On the empirical side, data sets typically include only a few types of objects. Often, there is just one. Data sets also typically include only a few types of relations between objects. Again, there is often just one.

Another reason that network models are easy to construct is that the construction of the model does not involve any attempt to capture patterns hidden in the data, and therefore does not involve data compression. Typically, once the data set has been cleaned, *every* object and relation gets represented. In this respect, network models are radically different from the compact, closed-form equations that have historically been viewed as the standard-bearer of scientific representation. These two observations about the construction of network models might lead one to think that such models must be superficial. They may look more like a trendy format for data summary than an innovative modeling strategy, capable of supporting profound scientific insight.

This last thought is mistaken, and it is the burden of this chapter to show why. The central thesis is that network models are capable of supporting profound insight into a surprisingly diverse range of phenomena. This view is supported by three case studies, selected to show that network models play an indispensable role in prediction, discovery, and explanation. Before getting into the case studies, the chapter provides a brief overview of the history that led to modern network modeling and introduces a few of the most common mathematical concepts. The chapter concludes with a discussion of the fact that network models are applicable to an enormously diverse range of empirical phenomena.

 DOI: 10.4324/9781003205647-45

This point has been emphasized by advocates of network modeling and has been used, at least occasionally, to support grandiose claims about the role of network models within the larger scientific enterprise. The account developed here is comparatively tempered, but not dismissive. It suggests that the trans-domain applicability of network models may sometimes offer us new and currently under-appreciated opportunities for scientific unification.

2. The emergence of modern network science

Modern network modeling emerged from the confluence of two historical research traditions, one in pure mathematics, and the other in social science. On the mathematical side, the paper, "On the Evolution of Random Graphs," by Paul Erdős and Alfred Rényi, introduced modern techniques for studying large graphs analytically (Erdős et al. 1960). In that paper, Erdős and Rényi imagine a large set of nodes, along with all of the possible graphs that can be constructed from that set, where a graph is simply a configuration of edges that connect the nodes. They prove that the set of all possible graphs with n nodes has several interesting properties. For example, they prove that, as n tends to infinity, the size of the largest connected subgraph follows a Poisson distribution. On the social science side, Mark Granovetter's paper, "The Strength of Weak Ties," (1973) showed how quantitative properties of social graphs could provide sociological insight. It showed, on the basis of both empirical data and hypothetical reasoning, that weak social ties play an outsized role in generating macroscopic sociological phenomena. The crux of his reasoning is that, unlike friends, mere acquaintances move in social circles different from one's own. As a result, acquaintances provide links to social groups that are both valuable and otherwise inaccessible. When it comes to finding a job, for example, acquaintances tend to be more advantageous than friends.

Modern network modeling can be viewed as a synthesis of the two research traditions that emerged, respectively, from these two papers. To see this, it helps to note some of the most salient differences between the two traditions. First, the sociological tradition used networks as a means of representing empirical data, while the mathematical tradition did not. Second, the sociological tradition focused primarily on networks with complex, non-random structures, while the mathematical tradition focused primarily on either random or lattice-like networks, both of which are more susceptible to mathematical analysis than complex graphs. Third, the mathematical tradition focused on networks that were large or infinite, while the sociological tradition, especially early on, focused on networks that were quite small (Granovetter's paper, for example, was based on data from just 54 people.)

Modern network analysis blends these two traditions together. It studies graphs that are large, but based on empirical data, and therefore finite. Moreover, most empirical networks are neither perfectly ordered nor perfectly random, and are, therefore, difficult to study using purely analytical techniques. To understand how large and complex networks behave under different parameter settings, computer simulations are required. Today, a large part of what is sometimes called *network science* involves the discovery of algorithms that can compute interesting properties of large complex graphs. Because many of these properties are probabilistic, one typically needs to study a whole ensemble of graphs, which is computationally demanding. It is therefore no accident that modern network modeling emerged only after the rise of cheap computing power.

The first papers to undertake this synthesis, which, in so doing, launched the modern era of network modeling, appeared between 1998 and 2000. The two most frequently cited are Watts and Strogatz (1998), in which the small-world model was introduced, and Barabási and Albert (1999) in which the so-called *BA preferential attachment model* was introduced. In 2005, a crucial and under-appreciated historical landmark in the development of network modeling was the release of an open-source Python library called NetworkX. NetworkX made it possible to convert lists and matrices into networks, compute common network properties, and visualize networks graphically (Hagberg et al. 2008). Once that software was released, scientists in many other fields began to use network analysis on their data, which in turn drove the development of new software for network analysis.

3. Common graph-theoretical concepts

Network models are based on the mathematics of graphs. A graph consists of a set of nodes and a set of edges, where an edge is just a two-element set of nodes (Trudeau 1976). Graphs are typically visualized as points and lines on a plane, but for the purposes of computing, a graph is represented as a type of matrix. The most common type of matrix for representing a graph is an adjacency matrix. In an adjacency matrix, both axes are defined by the set of nodes. If a direct connection exists between two nodes, their intersection is marked with a 1, indicating the presence of an edge; otherwise, it is marked with a 0.

$$A = \begin{matrix} 0 & 1 & 0 & 0 & 1 & 0 \\ 1 & 0 & 1 & 1 & 0 & 0 \\ 0 & 1 & 0 & 1 & 1 & 1 \\ 0 & 1 & 1 & 0 & 0 & 0 \\ 1 & 0 & 1 & 0 & 0 & 0 \\ 0 & 0 & 1 & 0 & 0 & 0 \end{matrix}$$

The same information can also be represented as an incidence matrix, which is a matrix defined by the nodes on one axis and the edges on the other. Incidence matrices are used less often than adjacency matrices but are favored for the representation of networks that are both large and sparse, because, in those cases, incidence matrices can be represented more compactly than adjacency matrices. (Where n is the number of nodes, and m is the number of edges, a sparse graph is one in which $n >> m$. An adjacency matrix has dimensions $n \times n$, which makes it larger than the corresponding incidence matrix with dimensions $n \times m$.)

The introduction sketched an intuitively appealing inference from the claim that networks are easy to construct to claim that they are superficial, or inferentially weak. One way to resist this cynical inference is to emphasize the distinction between *constructing* a network model and *using* it productively, once constructed. To use a network model to make inferences about the target system, you have to (i) choose the appropriate network properties to measure, and (ii) interpret the theoretical significance of those measurements. While the second of these two steps certainly *does* demand domain-specific empirical knowledge, it is less clear what sort of knowledge is required for the first step. A natural assumption would be that you need domain-specific knowledge of the target system in order to know which properties of the associated network representation are worth measuring. However, virtually all popular accounts of network science defend (or at least assert) the idea that, regardless of which empirical domain you are working in, the same set of network properties end up being important. This claim is both fascinating and puzzling, and it will be discussed in more detail below. Here, I only want to mention it as justification for the suggestion that one can understand a surprisingly large swath of network science modeling ideas on the basis of a rather small number of graph-theoretical concepts.

The following short list of network science concepts captures some of the most basic and most frequently used concepts. The characterizations are not rigorous definitions, but provide enough information to render the subsequent discussion accessible.

1 Node degree: the number of nodes with which a given node is connected.
2 Path length: the length of the shortest path that can be traversed between two nodes.
3 Clustering coefficient: a measure of how likely it is that there is an edge between nodes A and C, given that there is an edge between A and B, and another between B and C.
4 Small-worldness: the ratio of clustering coefficient to average path length.
5 Scale-free network: a network whose node degree distribution follows a power law.
6 Random graph: a graph in which the edges between nodes are determined by some random selection process.
7 Regular graph: a lattice-like graph in which every node has the same degree.

Although each of these properties is a property of a graph, not all of them can be found in books on graph theory per se. For an overview of network properties, as they pertain to network modeling, readers are advised to consult one of the many textbook treatments of network modeling ideas. Newman (2010) does a particularly good job of balancing ease of exposition with mathematical rigor.

4. Reasoning with networks

Each of the concepts listed above is exploited by the reasoning in the case studies below. Before turning to those, one source of potential confusion must be addressed. The reasoning in each case study draws not only on a graph and graph-to-data mapping but also on additional modeling apparatus. When additional modeling apparatus is required, a defender of the view that network models are superficial might say that the case studies described here fail to support the primary non-superficiality thesis, because the models in question are not *pure* network models. Their success, therefore, may have little to do with networks per se.

Although one can find examples of pure network models in the relevant sense (such as models that account for why author citation networks have the distributions they do), I do not discuss them here, since such pure network models have played a relatively minor role in the advancement of scientific knowledge over the past 25 years. More dramatic progress has been made by combining a graph theoretical representation with other kinds of modeling apparatus (for example, a system of differential equations.) The focus here on hybrid models, as we might call them, might prompt an objection of the following sort. One could *only* gather support for the thesis defended here (that network modeling supports profound forms of scientific inference) if one could first work out, with respect to any given inference, how to distinguish cleanly between the insight contributed by the network model, and the insight contributed by the other modeling apparatus.

This demand for a crisp criterion of model individuation is asking for too much. The claim that network modeling has made a substantive and distinctive contribution to science need not rely on specific criteria for counting network models. Adequate support for the claim can be provided simply by identifying network properties that are practically indispensable for novel forms of scientific inference. If at least some of those inferences can be described as profound, the central thesis defended in this chapter follows logically.

By formulating the central thesis in terms of properties rather than models, it can be reconciled with a wider variety of views about the nature of scientific models. In particular, a given collection of network properties can be viewed either as constitutive of a pure network model, which, as a contingent matter, was used in conjunction with another model *or* they can be viewed merely as a subset of elements within a larger, more multifaceted model. Both views are compatible with the central thesis of this chapter.

5. Discovery

One area in which network modeling has been used to make new discoveries is molecular biology. In a landmark paper, Spirin and Mirny (2003) undertook a network-based analysis of an existing, open-source database of protein-protein interactions in yeast, which were themselves detected by well-established experimental methods. Protein-protein interactions are biochemical interactions between proteins that allow them to function together as part of a molecular machine that accomplishes some cell function. Most mesoscale cell functions are carried out by a large family of proteins, not all of which engage in direct biochemical reactions with one another. In addition, many of the protein-protein interactions involved in any given mesoscale cell function simply have yet to be probed experimentally. For both reasons, there will often be proteins that play an important role in a given cell function, but which are not yet known to do so. Spirin and Mirny used network modeling to facilitate a new process for protein discovery that is radically more efficient than what was previously possible.

The network they constructed consisted of 3,992 nodes, each representing a protein type, and 6,500 edges, each representing a known protein-protein interaction. Their primary goal was to locate biologically significant clusters within this network. This is trickier than it sounds. Even the problem of identifying the single largest cluster in a graph is NP-hard, so developing efficient search algorithms is a non-trivial mathematical problem. Spirin and Mirny designed an algorithm to find the maximum of the function:

$$Q(m,n) = 2m / \left(n(n-1)\right)$$

where m is the number of interactions between n nodes. Q characterizes the density of the cluster. The algorithm uses a Monte Carlo procedure that starts with a set of nodes, selected at random, and then replaces members of that set, re-computing Q for each new set until it converges. They then selected all clusters with a Q value high enough to make it statistically significant. (Statistical significance is evidence that supports a rejection of the null hypothesis, which is itself typically formulated as the claim that the observation in question appeared by chance. In this setting, the operational meaning of "appeared by chance" is that it appears in a graph which is itself a member of an ensemble of graphs that was generated by a random graph construction procedure.)

With that done, Spirin and Mirny worked out which cell function the cluster of proteins contributes to. In each complex, at least some proteins were already known, and their functions were annotated in the open-source database. They hypothesized that the other proteins in the cluster would contribute to the same function; an inference strategy known as guilt-by-association.[2] This led to a suite of predictions about the functional role of proteins that were in the cluster, but not yet known to be involved in the cell function associated with that cluster.

The Spirin and Mirny paper counts as a significant contribution to scientific discovery for two reasons. First, their predictions radically reduced the space of proteins to be tested experimentally and thereby made it easier to choose experiments that were likely to have valuable results. The second reason is that, since it was first published in 2003, their predictions have been largely confirmed by experiments (Omranian et al. 2022). Moreover, the methods they developed for identifying protein complexes have been widely reused by other labs which have themselves made valuable discoveries with them.

Spirin and Mirny's work shows that network properties play a role in scientific discovery. There are good reasons to believe, furthermore, that the role they play is *practically ineliminable*. The space of possible protein-protein interactions is enormous. One could not practically perform the sort of experimental screening (such as two-hybrid screening) required to detect each possible interaction. In the absence of that brute force approach, one needs to make predictions about which proteins are likely to interact with each other from some set of known protein-protein interactions. Predictions of this sort can be divided into two classes: those that rely on theoretical knowledge of the proteins involved, and those that do not. If you go with the former class, you may get some predictive traction, but your predictions will be painstaking and slow. In the latter class, you make many predictions at scale. If you want to generate predictions at scale, it is necessary to represent the full suite of structural relations (interaction vs. no interaction) that characterize the pre-theoretical domain. To construct an uncompressed data representation of a suite of objects and structural relations is to construct a network. Therefore, if you want to make a large class of predictions about biologically significant protein clusters, network representation is practically ineliminable.

6. Prediction

One can hardly write about the use of network models in 2022 and fail to discuss their use in modeling the COVID-19 pandemic. One of the puzzling facts about the early phase of the COVID-19 pandemic was that in many countries, after an initial wave in which the infection rate grew exponentially, it continued to grow linearly, even though, according to standard epidemiological models, the probability of sustained linear growth is effectively zero. The models in question, commonly known as susceptible-immune-recovered (SIR) models, predict either exponential growth or exponential decay whenever the reproduction number R deviates even slightly from 1. The reproduction number can be defined as the expected number of secondary infections an infected person will cause, and it is rarely *precisely* equal to one. The fact that steady linear growth was observed over long periods of time suggests, therefore, that the SIR models were missing something important about COVID-19 dynamics.

Of course, epidemics are intrinsically difficult processes to predict. They are stochastic, they are influenced by many social and biological variables, and, at least in the early stages, they are non-linear. Consequently, one cannot expect high-precision predictions. Even when holding all parameter values constant, infection curves can look quite different from one run of the simulation to the next, and the total number of infected people can vary by a factor of two. Still, by casting the structure of the population as a network, it becomes possible to represent the critical degree of a population explicitly and use it to improve predictive traction.

Following the work of Pastor-Satorras and Vespignani (2001), Thurner et al. (2020) simulated the SIR model in a network environment. Nodes represent people, and edges between people represent physical proximity sufficient for viral transmission. Each person is represented as having a particular number of contacts per day to whom they could theoretically transmit the virus. That number is called the node *degree* and it varies from person to person, following a distribution that is designed to mimic the contact structure of real human populations. The network was based on empirical data, but Thurner et al. constructed their network by algorithmic means. Algorithmic construction allowed them to vary the parameters of the network systematically but also shouldered them with the burden of having to run model fitting tests to check how well the algorithmically generated model fits the empirical data. Of particular relevance are the facts that (i) during lockdown, the average degree D drops to near family size, and (ii) there will be some interactions between families, but these interactions will not take the form of giant hubs.

Intuitively, the higher the average degree, the more easily a virus will spread. In their simulations, Thurner et al. observed a qualitative shift in disease dynamics when D drops below a critical threshold. Above the threshold, the epidemic grows exponentially; below it, growth remains linear. This qualitative shift roughly captures the effect of lockdown policies during the summer of 2020. Once lockdown measures were in place, the contact structure of the population dropped to a value only slightly higher than the average family size. (The exact value of D at which this qualitative shift occurs is not a universal property of epidemics. It depends on the transmission rate of the virus, among other factors.)

Using this model, Thurner et al. managed to outperform extant predictions of the COVID-19 infection curves in both Austria and the USA. The choice of these two countries was significant because they differ so dramatically. Austria adopted strict lockdown policies early in the pandemic, while the USA introduced weaker lockdown policies later on. Despite these and many other differences between the two countries, Thurner et al. achieved this predictive improvement by choosing a value of D to fit empirical estimates of contact structure before and after lockdown measures were in place.

Crucially, this work deserves to be counted as a case of *prediction*, rather than *model fitting*, because the representative infection curves were captured without having to fit any other model parameters, all of which were chosen at the outset on the basis of measurement.

7. Explanation

At the neural level, an epileptic seizure is an episode of synchronized hyperexcitatory spiking activity. One of the puzzling things about epilepsy is that it is often caused by injuries that induce a substantial *loss* of neural connectivity. Intuitively, connectivity should *facilitate* hyperexcitability, since, when connectivity is high, there are more paths between the input activation and those neurons farthest from the input layer. In light of this intuition, a natural question is: why does a *loss* of connectivity lead to hyperexcitability? Answering this question is crucial to understanding why epileptic seizures occur.

The core of the answer to this question is that hyperexcitability is not due to the loss of connectivity itself, but to the new pattern of connectivity that emerges in the wake of that loss.

In other words, the answer depends on a change in the topological structure of the network. This idea was first suggested by Percha et al. (2005). Although that paper suggested the correct topological answer to our question, it was based on a small simulation of 144 neurons. As a result, it was unclear whether the simulation could justifiably be interpreted as a guide to post-injury epilepsy in humans. Dyhrfjeld-Johnson et al. (2007) published a radically more extensive model that confirmed and expanded the initial results.

The Dyhrfjeld-Johnson paper focused on the dentate gyrus, a part of the temporal cortex known both to be involved in the generation of seizures and to be unusually sensitive to injury. Dyhrfjeld-Johnson et al. built a nearly full-scale model of the dentate gyrus of the rat brain, with 50,000 neurons and over one billion connections. On top of each node in the graph, they built a compartmental model neuron, which captures the spiking behavior of neurons as a response to electrical input.

The dependent variable in this study is the degree of hyperexcitability in the network, which is defined as a function of (i) the proportion of the neurons in the network that get activated after a particular input, (ii) the length of the interval between initial activation and the activation of the last neuron to be activated, and (iii) the duration of the whole network activation, once achieved. The primary independent variable is the degree of small-worldness of the network topology.

How does getting hit on the head lead to an increase in the small-worldness of your dentate gyrus? The answer to this question is incomplete but interesting. Some cell types are more susceptible to injury than others. Hilar cells are both particularly susceptible to injury and highly connected. So, when these cells die out after injury, connectivity drops drastically. Soon afterward, granule cells (GCs) begin to form new connections at a greater rate than usual. Moreover, they form excitatory recurrent connections to other GCs, which, in the healthy brain is very rare (about 0.05% of all possible GC-GC pairs share a synaptic connection). Some of these GCs connect at extreme rates, in comparison with the expected level of connectivity, and thereby become network hubs. These facts are supported directly by physiological observation but are not themselves well-understood. So let us set aside the question of why GCs sprout new recurrent connections after the injury. Instead, we want to focus on what the new topology is like, and how that topology influences hyperexcitability in the spiking neuron model.

Two topological characteristics stand out. First, although the healthy dentate gyrus is already estimated to have the small-world property to some degree, the post-injury dentate gyrus has a very high degree of small-worldness, with a low-average path length, and, nevertheless, high-local connectivity. In this case, the path length dropped to an extremely low value: on average, two neurons are connected by less than three edges, even though there are 50,000 neurons, and despite the fact that connectivity in the post-injury model is only 4.7% of what it was previously. The topology leads to hyperexcitability gradually, until reaching a threshold. Until that threshold is reached, you get monotonic increases in hyperexcitability with small-worldness. Since other properties of the network are held fixed, only the topological property can explain the hyperexcitability. Robustness analysis reveals that the effect is stable. Under a parameter sweep, the link between small-worldness and hyperexcitability remains largely unchanged.

The Dyhrfjeld-Johnson model answers the following *why-question*: why do injuries to the temporal lobe increase susceptibility to epileptic seizures, even though they trigger substantial loss of connectivity? The answer is that (i) injury causes an increase in the small-worldness of the topology of the dentate gyrus, (ii) the increase in small-worldness promotes hyperexcitability, even in the absence of other physiological changes (iii) hyperexcitability is a state of increased susceptibility to epileptic seizure. For a full defense of the view that this case study deserves to be counted as an explanation, rather than as a mere description, it would be necessary to lay out one or more philosophical theories of explanation. There is no room for that project here, but one can find accounts of explanation congenial to the view in Rathkopf (2018), Kostic (2018), and Kostic and Khalifa (2021).

8. Network science is not superficial

In the introduction, two characteristics of the process of constructing a graph from empirical data were described. The first was that, once you have an appropriately structured data set, constructing a graph from the data does not require additional theoretical knowledge of the empirical domain. The second was that constructing a network model does not involve data compression. When you construct a network model, *all* the data gets recapitulated in graph-theoretic form. These two characteristics can give the misleading impression that network modeling is a superficial enterprise, in the sense that network models are likely to facilitate only rather shallow empirical inferences. The case studies above were selected to illustrate that this impression is incorrect. Network models are practically indispensable for certain kinds of scientific inference, some of which are profound. Here I will attempt to make the case for this claim more systematically.

Let us start with the subsidiary claim that network models are practically indispensable. The term "practically indispensable" refers to a weak form of necessity: it is not logically impossible to draw the conclusions at issue by means of some other modeling strategy, or by means of some other representational apparatus.[3] Rather, the claim is that, given the contingent constraints involved in real scientific practice, a network model of some kind is the only viable option. In the first case study, Spirin and Mirny used their network model to draw a host of conclusions about the functional contributions of various proteins. While it is logically possible that someone might have reached the same conclusions experimentally, the number of experiments required would be in the tens of millions. In the second case study, Thurner et al. used a network representation of the contact structure of a population under lockdown to improve predictions about the COVID-19 infection curve. In that case, the only alternative to network representation is to invoke the so-called mean-field assumption, which says that the probability of anyone coming into contact with anyone else is the same.[4] As Thurner et al. argue, however, the mean-field assumption breaks down under conditions of lockdown. In the third case study, Dyhrfjeld-Johnson et al. showed that the propensity for epilepsy in patients with head trauma is explained by the degree of small-worldness of the dentate gyrus. You cannot invoke the small-world property in an explanation without some form of network representation. In this case, therefore, the network representation is not only practically necessary but also logically necessary.

Furthermore, it does seem that the inferences enabled by network representations are at least sometimes profound. In each of the case studies, the inference enabled by appeal to network representation delivered a non-obvious answer to a substantive and important scientific question: (i) What is the function of *that* protein? (ii) why does COVID-19 last so long? (iii) Why do people sometimes get epilepsy after head injuries? If a scientific inference delivers a non-obvious and substantive answer to questions like these, the inference itself may be regarded as profound.

One might still suspect that there is another sense in which network modeling is superficial. Network modeling may appear, from an epistemological point of view, rather like a free lunch: it is substantial, but nevertheless undemanding. That is, although the incorporation of network representation into a given scientific enterprise may substantially enrich that enterprise, it does not demand any expertise. People from each discipline may find it useful to use network modeling, in much the same way they may find it useful to use elementary arithmetic, but the relevant techniques are so straightforward as to make claims of expertise in network science overblown. However, as Elek and Babarczy (2022) argue, network modeling has become a field in its own right, and one in which it is possible to gain expertise. This expertise is visible in our three case studies. Spirin and Mirny needed to know how to build an algorithm that identifies clusters satisfying quantitative criteria. The design of such algorithms demands considerable computational expertise, even though such expertise is not domain-specific empirical knowledge. In both the second and third case studies, the authors needed to integrate a static network model with a dynamical model, vary the topological properties of the network model systematically, and record the effects of that intervention on the dynamical model. To achieve that end, both studies went far beyond the simple task of constructing a graph from empirical data. Both studies involved the algorithmic construction of a graph, along with statistical analysis to check how well the algorithmically constructed graph fits the empirical data. In summary, we can identify at least four kinds of expert knowledge involved in network modeling.

1 Constructing graphs by algorithm and varying their properties systematically.
2 Devising algorithms to find empirically significant substructures in empirically constructed graphs.
3 Using model-fitting statistics to assess how well an algorithmically constructed graph fits an empirically constructed graph.
4 Integrating other representational apparatus into the network model.

Because network modeling is practically necessary for generating at least some profound scientific inferences, and because it is, increasingly, a field in which computational expertise can be accumulated, network modeling cannot reasonably be regarded as a superficial science.

9. Trans-domain applicability

Perhaps the most philosophically interesting thing about network modeling is the fact that the same network properties seem to crop up in many otherwise unrelated empirical domains. Network modelers often suggest that this fact has far-reaching implications. One of the leading voices in network modeling, Albert Barabási, says:

> A key discovery of network science is that the architecture of networks emerging in various domains of science, nature, and technology are similar to each other, a consequence of being governed by the same organizing principles. Consequently, we can use a common set of mathematical tools to explore these systems.
>
> *(Barabási 2016, 8)*

Is the similarity among observed network architectures really a consequence of the fact that networks across domains are governed by the same organizing principles? One of the most discussed principles in the network science literature is that all or most empirical networks have a scale-free distribution over node degree. That is, where k is the degree of a single node, the distribution over values of k is given by: $P(k) = k^{-\alpha}$, where the critical exponent α takes a value between 2 and 3.[5] There is a robust and quite technical debate about how broadly this power law relation actually applies to measured networks (Voitalov et al. 2019; Broido and Clauset 2019; Zhang et al. 2015; Newman 2005). The debate proceeds by gathering node-distribution estimates from many different data sets, fitting those estimates to a power law, and summarizing the results in a large table.

The curious thing about this debate is that little effort has been given toward determining the appropriate reference class. That is, if we think of the power law relation above as a first-order property of empirical systems, then there should be a way of expressing the generalization in terms of the logical schema: $\forall x(Fx \rightarrow Gx)$, where G refers to the property of satisfying the power law distribution. The reference class problem is simply that we should be able to say which property plays the role of F. As far as I can tell, there seem to be no hard conceptual boundaries for the class of systems that can be modeled as networks. If that impression is correct, property F may not exist. In that case, the frequency with which scale-free node degree distributions appear in nature is simply undefined, and the debate about the relative frequency of scale-free networks is conceptually muddled. Perhaps all that is needed to rectify the situation is a more careful exposition of the goal of gathering such data sets. Rather than framing the work as an attempt to capture the *frequency* of network properties in nature, it can be framed as an attempt to examine the many varieties of scale-free networks in nature, and then, to work out whether they have other interesting properties in common.

Thus far, I have talked about the trans-domain applicability of *networks* or *network models*. If we want to think more rigorously about the unit of scientific representation that has the capacity to be applied across empirical domains, these shorthand expressions are likely to be misleading. The term "network" is ambiguous between an object in nature and our representation of it. Arguably, the term "network model" refers to something that is not applicable across domains at all. This will depend on philosophical theories about what models are. As an example, however, consider the view that a network model consists of two parts: (i) a graph, and (ii) a mapping between the graph and the target system. If you conceive of models that way, then their individuation conditions are too fine-grained for them to be applicable across empirical domains. Instead, as suggested by Humphreys (2019), we should invoke the broader notion of a *computational template*. Computational templates are syntactic objects that come with an intended interpretation, but which are flexible enough to be applied to new phenomena. Their primary attraction is the fact that they serve as a point around which computational expertise can be gathered. As Richard Feynman likes to repeat in his famous lectures on physics, the same equations have the same solutions (Feynman 2010). Science is therefore easier if you can re-use equations whose solutions, whether computational or analytic, have been worked out by others.

In the case of network modeling, we should say that the unit of scientific representation that has the capacity for trans-domain applicability is the network template. Although this idea is new, there is a rich and growing literature on trans-domain modeling and computational templates which begins with Humphreys (2004), and includes many more recent articles. Of particular relevance to the discussion of networks Knuuttila and Loettgers (2016).

10. Generating conditions for networks

If we accept that the claims about network properties being realized in vastly different systems are both true and non-trivial, we will then naturally want to ask why this is the case. This sort of question might have many answers, but one important answer will point to the generating conditions for networks of that kind. If we can show that models of network generation can be described in terms of abstract conditions, and if we can get enough clarity about how to interpret those conditions in empirical terms, then perhaps we can show that different systems that appear radically distinct when described in terms of their more superficial empirical features are actually quite similar with respect to the fact that they both satisfy these abstract conditions. For example, it may be the case that citation networks of certain kinds can be explained by means of a preferential attachment model. It also may be the case that the distribution of city sizes can be explained by a similar preferential attachment model. We would then have a kind of model-based unification of the two processes that is grounded in an etiological mechanism, but which is nevertheless formulated at a level of abstraction that seems to leave implementation details far behind.

Preferential attachment, as formalized in the well-known BA model mentioned at the beginning, is only one of several models of network construction. Others include the Initial Attractiveness Model, Internal Links Model, Node Deletion Model, Accelerated Growth Model, and the Aging Model (for details see Barabási (2016)). All of these models of network generation are domain-general. So, although they could perhaps furnish an etiological explanation in terms of causes, they would presumably differ from mechanistic explanations, which are domain-specific. An interesting avenue for further research is to articulate the kind of explanation we have when we show that the process by which some empirical system was generated is well-described by one of these network generation models above. If we could gather a long list of such systems, we could group them in terms of their abstract generating model, and thereby broaden the unification base.

To conclude, network science shows that what you can learn about some phenomenon depends on the way you represent it. Most of the prominent examples of network science are cases in which we learn new facts about target systems that are composed of elements that are already familiar. Nevertheless, by choosing to represent these systems as networks, there is much to be learned that would have been inaccessible without the network representation.

Notes

1 Many thanks to Kareem Khalifa and Daniel Kostic for their insightful and detailed comments on an earlier draft of this chapter.
2 This aspect of the work is adroitly described by William Bechtel (2019).
3 Perhaps with the exception of the third case, which I discuss presently.
4 This is not to say that the style of network representation used by Thurner et al. is the only way to circumvent the mean-field assumption. For a review of alternative methods of incorporating graph-theoretic representation into epidemiological models, see Keeling et al. (2016).
5 The definition of a scale-free degree distribution is also somewhat contested. Incorporating the requirement that the degree distribution falls between 2 and 3 is what Broido et al. (2019) call a "strong definition" of the scale-free property.

References

Barabási, Albert-László. 2016. *Network Science*. Cambridge University Press.

Barabási, Albert-László, Réka Albert, and Hawoong Jeong. 1999. "Mean-field theory for scale-free random networks." *Physica A: Statistical Mechanics and its Applications* 272(1–2): 173–187.

Bechtel, William. 2019. "Analysing network models to make discoveries about biological mechanisms." *British Journal for the Philosophy of Science* 70(2): 459–484.

Bechtel, William and Robert C. Richardson. 2010. *Discovering Complexity: Decomposition and Localization as Strategies in Scientific Research*. MIT Press.

Broido, Anna D. and Aaron Clauset. 2019. "Scale-free networks are rare." *Nature Communications* 10(1): 1–10.

Dyhrfjeld-Johnsen, Jonas, Vijayalakshmi Santhakumar, Robert J. Morgan, Ramon Huerta, Lev Tsimring, and Ivan Soltesz. 2007. "Topological determinants of epileptogenesis in large-scale structural and functional models of the dentate gyrus derived from experimental data." *Journal of Neurophysiology* 97(2): 1566–1587.

Elek, Gábor and Eszter Babarczy. 2022. "Taming vagueness: The philosophy of network science." *Synthese* 200(2): 1–31.

Erdős, Paul and Alfréd Rényi. 1960. "On the evolution of random graphs." *Publications of the Mathematical Institute of the Hungarian Academy of Sciences* 5(1): 17–60.

Feynman, Richard, Robert Leighton, and Mathew Sands. 2010. *The Feynman lectures on physics; New millennium ed.* New York: Basic Books. Originally published 1963–1965.

Granovetter, Mark S. 1973. "The strength of weak ties." *American Journal of Sociology* 78(6): 1360–1380.

Hagberg, Aric. A., Daniel A. Schult, and Pieter J. Swart. 2008. "Exploring network structure, dynamics, and function using *Networkx*." In *Proceedings of the 7th Python in Science Conference*, edited by Gaël Varoquaux, Travis Vaught, and Jarrod Millman, 11–15. Pasadena, CA.

Humphreys, Paul. 2004. *Extending Ourselves: Computational Science, Empiricism, and Scientific Method*. Oxford University Press.

———. 2019. "Knowledge transfer across scientific disciplines." Studies in History and Philosophy of Science Part A 77: 112–119.

Huneman, Phillipe. 2010. "Topological explanations and robustness in biological sciences." *Synthese* 177(2): 213–245.

Keeling, Matt J., Thomas House, Alison J. Cooper, and Lorenzo Pellis. 2016. "Systematic approximations to susceptible-infectious-susceptible dynamics on networks." *PLOS Computational Biology* 12(12): e1005296.

Knuuttila, Tarja and Andrea Loettgers. 2016. "Model templates within and between disciplines: from magnets to gases–and socio-economic systems." *European Journal for Philosophy of Science* 6(3): 377–400.

Kostic, Daniel. 2018. "Mechanistic and topological explanations: an introduction." *Synthese*, 195(1): 1–10.

———. 2022. "Topological explanations: an opinionated appraisal." In *Scientific Understanding and Representation: Mathematical Modeling in the Life and Physical Sciences*, edited by Insa Lawler, Elay Shech, and Kareem Khalifa. Routledge.

Kostic, Daniel, and Kareem Khalifa. 2021. "The directionality of topological explanations." *Synthese* 199: 14143–14165.

———. 2022. "Decoupling topological explanations from mechanisms." *Philosophy of Science*, accepted manuscript, 1–39.

Newman, Mark E. 2005. "Power laws, pareto distributions, and Zipf's law." *Contemporary Physics*, 46(5): 323–351.

———. 2010. *Networks: An Introduction*. Oxford University Press.

Omranian, Sara, Zoran Nikoloski, and Dominik G. Grimm. 2022. "Computational identification of protein complexes from network interactions: present state, challenges, and the way forward." *Computational and Structural Biotechnology Journal* 20: 2699–2712.

Pastor-Satorras, Romualdo and Alessandro Vespignani. 2001. "Epidemic spreading in scale-free networks." *Physical Review Letters*, 86(14): 3200.

Percha, Bethany, Rhonda Dzakpasu, Michał Zochowski and Jack Parent. 2005. "Transition from local to global phase synchrony in small world neural network and its possible implications for epilepsy." *Physical Review E* 72(3): 031909.

Rathkopf, Charles. 2018. "Network representation and complex systems." *Synthese* 195: 55–78.

Spirin, Victor and Leonid A. Mirny. 2003. "Protein complexes and functional modules in molecular networks."*Proceedings of the National Academy of Sciences* 100(21): 12123–12128.

Thurner, Stefan, Peter Klimek and Rudolf Hanel. 2020. "A network-based explanation of why most Covid-19 infection curves are linear." *Proceedings of the National Academy of Sciences*, 117(37): 22684–22689.

Trudeau, Richard J. (1976). *Introduction to Graph Theory*. Dover Publishers.

Voitalov, Ivan, Pim van der Hoorn, Remco van der Hofstad, and Dmitri Krioukov. 2019. "Scale-free networks well done." *Physical Review Research* 1: 033034.

Watts, Duncan J., and Stephen H. Strogatz. 1998. "Collective dynamics of 'small-world' Networks." *Nature*, 393(6684): 440–442.

Zhang, Linjun, Michael Small, and Kevin Judd. 2015. "Exactly scale-free scale-free networks". *Physica A: Statistical Mechanics and Its Applications* 433: 182–197.

40
MODELS OF THE NERVE IMPULSE

Natalia Carrillo

1. Introduction

In the 1950s, Alan Hodgkin and Andrew Huxley published the results of a series of experiments on giant axons of squid and presented a mathematical model capable of reproducing such results (henceforth the Hodgkin–Huxley model). The model helped scientists who were trying to understand how excitable nerve cells (neurons) can transmit a signal from the dendrites (where it receives signals from other cells through synaptic connections with them) to the axonal terminals (where it establishes synaptic connections to other neurons). The mathematical model consists of a system of four differential equations known as the Hodgkin and Huxley equations. These equations became a milestone of mathematical biology—one of the first triumphs of formalization of biological phenomena, celebrated with the Nobel Prize of Physiology and Medicine in 1963.

What made this mathematical model such an achievement? Imagine you have two rooms. In one of them, you have scientists experimenting on a very large neuron (squids have these). In the other, you have scientists in front of four coupled equations. The experimenter stimulates the neuron with an electrical pulse at a particular temperature, with certain intensity. Someone shares these parameters with the modeler, who introduces the parameters in the system of equations and estimates the solutions to them. Figure 40.1 compares the measurements with the estimation of the solution to the mathematical equations.

How were these scientists able to produce equations that simulate the measurements in animal tissue so well? An answer to this question requires going through how Hodgkin and Huxley developed the model, but it also requires offering an interpretation of what the model does, epistemically speaking, which is a philosophical question. This chapter will examine these issues in the light of novel advances in neuroscience that challenge the status of the Hodgkin–Huxley model. The presentation proceeds as follows. The next section (Section 2) briefly presents the main ideas behind the Hodgkin–Huxley model. Afterward, in Section 3, the philosophical discussions on what makes the Hodgkin–Huxley model epistemically successful are presented. Most of those discussions take at face value that the ionic hypothesis resulting from the electrical approach to nerve impulse research (that the Hodgkin–Huxley model advanced) has been confirmed. However, some scientists have

 DOI: 10.4324/9781003205647-46

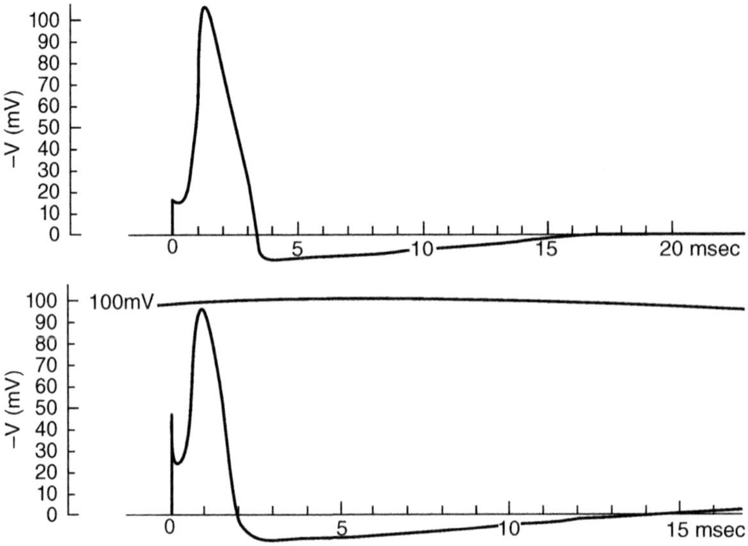

Figure 40.1 The measurements of transmembrane voltage in neurons compared with the voltage obtained by estimating the solution to the Hodgkin–Huxley model. Above: Solution of the Hodgkin and Huxley equations for a depolarization of 15 mV at 6°C. Below: Nerve impulse recording at 9.1°C. The vertical scale is identical in both curves, and the horizontal scales have been adjusted by a factor that is appropriate considering the difference in temperature. Voltage convention is that the resting voltage is zero. Image from Hodgkin and Huxley (1952).

challenged, in experimental and theoretical work, the central role of ion currents in explaining nervous excitability. The recalcitrant evidence is presented in Section 4, while Section 5 examines different alternatives to the electricity-centered approach to model the nerve impulse (also referred to as the *action potential*), all of which use thermodynamics. Section 6 examines the few philosophical discussions that examine the contrast of thermodynamic alternatives to the Hodgkin–Huxley model.

2. The Hodgkin and Huxley model and the ionic hypothesis

The research in nerve impulse generation and propagation involved, apart from Hodgkin and Huxley, also other physiologists like Cole, Curtis, and Katz, who in the first half of the 20th century performed many experiments in the newly discovered giant nerve cells of some squid's species. The axons (a long tubular section of every nerve cell) are so large (up to 1mm in diameter!) that the scientists were able to insert electrodes in them and measure the difference in electric potential between the inside and the outside of the nerve cell membrane (called transmembrane voltage).

The evidence collected from the experimentation in giant axons in the 1940s was combined with early attempts from the beginning of the 20th century to understand nerve impulse transmission in terms of electrochemical forces that affect the electric potential between the inside and the outside of the nerve cells. A very important element in the early stages of this development was the earlier discovery that free ions in a solution (as in the intracellular and extracellular fluid) are subject to electrical gradients (electrical attraction and repulsion) as well as chemical gradients (of diffusion). Thus, an electrical imbalance can produce a movement of ions, and a movement of ions can produce electrical effects. Already in 1890, Ostwald proposed that the electric potentials in nerves could be explained if nerve excitability was interpreted in terms of the electrochemical behavior of ions. Julius Bernstein exploited this idea in his famous "membrane theory" (Bernstein 1902). He proposed that there is a differential concentration of certain ions between the inside and the outside of the membrane. When the membrane is excited, he suggested, it "collapses" and the ions move freely toward their electrochemical equilibria, changing the electrical field in the vicinity of the membrane. These beginnings already suggested that it is the ions crossing the membrane that will explain the behavior, an idea I will refer to as "the ionic hypothesis."

When Hodgkin and Huxley were able to obtain more accurate measurements from the giant axons, they discovered that the voltage across the membrane did not disappear, as expected from Bernstein's membrane theory, and instead, the voltage is inverted, going from −70 to around +40 mV. The question then became what enables the nerve cell to (a) have an electrical imbalance across the membrane in the first place and (b) invert the electric potential when it is excited. In a very intense summer of research, Hodgkin and Huxley experimentally reexamined the role of ions in the electrical behavior of axons and were able to find good evidential support to the idea that excitation is not a collapse of the membrane but something more sophisticated. During excitation, there is an increased permeability to, first, potassium ions and, in a second moment, sodium ions, which together with an initial imbalance of more sodium inside the cell and more potassium outside the cell explains the voltage variations they were measuring.

In order to address the question of how these currents of ions could account for the observed voltage changes, neurophysiologists resorted to circuits that would be "equivalent" to the nerve membrane and the charges in its vicinity. Cole and Curtis (1938) and Hodgkin and Huxley (1952) depicted different kinds of electrical circuits where the electric charges would be distributed similarly to how they move across the membrane of squid neurons. Using the electric circuit they proposed, Hodgkin and Huxley were able to obtain a mathematical expression for the purported current across the membrane, as they derived the equations for the current in the circuit from the laws of electrodynamics. These equations and their derivations can be easily found in biophysics textbooks (e.g., Keener and Sneyd 2009, section 5.1). The system of equations comprising the Hodgkin–Huxley model does not have analytic solutions, but with numerical methods, the solutions can be approximated, which is how the graph presented in Figure 40.1 was produced. The graph shows the duration and intensity of the nerve membrane reaction to stimuli *in a specific point of the membrane*.[1] When Huxley approximated the solutions in order to see whether the model was in fact producing results that would be quantitatively equivalent to the experimental measurements, it took him around three weeks to calculate each curve by hand (e.g., the one displayed in Figure 40.1). With modern computational methods, the approximated solutions can be obtained in milliseconds. That allows the creation of simulators where one can perform virtual experimentation (i.e., quickly calculate the model's solutions for different parameter values representing different experimental conditions).

The Hodgkin–Huxley equations describe the current of ions across the membrane as the result of permeability changes. However, Hodgkin and Huxley did not know *how* the permeability of the membrane changed. Subsequent research focused on trying to figure out what changed the permeability of the membrane to ions. Is it active or passive transport? Are messenger molecules involved? The next two decades produced a lot of research addressing these questions, resulting in many interesting results like the crystallization of the potassium channel in the 90s (Doyle et al. 1998). These results together with patch clamp experiments supported the view that the ionic hypothesis was mostly correct (see Craver 2007).

3. Philosophical discussion on the epistemic value of the Hodgkin–Huxley model

The Hodgkin–Huxley model drew the attention of philosophers of science in the early 2000s. Being one of the most exemplary instances of mathematical representation of biological phenomena, the question of the epistemic contribution of the model was addressed. Is the model explaining how the nerve impulse generates and propagates, or is it merely reproducing the observed features of the phenomenon?

Weber (2008) answered this question by zooming in on the role of physical laws in electrical circuits that the Hodgkin–Huxley model employed (Kirchoff's laws, Ohm's law, etc.). According to Weber, the Hodgkin–Huxley model *is* explanatory, while he defended a thesis of "explanatory heteronomy": the explanatory generalizations in the Hodgkin–Huxley model are due to the laws of chemistry and physics, not of biology. Bogen (2008), in contrast, presented a non-explanatory account of the Hodgkin–Huxley equations. Paying attention to the use of analogical reasoning in the derivation of these equations, Bogen suggested that the laws function rather like *calculation tools*: "[Hodgkin and Huxley] applied Ohm's law to real-world quantities as if they were denizens of the equivalent circuit. In particular, they treated each ion permeability *as if* it varied with the membrane potential and current according to Ohm's law" (Bogen 2008, 1040). An intermediary position was postulated by Craver (2006; 2007). As part of the development of his mechanistic account of explanation, he argued that the Hodgkin–Huxley model is better understood as a how-possibly sketch of a mechanism that sustains nerve impulses. In his view, to gain the status of a how-actually explanation, the model would have to give an account of the "nuts and bolts" of the mechanism by which ions cross the nervous membrane.[2] Thus, Craver ascribed only *some* explanatory power to the Hodgkin–Huxley equations.

Later on, Levy (2014) picked up on Craver's claim that for the Hodgkin–Huxley model (or any mechanistic model) to explain, it would need to "account for all aspects of the phenomenon by describing how the component entities and activities are organized such that the phenomenon occurs" (Craver 2006, 374). Levy contested this completeness requirement, arguing that the explanatory achievement of the Hodgkin–Huxley model is in fact due to its *abstract* character (the completeness requirement was also more generally criticized by Knuuttila and Loettgers 2013 and Love and Nathan 2015). Levy argued that because the Hodgkin–Huxley model abstracts from the individual movement of ions, it can more generally account for the ionic currents—without having to open the "black box" of the mechanism of ion transport.[3] For Levy, the contribution of the model is due to its characterization of regularities at an *aggregative level:* "the discrete-gating picture relates whole-cell behavior to events at a lower level via aggregation: the system's total behavior is the sum of the behaviors of its parts" (Levy 2014, 15). He goes on to explain that such "aggregative abstraction" could be "truer to the mechanistic ideal, because it explains the relationship between lower-level mechanisms and higher-level ones" (20). While Levy thinks that the abstract interpretation he gives of the Hodgkin–Huxley equations still fits the mechanistic ideal, Colombo, Hartmann, and van Iersel (2015) have argued in favor of a more pluralistic approach to biological explanation. In their view, it is not necessary to commit to the real status of the entities and activities of a model to appreciate its explanatory value. Instead, they suggest treating the parts and activities of a mechanism as being defined (as opposed to discovered) in the modeling process.

4. Evidence in conflict with the electrical approach to nerve impulse research

The success of the ionic hypothesis that stemmed from the electrical approach put forth by Cole, Curtis, Hodgkin, Huxley, and many others suggests that a final explanation of the phenomenon of nervous excitability has been achieved. There is, however, a body of evidence that challenges such canonical explanation. The recalcitrant evidence involves the measurements of heat emission from the nerve cell while being excited that show heat being released in the initial phase of the action potential and reabsorbed in the second phase (Abbott et al. 1958). The issue is complex, but the basic idea is that one would expect a system like the one represented by the Hodgkin–Huxley model to dissipate more energy, since the movement of charges through a resistance always releases heat. However, the aforementioned measurements do not seem to fit this expectation.[4]

The conflicting evidence goes beyond the thorny issue of temperature measurements. Empirical research in the eighties showed that the nerve axons shorten (Tasaki and Iwasa 1980) and swell (Tasaki and Iwasa 1982; González-Pérez et al. 2016) when transmitting a nerve impulse. Also, findings that nerve impulses can be generated with mechanical stimulation can be traced back to the beginning of the 20th century in the work of Wilke (1912; see discussion in Drukarch et al. 2018). Finally, it has also been found that nerve impulses can be generated with cooling (see Heimburg and Jackson 2005 and references therein). It is not an easy task to explain the relation between mechanical effects and nervous transmission, nor the relation between temperature changes and excitability, on the basis of either the ionic hypothesis or the electrical approach to nerve impulse research more generally. The traditional explanation idealizes the membrane as a "sieve"—as a filter that allows some ions to cross and not others, and whose molecular features (the existence of voltage-sensitive protein ion channels) enable it to change its selective permeability when excited (see Ling 2007, for a philosophical discussion of the origin of the sieve idealizations, see Carrillo and Knuuttila 2022; 2023 and Carrillo and Martínez 2023). This abstract view of the membrane cannot easily relate mechanical or thermal changes of the membrane, to signal transmission.

Another issue is whether the membrane capacitance remains constant during the nerve impulse. Hodgkin and Huxley assumed that the membrane capacitance is constant during the transmission of a signal, and this assumption played an important role in their interpretation of voltage-clamp experiments (Hodgkin and Huxley 1952, 505). This assumption is supported by experiments, as Cole and Curtis (1939) showed that the membrane capacitance is around 1 $\mu F/cm^2$. These results have been criticized, however, for not accurately examining the capacitance during the transmission (Takashima 1979, 140; for discussion on the role of constant capacitance in the Hodgkin–Huxley model, see Carrillo and Knuuttila 2021).

5. Thermodynamical approaches to nerve impulse research

Considering the recalcitrant evidence of the Hodgkin–Huxley model, some scientists remained skeptical of the "electricity-centered" agenda the model promoted (see Drukarch et al. 2018). An early case is Ichiji Tasaki, who developed an alternative approach to nerve impulse research. Focusing on nerve excitation while paying attention to osmotic effects and phase transitions,[5] Tasaki challenged the widespread idea that the relevant cell barrier can be abstractly conceived as a "sieve," changing only its permeability. Instead, he and other scientists suggested that the membrane could change shape during excitation. This is how Teorell, a contemporary of Tasaki, formulated the idea:

> It is not likely that biological membranes are rigid; they may rather be distendable and (visco-) elastic [...] different layers in the composite membrane may have varying charge densities and hydraulic permeabilities [...]. It might perhaps be possible that this (membrane) structure can be subject to swelling or shrinkage
>
> *(Teorell 1962, 50–51)*

In the eighties, Tasaki invented the technique of intracellular perfusion and with it discovered that the macromolecular filamentous structure attached to the inside of the bilipid membrane (a part of the cytoskeleton) is essential for excitable behavior in neurons (Tasaki 1982, 160–162). He proposed that the filaments of macromolecules in the interior of the cell undergo phase transitions as part of the excitatory process, calling this the *macromolecular model of nerve excitability*. In this model, the relevant "barrier" between the inside and the outside of the cell includes the interior macromolecular structure of the axon in addition to the lipid bilayer already considered by Hodgkin, Huxley, and contemporaries (Tasaki 1982).

In his books, Tasaki underlines the need to recover tools from thermodynamics for the study of nerve impulse transmission. He thought that the discovery of the squid giant axon led too quickly to the entrenchment of a specific set of skills transferred from electrical engineering to nerve impulse research, and he lamented that this meant sidestepping other abilities that had been used in the beginning of the century to study nerve excitation. For instance, the deduction of Nernst's equation (in 1889), which the Hodgkin–Huxley model exploits to calculate the electrochemical equilibrium of sodium and potassium ions, involves calculations of thermodynamic potential (Gibbs free energy). Tasaki regrets that Nernst's equation is now used like a formula and that the ability to perform the calculations that allowed its derivation has been replaced with those of electric engineering (Carrillo and Martínez 2023).

A more recent effort to recover thermodynamical approaches to discuss nerve impulse transmission appears in the work of Thomas Heimburg and Andrew Jackson in many articles beginning in 2005. The Heimburg–Jackson model is based on findings that the isolated lipids of biomembranes display order–disorder (gel–fluid) transitions in physiological conditions. They propose that we look at the bilipid membrane not as a sieve but as a material that undergoes phase transitions, changing its mechanical and thermal properties. Ultimately, they suggest that lying at the base of the nerve impulse are such phase transitions that produce the much-discussed voltage variation but also changes in other variables like density, volume, temperature, and thickness.

The Heimburg-Jackson model agrees with Tasaki's in that it also claims that phase transitions are the basis of nervous transmission, but it differs with the macromolecular model in *what* undergoes the transition, since in the Heimburg-Jackson model it is the lipid barrier and not the macromolecules running along the inside of the bilipid layer that change state. The equations of the Heimburg-Jackson model assume that the membrane behaves like a fluid and the nerve impulse is a localized phase transition that forms a solitary wave (a soliton) in that fluid. Heimburg and Jackson calculated the possibility of soliton propagation in biomembranes and found that the conditions for this to happen are physiologically within range.[6] Solitary waves are conservative, which means they maintain their shape and velocity and do not annihilate or change shape when colliding with other waves, making them good transmitters of information. Heimburg and Jackson used hydrodynamic wave equations that have a solitary wave solution, and that solution is taken to model the nerve impulse transmission. This model predicts the speed of the wave transmission and can accommodate some of the evidence that the Hodgkin–Huxley model struggles with.

Perhaps the most interesting feature of the Heimburg–Jackson model is its potential to explain some known features of general anesthesia that had not been satisfactorily accounted for within the ionic framework suggested by the Hodgkin–Huxley model. The ionic hypothesis suggests that anesthetics act by blocking or otherwise affecting the transmembranal proteins that conform the voltage-sensitive ionic channels responsible for the voltage variation. But this explanation has proven to be at odds with some empirically established patterns. For instance, the Meyer–Overton rule establishes that a general anesthetic will be more potent the more liposoluble it is. But general anesthetics vary from simple atoms to complex molecules, which makes it hard to see what it is that makes them all have (qualitatively) the same anesthetic effect, and why the intensity of the effect might rely on how soluble these molecules or atoms are in lipids. In other words, there is no obvious structure-activity relationship for general anesthetics. Additionally, there seems to be a connection between general anesthesia and thermodynamic factors, since the anesthetic effect can be reversed with increased pressure (known as the "pressure reversal effect"). The Heimburg–Jackson model offers an alternative explanation to the effect of general anesthetics: by blending into the lipid membrane, anesthetics lower the freezing point of transitions in the biomembranes. As a result, more energy is required to generate a phase change. This accounts for the Meyer–Overton rule and pressure reversal effect by making use of the freezing point depression law (the one that establishes, e.g., that the melting temperature of water is lowered when salt is added to it).

The chapter has so far presented the Hodgkin–Huxley model and two alternatives that exploit the concept of phase transition to understand nerve impulses. What should we do with this diversity of models? How can we frame and understand their epistemic potential? A first observation is that the electrical and thermodynamic models are not so easily befriended, since there are clear differences with respect to what they expect experimentally and what they take to be the main explanatory units. The electricity-centered approach has led to the idea that most of the answers to why-questions involving nerve impulse transmission will be answered by referring to the characteristics of voltage-sensitive protein ion channels that open when the cell becomes excited. In turn, the thermodynamic models, while they do not claim that these protein channels do not do any work (a common misunderstanding in the discussion), they do not ascribe to them such a crucial role. Also, issues of what is expected to happen with heat along the signal transmission differ. Phase transitions are conservative processes, whereas current is dissipative (see below). In sum, the models disagree in important points leading to different explanations for nerve cell excitation. These issues appear to suggest that all these accounts cannot be right at the same time (as implied by the discussions in, e.g., Heimburg and Jackson 2005; Fox 2018). In light of these tensions, some scientists have attempted to develop models that integrate the mechanical with the electrical features of the nerve impulse in a single model. According to

Engelbrecht (2022), for instance, the nerve impulse can be understood as an "ensemble of waves," which should make it possible to integrate the Hodgkin–Huxley equations with the descriptions of the compression wave as described in the Heimburg–Jackson model. This proposal leads to problems, however, since there is the risk of generating logical inconsistencies in the resulting model. For instance, Holland et al. (2019) criticize Engelbrecht's approach for coupling equations that assume the membrane capacitance is constant (namely the Fitz-Hugh Nagumo[7] equations, a reduction of the Hodgkin–Huxley model) with equations that assume capacitance varies during the transmission of the nerve signal (Heimburg–Jackson equations).

An important underlying issue is how physics applies to biological phenomena. Whereas the statement that laws of physics must be obeyed by all matter, including biological matter, seems to be an obvious claim, not all scientific laws are equal. Einstein believed that thermodynamics' first principle (top-down) approach trumps constructive (bottom-up) approaches. According to Einstein's interpretation, constructive theories start from the constitutive details and derive the behavior from those constituents, whereas thermodynamics establishes relations between phenomenological variables that do not commit to specific micro-details. While constructive and first principle theories are not *in principle* in conflict, if inconsistencies were to be found between the descriptions coming from each of these theories, then the first principle constraints should rule out those proposed by the constructive theory. In recent works, it has been argued that Einstein's interpretation of thermodynamics suggests that thermodynamics offers the primary constraint, such that claims concerning the micromolecular process behind nerve impulse transmission should be reconsidered if they do not conform to the thermodynamical laws (Drukarch et al. 2021; Schneider 2021).

Very little philosophical discussion has appeared addressing the philosophical implications of the controversies between thermodynamical and electrical approaches to nerve impulse research. Notwithstanding, the diversity of models and the current discussions in neuroscience offer a unique possibility to re-examine important debates in the philosophy of science, touching important issues such as the reduction of biology to physics, the role of laws in explanations, what makes models explanatory, how scientists abstract and idealize and what that has to do with the diversity of models scientists produce. Some works along these lines include Holland et al. (2019), who following Giere suggest that the different models of the nerve impulse should not be understood as free-floating representations of the nerve impulse but as "tools" that are used for certain purposes by specific agents. This idea is articulated more carefully through the interpretation of these specific models as "epistemic artifacts" (a concept from Knuuttila 2011) in two articles (Carrillo and Knuuttila 2021; 2022). Using the artifactual account and focusing on the idealization of constant capacitance in the Hodgkin and Huxley model, the authors question the idea that the idealizations involved would have been any clear-cut distortions for the scientists themselves, as they are often thought by philosophers of science. Carrillo and Knuuttila instead suggest that some idealizations emerge from the artifactual context of the research practice, including experimental practices. Such idealizations are "holistic" in that they permeate whole research programs, so that challenging the assumptions undermines the research agenda (as is the case for constant capacitance in the electricity-centered approach to nerve impulse research).

Another set of articles discusses the tension between the ideal of constitutive explanation (that is foundational for many mechanistic accounts of explanation) and the thermodynamical approaches (Drukarch et al. 2021; Carrillo and Knuuttila, 2023). The issue is that whereas mechanistic explanation tends to require a description of constitutive entities and activities that exhibit the phenomenon, the very notion of phase transition only makes sense from a macroscopic point of view—the analysis starts from the macrolevel without considering the microlevel. For Heimburg (as well as for Einstein, as mentioned above), the macroscale approach offers an advantage when it comes to explaining many issues associated with the nervous impulse:

> The accepted model for nerve pulse propagation in biological membranes seems insufficient. It is restricted to dissipative electrical phenomena and considers nerve pulses exclusively as a microscopic phenomenon. A simple thermodynamic model that is based on the macroscopic properties of membranes allows explaining more features of nerve pulse propagation including the phenomenon of anesthesia that has so far remained unexplained.
>
> *(Heimburg 2010, 1)*

The discussion of whether microlevel details are necessary for a good explanation of nervous transmission relates to the aforementioned discussion of the relation between constructive and first principle theories.

Finally, some interesting lessons have been drawn from integrated philosophical and historical approaches. Carrillo and Martínez (2023) analyze the thermodynamical and electric modeling strategies from a historical perspective, tracing down the abstractions that structure the research programs leading to the different models of the nerve impulse. They trace different lineages of metaphors that are constitutive of the research programs and show how they relate to the material culture of the different scientific practices. The authors are then led to suggest a kind of pluralism that emerges from the different skills and artifacts used in scientific practices that then permeates the theoretical advancements leading to the diversity of models of the nerve impulse.

6. Conclusions

This chapter has contrasted different models of nerve impulse and has shown how they draw from different theoretical frameworks. The controversies among the models have been contextualized by putting them in their historical and disciplinary contexts. Importantly, the role of the ionic hypothesis in the electrical approach to model the nerve impulse has been exposed, and how the thermodynamical models challenge this hypothesis. Moreover, the earlier philosophical attempts to explain the epistemic power of the Hodgkin–Huxley model were contrasted to the few efforts that philosophically consider the alternative developments from scientists in terms of phase transitions. The discussion has brought forth several philosophical issues that await to be either addressed or further discussed, that were mentioned in the chapter, ranging from the reach of mechanistic explanation, the relation between physics and biology, the nature of idealizations and abstractions, to the epistemic value of modeling.

Acknowledgements

The author would like to thank Federico Faraci for reading an earlier version of this handbook entry and suggesting additional bibliography. This project received funding from the European Research Council under the European Union's Horizon 2020 Research and Innovation Programme (grant agreement no. 818772) and from the UNAM PAPIIT project number IN400422.

Notes

1 The signal also travels along the nerve cell's *axon*, the large tubular section of the cell that allows it to connect with distant neurons. In order to model the *transmission* of the signal, Wilfrid Rall later used the cable equation—an equation that was originally formulated to model the decay of telegraph (electric) signals in cables that covered long distances. The same principle applies to both models though, the idea is that the nerve cell behaves like a circuit, be it a circuit representing the transmembrane currents or a circuit representing the longitudinal currents.

2 Craver claimed that the explanation of the nervous impulse was not truly given until the proteins that form ionic channels across the membrane were discovered, thereby completing the explanatory sketch. As we see ahead, many scientists think that the evidence pointing to the role of protein ion channels is inconclusive. That goes against the idea that the Hodgkin–Huxley model plus such knowledge of protein ion channels counts as a how-actually model. The most recent empirical discussion has not been addressed by Craver or any of the other mechanists, however.

3 Schaffner (2008) had also highlighted the abstract/aggregative character of the Hodgkin and Huxley model in relation to its epistemic success, although he did not give it a mechanistic interpretation.

4 Some scientists have criticized the approach to nerve impulse as a conservative phenomenon (suggested by some thermodynamical models), based on measurements of temperature that show a slight release of heat (Meissner 2022, 51). However, the crux of the issue is whether the pulse is generated via a conservative or dissipative process (Drukarch et al. 2021), not whether the wave itself is completely conservative or not. In any case, advocates of the thermodynamical approach and even Hodgkin (1964, 70), consider that the measurements of heat are problematic for the Hodgkin–Huxley model.

5 Tasaki was able to articulate this novel perspective to understand nerve impulse transmission by recovering a different lineage of metaphors than those that served the electrical view (see Carrillo and Martínez 2023).

6 The Heimburg-Jackson model is empirically supported by experiments showing that synthetic membranes (e.g., black lipid membranes, DPPC and DPPA membranes) and lung surfactant display the kinds of properties required for traveling phase transitions to behave as they describe. Moreover, Heimburg, Jackson, and colleagues produced evidence showing permeability changes associated with phase transitions occurring in synthetic lipid membranes with no proteins (Laub et al. 2012).

7 The Fitz-Hugh Nagumo model is a simplification of the Hodgkin–Huxley equations, reducing the number of equations from 4 to 2, which allows for two-dimensional representation of the phase space. This model is also referred to as the Van der Pol - Bonhoeffer model.

References

Abbott, Bernard. C., Archibald Vivian Hill, and J. V. Howarth. 1958. "The positive and negative heat production associated with a nerve impulse." *Proceedings of the Royal Society B: Biological Sciences* 148(931): 149–187.

Bernstein, Julius. 1902. "Untersuchungen Zur Thermodynamik Der Bioelektrischen Ströme." *Pflüger Archiv Für Die Gesammte Physiologie Des Menschen Und Der Thiere* 92: 521–562.

Bogen, Jim. 2008. "The Hodgkin-Huxley equations and the concrete model: Comments on Craver, Schaffner, and Weber." *Philosophy of Science* 75(5): 1034–1046.

Carrillo, Natalia, and Tarja Knuuttila. 2021. "An artefactual perspective on idealization: Galvanic cells and electric circuits in nerve signal research." In *Models and Idealizations in Science: Fictional and Artefactual Approaches*, edited by Alejandro Cassini and Juan Redmond, 51–70. Switzerland: Springer Nature.

———. 2022. "Holistic idealization: An artifactual standpoint." *Studies in the History and Philosophy of Science* 91: 49–59. https://doi.org/10.1016/j.shpsa.2021.10.009

———. 2023. "Mechanisms and the problem of abstract models." *European Journal of Philosophy Science* 13: 27.

Carrillo, Natalia, and Sergio F. Martínez. 2023. "Scientific inquiry: From metaphor to abstraction." *Perspectives on Science* 31(2): 1–29.

Cole, Kenneth S., and Howard J. Curtis. 1938. "Transverse electric impedance of the squid giant axon." *Journal of General Physiology* 21(6): 757–765. https://doi.org/10.1085/jgp.21.6.757

———. 1939. "Electric impedance of the squid giant axon during activity." T*he Journal of General Physiology* 22(5): 649–670.

Colombo, Matteo, Stephan Hartmann, and Robert Van Iersel. 2015. "Models, Mechanisms and Coherence." *British Journal of Philosophy of Science* 66(1): 181–212.

Craver, Carl F. 2006. "When mechanistic models explain." *Synthese* 153: 355–376.

———. 2007. *Explaining the Brain - Mechanisms and the Mosaic Unity of Science*. Oxford University Press. New York.

Doyle, Declan A, J. Morais Cabral, Richard A Pfuetzner, Anling Kuo, Jacqueline M. Gulbis, Steven L. Cohen, Brian T. Chait, and Roderick Mackinnon. 1998. "The structure of the potassium channel: Molecular basis of K+ conduction and selectivity." *Science* 280 (April): 69–77.

Drukarch, Benjamin, Hannah A. Holland, Martin Velichkov, Jeroen J. G. Geurts, Pieter Voorn, Gerrit Glas, and Henk W. de Regt. 2018. "Thinking about the nerve impulse: A critical analysis of the electricity-centered conception of nerve excitability." *Progress in Neurobiology* 169(July): 172–185.

Drukarch, Benjamin, Micha M. M. Wilhelmus, and Shamit Shrivastava S. 2021. "The thermodynamic theory of action potential propagation: A sound basis for unification of the physics of nerve impulses." *Review Neuroscience* 33(3): 285–302. https://doi.org/10.1515/revneuro-2021-0094. PMID: 34913622.

Engelbrecht, Jüri, Kert Tamm, and Tanel Peets. 2022. "Physics shapes signals in nerves." *European Physical Journal Plus* 137: 696.

Fox, Douglas. 2018. "The Brain, Reimagined. Brain cells communicate with mechanical pulses, not electric signals." *Scientific American*, 318: 61–67.

Gonzalez-Perez, Alfredo, Lars D. Mosgaard, Rima Budvytyte, Edgar Villagran-Vargas, Andrew D.Jackson, Thomas Heimburg. 2016. "Solitary electromechanical pulses in lobster neurons." *Biophysical Chemistry* 216: 51–59. https://doi.org/10.1016/j.bpc.2016.06.005. Epub 2016 Jul 9. PMID: 27448851.

Heimburg, Thomas. 2010. "The physics of nerves. Translation from: Die Physik der Nerven." *Physik Journal* 2009 8(3): 33–39.

Heimburg, Thomas, and Andrew D. Jackson. 2005. "On soliton propagation in biomembranes and nerves." *Proceedings of the National Academy of Sciences* 102 (28): 9790–9795.

Hodgkin, Alan L. 1964. *The Conduction of the Nervous Impulse*. The Sherrington Lectures VII, Illinois: Charles C. Thomas Publisher.

Hodgkin, Alan L., and Andrew F. Huxley. 1952. "A quantitative description of membrane current and its application to conduction and excitation in nerve." *The Journal of Physiology* 117: 500–544.

Holland, Linda, Henk W. de Regt, and Benjamin Drukarch. 2019. "Thinking about the nerve impulse: The prospects for the development of a comprehensive account of nerve impulse propagation." *Frontiers in Cellular Neuroscience* 13: 1–12.

Keener, James, and James Sneyd. 2009. *Mathematical Physiology I: Cellular Physiology*. New York: Springer-Verlag.

Knuuttila, Tarja. 2011. "Modelling and representing: An artefactual approach to model-based representation." *Studies in History and Philosophy of Science Part A* 42(2): 262–271.

Knuuttila, Tarja, and Andrea Loettgers. 2013. "Synthetic modeling and mechanistic account: Material recombination and beyond." *Philosophy of Science* 80(5): 874–885.

Laub, Katherine R., Katja Witschas, Andreas Blicher, Søren B Madsen, Andreas Lückhoff, Thomas Heimburg. 2012. "Comparing ion conductance recordings of synthetic lipid bilayers with cell membranes containing TRP channels." *Biochimica et Biophysica Acta*. 1818(5): 1123–1134. https://doi.org/10.1016/j.bbamem.2012.01.014. Epub 2012 Jan 28. PMID: 22305677.

Levy, Arnon. 2014. "What was Hodgkin and Huxley's achievement?" *The British Journal for the Philosophy of Science* 65(3): 469–492.
Ling, Gilbert. 2007. "History of the membrane (pump) theory of the living cell from its beginning in mid-19th century to its disproof 45 years ago - though still taught worldwide today as established truth." *Physiological Chemistry and Physics and Medical NMR* 39(1): 1–68.
Love, Alan, and Marco J. Nathan. 2015. "The idealization of causation in mechanistic explanation." *Philosophy of Science* 82(5): 761–774. https://doi.org/10.1086/683263
Meissner, Scott. 2022. "Additional proposed tests of the soliton/wave-action potential model, and how the thermodynamic/theory-based philosophical approach abandons the scientific method" Preprints. https://doi.org/10.20944/preprints202208.0248.v1.
Schaffner, Kenneth. 2008. "Theories, models, and equations in biology: The heuristic search for emergent simplifications in neurobiology." *Philosophy of Science* 75(5): 1008–1021.
Schneider, Matthias F. 2021. "Living systems approached from physical principles." *Progress in Biophysics and Molecular Biology* 162: 2–25.
Takashima, Shiro. 1979. "Admittance change of squid axon during action potentials: Change in capacitive component due to sodium currents." *Biophysical Journal* 26(1): 133–142.
Tasaki, Ichiji. 1982. *Physiology and Electrochemistry Nerve Fibers*. Academic Press. New York.
Tasaki, Ichiji, and Kunihiko Iwasa. 1980. "Shortening of nerve fibers associated with propagated nerve pulse." *Biochemical and Biophysical Research Communications* 94(2): 716–720.
———. 1982. "Rapid pressure changes and surface displacements in the squid giant axon associated with production of action potentials." *The Japanese Journal of Physiology* 32: 69–81.
Teorell, Torsten. 1962. "Excitability phenomena in artificial membranes." *Biophysical Journal* 2: 27–52. https://doi.org/10.1016/s0006-3495(62)86947-1. PMID: 14039678.
Weber, Marcel. 2008. "Causes without mechanisms: Experimental regularities, physical laws, and neuroscientific explanation." *Philosophy of Science* 75:1008–1021.
Wilke, E. Atzler. 1912. "Experimentelle Beiträge zum Problem der Reizleitung im Nerven." *Pflügers Arch* 144(35): 430–446.

INDEX

Note: *Italic* page numbers refer to figures and page numbers followed by "n" denote endnotes.